CALCULUS

From Graphical, Numerical, and
Symbolic Points of View

CALCULUS

*From Graphical, Numerical, and
Symbolic Points of View*

- **ARNOLD OSTEBEE**

  St. Olaf College

- **PAUL ZORN**

  St. Olaf College

SAUNDERS COLLEGE PUBLISHING

Harcourt Brace College Publishers

Fort Worth   Philadelphia   San Diego   New York
Orlando   Austin   San Antonio   Toronto
Montreal   London   Sydney   Tokyo

This material is based upon work supported by the National Science Foundation under Grant No. USE-9053363

Any opinions, findings, and conclusions or recommendations expressed in this material are those of the authors and do not necessarily reflect the views of the National Science Foundation.

Derive is a registered trademark of Soft Warehouse, Inc.
HP-35 and HP-48 graphics calculators are registered trademarks of Hewlett-Packard Company.
IBM is a registered trademark of International Business Machines Corporation.
Macintosh is a registered trademark of Apple Computer, Inc.
Maple is a registered trademark of Waterloo Maple Software.
Mathematica is a registered trademark of Wolfram Research, Inc.
TI-81, TI-82, TI-83, TI-85, and TI-92 graphics calculators are trademarks of Texas Instruments, Inc.
Windows is a trademark of Microsoft Corporation.

Text Typeface: New Caledonia
Compositor: ETP/Harrison
Senior Associate Editor: Alexa Epstein
Managing Editor: Carol Field
Senior Project Editor: Nancy Lubars
Copy Editor: Mary Patton
Manager of Art and Design: Carol Bleistine
Art Director: Caroline McGowan
Art & Design Coordinator: Kathleen Flanagan
Cover Designer: Chazz Bjanes
Text Artwork: Techsetters, Inc.
Vice President of EDP: Tim Frelick
Manager of Production: Joanne Cassetti
Production Manager: Alicia Jackson
Product Manager: Nick Agnew

Cover Credit: Art Wolfe/© Tony Stone Images
Printed in the United States of America

ISBN 0-03-052659-0
Library of Congress Catalog Card Number: 96-70623

9012345 069 10 9876543

To Kay, Kristin, and Paul
To Janet, Anne, and Libby

About This Book: Notes for Instructors

This book aims to do exactly what its title suggests: present calculus from graphical, numerical, and symbolic points of view. In this preface we elaborate briefly on what this means.

Philosophy

Several common threads run through *Toward a Lean and Lively Calculus*, *Calculus for a New Century*, discussions on calculus at many professional meetings, and the NCTM *Curriculum and Evaluation Standards for School Mathematics*. First, there is a consistent call for *leaner and more conceptual courses*, driven by and focused on central ideas. Second, there is a realization that courses should *reflect modern technology both in content and in pedagogy*.

A Diagnosis.   Many calculus courses, we believe, slight the conceptual foundations of the subject and overemphasize routine techniques—formal differentiation, antidifferentiation, and convergence testing, for example. Analytic objects (such as integral, derivative, and convergence) are represented and manipulated only algebraically (i.e., via symbolic manipulation of explicit elementary functions). For example, textbooks often treat limits, derivatives, and integrals—all *analytic* objects—only as *algebraic* operations on *algebraic* functions. We try to take a broader view.

Whether one views calculus as an introduction to pure mathematics or as a foundation for applications (or both!), the conclusion is the same—concepts, not techniques, are truly fundamental to the course. Whatever uses they make of the calculus, students need more than a compendium of manipulative techniques. The *sine qua non* for a useful command of the calculus is a conceptual understanding that is deep and flexible enough to accommodate diverse applications.

A Prescription.   Our key strategy for improving conceptual understanding is combining, comparing, and moving among graphical, numerical, and algebraic "representations" of central concepts. This strategy pervades and unifies our exposition. Bringing graphical and numerical, as well as algebraic, viewpoints to bear on calculus ideas is the philosophical foundation of our text. By representing and manipulating calculus ideas and objects graphically, numerically, and algebraically, we believe that students gain a better, deeper, and more useful understanding.

Audience and Prerequisites

Our main strategy—combining graphical, numerical, and algebraic viewpoints to clarify concepts and make them concrete—aims explicitly at mainstream students, who especially need such help. We regard a more conceptual calculus as also more applicable. To use calculus ideas and techniques effectively, students must know what they are doing and why, not merely how. Thus, our text is appropriate for a general audience: mathematics majors, science and engineering majors, and non-science majors.

We expect that students have the "usual" precalculus preparation, including basic algebra and trigonometry and the rudiments of logarithm and exponential functions. However, we include reasonably complete reviews of all these subjects, with routine exercises, in the Appendices. Appendices are written in a style that supports independent reading and self-study; many instructors should be able to avoid explicit "coverage" of precalculus material in class time.

Annotated Table of Contents

Here is brief chapter-by-chapter information; for many more details, see the *Notes for Instructors*, described below.

Chapter 1 – Functions in Calculus   This chapter sets the stage by emphasizing graphical and numerical points of view, especially on functions. Functions are described by algebraic formulas, but also by graphs, by tables, and in words. The standard elementary functions are reviewed briefly, along with a variety of ways of producing new functions from old. The chapter reviews key precalculus concepts and foreshadows important calculus ideas, such as slope, concavity, and rates of change. (Appendices provide still more precalculus review for students who need it.)

Chapter 2 – The Derivative   The derivative function is first treated intuitively, in the parallel languages of slope and of rates of change. Next, a careful geometric study is made of functions and their derivatives, emphasizing relationships between the two. The second half of the chapter treats the derivative symbolically, via limits. Systematic symbolic calculation of derivatives begins in the next chapter.

Chapter 3 – Derivatives of Elementary Functions   Derivatives of the standard (algebraic and transcendental) elementary functions are first calculated, and their properties are studied. Antiderivatives are also introduced and used in simple calculations and applications. The second half of the chapter emphasizes combinatorial properties of functions and their derivatives—how the derivative of a built-up function (e.g., by composition) is formed from derivatives of its constituents. A final section treats inverse trigonometric functions.

Chapter 4 – Applications of the Derivative A flexible and varied collection of applications of the derivative, designed to allow easy rearrangements or omissions. Some topics (e.g., optimization) are standard; others (e.g., early differential equations, quadratic splines) are less usual. The chapter ends with some topics of a more analytic character, exploring continuity, differentiability, and their most important consequences, such as the mean value theorem. *Important Note: Chapter 4 is designed as a sampler; most instructors will want to cover only parts of it.*

Chapter 5 – The Integral The treatment generally mirrors that of the derivative in Chapter 2. The integral is first presented geometrically, as signed area. Analytic viewpoints follow, including a formal definition via approximating sums and a careful treatment of the fundamental theorem of calculus. Interpretations and applications of approximating sums foreshadow topics of Calculus II.

Chapter 6 – Finding Antiderivatives Students began finding antiderivatives early, in Chapter 3. Combined with that experience, this brief chapter completes a short introduction to rudiments of antidifferentiation. The last section refers both to printed tables and to technological aids.

Chapter 7 – Numerical Integration This chapter introduces, briefly but systematically, the idea of approximating integrals numerically—and of estimating the errors committed in doing so. Several alternative methods are presented and compared.

Chapter 8 – Using the Definite Integral Like Chapter 4, Chapter 8 offers a flexible variety of applications, this time of the definite integral. Geometric, physical, and economic applications are included. Choices and reordering are possible. An introductory section emphasizes what various applications have in common.

Chapter 9 – More Antidifferentiation Techniques This chapter continues and extends the themes of Chapter 6. Some instructors may choose to cover it before Chapter 8.

Chapter 10 – Improper Integrals This chapter both extends the idea of integral and foreshadows the treatment of infinite series in the next chapter.

Chapter 11 – Infinite Series Stressing analogies between improper integrals and infinite series, the chapter examines convergence and divergence numerically and graphically, as well as symbolically. The topic is traditionally difficult; technology is used to illustrate elusive ideas and make them concrete.

Chapter 12 – Differential Equations Differential equations are treated first graphically (using slope fields), then numerically (using Euler's method), and then symbolically (via separation of variables).

Chapter 13 – Polar Coordinates   This brief, readily transportable chapter stresses geometric understanding of an alternative coordinate system.

Chapter 14 – A First Look at Multivariable Calculus   The chapter offers a brief overview of derivatives and integrals of functions of several variables, stressing functions of two variables. The theme of linear approximation is stressed; technology is used to help students visualize the new objects. *Some instructors may prefer Chapter 14 as an alternative to the topics of infinite sequences and series.*

More on Content and Treatment

Several special features of our treatment cut across chapters, and deserve emphasis. For many more details and specific examples, see the *Notes for Instructors*, described below.

Foreshadowing   Important concepts and associated vocabulary often appear more than once—first informally and later in more detail. The geometric idea of concavity, for instance, is mentioned in Section 1.1, and then revisited several times. Instructors should know that it is *not* necessary to cover these ideas in full rigor at their first appearance.

Early Differential Equations   Differential equations are mentioned early in Chapter 2, introduced informally (but at some length) in Chapter 4, and then revisited more formally in Chapter 12. We believe that DE's are so basic and so important that students should see them early and often. Calculus I emphasizes the *idea* of a DE and its solutions; the (harder) question of finding solutions systematically is deferred to Calculus II.

Graphs, then Symbols   Derivatives and integrals are introduced in a parallel manner, with the graphical approach preceding the symbolic viewpoint. Thus Chapter 2 stresses the geometry of the derivative; the symbolic approach enters mainly in Chapter 3. Chapter 5 begins with a graphical definition of integral, and proceeds to the Riemann sum definition.

The Role of Technology

Combining graphical, numerical, and symbolic viewpoints in calculus can be forbiddingly time-consuming and distracting without technological assistance. With technology, these viewpoints become practically accessible; hence, we use technology freely.

Having said this, we emphasize that we do *not* intend to "automate" or "computerize" the calculus. We regard computers and graphing calculators strictly as instructional *tools*—albeit very powerful ones—to facilitate the crucial combination of graphical, numerical, and algebraic viewpoints. Our text exploits the capabilities of these tools to help students focus on the ideas that lie at the heart of the calculus.

We do not assume that students have any previous experience with using technology.

How Much Technology? What Kinds?   This text freely uses and refers to numerical and graphical computations, but it is independent of any particular technology. Any of the familiar high-level products—*Mathematica*®, *Maple*®, *Derive*®— are certainly sufficient, but so are many other forms of technology. Many graphing calculators (e.g., the TI-8X and 92, the HP-35 and 48) and special-purpose microcomputer software packages are also adequate.

Our use of technology is most conveniently described in terms of *functionalities*. The requirements for Calculus I and Calculus II differ somewhat:

Calculus I: Chapters 1–5 (the traditional content of Calculus I) draw freely on machine graphics; almost any up-to-date form will do. Most graphing calculators would suffice; so would almost any flexible microcomputer graphing program.

Calculus II: To make the best use of the remaining (Calculus II) material, students will require access to a modest level of numerical computation (mainly for estimating analytic quantities: integrals, series, etc.). Many microcomputer software packages provide the necessary functionality; so do programmable graphing calculators. In addition, access to simple symbolic operations (e.g., formal differentiation, Taylor series expansion) is desirable, but not strictly necessary.

Symbol Manipulation and Hand Computation

Although we emphasize concepts and use technology freely, we by no means ignore symbol manipulations in general, or hand computations in particular. We take the symbolic points of view of our title fully as seriously as the graphical and the numerical. We cover, for instance, the usual techniques for formal differentiation and antidifferentiation. Why do we do so?

That machines can do calculus manipulations does not render by-hand operations obsolete. Some manipulative practice and skill builds and supports conceptual understanding. Hand computation illustrates concepts concretely, builds "symbol sense" (the algebraic counterpart of numerical intuition) and an ability to estimate, and gives students a sense of mastery. It does *not* follow, though, that harder, more baroque computational problems are necessarily better or more useful; we deemphasize them. Both research results and our own experience suggest that diverting some time and attention to concepts does little, if anything, to reduce students' hand computational facility.

Distinctive Features of the Text

Our text differs significantly from standard treatments. Here are some of these differences, together with some of our assumptions, goals, and strategies.

Combining Graphical, Numerical, and Algebraic Viewpoints   Throughout the text we insist that students manipulate and compare graphical, numerical, and algebraic representations of mathematical objects. In studying functions, for example, students manipulate not only elementary functions but also functions presented graphically and tabularly. In the context of formal differentiation, exercises ask students to apply the chain rule to combinations of functions presented in various ways.

Graphical and numerical techniques, with error estimates, complement algebraic antidifferentiation methods. For series, routine convergence tests are emphasized less than finding—and defending—numerical limit estimates.

Concepts vs. Rigor   Proving theorems in full generality is less valuable, we think, than helping students understand concretely what theorems say, why they're reasonable, and why they matter. Too often, fully rigorous proofs address questions that students are unprepared to ask.

Still, we believe that introducing calculus students to the idea of proof—and to some especially important classical proofs—is essential. We prove major results, but emphasize only those that we believe contribute significantly to understanding calculus concepts. In examples and problems, too, we pay attention to developing analytic skills and synthesizing mathematical ideas.

Exercises   However clear its exposition, a textbook's problems generate most of students' mathematical activity and occupy most of their time. Through the problems we assign, we tell students concretely what we think they should know and do.

Routine drills and challenging theoretical problems are standard in calculus texts; ours includes many of the former and some of the latter. More distinctive, perhaps, are problems that fall between these poles:

- Problems that combine and compare algebraic, graphical, and numerical viewpoints and techniques.

- Problems that require "translation" among various representations and interpretations of calculus ideas (e.g., to interpret derivative information in terms of either slope or rate of change).

- Problems that use calculus as a *language* for interpreting and solving problems. Students are asked to translate problems into mathematical terms, solve these problems using the tools of calculus, and reinterpret mathematical results in the context of the original problem.

"Basic" vs. "Further" Exercises   For instructors' convenience, exercises are classified as either "Basic Exercises" or "Further Exercises." The former are, typically, (1) relatively straightforward; (2) conceptual in nature; and (3) focused on some important, basic principle. More routine exercises (e.g., symbol manipulation problems) are usually *not* named as "basic"—even though instructors should assign some of them. For more details and advice on assigning exercises, see the Instructor's Notes.

Strategies for Better Problem Solving   Emphasizing problem solving is nowadays *de rigueur* in calculus textbooks; ours is no exception. What, concretely, does such an emphasis entail in content and strategy?

Conceptual understanding, we believe, is the weakest link in students' ability to solve nontrivial problems. Thus, *the goal of better problem-solving skills is implicit in the goal of deeper conceptual understanding.* To that end:

- We use numerical and graphical methods both to improve students' understanding of concepts and to enlarge their kit of tools to solve problems. Students, we find, are surprisingly quick to master and apply elementary numerical and graphical methods, even error estimation.

- When technology is available, students are freed, indeed forced, to analyze the structure of a problem and plan a solution strategy. We emphasize general problem-solving strategies (reduction to more tractable subproblems, estimation, search for patterns, etc.) explicitly wherever we can.

- We provide a greater qualitative variety of exercises including problems that are posed more generally, problems that call for more synthesis, and problems that rely on a larger set of solution techniques. In this richer environment, we hope students will come to see mathematics as an open-ended, creative activity, not a rigid collection of recipes.

Teaching from the Text; Notes on Pedagogy

Our approach to the subject is non-traditional in some respects, e.g., in emphasizing graphs and geometry and in the nature of its exercises. Most teachers will find that using such a text successfully calls for some corresponding changes and adaptations in pedagogy. Many strategies have been used, including group homework assignments, takehome exams, writing assignments, computer labs, "gateway" tests for proficiency in differentiation and integration, and many others. What combination, if any, works best will depend on instructors' individual preferences and practical logistics.

No particular combination is ideal for everyone, but we believe that almost all instructors will benefit from looking closely at our *Notes for Instructors*, which is available either in physical form from Saunders College Publishing or electronically, in PostScript format, from our World Wide Web site: `http://www.stolaf.edu/ people/zorn/ozcalc`. The *Notes for Instructors* includes such features as these:

- Overviews of philosophy, strategies, and use of technology.

- Extensive pedagogical notes and suggestions, treating such topics as reading and writing mathematics, exercises, testing, pace and coverage, suggestions for use of classroom time, etc.

- Sample lesson plans.

- Section-by-section notes and comments, with suggested "basic" homework exercises from each section.

Supplements to the Text

The following aids for students and instructors are available from the publisher:

Student Answer Booklets.   These booklets, prepared by the authors, are available in two volumes, either separately or shrinkwrapped with the text. They contain answers to selected exercises.

Student Solutions Manuals.   Prepared by the authors, these manuals are available in two volumes, and contain complete solutions to all odd-numbered exercises.

Lecture Guide and Student Notes.   Prepared by Stephen M. Kokoska of Bloomsburg University of Pennsylvania, available in two volumes, these notes offer effective teaching aids that allow instructors and students to use identical material in class. Instructors can use these key examples, exercises, definitions, and theorems to prepare overhead transparencies. Ample room on each page allows students to add notes about the discussion, additional examples, and justification. Thus, instructors can prepare classroom materials with ease and students spend class time in discussion and group work rather than taking notes.

Calculus and Graphing Calculators.   Written by Joe May of North Hennepin Community College and Glen Van Brummelen of the King's University College. Part I introduces basic commands and features of the TI-82 and TI-85 graphing calculators, with numerous examples and exercises. Exploratory activities in Part II build on various applications and historical perspectives.

Calculus Explorations.   Three volumes are available: *Calculus Explorations Using Mathematica®*, by Allen Hibbard (Central College); *Calculus Explorations Using Maple®*, by Phoebe Judson (Trinity College); and *Calculus Explorations Using Derive®*, by Mary Kay Abbey (Montgomery College–Takoma Park).

These three technology supplements introduce students to a computer algebra system through meaningful calculus explorations. As students complete each lab activity, they learn the basic commands and discover (or reinforce) key (single-variable) calculus concepts through the numerous examples. The questions and examples in each lab encourage students to interpret, explain, and communicate their results. Each supplement offers over 20 lab activities to give instructors variety; the activities are ideal for group work or in-class discussion. Each lab begins with clearly stated goals; progressive questions help students check their understanding as they work. No previous knowledge of *Mathematica*, *Maple*, or *Derive* is required.

Instructors who adopt this text may also receive, free of charge, the following items.

Instructor's Resource Manual   The manual offers a variety of teaching tips and suggestions. Part I includes (i) advice on using technology, group work, homework, and testing, etc.; (ii) sample "lecture" plans; and (iii) section-by-section notes. Part II contains essays from several class-testers about their experiences. Part III includes several writing assignments.

Instructor's Solutions Manual   Available in two volumes, the manual contains complete solutions, prepared by the authors, for all exercises. (Answers to selected exercises are also available in the Student Answer Booklets.)

Test Bank   Prepared by Scott Inch of Bloomsburg University of Pennsylvania, this resource contains over 450 test questions, covering multiple viewpoints and levels of difficulty. Answers to all questions are included.

EXAMASTER+[TM] **Computerized Test Bank**   Available in Windows, Macintosh, and IBM formats, this resource combines the items of the Test Bank with a convenient features that allow instructors to prepare quizzes and examinations quickly and easily. Instructors can edit questions, add their own, administer tests over a computer network, and record student grades with the gradebook software.

Overhead Transparencies and Masters   These are available in two volumes and contain selected figures from the exposition and exercises. A selection of important figures is already printed on acetate, for direct viewing in the classroom. Additional useful figures are given on the masters, from which instructors may prepare their own transparencies.

Saunders College Publishing may provide complimentary instructional aids and supplements or supplement packages to those adopters qualified under our adoption policy. Please contact your sales representative for more information. If as an adopter or potential user you receive supplements you do not need, please return them to your sales representative or send them to

Attn: Returns Department
Troy Warehouse
465 South Lincoln Drive
Troy, MO 63379

Advice from You

We appreciate hearing instructors' comments, suggestions, and advice on this edition. Many of the suggestions received from users of earlier versions of this text have been incorporated in this version. Our physical and e-mail addresses are below.

Arnold Ostebee and Paul Zorn
Department of Mathematics
St. Olaf College
1520 St. Olaf Avenue
Northfield, Minnesota 55057-1098

e-mail: ostebee@stolaf.edu   zorn@stolaf.edu

Acknowledgments

This text owes its existence to (literally) countless professors, students, publishing company professionals, friends, advisors, critics, "competitors," family, and others. (These categories are not mutually exclusive!) It is a pleasure to acknowledge by name some—but, necessarily, only some—of the people who attended this text through its long gestation and birth.

Focus group participants, who offered useful early advice, include Wade Ellis, Jr., West Valley College; Gregory D. Foley, Sam Houston State College; Richard O. Hill, Jr., Michigan State University; Roger B. Nelsen, Lewis and Clark College; Patricia Roecklein, Montgomery College; Wayne Roberts, Macalester College; and Audrey Rose, Tulsa Junior College.

Various versions of the manuscript were meticulously reviewed, and many errors caught, by Mary Kay Abbey, Montgomery College–Takoma Park; Janet Andersen,

Hope College; Michael J. Bonanno, Suffolk Community College–Ammerman Campus; Matthew Brahm, University of Minnesota; Susanna Epp, DePaul University; Stephen Kuhn, University of Tennessee–Chattanooga; Rebecca Lee, Bowie State University; Herbert A. Medina, Loyola Marymount University; Edward S. Miller, Lewis–Clark State College; Jeffrey Ondich, Carleton College; John Polking, Rice University; John C. Peterson, Chattanooga State Technical Community College; Karen Saxe, Macalester College; Marsha Schoonover, Chattanooga State Technical Community College; Anita E. Solow, Grinnell College; and Larry Thomas, University of Virginia.

Accuracy of examples and exercises was carefully checked during production of preliminary and first editions by Eric Bibelnieks, University of Minnesota; Tracy Bibelnieks, University of Minnesota; Dan Kemp, South Dakota State University; Norm Loomer, Ripon College; Lewis Lum, University of Portland; John Peterson, Chattanooga State Technical Community College; Marsha Schoonover, Chattanooga State Technical Community College; and Kerry Wyckoff, Brigham Young University. All surviving errors are, of course, our own.

The Test Bank was checked for accuracy by Susan E. Fettes, SUNY–Oswego; and Art Richert, Southern College of Seventh-Day Adventists.

The overhead transparencies and masters were chosen by Tracy Bibelnieks, University of Minnesota; Peter Collins, Chattanooga State Technical Community College; and Jane Serbousek, Northern Virginia Community College–Loudoun Campus.

We are indebted, for many and different reasons, to members of Saunders College Publishing's professional staff, including Nancy Lubars, Senior Project Editor; Alicia Jackson, Production Manager; Caroline McGowan, Art Director; Nick Agnew, Product Manager; and Liz Wilchacky, Marketing Coordinator. To Alexa Epstein, Senior Associate Editor, our most special thanks are due, for her tireless support (extending even to attending several week-long summer workshops), unflagging encouragement, practical help, concrete advice, common sense, generous friendship, and (from time to time) exquisitely gentle prodding. Alexa's help has been simply indispensable and we thank her very much.

We have learned much from professional colleagues, here and around the country. Some are our teachers; some are departmental colleagues; some course-tested preliminary versions of these volumes at their own schools; some reviewed our grant proposals; some reviewed our project; some invited us to review theirs; some invited us to speak on panels or at workshops; some spoke on panels we organized; some helped us run our workshops; some are "calculus reform competitors"; some are advisers; some are critics. We list a small sample below, in alphabetical order. Regardless of category, we thank them all. Janet Andersen, Tracy Bibelnieks, Judith Cederberg, Caspar Curjel, Tom Dick, Ed Dubinsky, Wade Ellis, Bob Foote, Bonnie Gold, Michael Henle, Deborah Hughes Hallett, Paul Humke, Zaven Karian, John Kenelly, Harvey Keynes, Steve Kuhn, Richard Mercer, Lang Moore, John Peterson, Art Richert, Wayne Roberts, Don Small, David Smith, Keith Stroyan, Tom Tucker, Jerry Uhl, and Ted Vessey.

We thank St. Olaf College in general, and our departmental colleagues in particular, for their generous advice, support, good humor, and (when needed) forbearance during the many years of this project's development. Students (here and at many other institutions where preliminary versions of this book were course-tested) offered generous and useful advice.

We are grateful, too, for grants from the National Science Foundation (Grants CSI-8650912 and USE-9053363) and from the Fund for Improvement for Postsecondary Education (Grant G008642079). This support, extending over several years, has been essential to completing this project.

Our families, finally, deserve our deepest thanks. They have coped cheerfully with peculiar hours, extended absences, mental distraction, blizzards of paper, missed meals, and every other vagary that such a project entails. Without their love and sacrifice, we would never have begun—let alone completed—this project.

How to Use This Book: Notes for Students

All authors want their books to be *used*: read, studied, thought about, puzzled over, reread, underlined, disputed, understood, and, ultimately, enjoyed. So do we.

That might go without saying for some books—beach novels, user manuals, field guides, etc.—but it may need repeating for a calculus textbook. We know as teachers (and remember as students) that mathematics textbooks are too often read *backwards*: faced with Exercise 231(b) on page 1638, we've all shuffled backwards through the pages in search of something similar. (Very often, moreover, our searches were rewarded.)

A calculus textbook isn't a novel. It's a peculiar hybrid of encyclopedia, dictionary, atlas, anthology, daily newspaper, shop manual, *and* novel—not exactly light reading, but essential reading nevertheless. Ideally, a calculus book should be read in *all* directions: left to right, top to bottom, back to front, and even front to back. That's a tall order. Here are some suggestions for coping with it.

Read the Narrative Each section's narrative is designed to be read from beginning to end. The examples, in particular, are supposed to illustrate ideas and make them concrete—not just serve as templates for homework exercises.

Read the Examples Examples are, if anything, more important than theorems, remarks, and other "talk." We use examples both to show already-familiar calculus ideas "in action," and to set the stage for new ideas.

Read the Pictures We're serious about the "graphical points of view" mentioned in our title. The pictures in this book are not "illustrations" or "decorations." Pictures like this one—

Local but not global maxima and minima:
$$f(x) = x/2 + \sin x$$

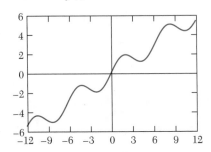

are everywhere, even in the middle of sentences. That's intentional: graphs are an important part of the language of calculus. An ability to think "pictorially"—as well as symbolically and numerically—about mathematical ideas may be the most important benefit calculus can offer.

Often they're put in margin notes, like this one.

Read with a Calculator and Pencil   This book is full of requests ◄ to check a calculation, sketch a graph, or "convince yourself" that something makes sense. Take these "requests" seriously. Mastering mathematical ideas takes more than reading; it takes doing, drawing, and thinking.

Read the Language   Mathematics is not a "natural language" like English or French, but it has its own vocabulary and usage rules. Calculus, especially, relies on careful use of technical language. Words like **rate**, **amount**, **concave**, **stationary point**, and **root** have precise, agreed-upon mathematical meanings. Understanding such words goes a long way toward understanding the mathematics they convey; misunderstanding the words leads inevitably to confusion. Whenever in doubt, consult the index.

Read the Appendices   The human appendix generally lies unnoticed—unless trouble starts, when it's taken out and thrown away. Don't treat *our* appendices that way. Though perhaps slightly enlarged, they're full of healthy matter: reviews of precalculus topics, help with "story problems," proofs of various kinds, even a graphical "atlas" of functions. Used as directed the appendices will help appreciably in digesting the material.

Read the Instructors' Preface (If You Like)   Get a jump on your teacher.

In short: *Read the book*.

A Last Note

Why study calculus at all? There are plenty of good practical and "educational" reasons: because it's good for applications; because higher mathematics requires it; because it's good mental training; because other majors require it; because jobs require it. Here's another reason to study calculus: because calculus is among our species' deepest, richest, farthest-reaching, and most beautiful intellectual achievements. We hope this book will help you see it in that spirit.

A Last Request

Last, a request. We sincerely appreciate—and take very seriously—students' opinions, suggestions, and advice on this book. We invite you to offer your advice, either through your teacher or by writing us directly. Our addresses appear below.

Arnold Ostebee and Paul Zorn
Department of Mathematics
St. Olaf College
1520 St. Olaf Avenue
Northfield, Minnesota 55057-1098

Contents

CALCULUS

From Graphical, Numerical, and Symbolic Points of View

Functions in Calculus

1.1 Functions, Calculus Style

Calculus is a branch of mathematical analysis—the study of functions. Functions describe relationships among varying quantities; calculus helps us discover and quantify these relationships. But we're getting ahead of ourselves.

Let's start from the beginning. What is a function? What kinds of functions are there? What can functions do? Informally speaking,

> *A function is a procedure* ▶ *for assigning a unique output to any acceptable input.*

I.e., a recipe.

Functions With Formulas

The most familiar functions are given by explicit algebraic formulas, such as these:

$$f(x) = x^3 - 4x + 3; \qquad g(x) = \frac{x^2 - 4}{x + 3}.$$

Functions like f and g are easy to write down, easy to plug numbers into, easy to draw graphs of, and (as we'll see later) easy to differentiate and antidifferentiate. ▶

We'll define these technical words soon.

Functions With or Without Formulas: A Sampler

Not every useful function comes equipped with, or even has, a convenient algebraic formula. Working with functions that are described in many ways—by graphs, by tables, in words, and so on—is an important theme of this course. In this section we sample some of the variety of possible functions and their uses.

EXAMPLE 1   Following are world population figures, in millions, over 20 centuries. [The data are from the Population Division, United Nations Secretariat.]

Human Population														
Year (C.E.)	0	1000	1500	1750	1800	1850	1900	1930	1940	1950	1960	1970	1980	1990
Number of people (millions)	300	310	500	790	980	1260	1650	2070	2300	2520	3020	3700	4450	5300

The table shows what we expect: Population grows faster and faster over time. *Plotting* the data shows this even more clearly:

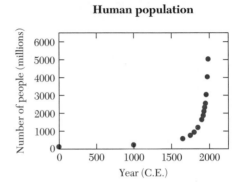

The preceding graph is a **scatter plot**, with one point per datum. Notice carefully how each point is plotted and how each axis is used.

Times are inputs; populations are outputs.

World population is certainly a function; corresponding to any time is a certain population. ◄   But does the population function have some convenient formula? An answer would be well worth having! ◄

What good would an explicit population formula be?

Later we'll describe ways of choosing mathematical formulas that "fit" observed data with reasonable accuracy. The following picture, for instance, is the graph of a mathematical function p, with formula

$$p(t) = 906e^{0.008(t-1800)}$$

superimposed on the data. The function appears to fit the data reasonably well.

Human population: fitting the data

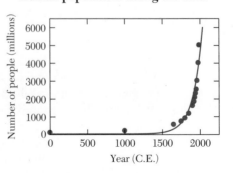

How and why we chose this particular formula for p doesn't matter here. ▶ The various constants—906, e, 0.008, 1800—involved in the formula are called **parameters**. Choosing parameter values appropriately is part of the process called **curve fitting**. ∎

We'll return to such matters at the end of this chapter.

E X A M P L E 2   For any real number t, let $l(t)$ be the outside temperature, in degrees Celsius, t hours after midnight, January 1, 1995, in Northfield, Minnesota. This rule defines a function: To an input t (a time), l assigns an output $l(t)$ (a temperature). We expect l to be continuous ▶ because temperature varies smoothly, not by leaps and bounds. However, we don't expect l to be given *exactly* by a simple algebraic formula that is valid for all time. ▶ Nevertheless, we might reasonably try to *estimate* l in some approximate or average sense. ∎

We'll define "continuous" rigorously later; here it's understood informally.

With such a formula we could predict the weather with perfect certainty.

No Formulas? Functions like l, without explicit formulas, may seem strange or artificial. In one sense the opposite is true: Most real-world functions are like the weather—messy, unpredictable, and imperfectly understood. The tame, well-understood functions we use to model natural phenomena are, from this point of view, comparatively artificial. The real surprise is how successfully the basic mathematical functions model, and even predict, real-world phenomena.

E X A M P L E 3   Information about a function n is given in the following table.

x	-3	-2	-1	0	1	2	3	4
$n(x)$	-4	-9	-1	0	1	8	3	-2

Our knowledge of n is, to say the least, incomplete. What's $n(100)$? For what inputs is n defined at all? Is the graph of n a smooth curve? Does n have some secret formula?

The table alone doesn't answer these questions. (Incomplete information is common in practice, e.g., for experimental data.) More information would help. If, say, the table records the outside temperature x hours after midnight on January 1, 1995, then (as in the preceding example) we'd expect $n(x)$ to be continuous and to accept noninteger inputs. To "model" the situation, we might look for an explicit function that agrees with the data in the table. The complicated-looking polynomial function

$$p(x) = -\frac{152}{105}x + \frac{5}{72}x^2 + \frac{683}{240}x^3 - \frac{11}{144}x^4 - \frac{33}{80}x^5 + \frac{1}{144}x^6 + \frac{1}{70}x^7$$

is one possibility; it happens to produce the values in the table. There's no guarantee, however, that $p(x)$ gives correct temperature information for any other values of x. (Experimentation quickly shows that $p(x)$ can't possibly tell temperature for all time. For instance, $p(-10) = -98{,}251$—an unlikely temperature even in Minnesota winter.) ∎

EXAMPLE 4   Let m be the function shown graphically as follows.

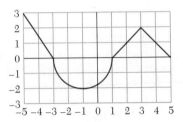

Some values of m, such as $m(-1) = -2$, are clear even without a formula. Other values of m are less clear; $m(-2)$, for example, is *about* -1.7, but the exact value cannot be determined from the graph.

As it happens, m can be represented algebraically, by a **piecewise-defined formula**. Because its graph consists of three line segments and a semicircle, the pieces are as follows: ◄

Check our work.

$$m(x) = \begin{cases} -\dfrac{3x}{2} - \dfrac{9}{2} & \text{if} \quad -5 \le x \le -3; \\ -\sqrt{4 - (x+1)^2} & \text{if} \quad -3 < x \le 1; \\ x - 1 & \text{if} \quad 1 < x \le 3; \\ -x + 5 & \text{if} \quad 3 < x \le 5. \end{cases}$$

With the formula we can now find $m(-2)$ exactly:

$$m(-2) = -\sqrt{4 - (-2+1)^2} = -\sqrt{3} \approx -1.732050808.$$

■

Patched-together Functions. Functions like m, defined "in pieces," may seem strange, but they are very useful in practice. Computer-drawn graphs, for example, are usually made up of many short line segments, or *arcs*—each of which can be described algebraically. The process of **splining** (joining properly chosen curves or line segments to form a single smooth curve) is based on elementary calculus ideas. Splines are used in computer-assisted design, engineering, and other applications.

The horizontal axis indicates time, not distance, so the graph is not a "picture" of the balloon's flight. Or could it be?

EXAMPLE 5   Hot days on the Kansas prairie generate strong updrafts and downdrafts. In the gondola of a hot air balloon, the altimeter needle swings. The function A (its graph follows) describes the altimeter readings over a certain period of time. Note the units: Time t is measured in minutes and altitude $A(t)$ in *hundreds* of feet above sea level. (East-central Kansas lies about 960 feet above sea level.) ◄

A balloon over the prairie: How high?

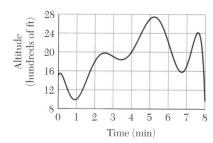

Reading Velocity from the Altitude Graph.   The graph of A tells a lot, directly, about altitude—when the balloon was highest, when it rose, when it fell, and so on. Consider another, equally interesting (for the passengers, probably more interesting) question:

How fast was the balloon rising or falling at time t?

In other words, how does the balloon's **vertical velocity** $V(t)$ vary with time? What can the altitude graph tell us about the vertical velocity? A lot:

Up or Down? The sign of $V(t)$ tells whether the balloon is rising or falling at time t. Altitude increases as the balloon rises, so $V(t) > 0$ when the balloon is rising (e.g., in the interval $4 < t < 5$), and $V(t) < 0$ when the balloon is falling (e.g., if $2.5 < t < 3.5$).

How Fast? At $t = 1.5$, altitude increases quickly relative to time; the balloon is soaring upward. By $t = 2.4$, the balloon has leveled off. Some ups and downs follow, but at $t = 7.8$ things look bad. Altitude is plunging, and so is the balloon.

Estimating Velocity The vertical velocity $V(t)$ clearly depends on how steeply the A-graph rises or falls at time t. At $t = 6$, for instance, the A-graph is dropping, but how fast? How, in other words, might we estimate a value for $V(6)$? One approach is to focus on a *small* interval containing $t = 6$, such as $5.5 \le t \le 6.5$. A close look at the graph shows that

$$A(5.5) \approx 27 \quad \text{and} \quad A(6.5) \approx 17 .$$

Therefore, over the 1-minute interval $5.5 \le t \le 6.5$, the balloon drops about 10 vertical units, or 1000 feet, and the **average velocity** over this interval is around -1000 feet per minute. Since the graph looks nearly straight over this short interval, the velocity at $t = 6$ should not be much different.

At the instant $t = 6$, the velocity is about -1000 feet per minute. In symbols: $V(6) \approx -10$. ▶

The unit of velocity is hundreds of feet per minute.

What Slope Means The vertical velocity $V(t)$ has an important graphical meaning: At any time t, $V(t)$ is the slope of the A-graph at time t. ▶ We just estimated that $V(6) \approx -10$. The graphical version of the same strategy runs like this: On a small interval near $t = 6$, the graph is almost straight. To estimate slope, therefore, any two nearby points should do.

Mathematical legalities arise in defining the slope of a curved graph. We'll consider them carefully later; here we can trust our intuition.

The line through $(5.5, 27)$ and $(6.5, 17)$, for instance, has slope

$$\frac{17 - 27}{6.5 - 5.5} = -10,$$

which agrees with our earlier estimate.

Plotting V   We can amass enough information to plot V by estimating the velocity $V(t)$ for *many* values of t. Doing so by hand would be tedious; we used a computer.

A balloon over the prairie: How fast?

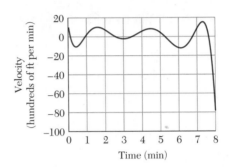

There's much to see in the new graph, but we'll limit ourselves to a final observation: At $t = 8$, things look bleak. ∎

Translation: Functions from Words

Mathematics is a language. Because real-world problems usually come phrased in "natural" languages—say French or English—solving such problems must start with translation into the special vocabulary and grammar of mathematics. From paragraphs of ordinary prose come functions, variables, equations, operations, and the like.

Translation from English to French is not mere word-for-word replacement. An excellent translation respects French grammar, syntax, and stylistic conventions as well as vocabulary. A well-translated text captures the essence of the original.

Natural languages are suited to expressing nuance, emotional tone, and even ambiguity. Mathematical language is more straightforward. It emphasizes simplicity and precision, even in matters of depth and subtlety. Nevertheless, the point just made about translation applies even more strongly to mathematical language: To translate is to express the *essence*, not merely the form, of the situation. Distilling the essence of a problem and expressing it mathematically are the first—and often the largest—steps toward a solution.

Calculus: The Language of Change.   The areas of mathematics that best describe and predict distinct sorts of phenomena differ. Phenomena of change are the special province of calculus. Calculus offers both language with which to *describe* changing quantities and rules with which to *predict* their behavior. Functions are the basic objects of calculus, so in this subject translation usually means translation into functional language.

EXAMPLE 6   To make a flimsy box from an ordinary $8.5'' \times 11''$ sheet of paper, cut squares from the corners and fold up the sides, as shown. (Cut on the heavy lines and fold on the dashed lines.) Which such box has the largest volume?

Boxmaking

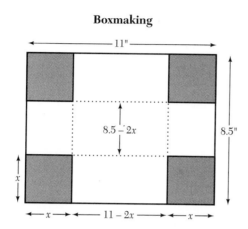

Solution   The volume V of the box depends only on the edge length—denoted by x—of the removed corners. In other words, V is a *function* of x. For any given x, V is easy to calculate algebraically:

$$V(x) = \text{length} \times \text{width} \times \text{depth} = (11 - 2x)(8.5 - 2x)x.$$

Notice that the formula makes sense in context only if all three factors are nonnegative—i.e., if $0 \le x \le 4.25$. In functional language, our problem is to find, among all feasible inputs x, the one that produces the largest possible output $V(x)$.

A graph ▶ shows how V varies with x:

Notice our choice of horizontal range.

Graph of $V(x) = (11 - 2x)(8.5 - 2x)x$: how volume varies

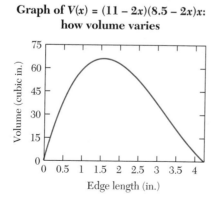

The maximum possible volume, apparently, is about 65 cubic inches; it occurs when $x \approx 1.6$ inches, i.e., for a box with dimensions $7.8'' \times 5.3'' \times 1.6''$.

By "zooming in" on the graph near $x = 1.6$, we can improve our estimates:

**Maximizing $V(x) = (11 - 2x)(8.5 - 2x)x$:
a closer look**

This sharper point of view suggests a maximum volume of about 66.1 cubic inches, achieved when $x \approx 1.59$ inches. (The dimensions then become $7.82'' \times 5.32'' \times 1.59''$.) With calculus methods, we can (and will) do better still. ◄

Just a bit better; for many practical purposes the results just found would be adequate.

New Functions from Old

Starting with "old" functions and building new ones is common throughout calculus. Here's a first example.

EXAMPLE 7   Start with the straight-line function $\ell(x) = x + 2$. Define a new function D, based on ℓ, as follows:

For any number x, let $D(x)$ be the distance from the origin $(0, 0)$ to the point $\big(x, \ell(x)\big)$ on the line ℓ.

Find a formula for $D(x)$; draw the graph.

Solution   Points on ℓ have the form $\big(x, \ell(x)\big) = (x, x + 2)$. By the distance formula for points in the plane, ◄ $D(x)$ has the formula

Appendix A has details on the distance formula.

$$D(x) = \sqrt{(x - 0)^2 + (x + 2 - 0)^2} = \sqrt{x^2 + x^2 + 4x + 4} = \sqrt{2x^2 + 4x + 4}.$$

Plotting $y = D(x)$ produces this graph:

Graph of $y = D(x)$

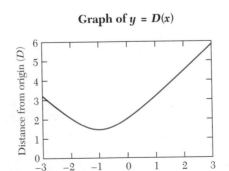

Notice the low spot on the graph of D; it seems to fall somewhere near $x = -1$. This low spot corresponds to the point on ℓ closest to the origin. ▶

■ *See the exercises at the end of this section for more details.*

Functions Approximating Other Functions

Sometimes it's useful to approximate one function with another.

E X A M P L E 8   (**A polynomial approximation to the square root function**)   The polynomial function

$$p(x) = \frac{5}{16} + \frac{15}{16}x - \frac{5}{16}x^2 + \frac{1}{16}x^3$$

turns out to approximate the square root function $f(x) = \sqrt{x}$ closely near $x = 1$. Graphs show what this means:

Approximating $\sqrt{x}$ with a polynomial

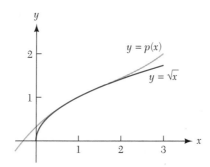

Near $x = 1$ the graphs are almost indistinguishable; elsewhere they pull apart. ▶ For example,

For $x < 0$, $f(x)$ is not even defined, so approximation is no longer an issue.

$$f(1.5) = \sqrt{1.5} \approx 1.22474 \qquad \text{and} \qquad p(1.5) \approx 1.22656$$

but

$$f(3) = \sqrt{3} \approx 1.73205 \qquad \text{and} \qquad p(3) = 2.$$

How we *chose* the polynomial p to approximate f must remain (for now) mysterious. Later in this book we'll learn to construct polynomials that approximate a wide variety of nonpolynomial functions. ■

> **Modern Times.**   Years ago students learned paper-and-pencil techniques for computing square roots. Today we use calculators. Still the question remains: How does a calculator compute roots, and to what accuracy? As an experiment, try to find a systematic way of estimating square roots with a calculator without using the square root key.

BASICS

1. (a) What does the approximate population function p given in Example 1 (page 2) predict for the year 2000?
   (b) What does it "predict" for the year 1000?
   (c) What does it "predict" for 1000 B.C.E.? (B.C.E, formerly known as B.C., means "before the Common Era.")
   (d) Are these predictions reasonable? Explain.

2. Use the algebraic representation of m in Example 4 (page 4) to compute the following:
   (a) $m(-4)$
   (b) $m(0)$
   (c) $m(2.3)$
   (d) $m(\pi)$

3. Find a piecewise-defined formula for the function whose graph follows.

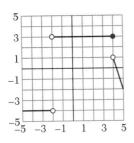

4. Let $\ell(x) = 2x + 1$.
   (a) Draw a graph of $\ell(x)$ for $0 \le x \le 5$.
   (b) Let $D(x)$ be the distance from the origin to the point $(x, \ell(x))$. Find a formula for $D(x)$.
   (c) Sketch a graph of $D(x)$ for $0 \le x \le 5$.

5. For any $t \ge 0$, let $A(t)$ be the area of the rectangle bounded above by the line $y = 3$, below by the x-axis, on the left by the y-axis, and on the right by the line $x = t$.
   (a) Write a formula for $A(t)$.
   (b) Draw a graph of $A(t)$ for $0 \le t \le 5$.

6. Let f be the function defined by $f(x) = x^2$. Define the function m as follows: For any $x \ne 0$, $m(x)$ is the slope of the line through $(0, 0)$ and $(x, f(x))$. For instance, $m(10) = 10$, the slope of the line through $(0, 0)$ and $(10, 100)$.
   (a) Find $m(-1)$, $m(-2)$, $m(1)$, $m(2)$, and $m(500)$.
   (b) Write a formula for $m(x)$ in terms of x.
   (c) Draw a graph of $m(x)$ for $-5 \le x \le 5$.

FURTHER EXERCISES

7. (a) At what time was the balloon in Example 5 (page 4) rising fastest? About how fast was it rising then?
   (b) At what time was the balloon in Example 5 falling fastest? About how fast was it falling then?
   (c) What happens at time $t = 8$?

8. Another possible way of depicting a balloon's flight graphically is to plot altitude vs. *horizontal position* instead of altitude vs. time. Since westerly (i.e., west-to-east) winds prevail on the American prairie, a balloon above Kansas often travels (ignoring vertical movement) due east. When this is so, it's natural to plot altitude against horizontal position, measuring the latter in miles east of a starting point. The resulting plot can be thought of as a "picture" of the balloon's flight.

   In this exercise we consider the relationship, if any, between these two types of graphs: altitude vs. time, as used in Example 5, and altitude vs. position, just discussed. In particular, can the two types of graphs possibly look the same?
   (a) Suppose that a *steady*, 60-mph westerly wind blows. Draw an altitude vs. position graph. How are the two graphs related?
   (b) Suppose that no horizontal wind blows. Draw a "picture" of the balloon's flight (i.e., an altitude-position graph). Do both types of graphs make sense?
   (c) What if an easterly wind blows? Draw a possible altitude-position graph. Does it look like an altitude-time graph?

9. When you drive an Ace Rental compact car x miles in a day, the company charges $f(x)$ dollars, where

   $$f(x) = \begin{cases} 30 & \text{if } 0 \le x \le 100; \\ 30 + 0.07(x - 100) & \text{if } x > 100. \end{cases}$$

   Describe Ace Rental's pricing policy in plain English. (Be sure to interpret the constants 30, 0.07, and 100 that appear in the pricing formula.)

Find a piecewise-defined formula for each function in Exercises 10 and 11; graphs are shown below. [**NOTE**: Each graph is made up of line segments and circular arcs.]

10.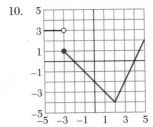

11.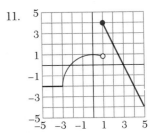

12. This problem refers to the functions ℓ and D in Example 7 (page 8).

    (a) Use a ruler to plot the straight-line function $\ell(x) = x + 2$ over the interval $-3 \le x \le 3$. Label the points $\big(3, \ell(3)\big)$ and $\big(-3, \ell(-3)\big)$ on your graph. (Use units of the same size—e.g., 1 cm—on both axes.)

    (b) For any x, $D(x)$ is the distance from the origin to the point $\big(x, \ell(x)\big)$. Use your graph and a ruler to estimate $D(3)$ and $D(-3)$. Then, for comparison, calculate $D(3)$ and $D(-3)$ using the formula.

    (c) Which point P on the line ℓ appears to be closest to the origin? Mark this point P with both its coordinates. How far is P from the origin? Relate your answer to the low point on the graph in Example 7.

    (d) We can use algebra to find point P exactly. Here's how. Completing the square under the radical sign gives $D(x) = \sqrt{2x^2 + 4x + 4} = \sqrt{2(x+1)^2 + 2}$. Use this fact to explain why $D(-1) = \sqrt{2}$ is the smallest possible value of D.

13. For any $t \ge 0$, let $A(t)$ be the area of the right triangle bounded above by the line $y = x$, below by the x-axis, and on the right by the line $x = t$.

    (a) Write a formula for $A(t)$.

    (b) Draw a graph of $A(t)$ for $0 \le t \le 5$.

14. Let $\ell(x) = 2x + 1$, and let $A(t)$ be the area of the trapezoid bounded by the x-axis, the y-axis, ℓ, and the line $x = t$.

    (a) Write a formula for $A(t)$.

    (b) Draw a graph of $A(t)$ for $0 \le t \le 5$.

15. Let f be the function defined by $f(x) = x^2$. Define the function g by this rule: $g(x)$ is the distance from $(0, 0)$ to $\big(x, f(x)\big)$. For instance, $g(10) = \sqrt{10{,}100}$, the distance from $(0, 0)$ to $(10, 100)$.

    (a) Find $g(0)$, $g(-1)$, $g(1)$, $g(2)$, and $g(500)$.

    (b) Write a formula for $g(x)$.

    (c) Draw a graph of $g(x)$ for $-5 \le x \le 5$.

16. Let f be the function defined by $f(x) = x^2$. Define the function j as follows: For any $x \ne 1$, $j(x)$ is the slope of the secant line through $(1, 1)$ and $\big(x, f(x)\big)$. For instance, $j(10) = 11$, the slope of the line through $(1, 1)$ and $(10, 100)$.

    (a) Find $j(2)$, $j(1.1)$, $j(1.01)$, $j(0.99)$, $j(0.9)$, and $j(0)$.

    (b) Write a formula for $j(x)$.

    (c) Draw a graph of $j(x)$ for $-2 \le x \le 2$.

17. Let f be the function defined by $f(x) = x^2$. Define the function k for any $x \ne 0$ by

$$k(x) = \frac{f(2 + x) - f(2 - x)}{2x}.$$

For instance,

$$k(0.1) = \frac{f(2.1) - f(1.9)}{0.2} = \frac{2.1^2 - 1.9^2}{0.2} = 4.$$

    (a) Find $k(-1)$, $k(-0.1)$, $k(0.1)$, $k(1)$, and $k(1000)$.

    (b) Write a formula for $k(x)$.

    (c) Draw a graph of $k(x)$ for $-5 \le x \le 5$.

18. Let f be the function $f(x) = x^2 + 3x - 4$ and ℓ the line that intersects the graph of f at the points $x = 2$ and $x = 2 + h$.

    (a) Find an equation of the line that intersects f at $x = -1$ and $x = 2$.

    (b) Find an algebraic expression for the slope of ℓ. [**NOTE:** The answer depends on h.]

    (c) Use your answer to part (b) to find an equation of the line ℓ.

    (d) Use your answer to part (c) to find an equation of the line that intersects f at $x = 2$ and $x = 2.003$.

    (e) Use your answer to part (c) to find an equation of the line that intersects the graph of f at $x = 2$ and $x = -1$.

19. Example 8 (page 9) describes one way to approximate the square root function, using a polynomial function. This problem is about a **piecewise-linear** approximation (i.e., a piecewise-defined function built from straight lines) of the square root function.

    (a) Draw a graph of $f(x) = \sqrt{x}$ for $0 \le x \le 9$. Mark the points $(0, 0)$, $(1, 1)$, $(4, 2)$, and $(9, 3)$ on your graph, then connect them with three line segments.

    (b) Let $g(x)$ be the function whose graph consists of the three line segments in part (a). Find a piecewise-defined formula for g.

    (c) Fill in the following table (round your answers to three decimal places).

x	0	0.5	0.9	3	4	6	8	8.9	8.99
$g(x)$									
$\sqrt{x}$									

20. Here's a highly simplified version of the U.S. federal income tax system. A taxpayer pays no federal tax on the first \$15,000 of annual income, 15% on the the next \$25,000 of income, and 28% on all earnings above \$40,000. Let x denote a taxpayer's annual income (in dollars), and let $T(x)$ denote the tax owed (in dollars) at income level x.

    (a) The tax function T is piecewise-linear—its graph consists of straight lines. Use this fact to draw the graph of $T(x)$ for $0 \le x \le 60{,}000$. [**HINT:** First calculate $T(x)$ for some convenient values of x.]

    (b) Give a piecewise-defined formula for $T(x)$.

    (c) A taxpayer earning, say, \$50,123 yearly would have to pay 28 cents more tax if he or she earned *one more dollar*. A taxpayer earning \$20,123 a year would need to pay only 15 cents tax on the next dollar of income.

In tax jargon, the first taxpayer has a **marginal tax rate** of 28%, and the second has a **marginal tax rate** of 15%. Let $MR(x)$ denote the marginal tax rate (as a decimal) at income level x. Give a piecewise-defined formula for $MR(x)$.

(d) Plot MR as a function of x for $0 \leq x \leq 60,000$. Describe in words how the graphs of T and MR are related.

21. (Exercise 20 continued.) Let $F(x)$ denote the fraction (written as a decimal) of a taxpayer's total annual income paid in federal tax. For instance, a taxpayer with income $40,000 pays $3750, so $F(40,000) = 3750/40,000 = 0.09375$—just over 9%.

(a) For all $x > 0$, $F(x) = T(x)/x$. Explain why.

(b) Write a piecewise-defined formula for $F(x)$. [**HINT**: Start with the piecewise-defined formula for $T(x)$.]

(c) Plot $F(x)$ for $0 < x \leq 60,000$. [**HINT**: A calculator may help you plot the separate pieces.]

(d) Could any taxpayer pay 25% of total income in taxes? Why or why not?

(e) Could any taxpayer pay 28% of total income in taxes? Why or why not?

22. An open box is made by cutting squares w inches on a side from the four corners of a sheet of cardboard that is 24 inches by 32 inches, and then folding up the sides.

(a) Express the volume of the box as a function of w. [**HINT**: Draw a picture.]

(b) Estimate the value of w that maximizes the volume of the box.

23. Coffee mugs can be manufactured for 60 cents each. At a price of 1 dollar each, 1000 mugs can be sold. For each penny by which the price is lowered, 50 more mugs can be sold. Let N be the number of coffee mugs manufactured (and sold!), and let x be the price reduction (in pennies).

(a) Explain why $N(x) = 1000 + 50x$.

(b) Explain why the profit on each mug sold is $(1 - 0.01x) - 0.60$ dollars.

(c) Find a function that relates total profit and x.

(d) What's the *best* price? [**HINT**: Graph the function from part (c).]

1.2 Graphs

Every equation in x and y has a **graph**—the set of ordered pairs (x, y) of real numbers that satisfy the equation. For example, the graph of any equation of the form

$$(x - a)^2 + (y - b)^2 = r^2$$

is a circle (with center (a, b) and radius r). Associating an equation (an algebraic object) to a graph (a geometric object) is a powerful tool for understanding both.

Graphs can be simple, complicated, or downright bizarre. Most graphs in calculus are of the everyday, $y = f(x)$ variety; almost every page of this book contains an example. Following, for a change, are some odder graphs.

EXAMPLE 1   Following are the graphs of three equations: $x^2 + y^2 = -1$, $x^2 + y^2 = 0$, and $x^2 + y^2 = 1$.

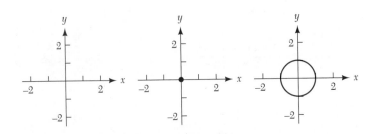

Is there some mistake? No; *no pair* (x, y) of real numbers satisfies the first equation, so the first graph is empty. Only the origin $(0, 0)$ satisfies the second equation, so its graph is a single point. The third graph plays no tricks. ■

Graphs of Functions

The function $f(x) = x^2$ corresponds in the obvious way to the equation $y = x^2$ and hence to the familiar parabolic curve. In the same way, every function f corresponds to an equation $y = f(x)$ and therefore to a graph. ▶

Not every equation (and therefore not every graph) corresponds to a function. Circles are obvious counterexamples: On a circle, most values of x correspond to two values of y.

> **Definition** The graph of a function f is the set of points (x, y) that satisfy the equation $y = f(x)$.

In other words:

The graph of a function is the set of all points of the form $(x, f(x))$.

The generic picture looks like this:

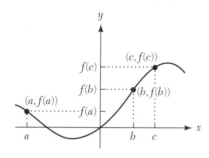

The graph of a function shows geometrically how the function varies over a range of inputs. Because calculus concerns precisely this question—how quantities vary with respect to each other—graphs play crucial roles throughout the subject.

Drawing the graph of a function given by an explicit formula, such as $p(x) = x^3 - 2x^2 + x - 2$, ▶ is an old, standard, and still important problem in elementary calculus. But other problems are just as important, especially now that machines draw graphs easily and accurately. ▶ Most important is to understand clearly what graphs say about functions and how they say it.

A graph of this function follows.

Not every machine-drawn graph is accurate. More on this subject in the next section.

Drawing graphs is sometimes useful, but more interesting is the connection between analytic properties of a function (e.g., whether it's increasing or decreasing) and geometric properties of its graph (e.g., whether it's rising or falling). Calculus tools unlock many such secrets.

Graphs—With and Without Formulas

Too much time and effort spent drawing graphs of "nice" functions with "nice" formulas can obscure an important fact:

Not every function has a nice formula—or any formula at all.

In real life, convenient formulas are the exception. In a sense, having a formula amounts to knowing *everything* about a function. In this imperfect world, little is that certain.

EXAMPLE 2   **(Function, formula, and graph)**   In the best possible circumstances we are given a function, its graph, and an explicit algebraic formula. Here, for instance, is a view of the graph of $p(x) = x^3 - 2x^2 + x - 2$. ◄

Technically speaking, we see only part of the graph. The full graph is an infinite object.

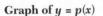

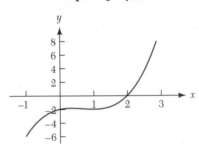

Graph of $y = p(x)$

Observe:

A y-intercept   The graph intercepts the y-axis at height -2. In symbols, $p(0) = -2$.

No breaks   The graph shown is free of "jumps" or "breaks"; it could be drawn manually, without lifting the pencil. In mathematical language, the function p is **continuous** on the domain shown. ◄

Later, we'll define continuity carefully.

An x-intercept   The graph's x-intercept is at $x = 2$, so $p(2) = 0$. Factoring the formula shows the same thing:

$$p(x) = x^3 - 2x^2 + x - 2 = (x - 2)(x^2 + 1).$$

(The factored form tells us something more: $p(x) = 0$ *only* if $x = 2$.)

The graph rises as we trace it from left to right.

Where Is p Increasing?   The graph of p rises from $x = 1$ to $x = 3$. ◄   In calculus language, the function p is said to be **increasing** for $1 < x < 3$. The picture ends at $x = 3$, so we can't tell from the graph alone what happens for larger values of x. It can be seen from the factored algebraic form of p, however, that $p(x)$ increases for *all* $x > 1$. (Soon we'll develop calculus methods to show the same thing more conveniently.)   ■

EXAMPLE 3   **(A bare graph: no formula, no context)**   Following is a portion of the graph of a garden-variety calculus function f. No formula for f is given. Two interesting points are labeled.

Graph of y = f(x)

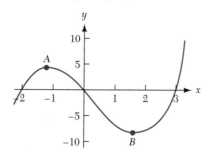

What can the graph tell us about the function f? What *can't* it say? Some important samples follow. Notice especially the words that appear in boldface; they'll reappear often.

- Because the point $(0, 0)$ lies on the graph, $f(0) = 0$. (Strictly speaking, we should say $f(0) \approx 0$, since graphical data are always approximate. Here we'll trust our eyes.)

- Judging from the graph, $f(1) \approx -6$. To know whether $f(1) = -6$ *exactly*, we would need more information.

- What's $f(10)$? The graph doesn't say. We might *guess* from the graph's general shape that $f(10)$ is some large positive number. Yet, from the graph alone, we can't be certain even that $f(10)$ exists.

- The points A and B correspond to **local maximum** and **local minimum values** ▶ of the function f. The coordinates of A—approximately $(-1.2, 4)$— imply that $f(-1.2) \approx 4$. Similarly, B's coordinates imply that $f(1.8) \approx -8$. Putting these data together, we would say (allowing for approximation errors) that f is *increasing* for $-2.5 < x < -1.2$, *decreasing* for $-1.2 < x < 1.8$, and *increasing* again for $1.8 < x < 3.5$.

  Informally, these are high and low spots on the graph. We'll define these terms formally in the next chapter.

- For which values of x is $f(x) < -5$? One approach is to draw the horizontal line $y = -5$. ▶ This line hits the graph of f at three different points—near $x = -2.4$, $x = 0.7$, and $x = 2.5$. Therefore, we would estimate that $f(x) < -5$ if either $x < -2.4$ or $0.7 < x < 2.5$.

  Do so now, with a ruler.

  Two cautions are necessary. First, every number mentioned is only approximate. Second, we have no idea how $f(x)$ behaves for x outside the interval shown.

- Later we'll define and study the slightly subtler idea of a graph's **concavity**. Concavity is a geometric property. The bowl of a spoon (in its usual position) is **concave up**; an umbrella (on a rainy day) is **concave down**. ▶ Clearly, the f-graph is concave down at point A and concave up at B. Somewhere between A and B, therefore, must be an **inflection point**, ▶ where the direction of concavity switches. Where does this happen?

  As rough synonyms, try concave up for "holds water" and concave down for "spills water."

  A close look shows that concavity changes somewhere between $x = 0$ and $x = 1$, but it's hard to be more precise. Soon, using calculus ideas, we'll see how to use the formula for p to locate inflection points precisely. ■

  Another term we'll define more precisely later.

EXAMPLE 4   **(Graphs in context)**   [This example was drawn from *Calculus in Context*, a project of the Five Colleges consortium.]   In certain disease epidemics, the affected population divides naturally into three groups: those *susceptible*, those *infected*, and those *recovered*. The epidemic's progress can be charted by observing how these populations vary over time. The following three graphs, labeled *S*, *I*, and *R*, show how the susceptible, infected, and recovered populations might vary over a 25-day period during a measles epidemic in a large school district.

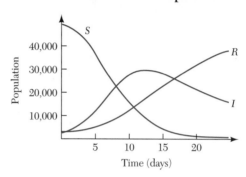

The course of an epidemic

Try to answer them for yourself before reading the answers.

With these graphs, we can answer many questions about the epidemic. ◄

1. *How many children were in each population on day 10?*

   Reading the height of each graph on day 10 shows (allowing for estimation errors):

   $$\text{Susceptible} \approx 13,000;$$
   $$\text{Infected} \approx 26,000;$$
   $$\text{Recovered} \approx 11,000.$$

2. *On which day is the infected population largest?*

   The *I*-graph, and hence the infected population, peaks around day 12. (In calculus language, *I* assumes a local maximum value at day 12.)

3. *When is the epidemic at its worst?*

   The answer depends on how we measure "badness." The number *I* of ill people is one reasonable measure. By this criterion the worst day is around day 12.

   Another possible measure of "badness" is the number of people who fall ill on a given day. In the situation at hand, to fall ill means to leave the susceptible population. From this point of view, the worst day occurs when the susceptible population drops most rapidly. A careful look at the *S*-graph suggests that this happens around day 7, where the graph points most steeply downward. (In calculus language, the day in question corresponds to an inflection point of the *S*-graph.)   ∎

EXAMPLE 5 **(Graphs from tabular data)** Suppose we know only what
this table tells about a function n: ▶

*The entries should look
familiar—see Section 1.1.*

x	-3	-2	-1	0	1	2	3	4
$n(x)$	-4	-9	-1	0	1	8	3	-2

How should we plot n? One problem is that many different functions agree with
n (i.e., have the same values) for $x = -3, -2, \ldots, 4$. In graphical language, many
different possible graphs pass through the eight data points.

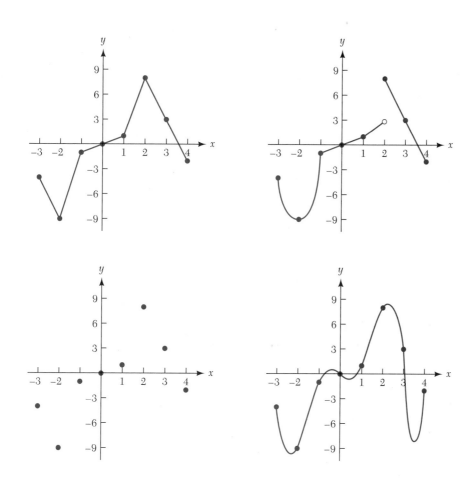

Is one of these graphs "right"? Without more information, we can't say. If we knew,
say, that the function n represented outdoor temperatures, hour by hour, on a cold
winter night, we might choose one as being more likely than the others. ▶

*Which one would you
choose?*

Notice, in particular, the **discontinuity** in the graph at upper right. (The hollow
dot indicates a "jump" at $x = 2$.) Physical experience says that air temperature varies
continuously over time, so this graph seems unlikely to tell temperature. ∎

New Functions from Old: Operations with Constants

Creating new functions—hence new graphs—from old is a recurring theme in calculus. There are many ways to "operate on" old functions to form new ones. The simplest such operations involve only **constant parameters** attached in various ways to an original function. From the ordinary sine function, for instance, we can produce mutants such as

$$m(x) = 42 + 31 \sin\left(\frac{2\pi}{52}x\right).$$

In Section 1.6 we'll discuss a real-life application of this function. Our present goal is more general: to understand how constant parameters affect *all* functions and their graphs.

Constant Parameters

Given an "old" function f and a constant a, we can create such new functions as

$$f(x) + a, \qquad f(x + a), \qquad f(ax), \qquad \text{and} \qquad af(x).$$

How do these new functions differ from plain old f? How is the particular value of a involved?

Additive Constants: Vertical and Horizontal Translations

Given a function f and a constant a, consider the new functions g and h defined by

$$g(x) = f(x) + a; \qquad h(x) = f(x + a).$$

If $a = 0$, then $f = g = h$.

Read this sentence again; it's important.

In each definition, a is an **additive constant**. ◄   Though similar in appearance, g and h are quite different: We form g from f by adding a to *outputs* of f; we form h by adding a to *inputs* of f. ◄

Here are graphs of g for several values of a:

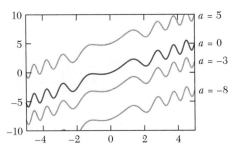

Vertical translations:
$y = f(x) + a$ **for various values of a**

The graphs show how the value of a affects g:

> *The graph of $y = f(x) + a$ is the result of shifting the f-graph a units upward.* ▶

If $a < 0$, the shift is downward.

In mathematics, moving an object without changing its shape is called **translation**; thus, a's effect is a **vertical translation**.

The relationship between f and h is different. Compare these graphs:

Horizontal translations:
$y = f(x + a)$ **for various values of a**

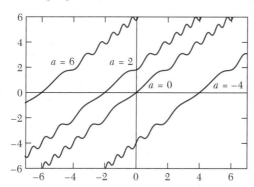

This time a causes a **horizontal translation**:

> *The graph of $y = f(x + a)$ is the result of shifting the f-graph a units to the left.*

Left, Not Right? Yes, Left. It might be surprising that the a in $f(x + a)$ moves the graph of f left, not right. To see why, consider the function $h(x) = f(x + 2)$. Then $h(-2) = f(0)$, $h(-1) = f(1)$, $h(0) = f(2)$, and so on. In effect, h has a two-unit head start on f: Whatever h does, f does two units later.

Multiplicative Constants: Stretching, Compressing, and Reflecting

Multiplicative parameters also make new functions from old. As before, we start with a function f and a constant a, and form two new functions:

$$g(x) = af(x); \qquad h(x) = f(ax).$$

Notice that g results from multiplying f's *outputs* by a; h comes from multiplying f's *inputs* by a. ▶

If $a = 1$, then $f = g = h$.

Vertical Stretching and Reflection.   Consider these graphs carefully:

Vertical stretching:
$y = af(x)$ **for various values of** a

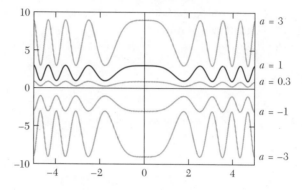

The graphs show the effect of a on $g(x) = af(x)$: Multiplying outputs from f by a causes **vertical stretching** by a factor of $|a|$ and, if $a < 0$, **reflection** about the x-axis.

Stretching the Language?   Our use of such terms as "stretch," "compress," and "reflect" deserves some comment.

- Stretching by a factor $a > 1$ is simplest. The graph of $y = 3f(x)$ (labeled $a = 3$ in the preceding figure), for instance, results from *tripling* the y-coordinate of each point on the f-graph.
- If $0 \leq a < 1$ (the case $a = 0.3$ is shown) stretching by a factor of a actually *decreases* vertical distances, flattening the f-graph toward the x-axis. In such situations we'll sometimes speak of **compressing** graphs.
- A negative multiplicative factor adds a reflection (about the x-axis) to the picture. The picture shows what happens for $a = -1$ and $a = -3$. In the simpler case, the graphs of f and $-f$ are mirror images. The graph of $-3f$, by comparison, requires *both* a reflection and a vertical stretch, with factor 3, starting from f.

Horizontal Stretching and Reflection.   The function $h(x) = f(ax)$ is formed by multiplying f's inputs by a. We just saw that multiplying *outputs* by a stretches the f-graph vertically. For similar reasons, multiplying *inputs* by a stretches or compresses the graph horizontally. The following graphs show h for several values of a.

$y = f(x); a = 1$ $y = f(0.5x); a = 0.5$

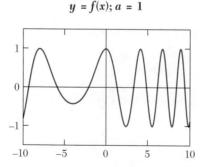
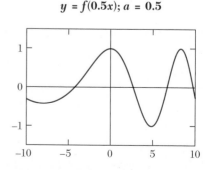

$$y = f(2x); \ a = 2 \qquad\qquad y = f(-x); \ a = -1$$

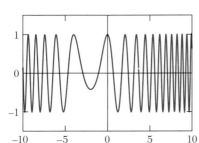 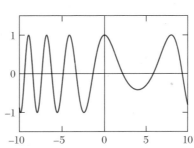

Notice the similarities to—but also the differences from—vertical stretching:

- The graphs of $f(x)$, $f(2x)$, and $f(0.5x)$ show that if $a > 0$, the graph of $f(ax)$ is the result of compressing the f-graph horizontally by a factor of a (or, alternatively, stretching it by a factor of $1/a$). ▶

- A negative value of a causes a reflection—this time, about the y-axis, The graph of $f(-2x)$, for instance, results from both compressing by a factor of 2 and reflecting about the y-axis.

Did you expect things to go the other way? If so, try this: If x measures time, then $f(x)$ is a quantity that varies with time. From this point of view, the parameter $a = 2$ in $h(x) = f(2x)$ makes $h(x)$ vary twice as fast. Anything f does, h does in half the time.

BASICS

1. The graph of a function f follows. Use the graph to complete the following table. (Estimate as necessary.)

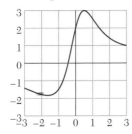

	x	$f(x)$
(a)	-2	
(b)	0	
(c)	1	
(d)		-1
(e)		2

2. Use the following sketch to answer each question.

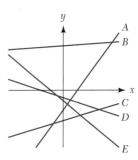

(a) Which lines have positive slope? Which have negative slope?

(b) Let m_A be the slope of line A, m_B be the slope of line B, and so on. Rank the numbers $m_A, \ldots, m_E$ in increasing order.

3. The graph of a function f follows. Several points of interest are labeled.

Graph of f

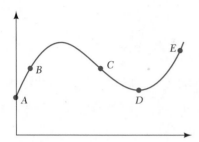

(a) Between which pairs of adjacent labeled points is f increasing? Decreasing?

(b) At which labeled point does f have an inflection point?

(c) Between which pairs of adjacent labeled points is f concave up?

(d) At which labeled point does f have a local minimum value?

(e) Between which pair of adjacent labeled points does f have a local maximum value?

Use the following graph of the function g to sketch a graph of each of the functions in Exercises 4–11.

Graph of g

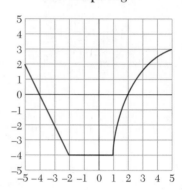

4. $f(x) = g(x) + 1$

5. $f(x) = g(x + 1)$

6. $f(x) = g(2x)$

7. $f(x) = 2g(x)$

8. $f(x) = -g(x)$

9. $f(x) = g(-x)$

10. $f(x) = -2g(-x)$

11. $f(x) = 3g(x - 2) + 1$

FURTHER EXERCISES

12. Let (x_1, y_1) and (x_2, y_2) be points on the line $y = -2(x - 1) + 3$. If $x_1 < x_2$, what can be said about y_1 and y_2? How do you know?

13. Let (x_1, y_1) and (x_2, y_2) be points on the line $y = 3(x + 1) - 2$. If $y_1 < y_2$, what can be said about x_1 and x_2? How do you know?

14. Let L_1 be the line through $(0, 0)$ and $(2, 1)$, and let L_2 be the line that passes through $(0, 0)$ and $(3, 7)$. For which values of m does the line $y = mx$ lie between the lines L_1 and L_2?

15. Suppose that the graph of a function g lies between a line A, with slope 2, and a line B, with slope -1, for all $x > 2$, as follows:

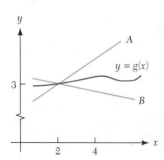

(a) Find an equation for line A.

(b) What is the height of line B at $x = 3$?

(c) Find numbers L and U such that $L < g(4) < U$.

Use the following graph of f for Exercises 16–24. [**NOTE:** There may be more than one possible answer to each exercise.]

Graph of f

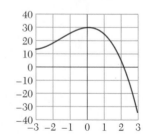

16. Find an interval over which $-10 \le f(x) \le 20$.

17. Is $f(x) \le 20$ if $-2 \le x \le -1$?

18. Find a number L such that $L \le f(x)$ if $-2 \le x \le 1$.

19. Find a number U such that $f(x) \le U$ if $-2 \le x \le 1$.

20. Find a number U such that $|f(x)| \le U$ if $-2 \le x \le 1$.

21. Find numbers L and U such that $L \le f(x) \le U$ if $2 \le x \le 3$.

22. Find a number U such that $|f(x)| \le U$ if $2 \le x \le 3$.
23. Find an interval over which $|f(x)| \le 30$.
24. Find an interval over which $|f(x) + 10| \le 20$.
25. Let $g(x) = \dfrac{x}{x^2 + 4}$. Find an interval on which $0.1 < g(x) < 0.3$.
26. Let $h(x) = 2x^3 - 4x^2 + 5x - 6$. Find an interval on which $|h(x) - 5| < 10$.
27. Let $f(x) = 1/(1 + x^2)$.
    (a) Does f have a maximum value? If so, find it. If not, explain why not.
    (b) Does f have a minimum value? If so, find it. If not, explain why not.

Suppose that $-5 \le f(x) < 11$ when $-3 < x \le 8$. In Exercises 28–35, indicate whether each statement *must* be true, *might* be true, or *cannot* be true. Justify each answer with a sentence or a sketch.

28. $f(x) > -6$ if $-3 < x \le 8$
29. $f(x) \le 100$ if $-3 < x \le 8$
30. $|f(x)| \le 53$ if $|x| < 2$
31. $f(2) = 11$ 33. $|f(0)| = 5$
32. $f(-3) = 11$ 34. $f(13) = -6$
35. $-4 < f(x) \le 9$ if $-3 < x \le 8$
36. Examine this sketch of a hemispherical water tank.

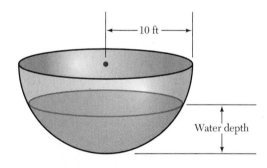

Let the letter D represent the depth of the water in the tank, and let V be the volume of the water in the tank. Regard volume as a function of water depth.
(a) What is $V(0)$?
(b) What is $V(10)$?
(c) Which of the graphs A–D could be the graph of $V(D)$? Explain your reasoning.

Graph A
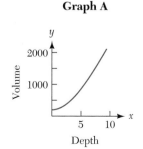
Depth

Graph B
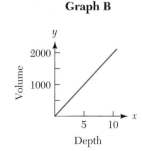
Depth

Graph C
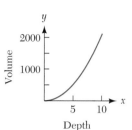
Depth

Graph D
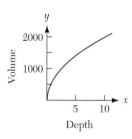
Depth

37. (a) Sketch a graph of temperature as a function of time that is consistent with the scenario described in the following sentence. "The child's temperature has been rising for the last 2 hours, but not as rapidly since we gave her the antibiotic an hour ago." (Let $t = 0$ correspond to the time when the sentence was written.)
    (b) What is the significance of the fact that the graph in part (a) is increasing on the interval $t = -2$ to $t = 0$?
    (c) What is the significance of the fact that the graph in part (a) is concave down on the interval $t = -1$ to $t = 0$?

In Exercises 38–40, sketch the graph of a function described by each of the following sentences. Briefly explain why your graph has the shape you sketched.

38. The cost of a new car continues to increase at an increasing rate.
39. The population is growing more slowly now than it was 5 years ago.
40. Upper Midwest Industries' profit growth is slowing.
41. The graph of a function f follows.

Graph of f
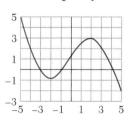

(a) Is f increasing on the interval $(-3, 1)$? Explain.
(b) Estimate the endpoints of the largest interval over which f is concave down.
(c) Estimate the x-value at which f has a local minimum value.

42. Sketch the graph of a function f that has *all* of these properties:
    (i) f is continuous on the interval $[-5, 5]$;
    (ii) f is concave up on $[-5, -1]$;
    (iii) f has a local minimum at $x = -2$;
    (iv) f has a local maximum at $x = 3$.
43. Can a function have more than one local maximum value? If not, why not? If so, give a graphical example.
44. The following figure shows the line $y = mx + b$.

    Sketch the following lines on a copy of the figure. (Be sure to identify each line.)
    (a) $y = mx - b$ (c) $y = -mx - b$
    (b) $y = -mx + b$
45. Let $g(x) = |3x + 2|$. Explain how to obtain the graph of g from the graph of the absolute-value function $f(x) = |x|$ by stretching, compressing, translating, and so on.
46. Let $f(x) = x^2$. For each of the following, (i) give a formula for $g(x)$, (ii) describe how to obtain a graph of g from a graph of f using translation, compression, stretching, and so on, and (iii) sketch a graph of g.
    (a) $g(x) = f(x) + 2$ (d) $g(x) = f(2x)$
    (b) $g(x) = f(x + 2)$ (e) $g(x) = f(-2x)$
    (c) $g(x) = 2f(x)$ (f) $g(x) = -2f(x)$
47. Let $f(x) = x^2$ and $g(x) = x^2 + 4x + 3$.
    (a) Complete the square to show that
        $g(x) = f(x + 2) - 1$.
    (b) Explain how to obtain the graph of g from the graph of f.
48. Let $f(x) = x^2$ and $g(x) = 2x^2 - 4x + 5$.
    (a) Complete the square to show that
        $g(x) = 2f(x - 1) + 3$.
    (b) Explain how the graphs of f and g are related.
49. Suppose that f is a function such that the slope of the secant line through the points $(2, f(2))$ and $(8, f(8))$ is -4.
    (a) Let $g(x) = f(x) + 3$. What is the slope of the secant line through the points $(2, g(2))$ and $(8, g(8))$? Justify your answer. [**HINT:** How are the graphs of f and g related?]
    (b) Let $h(x) = f(x + 3)$. What is the slope of the secant line through the points $(-1, h(-1))$ and $(5, h(5))$? Justify your answer. [**HINT:** How are the graphs of f and h related?]
    (c) Let $j(x) = f(2x)$. What is the slope of the secant

line through the points $(1, j(1))$ and $(4, j(4))$? Justify your answer.
    (d) Let $k(x) = f(-x)$. What is the slope of the secant line through the points $(-8, k(-8))$ and $(-2, k(-2))$? Justify your answer.
    (e) Let $m(x) = 5 - 3f(x)$. What is the slope of the secant line through the points $(2, m(2))$ and $(8, m(8))$? Justify your answer.
50. Suppose that g is a function such that the slope of the secant line through the points $(-3, g(-3))$ and $(5, g(5))$ is 4.
    (a) If $g(x) = f(x - 2)$, what is the slope of the secant line through the points $(-5, f(-5))$ and $(3, f(3))$? Justify your answer.
    (b) If $g(x) = f(2x)$, what is the slope of the secant line through the points $(-6, f(-6))$ and $(10, f(10))$? Justify your answer.
    (c) If $g(x) = 2f(x)$, what is the slope of the secant line through the points $(-3, f(-3))$ and $(5, f(5))$? Justify your answer.
    (d) If $g(x) = 4f(2x + 3) + 1$, what is the slope of the secant line through the points $(-3, f(-3))$ and $(13, f(13))$? Justify your answer.
51. Suppose that the minimum value of a function f is -7 and its maximum value is 3.
    (a) What is the maximum value of $g(x) = f(x) + 2$?
    (b) What is the minimum value of $g(x) = 2f(x) + 3$?
    (c) What is the maximum value of
        $g(x) = -3f(x + 2) + 5$?
    (d) What is the maximum value of $g(x) = |f(x)|$?
52. Suppose that the minimum value of f occurs at $x = 4$, the maximum value of f occurs at $x = -3$, and $-5 < f(x) < 2$ if $0 \le x \le 6$.
    (a) Where does the minimum value of $g(x) = 3f(x - 1) + 2$ occur? The maximum value?
    (b) Let g be the function defined in part (a). Find numbers a and b such that $-13 < g(x) < 8$ if $a \le x \le b$.
53. Suppose that the minimum value of f occurs at $x = 2$, the maximum value of f occurs at $x = 5$, and $-1 < f(x) < 3$ if $-4 \le x \le 6$.
    (a) Where does the minimum value of $g(x) = -3f(x) + 2$ occur? The maximum value?
    (b) Let g be the function defined in part (a). Find numbers L and U such that $L < g(x) < U$ if $-4 \le x \le 6$.
54. Suppose that the minimum value of f occurs at $x = -3$, the maximum value of f occurs at $x = 5$, and $-6 < f(x) < 4$ if $0 \le x \le 7$.
    (a) Where does the minimum value of $g(x) = 4f(2x + 1) - 3$ occur? The maximum value?
    (b) Let g be the function defined in part (a). Find a number U such that $|g(x)| < U$ if $1 \le x \le 15$.

55. Suppose that the minimum value of f occurs at $x = 4$, the maximum value of f occurs at $x = -5$, and $-1 < f(x) < 3$ if $-3 \le x \le 6$.
    (a) Where does the minimum value of $g(x) = -2f(4 - 3x) + 1$ occur? The maximum value?
    (b) Let g be the function defined in part (a). Find numbers a and b such that $-5 < g(x) < 3$ if $a \le x \le b$.

Suppose that f is increasing on the interval $(1, 10)$. In Exercises 56–69, indicate whether each statement *must* be true, *may* be true, or *cannot* be true. Justify each answer with a sentence or a sketch.

56. $f(8) < f(3)$.
57. $f(5) > 0$.
58. f is concave down on the interval $(1, 10)$.
59. $g(x) = f(x) - 25$ is increasing on the interval $(3, 7)$.
60. $g(x) = f(2x)$ is increasing on the interval $(1, 4)$.
61. $g(x) = 5f(x)$ is decreasing on the interval $(1, 2)$.
62. $g(x) = -3f(x)$ is decreasing on the interval $(1, 3)$.
63. $g(x) = f(x + 4)$ is increasing on the interval $(-2, 2)$.
64. $g(x) = -f(x)$ is decreasing on the interval $(1, 10)$.
65. $g(x) = -f(x)$ is increasing on the interval $(-10, -1)$.
66. $g(x) = |f(x)|$ is increasing on the interval $(1, 10)$.
67. $g(x) = f(-x)$ is decreasing on the interval $(1, 10)$.
68. $g(x) = f(-x)$ is increasing on the interval $(-10, -1)$.
69. $g(x) = f(|x|)$ is decreasing on the interval $(-10, -1)$.

Suppose that f is concave down on the interval $(1, 10)$. In Exercises 70–83, indicate whether each statement *must* be true, *may* be true, or *cannot* be true. Justify each answer with a sentence or a sketch.

70. $f(8) < f(3)$.
71. $f(5) > 0$.
72. f is increasing on the interval $(1, 10)$.
73. $g(x) = f(x) - 25$ is concave down on the interval $(3, 7)$.
74. $g(x) = f(2x)$ is concave up on the interval $(1, 4)$.
75. $g(x) = 5f(x)$ is concave down on the interval $(1, 2)$.
76. $g(x) = -3f(x)$ is concave up on the interval $(1, 3)$.
77. $g(x) = f(x + 4)$ is concave down on the interval $(-2, 2)$.
78. $g(x) = -f(x)$ is concave up on the interval $(1, 10)$.
79. $g(x) = -f(x)$ is concave down on the interval $(-10, -1)$.
80. $g(x) = |f(x)|$ is concave up on the interval $(1, 10)$.
81. $g(x) = f(-x)$ is concave up on the interval $(1, 10)$.
82. $g(x) = f(-x)$ is concave down on the interval $(-10, -1)$.
83. $g(x) = f(|x|)$ is concave up on the interval $(-10, -1)$.
84. If a function f is concave up on the interval $[a, b]$, then the line segment joining the points $\big(a, f(a)\big)$ and $\big(b, f(b)\big)$ lies above the graph of f over this interval.
    (a) Suppose that f is concave up on the interval $[a, b]$. Explain why
    $$f(a) + \frac{f(b) - f(a)}{b - a}(x - a) > f(x)$$
    for all x such that $a < x < b$.
    (b) Show that if f is concave up on the interval $[a, b]$, then
    $$\frac{f(x) - f(a)}{x - a} < \frac{f(b) - f(a)}{b - a}$$
    for all x such that $a < x < b$. [**HINT:** Use part (a).]
    (c) Suppose that g is concave down on an interval $[a, b]$. What is the relationship between the line segment joining the points $(a, g(a))$ and $(b, g(b))$ and the graph of g? [**HINT:** If g is concave down on an interval, $-g$ is concave up on that interval.]
85. Suppose that f is concave up on the interval $[0, 10]$, that $f(1) = 4$, and that $f(5) = 2$. [**HINT:** Use part (a) of Exercise 84.]
    (a) Explain why $f(3) < 3$.
    (b) Find a number U such that $f(4) < U$. Justify your answer.
86. Suppose that g is concave down on the interval $[0, 10]$, that $g(1) = 4$, and that $g(5) = 2$.
    (a) Explain why $g(3) > 3$.
    (b) Find a number L such that $g(4) > L$. Justify your answer.
87. Suppose that f is concave up on the interval $[a, b]$. Show that if $0 < t < 1$, then $f\big((1 - t)a + tb\big) < (1 - t)f(a) + tf(b)$. [**HINT:** Use part (a) of Exercise 84 and let $t = (x - a)/(b - a)$.]

1.3 Machine Graphics

Graphics—the ability to represent mathematical objects geometrically—is the most important advantage mathematical computing offers. Without computing, graphs can be tedious, if not impossible, to draw. Computing makes graphs and the insights they offer easily accessible. In this section we illustrate some of the power—and a few of the pitfalls—of computer graphics.

Graphical computing comes in many forms. Hardware ranges from handheld calculators to supercomputers. Software varies from simple plotting programs to sophisticated graphical environments. Plotting commands may be given by means of typed instructions, calculator keystrokes, mouse clicks, or in more exotic ways.

The *Maple* command

```
plot( sin(x), x=-5..5 );
```

produces a graph something like this:

A machine graph

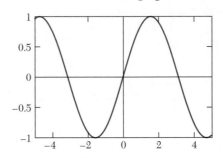

The *Mathematica* command

```
Plot[ Sin[x], {x,-5,5} ]
```

When plotting trigonometric functions in calculus, be sure your machine is set to radians, not degrees.

gives a similar result. (In both cases, the input variable x denotes radians, not degrees.) ◄

All such differences aside, several important ideas and terms apply across the board. Learn them well. They have to do not just with quirks of computing but with the essence of functions and graphs.

The Viewing Window

The sine curve shown earlier is incomplete. It shows what happens for $-5 \le x \le 5$, but the sine function undulates merrily on, in both directions, forever. Strictly speaking, we see only part of the sine graph, not the graph itself. We're not that fastidious—we'll usually say "graph," not "piece of the graph." But the point remains: Partial pictures are all we can expect. The full graph of a function may be infinitely long, so no finite picture can show all of it.

The term **viewing window** describes what *is* shown. In the preceding graph, for instance, the viewing window is the rectangle

$$\{ (x, y) \mid -5 \le x \le 5, \ -1 \le y \le 1 \}.$$

The square brackets in $[-5, 5]$ indicate a closed interval—one that contains its endpoints. See Appendix A for more on intervals and interval notation.

The next section has more details on domain and range.

(In **product notation** we'd denote the same rectangle by $[-5, 5] \times [-1, 1]$.) ◄

The viewing window is the "product" of two intervals, one horizontal and one vertical. We'll call them, respectively, **xrange** and **yrange**. In computer jargon, xrange and yrange are **plotting parameters**. (The term "xdomain" might be more accurate than "xrange"; alas, it's never used.) ◄

No Standard Jargon. The names "xrange" and "yrange", although suggestive, aren't standard; there *is* no standard terminology. Other common names include "hrange," "vrange," "inputrange," and "outputrange."

Six Views of the Sine Function

The appearance of a graph depends (sometimes startlingly) on the viewing window. Here are six views of—believe it or not—the same function:

Graph 1: $y = \sin(x)$
Window = [−30, 30] × [−5, 5]

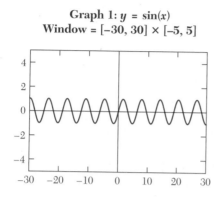

Graph 2: $y = \sin(x)$
Window = [−30, 30] × [−1, 1]

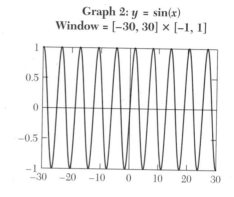

Graph 3: $y = \sin(x)$
Window = [−1, 1] × [−1, 1]

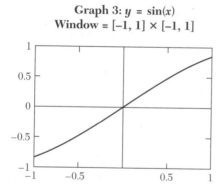

Graph 4: $y = \sin(x)$
Window = [−0.1, 0.1] × [−0.1, 0.1]

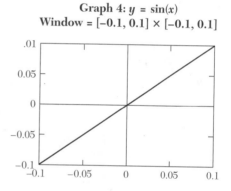

Graph 5: $y = \sin(x)$
Window = [3.13, 3.15] × [−0.1, 0.1]

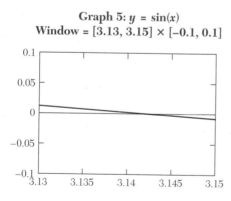

Graph 6: $y = \sin(x)$
Window = [3.13, 3.15] × [−0.01, 0.01]

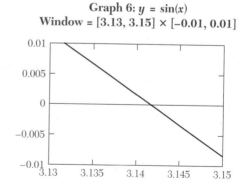

A graph's shape depends—strongly—on the viewing window. Therefore, every graphing device ▶ must somehow let the user choose the viewing window, whether with typed commands, mouse clicks, movable cross hairs, or some other device.

Calculator, computer, sharp stick, and so on.

Who Decides?   Some devices try to guess what the user wants to see. For example, in the *Maple* and *Mathematica* commands

```
plot( sin(x), x=-5..5 );
Plot[ Sin[x], {x,-5,5} ]
```

xrange is specified explicitly, but yrange is not; the programs somehow choose yrange. Leaving choices to the machine may be appropriate, especially for complicated functions. If the machine chooses poorly, the user can always intervene. Here's the bottom line: Whether specified explicitly or chosen automatically, the viewing window strongly affects the appearance of a graph.

Plotting Jargon

On paper, all six windows on the sine graph are about the same size and shape— about 1 inch by 1.75 inches. The differences reflect various choices of aspect ratio, center point, and so on. A few technical words are handy for describing such differences:

Scale: Measures the size of horizontal or vertical units on a graph. The horizontal and vertical scales on Graph 4 are ten times as large as those on Graph 3.

Zooming In: The result of *enlarging* the scale, by a **zoom factor**, in one or both axis directions. Graph 4 results from zooming in on Graph 3 with zoom factor 10 on each axis. (Horizontal and vertical zoom factors may differ.)

Zooming Out: The result of *shrinking* the scale in one or both axis directions. Graph 3 comes from zooming out from Graph 4; both zoom factors are 10.

Aspect Ratio: Measures the relative sizes of horizontal and vertical units. The viewing window $[-1, 1] \times [-3, 3]$ has aspect ratio 3:1—horizontal units are three times as large as vertical units. Aspect ratios matter especially when we plot such objects as circles, for which the relative sizes of units matter.

Sample Size; Connecting Dots

On any machine, the number of pixels limits the accuracy of a graph.

An unbroken curve has infinitely many points, but plotting devices can handle only finite data sets. ◄ Therefore, any physical graph can only approximate the "true" graph of a function. This limitation raises interesting metaphysical questions but is seldom serious for the everyday functions of elementary calculus. For most functions, 100 data points produce a passable graph. For really simple functions (lines, for example), even fewer data points suffice.

How to Lie with Computer Graphics

Don't try this at home. The policy applies only to software, not to hardware.

A good way to understand almost anything is to try to "break" it. ◄ It's fun and easy to fool even sophisticated plotting devices. Let's try.

Plotting devices use various strategies for drawing curves that look continuous. The simplest devices simply plot one data point per horizontal pixel. ▶ On machines with many pixels, this strategy is impractical; some way of "connecting the dots" is needed.

One "dot" on a computer screen represents a pixel, or picture element.

One simple strategy is to plot, say, 101 ▶ equally spaced points, join them with line segments, and hope for the best. This usually works fine, but not always. Look what happens to the sine graph in a huge viewing window:

Why 101? Because 100 equal subintervals have 101 endpoints.

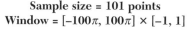

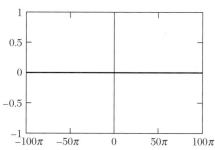

The graph is clearly wrong. ▶ The curve $y = \sin(x)$ actually oscillates *100 times* over the xrange shown. What went wrong?

The graph is just the x-axis. See it?

Nothing—the plotter did as it was told. The xrange interval, $[-100\pi, 100\pi]$, was cut into 100 equal pieces, and the sine function was sampled at the end of each piece. By unlucky accident, each sample point fell at a multiple of 2π, where the sine function is zero. Our plotting sample was too small and too regularly spaced to capture the function's complicated behavior. The moral should be clear:

Distrust large windows and ill-behaved functions. ▶

See page 352 for a truly nasty function.

Luckily, problems with sample size are uncommon in calculus applications. Many "smart" plotting devices recognize sticky situations and adjust sample sizes accordingly.

Cautions

Machine graphics should be viewed skeptically. Various surprises are possible; not all are the machine's fault. Here are some pitfalls to avoid.

Too Few Data Plotting functions with irregular or ragged graphs may require many, many data points. Using smaller windows (i.e., zooming in) sometimes helps.

Awkward Windows Let $f(x) = x^{10} - x$. By simple algebra, $f(0) = 0$ and $f(1) = 0$, so the graph of f crosses the x-axis at $x = 0$ and at $x = 1$. Does the machine-drawn graph agree? Here's a look. The machine chose its own yrange. ▶

The notation $1e + 07$ means 1×10^7, or 10 million.

Graph of $y = x^{10} - x$
Window = $[-5, 5] \times [-10^6, 10^7]$

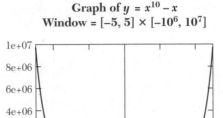

The graph shows disappointingly little near $x = 0$. The tenth power is at fault. Even for modest values of x, $f(x)$ is gigantic. To show everything, yrange must be correspondingly huge. As a result, all but the grossest behavior disappears. Here's the same function in a smaller window:

Graph of $y = x^{10} - x$
Window = $[-2, 2] \times [-10, 10]$

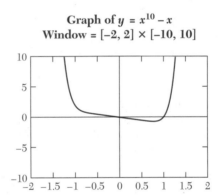

In this case, there is no really good choice of window.

Which graph is better depends—literally, in this case—on one's point of view. ◄

Scale Effects Different scales can give wildly different views of the same graph. Here are two views of the function $y = x + \sin(200x)/200$, the second a tenfold magnification of the first.

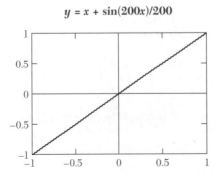

$y = x + \sin(200x)/200$

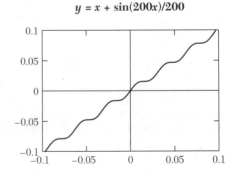

$y = x + \sin(200x)/200$

Notice the differences: The first graph looks almost straight; it's a slightly wiggly version of the line $y = x$. ▶ In the second graph, with smaller values of x, the rapid oscillations of $\sin(200x)/200$ produce a wiggly result. The moral: Viewing the same graph from different scales can show different aspects of a function's behavior.

Compared to x, the second summand, $\sin(200x)/200$, is small, so it has little effect.

Slippery Slopes   Shouldn't the line $y = 10x$ look steeper than the line $y = x$? ▶ Yet the following graphs look almost identical.

The lines have slopes 10 and 1, respectively.

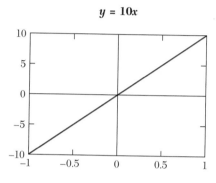

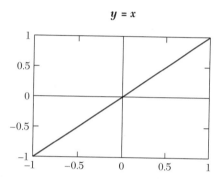

The apparent anomaly arises from different choices of yrange for the two graphs. Our plotting device chose yrange just large enough to fit the plotting data. ▶ Plotting both lines in the same window gives a very different view:

One curious effect of this policy: All lines with positive slope "look the same"; all go from lower left to upper right.

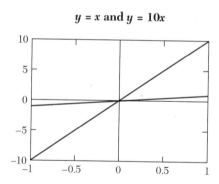

The moral: Watch vertical and horizontal scales carefully. Changing either can dramatically alter a graph's appearance.

Machine Graphs in Calculus

Graphs are crucial throughout this book. We'll draw some graphs by hand (mostly simple ones), but we'll leave most of that laborious work to machines. More important than drawing graphs is learning to read them and understand what they say about functions.

EXAMPLE 1   Find the maximum value, and where it occurs, of the function $f(x) = x \sin(x)$ on the interval $[0, 4]$.

Solution   Later we'll use a calculus concept, the derivative, to attack such problems. Here we'll simply use graphs. The ones that follow are captioned with the *Maple* commands that produce them.

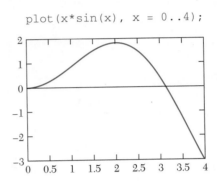

plot(x*sin(x), x = 0..4);

A maximum seems to occur around $x = 2$. Let's look more closely:

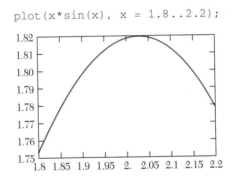

plot(x*sin(x), x = 1.8..2.2);

and still more closely:

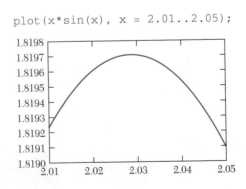

plot(x*sin(x), x = 2.01..2.05);

We conclude that the maximum value is about 1.8197; it occurs near $x = 2.03$. ∎

E X A M P L E 2   For x near 0, the function $p(x) = x - x^3/6$ is a good approximation to the sine function. ▶ *How well* does p approximate the sine function on $[-1, 1]$? On $[-2, 2]$? For which values of x can we be sure that $p(x)$ differs from $\sin(x)$ by less than, say, 0.001?

Later we'll discuss how to find simple functions, such as p, that approximate more complicated functions, such as the sine function.

Which is which? How can you tell?

S o l u t i o n   First we'll compare the graphs of $p(x)$ and $\sin(x)$ ▶ for x in the interval $[-2, 2]$.

Graphs of $y = \sin(x)$ and $y = x - x^3/6$

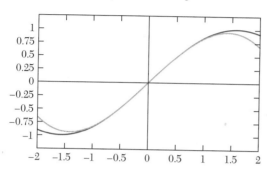

For $-1 \le x \le 1$ the approximation is good; there is little visible difference between the graphs. Things deteriorate when $|x| > 1$; by $x = \pm 2$, the graphs are far apart.

How might we *measure* how well one function approximates another on a given interval? One way is to find the largest difference, positive or negative, between the two functions on the given interval. On $[-1, 1]$, for example, we could plot the **error function** $p(x) - \sin(x)$:

Graph of $p(x) - \sin(x)$

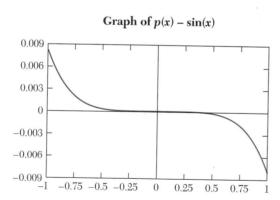

The graph shows that for x in $[-1, 1]$ the difference between the two functions is never more (in absolute value) than 0.009. In other words,

> On $[-1, 1]$, $p(x)$ approximates $\sin(x)$ *with error no worse than* 0.009.

Not bad; for x in this interval, we could almost safely forget the $\boxed{\text{SIN}}$ button on our calculators.

The remaining question asks: *For which values of x is* $|p(x) - \sin(x)| < 0.001$? Again a graph gives the answer:

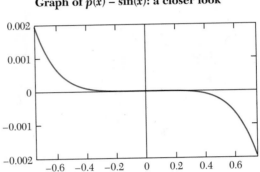

Graph of p(x) – sin(x): a closer look

Draw the horizontal lines $y = \pm 0.001$ *to see why.*

Conclusion: $p(x)$ approximates $\sin(x)$ with error less than 0.001 if $|x| < 0.65$. ◀
■

BASICS

NOTES: *The exercises in this section require some form of plotting technology. Almost any form will do; the main requirement is some ability to choose viewing windows (e.g., by zooming). When plotting trigonometric functions, be sure to use radians, not degrees.*

1. In this section we discussed the fact that $\sin(x) \approx p(x) = x - x^3/6$ for x near 0. Another, simpler, approximation of the sine function says that, for x near 0, $\sin(x) \approx \ell(x) = x$. Let's pursue the latter approximation a bit further.

   (a) Complete the following table, rounding your answers to five decimal places.

x	$\sin(x)$	$|x - \sin(x)|$	x	$\sin(x)$	$|x - \sin(x)|$
-1			1		
-0.5			0.5		
-0.1			0.1		
-0.01			0.01		

   (b) Earlier in this section we said, "On $[-1, 1]$, $p(x)$ approximates $\sin(x)$ with error no worse than 0.009." Does the same statement apply if $\ell(x)$ replaces $p(x)$? If not, what number could replace 0.009?

   (c) We also said: "$p(x)$ approximates $\sin(x)$ with error less than 0.001 if $-0.65 < x < 0.65$." Find an interval $(-a, a)$ on which $\ell(x)$ approximates $\sin(x)$ with error less than 0.001.

2. Use the following sketch to answer each question.

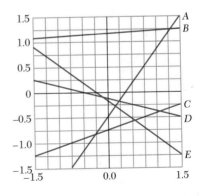

   (a) Do any lines have slope greater than 1? If so, which ones?

   (b) Where do lines A and C intersect? Justify your answer.

3. Use the following sketch to answer each question.
   [**HINT**: This is not the same graph as in Exercise 2.]

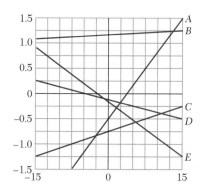

4. Use the following sketch to answer each question.
   [**HINT**: This is not the same graph as in Exercise 2.]

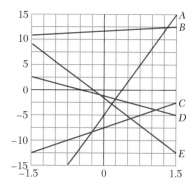

(a) Do any lines have slope greater than 1? If so, which ones?

(b) Where do lines A and D intersect? Justify your answer.

(a) Do any of the lines have slope greater than 1? If so, which ones?

(b) Where do lines C and D intersect? Justify your answer.

FURTHER EXERCISES

5. In each of the following parts, experiment to find a window in which the graph of $y = \sin(x)$ has (approximately!) the given shape or appearance. Express answers in the form $[a, b] \times [c, d]$.
   (a) A horizontal line.
   (b) A diagonal line from lower left to upper right.
   (c) A diagonal line from upper left to lower right.
   (d) A vertical line.
   (e) A rounded letter V.

6. Repeat Exercise 5, using the graph of $y = \cos x$.

7. Repeat Exercise 5, using the graph of $y = x^2$.

A number r is a **root** of a function f if $f(r) = 0$. In Exercises 8–11, use a grapher to locate the roots of each function in the interval $[-5, 5]$. (Report answers to two decimal places.)

8. $f(x) = x^2 - x$ 10. $f(x) = x - \cos x$

9. $f(x) = x - \sin x$ 11. $f(x) = 2^x - x^2$

12. For x near 0, the function $q(x) = 1 - x^2/2$ is a good approximation of the cosine function.
    (a) How well does q approximate the cosine function on $[-1, 1]$?

(b) Find an interval $(-a, a)$ on which $q(x)$ approximates $\cos x$ with error less than 0.001.

Graphs of most calculus functions "look straight at small scale." (Graphs 4, 5, and 6 on page 27 illustrate what this means for the sine function.) For each of the functions in Exercises 13–18, find a window centered at the given point in which the graph of the given function looks like a straight line.

13. $y = x^2$ at $(2, 4)$

14. $y = -x^3 + 3x^2 - 2$ at $(1, 0)$

15. $y = \cos(x)$ at $(0, 1)$

16. $y = \cos(x^2)$ at $(0, 1)$

17. $y = \cos(x^2)$ at $\left(5, \cos(25)\right)$

18. $y = \cos(x^2)$ at $\left(10, \cos(100)\right)$

For each function in Exercises 19–24, use graphics to estimate the maximum and minimum values of $f(x)$ when $-10 \le x \le 10$. (Round your answers to three decimal places.)

19. $f(x) = \sin(x)$ 22. $f(x) = x^2$

20. $f(x) = x + \sin(x)$ 23. $f(x) = 2^x$

21. $f(x) = \dfrac{x}{x - 11}$ 24. $f(x) = 2^x - x^2$

1.4 What *Is* a Function?

What exactly *is* a function? We've already seen many examples, given by formulas, by words, by graphs, and by tables. With these examples as fodder, it's time to ruminate on the *general* idea of function.

Five Examples

Any suitable definition of "function" covers a lot of ground. Functions appear in many guises; they can be described, formally or informally, in various ways. We'll refer repeatedly to five example functions at various levels of strangeness:

f $f(x) = x^2 + \dfrac{3}{x}$

g $g(t)$ = the world human population t years C.E.

h $h(u) = \begin{cases} 1 - u & \text{if } u \leq 0 \\ u^2 & \text{if } 0 < u < 2 \\ 42 & \text{if } u \geq 30 \end{cases}$

j The function given by this graph:

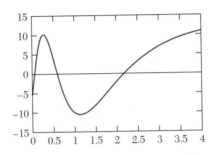

k The function given by this table:

n	-3	-2	-1	0	1	2	3	4
$k(n)$	-4	-9	-1	0	1	8	3	-2

These disparate functions have little in common except the property of *being* functions, to which we now turn.

The Definition At Last

> **Definition** A **function** is a rule for assigning to each member of one set, called the domain, one member of another set, called the range.

The definition uses three key words: "rule," "domain," and "range." We'll discuss each separately.

Rule

A function accepts an input and—somehow—assigns an output. The **rule** describes how this assignment is made. For our example functions:

f The rule is a simple, explicit, algebraic "recipe": *Given a real number x, compute $x^2 + 3/x$.* To 5, for example, f assigns the number $f(5) = 5^2 + 3/5 = 25.6$.

g To a number t—say, $t = 1992.23$—g assigns the world population at exactly that time, measured in years. Thus, imaginably, $g(1992.23) = 5{,}678{,}901{,}234$.

   Admittedly there's trouble here: World population is never known exactly, and certainly no convenient formula is available. Nevertheless, g makes perfectly good sense as a function. At any time t, there is one and only one population $g(t)$—whether or not we can compute it.

   Notice, finally, that although the rule for f tells how to *calculate* outputs, the rule for g only *describes* outputs.

h The rule for h involves "cases"; different inputs are treated differently. For example,

$$h(-5) = 1 - (-5) = 6$$

because the input -5 satisfies the first case of the definition, but

$$h(175) = 42$$

since the input 175 satisfies the third case.

   Functions defined using cases are called **piecewise-defined**, or **multiline, functions**. They arise naturally in computer graphics applications, where pieces of curves are fit together to form desired shapes. ▶

We will discuss this process, called splining, in Section 4.5. For a preview, glance at the pictures there.

j The rule for j—like everything else about j—is graphical: $j(x)$ *is the height of the graph at horizontal position x.* Thus, $j(3) \approx 7$, $j(1) \approx -10$, and $j(10)$ is not defined.

   As with any function defined by a graph, values of j can be read only approximately. Despite this inexactness, there's plenty to learn from graphs. Approximation occurs everywhere in the real world.

k We have only the table to use. Here is k's "rule": *For any n in the top row, k(n) is the entry below n.* Thus, $k(-2) = -9$, $k(2) = 8$, and $k(42)$ is undefined.

   Such a rule hardly seems to deserve the name. It gives no hint of *how* outputs are determined. Are we hiding some secret formula for k? No; sometimes a table contains all we know of a function. We might, with effort, find a formula or curve that "fits" tabular data, but deciding whether our formula is right would require more information. ▶

See, for instance, Example 5 in Section 1.2.

A Variable by Any Other Name. For most of our functions, we gave names—*x, t, u, n*—to the input variables. This was for convenience; the particular names don't matter. For example, the rules

$$f(x) = x^2 + \frac{3}{x} \quad \text{and} \quad f(z) = z^2 + \frac{3}{z}$$

define exactly the same function.

Domain

A function's **domain**, denoted by $\mathcal{D}$, is its set of acceptable inputs. For our example functions:

_f_   The algebraic rule $f(x) = x^2 + 3/x$ makes good sense unless $x = 0$, so the domain of f is the set $\mathcal{D} = \{x \mid x \neq 0\}$. In words: $\mathcal{D}$ is the set of real numbers x such that $x \neq 0$. ◄

_g_   The rule $g(t) =$ "the world human population t years C.E." makes sense for *any* real number. Hence, the domain of g can be taken as $(-\infty, \infty)$, the full set of real numbers.

     The inputs t for which $g(t)$ can be reasonably known or estimated are more restricted. Conceding something to reality, we might arbitrarily restrict our domain to the set $[-5000, 2100]$.

_h_   By its multiline definition, h accepts as input any real number u except those for which $2 \leq u < 30$. Therefore, the domain of h is the set $\{x \mid x < 2 \text{ or } x \geq 30\}$ or, in interval form, $(-\infty, 2) \cup [30, \infty)$. ◄

_j_   Acceptable inputs to j lie in the interval $[0, 4]$. (In "computerese," the domain is the xrange.)

_k_   Any function described by a table has a finite domain. ◄   The domain of k has just eight members: $\mathcal{D} = \{-3, -2, -1, 0, 1, 2, 3, 4\}$.

The vertical bar inside the set brackets means "such that." For more examples of such notation, see Appendix A.

The $\cup$ symbol means "union." See Appendix A for more details on set notation.

A table has only finitely many entries.

"Natural" Domains.   In practice, function domains are seldom stated explicitly. In this case ◄ the default is the **natural domain**—the set of inputs for which the rule makes sense. For a function such as f, given by an algebraic formula in x, the natural domain is the set of values of x for which the formula makes algebraic sense.

The most common case in calculus courses.

Range

A function's domain is its set of allowable inputs. The **range** is the corresponding set of outputs. In symbolic language,

    If f has domain $\mathcal{D}$, then the range of f, denoted by $\mathcal{R}$, is the set $\{f(x) \mid x \in \mathcal{D}\}$. ◄

The number 42 (picked at random) lies in the range of a function F ◄ if some input to F produces the output 42 (i.e., if for some number x, $F(x) = 42$). For example, 42 is in the range of the function $F(x) = \sqrt{x}$ but is not in the range of the function $G(x) = -x^2$. ◄

The expression $x \in \mathcal{D}$ means x is a member of the domain.

Why F and not f? Because we already used f.

Which input to F gives output 42? Why isn't 42 in the range of G?

Reading a Function's Range from Its Graph.   Defining a function's range is simple; *calculating* it—especially by hand—can be difficult. Graphs can help if viewing windows are chosen properly.

    The number 42 lies in the range of F, for instance, if and only if the horizontal line $y = 42$ hits the graph of F *somewhere*. For any function F,

    The range of F is the set of y values at which a horizontal line hits the graph of F.

Here's a more illuminating way of saying the same thing:

> *Let lights shine horizontally, from both left and right, on the graph of F.*
> *The range of F is the shadow thrown by the graph on the vertical axis.*

The following picture shows the shadow (or **horizontal projection**) cast by the sine graph on the vertical axis:

The shadow knows:
the range of $y = \sin x$ is $[-1, 1]$

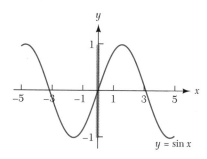

Finding Ranges of Our Example Functions.   We'll use both graphical and algebraic methods to calculate the ranges of our example functions.

f The domain of f is $\mathcal{D} = \{ x \mid x \neq 0 \}$, so the range of f is $\mathcal{R} = \{ x^2 + 3/x \mid x \in \mathcal{D} \}$. That's easy to write, but what, exactly, does $\mathcal{R}$ contain? Does $\mathcal{R}$ contain the number -2345.678 (chosen at random)?

Here's a graph:

Graph of $f(x) = x^2 + 3/x$

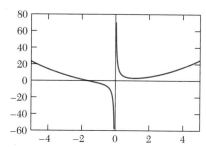

It suggests that *all* reals lie in the range: f seems to "hit" every real value at least once. ▶ Our random number -2345.678, in particular, seems to be the output for some input just to the left of zero. ▶ To estimate which

To see why, imagine the graph's shadow on the vertical axis.

Do you agree that inputs just to the left of zero produce large negative outputs?

x gives $f(x) = -2345.678$, let's look in another window.

Graph of $f(x)$: a closer look

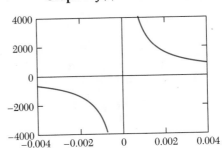

This graph suggests that $f(x) = -2345.678$ if $x \approx -0.0015$.

g By definition, g's range is the set of all world populations, for all time. This is clearly some set of nonnegative integers; saying much more is almost impossible.

h The domain of h has three pieces. To find the range, we'll decide what each domain piece contributes separately, and then assemble our answers.

Convince yourself of this; it's easy but important. A graph should help.

1. The output formula $1 - u$ holds if $u \leq 0$. As u varies from 0 through negative values, $1 - u$ covers the interval $[1, \infty)$. ◀ Thus, the range of h contains the interval $[1, \infty)$.

2. For $0 < u < 2$, $h(u) = u^2$. As u varies from 0 to 2, u^2 ranges from 0 to 4. Thus, the range of h contains the interval $(0, 4)$.

3. For $u \geq 30$, $h(u) = 42$. Thus, the third domain piece contributes the one-member set {42} to the range.

Putting the pieces together, we find that the range of h is

$$\mathcal{R} = [1, \infty) \cup (0, 4) \cup \{42\} = (0, \infty).$$

Notice that this union is redundant—the first two intervals overlap, and the "singleton" {42} adds nothing new. The redundancy reflects the fact that some members of the range are hit more than once.

From graphs alone we can only estimate.

j The j graph shows that the set of values assumed by j is the interval $[-11, 11]$. ◀

k The output set of k is the second row of k's table:

$$\{-4, -9, -1, 0, 1, 8, 3, -2\}.$$

The range of k, like its domain, is a finite set.

Domain and Range—a Machine's View.   Computers and calculators cannot escape issues of domain and range when plotting functions. Here's a machine-drawn plot of the function $p(x) = \sqrt{x^2 + 2x}$:

Graph of $p(x) = \sqrt{x^2 + 2x}$

The gap from $x = -2$ to $x = 0$ is there for good reason: The natural domain of p omits this interval. The picture also suggests (but, being finite, can't prove) that the range is $[0, \infty)$. With algebra, we could verify both of these observations.

The Definition of Function: Reprise

A function is a package with three chief parts: rule, domain, and range.

The rule can be given in various ways—as an algebraic formula, as a graph, as a table, as an English description, and so on. However it's given, the rule must not be ambiguous. To each element of the domain the rule associates exactly *one* element of the range.

The domain and range of a function f are, respectively, its input and output sets. The natural domain of f is the set of inputs for which f's rule makes sense. In calculus, both range and domain are usually sets of real numbers. ▶

Elsewhere in mathematics, domains and ranges may be sets of vectors, matrices, functions, complex numbers, or more exotic objects.

Odd Functions and Even Functions: Graphical Symmetry

Certain functions, like certain numbers. are said to be either odd or even. These properties, though defined symbolically, are easily recognized graphically. First, the symbolic definitions:

Definition A function f is **even** if $f(-x) = f(x)$ for all x in its domain; f is **odd** if $f(-x) = -f(x)$.

EXAMPLE 1   If $f(x) = x^2$, then

$$f(-x) = (-x)^2 = x^2 = f(x),$$

so f is even. If $g(x) = x^3$, then

$$g(-x) = (-x)^3 = -x^3 = -g(x),$$

so g is odd. The function $h(x) = x + x^2$ is neither even nor odd, because

$$h(-x) = -x + (-x)^2 = -x + x^2 \neq \pm h(x).$$ ■

Graphs and Symmetry

For many functions we can easily check evenness or oddness using algebraic manipulation. But what about the function $k(x) = \sin x$, which has no algebraic formula? To decide whether k is even, odd, or neither, let's look at the graph. To see how $k(x)$ and $k(-x)$ are related, we choose an xrange centered at $x = 0$:

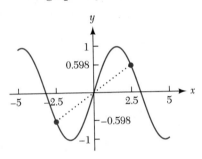

Symmetry about the origin:
graph of $y = \sin x$

The graph's **symmetry about the origin** means that for any input x, $k(-x) = -k(x)$. This is just what it means for k to be odd. If, say, $x = 2.5$, the bulleted points (points marked with dots) show that

$$k(2.5) = \sin(2.5) \approx 0.598 \qquad \text{and} \qquad k(-2.5) = \sin(-2.5) \approx -0.598.$$

In general,

See this section's exercises for more on odd and even functions.

A function f is odd if and only if the graph of f is symmetric about the origin. A function is even if its graph is symmetric about the y-axis. ◀

Periodic Functions: Repeating Patterns

See the preceding graph.

The sine function ◀ is called 2π-periodic because it repeats itself on every interval of length 2π. (So do the other basic trigonometric functions; we'll review them soon.) Any function is called **periodic** if it repeats itself on intervals of any fixed length.

Periodic Functions: The Graphical View

Here are graphs of three periodic functions, f, g, and h:

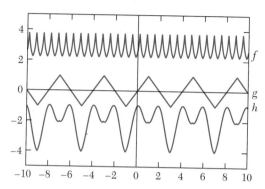

Graphs of three periodic functions

Each function's graph repeats some basic shape. How *often* a periodic function repeats itself depends on the **period**, i.e., the (horizontal) length of the basic repeating shape. It's easy to see from their graphs that g and h have period 4. Reading the period of the wigglier f is trickier. One way to avoid seasickness is to count that f completes 30 "cycles" from $x = -10$ to $x = 10$, an interval of length 20. Thus, f has period 20/30 = 2/3. ▶

This is a graphical approximation, but a good one.

Periodic Functions: The Symbolic View

Periodicity can be defined symbolically:

> **Definition** A function f is periodic if the equation
>
> $$f(x + P) = f(x)$$
>
> holds for all x in the domain of f and a positive number P. The smallest such number P is the period of f.

On the Graph If, say, $P = 3$, then the functional equation says that $f(x + 3) = f(x)$. What does this mean about the graph of f? The equation says that $f(1) = f(4)$, $f(1.5) = f(4.5)$, $f(2) = f(5)$, $f(2.5) = f(5.5)$, and so on. Thus, for any x the height of the graph at x is the same as the height of the graph at $x + 3$.

A Shift to the Left For any function f, the graph of $y = f(x + P)$ is the result of shifting the graph of $y = f(x)$ a distance of P units to the *left*. ▶ From this point of view, the periodicity equation $f(x + P) = f(x)$ means:

We discussed graphical translation in Section 1.2.

> *The graph of f doesn't change if it's shifted P units to the left.* ▶

Convince yourself of this for each preceding graph.

Repeats on Other Intervals Because the sine graph repeats itself on intervals of length 2π, it also repeats itself on intervals of length 4π, 6π, 8π, and so on. Remember, the period is the length of *the shortest* interval on which a function repeats.

EXAMPLE 2   Following are graphs of

$$f(x) = \frac{\sin x}{3} + 3 \quad \text{and} \quad g(x) = (\sin x)^2 + \sin(x).$$

Which is which? Why is each function 2π-periodic?

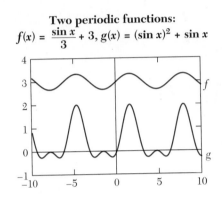

Two periodic functions:
$f(x) = \dfrac{\sin x}{3} + 3, \ g(x) = (\sin x)^2 + \sin x$

Solution   The upper graph is f—it's formed by compressing the sine curve vertically and translating it 3 units upward. The lower graph, therefore, is g.

Both functions appear to be 2π-periodic. So they are: They "inherit" the sine function's periodicity. A symbolic calculation shows how:

$$f(x + 2\pi) = \frac{\sin(x + 2\pi)}{3} + 3 = \frac{\sin x}{3} + 3 = f(x);$$

$$g(x + 2\pi) = \big(\sin(x + 2\pi)\big)^2 + \sin(x + 2\pi) = (\sin x)^2 + \sin x = g(x). \qquad \blacksquare$$

Functional Language: Names, Nicknames, Uses, and Abuses

Everyday speech is a rich mixture of formal and informal language, names and nicknames, careful descriptions and vague allusions. The common things of daily life—friends, places, college courses, and so on—are likeliest to attract nicknames, verbal shortcuts, and aliases.

In calculus, functions are the commonest objects of all. Mathematical language, true to form, offers various ways of talking about functions, some clearer and more precise than others. Some linguistic tips follow.

Function vs. Value.   It's sometimes important to distinguish between a function and the value of a function. The statement

Consider the function $f(x) = x^2$

may seem clear, but it blurs the distinction between f—a function—and $f(x)$—a value of f. It would be technically clearer (though less euphonious) to say ◄

We won't always be this fastidious.

Consider the function f defined by $f(x) = x^2$.

Composite Functional Expressions.   If f is defined by $f(x) = x^2$, then it's clear what is meant by such expressions as $f(3)$, $f(z)$, and $f(-13/2)$. **Composite expressions**, such as

$$f(x + 3), \qquad f(-x), \qquad \text{and} \qquad f\big(f(x)\big)$$

need more care.

To make sense of them, it helps to think of f as an **operator** or **procedure**: f accepts "anything" and returns its square. Thus,

$$f(x + 3) = (x + 3)^2 = x^2 + 6x + 9;$$

$$f(-x) = (-x)^2 = x^2;$$

$$f\big(f(x)\big) = \big(x^2\big)^2 = x^4.$$

Functions vs. Expressions.   The statement

The range of $x^2 - 6x$ is $[-9, \infty)$

is intelligible, but the *expression $x^2 - 6x$* is used—not quite correctly—to describe a function. A scrupulously correct version of the statement preserves this distinction:

Let f be the function defined by $f(x) = x^2 - 6x$. The range of f is $[-9, \infty)$.

Functions differ from expressions in subtle but important ways. Here are three ways:

- Not all functions are defined by symbolic expressions. Of the five functions that started this section, only two were defined symbolically. ▶   *Which two?*

- Several different symbolic expressions may define the same function. The following three expressions are equivalent:

$$x^2 - 6x; \qquad \frac{x^2 - 6x}{\sin^2 x + \cos^2 x}; \qquad \frac{x^4 - 6x^3 + x^2 - 6x}{x^2 + 1}.$$

  Any one of them could define f.

- Functions don't care about variable names; expressions do. For example,

$$x^2 - 6x, \qquad t^2 - 6t, \qquad \text{and} \qquad P^2 - 6P$$

  are different expressions, but all could define the same function.

Computers Know the Difference.   Computer programs insist on the distinction between functions and expressions. In *Maple*, for example, the command

```
f := x^2 - 6*x ;
```

defines f only as an expression. To define *f* as a *function*, we could type

```
f := x -> x^2 - 6*x;
```

The latter form isn't pretty, but it's explicit: f is defined as an *operator* that assigns to any input *x* the output $x^2 - 6x$.

BASICS

1. Indicate which of the following curves are graphs of functions.

   (a)
   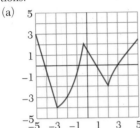

   (b)
   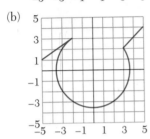

   (c)
   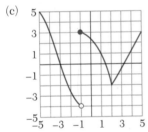

   (d)
   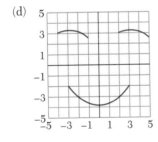

2. Let $f(x) = \begin{cases} 0 & \text{if } 0 \leq x < 1 \\ -2 & \text{if } 1 \leq x < 2 \\ x & \text{if } 2 \leq x \end{cases}$

   (a) Sketch a graph of f.
   (b) What is the domain of f?
   (c) What is the range of f?

Find the natural domain and the range of the functions in Exercises 3–6.

3. $g(x) = x^2 + x - 2$

4. $g(x) = \dfrac{x+2}{x-1}$

5. $f(x) = \dfrac{1}{(x-1)(x+2)}$

6. $h(x) = \sqrt{x+2}$

7. Indicate whether each of the following functions is even, odd, or neither.

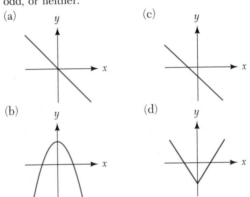

8. Suppose that f is an even function. Is the graph of f symmetric with respect to the y-axis? Justify your answer.

FURTHER EXERCISES

9. The **vertical line test** states, "A curve is the graph of a function only if each vertical line $x = k$ intersects the graph at most once." Explain why this test works.

10. (a) Sketch the solution curve of the equation $y^2 = x+4$.
    (b) Is the curve in part (a) the graph of a function? Justify your answer.

11. (a) Can the graph of a function have more than one x-intercept? Explain.
    (b) Can the graph of a function have more than one y-intercept? Explain.

In Exercises 12 and 13, sketch the graph of each piecewise-defined function; then find the domain and range of each.

12. $f(x) = \begin{cases} 0 & \text{if } x \leq -1 \\ 3x - 2 & \text{if } 1 < x < 1.5 \\ x^2 & \text{if } 1.5 \leq x < 2 \end{cases}$

13. $f(x) = \begin{cases} \dfrac{1}{2-x} & \text{if } x < 0 \\ \sqrt{1+4x^5} & \text{if } x > 1 \end{cases}$

Find the domain and range of each of the functions shown graphically in Exercises 14–17.

14.

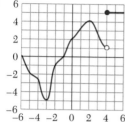

15.

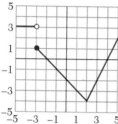

16.

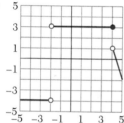

17.

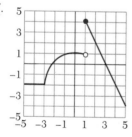

Find the natural domain and the range of each function in Exercises 18–25.

18. $f(x) = \sqrt{x^2 - 4}$

19. $r(x) = \dfrac{1}{\sqrt{(x+2)(1-x)}}$

20. $s(x) = \dfrac{1}{\sqrt{(x+2)(x-1)}}$

21. $t(x) = \sqrt{\dfrac{1-x}{x+2}}$

22. $h(x) = (x-2)(3-x)$

23. $j(x) = \sqrt{4 - x^2}$

24. $k(x) = (2x+5)^{1/3}$

25. $k(x) = (4 - 5x)^{2/3}$

26. Let $f(x) = \sqrt{x^4 + x^2} = \sqrt{x^2(x^2 + 1)}$ and $g(x) = x\sqrt{x^2 + 1}$.
    (a) Find the domain and range of f.
    (b) Find the domain and range of g.
    (c) Explain why $f \neq g$.

27. For each set that follows, find a function defined by a single algebraic expression that has the set as its natural domain. [**EXAMPLE:** The function $f(x) = \sqrt{(x-1)(x+3)}$ has the set $(-\infty, -3] \cup [1, \infty)$ as its domain.]
    (a) $\{x \mid x \neq -1 \text{ and } x \neq 3\}$
    (b) $[-1, 3]$
    (c) $[-1, \infty)$
    (d) $(-1, 3)$
    (e) $(-\infty, -1) \cup (3, \infty)$
    (f) $[-1, 3)$

28. Suppose that water is flowing into a cubical tank 10 feet on an edge.
    (a) Write a formula for V, the volume of water in the tank, as a function of d, the depth of water in the tank.
    (b) What is the domain of V? What is the range of V?
    (c) Write a formula for d as a function of V.

29. Let $f(x) = |x|$ and g be defined by the rule: "$g(x)$ is the slope of the graph of f at x."
    (a) Find the domain and range of f.
    (b) Explain why 0 is not in the domain of g.
    (c) Find the domain and range of g.
    (d) Find a symbolic expression for g. [**HINT:** Use a piecewise-defined function.]

30. Let $f(x) = x + |x|$ and g be defined by the rule: "$g(x)$ is the slope of the graph of f at x."
    (a) Find the domain and range of f.
    (b) Write f as a piecewise-defined function.
    (c) Find the domain and range of g.
    (d) Find a symbolic expression for g.

31. Let $f(x) = \dfrac{x}{|x|}$ and g be defined by the rule: "$g(x)$ is the slope of the graph of f at x."
    (a) Find the domain and range of f.
    (b) Write f as a piecewise-defined function.
    (c) Find the domain and range of g.
    (d) Find a symbolic expression for g.

In Exercises 32–37, indicate whether each function is even, odd, or neither. Justify your answers with algebra. [**HINT:** Start by looking at graphs of these functions.]

32. $f(x) = 3x^4 - 5x^2 + 17$

33. $f(x) = 12x^7 + 6x^3 - 2x$

34. $f(x) = |x|$

35. $f(x) = |x + 4|$

36. $f(x) = x^4 + |x^3| - 2$

37. $f(x) = 4x^3 - 7$

38. Suppose that f is an odd function whose domain includes zero. Explain why $f(0) = 0$ must be true. [**HINT:** $-0 = 0$.]

39. Can a function be both even and odd? [**HINT:** Consider constant functions (i.e., functions of the form $f(x) = c$).]

40. Each of the following graphs is a portion of the graph of an even function over the interval $[-5, 5]$. Complete these graphs.

(a)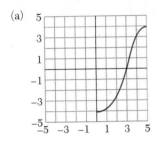

(b)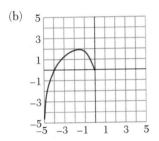

(c)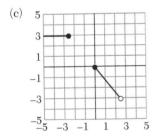

41. Which of the graphs in Exercise 40 could be a portion of the graph of an *odd* function over the interval $[-5, 5]$? Complete these graphs and explain why the others cannot be portions of the graph of an odd function.

42. Even functions are characterized by the fact that their graphs are symmetric about the y-axis. Are there functions whose graphs are symmetric about the x-axis? Explain.

43. Let $q(x) = ax^2 + bx + c$ where a, b, and c are constants.
    (a) Suppose that q is an even function. Show that $b = 0$.
    (b) Suppose that q is an odd function. What can be said about the values of the constants a, b, and c?

44. Suppose that the graph of a function f is symmetric with respect to the y-axis and that $f(3) = 1$. Evaluate $f(-3)$.

45. Suppose that f is an even function and that the point $(-3, 5)$ is on the graph of f. Find another point on the graph of f.

46. Suppose that f is an odd function and that $-12 \leq f(x) \leq 9$ when $x \geq 0$.
    (a) Is it possible that $f(-5) = 10$? Explain.
    (b) Is it possible that $f(-5) = -10$? Explain.

47. Suppose that f is an even function.
    (a) How are the graphs of f and $g(x) = f(|x|)$ related?
    (b) Is $h(x) = |f(x)|$ even, odd, or neither?
    (c) Is $k(x) = f(x) + |f(x)|$ even, odd, or neither?

48. Suppose that f is an odd function.
    (a) Is $g(x) = -f(x)$ even, odd, or neither? Justify your answer.
    (b) Is $h(x) = f(|x|)$ even, odd, or neither? Justify your answer.
    (c) Is $j(x) = |f(x)|$ even, odd, or neither?
    (d) Is $k(x) = f(x) + |f(x)|$ even, odd, or neither?

49. Let f be a function with domain R. Show that $g(x) = f(|x|)$ is an even function.

50. Suppose that f is a function with the following properties:
    (i) f is an even function;
    (ii) f is increasing over the interval $(2, 8)$;
    (iii) f is concave down over the interval $(3, 7)$; and,
    (iv) $4 \leq f(x) \leq 5$ when $1 \leq x \leq 6$.
    Indicate whether each of the following statements *must* be true, *may* be true, or *cannot* be true. Justify your answers. [**HINT:** Sketch a graph of f.]
    (a) f is increasing over the interval $(-8, -2)$.
    (b) f is concave down over the interval $(-7, -5)$.
    (c) $|f(-2)| < 9$
    (d) $-5 \leq f(-3) \leq -4$

51. Suppose that f is an odd function and that f is decreasing on the interval (a, b).
    (a) Is f increasing or decreasing over the interval $(-b, -a)$? Explain.
    (b) Is it possible that $a < 0 < b$? Explain.

52. Suppose that f is an odd function and that f is concave up on the interval (a, b).
    (a) Is f concave up or concave down over the interval $(-b, -a)$? Explain.
    (b) Is it possible that $a < 0 < b$? Explain.

53. Suppose that f is a periodic function with period 1/2 and that
$$f(2) = 5, \qquad f(9/4) = 2, \qquad \text{and} \qquad f(11/8) = 3.$$
    Evaluate each of the following:
    (a) $f(0)$ (d) $f(-3)$
    (b) $f(1/4)$ (e) $f(1000)$
    (c) $f(3/8)$ (f) $f(x + 5) - f(x)$

54. Suppose that f is a periodic function with period P and that g is defined by the rule: "$g(x)$ is the slope of the secant line through $(x, f(x))$ and $(x + 1, f(x + 1))$". Show that g is a periodic function with period P.

55. Suppose that f is a periodic function with period 2 and that f is increasing and concave down on the interval $[0, 1]$.
    (a) Is f increasing on the interval $[6, 7]$? Justify your answer.
    (b) Is it possible that f is concave down on the interval $[-3, -2]$? Justify your answer.
    (c) Suppose that f has a local maximum at $x = 13.5$. Where in the interval $[0, 2]$ does f have a local maximum?
    (d) Suppose that $-2 \le f(x) \le 5$ when $-4 \le x \le 3$. Explain why $-2 \le f(x) \le 5$ for all x.

56. Let f be a periodic function with period P.
    (a) Show that $f(x + 2P) = f(x)$.
    (b) What does part (a) mean graphically?

57. Suppose that f is a periodic function.
    (a) Is it possible that f is an even function? If so, sketch the graph of an even periodic function. If not, explain why not.
    (b) Is it possible that f is an odd function? If so, sketch the graph of an odd periodic function. If not, explain why not.
    (c) Is it possible that f is neither even nor odd? If so, sketch the graph of a periodic function that is neither even nor odd. If not, explain why not.

58. Suppose that f is a periodic function with period P and that $A \ne 0$ is a constant. Show that each of the following functions is periodic and find its period.
    (a) $g(x) = f(x) + A$ (c) $g(x) = Af(x)$
    (b) $g(x) = f(x + A)$ (d) $g(x) = f(Ax)$

1.5 A Field Guide to Elementary Functions

An **elementary function** is one built from certain legal basic elements, using certain legal operations. For example, the function

$$f(x) = \ln\left(\frac{\sin(2x)}{1 + 3x^4}\right)$$

is elementary because it is constructed (by addition, division, composition, and so on) ▶ from certain basic function elements: the sine and natural logarithm functions, powers of x, and constants.

We'll review these operations on functions in the next section.

Despite their name, elementary functions need not be trivial or simple-minded. Many different buildings can be built from standard sizes of brick. For the same reasons, elementary functions come in enormous variety.

What "standard" means for bricks is partly a practical matter ▶ but mainly a matter of choice. Our choice of functional building blocks is just as practical but much less arbitrary. The basic elementary functions of calculus have proved themselves useful throughout mathematics and its applications.

Bricks are twice as long as they are wide, not too heavy to lift, and so on.

A Pictorial Field Guide

This section contains brief descriptions, with pictures, of various elementary functions. Like any field guide, this one emphasizes identification, classification, and basic behavior rather than detailed study of its subjects. More details on each species of function met here can be found in later chapters or in the appendices.

Function Families

Like birds or plants, the building-block functions of calculus fall naturally into "families" (of which there are far fewer, fortunately, than there are for birds or plants!). We'll name just three.

Algebraic Functions Their rules involve only algebraic operations, such as these:

$$f(x) = x^2 + \frac{3}{x}; \qquad g(x) = \frac{x^2 + 3}{1 + \sqrt{x}}.$$

Exponential and Logarithm Functions Their rules involve exponentials or logarithms with various bases, such as these:

$$h(x) = 2^x; \qquad k(x) = e^x; \qquad m(x) = \log_2 x; \qquad n(x) = \ln x.$$

Trigonometric Functions The sine, cosine, and tangent functions, together with their descendants the cosecant, secant, and cotangent functions.

Hybrids

Like some plants, the basic calculus functions hybridize freely, combining in various ways to produce new elementary functions. How functions combine, and how their descendants' traits reflect their own, is an important theme throughout this book. In this section, our main concern is with the basic building-block functions themselves.

Algebraic Functions

An **algebraic function** is one defined using only the ordinary algebraic operations: addition, multiplication, division, raising to powers, and taking roots. Roughly speaking, algebraic functions are those that even the simplest calculator can handle. ◄ If, say, the function f is defined by the algebraic rule $f(x) = x - 3\sqrt{x+1}$, then checking with a calculator that

$$f(1.7) = 1.7 - 3\sqrt{2.7} \approx -3.229$$

requires nothing fancier than the arithmetic and square root keys.

$2.98 buys such a machine; trigonometric, exponential, and logarithm functions start a little higher.

Polynomials

Polynomials are the simplest algebraic functions; their rules require only multiplication and addition. ◄ (To review basic facts about linear, polynomial, and rational functions, see Appendix C.)

Positive integer powers of x can be found by repeated multiplication.

> **Definition** A **polynomial** in x is an expression that can be written in the form
>
> $$a_0 + a_1 x + a_2 x^2 + a_3 x^3 + \cdots + a_n x^n,$$
>
> that is, a sum of constant multiples of nonnegative integer powers of x.

EXAMPLE 1 The expressions

$$5x + 2, \qquad -x + 5x^{17}, \qquad \text{and} \qquad (x + 3)(\pi x^{17} + 7)$$

are all polynomials in x. ◄ None of the expressions

Use algebra to write the last example in standard form.

$$3 - \frac{1}{x}, \qquad \sqrt{x}, \qquad \text{and} \qquad \frac{x+3}{\pi x^{17} + 7},$$

is a polynomial. ∎

Polynomials as Functions. The polynomial expression $2x^3 + 5x$ corresponds naturally to the function $p(x) = 2x^3 + 5x$. Any function defined by a polynomial expression is called a **polynomial function**. In practice, the technical distinction between *polynomial* and *polynomial function* seldom matters; we'll use the terms interchangeably.

Domain and Range. Because polynomials involve only multiplication and addition, they accept all real numbers as inputs. In other words, every polynomial function has the same natural domain: all real numbers.

Polynomials' *ranges* vary drastically. The range of $p(x) = 10$ is $\{10\}$, a set with just one member. By contrast, the range of $q(x) = x^2 + 3$ is $[3, \infty)$, an infinite interval. The range of $r(x) = x^3 - 2x^2$ is not obvious from the formula, so we'll look at the graph. Here, together, are graphs of p, q, and r:

Polynomial graphs: $p(x) = 10$,
$q(x) = x^2 + 3$, $r(x) = x^3 - 2x^2$

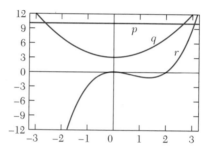

Blowing Up and Down The picture suggests that $r(x)$ "blows up" (tends to ∞) as $x \to +\infty$ and that $r(x)$ "blows down" (tends to $-\infty$) as $x \to -\infty$. ▶ This impression is correct; for any polynomial, the term with the *largest power of x* determines what happens as x tends to plus or minus infinity. ▶ (For $r(x)$, the x^3 term "dominates.") Thus, $r(x)$ assumes *all* real values, and its range is $(-\infty, \infty)$.

Plotting r in a larger window would show this even more clearly.

We'll show this soon, using limits.

Point of View The appearance of *any* graph depends strongly on how the plotting window is chosen. Polynomials are no exception. Here are two more views of the same three polynomials:

Graphs of p, q, and r: a closer look

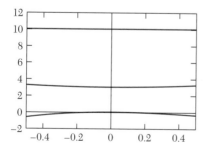

Graphs of p, q, and r: a farther view

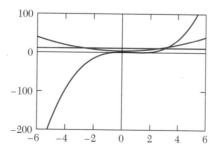

In the "farther view," the large vertical scale makes p, q, and r hard to distinguish near $x = 0$.

No Breaks Polynomial graphs are continuous and smooth everywhere—they have neither breaks or kinks. As we'll soon see, this reflects the fact that polynomials are among the simplest, best-behaved functions from the calculus point of view.

Lines and Linear Functions. The very simplest polynomials are constants, such as $p(x) = 10$. The simplest *interesting* polynomials are **linear functions**—polynomials that involve nothing higher than the first power of x. Each of the following expressions defines a linear function:

$$\ell_1(x) = 2x + 3; \qquad \ell_2(x) = 2(x - 3) + 3;$$

$$\ell_3(x) = 2x + 1; \qquad \ell_4(x) = -2x + 2.$$

Graphs of linear functions are, of course, lines:

Lines: $\ell_1(x) = 2x + 3$, $\ell_2(x) = 2(x - 3) + 3$,
$\ell_3(x) = 2x + 1$, $\ell_4(x) = -2x + 2$

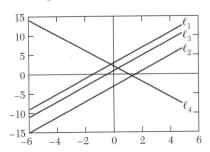

Each line's slope, y-intercept, and other important properties can be read from its equation.

Quadratics and Cubics: Characteristic Shapes. Graphs of **quadratic** and **cubic** polynomials—those with degree ◄ 2 and 3—have standard shapes, called **parabolas** and **cubic curves**. Here are three quadratics: ◄

The highest power of x with a nonzero coefficient.

Which is which?

Three quadratics: $q_1(x) = x^2$,
$q_2(x) = -4x^2 + 4$, $q_3(x) = 2x(x + 2)$

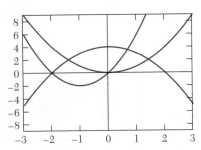

Here are three cubics: ▶

Which is which?

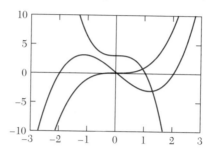

Three cubics: $c_1(x) = x^3$,
$c_2(x) = -3x^3 + 3$, $c_3(x) = x(x-2)(x+2)$

Polynomials in General: Many Shapes. Linear, quadratic, and cubic polynomials have characteristic shapes, but polynomial graphs in general can take almost any smooth, unbroken shape. Some shape restrictions do exist. For instance, a polynomial of degree n can have at most n **roots**; this means, graphically, that a polynomial graph can have at most n **x-intercepts**. (In particular, a polynomial function cannot be periodic.)

Rational Functions

Adding or multiplying two polynomials creates a new polynomial. ▶ *Dividing* two polynomials doesn't (usually) give another polynomial; it gives a rational function. Formally:

Is this really obvious? Think about it; try some examples.

> **Definition** A **rational function** is a function whose rule can be written as the quotient of two polynomials.

EXAMPLE 2   Each of the following functions is rational:

$$f(x) = \frac{1}{x}; \qquad g(x) = \frac{x^2}{x^2 + 3}; \qquad h(x) = \frac{2}{x} + \frac{3}{x+1}.$$

The function $m(x) = \sqrt{x} = x^{1/2}$ is not rational, because it involves a fractional power of x.

The function h may look suspicious. It's certainly the sum of two rational functions, but is h itself rational? The answer is yes, because we can rewrite $h(x)$ in the form

$$h(x) = \frac{2}{x} + \frac{3}{x+1} = \frac{2(x+1)}{x(x+1)} + \frac{3x}{x(x+1)} = \frac{5x+2}{x(x+1)}. \qquad \blacksquare$$

Graphs of Rational Functions

Rational functions, unlike polynomials, have an important new feature: the possibility of **horizontal** and **vertical asymptotes**.

EXAMPLE 3   Consider the rational function

$$r(x) = \frac{x^2}{x^2 - 1} = \frac{x^2}{(x-1)(x+1)}$$

and its graph:

Graph of $r(x) = x^2/(x^2 - 1)$

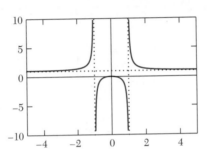

Asymptotes are straight lines toward which, "in the limit," a graph tends. We'll study asymptotes and their relations to limits in Chapter 2. Technicalities temporarily aside, it's already evident that the graph of r has three asymptotes: the vertical lines $x = -1$ and $x = 1$, and the horizontal line $y = 1$.

It's natural to expect trouble where a denominator vanishes. It may be less obvious from the formula how the function misbehaves.

Where the *vertical* asymptotes come from should be clear: They correspond to real roots of the denominator. ◄   The *horizontal* asymptote (the line $y = 1$) reflects r's "long run" behavior. It shows how r behaves for large positive and large negative inputs. Specifically, the asymptote shows that $r(x)$ approaches 1 as x approaches plus or minus infinity. Manipulating r's formula shows why:

$$r(x) = \frac{x^2}{x^2 - 1} = \frac{(x^2 - 1) + 1}{x^2 - 1} = 1 + \frac{1}{x^2 - 1}.$$

The graph of r falls toward the asymptote.

As x tends to plus or minus infinity, $1/(x^2 - 1)$ remains positive but tends to 0; hence, $r(x)$ approaches 1 from above. ◄   Tabulating values of r for inputs with large absolute values gives similar information:

Long-Run Values of $r(x) = x^2/(x^2 - 1)$				
x	10	100	1000	10,000	...
$r(x)$	1.01010101	1.00010001	1.00000100	1.00000001	...
x	−10	−100	−1000	−10,000	...
$r(x)$	1.01010101	1.00010001	1.00000100	1.00000001	...

■

Other Algebraic Functions

Polynomials and rational functions involve only the standard arithmetic operations: addition, subtraction, multiplication, division, and raising to integer powers. "Other" algebraic functions include ones such as

$$f(x) = \sqrt{x}, \qquad g(x) = \sqrt{x + \sqrt{x}}, \qquad \text{and} \qquad h(x) = \sqrt{\sqrt[3]{x^2 + 1} + x^{6/7}}$$

that involve nth roots. As we'll soon see, calculus methods and ideas apply just as readily to these functions as to polynomials and rational functions.

These "other" algebraic functions do pose one new problem: finding output values. To illustrate, compare the functions

$$m(x) = x + \frac{5x + 2}{x(x + 1)} \qquad \text{and} \qquad f(x) = \sqrt{x}.$$

It's easy to compute, even by hand, that $m(2) = 4$. Finding $f(2) = \sqrt{2}$ to any respectable numerical accuracy is harder, at least by hand. ▶ Even here, calculus can help. As we said in Section 1.1, ▶ calculus can help us choose polynomial functions that closely approximate the square root function.

A calculator has no trouble, of course.

See the graph on page 9.

Exponential and Logarithm Functions

Exponential and logarithm functions are among the most important and most practically useful functions in calculus. ▶ Understanding these functions requires, first of all, a familiarity with the algebra of logarithms and exponentials. Appendices E and F contain quick reviews of the necessary algebraic basics. Most basic of all are the following key facts. ▶

Section 1.7 contains a thoroughly practical application: using exponential functions to study how inflation affects college tuition.

If any are unfamiliar, review the appendices now!

> **How Logs and Exponentials Are Related** For a positive base b and real numbers x and y,
>
> $$y = \log_b x \iff x = b^y.$$
>
> This relation between logarithms and exponentials is crucial to understanding of either type of function. In particular, we'll use this relation to *define* logarithm functions using exponential functions.

> **Important Properties of Logs and Exponentials** We'll use these key properties often:
>
> $$b^x b^y = b^{x+y}; \qquad\qquad (b^x)^r = b^{xr};$$
> $$\log_b xy = \log_b x + \log_b y; \qquad\qquad \log_b(x^r) = r \log_b x.$$
>
> (**NOTE:** In each line, b must be positive. In the top two equations, x and y can take any real values. In the second two, x and y must be positive. For many more details, see the appendices.)

Exponential Functions

> **Definition** An **exponential function** is defined by an expression of the form
>
> $$f(x) = b^x,$$
>
> where b, the **base**, is a fixed positive number.

Each of the expressions

$$2^x, \qquad 4^t, \qquad \left(\tfrac{1}{2}\right)^z, \qquad \text{and} \qquad e^x$$

defines an exponential function. (The base is fixed; the exponent varies—hence the name.) *None* of the expressions

$$x^2, \qquad z^3, \qquad (w+1)^{15}, \qquad x^\pi$$

defines an exponential function, because each involves a fixed power of a variable base.

Base e. Among all exponential functions, the one with base e turns out to be especially useful and convenient. (The fact that e is usually the default base on scientific calculators attests to its special importance.) The number e is irrational; its decimal expansion begins $e = 2.71828182\ldots$. What makes base e special—for both exponential and logarithm functions—will become clearer soon, when we study derivatives.

The name is a bit unfortunate, because other functions with other bases are also exponential.

Parentheses are often omitted with such named functions as sin, cos, ln, and exp.

The function $f(x) = e^x$, with base e, is often called the **exponential function** ◄ or, sometimes, the **natural exponential function**. Alternative notations ◄ include

$$\exp(x), \qquad \exp x, \qquad \text{and} \qquad e^x.$$

Graphs of Exponential Functions. Graphs of several exponential functions follow. The rule for each is of the form b^x—only the base b varies. Points of special interest are bulleted.

Exponential functions with various bases

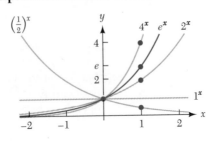

Observe:

Legitimate Bases   If $b > 0$, then the expression b^x makes sense for any real number x. ▶ Thus, any positive base b defines a sensible exponential function. (In practice, $b = 2$, $b = e$, and $b = 10$ are by far the most useful.)

Why not $b \le 0$? Appendix E gives details.

Domain and Range   Exponential functions accept all real inputs, so all have domain $(-\infty, \infty)$. The graphs show that unless $b = 1$, the range of the exponential function $f(x) = b^x$ is $(0, \infty)$.

Shapes   The shape of an exponential graph depends on the base b:

The greater the value of b, the faster b^x increases.

(If $b < 1$, b^x decreases. If $b = 1$, then b^x is a constant function.)

A Common Point   Every graph of the form $y = b^x$ passes through the point $(0, 1)$. In the language of functions:

If $f(x) = b^x$,   then $f(0) = b^0 = 1$.

The input $x = 1$ is also of interest. For any function $f(x) = b^x$, $f(1) = b^1 = b$. ▶

How do the graphs reflect these facts?

Exponential Functions Are Monotonic   All exponential functions $f(x) = b^x$ are **monotonic**, ▶ i.e., either nonincreasing everywhere or nondecreasing everywhere. In graphical language,

If $b > 1$, the graph of $y = b^x$ is everywhere rising. If $0 < b < 1$, the graph of $y = b^x$ is everywhere falling. ▶

Learn the word "monotonic"; it's convenient shorthand for a longer-winded phrase.

What happens if $b = 1$?

Faster and Faster   Perhaps the most important calculus property of exponential functions concerns *rates* of growth:

If $y = b^x$, then the rate of change of y is proportional to y itself.

This property of exponential functions is subtler than the others we've discussed. It can be seen only qualitatively in the graphs. The graph $y = 2^x$, for example, rises *faster* as it rises *higher*. We'll return to this important idea later, after studying rates of growth more carefully.

Exponential and Power Functions: They're Not the Same

The expressions 2^x and x^2 are typographically similar, but the resemblance ends there. They define entirely different functions, and their graphs show it:

2^x vs. x^2: which is which?

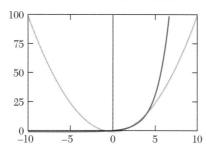

Note some differences:

Asymptotes   The graph of $y = x^2$ has no horizontal (or vertical) asymptotes. The graph of $y = 2^x$ has the line $y = 0$ as a horizontal asymptote, corresponding to the fact that $2^x \to 0$ as $x \to -\infty$. In fact, *every* exponential function $y = b^x$ (with $b > 0$ and $b \neq 1$) has the line $y = 0$ as a horizontal asymptote. By contrast, *no* power function x^b (with $b > 0$) has a horizontal asymptote. ◄

If $b < 0$, then $y = 0$ is a horizontal asymptote.

Comparing Rates of Growth   As $x \to \infty$, both

$$x^2 \to \infty \qquad \text{and} \qquad 2^x \to \infty.$$

However, 2^x blows up much, much faster than x^2. Graphs can only begin to suggest how quickly 2^x outruns x^2. To put the difference more dramatically, let $f(x) = x^2$ and $g(x) = 2^x$. Then $f(1000)$ dollars might buy a starter home in Hollywood, but $g(1000)$ is (more or less) the number of molecules in the universe. For even larger inputs, and larger values of b, the differences are even greater.

Logarithm Functions

This works for any positive base b except $b = 1$. Why not $b = 1$?

Notice the "mirror symmetry" in the table.

Exponential and logarithm functions come in pairs. An *exponential* function with base b corresponds to a *logarithm* function with the same base. ◄ The following table collects values of one such "matched pair" of functions, here with base $b = 2$. The values illustrate the correspondence between exponential and logarithm functions. ◄

Exponential and Logarithm Function Values								
Values of 2^x								
x	-2	-1	0	1	2	3	10	13.28771
2^x	1/4	1/2	1	2	4	8	1024	10,000
Values of $\log_2 x$								
x	1/4	1/2	1	2	4	8	1024	10,000
$\log_2 x$	-2	-1	0	1	2	3	10	13.28771

The formal definition makes the correspondence between logarithm and exponential precise:

Definition   The **logarithm function with base** b, denoted by

$$f(x) = \log_b x,$$

is defined by the condition

$$y = \log_b x \iff b^y = x.$$

Each of the expressions

$$\log_2 x, \qquad \log_4 x, \qquad \log_{1/2} x, \qquad \text{and} \qquad \log_e x$$

defines a logarithm function with some base b. The logarithm function with base e ▶ is called the **natural logarithm function**; in print and on calculator buttons, it's usually denoted by

Recall that $e \approx 2.718$.

$$\ln(x) \qquad \text{or} \qquad \ln x.$$

Graphs of Logarithm Functions. Graphs of logarithm functions with various bases follow. (Exponential functions with the same bases were plotted earlier; compare the pictures.) Interesting points are shown bulleted.

Logarithm functions with various bases

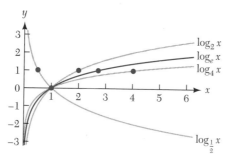

Not surprisingly, properties of logarithm function graphs reflect ▶ analogous properties of exponential function graphs.

Pun intended.

Legitimate Bases Any positive number b except $b = 1$ can be used as the base for a logarithm function. ▶ (In practice, as for exponentials, $b = 2$, $b = e$, and $b = 10$ are most useful.)

Why not $b = 1$? Because $1^y = 1$ for all $y > 0$. Appendix F has more details.

Domain and Range The graphs illustrate that for any suitable base b, the logarithm function accepts *positive* inputs; outputs take on *all* real values. Thus, every logarithm function has domain $(0, \infty)$ and range $(-\infty, \infty)$.

Growth Rates As for exponential functions, each logarithmic graph's shape depends on b. This time, the *greater* the value of b, the more *slowly* $\log_b x$ increases. (If $b < 1$, $\log_b x$ decreases.) Graphs of exponential functions rise faster and faster as x increases. Graphs of logarithm functions have the opposite property: They rise more and more slowly as x increases.

A Common Point The graph of every logarithm function passes through the point $(1, 0)$. ▶ In functional language,

Which point was common to exponentials?

$$\text{If } f(x) = \log_b x \qquad \text{then} \qquad f(1) = \log_b 1 = 0.$$

Similarly, for any $b > 0$, $f(b) = \log_b b = 1$.

Logarithm Functions Are Monotonic   Every logarithm function is strictly monotonic. ◄   If $b > 1$, the graph of $y = \log_b(x)$ *rises* everywhere. If $0 < b < 1$, the graph *falls* everywhere. ◄

That is, it is either strictly increasing everywhere or strictly decreasing everywhere.

There is no logarithm function for $b = 1$.

Exponential and Logarithm Functions as Inverses

The condition

$$y = \log_b x \iff x = b^y$$

means that a logarithm function and an exponential function with the same base are inverses of each other. ◄

We'll review the idea of inverse functions in the next section.

EXAMPLE 4   For any base b, $f(x) = b^x$ and $g(x) = \log_b x$ are inverse functions. What does this mean if $b = 10$ and $x = 100$? What if $x = 5$?

Solution   That the base 10 exponential and logarithm functions are inverses means that $y = \log_{10} x \iff x = 10^y$. For $x = 100$, this says that

$$y = \log_{10} 100 \iff 100 = 10^y.$$

Since the second equation holds only when $y = 2$, it follows that $\log_{10} 100 = 2$.
    What if $x = 5$? As any calculator knows, $\log_{10} 5 \approx 0.6989700043$. In this context, the inverse property says that $5 = 10^{0.6989700043}$. Your calculator should agree. ◄   ∎

Does it?

Inverse Functions: The Graphical View.   The base e exponential and logarithm functions exp and ln ◄   are inverses. The same holds for base 2. The following graphs show what this means geometrically.

Also known as e^x and $\log_e x$, respectively.

Logarithm and exponential functions are inverses

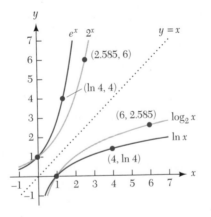

To make the picture look right, we used horizontal and vertical units of equal size.

The graphs of inverse functions are symmetric with respect to the line $y = x$. ◄
    The picture suggests, too, why the base e is special. Look carefully:

At the points $(0, 1)$ and $(1, 0)$ the graphs of $y = e^x$ and $y = \ln x$ are parallel to each other and to the line $y = x$.

The other two graphs don't have this property. At the moment, this property of e may seem unimportant, but it will soon prove surprisingly useful.

Logs or Exponentials: Which Are Simpler?   The equation

$$y = \log_b x \iff x = b^y$$

can be taken to define the function $\log_b(x)$ in terms of the intuitively simpler exponential function b^x. To see that $\log_4 1.38703096913 \approx 0.236$, for example, we could check that $4^{0.236} \approx 1.38703096913$.

  Does this approach to logarithms make sense? Are exponential functions really simpler than logarithm functions? In a strict logical sense, no: Knowing either implies knowing the other. Intuitively, though, exponentiation is probably simpler than finding logarithms, just as squaring is simpler than finding square roots.

  Defining logarithm functions via exponential functions is not the only possibility. The reverse strategy (to start with logarithms—defined without any reference to exponentials—and proceed to exponentials) is sometimes used. Although perhaps less intuitive, this approach can skirt troublesome technicalities.

Trigonometric Functions

In geometry, sines, cosines, tangents, and other trigonometric quantities are defined as certain ratios of the sides of right triangles. Here our point of view is different: We study sines, cosines, and other trigonometric objects as functions in their own right, connecting them only occasionally to right triangles. (For a review of trigonometric basics, see Appendix G.)

Why Trigonometric Functions Matter

The most important property of the trigonometric functions—sine, cosine, tangent, and so on—is their repetitive, or periodic, behavior. The famous undulating graphs of the sine and cosine functions reflect this wavelike character:

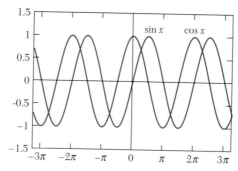

Graphs of $y = \sin x$ and $y = \cos x$

As always, the viewing window shows only part of the graph. In this case, what's shown is enough: The full graphs of the sine and cosine functions repeat the same pattern infinitely often.

Trigonometric Functions Defined: Circular Functions

Calling it the unit circle is a bit sloppy (many different circles have radius 1), but everyone does it.

Using u and v here saves x and y for later.

For calculus purposes, it's best to think of trigonometric functions in terms of circles, not triangles. The simplest circle, the **unit circle**, ◄ has radius 1 and center $(0, 0)$. In the uv plane, the unit circle's equation is $u^2 + v^2 = 1$. ◄

Imagine a point P moving *counterclockwise* around the unit circle, starting from the **east pole** $(1, 0)$. As P moves endlessly around the unit circle, many related quantities vary as well: the u-coordinate of P, the v-coordinate of P, the slope of the segment from P to the origin, and others. All the trigonometric functions can be derived from these varying quantities. (As a result, the sine and cosine functions are sometimes called **circular functions**.)

> **Definition** For any real number x, let $P(x)$ be the point reached by moving x units of distance counterclockwise around the unit circle, starting from $(1, 0)$. (If $x < 0$, go clockwise.) Then
>
> $$\cos(x) = u\text{-coordinate of } P(x);$$
>
> $$\sin(x) = v\text{-coordinate of } P(x).$$

It's worth many words; look carefully.

Here's the important picture: ◄

Defining sin x and cos x

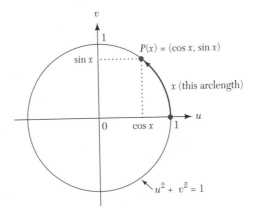

Simplest Properties of the Sine and Cosine Functions

Combining the definitions, the picture, the sine and cosine graphs, and a little reflection makes many properties of the sine and cosine functions evident.

> **Evident? Really?** Mathematicians use words like "evident," "clear," and "trivial" in peculiar ways. "Evident" does not mean "immediately obvious" or "This stuff is transparent to everyone with the slightest aptitude for mathematics; if it isn't clear to you, go away and never come back." "Evident" really means "not difficult to convince oneself of, with sufficient time and effort."

Domain and Range For any real number x, $\cos(x)$ and $\sin(x)$ are the coordinates of $P(x)$, the point x units of distance, measured counterclockwise, from $(1, 0)$. ▶ Thus, every real number x is a legal input; the domain of both the sine and cosine functions is $(-\infty, \infty)$. ▶

Recall: If $x < 0$, measure clockwise.

How do their graphs reflect this fact?

As our peripatetic point $P(x)$ circumnavigates the unit circle, its u- and v-coordinates fluctuate, but only between -1 and 1. Therefore, both functions share the same range: the closed interval $[-1, 1]$. In other words, for all x,

$$-1 \le \cos(x) \le 1 \quad \text{and} \quad -1 \le \sin(x) \le 1.$$

Famous Values The following table collects some important values of the sine and cosine functions. ▶

Check some!

Famous Sines and Cosines						
x	0	$\pi/2$	π	2π	$-\pi/2$	$\pi/4$
$P(x)$	$(1, 0)$	$(0, 1)$	$(-1, 0)$	$(1, 0)$	$(0, -1)$	$(\sqrt{2}/2, \sqrt{2}/2)$
$\cos(x)$	1	0	-1	1	0	$\sqrt{2}/2$
$\sin(x)$	0	1	0	0	-1	$\sqrt{2}/2$

Odd and Even It follows from the definition and the symmetry of the unit circle that the sine function is odd and the cosine function is even. ▶ In symbols,

How do their graphs show this?

$$\sin(-x) = -\sin(x) \quad \text{and} \quad \cos(-x) = \cos(x).$$

Trigonometric Functions Are 2π-Periodic For any real number x, the points $P(x)$ and $P(x + 2\pi)$ are identical. ▶ Therefore, both the sine and cosine functions repeat themselves, as the graphs show, on each interval of length 2π. In "math-speak," the sine and cosine functions are 2π-**periodic**. In symbols,

*Do you see why? [**HINT**: What's the circumference of the unit circle?]*

$$\sin(x + 2\pi) = \sin(x) \quad \text{and} \quad \cos(x + 2\pi) = \cos(x).$$

A Famous Trigonometric Identity For any real number x, the point $P(x) = (\cos(x), \sin(x))$ lies on the unit circle, so the coordinates of P must satisfy the defining equation $u^2 + v^2 = 1$. Therefore,

$$(\sin x)^2 + (\cos x)^2 = 1.$$

Equations like this one, that relate various trigonometric functions to each other, are called **trigonometric identities**. Many trigonometric identities exist; for examples, see Appendix G.

No Parentheses? Trigonometric, exponential, and logarithm expressions are often typeset without parentheses. For example,

$$\sin(x), \quad \ln(x), \quad \sin x, \quad \text{and} \quad \ln x$$

are all syntactically correct, but the last two are typographically standard. When confusion seems likely, we'll use parentheses.

No Algebraic Formulas   Might the sine and cosine functions have simple algebraic formulas? Alas, no: Trigonometric functions are known to be **transcendental**; i.e., they *can't* be written as algebraic combinations of powers, roots, sums, products, and so on. On the other hand, the trigonometric functions are far from mysterious; they've been studied, tabulated, and used for centuries. Even cheap calculators produce highly accurate values of these functions. How do they do it? Stay tuned.

Other Trigonometric Functions

Four additional trigonometric functions are defined in terms of the sine and cosine functions.

Definition For real numbers x,

$$\tan x = \frac{\sin x}{\cos x}; \qquad \cot x = \frac{\cos x}{\sin x};$$

$$\sec x = \frac{1}{\cos x}; \qquad \csc x = \frac{1}{\sin x}.$$

Pieces of the graphs of these functions follow. Pieces are enough: Like the sine and cosine functions, they repeat on every interval of length 2π. (In fact, the tangent and cotangent functions turn out to repeat even on intervals of length π. Appendix G explains why.)

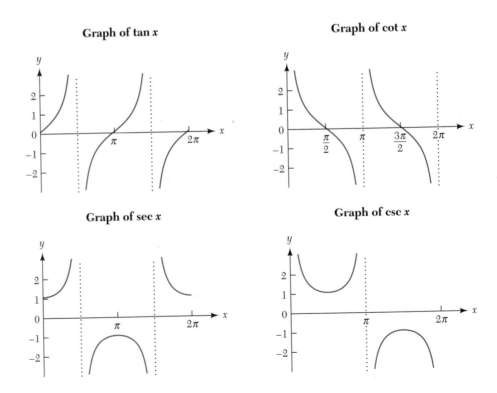

Graph of tan x   **Graph of cot x**

Graph of sec x   **Graph of csc x**

Properties of the "Other" Trigonometric Functions

As legitimate children of the sine and cosine functions, the tangent, cotangent, secant, and cosecant functions inherit certain properties from their parents, but have their own personalities too.

Domains and Ranges Because the sine and cosine functions are defined for *all* real numbers, the other trigonometric functions are also defined for all real numbers except those at which a denominator is 0. The domain of the cosecant function, for instance, consists of all real numbers that are not integer multiples of π.

As their graphs show, the tangent and cotangent functions have range $(-\infty, \infty)$—all real numbers are possible outputs. Outputs from the secant and cosecant functions, by contrast, must have absolute value at least 1. In interval language, the secant and cosecant functions have range $(-\infty, -1] \cup [1, \infty)$.

Asymptotes The tangent, cotangent, secant, and cosecant functions all have vertical asymptotes. Appendix G discusses the asymptotes of the tangent function in detail.

Functional Properties The following properties of the "other" trigonometric functions follow directly from their definitions. ▶ Each holds for all inputs x in the given function's domain.

$$|\sec x| \geq 1; \qquad\qquad |\csc x| \geq 1;$$

$$\tan(-x) = -\tan x; \qquad\qquad \sec(-x) = \sec x.$$

Convince yourself carefully of each one, using only the definitions and the properties of the sine and cosine functions. Then observe how the graphs reflect each property.

The Tangent Function, the Unit Circle, and Slope. Like the sine and cosine functions, the tangent function has an important and pleasing geometric interpretation in terms of the unit circle:

$$\tan x = \frac{\sin x}{\cos x} = \text{slope of the line from the origin to } P(x).$$

Interpreting the tangent function as slope shows, for instance, why $\tan(\pi/4) = 1$, and why $\tan x \to \pm\infty$ as $x \to \pi/2$.

BASICS

In Exercises 1–8, find the domain and range of each function.

1. $f(x) = e^{x^2}$
2. $f(x) = \ln(x^2)$
3. $f(x) = \ln(x^3 + 1)$
4. $f(x) = \exp\left((x - 4)^2 - 1\right)$
5. $f(x) = \sin(3x - 1)$
6. $f(x) = 1 + 2\cos(3x)$
7. $f(x) = \sqrt{1 + \cos x}$
8. $f(x) = (1 + \sin x)^{-1/3}$

In Exercises 9–16, indicate whether the function is even, odd, or neither. If a function is even or odd, show this using the facts that sine is an odd function and cosine is an even function. [**HINT:** Start by examining a graph of the function.]

9. $f(x) = e^{\cos x}$
10. $f(x) = 1 + \cos x$
11. $f(x) = x + \sin x$
12. $f(x) = x \sin x$
13. $f(x) = x \cos x$
14. $f(x) = |\cos x|$
15. $f(x) = |\sin x|$
16. $f(x) = \sin^2 x$

In Exercises 17–22, find the period of each of the following functions.

17. $f(x) = \cos x - 2$
18. $f(x) = e^{\sin x}$
19. $f(x) = \sin(2x)$
20. $f(x) = \sin(x/2)$
21. $f(x) = \sin(2\pi x)$
22. $f(x) = \cot x$

23. Do the points shown in the following table lie on a line? If so, find an equation of the line. Justify your answers.

x	5.2	5.3	5.4	5.6	5.9	6.5
y	27.8	29.2	30.6	33.4	37.6	46.0

24. Do the points shown in the following table lie on a line? If so, find an equation of the line. Justify your answers.

x	0	1	3	6	10	15
y	0.000	0.785	1.249	1.406	1.471	1.504

FURTHER EXERCISES

Each of the functions graphed in Exercises 25–28 is described by a symbolic expression having *one* of the following forms, where A and B are real numbers.

(i) $f(x) = Ax + B$
(ii) $f(x) = Ax^2 + B$
(iii) $f(x) = A\cos x + B$
(iv) $f(x) = \cos(Ax) + B$
(v) $f(x) = A\sin x + B$
(vi) $f(x) = \sin(Ax) + B$
(vii) $f(x) = 3^{(x+A)} + B$
(viii) $f(x) = \log_2(x + A) + B$

Decide which expression corresponds to each function and find appropriate numerical values for the parameters A and B. Explain each choice in a sentence or two.

(a)

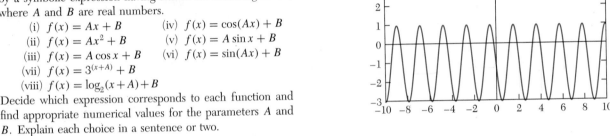

(b)

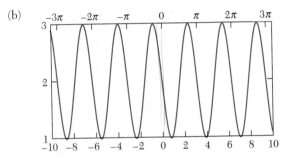

25.

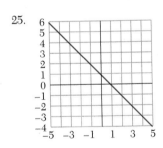

26.

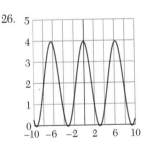

27.

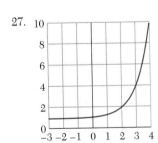

28.

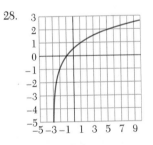

29. Each of the following graphs is of a function of the form $f(x) = A\sin(Bx) + D$. For each graph, identify values of the parameters A, B, and D.

30. Each of the following graphs is of a function of the form $g(x) = A\cos(x + C) + D$. For each graph, identify values of the parameters A, C, and D.

(a)

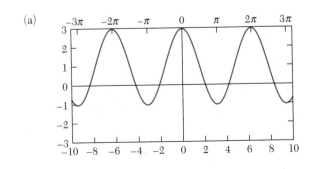

(b)

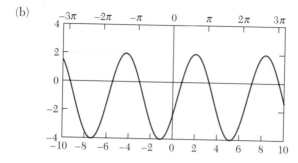

31. (a) Give an example of a polynomial of even degree that defines an even function.
    (b) Does *every* polynomial of even degree define an even function? Why or why not?

32. Let $f(x) = e^x$ and $g(x) = 10^x$.
    (a) What is the domain of g?
    (b) Show that there is a number k such that $g(x) = f(kx)$.
    (c) Explain how to obtain the graph of g from the graph of f.

33. Let $f(x) = \ln x$ and $g(x) = \log_{10} x$.
    (a) What is the domain of g?
    (b) Show that there is a number k such that $g(x) = kf(x)$.
    (c) Explain how to obtain the graph of g from the graph of f.

34. Let $f(x) = \ln(x^3)$ and $g(x) = \ln(x^2)$.
    (a) What is the domain of f?
    (b) Explain how to obtain the graph of f from the graph of $y = \ln x$.
    (c) What is the domain of g?
    (d) Explain how to obtain the graph of g from the graph of $y = \ln x$.

35. Let $f(x) = \ln(2 + \sin x)$.
    (a) Evaluate $f(0)$, $f(\pi/6)$, and $f(\pi/2)$ using a calculator.
    (b) What are the domain and range of f?
    (c) Is f even, odd, or neither? Justify your answer.
    (d) Is f periodic? If so, find its period. If not, explain why not.
    (e) Find the maximum and minimum values of f.

36. Let $f(x) = Ae^{kx}$, where $A > 0$ and k are constants. Show that the graph of $y = \ln(f(x))$ is a straight line.

37. Let $g(x) = Ax^k$, where $A > 0$ and k are constants. Show that the graph of $y = \ln(g(e^x))$ is a straight line.

38. Suppose that $y = 20x + 1$ is an equation of the line that intersects the curve $y = b^x$ at $x = 0$ and $x = 4$.
    (a) Find the number b.
    (b) What is the slope of the line that intersects the curve $y = \log_b x$ at $x = 1$ and $x = 81$?

39. Use properties of the sine and cosine functions to show that $\sec x$ is an even function.

40. Use properties of the sine and cosine functions to show that $\tan x$ is an odd function.

In Exercises 41–46, use properties of the trigonometric functions to find the period of each function. [**HINT:** $\cos(x + \pi) = -\cos x$; $\sin(x + \pi) = -\sin x$.]

41. $f(x) = |\sin x|$
42. $f(x) = 2|\cos(3x)|$
43. $f(x) = \cos^2 x$
44. $f(x) = \cos x \sin x$
45. $f(x) = \cos(|x|)$
46. $f(x) = \sin(|x|)$

47. Use properties of the cosine function to explain why $|\sec x| \geq 1$ for every x in the domain of the secant function.

48. Use properties of the sine function to explain why $|\csc x| \geq 1$ for every x in the domain of the cosecant function.

49. Explain why $\dfrac{1}{3} \leq \dfrac{1}{2 + \cos x} \leq 1$ for all x.

50. Show that $0.3 < \dfrac{1}{1 + \tan x} \leq 1$ if $0 \leq x \leq 1$.
    [**HINT:** $\tan x$ is increasing on the interval $[0, 1]$ and $\tan(\pi/3) = \sqrt{3} < 2$.]

51. Find a number U such that $|1 + \sec x| \leq U$ if $2 \leq x \leq 4$.

52. Each of the following statement shows that the sine function is *not* a polynomial. Explain why.
    (a) The equation $\sin x = 0$ has infinitely many solutions ($x = n\pi$ is a solution for every $n \in \mathbb{Z}$).
    (b) The sine function is bounded below by -1 and above by 1 (i.e., for every x, $-1 \leq \sin x \leq 1$).

53. Suppose that f is a linear function (i.e., $f(x) = ax + b$, where a and b are constants).
    (a) Show that $f(5) - f(3) = f(10) - f(8)$.
    (b) Find a number c such that $f(4) - f(1) = f(c) - f(-9)$.
    (c) Show that $3(f(2) - f(-1)) = f(10) - f(1)$.
    (d) Assume that $h \neq 0$ is a constant. Show that the value of the expression

    $$\frac{f(x + h) - f(x)}{h}$$

    does not depend on either x or h. What is the geometric significance of this expression?

54. Suppose that f is an exponential function (i.e., $f(x) = Ab^x$, where A and $b > 0$ are constants).
    (a) Explain why $f(5)/f(3) = f(10)/f(8)$.
    (b) Find a number a such that $f(4)/f(1) = f(a)/f(-9)$.
    (c) Explain why $\left(f(2)/f(-1)\right)^3 = f(10)/f(1)$.
    (d) Show that the value of the expression $f(x + h)/f(x)$ does not depend on x.
    (e) The value of the expression in part (d) does depend on h. Show that the value of the expression

    $$\frac{\ln\left(f(x + h)/f(x)\right)}{h}$$

    does not depend on h. (Assume $h \neq 0$.)

55. The points listed in the following table lie on the graph of an exponential function f.

x	1.5	2.0	3.0	5.0	8.0
y	8.075	8.470	9.317	11.273	15.005

(a) Evaluate $f(12)$. [**HINT:** See the previous exercise.]
(b) Evaluate $f(0)$.
(c) Evaluate $f(1)$.
(d) Find a symbolic expression of the form Ab^x for f.
(e) Find a symbolic expression of the form Ae^{kx} for f.

56. Suppose that the line $y = 2$ is a horizontal asymptote of the function f.
(a) Explain why the line $y = 5$ is a horizontal asymptote of the function $g(x) = f(x) + 3$.
(b) Find the equation of a horizontal asymptote of the function $g(x) = f(x + 1)$.
(c) Find the equation of a horizontal asymptote of the function $g(x) = 3f(x + 4) + 1$.

57. Suppose that the line $x = -3$ is a vertical asymptote of the function f.
(a) Explain why the line $x = -3$ is a vertical asymptote of the function $g(x) = f(x) + 3$.
(b) Find the equation of a vertical asymptote of the function $g(x) = f(x + 1)$.
(c) Find the equation of a vertical asymptote of the function $g(x) = 3f(x + 4) + 1$.

58. Let $f(x) = \dfrac{\sqrt{x} - 3}{x - 9}$ and $g(x) = \dfrac{1}{\sqrt{x} + 3}$.

(a) Show that $f(x) = g(x)$ for all $x \geq 0$ such that $x \neq 9$.
(b) Find the domain and range of f.
(c) Find the domain and range of g.
(d) How are the graphs of f and g related?
(e) Does the graph of f have a horizontal asymptote? If so, find an equation for this asymptote.
(f) Does the graph of f have a vertical asymptote? If so, find an equation for this asymptote.

59. Let $f(x) = \dfrac{x - 3}{x^2 - 9}$ and $g(x) = \dfrac{1}{x + 3}$.
(a) Show that $f(x) = g(x)$ for all x such that $x \neq \pm 3$.
(b) Find the domain and range of f.
(c) Find the domain and range of g.
(d) How are the graphs of f and g related?
(e) Does the graph of f have a horizontal asymptote? If so, find an equation for this asymptote.
(f) Does the graph of f have a vertical asymptote? If so, find an equation for this asymptote.

60. Find a formula for a rational function that has a vertical asymptote at $x = 1$ but has no horizontal asymptote. [**HINT:** There are many possible answers. Check your answer by graphing.]

61. Find a formula for a rational function that has the line $y = 1$ as a horizontal asymptote but has no vertical asymptotes. [**HINT:** There are many possible answers. Check your answer by graphing.]

62. Find a formula for a rational function that has vertical asymptotes at $x = -1$ and $x = 2$ and that has the line $y = 3$ as a horizontal asymptote. [**HINT:** There are many possible answers. Check your answer by graphing.]

1.6 New Functions from Old

Operations

Ordinary addition, as in $2 + 3 = 5$, is an **operation on numbers**: Two old numbers are operated on to produce a "new" number. Subtraction, multiplication, division, and negation ◄ are also operations on numbers. Each begins with one or more old numbers and—somehow or other—produces a new number.

Multiplication by −1.

Operations on functions are similar to operations on numbers. Addition, for instance, makes sense for functions: The function $h(x) = x + \cos x$ is, in a natural way, the *sum* of the functions $f(x) = \cos x$ and $g(x) = x$.

A mathematical operation starts with one or more objects of some type and produces a new object *of the same type*. In particular:

An operation on functions accepts one or more functions as input and produces a new function as output.

Operations on Functions: The Machine View

Schematic diagrams like these

represent functions as input–output machines; they accept inputs and produce outputs. ▶   From this "mechanical" point of view, function operations hook simple
machines together to produce fancier machines.

*The first function name is
generic; the second refers to
the squaring function.*

> **A Literal Function Machine.** The boxes that represent functions in our machine
> diagrams resemble the function keys on a scientific calculator. This is a happy coincidence,
> because calculator function keys, such as $\boxed{\text{SIN}}$, nicely illustrate the idea of a function machine.
> Pressing $\boxed{\text{SIN}}$ takes an input, does something to it, and returns an output.

Adding Machines.   The following diagram gives a machine view of function addition: Two function machines, f and g, are linked in parallel to form a new function
machine, h:

Building a new function machine: addition

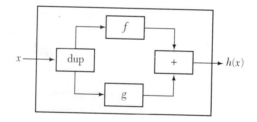

The first box duplicates an input x, because one copy is needed for f, and one for g.
The last box adds its two inputs. The large outer box represents the new function,
h. Its presence emphasizes that the sum of two functions is another function.
   The definition of function addition makes precise what the picture depicts. ▶

*The other algebraic
operations have similar
definitions and similar
machine diagrams. We'll
handle them less formally.*

> **Definition** Let f and g be two functions. Their sum, denoted by
> $f + g$, is the new function whose rule is
>
> $$(f + g)(x) = f(x) + g(x).$$

Algebraic Operations on Functions

Algebraic operations—addition, subtraction, multiplication, and division—on symbolically defined functions are simple and familiar. The following table illustrates
various operations applied to the functions f and g defined by $f(x) = \cos x$ and
$g(x) = x$.

Algebraic Operations on Functions			
Operation	**Output**	**Rule**	**Domain**
Addition	$f + g$	$\cos(x) + x$	$(-\infty, \infty)$
Subtraction	$f - g$	$\cos(x) - x$	$(-\infty, \infty)$
Multiplication	$f \cdot g$	$x \cos(x)$	$(-\infty, \infty)$
Division	f/g	$\cos(x)/x$	$\{x \mid x \neq 0\}$

The table contains few surprises, but note the following.

Outputs Are Functions   Each output is a new function; its rule appears in the third column.

Domains   The natural domain of f/g is smaller than the domains of f and g, because the quotient makes sense only if $g(x) \neq 0$. Sometimes, as the next example shows, domains need closer attention.

Algebraic Operations on "Other" Functions.   Not all functions have symbolic formulas. With due care taken, however, algebraic operations still make sense.

EXAMPLE 1   Let f be defined by the formula $f(x) = x^2 + 3/x$, and let g be defined by the table

x	-3	-2	-1	0	1	2	3	4
$g(x)$	-4	-9	-1	2	0	8	3	-2

Find the natural domain and some output values of f/g.

Solution   The expression $f(x)/g(x)$ makes sense if and only if both numerator and denominator are defined and $g(x) \neq 0$. If, say, $x = 3$, then

$$\frac{f}{g}(3) = \frac{f(3)}{g(3)} = \frac{10}{3}.$$

However,

$$\frac{f}{g}(2.3) = \frac{f(2.3)}{g(2.3)}, \qquad \frac{f}{g}(0) = \frac{f(0)}{g(0)}, \qquad \text{and} \qquad \frac{f}{g}(1) = \frac{f(1)}{g(1)}$$

Why? Convince yourself.   are all undefined—each for a different reason. ◀   In fact, $f(x)/g(x)$ is defined only for x in the finite set $\{-3, -2, -1, 2, 3, 4\}$. Thus, the domain of f/g is smaller than either the domain of f or the domain of g. ∎

Composition

Let f and g be the functions defined by $f(x) = x^2$ and $g(x) = \cos x$. The new functions h and k defined by

$$h(x) = \cos(x^2) \qquad \text{and} \qquad k(x) = (\cos x)^2$$

are constructed from f and g by **composition**, i.e., by following one function with the other. More precisely:

> **Definition** Let f and g be functions. The **composition** of f and g, denoted $f \circ g$, is the function defined by the rule
>
> $$(f \circ g)(x) = f\big(g(x)\big).$$

For the preceding functions f, g, h, and k,

$$(f \circ g)(x) = f\big(g(x)\big) = f(\cos x) = (\cos x)^2 = k(x);$$

$$(g \circ f)(x) = g\big(f(x)\big) = g\left(x^2\right) = \cos(x^2) = h(x).$$

Notice the following properties of composition:

Order Matters   If $f(x) = \sqrt{x}$ and $g(x) = x + 5$, then

$$(f \circ g)(x) = f(x + 5) = \sqrt{x + 5};$$

$$(g \circ f)(x) = g\left(\sqrt{x}\right) = \sqrt{x} + 5.$$

The two functions $f \circ g$ and $g \circ f$ are different. For instance, $(f \circ g)(4) = \sqrt{9} = 3$, while $(g \circ f)(4) = \sqrt{4} + 5 = 7$.

Domains   A legal input x to $f \circ g$ must first be a legal input to g. To keep things moving, the output $g(x)$ must also be a legal input to f. ▶   Thus, the domain of $f \circ g$ is a subset of the domain of g. For the preceding functions, the domains of $f \circ g$ and $g \circ f$ are, respectively, the intervals $[-5, \infty)$ and $[0, \infty)$.

The following machine diagram illustrates the idea: A legal input to $f \circ g$ must pass successfully through the whole contraption.

Composition: Function Machines in Series

The machine picture of composition is especially simple:

Mechanical composition

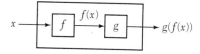

The machines f and g are linked "in series"; the output of f is "piped" through g.

Composing Functions Defined by Formulas: Substitution

Composing functions defined by explicit symbolic formulas, such as $f(x) = x + 2$ and $g(x) = \sin x$, is a straightforward matter:

$$(f \circ g)(x) = \sin(x) + 2; \qquad (g \circ f)(x) = \sin(x + 2).$$

Like f and g themselves, the composite functions $f \circ g$ and $g \circ f$ have explicit formulas. Finding these formulas amounts to substituting one symbolic expression into another.

Composing "Other" Functions

The following examples illustrate various ways of composing functions that are not given by formulas. The common thread is the idea of composition itself.

EXAMPLE 2 **(Composing with a tabular function)** Let $f(x) = x + 2$; let g be defined by the table

x	-3	-2	-1	0	1	2	3	4
$g(x)$	-4	-9	-1	2	0	8	3	-2

Describe the functions $f \circ g$ and $g \circ f$; describe their domains.

Solution Because $f(x) = x + 2$,

$$(f \circ g)(x) = f(g(x)) = g(x) + 2 \quad \text{and} \quad (g \circ f)(x) = g(f(x)) = g(x + 2).$$

The formulas permit calculations such as these:

$$(f \circ g)(2) = g(2) + 2 = 8 + 2 = 10;$$

$$(g \circ f)(2) = g(2 + 2) = g(4) = -2.$$

But not just *any* input works;

$$(f \circ g)(10) = g(10) + 2$$

is undefined. The problem is with domains. Although f accepts any real number as input, g has a tiny domain—the finite set $\{-3, -2, \ldots, 3, 4\}$. The effect is to restrict sharply the domains of both $f \circ g$ and $g \circ f$.

Which inputs pass safely through the composite functions $f \circ g$ and $g \circ f$? Because only g presents an obstruction, the domain of $f \circ g$ is the full domain of g.

The domain of $g \circ f$ is different. Because $(g \circ f)(x) = g(x + 2)$, an input x survives the trip through f and g only if $x + 2$ is in the domain of g, i.e., only if $x \in \{-5, -4, \ldots, 2\}$.

The following table gives full details on both $g \circ f$ and $f \circ g$; gaps in the table reflect gaps in the respective domains.

Composition with a Tabular Function									
x	-5	-4	-3	-2	-1	0	1	2	3	4
$g(x)$			-4	-9	-1	2	0	8	3	-2
$(f \circ g)(x)$			-2	-7	1	4	2	10	5	0
$(g \circ f)(x)$	-4	-9	-1	2	0	8	3	-2		

■

Both f and g actually have fairly simple formulas. Try to guess them if you like, but the graphs are the point here.

EXAMPLE 3 **(Graphical composition)** Let f and g be the following functions: ◀

Composing graphical functions

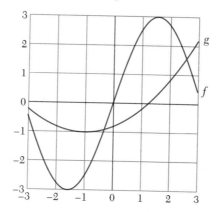

Find approximate values of the composite function $f \circ g$ for integer inputs in $[-3, 3]$; use them to plot $f \circ g$. Is there an x for which $(f \circ g)(x) = 0$?

Solution   By definition, $(f \circ g)(3) = f(g(3))$. Reading approximate values ▶ from the graphs gives

$$(f \circ g)(3) = f(g(3)) \approx f(2.2) \approx 2.4.$$

Check these estimates for yourself.

Similar graphical estimates produced this table of values:

Composing Graphical Functions							
x	-3	-2	-1	0	1	2	3
$(f \circ g)(x)$	-0.6	-2.2	-2.5	-2.2	-0.6	2.2	2.4

With enough such data, we can plot $f \circ g$ as follows (bold curve). The tabulated data points are bulleted.

Composing graphical functions

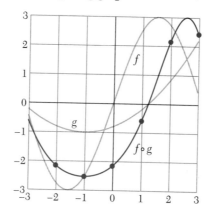

EXAMPLE 4   **(Composing functions described in words)**   Motoring across Montana from Billings to Missoula, one's *altitude* (measured in feet above sea level) varies with *distance* traveled (in miles from Missoula). Distance, in turn, varies with *time* (in hours since departure), so altitude also varies with time. Interpret the situation in the language of functions and composition.

Solution   Let A, d, and t denote altitude, distance, and time. Altitude is a function—say f—of distance; distance is another function—say, g—of time. In function notation,

$$A = f(d) \qquad \text{and} \qquad d = g(t).$$

The composite function

$$A(t) = (f \circ g)(t) = f(g(t))$$

represents altitude (in feet) as a function of time (in hours).   ∎

The Identity Function

The numbers 0 and 1 are **identities** for (ordinary) addition and multiplication, respectively, because

$$a + 0 = 0 + a = a \qquad \text{and} \qquad a \cdot 1 = 1 \cdot a = a$$

for any real number a.

We'll assume that all functions here have sets of real numbers as their domain and range. This does no harm and skirts distracting technicalities.

A similar notion of identity applies for composition. Composition is an operation on functions, ◄   so an identity for composition must also be a function. The identity function, denoted by I, has a simple definition: ◄

As a function machine, I is just a hollow tube; inputs pass through unchanged.

> **Definition** The **identity function** for composition is defined by the rule
>
> $$I(x) = x.$$

To deserve the name "identity function," I should satisfy the equations $f \circ I = I \circ f = f$. It does; for any input x,

$$(f \circ I)(x) = f(I(x)) = f(x) \qquad \text{and} \qquad (I \circ f)(x) = I(f(x)) = f(x).$$

In words:

Composing any function f with I, in either order, gives f.

Inverse Functions

But all the meanings are related—in each case, one object is the inverse of another with respect to a certain operation.

The word "inverse" has various meanings in mathematics. ◄   The numbers 3 and -3 are additive inverses because their sum is 0, the additive identity. Similarly, 3 and 1/3 are multiplicative inverses because their product is 1, the multiplicative identity.

Beware: $f^{-1}(x) \neq 1/f(x)$.

The term "inverse function" has yet another meaning. Two functions f and g are inverses if *composing* them—in either order—gives the identity function. Thus the inverse of a function f is another function, denoted by f^{-1}, ◄   that undoes the effect of the original function f.

The following examples illustrate the idea of inverse function in the best and worst possible cases.

EXAMPLE 5 If $f(x) = x + 3$, then $f^{-1}(x) = x - 3$. ▶ For any input x,

$$\left(f^{-1} \circ f\right)(x) = f^{-1}(x+3) = (x+3) - 3 = x = I(x);$$

$$\left(f \circ f^{-1}\right)(x) = f(x-3) = (x-3) + 3 = x = I(x).$$

f adds three; f^{-1} takes it away.

In other words, $f^{-1} \circ f = I = f \circ f^{-1}$. As required, both compositions give the identity function.

■

EXAMPLE 6 Let f be the constant function $f(x) = 3$. This function has no inverse. The problem is that for every function g, $f \circ g$ is still a constant function; ▶ thus $f \circ g$ cannot be the identity function I. Specifically, if g is any function and x any input, then

Composition with a constant function always produces a constant function. Do you see why?

$$(f \circ g)(x) = f(g(x)) = 3.$$

■

Here is the scrupulously correct definition:

Definition Let f and g be functions. If the equations

$$(f \circ g)(x) = x \qquad \text{and} \qquad (g \circ f)(x) = x$$

hold for all x in the domains of g and f, respectively, then f and g are **inverse functions**. In this case, we write $g = f^{-1}$ (and $f = g^{-1}$).

Machine Views Viewing f and g as function machines, the definition says this:

Inverses

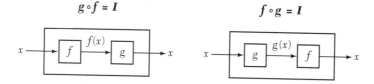

In effect, the machine g is really the machine f running in reverse.

Famous Families: Logs and Exponentials Logarithm and exponential functions with the same base are the most important and useful examples of inverse functions. ▶ Indeed, we *defined* logarithms as inverses of exponentials. In base e, for instance, the natural logarithm function is defined by the condition

We reviewed these functions in Section 1.5. See the graphs on page 60. See also Appendix F.

$$y = \ln x \iff x = e^y.$$

Similarly, in base 10, $y = \log_{10} x \iff x = 10^y$. ▶

Notice how log and exponential keys appear in "inverse" positions on a calculator keypad.

Equivalent Conditions If f and g are inverse functions, then

$$f(a) = b \iff g(b) = a,$$

for every a in the domain of f.

Domains and Ranges If f and g are inverses, then inputs to g are outputs from f, and vice versa. Therefore, the domain of f is the range of g, and the range of f is the domain of g.

Which Functions Have Inverses?

Does every number have an additive inverse? A multiplicative inverse?

We saw earlier that the constant function $f(x) = 3$ has no inverse function. ◄ Which functions *do* have inverses?

One-to-One Functions.   Every function f, invertible or not, sets up a correspondence between its domain and its range; to an input x from the domain, f assigns $f(x)$, one member of the range. An inverse function—if one exists—reverses this correspondence. To be invertible, therefore, f must be **one-to-one**, i.e., each output must correspond to only one input. A constant function fails this test miserably: it sends *every* input to the same output.

Recognizing Invertibility Graphically.   Graphically speaking, one-to-one functions are those that pass the **horizontal line test**:

No horizontal line crosses the graph more than once.

The test guarantees that every possible output y corresponds to no more than one input x.

EXAMPLE 7   The functions $f(x) = x^2$ and $h(x) = 2x + 3$ are shown graphically. Is either function invertible? If so, find its inverse.

Not one-to-one: $f(x) = x^2$ One-to-one: $h(x) = 2x + 3$

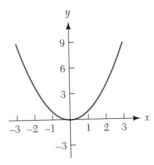
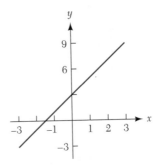

The graph of f fails the horizontal line test at $y = 4$, for example.

Solution   The function f is not one-to-one on the interval $[-3, 3]$, ◄   so it's not invertible on the domain $[-3, 3]$. The function h is one-to-one and therefore invertible.

With algebra we can find an explicit formula for h^{-1}: ▶

Check each step.

$$y = h^{-1}(x) \iff x = h(y) \iff x = 2y + 3 \iff y = h^{-1}(x) = \frac{x-3}{2}.$$

∎

Inversion and Symmetry

The definition of inverse function, interpreted graphically, says this:

> *A point* (a, b) *lies on the graph of* f *if and only if* (b, a)—*the reflection of* (a, b) *across the line* $y = x$—*lies on the graph of* f^{-1}.

Graphs of h and h^{-1} from the preceding example illustrate what this means:

Graphs of h and h^{-1}

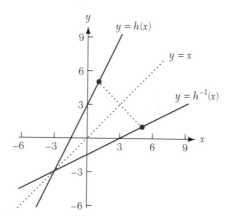

The two graphs are symmetric about the line $y = x$: Every point on either graph is the mirror image of a point on the other. One symmetric pair of points, $(1, 5)$ and $(5, 1)$, is bulleted. ▶

The graphs look symmetric because equal scales are used on the two axes.

Restricting Domains to Make Functions Invertible

Is the function $f(x) = x^2$ invertible? Earlier we said no, judging from the graph over the interval $[-3, 3]$. But isn't $g(x) = \sqrt{x}$ an inverse? Resolving this quandary requires a closer look at domains.

EXAMPLE 8   Make careful sense of the functions $f(x) = x^2$ and $g(x) = \sqrt{x}$ as inverses.

It's "two-to-one." Do you understand why?

Solution   The function $f(x) = x^2$ is not one-to-one ▶ on its natural domain—the full set of real numbers. However, f *becomes* one-to-one if its domain is restricted to nonnegative inputs. With this restriction, f *is* invertible and has the expected inverse $g(x) = \sqrt{x}$, as the following graphs show. ▶

Check algebraically, too, that $(f \circ g)(x) = x = (g \circ f)(x)$ *for all appropriate values of* x.

Inverses: $f(x) = x^2$ and $g(x) = \sqrt{x}$

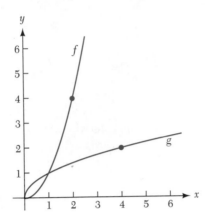

The graphs of f and g are, as expected, mirror images through the line $y = x$. The bulleted points, for instance, are such mirror images; they show that $f(2) = 4$ and

This is expected, because f and g are inverses.

$g(4) = 2$. ◄   The conclusion: $f(x) = x^2$ is invertible on the domain $[0, \infty)$; its inverse there is $g(x) = \sqrt{x}$.   ■

A Moral about Domains.   For most calculus purposes, it's enough to treat domains informally. For most functions the **natural domain** (the set of all inputs for which the function's rule makes sense) is clear from context and seldom requires explicit attention.

The question of finding inverses is an exception to this rule. As the preceding example showed, a function may well be invertible on one domain but not on another. The practical moral is to take care with domains when working with functions and their inverses.

BASICS

1. (This problem is related to Example 1 on page 70.) Let f be defined by the formula $f(x) = x^2 + 3/x$, and let g be defined by the following table.
   (a) Fill in as many entries in the following table as possible. (A few gaps are inevitable.)

x	-3	-2	-1	0	1	2	3	4
$f(x)$	8				4		10	
$g(x)$	-4	-9	-1	2	0	8	3	-2
$(f+g)(x)$								
$(f \cdot g)(x)$								
$(f/g)(x)$								
$(f \circ g)(x)$								

   (b) Briefly explain each gap in the table.
   (c) State the domains and ranges of $f + g$, $f \cdot g$, f/g, and $f \circ g$.

2. The functions f and g shown in Example 3, page 72, are defined by the formulas $f(x) = a \sin x$ and $g(x) = (x + b)^2/5 + c$ for certain constants a, b, and c.
   (a) Use the graphs in Example 3 to find numerical values of a, b, and c.
   (b) Use the formulas for f and g to find values for $(f \circ g)(-3)$, $(f \circ g)(-2)$, $(f \circ g)(-1)$, $(f \circ g)(0)$, $(f \circ g)(1)$, $(f \circ g)(2)$, and $(f \circ g)(3)$. (Approximate values are tabulated in Example 3.)
   (c) Use the formulas you found for f and g in part (a) to find a formula for $f \circ g$. Use a calculator or computer to plot $f \circ g$ in the window $[-5, 5] \times [-3, 3]$. (Compare your plot with the one in Example 3.)

In Exercises 3–14, let $f(x) = \sin x$, $g(x) = x^2$, and $h(x) = x + 3$. Write each of the following functions as a composition of one or more of the functions f, g, and h.

3. $j(x) = \sin(x + 3)$
4. $j(x) = (x + 3)^2$
5. $j(x) = \sin^2 x$
6. $j(x) = \sin(x^2)$
7. $j(x) = x^2 + 3$
8. $j(x) = 3 + \sin x$
9. $j(x) = \sin^2(x + 3)$
10. $j(x) = \sin\big((x + 3)^2\big)$
11. $j(x) = \sin(x^2) + 3$
12. $j(x) = \big(\sin(x + 3)\big)^2$
13. $j(x) = x^4$
14. $j(x) = x + 6$

In Exercises 15–22, write each function as the composition of two (or more) simpler functions. For example, the function $h(x) = (x + 1)^2$ can be written as $h(x) = (f \circ g)(x)$ where $f(x) = x^2$ and $g(x) = x + 1$.

15. $h(x) = \big(x^3 - 4x + 5\big)^6$
16. $h(x) = \sqrt{(x + 1)^3}$
17. $h(x) = \ln(2 + \sin x)$
18. $h(x) = e^{x^2}$
19. $h(x) = \sqrt{1 + e^x}$
20. $h(x) = \big|x^2 - 4\big|$
21. $h(x) = \sin\big(x^2 - 4x + 4\big)$
22. $h(x) = \cos^2 x + 4\cos x + 4$
23. Consider the function $f(x) = x^2$ restricted to the domain $(-\infty, 0]$. Since f is one-to-one on this interval there is an inverse function f^{-1}.
    (a) Sketch graphs of both f and f^{-1} on the same axes. [**HINT:** Reflect the graph of f across the line $y = x$.]
    (b) What's the formula for f^{-1}?
    (c) What are the domain and range of f? What are the domain and range of f^{-1}? How do the graphs of f and f^{-1} reflect this?
24. Suppose that the point $(-3, 5)$ is on the graph of f. Find a point on the graph of f^{-1}.
25. Suppose that $f^{-1}(2) = 3$. Find a point on the graph of f.
26. Suppose that f is a nonconstant linear function (i.e., $f(x) = ax + b$ and $a \neq 0$).
    (a) Explain why f^{-1} exists.
    (b) Show that f^{-1} is a linear function.
    (c) How are the slopes of f and f^{-1} related?
    (d) Why is the assumption $a \neq 0$ needed in part (a)?

FURTHER EXERCISES

27. Let $f(x) = x^3$ and $g(x) = \sqrt{x + 1}$.
    (a) What is the domain of $f + g$?
    (b) What is the domain of $f \cdot g$?
    (c) What is the domain of $f \circ g$?
    (d) What is the domain of $g \circ f$?
    (e) What is the domain of $g \circ g$?
    (f) What is the domain of f/g?
    (g) What is the domain of g/f?
28. Let $f(x) = \sqrt{4 - x}$ and $g(x) = x^2$. What are the domain and range of $f \circ g$?
29. Let $f(x) = \sqrt{x - 2}$ and $g(x) = -x$.
    (a) What are the domain and range of $f \circ g$?
    (b) What are the domain and range of $g \circ f$?
30. Let $h(x)$ be the concentration of oxygen (in atoms/cm^3) in the air x meters above sea level, and let $g(x)$ be the number of breaths per minute an average person has to take when the concentration of oxygen in the air is x atoms/cm^3. Which of the following tells something useful: $g \circ h$ or $h \circ g$? What does it tell?
31. Suppose that f is an odd function, g is an even function, and h is a function that is neither even nor odd, all with domain $\mathbb{R}$.
    (a) Let $k = g \circ f$. Is k even, odd, or neither? Explain.
    (b) Let $j = h \circ g$. Is j even, odd, or neither? Explain.
    (c) Let $m = f \circ f$. Is m even, odd, or neither? Explain.
32. Let $f(x) = ax + b$ and $g(x) = cx + d$. What values of the parameters a, b, c, and d make each of the following statements true?
    (a) $f \circ g = g \circ f$
    (b) $f \circ g = f$
    (c) $f \circ g = g$
33. Let f, g, and h be functions defined for all real numbers.
    (a) Find functions f, g, and h for which $f \circ (g + h) \neq f \circ g + f \circ h$.
    (b) Find functions f, g, and h for which $f \circ (g + h) = f \circ g + f \circ h$.
34. Let $f(x) = 1$ when x is irrational, and $f(x) = 0$ when x is rational. For which values of x, if any, is $(f \circ f)(x) = 0$?
35. This problem is about inverting the function $f(x) = \sin x$.
    (a) Explain why f is *not* invertible on its full domain, $(-\infty, \infty)$.
    (b) Explain why f *does* have an inverse function f^{-1} when f is restricted to the domain $[-\pi/2, \pi/2]$.
    (c) On one set of axes, carefully draw graphs of f (restricted to the domain $[-\pi/2, \pi/2]$) and f^{-1}. [**HINT:** Exploit symmetry to draw f^{-1}.]
    (d) The function f^{-1} you drew in part (c) is known as the **arcsine function**. It's usually written either $\arcsin(x)$ or $\sin^{-1}(x)$. Use a calculator or computer to plot the sine and arcsine functions together in the viewing window $[-\pi/2, \pi/2] \times [-1, 1]$. (We'll study the arcsine and other "inverse trigonometric functions" carefully later in this book.)

36. Let f be the function shown graphically as follows.

Graph of f

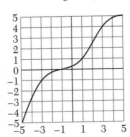

(a) Explain why f has an inverse.
(b) What are the domain and range of f^{-1}?
(c) Fill in the following table. (Because f is given graphically, your answers will be estimates.)

x	−3	0	2
$f^{-1}(x)$			
$f^{-1}(x+2)$			
$f^{-1}(x)+2$			
$2f^{-1}(x)$			
$f^{-1}(2x)$			

37. Let f be the function graphed below. (The graph is made up of a line segment and a quarter-circle.)

Graph of f

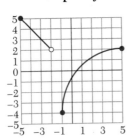

(a) What is the domain of f^{-1}?
(b) Evaluate $f^{-1}(-4)$, $f^{-1}(0)$, $f^{-1}(2)$, and $f^{-1}(3)$.
(c) Find a formula for $f^{-1}(x)$.

In Exercises 38–49, find the domain, the range, and a symbolic expression for the inverse of each function. Also, sketch graphs of f and f^{-1} on the same axes. [**HINT:** In some cases you will need to restrict the domain of the given function to make it invertible.]

38. $f(x) = x^{1/3}$
39. $f(x) = -\sqrt{x}$
40. $f(x) = x^3 + 1$
41. $f(x) = 1/x$
42. $f(x) = 1/x^2$
43. $f(x) = 2/(3 + 4x)$
44. $f(x) = (x^2 + 1)^2$
45. $f(x) = x/(x^2 - 4)$
46. $f(x) = x/(x - 1)$
47. $f(x) = \sqrt{1 - x^2}$
48. $f(x) = e^{x^2}$
49. $f(x) = \ln\big(x/(x + 1)\big)$

50. Suppose that f is a periodic function with domain R. Explain why f does not have an inverse.
51. Sketch the graph of a function that is concave down and whose inverse is concave up.
52. If the domain of $f(x) = \sin(1/x)$ is taken to be an interval of the form $(0, r)$, where r is a positive real number, then f^{-1} does not exist. Explain why.
53. Let g be an odd function. Some values of g are shown in the following table. Fill in as many of the missing entries as possible.

x	−5	−2	1	3	4
$g(x)$	6	3	2	−1	−5
$g^{-1}(x)$					

54. Suppose that f and f^{-1} both have domain $(-\infty, \infty)$ and that f is an odd function. Show that f^{-1} is also an odd function.
55. Suppose that f is an even function, that the domain of f is $(-\infty, \infty)$, and that $f(0) = 1$. Explain why the domain of f^{-1} cannot be $(-\infty, \infty)$.
56. Suppose that $1 \le f(x) \le 5$ if $-2 \le x \le 3$. Is it possible that $f^{-1}(0) = 0$? Explain.
57. The graph of a function g is the reflection about the line $y = x$ of the function $f(x) = x^3 - 2$. Find a formula for $g(x)$.
58. Let $f(x) = e^{2x}$ and $g(x) = f^{-1}(x)$. Evaluate $f\big(g(\ln 2)\big)$.
59. (a) Suppose that the function f has an inverse and that $g(x) = f(x + a)$. Show that $g^{-1}(x) = f^{-1}(x) - a$.
    (b) Use the result in part (a) to find an expression for f^{-1} if $f(x) = \ln(x + 1)$.
60. (a) Suppose that the function f has an inverse and that $h(x) = f(ax)$. Show that if $a \ne 0$, then $h^{-1}(x) = f^{-1}(x)/a$.
    (b) Use the result in part (a) to find an expression for f^{-1} if $f(x) = e^{2x}$.
61. Suppose that $h = f \circ g$ where f and g are invertible functions.
    (a) Show that $h^{-1} = g^{-1} \circ f^{-1}$.
    (b) Use the result in part (a) to obtain the result stated in part (a) of Exercise 59.
    (c) Use the result in part (a) to obtain the result stated in part (a) of Exercise 60.
62. Find all values of the parameter k for which the function $f(x) = kx + |x|$ with domain $(-\infty, \infty)$ is invertible.

1.7 Modeling with Elementary Functions

Why do the basic elementary function families—algebraic, exponential, and trigonometric—deserve their starring roles in calculus? What makes these particular functions uniquely valuable and important?

One answer is entirely practical: Elementary functions can be used to model real-world phenomena. We'll illustrate with several examples.

What Is Mathematical Modeling?

To *model* a phenomenon—the U.S. Consumer Price Index, snowshoe hare and fox populations in the Arctic, gravitational forces on spacecraft, the shape of the Golden Gate Bridge, heating requirements in Minnesota, demand for hard red Montana wheat, risks and benefits of birth control drugs, varying weights of 1-pound loaves, the progress of AIDS, Big Mac prices over the years—mathematically is to *describe* or *represent* the phenomenon quantitatively, in mathematical language. All mathematical models have two important features:

Simplification   No model, mathematical or not, reproduces every property or aspect of the thing modeled. Every model necessarily *simplifies*, to some extent, the reality it describes. A useful model is simple enough to permit calculations but detailed enough to capture the essence of the phenomenon modeled.

Prediction   A useful mathematical model must do more than describe already observed behavior. It should also help us predict or understand the behavior.

The process of mathematical modeling involves three steps:

Description   Describing a phenomenon in mathematical language, using mathematical ingredients: functions, variables, equations, etc.

Deduction   Deducing mathematical consequences of the mathematical description.

Interpretation   Interpreting mathematical results in the language of the phenomenon under study.

Polynomials, Rational Functions, and the Physical World

Polynomials and rational functions are among the simplest functions, because they're built using only the familiar arithmetic operations. ▶ Despite their simplicity, these functions nicely model a surprising number and variety of real-world physical phenomena.

Addition, multiplication, subtraction, and division.

Parabolic Trajectories.   What sort of path (or **trajectory**) does a projectile (a baseball, say, or a high jumper) describe during its flight from the ground into the air and back again? The general shape—some sort of arch—is clear enough, but how can we describe the trajectory mathematically? Is it a parabola? Part of a circle? One arch of a sine curve? Here are curves of all three types:

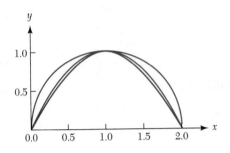

**A semicircle, a parabola, and a sine curve:
is one a trajectory?**

*Actually, one is less
convincing than the other
two. For more information,
see this section's exercises.*

*For example, ignoring wind
resistance, glider effects, and
other stray forces.*

All three curves look possible at first glance, ◄   but only one is correct. It's a striking fact of mathematical physics that under ideal conditions ◄   a projectile influenced only by the force of gravity follows a *parabolic* trajectory. Later, using calculus tools, we'll be able to explain exactly why this fact holds. For now we'll assume it, in the following form.

Fact Suppose that a projectile leaves the origin with initial velocity v_0, aimed along the line $y = mx$, and influenced only by gravity. The projectile's trajectory is a parabola, given by the quadratic equation

$$y = mx - \frac{g}{2v_0^2}(1 + m^2)x^2$$

(g is the [constant] acceleration due to gravity).

We'll use the formula in several ways. First note several remarks:

Initial Angle The projectile's line of aim can be described either by its slope m (as above) or by the angle θ the line makes with the positive x-axis. In this case, $m = \tan\theta$, ◄   and the trajectory equation becomes

*Is it clear that $m = \tan\theta$?
For more details see the
exercises at the end of this
section.*

$$y = x\tan\theta - \frac{g}{2v_0^2}(1 + \tan^2\theta)x^2.$$

Units The preceding formulas work only with consistently chosen units of measurement. If x and y (which measure horizontal and vertical distance) are measured in feet and v_0 is measured in feet per second, then $g = 32$ feet per second per second. If, instead, x and y are measured in meters and v_0 in meters per second, then the *downward* acceleration g is (approximately) 9.8 meters per second per second.

Up or Down? Gravity causes *downward* acceleration—that's why the downward acceleration g is *positive* ($g \approx 9.8$ in metric units). We might have said, equivalently, that the *upward* acceleration is −9.8 meters per second per second.

EXAMPLE 1   A pitcher standing at the origin throws baseballs along the positive x-axis, always with initial velocity 128 feet per second ▶ but aiming along various lines $y = mx$. Describe the possible trajectories.

About 87 miles per hour—typical in the major leagues.

Solution   In present units, $v_0 = 128$ and $g = 32$, so a trajectory with initial slope m has the form ▶

Check to make sure we've worked out the constants correctly.

$$y = mx - \frac{1+m^2}{1024}x^2 = -\frac{1+m^2}{1024}x\left(x - \frac{1024m}{1+m^2}\right).$$

Here are trajectory curves for many values of m: ▶

The equations make sense in context only when $y \geq 0$, of course.

Baseball trajectories

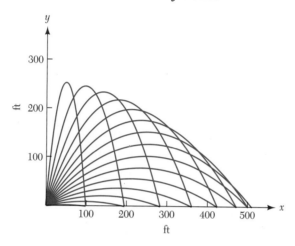

Observe:

Various Shapes   Every trajectory parabola begins at the origin. Depending on its initial slope m, a trajectory may be either tall and narrow or short and squat. As the picture shows, balls that fly farthest have trajectories somewhere between these two extremes. In the next example we'll discuss how to maximize the ball's horizontal range.

Back to Earth   When a ball hits the ground, its trajectory function is zero. As the preceding formulas show, ▶ each trajectory parabola has two roots: $x = 0$, where the ball starts, and

Check our work.

$$x_{\max} = 1024\frac{m}{1+m^2}$$

where the ball lands. (If, say, $m = 1$, then $x_{\max} = 1024/2 = 512$ feet.)

At the Peak   Knowing $x_{\max}$ lets us find $y_{\max}$, the ball's height at its peak. The symmetry of parabolas ▶ means that every trajectory peaks exactly halfway from $x = 0$ to $x = x_{\max}$, i.e., at $x_{\text{middle}} = 512m/(1+m^2)$. This value of x gives the ball's maximum height for any given m:

A parabola is symmetric about the vertical line through its vertex.

$$y_{\max} = 256\frac{m^2}{1+m^2}.$$ ■

EXAMPLE 2   Consider again the situation of Example 1. With what initial slope m should the ball be thrown to rise as high as possible? What initial slope gives the maximum possible horizontal range?

And we'll hope that our calculations agree ...

Solution   We'll start with common sense. ◄   Intuition says that, for a ball to go as high as possible, it should be thrown straight up. To throw as *far* as possible, we want a trajectory that's neither too tall nor too flat: setting $m = 1$ (or, equivalently, $\theta = 45$ degrees) seems a reasonable compromise.

To check these guesses algebraically, we consider two functions of the initial slope m. For any given slope $m \geq 0$, let $f(m)$ be the peak *height* attained, and $g(m)$ the horizontal *distance* traveled. As we showed above, f and g can be written

Both f and g are rational functions of m.

as functions of m: ◄

$$f(m) = 256\frac{m^2}{1 + m^2}; \qquad g(m) = 1024\frac{m}{1 + m^2}.$$

Plotting f and g suggests answers to our questions:

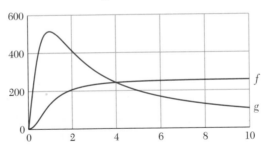

Graphs of $f(m) = 256m^2/(1 + m^2)$
and $g(m) = 1024m/(1 + m^2)$

Note the horizontal asymptote.

As its graph suggests, ◄   the height function f increases with m. Specifically, as $m \to \infty$,

$$256\frac{m^2}{1 + m^2} \to 256.$$

See this section's exercises for an algebraic approach to the same problem.

Do real baseball players obey this rule? Do shot-putters?

In baseball terms, as the slope m gets larger and larger (and the line of aim approaches the vertical), the ball's maximum height approaches 256 feet.

The g-graph suggests that (as we guessed) the range function peaks at or near $m = 1$, where $g(1) = 512$ feet. Using ideas from calculus, we'll soon be able to show this rigorously by calculation. Until then, graphical evidence will do. ◄   In baseball terms: *For maximum range, aim the ball 45 degrees above the horizon.* ◄

■

Exponential Functions, College Tuition, and Growth with Interest

Exponential functions model phenomena that grow, like undisturbed bank accounts, with "interest." A fable will illustrate.

Ingrid and Eric.   Ingrid was born on Labor Day, 1991. Her family is poor, but her wealthy Uncle Eric is determined that Ingrid will someday enroll at his alma mater, St. Lena College. To this end, Eric resolves to pay Ingrid's first-year comprehensive fee at St. Lena. (After a year, Eric thinks, college students should fend for themselves.) If all goes well, Ingrid will enter St. Lena on Labor Day, 2010, her 19th birthday.

Eric's Problem.   Uncle Eric, ever prudent, wants to deposit the money now, at Ingrid's birth. Tax-sheltered 19-year bonds that pay 8% annually are available. Eric wants to cover Ingrid's first-year fee, but not trips to Fort Lauderdale or other foolishness. How much should he deposit now? Would $14,300—the 1991 fee at St. Lena—be enough?

The crucial question is St. Lena's fee in the year 2010. Uncle Eric can't hope to know it 19 years ahead, but perhaps he can *estimate* it on the basis of historical data. He starts by collecting information on St. Lena's annual fee from 1961, when the fee was $1575, through 1992, when it will be $15,400. Looking for a pattern, he plots the data. (For convenience, he lets $t = 0$ correspond to the year 1960.)

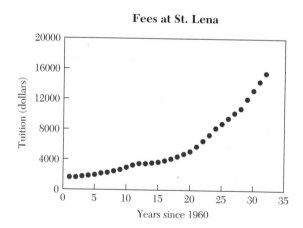

Fees at St. Lena

"Hmmm," thinks Eric. "Looks exponential to me. Might some exponential-type formula fit these data? If so, I could use the formula to estimate St. Lena's fee in 2010."

Uncle Eric Fits a Curve.   Curious, Uncle Eric plots functions of the form

$$f(t) = ae^{bt}$$

for various constants a and b, along with the fee data. Rusty on theory, he relies on trial and error. Eventually he settles on the function

$$f(t) = 1350e^{0.07606t},$$

based on this graph:

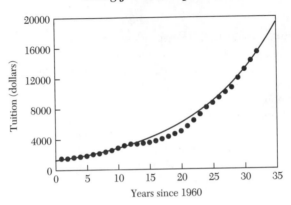

**Fees at St. Lena:
fitting $y = 1350 \exp(0.07606t)$**

The year 2010 corresponds to $t = 50$.

"A nice fit," thinks Eric. "If this keeps up, in 2010 ◄   I'd expect St. Lena to charge about

$$f(50) = 1350\, e^{0.07606 \cdot 50} \approx 60{,}527.91.$$

The cost of books in 2010 we leave to the reader's imagination.

What the heck—I'll make it a round $60,550; the kid will need something for books." ◄

Uncle Eric Plans Ahead.   How much should Eric deposit for Ingrid *now?*

Eight percent interest compounded annually means that on each Labor Day Ingrid's old balance will be multiplied by 1.08. In functional language, let $P(t)$ denote the value of the deposit after t years; then the initial deposit—still to be determined—is $P(0)$. Here's what happens over time:

$$P(1) = P(0) \cdot 1.08;$$

$$P(2) = P(1) \cdot 1.08 = P(0) \cdot 1.08^2;$$

$$P(3) = P(2) \cdot 1.08 = P(0) \cdot 1.08^3;$$

$$\vdots$$

$$P(t) = P(t-1) \cdot 1.08 = P(0) \cdot 1.08^t.$$

Now we can solve for $P(0)$, the initial deposit. We want $P(19) = P(0) \cdot 1.08^{19} = 60{,}550$, so

$$P(0) = \frac{60{,}550}{1.08^{19}} \approx 14{,}030.17.$$

How It Ended, and the Moral.   Uncle Eric put $14,030.17—not quite this year's fee—into 19-year bonds. Baby Ingrid, although still an infant, already looks like college material.

The fable hints at an important—the most important—use of exponential functions: modeling the behavior of quantities that grow (or shrink) under the influence of "interest." Financial quantities most familiarly exhibit such behavior, but so do others: populations, radioactive masses, and many more. ▶

We'll revisit exponential growth in Chapter 4.

> **Do Models Describe or Predict?**   With a little mathematics and lots of hindsight, we could fit the *past* behavior of the St. Lena fee as well as we like. For example, it's possible (but messy) to find a polynomial function, of degree 30 or so, whose graph hits every one of the 31 points Uncle Eric plotted. Such a function might be said to describe the past, but must it also accurately predict the future?
>
> No! How accurately *any* mathematical model can be expected to predict behavior is a difficult and subtle question. The answer depends on many factors, some nonmathematical: the nature and complexity of the phenomenon being modeled, the reliability and validity of the observed data, and other factors.

Trigonometric Functions: Modeling Weather

Minnesota weather is famous for its wild swings. Averaged over many years, however, even the weather becomes surprisingly predictable. Weather phenomena are by nature periodic, so it's natural to use trigonometric functions in weather models.

Temperature in Minneapolis.   Plotting average weekly temperatures in Minneapolis ▶ produces a familiar shape:

From U.S. Weather Service data. Notice the horizontal scale; the "year" starts around March 21.

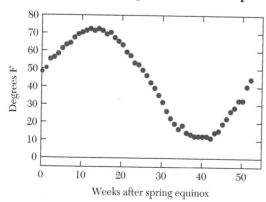

Average weekly temperature in Minneapolis

Weeks after spring equinox

Although the plot resembles a sine curve, the unembellished formula $y = \sin x$ clearly doesn't fit the data:

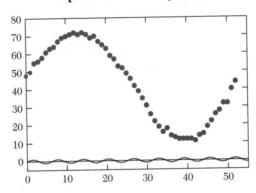

Fitting the data: start with sin x
Temperature data and y = sin x

Unpromising as the preceding graph may seem, let's persevere. Weather varies periodically (in an average sense, at least), so it's reasonable to use some **sinusoidal function**

$$a \sin(bx + c) + d$$

to model our data. Our problem, then, is to choose appropriate values for a, b, c, and d.

A close look at the temperature scatter plot tells a lot.

A Vertical Center The temperature data are vertically centered around 42 degrees Fahrenheit. Therefore, an appropriate sinusoid should be lifted 42 units up. We'll try $d = 42$.

There's no shame in the word "about"— mathematical modeling is an inexact science!

High and Low Temperatures The maximum and minimum temperatures— 72 and 11 degrees, respectively—suggest a vertical variation of about ◀ 31 degrees Fahrenheit in each direction. This argues for stretching the sine function vertically with factor $a = 31$.

Stretching the Period The period for temperature data is 52 weeks; the sine function's period is 2π. To match things up, we must stretch the sine curve horizontally, with factor $52/(2\pi)$. Setting $b = 2\pi/52 \approx 0.12083$ has this effect.

No Horizontal Shift The scatter plot suggests that no horizontal shift is needed: the sinusoid seems to start at the left. Thus, we set $c = 0$. ◀

We arranged this by starting our 1-year period in spring. This does no harm. Had we insisted on an ordinary calendar year we would have needed a nonzero horizontal shift.

Combining these observations argues for selecting the sinusoid

$$y = 31 \sin \left(\frac{2\pi}{52} x \right) + 42$$

as our best guess to fit the data. The remaining plots show the effects of adding these parameters one by one to the basic sine function.

Setting $d = 42$ produces a vertical shift:

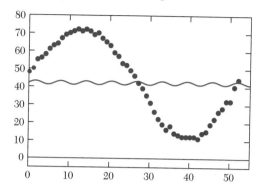

Fitting the data: a vertical translation
Temperature data and $y = \sin x + 42$

The graph is far too flat. Setting $a = 31$ turns up the "amplitude":

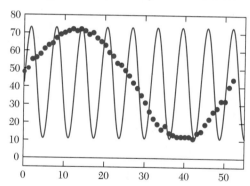

Fitting the data: increase amplitude
Temperature data and $y = 31 \sin x + 42$

Now it's wiggling too fast. Including $b = 2\pi/52$ stretches the period:

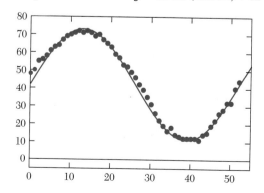

Fitting the data: make the period 52
Temperature data and $y = 31 \sin (2\pi x/52) + 42$

This graph is much better.

EXAMPLE 3   Predict the mean temperature in Minneapolis on December 31.

Solution   December 31 falls 40 weeks and 5 days (i.e., 40.71 weeks) after the spring equinox. The formula predicts a mean temperature of

$$42 + 31 \sin \left(\frac{2\pi}{52} \cdot 40.71 \right) \approx 11.7°\text{F} . \qquad \blacksquare$$

BASICS

1. Consider again the three curves in the picture on page 82. (The curves are candidates for trajectories of an ideal projectile.)

   (a) The three curves have three points in common. Assign these points the coordinates $(0, 0)$, $(1, 1)$, and $(2, 0)$, respectively. In this system, the three curves have equations $y = \sqrt{1 - (x - 1)^2}$, $y = 1 - (x - 1)^2$, and $y = \sin(\pi x / 2)$. Which curve has which equation? How do you know?

   (b) The angle a trajectory curve makes with the ground (i.e., the horizontal axis) is the angle at which the projectile is "fired." Estimate these angles (in radians) for the three curves shown. Using the result, explain why the semicircle cannot be a trajectory curve.

   (c) The Fact box on page 82 says that among the three candidate curves, only the quadratic curve $y = 1 - (x - 1)^2 = 2x - x^2$ is a trajectory. Suppose that this curve *is* a trajectory and that the x and y units represent feet (so the projectile travels 2 feet horizontally and has a maximum height of 1 foot). Find the initial slope m and the initial velocity v_0. (Use appropriate units throughout; time is measured in seconds.)

   (d) Repeat part (c), but assume now that the x and y units represent meters (so that the projectile travels 2 meters horizontally and has a maximum height of 1 meter; time is still measured in seconds). How do your answers compare with those in the previous part?

2. We claimed in this section that if θ is the angle from the positive x-axis to the line $y = mx$, then $m = \tan \theta$. Explain why, using the right triangle with vertices $(0, 0)$, $(1, 0)$, and $(1, m)$.

3. All parts of this exercise refer to the situation of Example 1 (page 83). In particular, assume throughout that a ball thrown from the origin with initial slope m and initial velocity 128 feet per second follows the trajectory curve

$$y = mx - \frac{1 + m^2}{1024} x^2 .$$

   (a) Show that the ball lands at $x = \dfrac{1024m}{1 + m^2}$ feet. (The ball lands where $y = 0$.)

   (b) Let m be given. At what value of x is the ball highest? (Since the trajectory is a parabola, the vertex occurs halfway between the x-intercepts.)

   (c) Let m be given. How high is the ball at its peak? (Use the preceding result.)

   (d) Use results of part (c) to complete the following table:

m	0.1	0.5	1.0	2.0	10.0	
Peak height						255.9

   (e) A ball thrown straight up with initial velocity 128 feet per second rises 256 feet. Give an informal reason why, using results from part (d).

4. In Example 2 (page 84) we claimed that if $g(m) = 1024m/(1 + m^2)$, then $g(1) = 512$ is the largest possible value of g. Equivalently, $1/2$ is the largest possible value of $m/(1 + m^2)$. Prove the latter claim by justifying each of the following steps.

$$0 \le (1 - m)^2 \implies 0 \le 1 - 2m + m^2$$

$$\implies 2m \le 1 + m^2$$

$$\implies \frac{m}{1 + m^2} \le \frac{1}{2} .$$

5. **(The danger zone)**   A curious bird flies over the baseball field of Example 1 (page 83) . The bird wants to see what's happening but also wants to avoid collisions with thrown balls. How high must the bird fly to be safe? (If you find this scenario unconvincing, then try this: What

shape should the stadium's roof have to avoid being hit by thrown balls?) The answer depends, of course, on x: The bird's danger zone is tall for x near zero but gets shorter as x takes progressively larger positive values; for sufficiently large x, the bird is safe even on the ground. The bird's danger zone is defined by still another curve. Here's the picture:

Baseball trajectories and their upper envelope

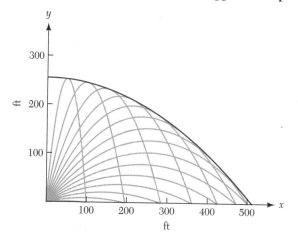

The curve on top is called the **upper envelope** of our family of baseball trajectories. In this exercise we'll study the upper envelope curve.

(a) We said in Example 2 (page 84) that (i) a ball thrown straight upward at 128 feet per second rises to 256 feet and (ii) the maximum horizontal range for any trajectory is 512 feet. Use these facts to label the x- and y-intercepts of your upper-envelope curve.

(b) The upper-envelope curve looks like half a parabola, but is it really parabolic? If so, its equation must be of the form $y = a - bx^2$ for some constants a and b. Use this fact and results from part (a) to find numerical values for a and b.

(c) Using your results from the previous part, plot the parabola $y = a - bx^2$ and the trajectory curves for $m = 0.5$, $m = 1.0$, $m = 2.0$, and $m = 4.0$—all on

the same axes. Use the window $[0, 600] \times [0, 300]$. Does your picture suggest that you've chosen the upper envelope curve correctly? Why or why not?

6. Exercise 5 gives convincing graphical evidence that the upper-envelope curve for the baseball trajectories is (like the trajectories themselves) a parabola. A rigorous proof of this fact requires some careful algebraic manipulation. Provide such a proof, using the following outline.

(a) Let $q(t) = at^2 + bt + c$ be a quadratic polynomial in t; a, b, and c are constants, with $a \neq 0$. Show that $q(t)$ can be written in the form

$$q(t) = a \left(t + \frac{b}{2a} \right)^2 + \left(c - \frac{b^2}{4a} \right).$$

Conclude from this (explain your reasoning) that the parabola $y = q(t)$ has its vertex at $t = -b/2a$.

(b) Let u be the function whose graph is the upper envelope for the baseball trajectories. We want to show that for any fixed x_0 between 0 feet (the minimum horizontal range) and 512 feet (the maximum range), $u(x_0)$ is given by a quadratic expression in x_0. (We'll use x_0 rather than x to emphasize that, for now, x_0 is a fixed number.) Let $h(m)$ denote the height at x_0 of the trajectory curve with initial slope m. Explain the statement:

$u(x_0)$ is the maximum possible value of $h(m)$.

(c) Show that, for $m > 0$,

$$h(m) = mx_0 - (1 + m^2)\frac{x_0^2}{1024}$$
$$= -\frac{x_0^2}{1024}m^2 + x_0 m - \frac{x_0^2}{1024}.$$

(d) Part (c) shows that $h(m)$ is a quadratic function of m. Since the coefficient of m^2 is negative, h takes a maximum value at its vertex. Show that this maximum occurs at $m = 512/x_0$ and that the maximum value is $256 - x_0^2/1024$.

(e) Conclude that $u(x_0) = 256 - x_0^2/1024$, so the upper envelope is indeed a parabola.

7. According to Uncle Eric's model, St. Lena's tuition doubles every X years. What is X? Why?

1.8 Chapter Summary

This chapter is a general introduction and fly-by overview of some of the objects, ideas, uses, and techniques of calculus.

Functions: More than Formulas.   The most important idea in this chapter is that of **function**. We expect most readers are familiar, from precalculus experience, with functions such as

$$f(x) = x^2 + \frac{3}{x} + \sqrt{x}$$

that are defined by formulas. Such functions will indeed play important roles in this book. More characteristic of the chapter, however, are functions defined in other ways: by graphs, by tables, and in words. An important goal of this book is to present calculus ideas from graphical and numerical (as well as algebraic) points of view. In real-world practice, functions seldom appear ready-made as formulas.

Graphs: Pictures of Functions.   Formulas, even simple ones, give little sense of how a function behaves for various inputs. Graphs, by contrast, are ideally suited for this purpose. A graph gives a function a geometric, not just symbolic, meaning. Best of all, graphs show how functions vary—when they rise, when they fall, when they're concave up or concave down, and so on. All of these properties are crucially important in calculus itself, but even more so in applications of the subject.

Machine Graphics.   With computers and calculators, drawing accurate graphs becomes easy. What remains is the good stuff: interpreting what graphs say. Doing so is an unending refrain throughout this book.

Using machines effectively to plot functions requires some care (things can go wrong) and some familiarity with newfangled jargon: **xrange**, **yrange**, **plotting windows**, **aspect ratio**, etc. Section 1.3 discussed such niceties.

Functions, Formally.   Truly understanding the idea of function means understanding its three main parts: **rule**, **domain**, and **range**. Section 1.4 treated such things, with many examples. The variety is important: Seeing function ideas and terminology in various settings helps clarify what they mean in *any* setting.

A Field Guide to Elementary Functions.   Section 1.5 introduced the cast of leading functional characters in calculus. Many of the functions we'll treat in calculus are **elementary**—that is, built by combining members of the leading function families: algebraic, trigonometric, exponential, and logarithm functions. We met these function families briefly in Section 1.5. More details about each family appear in the appendices and in later sections.

New Functions from Old.   Section 1.6 described various ways of combining old functions to form new ones. Algebraic combinations—sums, differences, products, and quotients—are the simplest; **composition** and **inversion**, perhaps less familiar, are equally important in calculus.

Modeling with Elementary Functions.   The basic elementary function families—algebraic, exponential, and trigonometric—deserve their starring roles in calculus because they're practically useful, both mathematically and in applications. Section 1.7 described several uses of elementary functions in modeling real-world phenomena.

2

The Derivative

2.1 Amount Functions and Rate Functions: The Idea of the Derivative

Introduction

Calculus is the mathematics of change. The immense practical power of calculus derives from its ability to describe—and predict—the behavior of changing quantities. Falling apples, orbiting spacecraft, growing populations, decaying radioactive materials, rising consumer prices—all can be modeled through the calculus of change.

The calculus tool that makes all this possible is called the derivative function. ▶ In coming chapters we'll spend a lot of time and effort studying how to define derivative functions formally and how to calculate them symbolically. In the process, various subtleties and puzzles will inevitably arise. Nevertheless, the basic idea behind derivatives is quite simple: *For a given function* f, *the derivative function,* denoted by f', ▶ *tells the rate of change of* f. This principle, suitably interpreted, lies behind all our calculations and applications of the derivative.

The name isn't very descriptive, unfortunately. There are many ways to derive one function from another.

Read: "eff prime."

Whose Idea? No one person invented the derivative. Related ideas and methods appear throughout mathematical history, spanning at least 2000 years. The modern European development of derivative happened largely in the 17th and 18th centuries, ranging from Pierre de Fermat (1601–1665; French) through Isaac Newton (1642–1727; English) and Gottfried Leibniz (1646–1716; German)—generally considered the cofounders of modern calculus—to Leonhard Euler (1707–1783; Swiss) and Joseph-Louis Lagrange (1736–1813; French). Lagrange, building on Euler's idea of function, may have been first to use both the phrase "derivative function" ▶ and the prime symbol to denote it.

In French, fonction dérivée.

About This Chapter. This chapter is about the concept of the derivative function. The relationship between f and f' is the main theme; we'll study what it means for f' to be the rate function derived from f, what each function says about the other and especially how graphs of f and f' are related. Formal definitions appear toward the end of the chapter, and symbolic computational techniques after that.

Well-Behaved Functions.   It must be admitted that functions f exist for which the derivative function f' is either somehow ill-behaved or, worse, doesn't exist at all. ◀   In this chapter, though, we assume that (unless otherwise noted) our functions are well-behaved—i.e., that a given function f has a derivative f' and that both f and f' have smooth, unbroken graphs. Later, as we develop new language and tools, we'll study more carefully some of what can go wrong.

We'll see some examples soon.

The Derivative as a Rate Function

Any function f can be used to build new functions, derived in one way or another from f. Starting with $f(x) = \sin(x)$, for instance, we can concoct all sorts of new functions. The rules

$$g(x) = \sin(2x), \qquad h(x) = \sin(x + \pi/2), \qquad \text{and}$$

$$k(x) = \frac{\sin(x + 0.1) - \sin x}{0.1}$$

all define "relatives" of f; the possibilities are endless.

Among all the possible functions one might derive from a given function f, the derivative function f' is unquestionably most important. Here's an informal description:

> **Definition** (The Derivative as Rate Function) Let f be any function. The new function f', called the **derivative** (or **rate function**) of f, is defined by the rule
>
> $$f'(x) = \text{instantaneous rate of change of } f \text{ at } x.$$

(We'll use "derivative" and "rate function" interchangeably to describe f'; when we do, we'll call f an **original function** or an **amount function**.)

We'll revisit and refine our definition of derivative in later sections. ◀   First, however, we'll try out the ideas of rate function and amount function "on the road."

We'll have to say, for instance, exactly what "instantaneous" means, mathematically.

Rates, Amounts, and Cars: The Prime Example

Moving cars provide the most familiar setting for comparing rates and amounts; cars nicely illustrate most of the basic ideas and even some of the subtleties. Here's a typical situation, stripped to essentials:

At noon on October 1, 1995, a car enters Interstate 94, heading east, from the State Capitol on-ramp in Bismarck, North Dakota. For several hours, the car moves—sometimes east, sometimes west—at varying speeds.

To put this scenario in functional language, let's define some symbols:

t   the elapsed time, measured in hours, since a reference time $t = 0$

P(t)   the car's position at time t, measured in miles *east* of a reference point at time t

V(t)   the car's *eastward* velocity at time t, measured in miles per hour (mph)

Here are possible graphs of the position and velocity functions:

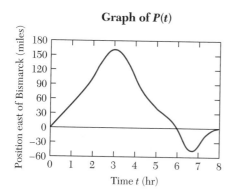

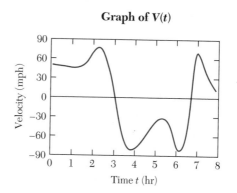

There's a lot to see: ▶

Watch for several uses of the prime notation.

Points of Reference.   Choices of a reference time and a reference point were entirely up to us. We could have used, say, the moment in the year 800 when Charlemagne was crowned Emperor of the West as our reference time, and downtown Glendive, Montana, as our reference point. But it seems natural to choose noon on October 1, 1995, and the State Capitol on-ramp in Bismarck.

Position and Velocity as Functions of Time.   Once we've specified the necessary reference points and units of measurement, P and V become usable functions of t; at any sensible input time t, the car has a well-defined position $P(t)$ and velocity $V(t)$. For example, the statements

$$P(2) = 100 \quad \text{and} \quad V(2) = 70$$

mean that at 2 P.M. the car is 100 miles *east* of Bismarck, moving *east* at 70 mph.

Amount vs. Rate.   It's natural to think of P as an amount function and V as the associated rate function. ▶   For any time t, $V(t)$ is the *instantaneous rate of change of position with respect to time*. In symbols, $V(t) = P'(t)$, or simply $V = P'$.

The units—miles and miles per hour, respectively—support this view.

Positive and Negative Values.   Both P and V can take either positive or negative values. The statements

$$P(6.5) = -40 \quad \text{and} \quad V(6.5) = -60$$

mean that at 6:30 P.M. the car is 40 miles *west* of Bismarck, traveling *west* at 60 mph. ◄   When $P(t) > 0$, the car is *east* of the State Capitol. When $V(t) > 0$, the car is headed *eastward*. ◄

Position vs. Distance, Velocity vs. Speed.   Our functions describe the car's position and velocity. Why not use the more familiar terms "distance" and "speed"? The difference is technical but nevertheless important. In physical language, distance and speed connote *nonnegative* quantities. ◄   Because position and velocity can take either positive or negative values, they can convey information about direction (in this case, east or west).

What the Speedometer Tells; What It Doesn't.   Speedometers can't tell direction, so a speedometer reading of 70 mph at $t = 4$ means that $V(4) = \pm 70$. ◄   The speedometer reading, moreover, is an instantaneous speed. It means that if the car maintains the same speed for an hour, then it covers 70 miles.

Acceleration—Another Rate.   Velocity, like position, has its own rate function. In physical language, this new rate function—the rate of change of velocity with respect to time—is called acceleration. Like position and velocity, the acceleration varies with time; in symbols: $A(t) = V'(t)$. Acceleration, too, can take either positive or negative values. The statement $A(2) > 0$, for instance, means that at $t = 2$ the car's eastward velocity is *increasing*. Similarly, $A(1) < 0$ implies that the car's eastward velocity is *decreasing*. ◄

Velocity—Rate or Amount? It Depends.   Earlier we viewed V as the *rate* function derived from P. Now our point of view has changed: When we consider acceleration, velocity becomes the associated *amount*.

Here's the moral and some advice: *Any function can be thought of either as a rate or as an amount, depending on the situation.* To avoid confusion, think of rate functions and amount functions together, as inseparable pairs.

Units of Velocity and Acceleration.   Every rate, properly understood, is the rate of change of something with respect to something else. The proper units for measuring rates depend both on the quantities that vary and on the units used to measure those quantities.

In the current setting, velocity is the rate of change of position (measured in miles) with respect to time (measured in hours), so the unit of velocity is miles per hour. Acceleration is the rate of change of velocity with respect to time, so its appropriate unit of acceleration is miles per hour per hour—the ratio of the unit of velocity to the unit of time. ◄

Rate Functions in Other Contexts

The ideas of amount function and rate function are intuitively natural when applied to physical quantities, such as distance and velocity, but the ideas of rate and amount make good sense in any context. *Any* function f can be thought of as an amount function: To an input t, f assigns the amount $f(t)$. ◄   How one interprets f' as a rate depends on context. ▶ If $f(t)$ is the position of a moving object at time t, then $f'(t)$ is the velocity at time t, i.e., the rate of change of position with respect to time. If $f(t)$ represents altitude, then $f'(t)$ represents *rate of ascent*, i.e., the rate of change of altitude with respect to time. If $f(t)$ represents the varying value of a bank deposit, then $f'(t)$ represents the rate, at time t, at which money flows into the account. If $f(x)$ tells how many gallons of gasoline are consumed in covering x miles, then $f'(x)$ represents the varying rate of consumption, in gallons per mile. ▶

The units of f' depend on the situation, too.

Even if f is a "pure" mathematical function—unattached to cars, balloons, money, or anything else—the idea of a rate function still makes good sense. To see how, notice that in each instance above, f' measured how fast the *output* from f increased with respect to the *input* to f. The same idea applies to "pure" mathematical functions: If f is any function, and $y = f(x)$, then $f'(x)$ represents the rate, at x, at which y increases with respect to x. For instance, the statement $f'(3) = 5$ means, that if $x \approx 3$, increasing x by a small amount produces about five times as much increase in $f(x)$.

Here we use gallons per mile—not the more familiar miles per gallon.

The Derivative Graphically: f′ as a Slope Function

What does the derivative mean *graphically*? How do the graphs of f and f' reflect the amount–rate relationship?

Slope is the key idea that links graphs and rates. By definition, slope is a rate of change, i.e., the *ratio* ▶ of two changes, Δy and Δx. Specifically, the slope of the f-graph at any point $(x, f(x))$ is the instantaneous rate of change of f with respect to x. In a nutshell:

"Rate" and "ratio" have common roots.

> **Definition** (**The Derivative as a Slope Function**) Let f be any function. The derivative function f' is given by the rule
>
> $$f'(x) = \text{slope of the } f\text{-graph at } x.$$

Like our first, informal definition of derivative, this one needs—and will get—plenty of refinement and discussion. ▶   First let's see what the definition means.

We'll have to say, for instance, precisely how to calculate a graph's "slope at a point."

Derivatives of Linear Functions

It's especially easy to find derivatives of linear functions.

EXAMPLE 1   Let f be the linear function

$$f(x) = 3x + 1.$$

Explain, graphically and in terms of rates, why the derivative is the *constant* function

$$f'(x) = 3.$$

Solution   The graph of f is a straight line, so its slope—at *any* value of x—is 3. In symbols, $f'(x) = 3$ for all x. Here are graphs of f and f':

Graph of $f(x) = 3x + 1$

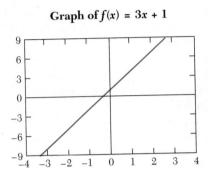

Graph of $f'(x) = 3$

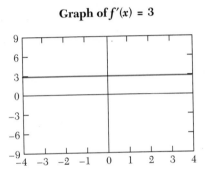

That f is linear with constant slope 3 means that the f-graph rises 3 units for every unit it runs. In rate language, f increases everywhere at the rate of 3 y-units per x-unit.   ∎

The example illustrates a general fact:

> **Fact** For any linear function $f(x) = ax + b$, the derivative (rate) function is the constant function $f'(x) = a$.

The Slope of a Graph at a Point: Tangent Lines

For straight lines, the idea of slope is simple: A line has only one slope, and any two points suffice to compute it. But what about a *curved* graph—how can we measure *its* slope at a point? Indeed, does a curve even *have* slope at a point?

Defining and finding the slope of a curve at a point are key ideas (perhaps *the* key ideas) of calculus. In this section we take a first graphical look; in the next section we'll return to the problem in more detail.

The slope of a curve at a point P can be thought of as the slope of a certain line through P (called the **tangent line at P**) that "fits" the curve at P. Equivalently, the tangent line at P "points in the direction of the curve." ◄   In car talk:

The words in quotation marks are an intuitive description. Formal definitions come later in this chapter.

The curve is a highway, seen from above. A car, headlights blazing through fog, follows the road. When the car is at point P, its headlights point in the direction of the tangent line at P. ▶

If the driver lost control at P, the car would (literally) go off on a tangent.

Using these intuitive ideas, one can usually draw tangent lines with reasonable accuracy and estimate their slopes. Here are short segments of tangent lines at several points along a curve:

Examples of tangent lines

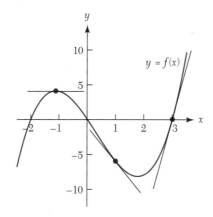

The leftmost tangent segment, at about $x = -1.1$, looks approximately horizontal. In function notation: $f'(-1.1) \approx 0$. Estimating slopes of the other two tangent segments suggests that $f'(1) \approx -5$ and $f'(3) \approx 15$. ▶

Do these estimates look reasonable to you?

Relating Amount and Rate Functions

How are an amount function f and its rate function f' related to each other? What can either function say about the other? These questions are at the heart of calculus. We'll revisit them often, in graphical, numerical, and symbolic forms. The rest of this section begins the discussion.

The Racetrack Principle: Fast Dogs Win

Here's a safe bet:

> *If dogs f and g start together and dog g always runs faster than dog f, then dog g always leads and therefore wins the race.*

Credit for this terminology goes to H. Porta and J. J. Uhl, University of Illinois.

This so-called **racetrack principle**, ▶ appropriately rephrased, is surprisingly useful. The first step is to interpret it as a property of rate and amount functions. ▶

f′ tells the velocity of dog f.

Here's one way to do so:

Fact (**The Racetrack Principle**) Let f and g be functions defined for all x in $[a, b]$, and suppose that $f(a) = g(a)$. If

$$f'(x) \leq g'(x)$$

for all x in $[a, b]$, then

$$f(x) \leq g(x)$$

for all x in $[a, b]$.

Observe these fine points.

Fair Start The requirement that $f(a) = g(a)$ means that f and g "start together" at $x = a$.

Rates of Growth The condition that $f'(x) \leq g'(x)$ means that f "grows no faster" than g for $a \leq x \leq b$.

Never Ahead The conclusion means that f "never outruns" g for $a \leq x \leq b$.

In the strict version, $<$ replaces $\leq$.

Strict Inequalities A strict version ◄ of the racetrack principle also holds: If $f(a) = g(a)$ and $f'(x) < g'(x)$ for x in $[a, b]$, then $f(x) < g(x)$ for every x in $(a, b]$.

The racetrack principle is easiest to use when either f' or g' is a constant function. In that case, we can compare f or g to a linear function.

EXAMPLE 2   Let f be a function such that (i) $f(0) = 0$; (ii) $f'(x) \leq 10$ if $x \geq 0$. Show that $f(1) \leq 10$.

Solution   Informally speaking, (ii) means that, for $x \geq 0$, $y = f(x)$ increases at a rate no greater than 10 y-units per x-unit. Equivalently, the slope of the f-graph doesn't exceed 10. Thus, as x runs 1 unit (from $x = 0$ to $x = 1$), y rises no more than ten units (from $y = 0$ to $y = 10$), so $f(1) \leq 10$. ◄

A dog with a top speed of 10 meters per second can't run more than 10 meters in one second.

The racetrack principle restates these ideas a little more formally. To apply it, let g be the linear function $g(x) = 10x$. (We chose this particular g because it has constant slope 10, so $g'(x) = 10$, and because $g(0) = 0 = f(0)$.) Now all the ingredients are ready. Since $f(0) = g(0)$ and $f'(x) \leq 10 = g'(x)$, the racetrack principle guarantees that for $x \geq 0$, $f(x) \leq g(x) = 10x$. Setting $x = 1$ gives $f(1) \leq g(1) = 10$. ∎

More Safe Bets: Racetrack Variants

The basic racetrack principle—that faster dogs pull ahead—can be restated in several other useful forms by changing the conditions slightly. In the three variants that follow, we no longer assume that dogs f and g start together. Each principle appears first in words, then in symbolic notation. ◄

Does each word version seem reasonable? Do the symbols really say the same thing?

C represents the constant distance between the dogs.

Equal Speeds, Equal Distances In words: If two dogs run equally fast from time a to time b, then the distance between them remains constant. In symbols: If $f'(x) = g'(x)$ for $a \leq x \leq b$, then for some constant C, $f(x) - g(x) = C$ for $a \leq x \leq b$. ◄

Faster Means Farther In words: If dog f runs no faster than dog g, then dog f runs no farther than dog g from time a to time x—regardless of where each dog starts. In symbols: If $f'(x) \leq g'(x)$ for $a \leq x \leq b$, then

$$f(x) - f(a) \leq g(x) - g(a).$$

for $a \leq x \leq b$. (Note that $f(x) - f(a)$ and $g(x) - g(a)$ represent, respectively, the net changes in f and g from a to x. In racing terms, $f(x) - f(a)$ and $g(x) - g(a)$ are the net distances covered by dogs f and g from time a to time x.)

Photo Finish In words: If dogs f and g finish together, then the faster dog cannot have been ahead at any time. In symbols: Suppose that $f'(x) \leq g'(x)$ for $a \leq x \leq b$ and $f(b) = g(b)$. Then, for $a \leq x \leq b$,

$$f(x) \geq g(x).$$

EXAMPLE 3 Suppose that $f(0) = 3$ and $f'(x) \leq 2$ for all x. What can be said about the values $f(2)$ and $f(-4)$? Interpret the result graphically.

Solution We apply the racetrack principle with $g(x) = 2x + 3$. We chose this g because (i) $g(0) = f(0) = 3$; and (ii) $g'(x) = 2 \geq f'(x)$. ▶ To estimate $f(2)$ we use the ordinary racetrack principle on the interval $[0, 2]$. It says that

Why is $g'(x) = 2$? Because g defines a line with slope 2.

$$f(0) = g(0) \quad \text{and} \quad f'(x) \leq g'(x) \implies f(2) \leq g(2) = 7.$$

To estimate $f(-4)$, we'll use the photo-finish version on the interval $[-4, 0]$. It says that

$$f(0) = g(0) \quad \text{and} \quad f'(x) \leq g'(x) \implies f(-4) \geq g(-4) = -5.$$

All of this can be understood geometrically. Here are possible graphs of f and g:

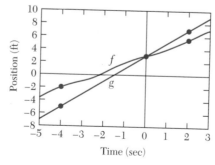

Racing functions: graphs of f and g

As the picture shows, f and g agree at $x = 0$, and f never rises faster than g. As a result, says the racetrack principle, f lies above g before $x = 0$, and below g after $x = 0$. ▶

Compare, for example, the points at $x = -4$ and $x = 2$.

Derivatives in Action: Using Rate and Amount Functions

Many real-world phenomena naturally involve rate and amount functions; derivatives are just what's needed to describe and predict their behavior. We end this section with some examples of derivatives in action. As before, we defer calculations until later sections.

Free Fall Without Air Resistance

A moving object is said to be in **free fall** if it is affected only by gravity. Other possible forces—air resistance, engine thrust, etc.—are assumed to play no role.

Gravity exerts the same downward acceleration on all falling objects regardless of their size, weight, initial velocity, or other characteristics. (For objects near the surface of the earth, this acceleration is essentially constant.) Galileo Galilei (1564–1642) is supposed to have discovered this fact by tossing a variety of objects from the Leaning Tower of Pisa and noticing that all landed at about the same time.

Can Air Resistance Be Ignored? In practice, air resistance matters. Parachutists depend on it for survival; cannonballs fall faster than, say, live chickens. *What* Galileo dropped isn't known; indeed, the whole Pisa story seems unlikely, with all those tourists milling around below. For physical reasons, air resistance is usually negligible for small, dense objects moving at moderate speeds, as when a coin falls to the floor.

That is, v is the rate function of h; a is the rate function of v.

A falling object has, at any time t, a certain height $h(t)$, an upward velocity $v(t)$, and an upward acceleration $a(t)$. Given appropriate measurement scales for time and distance, h, v, and a are all functions of t, and the derivative relations $v = h'$ and $a = v'$ hold. ◄ Galileo's discovery that an object in free fall has constant downward acceleration says symbolically that $a(t) = -g$ or, equivalently, that

$$v'(t) = -g$$

The g stands for gravity.

for some positive constant g. ◄ (The negative sign is there because the object accelerates downward.) In English units, $g \approx 32$ feet per second per second; in metric units, $g \approx 9.8$ meters per second per second.

We reasoned similarly earlier, when we said that if a function is linear, then its derivative is constant.

$v_0 > 0$ means that the object rises initially, perhaps by having been thrown upward.

Formulas for Distance and Velocity.   In the late 1600s Isaac Newton used Galileo's discovery to derive his laws of motion. Newton reasoned that if (as in free fall) the acceleration function $a(t)$ is *constant*, then the velocity function must be linear. ◄

Suppose that at time $t = 0$ seconds an object is h_0 feet above ground level, and has *upward* velocity v_0. ◄ Newton's laws of motion imply rather simple formulas for $h(t)$ and $v(t)$, the functions for altitude (in feet) and (vertical) velocity (in feet per second):

$$h(t) = h_0 + v_0 t - 16t^2; \qquad v(t) = v_0 - 32t.$$

With these formulas ▶ we can answer many questions related to free fall.

Consider: Why are the minus signs present?

EXAMPLE 4   Could a major league pitcher throw a ball to the 195-foot ceiling of the Hubert H. Humphrey Metrodome, home of the Minnesota Twins?

Solution   Not likely. A pitcher throws a fast ball with an initial horizontal velocity of around 95 mph, or 140 ft/sec. Throwing straight up is awkward, so let's assume a smaller initial *vertical* velocity, say 100 ft/sec. If the ball starts from an initial height of 6 feet, then the preceding formulas say that

$$h(t) = 6 + 100t - 16t^2; \qquad v(t) = 100 - 32t.$$

Height is greatest at the instant when upward velocity is zero, ▶ that is, when

Why?

$$v(t) = 100 - 32t = 0 \iff t = \frac{100}{32}.$$

At that time, ▶

Check the calculation.

$$h\left(\frac{100}{32}\right) = 162.25 \text{ ft,}$$

lower than the Metrodome's ceiling. Graphs of h and v follow.

Graph of $h(t) = 6 + 100t - 16t^2$

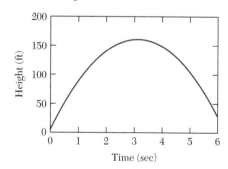

Graph of $v(t) = 100 - 32t$

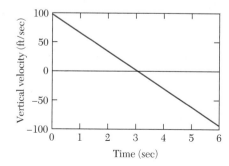

Differential Equations

Any equation that involves derivatives is called a **differential equation (DE).** The equations

$$f'(x) = 3; \qquad v'(t) = -g + kv(t)^2; \qquad f'(x) = f(x)$$

are all DEs. A **solution** to a differential equation is a *function* that satisfies the differential equation. ◄ (As we saw, $f(x) = 3x + 1$ is one solution to the first differential equation. We'll see solutions to the others later.)

NOTE: A solution to a DE is a function, not a number.

Many of the most important applications of the calculus—especially physical applications—involve posing problems as differential equations and then finding and using solutions.

Falling with Air Drag

It's a fact of physics that when small, dense objects (such as baseballs) fall at moderate speeds, the deceleration due to air drag is (approximately) proportional to the *square* of the velocity. In derivative notation, this fact says that the total acceleration is

$$v'(t) = -g + kv(t)^2$$

for some positive proportionality constant k. (The first term, $-g$, represents the downward acceleration due to gravity.) The numerical value of k depends on physical factors—smoothness of the object, air density, and so on. For a falling baseball, the value $k \approx 0.003$ is reasonable.

EXAMPLE 5   On an outing with Dad, Bud and Sis play catch on the observation deck of the IDS Tower in Minneapolis. To Dad's horror their baseball slips over the edge, squarely beaning a lawyer who is on a lunchtime stroll along the Nicollet Mall, 600 feet below. What happens next?

Solution   Thanks to air drag, Dad is sued for personal injury, not negligent homicide. Here's why.

Note that $h(0) = 600$; what does this mean in the present context?

Without air drag, Newton's free-fall model applies, so ◄

$$h(t) = 600 - 16t^2,$$
$$h'(t) = v(t) = -32t, \qquad \text{and}$$
$$v'(t) = -32.$$

Taking air drag into account leads to another differential equation:

$$v'(t) = -32 + 0.003v(t)^2.$$

Solving this differential equation symbolically is hard. We won't try, but we'll show graphs of the functions v and h that apply if air drag is taken into account. In the following pictures, the solid curves show the effect of air drag; the dotted curves ignore it. Notice, for example, the difference between the two velocity graphs (in the right-hand picture). Ignoring air drag, the ball accelerates at a constant rate. Taking account of air drag gives a more realistic result: as the ball gathers speed, its acceleration tapers toward zero.

Effects of air drag on position

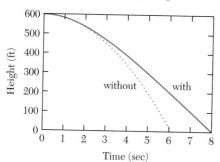

Effects of air drag on velocity

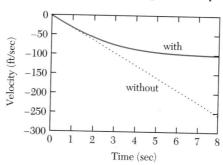

The two velocity curves show the advantage (to both Dad and the lawyer) of air drag. *With* air drag, the ball hits the lawyer at about 100 feet per second—an unpleasant blow, certainly, but a slow pitch by major league standards. *Without* air drag, the ball would be traveling more than twice as fast when it landed. ▶   ■

Of course, it might also then miss the strolling lawyer.

Understanding the Universe. The idea of derivative is among the most important and powerful in all of mathematics—indeed, in all of human history. This concept, more than any other, distinguishes calculus from other branches of mathematics and accounts for the great power of the calculus to describe, and therefore predict, physical phenomena.

Planetary motion is one such phenomenon. In the early 1600s Johannes Kepler discovered (by direct observation) that the orbit of Mars is an ellipse; he suggested the same about other planets. Isaac Newton, using his early theory of calculus, demonstrated all this (and more) mathematically toward the end of the 1600s. As George Simmons put it,[a]

Newton's enormous success revived and greatly intensified the almost-forgotten Greek belief that it is possible to understand the universe in a rational way. This new confidence in its own intellectual powers permanently altered humanity's perception of itself, and over the past 300 years almost every department of human life has felt its consequences.

[a]*Calculus Gems: Brief Lives and Memorable Mathematics*, McGraw-Hill, 1992, pp. 334–335.

BASICS

1. An airplane takes off from one airport and, 45 minutes later, lands at another airport 300 miles away. Let t represent the time since the airplane took off, $x(t)$ the *horizontal* distance traveled, and $y(t)$ the *altitude* of the airplane.
   (a) What is the physical meaning of $x'(t)$? Sketch plausible graphs of x and x'.
   (b) What is the physical meaning of $y'(t)$? Sketch plausible graphs of y and y'.

2. The graph of a function f is shown on the right. Rank the values of $f'(-3)$, $f'(-2)$, $f'(0)$, $f'(1)$, and $f'(4)$ in increasing order.

Graph of f

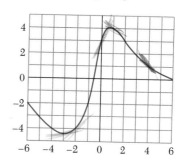

3. Suppose that $f(x) = x^3 - 5x^2 + x - 1$ and that $g(x) = x^3 - 5x^2 + x + 4$. Explain why $f'(x) = g'(x)$ for every x. [**HINT**: How are the graphs of f and g related?]

4. Let $f(x) = 5$, $g(x) = x$, and $h(x) = -2x$. Evaluate $f'(11)$, $g'(-53)$, and $h'(37)$.

5. Let $f(x) = 3x + 2$ and $g(x) = 3x - 4$.
   (a) Evaluate $f'(0)$ and $g'(0)$.
   (b) Give a graphical explanation of the fact that $f'(x) = g'(x)$ for all values of x.
   (c) Find a function h such that $h(1) = 2$ and $h'(x) = f'(x)$ for all values of x.

6. Suppose that $f(x) = x - 2$ and $g(x) = 4x + 1$.
   (a) Plot graphs of f and g on the same axes.
   (b) Show that $f'(x) \le g'(x)$ for all values of x.
   (c) Show how the racetrack principle implies that $f(x) \le g(x)$ for all $x \ge -1$.
   (d) What does the racetrack principle say about the relationship between the numbers $f(x)$ and $g(x)$ when $x \le -1$?

FURTHER EXERCISES

7. Interpret each sentence below as a statement about a function and its derivative. In each case, indicate clearly what the function is, what each variable means, and appropriate units.
   (a) The price of a product decreases as more of it is produced.
   (b) The increase in demand for a new product decreases over time.
   (c) The work force is growing more slowly now than it was five years ago.
   (d) Health care costs continue to rise, but at a slower rate than three years ago.
   (e) During the past two years, the United States has continued to cut its consumption of imported oil.

8. Let $G(v)$ be the number of miles per gallon that a vehicle gets as a function of its speed, v, in miles per hour. Interpret the statement $G'(35) = 0.4$ in plain English using terms such as gas mileage and speed.

9. A company must budget for research and development of a new product. Let x represent the amount of money invested in R&D and T be the time until the product is ready to market.
   (a) Give reasonable units for T and x. What is T' in these units?
   (b) What economic interpretation can be given to T'?
   (c) Would you expect T' to be positive or negative? Explain.

10. The height of a ball (in feet) t seconds after it is thrown is given by $h(t) = -16t^2 + 38t + 74$; its (upward) velocity at time t is $v(t) = -32t + 38$.
    (a) From what height was the ball thrown?
    (b) What was the ball's initial velocity? Was it thrown up or down? How can you tell?
    (c) Was the ball's height increasing or decreasing at time $t = 2$?
    (d) At what time did the ball reach its maximum height? How high was it then?
    (e) How long was the ball in the air?

11. Suppose that $g(x) = f(x) + 3$ and that $f'(x)$ exists for all x.
    (a) Explain how the graphs of f and g are related.
    (b) How is the graph of g' related to the graph of f'? Explain.
    (c) If $f'(1) = 5$, what is $g'(1)$?
    (d) Let $h(x) = f(x) + k$ where k is a constant. How is the graph of h' related to the graph of f'?

12. Let $f(x) = \sin x$ and $g(x) = \cos x$. Explain why the graph of g' is the graph of f' shifted $\pi/2$ units to the left. [**HINT**: $\cos x = \sin(x + \pi/2)$.]

13. Let $g(x) = f(x + 3)$ and suppose that $f'(x)$ exists for all x.
    (a) Explain how the graphs of f and g are related.
    (b) How is the graph of g' related to the graph of f'? Explain.
    (c) If $f'(2) = 4$, what is $g'(-1)$?
    (d) Let $h(x) = f(x + k)$ where k is a nonzero constant. How is the graph of h' related to the graph of f'?

14. Suppose that $g(x) = 3f(x)$ and that $f'(x)$ exists for all x.
    (a) Explain how the graphs of f and g are related.
    (b) How is the graph of g' related to the graph of f'? Explain.
    (c) If $f'(-1) = 2$, explain why $g'(-1) = 6$.
    (d) Let $h(x) = kf(x)$ where $k > 0$ is a constant. How is the graph of h' related to the graph of f'?
    (e) Let $j(x) = kf(x)$ where $k < 0$ is a constant. How is the graph of j' related to the graph of f'?

15. Suppose that f is periodic with period 5 and that $f'(x)$ exists for all x. Explain why f' is also a periodic function with period 5.

16. Suppose that $f'(3) = -2$.
    (a) Assume that f is an *even* function. Explain why $f'(-3) = 2$. [**HINT**: Give a graphical argument.]
    (b) Now assume that f is an *odd* function. What is $f'(-3)$?

17. Suppose that $f'(x) = 1$ for all x. Find an algebraic expression for $f(x)$ assuming that
    (a) $f(0) = 2$.
    (c) $f(0) = -2$.
    (b) $f(1) = 2$.
    (d) $f(-1) = -2$.

18. Suppose that $f(a) = b$ and $f'(x) = c$ for all x. Find an algebraic expression for $f(x)$.

19. Suppose that f and g are defined on the interval $[a, b]$, that $f(a) = g(a)$, and that $f'(x) \le g'(x)$ for all x in $[a, b]$.
    (a) Use car talk to explain why $f(b) \le g(b)$.
    (b) Suppose that $f(b) = g(b)$. What does this imply about f' and g'?

20. Suppose that f and g are defined on the interval $[a, b]$, that $f(a) = g(a)$, and that $f'(x) \le g'(x)$ for all x in $[a, b]$. Use the basic racetrack principle to show that $f(x) - f(a) \le g(x) - g(a)$ for all x in $[a, b]$.

21. Suppose that $f(1) = -2$ and $f'(x) \le 3$ for all x in $[-10, 10]$.
    (a) Show that $f(2) \le 1$.
    (b) Show that $f(5) \le 10$.
    (c) Show that $f(0) \ge -5$.
    (d) Is $f(9) < 31$? Justify your answer.
    (e) Is $f(-5) > -23$? Justify your answer.
    (f) Could $f(4) = 8$? Justify your answer.
    (g) Could $f(-7) = -25$? Justify your answer.
    (h) Could $f(8) = -25$? Justify your answer.
    (i) Let $g(x) = 4x + 3$. Compare the numbers $f(7)$ and $g(7)$.

22. Suppose that $f(1) = 3$ and $f'(x) \ge 2$ for all x in $[0, 5]$.
    (a) Show that $f(4) \ge 9$.
    (b) Show that $f(0) \le 1$.
    (c) Let $g(x) = x + 1$. Compare the numbers $f(3)$ and $g(3)$.

23. Suppose that $f(2) = 1$ and $-4 \le f'(x) \le 3$ for all x in $[-10, 10]$.
    (a) Find upper and lower bounds on the value of $f(6)$.
    (b) Find upper and lower bounds on the value of $f(-5)$.

24. Suppose that $f(1) = 2$ and $f'(x) \le -3$ for all x in $[-5, 5]$.
    (a) What does the racetrack principle imply about the value of $f(4)$?
    (b) What does the racetrack principle imply about the value of $f(-2)$?

25. Suppose that $f(0) = 2$ and $|f'(x)| \le 1$ for all x in $[-5, 5]$.
    (a) Find upper and lower bounds on the value of $f(1)$.
    (b) Find upper and lower bounds on the value of $f(-3)$.

26. Suppose that $f'(x) \le 0$ for all x in $(2, 7)$. Explain why $f(3) \ge f(6)$.

27. Suppose that $f'(x) > 0$ for all x in $(2, 7)$. Explain why $f(4) < f(5)$.

28. The graph of the derivative of a function f is shown below. Use the graph of f' and the racetrack principle to answer the following questions about f. [**NOTE:** The graph of f is not shown.]

Graph of f'

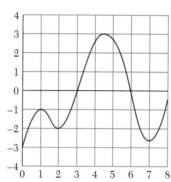

(a) Explain why $f(5) - f(4) \le 2$.
(b) Is $-6 < f(2) - f(0) < -2$? Justify your answer.
(c) Show that $-5 \le f(2) - f(0)$.
    [**HINT:** $f(2) - f(0) = \big(f(2) - f(1)\big) + \big(f(1) - f(0)\big)$.]
(d) Which is larger: $f(3)$ or $f(6)$? Justify your answer. [**HINT:** Use part (a).]
(e) Which is larger: $f(0)$ or $f(3)$? Justify your answer.

29. The graph of the derivative of a function g is shown below. Use the graph of g' and the racetrack principle to answer the following questions about g. [**NOTE:** The graph of g is not shown.]

Graph of g'

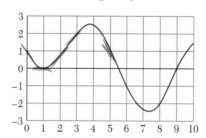

(a) Is $g(2) < g(5)$? Justify your answer.
(b) Is $g(6) > g(8)$? Justify your answer.
(c) Rank the five numbers 0, 1, $g(2) - g(1)$, $g(8) - g(7)$, and $g(9) - g(8)$ in increasing order.
(d) Rank the four numbers $g(0)$, $g(1)$, $g(3)$, and $g(5)$ in increasing order.

30. Suppose that $f(-1) = -2$ and that f' is the function shown below. Use the racetrack principle to answer the following questions about f. [**NOTE:** The graph of f is not shown.]

Graph of f'

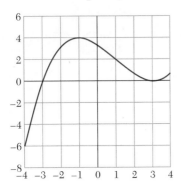

(a) Could $f(3) = -6$? Justify your answer.

(b) Could $f(3) = 2$? Justify your answer.

(c) Show that f achieves its largest value in the interval $[-4, 4]$ at $x = 4$. [**HINT:** Show that $0 < f(-4) - f(-3) < 6$ and $f(1) - f(-2) > 6$.]

(d) Where in the interval $[-4, 4]$ does f achieve its smallest value?

31. Suppose that k is a real number and that f is a function such that $f'(x) \leq k$ for all x in an interval $[a, b]$. Show that $f(x) \leq k(x - a) + f(a)$ for all x in $[a, b]$.

32. The converse of the racetrack principle says that if $f(a) = g(a)$ and $f(x) \leq g(x)$ for all x in $[a, b]$, then $f'(x) \leq g'(x)$ for all x in $[a, b]$. Show that this statement is not true in general by giving graphical examples of functions f and g that satisfy the hypotheses of the statment but not its conclusion.

33. (a) Suppose that $f'(x) \leq 0$ for all x in $[a, b]$ and $f(a) = c$. Use the racetrack principle to show that $f(x) \leq c$ for all x in $[a, b]$.

(b) Suppose that $f'(x) \geq 0$ for all x in $[a, b]$ and $f(a) = c$. Use the racetrack principle to show that $f(x) \geq c$ for all x in $[a, b]$.

(c) Use parts (a) and (b) to prove that if $f'(x) = 0$ for all x in $[a, b]$, then f is a constant function on the interval $[a, b]$.

34. Suppose that $f'(x) = g'(x)$ for all x in $[a, b]$. Use the previous problem to show that there is a constant C such that $g(x) = f(x) + C$ for all x in $[a, b]$. [**HINT:** If $h(x) = g(x) - f(x)$, then $h'(x) = 0$ for all x in $[a, b]$.]

2.2 Estimating Derivatives: A Closer Look

Writing $x = a$ helps "fix" our attention on a single point of the domain.

In the preceding section we described the derivative function in terms of rates and slopes. Given a function f and an input value $x = a$, ◀

$f'(a)$ is the instantaneous rate of change of f with respect to x at $x = a$.

In graphical language,

$f'(a)$ is the slope of the f-graph at $x = a$.

This brief section is—literally—a closer look at the derivative at a point. By "zooming in" on a function or its graph near an input value $x = a$, we can estimate numerical values of f'; we can also see more clearly what the derivative means, both graphically and numerically.

Estimating $f'(a)$: The Slope of a Graph at a Point

For a line, finding slope is a routine matter. For a curve, the idea of slope at a point is a little subtler, but a lot more interesting.

Tangent Lines

Remember the headlights-through-fog analogy? See page 99.

The slope of a curve at a point P can be thought of as the slope of the curve's tangent line at P, i.e., the straight line through P that, roughly speaking, "points in the direction of the curve" at P. ◀ This description, although vague by strict mathematical standards, is perfectly adequate to illustrate the idea. For nearly all calculus functions, moreover, it leads reliably to good derivative estimates.

EXAMPLE 1   Let $f(x) = x^2$. Use tangent lines to estimate $f'(1)$ and $f'(-1.5)$. Interpret the results as rates of change.

Solution   Here's the graph of f, together with reasonable candidates ▶ *Do you agree that they're* for tangent lines at $x = 1$ and $x = -1.5$:   *reasonable?*

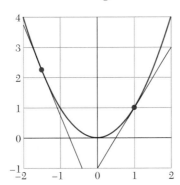

Graph of $f(x) = x^2$ and two tangent lines

We can now estimate the slope of each tangent line directly from its graph. The line at $x = 1$, for instance, seems to rise 2 units as it runs 1, so its slope is 2. Similarly, the tangent line at $x = -1.5$ seems to have a slope around -3. ▶ We conclude   *Check our estimates.* that

$$f'(1) \approx 2; \qquad f'(-1.5) \approx -3.$$

In the language of rates of change, the results mean that (i) at $x = 1$, y is increasing twice as fast as x, and (ii) at $x = -1.5$, y *decreases* three times as fast as x *increases*. ■

Zooming In to Estimate Slope

Drawing a tangent line is one way to estimate the graph's slope at a point. The strategy works because, near the point in question, the graph and the tangent line are almost indistinguishable.

Another strategy—often more convenient, given plotting technology—is to "zoom in" on the point in question, perhaps repeatedly, until the graph *itself* looks like a straight line. Once the graph looks straight, it's easy to estimate the slope.

EXAMPLE 2   Let $f(x) = x^2$, as before. Estimate $f'(1)$ by zooming. Use the result to find an equation for the tangent line at $x = 1$.

Note the plotting windows.

S o l u t i o n   Zooming in successively on the f-graph near $(1, 1)$ gives these results: ◄

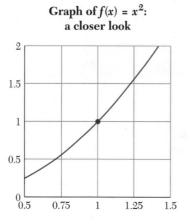

Graph of $f(x) = x^2$:
a closer look

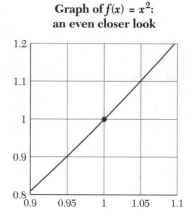

Graph of $f(x) = x^2$:
an even closer look

Check the estimate; use the grid.

Sketch this line into the two preceding pictures.

In the last picture, the graph looks virtually straight; its slope seems to be about 2. ◄ The tangent line in question has slope 2 and passes through $(1, 1)$; its equation, therefore, is $y = 2(x - 1) + 1$. ◄   ■

When Zooming Works: Local Linearity.   Remarkably, the just-illustrated strategy of zooming in to estimate slope almost always works. Zooming in on the graph of almost any calculus function f, at almost any point $(a, f(a))$, eventually produces what looks like a straight line with slope $f'(a)$. A function with this property is sometimes called **locally linear** (or **locally straight**) at $x = a$. ◄   Local linearity says, in effect, that f "looks like a line" near $x = a$ and therefore has a well-defined slope at $x = a$.

These aren't formal definitions, just descriptive phrases.

Example 2 showed that $f(x) = x^2$ is locally linear at $x = 1$. Now let's try zooming in on a trigonometric function.

EXAMPLE 3   Is $f(x) = \sin x$ locally linear at $x = 1.4$? What does the answer say about f'?

S o l u t i o n   Here are two views of the sine graph near the target point $P = (1.4, \sin 1.4) \approx (1.4, 0.9854)$:

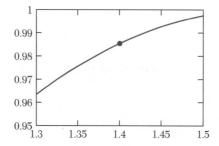

The sine graph near $P = (1.4, \sin 1.4)$

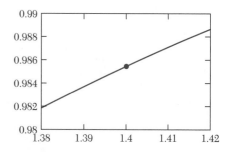

Zooming in near $P = (1.4, \sin 1.4)$

In the second window, the sine graph is almost straight. To estimate the slope at P we can use any two points from the graph. Using $x = 1.38$ and $x = 1.42$, we find (approximately, of course) a slope of 0.17. ▶   In function notation, $f'(1.4) \approx 0.17$.

Check for yourself. Your estimate may differ in the second decimal place.

When Zooming Fails: Local Nonlinearity.   Not every function is locally linear everywhere. A kink in a graph can't be smoothed out by repeated zooming in. At such a point the derivative doesn't exist, and our model of slope breaks down.

EXAMPLE 4   Let $f(x) = |x|$. Does $f'(0)$ exist? Does $f'(a)$ exist if $a \neq 0$?

Solution   Here's the result of zooming near the origin:

Graph of $f(x) = |x|$

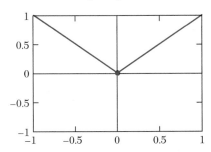

Graph of $f(x) = |x|$: a closer look

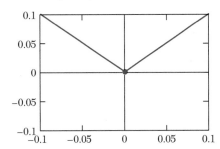

There's no mistake in the pictures. Zooming in had no effect on the kink at the origin. Zooming again would make no difference; that kink is there to stay, and $f'(0)$ doesn't exist.

For $a \neq 0$, on the other hand, the graph is already straight (with slope ± 1), so there's no problem with derivatives:

$$f'(a) = 1 \quad \text{if } a > 0; \qquad f'(a) = -1 \quad \text{if } a < 0.$$

In words: $f(x) = |x|$ is not locally linear at $x = 0$, but is locally linear everywhere else.

Zoom Factors and Distortion.   Throughout this section, "zooming in" means "zooming in with equal zoom factors in both directions." Requiring equal zoom factors is more than a technicality. Zooming repeatedly with *different* zoom factors can so distort the picture that all information is lost.

Consider, for example, the kink at $x = 0$ on the graph $y = |x|$, as in Example 4. Zooming in repeatedly—but *only* in the x-direction (e.g., with zoom factors 4 in x and 1 in y)—gradually flattens the vee. Eventually, the graph looks horizontal. Try it.

Plotting the Rate Function

Because our slope results are estimates, the plot is only approximate. But so is every plot.

Zooming in or drawing tangent lines to the graph of f lets us estimate $f'(a)$ for various values of a. By pooling such information we can *plot* the derivative function. ◄ The results are sometimes surprising.

EXAMPLE 5   For $f(x) = x^2$, plot $f'(x)$. Then guess a formula for $f'(x)$.

About the y-axis.

Solution   In Example 1 we estimated that $f'(1) = 2$ and $f'(-1.5) = -3$. From the symmetry ◄ of the graph $y = x^2$, a few more values come easily:

x	-1.5	-1	0	1	1.5
$f'(x)$	-3	-2	0	2	3

Our guess is correct. Soon we'll see exactly why, by symbolic calculation.

Plots of these points suggest that the graph of f' is a straight line with slope 2, i.e., that $f'(x) = 2x$. ◄

Graph of f'

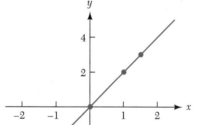

What's Natural about e^x?   Applying the same technique to the natural exponential function produces a real surprise—one that helps explain why the adjective "natural" fits.

EXAMPLE 6   Let $f(x) = e^x$. Use tangent lines to plot f'; then guess a formula for f'.

Solution   Here's a graph of f (shown dotted); at several points, small segments of tangent lines (shown solid) are drawn:

Tangent segments to the graph of
$$f(x) = e^x$$

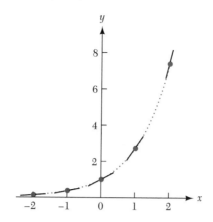

Using a ruler or any convenient plotting technology, one can estimate the slope of each tangent segment. Several such results (rounded to one decimal place) are tabulated below. ▶

Check some table entries yourself. Are they plausible?

The Derivative of e^x—Graphical Estimates					
x	−2	−1	0	1	2
y	0.135	0.368	1.000	2.718	7.389
Slope at (x, y)	0.1	0.4	1.0	2.7	7.4

Plotting the slope data (with a smooth curve through the points) gives a striking result:

The derivative of $f(x) = e^x$

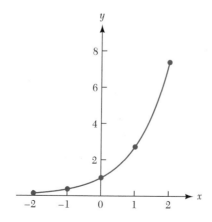

We'll prove this symbolically later.

Judging from the graphs, f and f' are identical! They *are* identical: ◀

If $f(x) = e^x$, then $f'(x) = e^x$. *The natural exponential function is its own derivative.*

In other words: $f(x) = e^x$ solves the **differential equation** $f'(x) = f(x)$. This remarkable property makes the natural exponential function one of the most important and useful in all of mathematics. ■

BASICS

1. The graph of a function f is shown below.

Graph of f

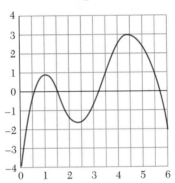

(a) Fill in the missing entries in the table below.
    [**NOTE**: Because f is given graphically, your answers will be estimates. However, your answers should approximate the exact values within ±0.5.]

x	−2	−1	0	1
$f'(x)$	6.0			−2.0
x	2	2.5	3	4
$f'(x)$				3.5

(b) Use your answers to part (a) to sketch a graph of f'.

2. The graph of a function f is shown below.

Graph of f

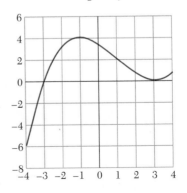

(a) Check the entries in the table below and replace any that are wrong with reasonable estimates.

x	1	2	3	4	5	6
$f'(x)$	0	2	6	3	−2	42

(b) Sketch a graph of f'.

3. The graph of the derivative of a function f appears below. [**NOTE**: The graph of f is not shown.]

Graph of f'

(a) Suppose that $f(1) = 5$. Find an equation of the line tangent to the graph of f at $(1, 5)$.

(b) Suppose that $f(-3) = -6$. Find an equation of the line tangent to the graph of f at $(-3, -6)$.

FURTHER EXERCISES

4. Graphs of a function f and its derivative f' are shown below.

Graph of f

Graph of f'

(a) Draw short segments of the lines tangent to f at $x = -1$, $x = 0$, $x = 1$, $x = 2$, $x = 3$, and $x = 3.5$ on the graph of f.
(b) Estimate the slope of each of the tangent lines you drew in part (a).
(c) Are your answers in part (b) consistent with the graph of f' shown above? Explain.
(d) Find equations for the lines tangent to f at $x = -1$, $x = 0$, $x = 1$, $x = 2$, and $x = 3$.

5. The graph of a function f is shown below.

Graph of f

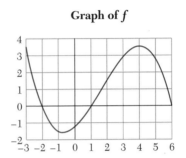

(a) Fill in the missing entries in the table below.

x	-3	-2	-1	0	1
$f'(x)$	-4.9			0.8	
x	2	3	4	5	6
$f'(x)$		1.1			-4.0

(b) Sketch a graph of f'.

6. The graph of a function f is shown below.

Graph of f

(a) Estimate $f'(-5)$, $f'(-3)$, $f'(-1)$, $f'(2)$, and $f'(4)$.
(b) Sketch a graph of f'.

7. Let f be the function whose graph is shown below.

Graph of f

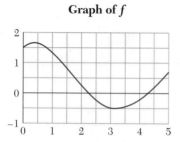

(a) Estimate $f'(0)$, $f'(1)$, $f'(2)$, $f'(3)$, $f'(4)$, and $f'(5)$.

(b) Use your results from part (a) to sketch a graph of the function f'.

8. Let $f(x) = x^2$.
   (a) Use zooming to estimate $f'(0)$, $f'(1/2)$, $f'(1)$, $f'(2)$, and $f'(3)$.
   (b) Use your results from part (a) to sketch a graph of f' over the interval $[-3, 3]$.   [**HINT:** Why is $f'(-1) = -f'(1)$?]
   (c) Use your results from parts (a) and (b) to guess an algebraic formula for $f'(x)$.

9. Let $f(x) = \sqrt{x}$.
   (a) Use zooming to estimate $f'(1/4)$, $f'(1)$, $f'(9/4)$, $f'(4)$, $f'(25/4)$, and $f'(9)$.
   (b) Use your results from part (a) to sketch a graph of f' over the interval $[0, 9]$.
   (c) Why is $f'(-1)$ undefined? Why is $f'(0)$ undefined? What is the domain of f'?   [**HINT:** Look at a graph of f.]
   (d) Use your results from part (a) to guess an algebraic formula for $f'(x)$.   [**HINT:** Evaluate $1/f(x)$ at $x = 1/4$, $x = 1$, $x = 9/4$, $x = 4$, $x = 25/4$, and $x = 9$.]

10. Let $f(x) = x^3$ and $g(x) = x^{1/3}$.
   (a) Use zooming to estimate $f'(0)$, $f'(1/2)$, $f'(1)$, $f'(3/2)$, and $f'(2)$.
   (b) Use your results from part (a) to sketch a graph of f' over the interval $[-2, 2]$.   [**HINT:** Why is $f'(-1) = f'(1)$?]
   (c) Use your results from parts (a) and (b) to guess an algebraic formula for $f'(x)$.
   (d) Estimate $g'(-1)$ and $g'(1)$.
   (e) Why is $g'(0)$ undefined? What is the domain of g'? What is the domain of g?
   (f) It can be shown that $f'(4) = 48$. Use this fact to find $g'(64)$. [**HINT:** How are the graphs of f and g related? The point $(4, 64)$ is on the graph of f and the point $(64, 4)$ is on the graph of g.]

11. Let $f(x) = 1/x$.
   (a) Use zooming to estimate $f'(1/4)$, $f'(1/2)$, $f'(1)$, $f'(2)$, and $f'(3)$.
   (b) Use your results from part (a) to sketch a graph of f' over the intervals $[-3, 0)$ and $(0, 3]$.   [**HINT:** Why is $f'(-1) = f'(1)$?]
   (c) Use your results from parts (a) and (b) to guess an algebraic formula for $f'(x)$.

12. Let $f(x) = e^x$.
   (a) Use zooming to estimate $f'(-1)$, $f'(0)$, $f'(1)$, $f'(1.5)$, and $f'(2)$.
   (b) Use your results from part (a) to sketch a graph of f' over the interval $[-1, 2]$.
   (c) Use your results from part (a) to guess a formula for $f'(x)$. [**HINT:** Evaluate $f(x)$ at $x = -1$, $x = 0$, $x = 1.5$, and $x = 2$.]

13. Let $f(x) = \ln x$.
   (a) Use zooming to estimate $f'(1/10)$, $f'(1/5)$, $f'(1/2)$, $f'(1)$, $f'(2)$, and $f'(5)$.
   (b) Use your results from part (a) to sketch a graph of f' over the interval $[0, 5]$.
   (c) Use your results from part (a) to guess a formula for $f'(x)$.

14. Let $f(x) = \sin x$.
   (a) Use zooming to estimate $f'(0)$, $f'(\pi/6)$, $f'(\pi/4)$, $f'(\pi/3)$, $f'(\pi/2)$, and $f'(\pi)$.
   (b) Use the symmetry of the sine function and your answers from part (a) to estimate $f'(2\pi/3)$, $f'(3\pi/4)$, and $f'(5\pi/6)$. Check your answers using zooming.
   (c) Use your results from parts (a) and (b) to guess an expression for $f'(x)$.

15. Let $f(x) = \cos x$.
   (a) Use zooming to estimate $f'(0)$, $f'(\pi/6)$, $f'(\pi/4)$, $f'(\pi/3)$, $f'(\pi/2)$, and $f'(\pi)$.
   (b) Use your results from part (a) to guess an expression for $f'(x)$.

16. (a) Zoom in on the graph of $y = e^x$, near the point $(0, 1)$, until the graph looks like a straight line. What's the slope of this line?
   (b) Zoom in on the graph of $y = \ln x$, near the point $(1, 0)$, until the graph looks like a straight line. What's the slope of this line?

17. The previous exercise explains (partly) why base e is special. The italicized statement on p. 60 holds *only* for exponential and logarithm functions with base e. For other bases, a slightly different relation holds between the slope of $y = b^x$ at $(0, 1)$ and the slope of $y = \log_b x$ at $(1, 0)$. Let's "discover" this relationship for $b = 10$.
   (a) Zoom in on the graph of $y = 10^x$ near the point $(0, 1)$ until the graph looks like a straight line. (This is the "tangent line" to the graph of $y = 10^x$ at $(0, 1)$.) Find the slope as accurately as possible.

(b) Zoom in on the graph of $y = \log_{10} x$ near the point $(1, 0)$ until the graph looks like a straight line. (This is the "tangent line" to the graph of $y = \log_{10} x$ at $(1, 0)$.) Find the slope as accurately as possible.

(c) How are the slopes in parts (a) and (b) related? What "variant" of the italicized statement referred to above holds in this situation?

18. In Section 1.5 we claimed:

If $y = b^x$, then the rate of change of y is proportional to y itself.

In this exercise we explore this claim.

(a) By zooming in, estimate the slope of the tangent line to the graph of $y = e^x$ at $x = 0$, $x = 1$, and $x = 2$. Are your results consistent with the above statement? With what constant of proportionality?

(b) By zooming in, estimate the slope of the tangent line to the graph of $y = 10^x$ at $x = 0$, $x = 1$, and $x = 2$. Are your results consistent with the above statement? With what constant of proportionality?

19. (a) Sketch the graph of a function f that is consistent with these data:

x	-2	-1	1	2
$f(x)$	1	-1	-1	2
$f'(x)$	-3	0	-1	-2

(b) Sketch the graph of a function g that is consistent with these data:

x	-2	-1	1	2
$g(x)$	1	-1	-1	2
$g'(x)$	1	0	-1	-2

20. The line tangent to f at $(5, 2)$ passes through the point $(0, 1)$. Find $f(5)$ and $f'(5)$. Justify your answers.

21. Suppose that the line tangent to the curve $y = f(x)$ at $x = 3$ passes through the points $(-2, 3)$ and $(4, -1)$. Find $f(3)$ and $f'(3)$.

22. The line tangent to the curve $y = f(x)$ at $x = 2$ has slope -1 and crosses the x axis at $x = 5$. Find $f(2)$ and $f'(2)$.

23. Suppose that the line tangent to the curve $y = f(x)$ at $x = -2$ is described by the equation $y = 4x + 3$. Find $f(-2)$ and $f'(-2)$.

Exercises 24–32 concern a function f whose derivative, f', is the function shown below. [**NOTE:** The graph of f is not shown.]

Graph of f'

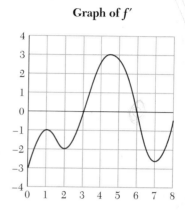

24. Explain why each of the following *cannot* be the line tangent to the curve $y = f(x)$ at $x = 1$: (i) $y = x + 1$; (ii) $y = 3$; (iii) $y = -2x - 1$; (iv) $y = 2x$.

25. Which of the following could be the line tangent to the curve $y = f(x)$ at $x = 6$: (i) $y = -5$; (ii) $y = 6(x + 1)$; (iii) $6y = 2 - 3x$; (iv) $y = -6x$; (v) $y = 6x + 1$?

26. Find the equation of a line parallel to the line tangent to the curve $y = f(x)$ at $x = 4.5$.

27. Suppose that $f(0) = 1$. Explain why $y = -2x + 3$ is *not* the line tangent to the curve $y = f(x)$ at $x = 2$.

28. Suppose that $f(0) = 1$. Explain why $y = -2x - 1$ is *not* the line tangent to the curve $y = f(x)$ at $x = 2$.

29. Let $\ell(x) = f'(5)(x - 5) + f(5)$.
   (a) Is $\ell(6)$ greater than, less than, or equal to $f(6)$? Justify your answer.
   (b) Is $\ell(8)$ greater than, less than, or equal to $f(8)$? Justify your answer.

30. Let $g(x) = -3x + f(0)$.
   (a) Explain why $g(1) < f(1)$.
   (b) Is $g(2) < f(2)$? Justify your answer.
   (c) Is $g(8) < f(8)$? Justify your answer.

31. Let ℓ be the line tangent to f at $x = 1$.
   (a) Is $\ell(2) \le f(2)$? Justify your answer.
   (b) Is $\ell(0) \le f(0)$? Justify your answer.

32. Let ℓ be the line tangent to f at $x = 7$.
   (a) Is $\ell(8) \le f(8)$? Justify your answer.
   (b) Is $\ell(6) \le f(6)$? Justify your answer.
   (c) Is $\ell(2) \le f(2)$? Justify your answer.

33. The graph of the derivative of a function g is shown below. [**NOTE:** The graph of g is not shown.]

Graph of g'

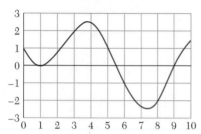

(a) Suppose that $g(3) = 11$. Find an equation of the line tangent to g at the point $(3, 11)$.
(b) Let ℓ be the line tangent to g at $x = 2$. Is $g(5) \leq \ell(5)$? Is $g(1) \leq \ell(1)$? Justify your answers.
(c) Let h be the line tangent to g at $x = 6$. Explain why $g(x) \leq h(x)$ if $6 \leq x \leq 8$.
(d) Let h be the line tangent to g at $x = 6$. How do the values of $g(x)$ and $h(x)$ compare if $0 \leq x \leq 6$? Justify your answer.

34. Suppose that $f(2) = 4$ and $f'(x) = \sqrt{x^3 + 1}$ for $x \geq -1$.
(a) Find an equation for the line tangent to f at $x = 2$.
(b) Let ℓ be the tangent line from part (a). Does $\ell(0)$ overestimate or underestimate $f(0)$? Justify your answer.
(c) Does $\ell(3)$ overestimate or underestimate $f(3)$? Justify your answer.

For each pair of functions in Exercises 35–40, the constant A can be chosen so that $f'(x) = Ag(x)$ for every x in the domain of g. Find this value of A. [**HINT:** Examine graphs of f and g.]

35. $f(x) = x^3$, $g(x) = x^2$
36. $f(x) = \sqrt{x}$, $g(x) = x^{-1/2}$
37. $f(x) = \sin x$, $g(x) = \cos x$
38. $f(x) = \cos x$, $g(x) = \sin x$
39. $f(x) = e^x$, $g(x) = e^x$
40. $f(x) = \ln x$, $g(x) = 1/x$

2.3 The Geometry of Derivatives

The geometric relationship between a function f and its derivative f' is easy to state:

> *For any input a, $f'(a)$ is the slope of the line tangent to the f-graph at $x = a$.*

This statement—innocuous as it appears—is one of the most important in this book. The next two sections (and much of this book) explore its meaning and implications.

Graphs of *f* and *f'*: An Extended Example

But don't peek.

Corresponding points (e.g., A and A') have the same x-coordinate.

Graphs of a function f and its derivative f' follow. (For the moment, no formulas are given—or needed. We'll return to these functions symbolically at the end of the next section.) ◄   Three interesting points are labeled on each. ◄

Graph of f

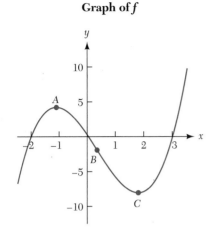

Graph of f'

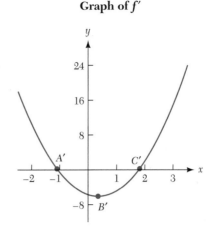

A First Look

We've only asserted, not proved, that f' has the graph shown. Tools to be developed in later sections will let us *prove* this; here we'll just assume it.

The f'-graph tells how slopes of tangent lines to the f-graph behave. At B, for instance, the f-graph seems ▶ to have a slope around -6; for this reason, the f'-graph has height -6 at B'.

Check carefully for yourself.

First let's observe several straightforward geometric relationships between the two graphs, introducing some useful new terminology in the process.

The Sign of f' The graph of f rises to the left of A, falls between A and C, and rises again to the right of C. The sign of $f'(x)$ tells whether the line tangent to the f-graph at x points up or down. At A' and C', ▶ f' changes sign. Thus, at A and C, f itself changes direction.

A' and C' correspond to roots of f'.

In the Long Run The graph suggests that f blows up to the right of C and blows down to the left of A. In geometric terms, as $x \to \infty$, the graph of $f(x)$ rises more and more steeply. What happens over the same intervals on the f'-graph? To the right of C', f' is positive and increasing. To the left of A', f' is positive and decreasing.

Stationary, Maximum, and Minimum Points The points A and C, with approximate coordinates $(-1.1, 4)$ and $(1.8, -8)$, ▶ where the f-graph is horizontal, are obviously of interest. They mark, respectively, high and low points of the graph. The situation looks clear, but to avoid later confusion, it will pay to be extremely picky with language here—especially about the distinction between inputs to f (called points) and outputs from f (called values).

We'll improve these guesses later by symbolic methods.

Here, in full detail, is the situation at $A \approx (-1.1, 4)$. The domain point $x = -1.1$ is called a local maximum point of f; the corresponding output—$f(-1.1) \approx 4$—is called a local maximum value of f. At C ▶ the situation is similar: $x = 1.8$ is a **local minimum point** of f, and $f(1.8) \approx -8$ is the corresponding **local minimum value** of f. The

$C \approx (1.8, -8)$.

Careful definitions of all these terms will follow.

x-coordinates of both *A* and *C* are called **stationary points** of *f*. ◄   (We say local rather than global because elsewhere in its domain *f* may assume larger or smaller values.) The corresponding points *A'* and *C'* occur where the *f'*-graph crosses the *x*-axis (i.e., at roots of *f'*).

Any More Stationary Points?   Are *A* and *C* the only stationary points of *f*? The picture doesn't say. It shows only the interval $-2.5 \leq x \leq 3.5$; additional stationary points could lie outside this interval.

Recall: "Holds water" and "spills water" are rough synonyms for "concave up" and "concave down."

Concavity and Inflection   The point *B*, near $x = 0.3$, is an **inflection point** of *f*: At *B*, the *f*-graph's **direction of concavity** changes, from concave down to concave up. ◄   The point *B* has another special geometric property: At *B*, the graph of *f* points most steeply downward.

Locating B or B' precisely by magnifying either graph doesn't work very well. Viewed closely, both graphs look like straight lines.

The corresponding point on the *f'*-graph, *B'*, is easier to see; it's a local minimum point. Later we'll use this property and some algebra to find the *exact* location of *B*. ◄

What *f'* Says about *f*

Interpreting the derivative function *f'* in terms of the slopes of tangent lines on the *f*-graph has many important geometric implications. We summarize several below.

Nicely Behaved Functions.   Many ideas in this section depend on both *f* and *f'* being reasonably "nice," or "well behaved." "Nice enough" means, roughly that neither *f* nor *f'* displays sudden jumps or breaks. Any function *f* whose derivative function *f'* is continuous (i.e., without breaks) is nice enough for us. Fortunately, most calculus functions *are* nice in this sense. As we proceed we'll see more of what nice means and why it matters.

Increasing or Decreasing?

From left to right!

A function *f* **increases** where its graph rises ◄   and **decreases** where its graph falls. The following definition captures these natural ideas in analytic language.

Definition Let *I* denote the interval (a, b). A function *f* is **increasing** on *I* if

$$f(x_1) < f(x_2)^* \qquad \text{whenever} \qquad a < x_1 < x_2 < b.$$

f is **decreasing** on *I* if

$$f(x_1) > f(x_2) \qquad \text{whenever} \qquad a < x_1 < x_2 < b.$$

In these definitions, ≤ and ≥ replace < and >.

(If $f(x_1) \leq f(x_2)$ for $x_1 < x_2$, then *f* is **nondecreasing** on *I*; if $f(x_1) \geq f(x_2)$ for $x_1 < x_2$, then *f* is **nonincreasing** on *I*. ◄   In either case, *f* is **monotonic** on *I*.)

The derivative detects whether a function increases or decreases on an interval I.

Fact If $f'(x) > 0$ for all x in I, then f increases on I. If $f'(x) < 0$ for all x in I, then f decreases on I.

This fact certainly *sounds* reasonable. To say that $f'(x) > 0$ means that the tangent line to the f-graph at x points upward. With any luck, so should the f-graph itself. Racetrack ▶ language makes the fact seem even simpler: If dog f has positive velocity, then dog f always runs *forward*.

See page 99 for more on the racetrack principle.

Plausible as the fact sounds, proving it rigorously requires one of calculus's "big theorems": the mean value theorem (we'll see it later).

Behavior at a Point.   Functions are often said to increase or decrease *at a point*. To say, for example, that f increases *at* $x = 3$ means that f increases on some interval—perhaps a small one—containing $x = 3$. Now we can restate the preceding fact ▶ as follows:

Here, too, we're assuming that both f and f' are free of discontinuities and other naughty behavior.

Fact If $f'(a) > 0$, then f is increasing at $x = a$. If $f'(a) < 0$, then f is decreasing at $x = a$.

At $x = 1$, for instance, $f'(1) < 0$. The Fact says—and the picture agrees—that at $x = 1$ the f-graph is decreasing.

Derive Carefully.   The previous fact's **converse** reads like this: *If f is increasing at $x = a$, then $f'(a) > 0$.* Tempting as it is, the converse is false. The next example shows why.

EXAMPLE 1   Where is the function $f(x) = x^3$ increasing? Where is $f'(x) > 0$? What happens at $x = 0$?

Solution   As its graph suggests,

Graph of $f(x) = x^3$

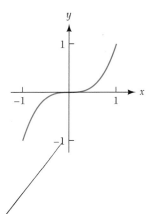

Imagine tangent lines.   the function $f(x) = x^3$ increases *everywhere*. The graph suggests, too, ◄ that $f'(x) > 0$ if $x \neq 0$, but that $f'(0) = 0$.   ■

The Derivative of an Increasing Function.   What *can* be said about f' where f increases? This:

> **Fact** If f increases at $x = a$, then $f'(a) \geq 0$; if f decreases at $x = a$, then $f'(a) \leq 0$.

The derivative of an increasing function must be *nonnegative*, but not necessarily *strictly positive*.

Maximum Points, Minimum Points, and Stationary Points: Formal Definitions

See page 119.
Look carefully to see why.   The point A on the first graph in this section ◄ corresponds to a local, but not global, maximum value of f. ◄ Successful use of such technical language requires clear definitions.

> **Definition** Let f be a function and x_0 a point of its domain.
> - x_0 is a **stationary point** of f if $f'(x_0) = 0$.
> - x_0 is a **local maximum point** of f if $f(x_0) \geq f(x)$ for all x in some open interval containing x_0. The number $f(x_0)$ is a **local maximum value** of f.
> - x_0 is a **global maximum point** of f if $f(x_0) \geq f(x)$ for all x in the domain of f. The number $f(x_0)$ is the **global maximum value** of f.
>
> (Local and global **minimum points** are defined similarly.)

Several linguistic subtleties deserve mention.

Points versus Values Local and global maximum *points* are inputs to a function; local and global maximum *values* are the corresponding outputs. The difference, although important, is sometimes blurred in informal speech. (The first sentence of the next paragraph is impeccably correct on this score.)

Local versus Global At a *local* maximum point, a function f may or may not assume a *global* maximum value. Minimum points, too, may be local, global, or both.

Extreme Values The word **extremum** (plural extrema), which means "either maximum or minimum value," is a convenient shorthand. We'd say, for instance, that for the function f at the beginning of this section, both points A and C correspond to extrema of f.

Finding Maximum and Minimum Points

Geometric intuition says that at a local maximum point or local minimum point, a smooth graph must be "flat." ▶ More succinctly:

See A and C on the preceding graphs.

> **Fact** On a smooth graph every local maximum or local minimum point x_0 is a stationary point—i.e., a root of f'.

This fact has immense practical value. To find maximum and minimum values of f, the fact says, we can limit our search to roots of f'. Each such root is a stationary point and therefore, possibly, a maximum or minimum point (but not a sure thing—a stationary point may be only a "flat spot" in the graph). ▶

The next example—an important one—shows how to sort out all the possibilities.

See the previous example: $f(x) = x^3$ has a "flat spot"—but not a maximum or minimum—at $x = 0$.

EXAMPLE 2   The graph of a function f' appears as follows; the f-graph is not shown (for now). Three points of interest are bulleted. Where, if anywhere, does f have local maximum or local minimum points? Why?

Graph of f'

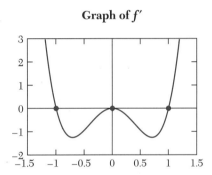

Solution   The three bullets on the graph—at $x = -1$, $x = 0$, and $x = 1$—represent roots of f' and therefore correspond to stationary points of f. What type of stationary point is each one: a local maximum, a local minimum, or just a flat spot? The key to deciding is to check the sign of f' just before and just after each stationary point. We take each root of f' in turn.

At $x = -1$ Just before (i.e., to the left of) $x = -1$, $f'(x) > 0$. Therefore (by an earlier Fact), f increases until $x = -1$. Just after $x = -1$, $f'(x) < 0$, so f decreases immediately after $x = -1$. This means that f has a local maximum at $x = -1$.

At $x = 1$ Consider values of x near $x = 1$. The graph shows that if $x < 1$, $f'(x) < 0$; if $x > 1$, $f'(x) > 0$. Thus, reasoning as above, f decreases before $x = 1$ and increases after $x = 1$. This means that f has a local minimum at $x = 1$.

At $x = 0$ This time the graph shows that $f'(x) < 0$ on both sides of $x = 0$. Thus, f must decrease before *and* after $x = 0$, so $x = 0$ is neither a maximum nor a minimum point, but just a flat spot in the f-graph.

Here, at last, is a possible f-graph. It agrees with everything we said.

Graph of f

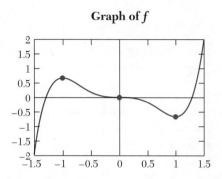

Let's summarize what we know about stationary points.

Fact (First Derivative Test) Suppose that $f'(x_0) = 0$.

- If $f'(x) < 0$ for $x < x_0$ and $f'(x) > 0$ for $x > x_0$, then x_0 is a local minimum point.
- If $f'(x) > 0$ for $x < x_0$ and $f'(x) < 0$ for $x > x_0$, then x_0 is a local maximum point.

The next example shows what happens on a crooked graph.

EXAMPLE 3   The function $f(x) = |x|$ has a local (and global) minimum point at $x = 0$, but not a stationary point. ◄   The graph shows why:

At $x = 0$ this function is not nice in the sense discussed earlier.

Graph of $f(x) = |x|$:
a local minimum at $x = 0$

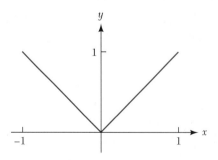

At $x = 0$ the graph has a kink and therefore *no* tangent line and no well-defined slope. In functional language, $f'(0)$ does not exist.   ∎

Concave Up or Concave Down?

So far we've described concavity and inflection points informally, in graphical language. Here's a more formal, analytic definition:

> **D e f i n i t i o n** The graph of f is **concave up** at $x = a$ if the slope function f' is increasing at $x = a$. The graph of f is **concave down** at $x = a$ if f' is decreasing at $x = a$. Any point at which a graph's direction of concavity *changes* is an **inflection point** of the graph.

The graphs of f and f' on page 119 ▶   illustrate all of this.

Look again.

Concave Down   To the left of B, the f-graph is concave down; to the left of B', the f'-graph is decreasing.

Concave Up   To the right of B, the f-graph is concave up; to the right of B' the f'-graph is increasing.

An Inflection Point B is an inflection point of f; f' has a minimum point at B'.

Finding Inflection Points from the Graph of f′

The direction of concavity of the graph of f depends, as the definitions show, on whether f' increases or decreases. An inflection point occurs wherever f' *changes direction*—i.e., wherever f' has a local minimum or a local maximum.

E X A M P L E 4   Discuss concavity of the sine function. Find all inflection points; explain the results in derivative language.

S o l u t i o n   It's a fact—we'll show it carefully later—that if $f(x) = \sin x$, then $f'(x) = \cos x$. The graphs certainly make the claim plausible: ▶

Do you agree? Are the roots of f' where they should be?

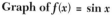

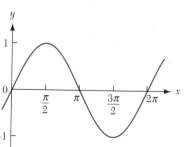

Graph of $f(x) = \sin x$

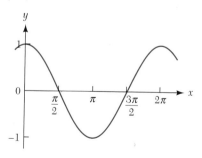

Graph of $f'(x) = \cos x$

Notice:

Stationary Points f has stationary points (a local ▶   maximum point and a local minimum point) $x = \pi/2$ and at $x = 3\pi/2$—exactly the roots of f'.

Global, too, in this case.

Increasing or Decreasing? f increases on the intervals $(0, \pi/2)$ and $(3\pi/2, 2\pi)$; on the same intervals, f' is positive.

Concavity f is concave down on $(0, \pi)$—where f' decreases—and concave up on $(\pi, 2\pi)$—where f' increases. In fact, f illustrates every possible combination of increasing/decreasing and concavity behavior.

Inflection Points f has an inflection point at each multiple of π—precisely where f' assumes a local maximum or local minimum.   ■

A Sample Problem

Here's a problem in the spirit of this section.

EXAMPLE 5   The graph of the derivative of a function f is shown.

Graph of f'

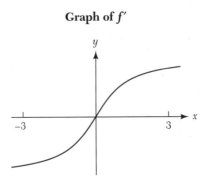

a. The equation $f(x) = 0$ can have no more than two solutions in the interval $[-3, 3]$. Explain why.

b. Explain why f cannot have two roots in the interval $[0, 3]$.

c. Suppose that $f(0) = 1$. How many solutions does the equation $f(x) = 0$ have in the interval $[-3, 3]$? Explain.

Solution   Because f' is negative on the interval $(-3, 0)$ and positive on the interval $(0, 3)$, f itself must decrease on $(-3, 0)$ and increase on $(0, 3)$. Thus f has a local minimum at $x = 0$ and no other stationary points. Therefore:

a. The graph of f is U-shaped, so it can intersect the interval $[-3, 3]$ in at most two places.

b. The graph of f increases on the interval $(0, 3)$, so it can cross the x-axis no more than once on that interval.

c. If $f(0) = 1$, then f has no roots in the interval $[-3, 3]$, because f has its minimum value at $x = 0$.   ■

BASICS

1. The graph of a function f is shown below. Several points are labeled.

Graph of f

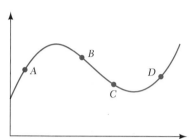

(a) At which labeled points is f' positive? Explain.
(b) Between which pairs of labeled points does f have a stationary point?
(c) Between which pair of labeled points does f have an inflection point?
(d) Between which pair of labeled points is f' increasing?
(e) Between which pair of labeled points does f' achieve its minimum value?

2. The graph of a function f is shown below.

Graph of f

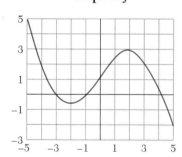

(a) On which intervals is f' negative? Positive?
(b) On which intervals is f' increasing? Decreasing?
(c) Where does f' achieve its maximum value? Estimate this value of f'.
(d) Where does f' achieve its minimum value? Estimate this value of f'.
(e) Sketch a graph of f'.   [**NOTE:** Your sketch should be consistent with your answers to parts (a)–(d).]

3. The graph of a function f is shown below

Graph of f

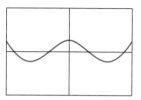

Which of the following figures could be the graph of f'? For each graph that you did *not* select, explain why it can't be the graph of f'.

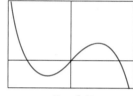

Graph I

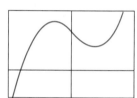

Graph II

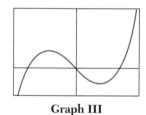

Graph III

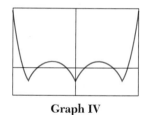

Graph IV

4. The graph of the derivative of a function g is shown below. Use the graph of g' to answer the following questions about g.   [**NOTE:** The graph of g is not shown.]

Graph of g'

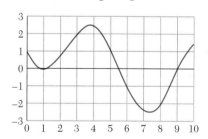

(a) Where does g have stationary points?
(b) Where does g have local maxima? Local minima?
(c) The graph of g' has a local maximum at $x = 3.8$ and a local minimum at $x = 7.4$. What do these facts say about the graph of g?

(d) Is g concave up or concave down at $x = 5$? At $x = 8$? Justify your answers.

(e) Suppose that $g(0) = 0$. Sketch a graph of g.

5. The graph of the derivative of a function f appears below. [**NOTE:** The graph of f is not shown.]

Graph of f'

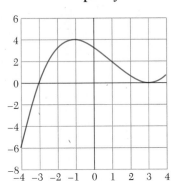

(a) Where does f have stationary points?

(b) Where does f have local maxima? Local minima? Points of inflection?

(c) Is f concave up or concave down at $x = 1$? Justify your answer.

(d) Sketch the graphs of *two* functions that could be f.

6. Suppose that the derivative of a function f is $f'(x) = x + 2$.

(a) On which intervals, if any, is f increasing?

(b) At which values of x, if any, does f have a local maximum? A local minimum?

(c) On which intervals, if any, is f concave down?

(d) Sketch a graph of f.

Repeat the previous exercise for the derivative functions in Exercises 7–18. (Report approximate answers to two decimal places.)

7. $f'(x) = (x - 2)(x + 4)$

8. $f'(x) = (x - 2)^2(x + 4)$

9. $f'(x) = (x + 1)^2(x - 2)$

10. $f'(x) = (x^2 + 9)(x + 6)$

11. $f'(x) = (x^2 - 9)(x + 3)$

12. $f'(x) = \dfrac{1}{x^2 + 1}$

13. $f'(x) = \dfrac{x - 2}{x^2 + 5}$

14. $f'(x) = \dfrac{x + 1}{\sqrt{x^2 + 1}}$

15. $f'(x) = \dfrac{x^2 + 2}{\sqrt{x^4 + 8}}$

16. $f'(x) = e^{-x^2}$

17. $f'(x) = (9 - x^2)e^{-x/4}$

18. $f'(x) = \ln(1 + x^2)$

FURTHER EXERCISES

19. In the left-hand column below are graphs of several functions. In the right-hand column—in a different order—are graphs of the associated *derivative* functions. Match each function with its derivative. [**NOTE:** The scales on the graphs are not all the same.]

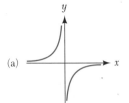

(a)

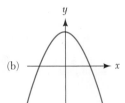

(b)

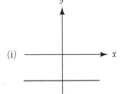

(i)

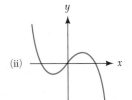

(ii)

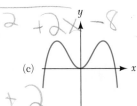

(c)

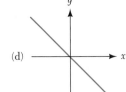

(d)

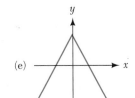

(e)

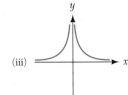

(iii)

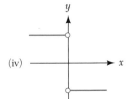

(iv)

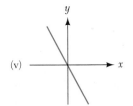

(v)

20. Sketch a graph of the derivative of each function labeled (i)–(v) in the right column of the preceding problem.

21. For each function labeled (a)–(e) in the left column of Exercise 19, sketch the graph of a function whose *derivative* is the function shown.

22. Graphs of f, f', and g (not the derivative of f) are shown below. Which is which? Explain how you can tell.

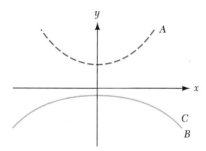

23. Graphs of f, f', and g (not the derivative of f) are shown below. Which is which? Explain how you can tell.

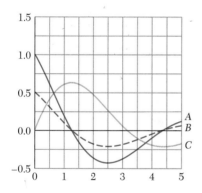

24. (a) Sketch the graph of a function f for which $f(x) > 0$ and $f'(x) > 0$ for all x.
    (b) Sketch the graph of a function g for which $g(x) > 0$ and $g'(x) < 0$ for all x.
    (c) Give formulas for functions f and g with these properties.

25. (a) Sketch the graph of a function f for which $f(x) < 0$ and $f'(x) > 0$ for all x.
    (b) Sketch the graph of a function g for which $g(x) < 0$ and $g'(x) < 0$ for all x.
    (c) Give formulas for functions f and g with these properties.

26. The graph of the derivative of a function f is shown below. Use the graph of f' to answer the following questions about f. [**NOTE:** The graph of f is not shown.]

Graph of f'

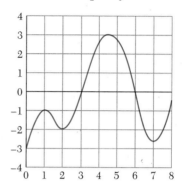

(a) On which intervals is f increasing? Decreasing?
(b) Where does f have a stationary point?
(c) Where does f have a local maximum? A local minimum?
(d) On which intervals is f concave up? Concave down?
(e) Where does f have a point of inflection?
(f) Where does f achieve its maximum value on the interval $[0, 2]$? Its minimum value?
(g) Where does f achieve its maximum value on the interval $[3, 6]$? Its minimum value?
(h) Assume that $f(0) = 0$. Sketch a graph of f. (Your graph need only have the right general shape.)
(i) How does your answer to part (h) change if $f(0) = 5$?

27. Suppose that $f(3) = -1$ and that f' is the function graphed in the previous exercise.
    (a) Let ℓ be the line tangent to f at the point $(3, -1)$. Find an equation for ℓ.
    (b) Explain why ℓ lies below the curve $y = f(x)$ near $x = 4$.
    (c) Does ℓ lie above or below the curve $y = f(x)$ near $x = 2$? Justify your answer.

28. The graph of the *derivative* of a function f is shown below.

Graph of f'

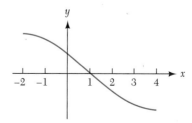

(a) Where in the interval $[-2, 4]$ is f increasing? Where is f decreasing? Why?

(b) Where in $[-2, 4]$ does f have a stationary point? Is it a local maximum, a local minimum, or neither?

(c) Describe in words the general shape of the f-graph over the interval $[-2, 4]$.

29. Let $f(x) = e^{\sin x}$. How many roots does f' have in the interval $[0, 2\pi]$? Where are they located? Justify your answers.

30. Suppose that the following information is known about a function f:
    (i) f is positive on $[-7, 8]$
    (ii) f is increasing on $[-2, 3]$
    (iii) f is decreasing on $(4, 10)$
    (iv) f' is decreasing on $[-3, 6]$
    (v) $f'(4) = 0$
    (vi) f has an inflection point at $x = 7$

    (a) Explain why f has a local extremum at $x = 4$. Is it a local maximum or a local minimum?
    (b) Where is f concave down?
    (c) Where is f' nonnegative?
    (d) Rank the four values $f'(4)$, $f(5)$, $f'(6)$, and $f(7)$ in increasing order.
    (e) Sketch the graph of a function f with these properties.

2.4 The Geometry of Higher-Order Derivatives

As before, "nice" means that all the derivatives in question exist and have continuous graphs.

"Differentiate" means "find the derivative of."

The derivative of a nice function f is another nice function, f'. ◄   Given the many uses of f', the next step is natural: Differentiate ◄   again. The resulting function, denoted by f'', is called the **second derivative** of f. Like f', the second derivative f'' carries useful information about f. What information? How do we read it?

Prime Advice.   With several functions (f, f', f'', etc.) in the picture, it's easy to lose one's bearings. One organizing principle helps:

> *The derivative of any function tells whether, and how fast, that function increases or decreases.*

We defined concavity in terms of f' in the last section.

If, say, $f''(a) > 0$, then f' is *increasing* at $x = a$. Therefore, it follows that f must be concave up at $x = a$. ◄

What f″ Says about f—Concavity Revisited

The graph of f is concave up or down at $x = a$ depending on whether f' is increasing or decreasing at $x = a$. The preceding principle, applied to f', reads like this:

> *If $f''(a) > 0$, then f' increases at $x = a$. If $f''(a) < 0$, then f' decreases at $x = a$.*

This leads to a useful test for concavity:

F a c t The graph of f is concave up where $f'' > 0$ and concave down where $f'' < 0$.

(The fact gives no information if $f''(a) = 0$. In that case, as we'll see, f may be concave up, concave down, or neither.)

EXAMPLE 1   Interpret the concavity of the sine function using the second derivative.

Solution   If $f(x) = \sin x$, then $f'(x) = \cos x$ and $f''(x) = -\sin x$. (We'll derive both of these formulas carefully later in the chapter. For the moment, we'll assume that they hold.) Following are graphs of f and f'' on the same axes.

Graphs of f and f''

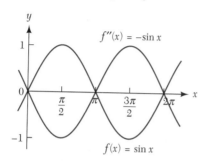

It's easy to see ▶ that the advertised relationship holds: f is concave up precisely where $f'' > 0$ and is concave down where $f'' < 0$. ∎   *See for yourself.*

> **Wave Claims.**   We claimed earlier, on graphical evidence alone, that if $f(x) = \sin x$, then $f'(x) = \cos x$. We also claimed that if $g(x) = \cos x$, then $g'(x) = -\sin x$. Both claims require careful proof.
>
> Notice, though, that the two claims are essentially the same. Since the graphs of $f(x)$ and $g(x)$ differ only by a horizontal translation, so must the graphs of their derivatives.

Finding Inflection Points

A function f has an inflection point wherever the *sign* of f'' changes. It follows, therefore, that for nice functions,

> *Every inflection point of f occurs at a root of f''.*

On the sine graph, for instance, an inflection point occurs at every root of f''.

Derive Carefully, Again.   The converse of the preceding statement is false:

> $f''(a)$ *may be zero even if no inflection occurs at* $x = a$.

The next example illustrates why: If $f''(a) = 0$, then a is a stationary point of f' but not necessarily a local maximum or local minimum.

The functions shown have simple formulas: $f(x) = x^4$, $f'(x) = 4x^3$, and $f''(x) = 12x^2$. We "hid" the formulas here in the margin because they're not necessary to this problem.

EXAMPLE 2   Graphs of a function f and its derivatives f' and f'' follow. ◄ Discuss concavity and inflection points of f.

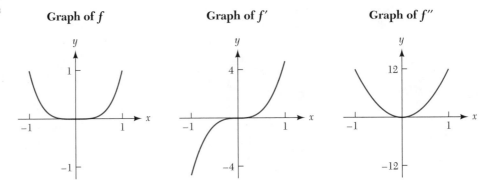

Graph of f   **Graph of f'**   **Graph of f''**

S o l u t i o n   Although $f''(0) = 0$, $(0, 0)$ is *not* an inflection point. On the contrary, the graph of f is concave up everywhere, because f' is everywhere increasing.

Notice too how the shape of the f-graph reflects the fact that $f(0) = f'(0) = f''(0) = 0$. If *several* derivatives of a function are zero at a point $x = a$, the graph of f looks flat near $x = a$. ◄   ■

The greater the number of derivatives that vanish at $x = a$, the flatter the graph.

All Together Now: Maximum Points, Minimum Points, Concavity, and the Second Derivative

A stationary point on the graph of a function f may be a local maximum point, a local minimum point, or neither. If an accurate graph of f is available, it's usually obvious which case applies. If not, concavity often helps us distinguish among these alternatives. The idea is simple:

> *At a local minimum point, the graph of f is concave up.*
> *At a local maximum point, the graph of f is concave down.*

The following claims are all immediate consequences. Together, they're known as the second derivative test.

Fact (Second Derivative Test) Suppose that $f'(a) = 0$.

If $f''(a) > 0$, then f has a local minimum at $x = a$.

If $f''(a) < 0$, then f has a local maximum at $x = a$.

If $f''(a) = 0$, anything can happen: f may have a local maximum, a local minimum, or neither at $x = a$, and the f-graph may be concave up, be concave down, have an inflection point, or have none of these properties.

If—but only if—appropriate care is taken, the second derivative test can be a convenient tool for finding maxima and minima of functions.

Stirring in the Formulas

We return, finally, to the functions f, f', and f'' we first met at the beginning of the preceding section—this time with algebraic formulas added. Here are all three formulas:

$$f(x) = x^3 - x^2 - 6x; \qquad f'(x) = 3x^2 - 2x - 6; \qquad f''(x) = 6x - 2.$$

Here are all three graphs:

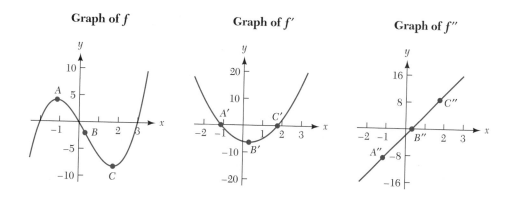

Graph of f   **Graph of f'**   **Graph of f''**

Combining graphs *and* formulas, we can sharpen and reinterpret our earlier observations:

What the Derivative Says Both the graph and the algebraic definition of f' suggest that as $x \to \infty$, $f'(x) \to \infty$. Therefore, as $x \to \infty$, the slope of the f-graph is (a) positive and (b) increasing. Equivalently, the f-graph itself is (a) rising and (b) rising faster and faster.

Stationary Points The f-graph has horizontal tangent lines at the stationary points A and C. According to the derivative formula, stationary points occur at values of x for which

$$f'(x) = 3x^2 - 2x - 6 = 0$$

By great (and, in some respects, atypical) good luck, $f'(x)$ happens to be a quadratic polynomial. By the quadratic formula, the roots of f' fall at $x = (1 \pm \sqrt{19})/3$. In decimal form, the roots are $x_1 \approx -1.1196$ and $x_2 \approx 1.7863$.

How Many Stationary Points? The preceding calculation shows, too, that f' has *only* two roots. Hence, A and C are the only stationary points of f.

Finding the Inflection Point The inflection point B occurs where f' takes its minimum value, somewhere near $x = 0.3$. Where *exactly* is B? The formula for f'' answers the question: A root of f'' occurs at $x = 1/3$, where

$$f''(x) = 6x - 2 = 0.$$

As its graph shows, f'' changes sign at $x = 1/3$, so f changes its direction of concavity there. ▶

The formula for f'' shows a little more: f'' changes sign only at $x = 1/3$. Therefore, f changes concavity only once.

Morals

We have mentioned a large, even bewildering, variety of connections among f, f', and f''. The reader may be tempted to memorize complicated lists of cases and subcases, particularly as they apply to concavity, inflection points, and the second derivative. That would be a mistake. A better strategy is to learn and understand the single organizing principle behind this welter of information:

> *The derivative function f' tells whether, and how fast, the original function f increases or decreases.*

A Sample Problem

The following example problem gives some idea of the variety of uses to which derivatives and higher derivatives can be put.

EXAMPLE 3   The graph of f', the derivative of a certain function f, follows. (The f-graph is not shown!)

Graph of f'

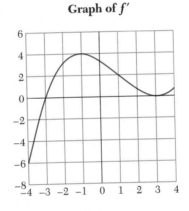

a. Estimate $f''(-3)$.

b. Where is f concave up? Concave down?

c. Where does f have inflection points?

Solution   Looked at properly, the graph answers all our questions. It's crucial to remember always that the picture shows f', not f or f''.

a. By definition, $f''(-3) = (f')'(-3)$, i.e., the slope of the graph above at $x = -3$. A careful look reveals that this slope is about 5.

b. Since $f''(-3) > 0$, f is concave up at $x = -3$. More generally, f is concave up wherever $f'' > 0$, i.e., where f' is increasing. The picture shows that this occurs for $-4 \le x \le -1$ and for $3 \le x \le 4$. For the same reason, f is concave down for $-1 \le x \le 3$.

c. The function f has an inflection point wherever its direction of concavity changes. Part (b) shows that this happens at $x = -1$ and at $x = 3$. ∎

BASICS

1. The graph of a function f appears below.

Graph of f

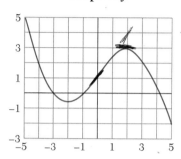

(a) For which values of x, if any, is $f''(x)$ negative? Positive? Zero?
(b) Rank the four numbers $f'(0)$, $f(2)$, $f'(2)$, and $f''(2)$ in increasing order.

2. The graph of a function f appears on the right. [**NOTE:** f has points of inflection near $x = 0.82$ and near $x = 1.93$; f has stationary points near $x = 0.52$, $x = 1.19$, and $x = 2.42$.]

Graph of f

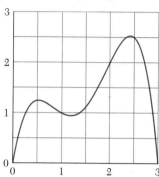

(a) Where is f'' positive? Negative? Zero?
(b) Rank the four values 0, $f''(0.5)$, $f''(1.2)$, and $f''(2.4)$ in increasing order.

3. Water flows at a constant rate into a large tank that is a cylinder with a circular base. Let $V(t)$ be the volume of water in the tank at time t and $H(t)$ be the height of the water in the tank at time t. Give a physical interpretation of each of the following expressions and indicate whether each is positive, negative, or zero at the time t^* when the tank is one-quarter full:

(a) $V'(t)$ (c) $V''(t)$
(b) $H'(t)$ (d) $H''(t)$

FURTHER EXERCISES

4. The graph of the second derivative of a function g is shown below. Use this graph to answer the following questions about g and g'.

Graph of g''

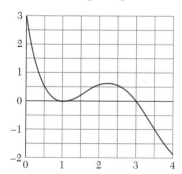

(a) Where is g concave up?
(b) Where does g have points of inflection?
(c) Rank the four numbers $g'(0)$, $g'(1)$, $g'(2)$, and $g'(3)$ in increasing order.
(d) Suppose that $g'(0) = 0$. Is g increasing or decreasing at $x = 2$? Justify your answer.
(e) Suppose that $g'(0) = -4$. Is g increasing or decreasing at $x = 2$? Justify your answer.
(f) Suppose that $g'(1) = -1$. Is g increasing or decreasing at $x = 2$? Justify your answer.

5. Suppose that $g' = f$ where f is the function defined in Exercise 2 above.

(a) Estimate $g''(0)$, $g''(0.5)$, $g''(1)$, $g''(1.5)$, $g''(2)$, and $g''(3)$.
(b) Sketch a graph of g'' over the interval $[0, 3]$.

6. In the left-hand column below are graphs of several functions. In the right-hand column—in a different order—are graphs of the *second* derivatives of these functions. Match each function with its second derivative function. [**NOTE:** The scales on the graphs are not all the same.]

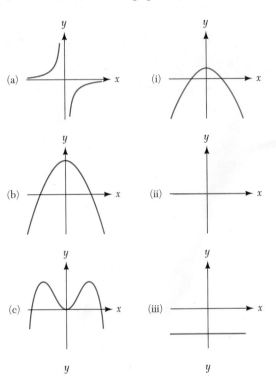

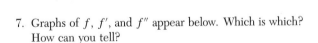

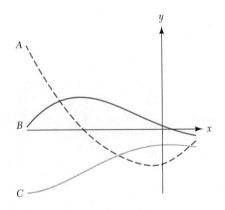

7. Graphs of f, f', and f'' appear below. Which is which? How can you tell?

8. Graphs of f, f', and f'' appear below. Which is which? How can you tell?

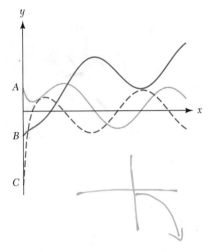

9. Graphs of f, f', f'', and g (not a derivative of f) are shown below. Which is which? How can you tell?

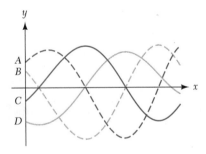

10. Suppose that the functions f, f', and f'' are continuous on $(-\infty, \infty)$ and that f has the following properties:
    (i) f is negative on $(-10, 1)$ and positive on $(3, 10)$.
    (ii) f is increasing on $(-\infty, 5)$ and decreasing on $(5, \infty)$.
    (iii) f is concave up on $(-\infty, 2)$ and concave down on $(2, \infty)$.
    Rank the five values $f'(0)$, $f'(1)$, $f'(2)$, $f''(2)$, and $f'(8)$ in increasing order.

11. Repeat Problem 3 assuming that the tank is a sphere.

12. Repeat Problem 3 assuming that the tank is an inverted cone (i.e., a cone with its vertex pointing down).

13. Interpret each sentence below as a statement about a function and its derivatives. In each case, indicate clearly what the function is and what each symbol means.
    (a) The child's temperature is still rising, but the penicillin seems to be taking effect.
    (b) The cost of a new car is increasing at an increasing rate.
    (c) Upper Midwest Industries' profit growth slows.

14. Let $U(t)$ be the number of people unemployed in a country t months after the election of an economically conservative president. Translate each of the following facts about the graph of $U(t)$ into statements about the unemployment situation. (Don't forget to specify units in your answers.)
    (a) The y-intercept of $U(t)$ is 2,000,000.
    (b) $U(2) = 3,000,000$.
    (c) $U'(20) = 10,000$.
    (d) $U''(36) = 800$ and $U''(36) = 0$.

15. (a) Sketch the graph of a function f for which $f(x) > 0$, $f'(x) > 0$, and $f''(x) > 0$ for all x.
    (b) Sketch the graph of a function f for which $f(x) > 0$, $f'(x) < 0$, and $f''(x) > 0$ for all x.

16. (a) Sketch the graph of a function f for which $f(x) < 0$, $f'(x) < 0$, and $f''(x) < 0$ for all x.
    (b) Sketch the graph of a function f for which $f(x) < 0$, $f'(x) > 0$, and $f''(x) < 0$ for all x.

17. Is there a function f for which $f(x) > 0$, $f'(x) < 0$, and $f''(x) < 0$ for all x? If so, sketch the graph of such a function. If not, explain why no such function exists.

18. Is there a function f for which $f(x) < 0$ and $f''(x) > 0$ for all x? If so, sketch the graph of such a function. If not, explain why there is no function with these properties.

19. Assume that f is a continuous function defined on the closed interval $[-3, 3]$ such that $f(-3) = 4$ and $f(3) = 1$. Furthermore, assume that f' and f'' are continuous on $(-3, 3)$ and that the information in the table below is known about these functions.

x	$-3 \le x < -1$	-1	$-1 < x < 0$	0
$f'(x)$	$+$	0	$-$	$-$
$f''(x)$	$-$	$-$	$-$	0
x	$0 < x < 1$	1	$1 < x \le 3$	
$f'(x)$	$-$	0	$-$	
$f''(x)$	$+$	0	$-$	

    (a) Sketch the graph of a function that has the given properties.
    (b) What are the x-coordinates of the points where f

achieves its maximum and minimum values on the interval $[-3, 3]$? Justify your answer.

20. Sketch the graph of a continuous function g that has *all* of the following properties:
    (i) The domain of g is the interval $[-3, 5]$.
    (ii) $g''(x) < 0$ for all $x \in (-2, 1)$
    (iii) $g''(x) > 0$ for all $x \in (1, 2)$
    (iv) $g'(-2) > 0$
    (v) $g'(4) > 0$
    (vi) $g'(1) = 0$ but $g(1)$ is neither the maximum nor the minimum value of g over the interval $[-2, 4]$

21. Sketch the graph of a continuous function that satisfies the following conditions:
    (i) $f'(x) < 0$ for all $x \ne 4$
    (ii) $f'(4)$ does not exist
    (iii) $f''(x) < 0$ for all $x < 4$
    (iv) $f''(x) > 0$ for all $x > 4$

22. It can be shown that if f is an even function, then f' is an odd function, and that if f is an odd function, then f' is an even function. Use these facts to show that if f is an even function, then f'' is an even function.

23. Suppose that $1 \le f''(x) \le 3$ if $2 \le x \le 4$, that $f'(2) = 5$, and that $f(2) = -6$.
    (a) Explain why $f'(3) \le 8$.
    (b) Use part (a) to show that $f(3) \le 2$.
    (c) Explain why $f'(3) \ge 6$.
    (d) Find upper and lower bounds on $f'(4)$.
    (e) Find upper and lower bounds on $f(4)$.

24. Suppose that the functions f, f', and f'' are continuous on $(-\infty, \infty)$, that $f'(2) = -1$, and that $0.1 < f''(x) < 0.5$ if $2 < x < 4$. Find upper and lower bounds on the value of $f'(3)$. [**HINT:** Apply the racetrack principle.]

25. Suppose that the line $y = 2x - 1$ is tangent to the graph of f at $x = 3$ and that $f''(x) > 0$ if $0 < x < 5$.
    (a) Find $f(3)$ and $f'(3)$.
    (b) Explain why $7 < f(4)$ must be true.
    (c) Find a lower bound on the value of $f(2)$.

26. Suppose that $f(2) = 5$, $f'(2) = 2$, and that f is concave up everywhere.
    (a) How many roots could f have? What can you say about the location of these roots?
    (b) Find a lower bound on the value of $f(0)$.
    (c) Suppose that $f(-2) = -1$. Show that $f(0) < 3$.

27. Is the following statement true or false? Justify your answer.

    If $f(2) = 3$, $f'(2) = 1$, and $f''(2) = 0$, then the graph of $f(x)$ has an inflection point at $(2, 3)$.

28. Suppose that $f'(c) = f''(c) = 0$, but $f'''(c) > 0$. Does f have a local minimum, a local maximum, or an inflection point at c? Explain.

2.5 Average and Instantaneous Rates: Defining the Derivative

In earlier sections we introduced and studied derivatives informally, interpreting them as rates or slopes. In this section we take a more formal point of view.

Given an original function f and an input $x = a$, what exactly do we mean, mathematically, by $f'(a)$? We've thought of $f'(a)$ as the slope of f at $x = a$ or as the instantaneous rate of change of y with respect to x at $x = a$, but important questions remain: What, exactly, is the *instantaneous* rate of change of a function at a point? Exactly how is the tangent line to a graph at a point defined? Most important of all:

> How can we use a formula for f to calculate—not just estimate—values of f'?

The Rate-of-Change Problem

In this no-frills model, drivers calculate speed from odometer readings.

Exactly at midnight, Professor X drives away in her brand-new Göttingen automobile. The odometer reads zero. There is no speedometer. ◄ Consulting her odometer and wristwatch, Professor X can find, for any time t (in hours), her total distance $D(t)$ (in miles) traveled since midnight. She might even plot her readings, as we did:

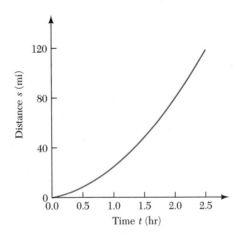

Professor X's trip: graph of $D(t)$

How can Professor X compute her speed at time $t = 2$, when she expects to reach a notorious speed trap? Reading the odometer at $t = 2$ won't help. It may say that she's traveled, say, 80 miles in 2 hours and hence that her *average* speed was 40 mph for the trip; but what about her speed at the *instant* $t = 2$?

She's right: The inexpensive Göttingen responds slowly both to throttle and to brake.

Professor X reasons that her *average* speed over a short interval—say $t = 1.99$ to $t = 2.00$—should be about equal to her *instantaneous* speed at any time during the same interval. ◄ The odometer reports 0.6985 miles of travel during the interval.

"Hence," ▶ thinks Professor X, "my average speed over the short period must have been

Professors say "hence," even to themselves.

$$\frac{0.6985 \text{ miles}}{0.01 \text{ hours}} = 69.85 \text{ mph.}$$

I'd better slow down."

The Tangent-Line Problem

The tangent-line problem asks:

Given a curve C and a point P on C, describe the line tangent to C at P.

EXAMPLE 1   Which line is tangent to the graph of $f(x) = x^2 + x$ at $x = 2$?

Solution   Here's part of the graph, with a plausible tangent line at the point $P = (2, 6)$.

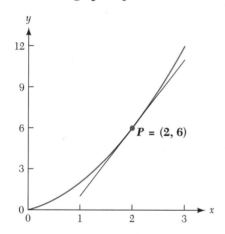

A tangent line to the graph of $y = x^2 + x$

From the graph alone we can estimate that the tangent line L has slope 5. ▶ If we're right, then L has equation $y = 5(x - 2) + 6$.

Check this for yourself.

Finding the slope of L *exactly* is complicated by the fact that we know for certain only one point on L. To start, we might look at a line L' through *two* nearby points on the graph, such as those at $x = 2$ and $x = 2.01$. We expect that L and L' will have similar slopes. Knowing *two* points on L' means that its slope is easy to find:

$$\frac{\Delta f}{\Delta x} = \frac{f(2.01) - f(2)}{2.01 - 2} = \frac{6.0501 - 6}{0.01} = 5.01.$$

This numerical evidence further supports our guess that $f'(2) = 5$. ▶   ■

Because of the shape of the graph near $x = 2$, the slope of L' slightly overestimates the slope of L.

Sharpening the Tangent-Line Problem

We stated the tangent-line problem in intuitive but slightly vague terms. What does it mean for a curve to be "given," and how should we "describe" the line tangent to C at P?

Given by Equations   Curves are often given as equations (e.g., $x^2 + y^2 = 4$) or as functions (e.g., $f(x) = x^2 + x$). Some curves are given *only* graphically or numerically, without any algebraic formula attached. In that case tangent lines can be found only approximately. If a curve C *has* an exact formula, we'd like to use it to find an exact formula for the line tangent to C at P.

Wanted: Slope   A line L is completely determined by its slope and a point P on L. In the tangent-line problem, a point P is given from the start, so "finding" L means, in effect, finding the *slope* of L.

Estimating Slope   If a curve C has a tangent line at P, we can estimate its slope either by hand—i.e., by drawing a line that fits C at P—or by zooming in—i.e., by viewing C near P in a plotting window small enough that C appears straight.

Draw a picture to convince yourself.

No Tangent Line   Not every curve C has a tangent line at a given point P. For example, the graph of $y = |x|$ has a sharp corner ◀   at $P = (0, 0)$, so no single straight line can be said to fit the graph at P.

EXAMPLE 2   We guessed earlier that the line L with equation $y = 5(x - 2) + 6$ is tangent to C, the graph of $f(x) = x^2 + x$, at $x = 2$. Does zooming support this guess? How?

Solution   Here are graphs of C and L at four scales:

C and L: window [–2, 6] × [–18, 30]

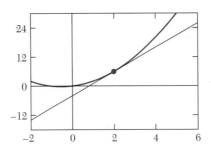

C and L: window [1, 3] × [0, 12]

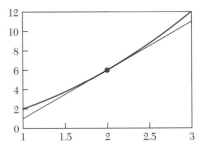

C and L: window [1.5, 2.5] × [3, 9]

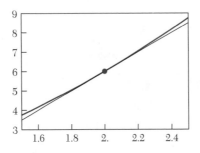

C and L: window [1.9, 2.1] × [5.4, 6.6]

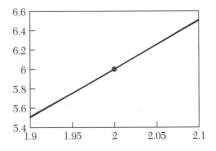

The graphs show that C, viewed close up, does resemble L, as claimed. In particular, because L has slope 5, so does the curve C at the point $(2, 6)$. ■

The Problems Solved: Secant Lines and Tangent Lines, Average Rates and Instantaneous Rates

Graphical methods (zooming until straight, drawing tangent lines by hand, and so on) give simple and useful *estimates* of slopes and rates. For more precision, and to get deeper under the skin of the derivative, we need clear, concise definitions, in mathematical language, for instantaneous rates and for slopes of tangent lines. The concept of limit turns out to be just the tool we need. ▶

We'll solve our two problems in parallel. First we pose them explicitly.

We'll see much more on limits in the next two sections.

> **The Tangent-Line Problem** Let $y = f(x)$ be a function. Find the slope of the line tangent to the graph of f at $x = 2$.
>
> **The Instantaneous-Speed Problem** Let the function $s = D(t)$ give the distance, in miles, traveled by a car up to time t, in hours. Find the instantaneous speed at time $t = 2$.

Secant Lines, Tangent Lines, and Limits

Given two points P and Q on a curve C, we can draw the **secant line** through P and Q. If P and Q are close together, it's reasonable to expect the secant line they determine to resemble the line tangent to C at P. ▶

Finding slopes of tangent lines poses a challenge, but finding slopes of secant lines is easy. If C is the graph of a function $y = f(x)$, and $P = (a, f(a))$ and $Q = (b, f(b))$ are points on the graph, then, as the following picture shows,

Before reading on, draw some curves and secant lines to convince yourself of this easy but important idea.

The secant line from $x = a$ to $x = b$

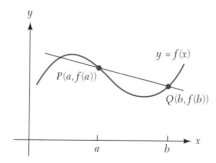

the slope of the secant line from P to Q (i.e., from $x = a$ to $x = b$) is

$$\frac{\Delta f}{\Delta x} = \frac{f(b) - f(a)}{b - a}.$$

Earlier we calculated that the secant line to $f(x) = x^2 + x$ from $x = 2$ to $x = 2.01$ has slope 5.01. This result suggests our strategy: We'll take the slope of the tangent line to be the limit of slopes of secant lines over smaller and smaller intervals.

What Is a Limit?   The **limit** is a mathematical tool used to describe how a function $f(x)$ behaves for x *near* a specific value $x = a$. We'll discuss limits carefully in the next two sections. In this section we treat limits informally, in context.

EXAMPLE 3   Compute slopes of secant lines to the graph of $f(x) = x^2 + x$ over small intervals near $x = 2$, and interpret the results in limit language. Draw (and defend) a conclusion about $f'(2)$.

Solution   Results for various small intervals appear below:

Slopes of Secant Lines over Small Intervals					
Interval	$[1.99, 2]$	$[1.999, 2]$	$[2, 2.001]$	$[2, 2.01]$	$[2, 2.1]$
Slope	4.99	4.999	5.001	5.01	5.1

Only a mathematician could doubt that these slopes approach 5.

To put the situation in limit language, consider the interval from $x = 2$ to $x = 2 + h$. ◄   Then the secant line from $x = 2$ to $x = 2 + h$ has slope

We think of h as a small (positive or negative) increment.

$$\frac{\Delta f}{\Delta x} = \frac{f(2+h) - f(2)}{(2+h) - 2} = \frac{f(2+h) - f(2)}{h}.$$

The slope of the tangent line at $x = 2$ is the limit as h tends to zero of this ratio. In symbols,

$$f'(2) = \lim_{h \to 0} \frac{f(2+h) - f(2)}{h}.$$

At last we can put this problem out of its misery, by calculating the slope of the tangent line at $x = 2$:

$$\lim_{h \to 0} \frac{f(2+h) - f(2)}{h} = \lim_{h \to 0} \frac{(2+h)^2 + (2+h) - 6}{h}$$

$$= \lim_{h \to 0} \frac{5h + h^2}{h}$$

$$= \lim_{h \to 0} (5 + h)$$

$$= 5. \qquad \blacksquare$$

(The preceding limit calculation should be, for the moment, intuitively reasonable. We'll study the algebra of limits carefully in the next two sections.)

Average Speeds, Instantaneous Speeds, and Limits

The rate-of-change problem, for Professor X, is to convert information about distance traveled to information about instantaneous speed. Its solution closely mimics that for the tangent-line problem.

Average speed over an interval of time is easy to compute. As Professor X knows,

$$\text{average speed} = \frac{\text{distance traveled}}{\text{time elapsed}}.$$

The remaining problem is to connect average speeds (over intervals) with instantaneous speeds. Again, limits provide the missing link: The instantaneous speed at time t_0 is the limit of average speeds over small time intervals that contain t_0. More succinctly put:

Consider the interval from $t = 2$ to $t = 2 + h$ hours. ▶ *The average speed from $t = 2$ to $t = 2 + h$ is*

Again, h represents a small positive or negative increment.

$$\frac{\Delta D}{\Delta t} = \frac{D(2 + h) - D(2)}{(2 + h) - 2} = \frac{D(2 + h) - D(2)}{h}.$$

Therefore, the instantaneous speed at $t = 2$ is

$$\lim_{h \to 0} \frac{D(2 + h) - D(2)}{h}.$$

We confess that, all along, we've had the function $D(t) = 10t + 15t^2$ secretly in mind for Professor X's trip. Finding her speed at $t = 2$ hours is now a simple calculation: ▶

But check each step!

$$\lim_{h \to 0} \frac{D(2 + h) - D(2)}{h} = \lim_{h \to 0} \frac{10(2 + h) + 15(2 + h)^2 - 80}{h}$$

$$= \lim_{h \to 0} \frac{70h + 15h^2}{h}$$

$$= \lim_{h \to 0} (70 + 15h)$$

$$= 70.$$

At *exactly* 2:00 A.M., Professor X's Göttingen was doing *exactly* 70 mph.

Derivative: The Formal Definition

The tangent-line and rate-of-change problems turned out to be identical. Both problems—and many others—lead to limits of a common form. This form appears in the following definition.

> **Definition** Let f be a function defined near and at $x = a$. The **derivative of f at $x = a$**, denoted by $f'(a)$, is defined by the limit
> $$f'(a) = \lim_{h \to 0} \frac{f(a + h) - f(a)}{h}.$$

The most important in all of calculus.

This important definition ◄ deserves some amplification:

Must $f'(a)$ Exist? In a word, no. The limit in the definition may or may not exist. If it *does* exist, we say that f is differentiable at $x = a$ or that f has a derivative at $x = a$. Most of the functions we treat in calculus are differentiable wherever they make sense. The absolute value function $g(x) = |x|$ is one exception. The kink at $x = 0$ means that g has no tangent line—and hence no derivative—at $a = 0$. (We'll see in the next section why the limit that defines $g'(0)$ does not exist.)

Dividing by Zero? The limit in the derivative definition *always* involves a denominator tending to zero; evaluating it always involves some trickery or other. In practice, things often work out as in the preceding examples: After some algebra in the numerator, the troublesome h in the denominator goes away. ◄

But only if the algebra is done carefully.

Other Forms The limit in the derivative definition can be written in other equivalent forms. Here are several popular ones:

$$\lim_{h \to 0} \frac{f(a + h) - f(a)}{h} \qquad\qquad \lim_{\Delta x \to 0} \frac{f(a + \Delta x) - f(a)}{\Delta x}$$

$$\lim_{x \to a} \frac{f(x) - f(a)}{x - a} \qquad\qquad \lim_{\Delta x \to 0} \frac{\Delta f}{\Delta x}$$

Depending on f, one form may be simpler or easier to manipulate than another, but all convey the same information.

Difference Quotients, Average Rates of Change, and Their Limits

The quotients in the preceding limits are all instances of a general idea:

> **Fact** For any function f, the ratio
> $$\frac{f(a + h) - f(a)}{h}$$
> is called a **difference quotient**. It measures the *average rate of change* of the function f from $x = a$ to $x = a + h$.

To understand what it tells, let's consider the various parts of the difference quotient separately

The Denominator The difference

$$h = (a + h) - a = \Delta x$$

represents a change, or increment, in the input variable, from $x = a$ to $x = a + h$. (The change may be either an increase or a decrease.)

The Numerator As x moves from $x = a$ to $x = a + h$, the output $f(x)$ changes too. The numerator

$$\Delta f = f(a + h) - f(a)$$

measures this change. Like Δx, Δf may have either sign.

The Quotient The difference quotient is the *ratio* of two changes Δf and Δx, so it represents a *rate* of change of $f(x)$ with respect to x. Because the changes occur over an interval in x, the quotient represents an *average* rate. The sign of $\Delta f / \Delta x$ tells whether x and $f(x)$ change in the same direction. If, say, $\Delta f / \Delta x < 0$, then an increase in x produces a decrease in $f(x)$.

From Average to Instantaneous The difference quotient is an average rate of change over the interval $[a, a + h]$. Therefore, the limit

$$\lim_{h \to 0} \frac{f(a + h) - f(a)}{h}$$

is the *instantaneous* rate of change at the instant $x = a$.

At Any Rate. In everyday speech, the words "rate" and "instantaneous" suggest quantities varying in time. "Rate of change" should be understood more generally in calculus. If f is *any* function—perhaps having nothing to do with time—the difference quotient $\Delta f / \Delta x$ and its limit are still thought of as rates; they measure how fast outputs from f change relative to inputs x.

A case could be made for replacing the customary phrase "rate of change"—which seems to apply equally well to increasing and decreasing quantities—with "rate of increase." "Rate of increase" makes the bookkeeping a little simpler: If $\Delta f / \Delta x > 0$, the rate of increase of $f(x)$ with respect to x is positive; this means that $f(x)$ increases as x does. If $\Delta f / \Delta x < 0$, the rate of increase is negative; $f(x)$ decreases as x increases.

Difference Quotients and Derivatives: What They Mean in Various Settings

Any function f has many possible interpretations. Our old favorite $f(x) = x^2 + x$ could represent the weight in ounces of a kitten x days after birth, the area of a rectangle with variable sides, or almost anything else. Any particular interpretation of f leads to a corresponding interpretation of the difference quotient. Several examples follow. In every case the difference quotient and derivative are rates of change. Notice especially the various units of measurement: Choices of units for x and y determine appropriate units for the difference quotient and the derivative.

Slopes   On any graph $y = f(x)$, the difference quotient represents the *slope of the secant line* from $x = a$ to $x = a + h$. The derivative $f'(a)$, as the limit of slopes of secant lines, represents the *slope of the tangent line* at $x = a$.

Velocities   If $y = f(x)$ represents the position of a moving object at time x, then the difference quotient $\Delta f / \Delta x$ represents *average* velocity over the interval from $x = a$ to $x = a + h$. The derivative $f'(a)$, as a limit of average velocities, represents *instantaneous* velocity at time a. ◄

Recall that speed is the absolute value of velocity.

Gas Mileage   Let y be the total distance (in miles) traveled by Professor X on x gallons of fuel. Then y is a function—say, g—of x. (Even the Göttingen has a gas gauge.) The professor takes these readings:

x (gallons)	0	1.5	1.9	1.99	2
y (miles)	0	43.2725	55.4993	58.3969	58.7206

A difference quotient, such as

$$\frac{g(2) - g(0)}{2 - 0} = \frac{58.7206 \text{ miles}}{2 \text{ gallons}} = 29.3603 \text{ miles per gallon,}$$

represents the *average* rate of gas consumption over the interval in which the first two gallons of gas were consumed. The *instantaneous* rate of gas consumption varies from moment to moment. To find the instantaneous rate at $x = 2$, we need to find a limit. Without an explicit formula for g, we can only estimate. The following table shows average rates for various intervals in x.

Interval	[0, 2]	[1.5, 2]	[1.9, 2]	[1.99, 2]
Average Miles per Gallon	29.3603	30.8962	32.213	32.37

A sensible guess, then, is that at the moment when $x = 2$, gas mileage was about 32.4 mpg.

Units   What units make sense for difference quotients and derivatives? It depends on the units chosen for x and y. For Professor X's speed we had $s = D(t)$, with s in miles and t in hours; the natural rate unit, therefore, is miles per hour (i.e., the quotient miles/hours). For fuel consumption we used miles and gallons as units, so rates were measured in miles per gallon. Had Professor X chosen another vehicle and other units of distance and fuel (e.g., horse, leagues, and bales) the proper units for difference quotient and derivative would have been different (e.g., leagues per bale).

The Tangent-Line Problem in Historical Perspective

The problem of finding a straight line that fits a curve at a given point is very old. For the Greeks the problem had less to do with description—how to *define* the tangent line rigorously—than with construction—how, given a specific curve, to use ruler, compass, and other tools to *draw* tangent lines. For a circle, constructing the tangent line L in the Greek sense is easy: L is the line perpendicular to the radius at P. The Greeks could construct tangent lines for circles, parabolas, and a few other curves of classical geometry. In his treatise *On Spirals* (ca. 250 B.C.E.) Archimedes constructed tangent lines to certain spirals. The following figure illustrates the situation for circles and spirals.

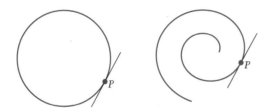

If C is a circle, the tangent line L at P happens to be the only line through P that touches ► the circle only at P. This touches-only-at-P criterion works (i.e., identifies a line which deserves to be called "tangent") for some curves but not for many others we meet in modern calculus courses. For the curve $y = \sin x$, for example, most tangent lines cross the curve at several points.

Tangere is Latin for "to touch."

The Greeks seem to have taken the touching criterion as their *definition* of tangent line. That they did so is a bit surprising, because even for the Archimedean spiral the criterion fails: Every tangent line eventually crosses the spiral. Perhaps the Greeks accepted such an inadequate definition because they were less interested in defining tangent lines than in constructing them.

Finding the "right" definition of tangent line is slightly subtle. In one sense, the tangent line is anything we say it is; we can define it any way we like. We might, for example, foolishly define the line tangent to C at P to be the horizontal line through P. This definition is not so much logically incorrect as useless: A tangent line by this definition would tell us nothing about the direction of C near P. A good definition must be both simple enough to work with and precise enough to capture the essence of the idea in question.

A good mathematical definition starts with a general intuition and makes it precise. For tangent lines we have plenty of geometric intuition to build on. Given a curve and a point on it, it is usually easy to draw a tangent line and agree that it's correct (allowing duly for imperfect pencils, rulers, eyesight, and so on). The challenge is to describe in *mathematical language* the line we would draw anyway.

The right definition of tangent line involves a limit. In retrospect, this conclusion seems obvious, but it was arrived at only after a long historical process. Although the question goes back to ancient times, not until the work of Cauchy in the 1820s—100 years after Newton's death—were limits in the present-day sense known. That is not to say that nothing happened in the meantime. By 1630, Fermat had essentially the modern idea of tangent line. He used his idea to solve problems like those that appear in calculus courses today.

BASICS

1. Amy traveled from Chicago to Milwaukee. Due to road construction, she drove the first 10 miles at a constant 20 mph. For the next 30 miles she drove at a constant 60 mph. She stopped at a rest area for a 10-minute snack, then drove the last 45 miles at a constant 45 mph.
   (a) Plot Amy's distance along the road from her starting point as a function of time.
   (b) Plot Amy's velocity as a function of time.
   (c) What was Amy's average speed for the trip (including the stop at the rest area)?

2. If $f(1) = 2$ and the average rate of change of f from $x = 1$ to $x = 5$ is 3, what's $f(5)$?

3. If $f(2) = 1$ and the average rate of change of f from $x = -3$ to $x = 2$ is 4, what's $f(-3)$?

4. True or false? Justify your answers.
   (a) Velocity is the rate of change of position with respect to time.
   (b) If the average rate of change of a function over an interval is zero, then the function must be constant.
   (c) If the average rate of change of a function over every interval is 17, then the graph of the function is a straight line.

5. The graph below shows how the price of a certain stock varied over a recent trading day.

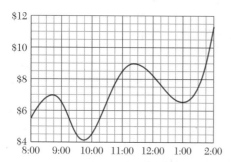

   (a) For each time interval below, find the total change in price and the average rate of change of price. (Be sure to indicate the units used to measure these quantities.)
(i) 8:00 to 11:00	(iii) 9:30 to 2:00
(ii) 9:00 to 1:00	(iv) 11:00 to 1:00
   (b) Estimate the instantaneous rate of change of the stock's price at each of the following times. (Be sure to indicate units with your answers.)
(i) 9:15 A.M.	(iii) 12:15 P.M.
(ii) 10:30 A.M.	(iv) 1:45 P.M.

6. The position $p(t)$ of an object at time t (in seconds) along a line with units marked in meters is given by $p(t) = 3t^2 + 1$.
   (a) Find the change in position of the object between time $t = 1$ and $t = 3$.
   (b) Find the average velocity of the object between time $t = 1$ and $t = 3$.
   (c) On one set of axes, sketch a graph of $p(t)$ and the secant line from $t = 1$ to $t = 3$. Find an equation of the secant line.
   (d) Find the average velocity of the object between time $t = 1$ and time $t = 1+h$. What are the *units* of your answer?
   (e) Find the instantaneous velocity of the object at time $t = 1$.
   (f) Find an equation of the line tangent to the graph of $p(t)$ at $t = 1$. Add this line to the sketch you made in part (c) above.

7. For each function f below, use a difference quotient with $h = 0.001$ to estimate the value of f' at the given point.
   (a) $f(x) = x^2 - 3x$ at $x = 3$
   (b) $f(x) = x^{1/3}$ at $x = 2$
   (c) $f(x) = x^{-2}$ at $x = 1$
   (d) $f(x) = \sin x$ at $x = 0$
   (e) $f(x) = \cos x$ at $x = 0$
   (f) $f(x) = \cos(3x)$ at $x = \pi/2$
   (g) $f(x) = 2^x$ at $x = 0$
   (h) $f(x) = \ln x$ at $x = 1$

8. For each of the following functions, use a graph of
   $$g(\Delta x) = \frac{f(1 + \Delta x) - f(1)}{\Delta x}$$
   over a small interval centered at $\Delta x = 0$ to estimate $f'(1)$.
(a) $f(x) = x^3$	(c) $f(x) = \sin(\pi x)$
(b) $f(x) = x + (1/x)$	(d) $f(x) = e^x$

9. Suppose that f is a function for which $f'(2)$ exists. Use the values of f given below to estimate $f'(1.99)$, $f'(2)$, $f'(2.01)$, and $f'(2.1)$. Explain how you obtained your estimates.

x	1.9	1.99	1.999	2.0
$f(x)$	25.34	33.97	34.896	35

x	2.001	2.01	2.1	
$f(x)$	35.104	36.05	46.18	

10. Suppose that f is a function for which $f'(2)$ exists. Use the values of f given at right to estimate $f'(1.9)$, $f'(2)$, and $f'(2.02)$.

x	1.9	1.97	2.0	2.02	2.2
$f(x)$	6.6	6.905	7	7.059	7.5

FURTHER EXERCISES

11. On a 50-minute trip, a car travels for 20 minutes with an average velocity of 30 mph and then for 30 minutes with an average velocity of 50 mph. Find the total distance traveled and the average velocity over the entire trip.

12. (a) A car travels for 30 miles with an average velocity of 40 mph and then for another 30 miles with an average velocity of 60 mph. What is the average velocity of the car for the entire trip?
    (b) Another car travels for 30 minutes at 40 mph and then for 30 minutes at 60 mph. Find the average velocity over the 1-hour time period.
    (c) A car is to travel 2 miles. It went the first mile at an average velocity of 30 mph. The driver wishes to average 60 mph for the entire 2-mile trip. Is this possible? Explain.

13. The number of deer in a forest at time t years after the beginning of a conservation study is shown on the graph below.

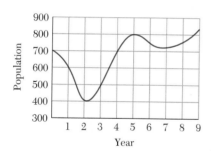

Year

(a) Over which two of the following periods did the population of deer decline at an average rate of 50 deer per year?
    (i) 1–2   (ii) 1–3   (iii) 1–4   (iv) 2–3   (v) 5–6
(b) Approximately how fast was the deer population increasing (or decreasing) at time $t = 1.5$? At time $t = 3$? (Include appropriate units.)

14. Let $f(x) = \sqrt{x}$. Tabulate $f(x)$ at $x = 1.998$, 1.999, 2.000, 2.001, and 2.002 and use your results to answer the following questions. (Round numerical values of $f(x)$ to seven decimal places.)
    (a) Explain how these tabulated values show that f is "locally linear" near $x = 2$.
    (b) Estimate $f'(1.999)$, $f'(2)$, and $f'(2.001)$ using the tabulated values of f.
    (c) Does $\big(f(2.001) - f(2)\big)/0.001$ overestimate or underestimate the exact value of $f'(2)$? Why? [**HINT:** Look at a graph of $f(x)$ near $x = 2$.]
    (d) Estimate $f''(2)$ using the results from part (b).

15. Repeat the previous problem for each of the following functions.
    (a) $f(x) = x^2$ (c) $f(x) = e^x$
    (b) $f(x) = \sin x$ (d) $f(x) = \ln x$

16. Let $f(x) = |x|$.
    (a) Sketch a graph of f over the interval $[-5, 5]$.
    (b) Explain why $f'(x) = \begin{cases} -1, & \text{if } x < 0 \\ 1, & \text{if } x > 0. \end{cases}$
    (c) Explain why $f'(0)$ doesn't exist.

17. Let $f(x) = 2^x$.
    (a) Find the average rate of change of f between $x = 0$ and $x = 0.01$. Use this result to estimate $f'(0)$. Explain why this estimate is too large.
    (b) Find the average rate of change of f between $x = -0.01$ and $x = 0$. Use this result to estimate $f'(0)$. Explain why this estimate is too small.
    (c) Find the average rate of change of f between -0.01 and 0.01. Use this result to estimate $f'(0)$.
    (d) Why is the estimate of $f'(0)$ found in part (c) more accurate than the estimates computed in parts (a) and (b)?
    (e) Describe a method based on part (c) for estimating $f'(a)$ from tabulated data.

18. Each of the following limits is $f'(a)$ for some function f and some number a. In each case, identify f and a.
    (a) $\displaystyle \lim_{x \to 4} \frac{\sqrt{x} - 2}{x - 4}$ (c) $\displaystyle \lim_{u \to \pi/4} \frac{\cos u - \sqrt{2}/2}{u - \pi/4}$
    (b) $\displaystyle \lim_{h \to 0} \frac{5^{2+h} - 25}{h}$ (d) $\displaystyle \lim_{s \to 8} \frac{\log_2 s - 3}{s - 8}$

2.6 Limits and Continuity

We broached the subject of limits in the preceding section, in defining the derivative. In this section we study limits in their own right. Then we use limits to define precisely another important concept: continuity of a function at a point of its domain.

The Basic Idea; Simplest Examples

The general idea of limit is simple. Suppose that f is a function defined near, but not necessarily *at*, $x = a$. To say that

$$\lim_{x \to a} f(x) = L$$

means, informally, that $f(x)$ approaches L as x approaches a. In other *symbols*,

$$f(x) \to L \qquad \text{as} \qquad x \to a.$$

In other *words*,

> *As inputs to f approach a, outputs from f approach L.*

In still other words,

$$f(x) \approx L \qquad \text{whenever} \qquad x \approx a.$$

Intuitive ideas aren't formal definitions. We'll give a definition soon. First, here are some examples.

EXAMPLE 1   What is $\lim_{x \to 3} x^2$? Why?

Solution   The answer is no surprise:

$$\lim_{x \to 3} x^2 = 9.$$

The squaring function $f(x) = x^2$ is well behaved near $x = 3$. In particular, $f(x) \approx 9$ whenever $x \approx 3$. (In fact, $f(x)$ *equals* 9 when $x = 3$, but that's irrelevant to the limit.)   ■

EXAMPLE 2   The statement

$$\lim_{x \to 3} \frac{x^2 - 9}{x - 3} = 6$$

is about the limit of a certain function at a certain point. Which function? Which point? What does the statement say? Is it true?

Solution   The function—we'll call it g—is defined by the rule $g(x) = (x^2 - 9)/(x - 3)$. The point in question is $x = 3$.

Why not?

Notice that $g(x)$ is not defined ◄ at the interesting point $x = 3$. (Everywhere else, the rule for g makes good sense.) The limit statement says that, although g has no ordinary *value* at $x = 3$, it does the next best thing: $g(x)$ approaches 6 as x approaches 3. In symbols,

$$g(x) \to 6 \qquad \text{as} \qquad x \to 3.$$

That's what the statement says. Is it true? The formula for g isn't immediately helpful, precisely because $g(3)$ isn't defined. The graph of g helps:

Graph of $g(x) = (x^2 - 9)/(x - 3)$

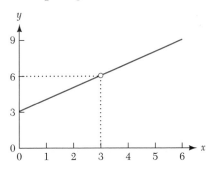

It shows convincingly that the limit statement is true. Despite the gap ▶ at $x = 3$, $g(x)$ does approach 6 as x approaches 3, from either right or left.
    Numerical evidence points to the same conclusion:

How does the gap appear on the graph?

x	2.9	2.99	2.999	2.9999	...	3.0001	3.001	3.01	3.1
$g(x)$	5.9	5.99	5.999	5.9999	...	6.0001	6.001	6.01	6.1

Observe, finally, that the graph of g (except at $x = 3$) looks suspiciously like the line $y = x + 3$. A little algebra shows why: If $x \neq 3$,

$$g(x) = \frac{x^2 - 9}{x - 3} = \frac{(x - 3)(x + 3)}{x - 3} = x + 3.$$

(We repeat: The last equation holds only for $x \neq 3$—but that's good enough, because we're interested in $g(x)$ only for values *near* $x = 3$.) Now it's clear that

$$\lim_{x \to 3} g(x) = \lim_{x \to 3} (x + 3) = 6. \qquad \blacksquare$$

Why Limits Matter: Functions with Gaps

The simplest limit expressions (e.g., $\lim_{x \to 3} x^2 = 9$ and $\lim_{x \to 3}(x + 3) = 6$) illustrate what limits are, but not why they matter. Judging only from such examples, limits may seem to be much ado about little. That impression is wrong. Limits do matter in calculus. One reason is that functions with gaps in their domains (such as g in Example 2) arise often in calculus. Finding derivatives is the most common and most important way in which such functions arise.

EXAMPLE 3   Let $f(x) = x^2$. Use the limit definition to find the derivative $f'(3)$. What does the result mean graphically?

Difference quotients can be written in various forms. See page 144.

Solution   By definition, the derivative is a limit of difference quotients. We write ours like this: ◀

$$f'(3) = \lim_{x \to 3} \frac{f(x) - f(3)}{x - 3} = \lim_{x \to 3} \frac{x^2 - 9}{x - 3}.$$

That's the limit we found in Example 2, so $f'(3) = 6$. In graphical terms: *At $x = 3$ the curve $y = x^2$ has slope 6.*   ■

Limits in Theory

Functions with gaps or other anomalies in their domains arise in connection with instantaneous phenomena, including tangent lines and rates of change. Example 3 showed how the concept of limit permits us to handle such functions.

With a satisfactory theory of limits, based on precise definitions, mathematicians can make sense of—and calculate reliably with—mysterious-seeming concepts such as "infinitely small quantities." Without such theory as footing, we would remain mired (as mathematicians did for hundreds of years) in the vague, metaphorical—sometimes even mystical—language of "fluxions," "fluents," and "indivisibles."

> **Seeing Ghosts.**   A precise idea of limit is surprisingly subtle. Developing a rigorous but usable definition of limit was an important historical step toward putting calculus on a sound logical footing.
>
> The modern idea of limit stems from the work of the French mathematician Augustin-Louis Cauchy, around 1820—over 150 years after Isaac Newton and Gottfried Leibniz developed calculus. Newton's and Leibniz's own understanding of limit could not have been, by modern standards, rigorous. Sensing this, Bishop Berkeley satirized the intellectual pretensions of mathematicians in his 1734 philosophical treatise *The Analyst: A Discourse Addressed to an Infidel Mathematician*, which ridiculed Newton's limits as "ghosts of departed quantities."

Limits in Practice

Most of the limits encountered in calculus are relatively simple and straightforward. Combining algebraic, graphical, and numerical tools makes them even more so. The rest of this section aims to support this claim.

EXAMPLE 4   Let $f(x) = \sin x$. What limit defines $f'(0)$? What does the limit mean graphically? Find the limit.

Solution   The derivative $f'(0)$ is the slope of the tangent line to the curve $y = \sin x$ at $x = 0$. As we saw in the last section, this slope is the limit of slopes of secant lines over small intervals starting (or ending) at $x = 0$. To investigate that limit, we'll define a new function, g, by the rule

$$g(t) = \text{slope of secant line joining } (0, 0) \text{ and } (t, \sin t).$$

The rule makes sense for any $t \neq 0$, as the next picture shows.

Defining the function g

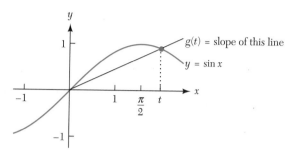

Even though $g(t)$ is not defined at $t = 0$, we'll find the limit $\lim\limits_{t \to 0} g(t)$.

Although we defined g in words, finding an explicit formula isn't hard. For $t \neq 0$,

$$g(t) = \frac{\Delta y}{\Delta x} = \frac{\sin t}{t}.$$

We want to find

$$\lim_{t \to 0} g(t) = \lim_{t \to 0} \frac{\sin t}{t}.$$

Because $g(0)$ isn't defined, we examine values of $g(t)$ for t near 0. Here are some numerical results, rounded to five decimals: ▶

Check some entries with a calculator; use radians.

t	-0.1	-0.01	-0.0001	...	0.0001	0.01	0.1
$g(t)$	0.99833	0.99998	1.00000	...	1.00000	0.99998	0.99833

The numbers suggest, rightly, that the limit is 1. ▶ An airtight proof would take more work. We won't give one here, but see Appendix H.

The answer means that the line $y = x$, with slope 1, is tangent to the sine curve at $x = 0$. ▶

■

For graphical evidence, plot $g(t)$ for t near 0.

Draw this line on the preceding graph to understand what tangency means.

The Importance of Being Algebraic. We calculated that

$$\lim_{x \to 3} \frac{x^2 - 9}{x - 3} = \lim_{x \to 3} (x + 3) = 6.$$

Won't something similar work with $\lim_{x \to 0} \sin(x)/x$?

No. The difference is that $x^2 - 9$ is an *algebraic* expression and so permits factoring, division, and other algebraic operations. The numerator $\sin x$, by contrast, is not algebraic, so ordinary factoring and division aren't possible.

No Limit. A function f may not *have* a derivative at a point of its domain. The absolute value function $f(x) = |x|$, for example, has a kink in its graph at $x = 0$. By rights, therefore, $f'(0)$ shouldn't exist. The next example shows exactly why it doesn't.

Here, again, is our favorite "not-nice" function.

EXAMPLE 5   Let $f(x) = |x|$. ◄   Show, using limits, that $f'(0)$ doesn't exist.

Solution   By definition,

$$f'(0) = \lim_{x \to 0} \frac{f(x) - f(0)}{x - 0} = \lim_{x \to 0} \frac{|x|}{x},$$

if the limit exists.

To see that the limit doesn't exist, consider the function h defined by $h(x) = |x|/x$. Does $\lim_{x \to 0} h(x)$ make sense? A moment's thought reveals that $h(x) = 1$ if $x > 0$, $h(x) = -1$ if $x < 0$, but $h(0)$ is undefined. The graph agrees:

Graph of $h(x) = |x|/x$

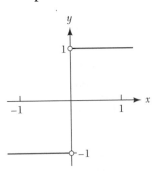

The limit $\lim_{x \to 0} h(x)$ does not exist; there is no *single* number to which $h(x)$ tends as x tends to 0. The problem is that, as $x \to 0$ from the *left*, $h(x) \to -1$, but as $x \to 0$ from the *right*, $h(x) \to 1$. Because the limit doesn't exist, neither does $f'(0)$.   ∎

Definitions of Limit—Informal and Precise

Let L be any finite number. The limit expression

$$\lim_{x \to a} f(x) = L \tag{2.1}$$

means, informally, that $f(x)$ approaches L as x approaches a. To be precise, we'll say exactly what we mean by Equation 2.1, first informally and then rigorously.

Equation 2.1 has two types of ingredients:

- real numbers a and L;
- a function f defined for all x near $x = a$, but not necessarily *at* $x = a$.

> **Definition** (Limit, Informally)   If $f(x)$ tends to a single number L as x approaches a, then $\lim_{x \to a} f(x) = L$.

Definition (**Limit, Formally**) Suppose that for every positive number ϵ (epsilon), no matter how small, there is a positive number δ (delta) so that

$$|f(x) - L| < \epsilon \quad \text{whenever} \quad 0 < |x - a| < \delta.$$

Then

$$\lim_{x \to a} f(x) = L.$$

Observe:

Estimating Limits If $\lim_{x \to a} f(x) = L$, then $f(x) \approx L$ if $x \approx a$. This means that we can investigate limits numerically (by evaluating $f(x)$ for x near a) and graphically (by graphing $f(x)$ for x near a). ▶

We took both of these approaches above.

A Gap at a? No Problem It's an important fact that $\lim_{x \to a} f(x)$ may exist *even if $f(a)$ is undefined*. Indeed, $f(a)$ is *usually* undefined in interesting cases, such as these:

$$\lim_{x \to 4} \frac{\sqrt{x} - 2}{x - 4}; \qquad \lim_{x \to 4} \frac{x^2 - 16}{x - 4}; \qquad \lim_{x \to 0} \frac{\cos x - 1}{x}.$$

None of these functions is "defined at a." In a sense, the problem of limits is to *find* a suitable value for $f(a)$—if one exists.

No Guarantees Even if $f(a)$ is defined, there is no guarantee that $\lim_{x \to a} f(x) = f(a)$. The limit definition "doesn't see" $f(a)$. Whether or not $f(a)$ exists, the definition refers only to values of x for which $|x - a| > 0$—i.e., $x \neq a$. (See the next example.)

From Greek to English In words, the formal definition of limit says:

> *We can ensure that $f(x)$ differs from L by less than any prearranged amount ϵ by requiring that x be within δ of a.*

The full idea is subtle, but the following "generic" picture gives some idea of how ϵ and δ are related.

**Epsilon and delta:
the generic picture**

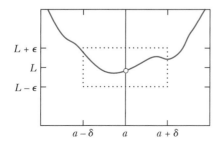

Graphically, the formal definition means this:

> *For any $\epsilon > 0$ we can choose a viewing window centered at (a, L), of width 2δ and height 2ϵ, from which the f-graph never escapes out the top or bottom, except possibly at $x = a$.*

We'll Be Informal   In elementary calculus—and in this book—informal views of the limit are almost always adequate. In more advanced work (e.g., in "real analysis"), a rigorous understanding of limits is essential.

Airtight Proofs.   Careful proofs of limit statements are essential at higher levels of mathematical analysis; inevitably, such proofs involve careful balancing of epsilons and deltas. We are more interested in the *concept* of limit, and how to use it, than in formal proofs. We'll discuss some theoretical properties of limits, but mainly as aids to concrete computation.

Variations on a Theme: One-Sided Limits

The most basic and most important limit idea is the one just studied: the (two-sided) limit of a function at a point. However, several variants on the same idea arise often enough to deserve mention: "one-sided" limits and limits that involve the ∞ symbol. ◄   Slightly different definitions apply to these variants. Informal versions will be enough: ◄

We'll see the latter in the next section.

Notice the unusual use of "+" and "−" as superscripts.

Definition (Right-Hand Limit) $\lim\limits_{x \to a^+} f(x) = L$ means that $f(x) \to L$ as $x \to a$ from the right.

Definition (Left-Hand Limit) $\lim\limits_{x \to a^-} f(x) = L$ means that $f(x) \to L$ as $x \to a$ from the left.

EXAMPLE 6   In Example 5 we saw (e.g., from the graph) that if $h(x) = |x|/x$, then $h(x)$ tends either to 1 or to -1 depending on whether x tends to 0 from the right or from the left. In one-sided limit notation,

$$\lim_{x \to 0^-} h(x) = -1 \qquad \text{and} \qquad \lim_{x \to 0^+} h(x) = 1. \qquad \blacksquare$$

Because these one-sided limits aren't equal, we concluded that the ordinary limit $\lim_{x \to 0} h(x)$ does not exist. There's a useful principle lurking here:

Fact The ordinary limit $\lim\limits_{x \to a} f(x)$ exists if and only if both corresponding one-sided limits exist, and are equal. In symbols,

$$\lim_{x \to a} f(x) = L \iff \lim_{x \to a^+} f(x) = L = \lim_{x \to a^-} f(x).$$

Continuity

To say that f is **continuous** on an interval I means, roughly speaking, that the graph of f is "unbroken" for x in I. This intuitive idea, although not a formal definition, works reliably enough for most calculus functions.

EXAMPLE 7   Each of the following functions is continuous on the full set $(-\infty, \infty)$ of real numbers:

$$f(x) = x^2; \qquad g(x) = \sin x; \qquad h(x) = \frac{x^2}{x^2+1}; \qquad k(x) = |x|.$$

(The graph of k has a kink at $x = 0$, but that doesn't destroy continuity.)

By contrast, $\sqrt{x}$ is continuous on the interval $[0, \infty)$. The function $m(x) = 1/x$ is continuous on $(0, \infty)$ and on $(-\infty, 0)$ but not on any interval that contains 0. ■

EXAMPLE 8   Following is a function n, shown graphically. Where is n continuous?

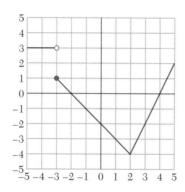

Solution   The only discontinuity occurs at $x = -3$. In particular, n is continuous on any interval that omits $x = -3$. ■

Continuity Defined: Limits Again

Graphical intuition is helpful, but a precise definition of continuity cannot depend on pictures. ▶ The idea of limit turns out to be the right one to express, in analytic terms, the geometric idea of "unbrokenness" at a point. Here's the formal definition:

One problem is that drawing accurate pictures is difficult. In higher dimensions, things are even worse.

> **Definition** Let f be a function defined on an interval I containing $x = a$. If
>
> $$\lim_{x \to a} f(x) = f(a),$$
>
> then f is continuous at $x = a$. (If a is the left or right endpoint of I, then a right-hand limit or a left-hand limit replaces the preceding two-sided limit.) To say that f is continuous on I means that f is continuous at each point of I (including endpoints of I, if any).

The formal definition illustrates the important theoretical role played by limits in calculus. In practice, however, the intuitive idea of continuity as unbrokenness is usually adequate. We'll content ourselves with several brief remarks.

Continuous Where?   A function f may be continuous on one set but discontinuous on another set. Thus, the common phrase "f is continuous" is technically ambiguous—it doesn't say *where* f is continuous. Here's the rule of thumb:

> *Unless otherwise qualified, "continuous" means "continuous throughout the domain of f."*

No Surprises   The definition says that f is continuous at a if, whenever $x \to a$, $f(x) \to f(a)$. In other words, inputs near a produce outputs near $f(a)$. In effect, this means that f behaves "predictably" at $x = a$.

Continuity Is the Rule   As a rule, the elementary functions of calculus are continuous wherever they're defined. For example, $f(x) = \sqrt{x-1}$ is continuous for $x \geq 1$, i.e., on the natural domain of f. Similarly, every polynomial function $p(x)$ is continuous on the full set of real numbers. (Proofs of such facts are important in higher analysis; in this book we'll deemphasize them.)

Easy Limits   Finding limits of continuous functions is as easy as can be. If f is continuous at $x = a$, then—by the very definition of continuity—

$$x \to a \implies f(x) \to f(a).$$

In words: The *limit* of f as x approaches a is the *value* of f at a.

EXAMPLE 9   Let $k(x) = x^2 + 3x + 5$. Find $\lim_{x \to 2} k(x)$.

Solution   The problem looks easy; neither the formula nor the graph of k suggests any surprises near $x = 2$.

Graph of $k(x) = x^2 + 3x + 5$

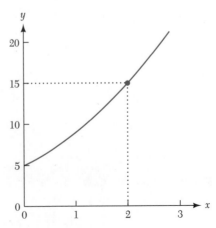

The obvious guess,

$$\lim_{x \to 2} k(x) = k(2) = 15,$$

is correct, precisely because k is continuous at $x = 2$. ▶

As we said earlier, a polynomial function is continuous everywhere.

Discontinuity: Three Possible Problems

A function f is **discontinuous** at $x = a$ if—for any reason—the limit condition in the definition doesn't hold. Three things can go wrong:

1. $f(a)$ may not be defined; this happens for $f(x) = 1/x$ at $x = 0$.

2. The limit $\lim_{x \to a} f(x)$ may not exist, this happens at $x = -3$ for the function n in Example 8.

3. The limit $\lim_{x \to a} f(x)$ may exist but differ from $f(a)$.

The next example illustrates the last form of pathology.

EXAMPLE 10   Consider the function f ▶ defined by

$$f(x) = \begin{cases} x & \text{if } x \neq 0 \\ 1 & \text{if } x = 0 \end{cases}.$$

What's $\lim_{x \to 0} f(x)$?

Plot this function by hand to understand the example.

Solution   The fact that $f(0) = 1$ makes no difference to the limit. Because $f(x) = x$ for $x \neq 0$, we have

$$\lim_{x \to 0} f(x) = \lim_{x \to 0} x = 0.$$

In a sense, f has the wrong value at $x = 0$; for this reason, f is called discontinuous at $x = 0$. ■

Limits and Computer Graphics. Most plotting devices can successfully plot a function over an interval containing one suspicious point. They will, for instance, draw reasonable pictures for $h(x) = |x|/x$ on $[-1, 1]$ even though h isn't defined at $x = 0$. If a device "complains" or produces nonsense, one can still use intervals that avoid the problem point. For $h(x) = |x|/x$ we might try graphs over $[-1, -0.01]$ and $[0.01, 1]$.

Sample Problems

Here are two problems that combine ideas from this section.

EXAMPLE 11   Let $f(x) = \begin{cases} bx^2 + 1 & \text{if } x < -2 \\ x & \text{if } x \geq -2 \end{cases}.$

What value of b makes f continuous at $x = -2$?

Solution   To make f continuous at $x = -2$, we need

$$\lim_{x \to -2} f(x) = f(-2) = -2.$$

Viewing the two lines of the definition separately, we see that

$$f(x) \rightarrow 4b + 1 \qquad \text{as} \qquad x \rightarrow -2^-;$$
$$f(x) \rightarrow -2 \qquad \text{as} \qquad x \rightarrow -2^+.$$

For the two one-sided limits to be equal, we need $4b + 1 = -2$, or $b = -3/4$. ∎

EXAMPLE 12   Interpret $\lim\limits_{h \to 0} \dfrac{e^h - 1}{h}$ as a derivative; estimate its value.

Check this equation.

Solution   If $f(x) = e^x$, then ◀

$$\lim_{h \to 0} \frac{e^h - 1}{h} = \lim_{h \to 0} \frac{f(h) - f(0)}{h}.$$

We took one in Example 6, Section 2.2. Look back or plot the curve yourself.

Thus, the given limit is (by the definition of derivative) the slope of the curve $y = e^x$ at $x = 0$. A close look at this curve ◀ suggests that this slope is 1. ∎

BASICS

1. Let $f(x) = \begin{cases} 1, & \text{if } x \text{ is an integer} \\ 2, & \text{if } x \text{ is not an integer.} \end{cases}$

   (a) Draw a graph of f over the interval $[0, 5]$.
   (b) Evaluate $\lim\limits_{x \to 4} f(x)$.
   (c) Evaluate $\lim\limits_{x \to 5/2} f(x)$.
   (d) For which values of a does $\lim\limits_{x \to a} f(x)$ exist? Why?

Let f be the function whose graph is shown below. Using the graph, evaluate each of the limits in Exercises 2–15 or explain why the limit does not exist. [**NOTE:** $f(1) = 2$.]

Graph of f

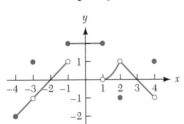

2. $\lim\limits_{x \to 0} f(x)$
3. $\lim\limits_{x \to -3} f(x)$
4. $\lim\limits_{x \to -1} f(x)$
5. $\lim\limits_{x \to 2^-} f(x)$

6. $\lim\limits_{x \to 2^+} f(x)$
7. $\lim\limits_{x \to 2} f(x)$
8. $\lim\limits_{x \to -2^-} f(x)$
9. $\lim\limits_{x \to 1^+} f(x)$

10. $\lim\limits_{x \to -4^+} f(x)$
11. $\lim\limits_{x \to 4^-} f(x)$
12. $\lim\limits_{x \to 3} f'(x)$
13. $\lim\limits_{x \to -1^+} f'(x)$
14. $\lim\limits_{x \to -1^-} f'(x)$
15. $\lim\limits_{x \to -3} f'(x)$

16. Over which intervals is f continuous?

Let g be the function whose graph is shown below. Using the graph, evaluate each of the limits in Exercises 17–26 or explain why the limit does not exist.

Graph of g

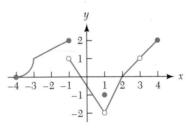

17. $\lim\limits_{x \to -4^+} g(x)$
18. $\lim\limits_{x \to -3} g(x)$
19. $\lim\limits_{x \to -2} g(x)$
20. $\lim\limits_{x \to -1^+} g(x)$
21. $\lim\limits_{x \to -1} g(x)$

22. $\lim\limits_{x \to 0^-} g(x)$
23. $\lim\limits_{x \to 1} g(x)$
24. $\lim\limits_{x \to 2} g(x)$
25. $\lim\limits_{x \to 3} g(x)$
26. $\lim\limits_{x \to 4^-} g(x)$

27. Over which intervals is g continuous?

28. Let $s(t)$ be the slope of the secant line from $x = 3$ to $x = t$ on the graph of the function $f(x) = x^2 + x$.
    (a) Find a formula for $s(t)$.
    (b) Find $\lim_{t \to 3} s(t)$.

(c) What does your answer to part (b) imply about the graph of f? Draw graphs of f and the line tangent to f at $x = 3$ over the interval $[2, 4]$.

FURTHER EXERCISES

29. Suppose that f is continuous at $x = 1$ and that $f(1) = -2$. Evaluate $\lim_{x \to 1} f(x)$.

30. Suppose that f is continuous at $x = 3$ and that $\lim_{x \to 3} f(x) = 17$. Indicate which of the following statements about f *must* be true, which *might* be true, and which *cannot* be true. Justify your answers.
    (a) 3 is in the domain of f.
    (b) $f(3) = 17$.
    (c) $\lim_{x \to 3^-} f(x) = 17$.

31. Let $f(x) = \begin{cases} \frac{\sin x}{x} & \text{if } x \neq 0 \\ 1 & \text{if } x = 0. \end{cases}$
    Explain why f is continuous at $x = 0$.

32. Let $g(x) = \begin{cases} \frac{\sin x}{|x|} & \text{if } x \neq 0 \\ 1 & \text{if } x = 0. \end{cases}$
    Is g continuous at $x = 0$? Justify your answer.

33. Let $h(x) = \begin{cases} \frac{\sin x}{\sqrt{1 - \cos^2 x}} & \text{if } x \neq 0 \\ 1 & \text{if } x = 0. \end{cases}$
    Is h continuous at $x = 0$? Justify your answer.

34. Let $f(x) = \begin{cases} ax + 1 & \text{if } x < 2 \\ x^2 & \text{if } x \geq 2. \end{cases}$
    Find the value of a for which $\lim_{x \to 2} f(x)$ exists.

35. Suppose that the function f is continuous at $x = 2$ and that f is defined by the rule
    $$f(x) = \begin{cases} ax^2 + 3 & \text{if } x < 2 \\ 3x - 5 & \text{if } x \geq 2. \end{cases}$$
    (a) Find the value of a.
    Evaluate each of the following limits.
    (b) $\lim_{x \to 3} f(x)$
    (c) $\lim_{x \to 0} f(x)$
    (d) $\lim_{x \to 2} f(x)$
    (e) $\lim_{x \to 1} f(x)$

36. Let $g(x) = \begin{cases} x^2 + a & \text{if } x < -1 \\ x + a^2 & \text{if } x = -1 \\ a - x & \text{if } x > -1. \end{cases}$
    (a) Find *all* values of a for which $\lim_{x \to -1} g(x)$ exists.
    (b) Find *all* values of a for which g is continuous at $x = -1$.

37. Let $f(x) = \begin{cases} x^2 - 2 & \text{if } x \leq b \\ x & \text{if } x > b. \end{cases}$
    Find *all* values of b for which $\lim_{x \to b} f(x)$ exists.

Each of the limits in Exercises 38–47 is $f'(a)$ for some function f and some number a. Identify f and a, then use a calculator or grapher to estimate the value of the limit.

38. $\lim_{h \to 0} \dfrac{\tan h}{h}$

39. $\lim_{h \to 0} \dfrac{\cos h - 1}{h}$

40. $\lim_{h \to 0} \dfrac{\sin(3h)}{h}$

41. $\lim_{x \to \pi} \dfrac{\sin x}{x - \pi}$

42. $\lim_{x \to \pi/2} \dfrac{\cos x}{x - \pi/2}$

43. $\lim_{x \to 16} \dfrac{\sqrt[4]{x} - 2}{x - 16}$

44. $\lim_{x \to 1} \dfrac{2^x - 2}{x - 1}$

45. $\lim_{h \to 0} \dfrac{\sqrt{9 + h} - 3}{h}$

46. $\lim_{h \to 0} \dfrac{\sin(\pi/6 + h) - 1/2}{h}$

47. $\lim_{h \to 0} \dfrac{\ln(1 + h)}{h}$

48. Suppose that $\lim_{x \to a} f(x) = L$. Explain why $\lim_{h \to 0} f(a + h) = L$.

49. Suppose that $\lim_{h \to 0} \frac{f(a+h) - f(a)}{h} = L$. Explain why $\lim_{x \to a} \frac{f(x) - f(a)}{x - a} = L$.

50. Suppose that g is the linear function $g(x) = mx + b$.
    (a) Show that $g'(a) = \lim_{h \to 0} \frac{g(a+h) - g(a)}{h} = m$ for every value of a.
    (b) Interpret the result in part (a) in terms of slopes and rates of change.

In Exercises 51–56 use the graphs of the functions f (see page 160) and g (see page 160) to evaluate each limit. If a limit does not exist, explain why.

51. $\lim_{x \to 0} \dfrac{f(x) - 2}{x}$

52. $\lim_{x \to 3} \dfrac{f(x)}{x - 3}$

53. $\lim_{x \to -1^+} \dfrac{f(x)}{x + 2}$

54. $\lim_{h \to 0} \dfrac{g(-2 + h) - 3/2}{h}$

55. $\lim_{h \to 0^-} \dfrac{g(-1 + h) - 2}{h}$

56. $\lim_{h \to 0} \dfrac{g(2 + h)}{h}$

57. Let $f(x) = x^2$. Carefully explain each step in the following evaluation of $f'(1)$:
    (a) $f'(1) = \lim_{x \to 1} \dfrac{x^2 - 1}{x - 1}$
    (b) $= \lim_{x \to 1} (x + 1)$
    (c) $= 2$.

58. Let $f(x) = x^2$. Carefully explain each step in the following evaluation of $f'(3)$:
    (a) $f'(3) = \lim_{h \to 0} \dfrac{(3 + h)^2 - 9}{h}$
    (b) $= \lim_{h \to 0} \dfrac{6h + h^2}{h}$
    (c) $= \lim_{h \to 0} (6 + h)$
    (d) $= 6$.

In Exercises 59–65, use the limit definition of the derivative to compute $f'(4)$ for each function.

59. $f(x) = 3$
60. $f(x) = 2x + 1$
61. $f(x) = x^2 - 3x$

62. $f(x) = \sqrt{x}$
63. $f(x) = x^{-1}$
64. $f(x) = 3x^{-2}$

65. $f(x) = x^{-1/2}$

66. Suppose that $f(x + y) = f(x) + f(y) + 3xy$ for all real numbers x, y, and that $\lim_{h\to 0} f(h)/h = 42$.

   (a) Explain why $f(0) = 0$.   [**HINT**: Continuity of f *cannot* be assumed.]

   (b) Find a symbolic expression for $f'(x)$.

67. Suppose that $f'(2) = 3$. Find $\lim_{h\to 0} \dfrac{f(2-h) - f(2)}{h}$.

68. Each of the following limits has the form

$$\lim_{x\to a} f(x) = L.$$

For each, use a grapher to find a value of δ that guarantees that $|f(x) - L| < 0.001$ when $|x - a| < \delta$.

   (a) $\lim_{x\to 2} 3x = 6$

   (b) $\lim_{x\to 0} \dfrac{1 - \cos x}{x} = 0$

   (c) $\lim_{x\to 1} \dfrac{x+1}{x+3} = 0.5$

   (d) $\lim_{x\to 0} \dfrac{1 - \cos(2x)}{x^2} = 2$

In Exercises 69–73, use the value of δ given to prove the limit. (Assume $\epsilon < 1$.)

69. $\lim_{x\to 5} x = 5;$ $\delta = \epsilon$

70. $\lim_{x\to 4} x^2 = 16;$ $\delta = \epsilon/10$

71. $\lim_{x\to 2} 6 = 6;$ $\delta = 0.2$

72. $\lim_{x\to 3} \dfrac{x^2 - 9}{x - 3} = 6;$ $\delta = \epsilon$

73. $\lim_{x\to 0} |x| = 0;$ $\delta = \epsilon$

74. Use the formal definition of limit to show that $\lim_{x\to 1} 0.99999x \ne 1$.

75. Let $f(x) = \begin{cases} x^2 + 1 & \text{if } x \le 0 \\ -x & \text{if } x > 0. \end{cases}$

   Use the formal definition of limit to show that f is not continuous at $x = 0$.   [**HINT**: $f(0) - f(x) > 1$ when $x > 0$.]

76. Let f be the function defined in the previous problem.

   (a) Show that for every positive number δ, no matter how small, there is a positive number ϵ such that $|f(x) - 1| < \epsilon$ whenever $0 < |x| < \delta$.

   (b) Why doesn't the result in part (a) show that f is continuous at $x = 0$?   [**HINT**: Compare part (a) with the formal definition of limit.]

2.7 Limits Involving Infinity; New Limits from Old

In the preceding section we defined limits, explored several examples and extensions of the idea, and then applied limits to define continuity. This section continues the limit theme. First we define another variant on the basic limit idea—limits that involve the ∞ symbol—and interpret results graphically in terms of asymptotes. Then we consider how to combine simple, known limits to obtain new, more challenging limits.

Limits at Infinity, Infinite Limits, and Asymptotes

Limits at infinity and infinite limits (they're different!) describe the behavior of functions when either inputs or outputs increase or decrease "without bound." Informal definitions and examples will suffice to give the idea.

> **Definition** (Limit at Infinity) $\lim_{x\to\infty} f(x) = L$ means that $f(x) \to L$ as $x \to \infty$ (i.e., as x increases without bound).

> **Definition** (Infinite Limit) $\lim_{x\to a} f(x) = \infty$ means that $f(x) \to \infty$ (i.e., $f(x)$ "blows up") as $x \to a$.

For instance, limits as $x \to -\infty$, and one-sided infinite limits.

Additional variants exist. ◄   Defining them all separately would be tedious. What each means is usually clear from context.

EXAMPLE 1   Discuss the meaning of each of the following limit statements:

$$\lim_{x \to \infty} \frac{1}{x} = 0; \qquad \lim_{x \to 0} \frac{1}{x^2} = \infty;$$

$$\lim_{x \to 0^-} \frac{1}{x} = -\infty; \qquad \lim_{x \to \infty} \sin x \text{ does not exist.}$$

S o l u t i o n   The limit $\lim_{x \to \infty} \sin x$ does not exist because the sine function oscillates forever, never settling down near any single value. The following graphs should help the reader interpret the first three limits. ▶

What other limit statements do the graphs suggest?

Graph of $y = 1/x$ **Graph of $y = 1/x^2$**

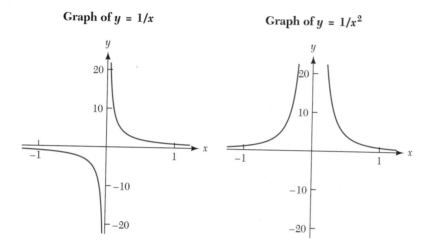

Notice especially the graphical meaning of each infinite limit and limit at infinity. Each one corresponds to an asymptote, either vertical or horizontal. We'll refine this observation in what follows. ∎

Asymptotes and Limits

The connections among asymptotes, limits at infinity, and infinite limits are easily summarized:

Horizontal Asymptotes; Limits at Infinity The horizontal line $y = L$ is an asymptote to the graph of f if (and only if) f approaches L as $x \to \pm\infty$—that is, if

$$\lim_{x \to \infty} f(x) = L \qquad \text{or} \qquad \lim_{x \to -\infty} f(x) = L.$$

Vertical Asymptotes; Infinite Limits The vertical line $x = a$ is an asymptote to the graph of f if (and only if) f blows up (or blows down) at a—that is, if

$$\lim_{x \to a^-} f(x) = \pm\infty \qquad \text{or} \qquad \lim_{x \to a^+} f(x) = \pm\infty.$$

Limits and Asymptotes of Rational Functions

Asymptote information is useful for any function. Horizontal asymptotes reflect long-term behavior; vertical asymptotes reveal sudden spikes and other anomalies. For polynomials and rational functions (quotients of polynomials) the story is especially simple.

Can Polynomials Have Asymptotes?   Polynomials do not have vertical asymptotes. A vertical asymptote occurs at a *finite* value of x, near which a function's value is unbounded. Since polynomials have the pleasant form

$$p(x) = a_0 + a_1 x + \cdots + a_n x^n,$$

free of troublesome denominators, polynomials cannot blow up (or down) except as x approaches $\pm\infty$.

Draw the graph of $p(x) = 3$ to convince yourself.

Only a constant polynomial can have a *horizontal* asymptote. If, say, $p(x) = 3$, then the p-graph is itself horizontal—it is its own asymptote. ◄ Nonconstant polynomials cannot have horizontal asymptotes. Moreover:

> **Fact** If $p(x)$ is a nonconstant polynomial, then
> $$\lim_{x \to \pm\infty} |p(x)| = \infty.$$

We won't prove this fact formally, but we'll illustrate why it holds.

EXAMPLE 2   Let $p(x) = 2x^3 - 53x^2 - 123x$. How does p behave as $x \to \pm\infty$?

Solution   It's clear at a glance that as $x \to \infty$, $2x^3 \to \infty$, $-53x^2 \to -\infty$, and $-123x \to -\infty$. But what about the sum?

Convince yourself of each step. Watch the last step in each operation carefully—we'll say more soon about algebra with the ∞ symbol.

The trick is to *factor out the largest power of x:* ◄

$$\lim_{x \to \infty} \left(2x^3 - 53x^2 - 123x \right) = \lim_{x \to \infty} x^3 \left(2 - \frac{53}{x} - \frac{123}{x^2} \right)$$

$$= \lim_{x \to \infty} x^3 \cdot \lim_{x \to \infty} \left(2 - \frac{53}{x} - \frac{123}{x^2} \right)$$

$$= \infty \cdot 2 = \infty.$$

Similarly,

$$\lim_{x \to -\infty} \left(2x^3 - 53x^2 - 123x \right) = \lim_{x \to -\infty} x^3 \cdot \lim_{x \to -\infty} \left(2 - \frac{53}{x} - \frac{123}{x^2} \right)$$

$$= -\infty \cdot 2 = -\infty.$$

More on this fact soon.

(In each calculation we used the fact that the limit of a product is the product of the limits. ◄   ) ∎

Algebra with the Infinity Symbol.   In Example 2 we breezed past two possibly suspicious claims: $\infty \cdot 2 = \infty$ and $-\infty \cdot 2 = -\infty$. Do they really make sense? Since ∞ and $-\infty$ are not numbers, their appearance in equations should always raise flags of caution. Here, fortunately, we're OK—neither $\infty \cdot 2$ nor $-\infty \cdot 2$ is ambiguous. The following table collects some rules of etiquette for the use of the infinity symbol.

Good and Bad Uses of ∞	
Expression	**What It Means; Remarks**
$1/\infty = 0$	Usually OK, but if we really mean, say, $\lim_{x \to \infty} 1/x = 0$, why not say so?
$3 \cdot \infty = \infty$	OK as mathematical shorthand. It means: If any quantity increases without bound, so does three times that quantity.
$\infty + \infty = \infty$	OK again. It means: If two quantities increase without bound, so does their sum.
$1/0 = \infty$	Wrong. Division by 0 is not defined for real numbers. Worse, $1/x \to \infty$ as $x \to 0$ from the right, but $1/x \to -\infty$ as $x \to 0$ from the left.
$\infty - \infty = 0$	Wrong again. For example, as $x \to \infty$, $x^3 \to \infty$ and $x^2 \to \infty$, but $x^3 - x^2 \to \infty$.
$\infty/\infty = 1$	Wrong. For example, as $x \to \infty$, $x^3 \to \infty$ and $x^2 \to \infty$, but $x^3/x^2 = x \to \infty$.

Horizontal Asymptotes of Rational Functions: Appeal to Higher Powers.

Finding horizontal asymptotes of rational functions means finding limits at infinity. A simple algebraic strategy applies:

> *To find limits at infinity of rational functions, factor out the highest powers of x from numerator and denominator.*

Let's apply it:

EXAMPLE 3   Consider the three rational functions

$$r(x) = \frac{2x^3 - x}{x^2 - 1}, \qquad s(x) = \frac{2x^3 - x}{x^3 - 1}, \qquad \text{and} \qquad t(x) = \frac{2x^3 - x}{x^4 - 1}.$$

Find all horizontal asymptotes.

Solution   The graphs follow soon, but we won't peek. Instead we'll use the preceding factoring strategy:

$$r(x) = \frac{2x^3 - x}{x^2 - 1} = \frac{x^3(2 - 1/x^2)}{x^2(1 - 1/x^2)} = x \cdot \frac{2 - 1/x^2}{1 - 1/x^2};$$

$$s(x) = \frac{2x^3 - x}{x^3 - 1} = \frac{x^3(2 - 1/x^2)}{x^3(1 - 1/x^3)} = \frac{2 - 1/x^2}{1 - 1/x^3};$$

$$t(x) = \frac{2x^3 - x}{x^4 - 1} = \frac{x^3(2 - 1/x^2)}{x^4(1 - 1/x^4)} = \frac{1}{x} \cdot \frac{2 - 1/x^2}{1 - 1/x^4}.$$

Now we can see that as $x \to \infty$,

$$r(x) \to \infty, \qquad s(x) \to 2, \qquad \text{and} \qquad t(x) \to 0. \blacktriangleright$$
Convince yourself.

As $x \to -\infty$,

$$r(x) \to -\infty, \qquad s(x) \to 2, \qquad \text{and} \qquad t(x) \to 0.$$

Thus, s and t have *horizontal* asymptotes, at $y = 2$ and $y = 0$, respectively; r has none. ∎

Vertical Asymptotes of Rational Functions: Look for Roots. If a rational function has vertical asymptotes, they are located at roots of the denominator.

EXAMPLE 4   Let r, s, and t be as before. Find all vertical asymptotes.

Solution   A rational function has a vertical asymptote at any position where the denominator is zero and the numerator isn't. (If both numerator and denominator are zero at the same place, they have a common factor; for present purposes, it can be canceled.) To locate the asymptotes, factor the denominator. ◄   For r, s, and t:

Factoring polynomials of high degree can be hard or impossible. Authors usually choose polynomials that are easy to factor. So will we.

$$r(x) = \frac{2x^3 - x}{x^2 - 1} = \frac{2x^3 - x}{(x + 1)(x - 1)};$$

$$s(x) = \frac{2x^3 - x}{x^3 - 1} = \frac{2x^3 - x}{(x - 1)(x^2 + x + 1)};$$

$$t(x) = \frac{2x^3 - x}{x^4 - 1} = \frac{2x^3 - x}{(x^2 - 1)(x^2 + 1)} = \frac{2x^3 - x}{(x - 1)(x + 1)(x^2 + 1)}.$$

The factored denominators show that, for each function, $x = \pm 1$ are the suspicious values, or singular points. Since the numerator $2x^3 - x$ is not zero at $x = \pm 1$, it follows that r and t have vertical asymptotes at $x = 1$ and at $x = -1$; s has a vertical asymptote at $x = 1$.

On either side of a vertical asymptote, a function may tend either to ∞ or to $-\infty$. What happens at each of the preceding asymptotes? Graphs make the answers obvious, but even without graphs, the question is easy to answer. We need only determine the *sign* of each function just to the right and just to the left of each asymptote.

For $r(x)$, just to the left of $x = -1$, we see that

$$2x^3 - x < 0, \qquad x + 1 < 0, \qquad \text{and} \qquad x - 1 < 0.$$

r blows down to $-\infty$ as $x \to -1$ from the left.

Therefore, $r(x) < 0$, so $\lim_{x \to -1^-} r(x) = -\infty$. ◄   A similar sign check reveals that r blows *up* to ∞ as $x \to -1$ from the *right*. ◄

Can you mimic this analysis for the functions s and t?

We've waited long enough. Here are the graphs:

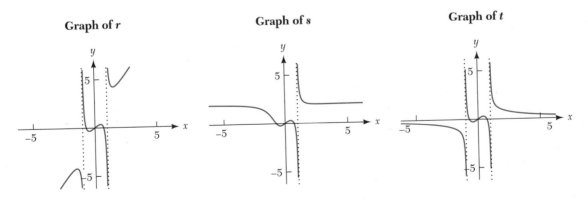

Graph of r   **Graph of s**   **Graph of t**

Asymptotes of Other Algebraic Functions

Like rational functions, other algebraic functions may have asymptotes, with exactly the same connection to infinite limits and limits at infinity. Calculating these limits algebraically may take extra ingenuity.

EXAMPLE 5   The algebraic function defined (for $x > 0$) by $f(x) = \sqrt{x^2 + x} - x$ has a horizontal asymptote. Find it.

Solution   How $f(x)$ behaves as $x \to \infty$ isn't obvious from the formula, ▶ but the following table of values argues persuasively for a limit of 1/2. ▶

Is it? Why or why not?

So would a graph; draw one.

As $x \to \infty$, $\sqrt{x^2 + x} - x \to 1/2$								
x	1	2	4	8	16	32	64	128	256
$f(x)$	0.4142	0.4495	0.4721	0.4853	0.4924	0.4962	0.4981	0.4990	0.4995

With algebra we'll clinch our case. The first step is a trick: multiplying top and bottom by the conjugate.

$$\lim_{x \to \infty} (\sqrt{x^2 + x} - x) = \lim_{x \to \infty} \frac{(\sqrt{x^2 + x} - x)(\sqrt{x^2 + x} + x)}{\sqrt{x^2 + x} + x}$$

$$= \lim_{x \to \infty} \frac{x}{\sqrt{x^2 + x} + x}$$

$$= \lim_{x \to \infty} \frac{x}{x\left(\sqrt{1 + 1/x} + 1\right)}$$

$$= \lim_{x \to \infty} \frac{1}{\sqrt{1 + 1/x} + 1} = \frac{1}{2}. \qquad ∎$$

New Limits from Old

When calculating limits, we often take a few basic limits to be obvious and try to reduce more complicated limits to combinations of these simpler ones. The next example illustrates that process.

EXAMPLE 6   Find $\lim\limits_{x \to \infty} \dfrac{x^2 + 3}{2x^2 - 5}$.

Solution   With a graph it would be easy to guess that the limit is 1/2. Algebra makes the case airtight: ▶

The key step is dividing numerator and denominator by x^2.

$$\lim_{x \to \infty} \frac{x^2 + 3}{2x^2 - 5} = \lim_{x \to \infty} \frac{x^2 \cdot (1 + 3/x^2)}{x^2 \cdot (2 - 5/x^2)}$$

$$= \lim_{x \to \infty} \frac{1 + 3/x^2}{2 - 5/x^2}$$

$$= \frac{\displaystyle \lim_{x \to \infty} 1 + \lim_{x \to \infty} 3/x^2}{\displaystyle \lim_{x \to \infty} 2 - \lim_{x \to \infty} 5/x^2}$$

$$= \frac{1 + 0}{2 - 0} = 1/2 \, . \qquad \blacksquare$$

We used several plausible facts in the preceding calculations: that $\lim_{x \to \infty} 3/x^2 = \lim_{x \to \infty} 5/x^2 = 0$, that the limit of a sum or quotient is the sum or quotient of the limits, and so on.

The following theorem justifies combining old limits to find new ones. Each part tells how limits behave with respect to some algebraic operation on functions.

Theorem 1 (Algebra with Limits)   Suppose that

$$\lim_{x \to a} f(x) = L \qquad \text{and} \qquad \lim_{x \to a} g(x) = M,$$

where L and M are finite numbers. Let k be any constant. Then

(i) $\displaystyle \lim_{x \to a} k f(x) = kL$;

(ii) $\displaystyle \lim_{x \to a} [f(x) + g(x)] = L + M$;

(iii) $\displaystyle \lim_{x \to a} f(x) \cdot g(x) = L \cdot M$;

(iv) $\displaystyle \lim_{x \to a} \frac{f(x)}{g(x)} = \frac{L}{M}$ (if $M \neq 0$).

Remarks on the Theorem.

Why They Hold   Each rule is intuitively plausible, but rigorous proofs would require the ϵ–δ definition. We'll skip the proofs here, but a typical proof of this kind appears in Appendix I.

Limits That Involve Infinity   All the rules hold for limits at infinity (i.e., when a is $\pm\infty$). They also hold for infinite limits (i.e., when $L = \pm\infty$ and/or $M = \pm\infty$), but only if appropriate care is taken with uses of the infinity symbol, as described in the preceding table.

"Obvious" Limits   Using these algebraic rules, we can find many new limits—provided that we know some old limits. To get such a foothold, we'll take as obvious ◄   limits such as these:

To be fully rigorous we'd have to prove even these limits.

"Obvious" Limits	
$\displaystyle \lim_{x \to 0} |x| = 0$	
$\displaystyle \lim_{x \to a} x = a$	for any real number a
$\displaystyle \lim_{x \to \infty} x^n = \infty$	for any $n > 0$
$\displaystyle \lim_{x \to \infty} \frac{1}{x^n} = 0$	for any $n > 0$

New Continuous Functions from Old: Algebraic Combinations

If f and g are continuous functions, it sounds reasonable that such algebraic combinations as $f + g$, f/g, and $3f - 2g$ should also define continuous functions. As a rule, they do. Because continuity is defined via limits, ▶ it's not surprising that properties of limits are reflected in properties of continuous functions. The next theorem gives the fine print.

See the previous section!

> **Theorem 2** (Algebra with Continuous Functions) Suppose that both f and g are continuous at $x = a$. Then each of the functions
>
> $$f + g, \qquad f - g, \qquad \text{and} \qquad fg$$
>
> is also continuous at $x = a$. If $g(a) \neq 0$, then f/g is continuous at $x = a$.

Proof. Each claimed property of continuous functions follows directly from the analogous property of limits. The first claim, for instance, goes like this:

$$\lim_{x \to a}(f+g)(x) = \lim_{x \to a} f(x)+g(x) = \lim_{x \to a} f(x)+\lim_{x \to a} g(x) = f(a)+g(a) = (f+g)(a).$$

The result is just what's needed to show that $f + g$ is continuous at $x = a$.

Composition, Limits, and Continuity

Limits often involve functions built up by composition, so it's helpful to know how the operations of composition and taking limits interact. First let's discuss some straightforward examples.

EXAMPLE 7 Find and discuss each of the following limits. ▶

Guess all three limits before reading on. With any luck you'll guess correctly.

$$L_1 = \lim_{x \to 0} \ln(\cos x); \qquad L_2 = \lim_{x \to 0} \ln\left(\frac{\sin x}{x}\right); \qquad L_3 = \lim_{x \to \infty} \exp\left(\frac{x+1}{x+2}\right).$$

Solution The first limit is easiest:

As $x \to 0$, $\cos x \to \cos 0$, so $\ln(\cos x) \to \ln(\cos 0) = \ln 1 = 0$.

Why did things work so well? Because both functions involved are continuous at the points that matter: As $x \to 0$, $\cos x \to \cos 0$ *because the cosine function is continuous at $x = 0$; as $\cos x \to 1$, $\ln(\cos x) \to \ln 1$ because the natural logarithm function is continuous at $x = 1$.*

To find L_2, we reason similarly. Recall, first, that as $x \to 0$, $(\sin x)/x \to 1$. ▶ Therefore:

We've used this fact repeatedly. See, for example, Example 4, page 152, for graphical and numerical evidence.

As $x \to 0$, $\dfrac{\sin x}{x} \to 1$, so $\ln\left(\dfrac{\sin x}{x}\right) \to \ln 1 = 0$.

Continuity helps again: Because $\ln$ is continuous at $x = 1$,

$$\frac{\sin x}{x} \to 1 \Longrightarrow \ln\left(\frac{\sin x}{x}\right) \to \ln 1.$$

Convince yourself of the first implication.

The limit L_3 is "at infinity," but the same sort of reasoning applies as for L_1 and L_2: ◀

$$x \to \infty \Longrightarrow \frac{x+1}{x+2} \to 1 \Longrightarrow \exp\left(\frac{x+1}{x+2}\right) \to \exp(1) = e.$$

Again, continuity is key: The second implication holds because the exponential function is continuous at $x = 1$. ∎

Limits of Composite Functions: A General Principle. One basic principle applies to all three limits in Example 7. We'll state it in general terms. ◀ (The formal proof, which we omit, appeals to the ϵ–δ definition of limit.)

In the following Fact, the constant a may be ±∞.

> **Fact** (Limit of a Composite Function) Let f and g be functions, such that $\lim_{x \to a} g(x) = b$ and f is continuous at $x = b$. Then
>
> $$\lim_{x \to a}(f \circ g)(x) = \lim_{x \to a} f(g(x)) = f(b).$$

We used this fact repeatedly in Example 7.

The precise version says exactly where f and g must be continuous.

Composing Continuous Functions.   Recall that, by definition, a function f is continuous at a point $x = a$ if and only if $\lim_{x \to a} f(x) = f(a)$. ◀ With this definition and the preceding Fact, it's easy to show that if f and g are continuous functions, then so is $f \circ g$. More precisely: ◀

> **Theorem 3** Suppose that g is continuous at $x = a$ and f is continuous at $g(a)$. Then $f \circ g$ is continuous at $x = a$.

Proof.   If $x \to a$, $g(x) \to g(a)$, because g is continuous at a. As $g(x) \to g(a)$, $f(g(x)) \to f(g(a))$, because f is continuous at $g(a)$. Putting everything together,

$$x \to a \Longrightarrow (f \circ g)(x) \to (f \circ g)(a).$$

That's exactly what we needed to show.   ∎

Limits and Inequalities: The Squeeze Principle

Algebraic limit rules don't help with limits such as

$$\lim_{x \to 0} x \cdot \sin\left(\frac{1}{x}\right) \qquad \text{and} \qquad \lim_{x \to \infty} \frac{\sin x}{x}.$$

The product-of-limits rule doesn't apply to the first limit, for example, because the second factor, $\sin(1/x)$, has no limit as $x \to 0$. With graphs the limits are easy to guess, but let's try to compute them directly.

Squeezing Inequalities. The idea is to use carefully chosen inequalities to "squeeze" the function of interest between two simpler functions. For the first of the preceding limits, the key inequalities say that, for any x,

$$-|x| \leq x \sin\left(\frac{1}{x}\right) \leq |x|.$$

A graph shows convincingly that the inequality is valid:

Finding a limit by squeezing

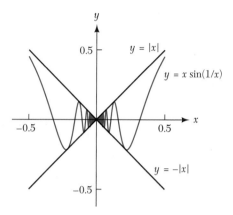

As $x \to 0$, both the left and the right sides of this three-term inequality tend to 0. Squeezed between $-|x|$ and $|x|$, $x \sin(1/x)$ has only one place to go:

$$x \sin\left(\frac{1}{x}\right) \to 0 \quad \text{as} \quad x \to 0.$$

There's a general theorem here. The preceding graph makes it reasonable; we omit the formal proof.

Theorem 4 (**The Squeeze Principle**) Suppose that

$$f(x) \le g(x) \le h(x)$$

for all x near $x = a$. If $f(x) \to L$ and $h(x) \to L$ as $x \to a$, then $g(x) \to L$ as $x \to a$.

Note:

- The theorem is good for more than computation. It also guarantees that $\lim_{x \to a} g(x)$ exists—no matter how crazily g careens between f and h.
- The squeeze principle is conceptually easy; the hard part, in practice, is to choose the "bounding functions" f and h properly.

EXAMPLE 8   Explain why $\displaystyle \lim_{x \to \infty} \frac{\sin x}{x} = 0$.

Solution   The inequality

$$-\frac{1}{x} \le \frac{\sin x}{x} \le \frac{1}{x}$$

holds ▶ for all $x > 0$. Because the left and the right sides tend to 0 as $x \to \infty$, so must the middle expression.   ■

Convince yourself; use the fact that for all x,
$$-1 \le \sin x \le 1.$$

Finding Limits Graphically and Numerically

If $\lim_{x \to a} f(x) = L$, then $f(x) \approx L$ whenever $x \approx a$. Therefore, in order to guess or estimate L, it's natural to look at values of f near $x = a$. We can do this either numerically, by plugging in numbers, or graphically, by plotting f near $x = a$. Each strategy has its advantages. Numerical outputs may be more accurate, but graphs show better how f varies. With technology, both approaches are almost painless.

Can you guess these limits using only mental gymnastics?

Check some with a calculator.

EXAMPLE 9   Let $f(x) = x \sin(1/x)$.   Discuss ◄ $\lim_{x \to 0} f(x)$ and $\lim_{x \to \infty} f(x)$.

Solution   We saw earlier, by squeezing, that the first limit is 0. Numerical values of f near $x = 0$ (rounded to 4 decimal places) agree: ◄

x	−0.1	−0.01	−0.001	...	0.001	0.01	0.1
$f(x)$	−0.0544	−0.0051	0.0008	...	0.0008	−0.0051	−0.0544

To investigate $\lim_{x \to \infty} f(x)$ we'll consider large values of x, first graphically

Graph of $y = x \sin(1/x)$ for large x

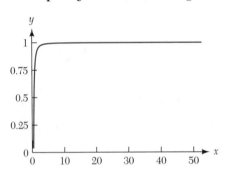

and now numerically, to six decimal places:

x	10	100	1000	10000
$f(x)$	0.998334	0.999983	1.000000	1.000000

Again the conclusion is irresistible: $\lim_{x \to \infty} f(x) = 1$.   ■

EXAMPLE 10   Let $g(x) = \sin(1/x)$. Discuss $\lim_{x \to 0} g(x)$.

Solution   This time both graph ▶   and numbers look suspiciously chaotic:

The graph shown is of poor quality. What's wrong with it? Why is a good graph hard to find? Would a smaller window help?

A (poor quality) graph of y = sin(1/x)

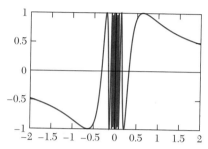

x	−0.1	−0.05	−0.025	...	0.001	0.0025	0.0075
$g(x)$	0.544	−0.913	−0.745	...	0.827	−0.851	0.983

In fact, the limit doesn't exist. As $x \to 0$ from the right, $1/x \to +\infty$, so $\sin(1/x)$ oscillates faster and faster between −1 and 1.   ∎

BASICS

1. Let f and g be the functions whose graphs are shown below. Use the graphs to evaluate the following limits. If a limit doesn't exist, explain why. [**NOTE:** $f(1) = 2$.]

Graph of f   **Graph of g**

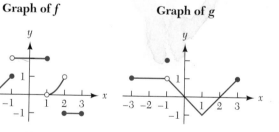

(a) $\lim\limits_{x \to 0} \big(f(x) + g(x)\big)$ (e) $\lim\limits_{x \to 0} \dfrac{f(x)}{g(x)}$

(b) $\lim\limits_{x \to 2} \big(f(x) + g(x)\big)$ (f) $\lim\limits_{x \to 0} \big(f(x) \cdot \cos x\big)$

(c) $\lim\limits_{x \to 1} \big(f(x) \cdot g(x)\big)$ (g) $\lim\limits_{x \to -2} x^2 g(x)$

(d) $\lim\limits_{x \to 2} \big(f(x) \cdot g(x)\big)$

2. Let f and g be the functions whose graphs are shown below. Use the graphs to evaluate the following limits. If a limit doesn't exist, explain why.

Graph of f   **Graph of g**

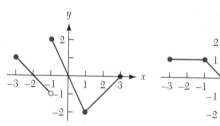

(a) $\lim\limits_{x \to -2} \left(\dfrac{f(x)}{g(x)}\right)$ (c) $\lim\limits_{x \to 0} \left(\dfrac{f(x)}{g(x)}\right)$

(b) $\lim\limits_{x \to 2} \left(\dfrac{f(x)}{g(x)}\right)$ (d) $\lim\limits_{x \to 1} \left(\dfrac{f(x)}{g(x)}\right)$

3. Use a calculator to estimate each of the following limits numerically.

(a) $\lim\limits_{x \to \infty} \dfrac{3}{x^2}$ (f) $\lim\limits_{x \to 0} \sin\big(\sin x\big)$

(b) $\lim\limits_{x \to \infty} \dfrac{5}{\sqrt{x}}$ (g) $\lim\limits_{x \to \infty} \dfrac{2^x}{x^2}$

(c) $\lim\limits_{x \to \infty} \dfrac{x^2 + 1}{x}$ (h) $\lim\limits_{x \to \infty} \dfrac{x^2}{2^x}$

(d) $\lim\limits_{x \to 0^+} \dfrac{\sqrt{x}}{\sin x}$ (i) $\lim\limits_{x \to -\infty} \dfrac{2^x}{x^2}$

(e) $\lim\limits_{x \to 0^+} \big(x \log_{10} x\big)$ (j) $\lim\limits_{x \to 0^+} x^x$

4. Use a grapher to guess whether each of the following limits exist. Estimate values of those that do exist.

   (a) $\lim\limits_{x\to\infty} \dfrac{2000x^2}{1000x^2 + \sin x}$

   (b) $\lim\limits_{x\to0} \dfrac{1 - \cos^2(3x)}{x^2}$

   (c) $\lim\limits_{x\to0} \cos\left(\dfrac{1 - \cos x}{x}\right)$

   (d) $\lim\limits_{x\to\infty} \dfrac{3^x - 3}{2^x - 2}$

   (e) $\lim\limits_{x\to-\infty} \dfrac{3^x - 3}{2^x - 2}$

5. Evaluate

   (a) $\lim\limits_{x\to\infty} \sin\left(\dfrac{\pi}{x}\right)$.

   (b) Evaluate $\lim\limits_{x\to0} \sin\left(\dfrac{\pi}{x}\right)$.

   (c) Evaluate $\lim\limits_{x\to\infty} \exp\left((3x^2 + x)/2x^3\right)$.

6. Suppose that $\lim\limits_{x\to1} f(x) = 5$ and $\lim\limits_{x\to1} g(x) = -2$. Use this information to evaluate each of the following limits.

   (a) $\lim\limits_{x\to1}\big(f(x) + g(x)\big)$

   (b) $\lim\limits_{x\to1}\big(g(x) - f(x)\big)$

   (c) $\lim\limits_{x\to1} \dfrac{f(x) + 2}{f(x)g(x)}$

FURTHER EXERCISES

7. Let $f(x) = \sqrt[3]{x}$. Explain why $\lim\limits_{x\to0} f'(x) = \infty$.

8. Suppose that f is a (nonconstant) periodic function. Explain why $\lim\limits_{x\to\infty} f(x)$ does not exist.

9. Sketch the graph of a function f that has *all* of the following properties. (Be sure to label axes and indicate units on each axis.)

   (i) f has domain $(-3, \infty)$

   (ii) f has range $(-\infty, 2)$

   (iii) $f(-1) = 0$; $f(0) = 0$; $f(5) = 0$

   (iv) $\lim\limits_{x\to-3^+} f(x) = -\infty$

   (v) $\lim\limits_{x\to2} f(x) = -3$

   (vi) $\lim\limits_{x\to\infty} f(x) = 2$

10. Sketch the graph of a function f that has *all* of the following properties:

    (i) f has domain $\mathbb{R}$

    (ii) f has range $(-3, \infty)$

    (iii) $\lim\limits_{t\to2} f(t) = -2$

    (iv) $\lim\limits_{t\to\infty} f(t) = -3$

    (v) $\lim\limits_{t\to-\infty} f(t) = 3$

11. Let P be the (fixed) point $(0, 1)$, let Q be a (moveable) point on the x-axis, and let ℓ be the line joining P and Q. In this exercise we'll explore how the *slope* of ℓ behaves as Q slides back and forth along the x-axis.

    (a) Suppose that Q slides to the *right* along the x-axis. How does the slope of ℓ behave? Does this slope approach a limit as Q moves farther and farther to the right? What limit? Give a geometric reason for your answer—no calculation is needed. (Draw a picture to see what's going on.)

    (b) Redo the previous part algebraically: First find a symbolic expression for the slope in question, then find an appropriate limit. [**HINT:** Q has coordinates $(x, 0)$.]

    (c) Repeat parts (a) and (b) but assuming this time that Q slides farther and farther to the *left*.

    (d) Suppose that Q slides *toward the origin* from the right. Does the slope of ℓ approach a limit? Justify your answer with an appropriate limit calculation.

12. Let O be the origin $(0, 0)$, let Q be a (moveable) point on the line $y = 2x + 3$, and let ℓ be the line joining O and Q. This exercise explores how the *slope* of ℓ behaves as Q moves back and forth along the line $y = 2x + 3$, while O remains fixed.

    (a) Suppose that Q slides to the *right* along the line $y = 2x + 3$. How does the slope of ℓ behave? Does this slope approach a limit as Q moves farther and farther to the right? What limit? Give a geometric reason for your answer—no calculation is needed. (Draw a generic picture to see what's going on.)

    (b) Redo the previous part algebraically: First find a symbolic expression for the slope in question, then find an appropriate limit at infinity. [**HINT:** Q has coordinates $(x, 2x + 3)$.]

    (c) Does it make any difference in the two parts above if Q slides farther and farther along ℓ to the *left*? Explain your answer.

13. *This is a more general and open-ended version of Exercises 11 and 12. One or both of them should be done first.* Let P be any point in the plane, ℓ_1 any line, and Q any point on ℓ_1; assume that $P \neq Q$. Let ℓ_2 be the line through P and Q. This exercise explores what happens to the slope of ℓ_2 as Q moves to and fro along ℓ_1. [NOTE: No algebraic formulas or formal limit calculations are needed until the last part. In all other parts, just use pictures.]

    (a) Suppose that P happens to lie on ℓ_1. How does the slope of ℓ_2 behave as Q moves along ℓ_1? (Draw a picture to guess the answer—no calculation is needed.)

    (b) Suppose that ℓ_1 is a *vertical* line. (Don't assume that P lies on ℓ_1.) How does the slope of ℓ_2 behave as Q moves higher and higher along ℓ_1? Does the slope approach a limit? What limit? [**HINT:** Does it matter whether P lies to the left or right of ℓ_1?]

    (c) Suppose that ℓ_1 is a *horizontal* line. (Don't assume that P lies on ℓ_1.) How does the slope of ℓ_2 behave as Q moves farther and farther to the right along ℓ_1? Does the slope approach a limit? What limit?

(d) Suppose that ℓ_1 is an *oblique* line (i.e., with slope $m \neq 0$). (Don't assume that P lies on ℓ_1.) How does the slope of ℓ_2 behave as Q moves farther and farther to the right along ℓ_1? Does the slope approach a limit? What limit?

(e) Use limits at infinity to verify as many of your answers above as possible. To do so, write P in the form (a, b) and ℓ_1 either in the form $y = mx + k$ or in the form $x = c$.

14. Show that there is only one value of $k \geq 1$ for which $\lim\limits_{x \to 0} \dfrac{\sin(2 \sin x)}{x^k}$ is a real number. Find this value of k and evaluate the limit.

15. Let $p(x) = (x + 1)(2 - x)$ and $q(x) = x^{17} - 3x^{14} + 39$. Evaluate each of the following:

(a) $\lim\limits_{x \to \infty} p(x)$ (c) $\lim\limits_{x \to \infty} q(x)$

(b) $\lim\limits_{x \to -\infty} p(x)$ (d) $\lim\limits_{x \to -\infty} q(x)$

16. Suppose that $r(x)$ is a polynomial of degree n such that $\lim\limits_{x \to \infty} r(x) = -\infty$ and $\lim\limits_{x \to -\infty} r(x) = -\infty$. Can you tell if n is even or odd? Explain.

17. Let h be a function that has a vertical asymptote at $x = 3$ and the properties $\lim\limits_{x \to -\infty} h(x) = \infty$ and $\lim\limits_{x \to \infty} h(x) = -1$.

(a) Must h have a horizontal asymptote? If so, what is it?

(b) Could h be a rational function? Why or why not?

18. Any rational function $r(x)$ behaves like ax^n for some real number a and some integer n in the sense that

$$\lim\limits_{x \to \infty} \frac{r(x)}{ax^n} = 1.$$

For example, $k(x) = 2x^3 / (x^2 + 1)$ behaves like $2x$ ($a = 2$ and $n = 1$) for large values of x because

$$\lim\limits_{x \to \infty} k(x) = \lim\limits_{x \to \infty} \frac{2x^3}{x^2 + 1}$$
$$= \lim\limits_{x \to \infty} \left(2x \cdot \frac{x^2}{x^2 + 1} \right)$$

and $\lim\limits_{x \to \infty} x^2 / (x^2 + 1) = 1$.

Find the appropriate values of a and n for each of the following rational functions. Then graph the function and ax^n on the same axes for large values of x (e.g., on the interval $[50, 100]$).

(a) $f(x) = \dfrac{4x^2 + 3x}{3x^4 + 2x^2 - 1}$

(b) $g(x) = \dfrac{3x^4 - 2x^3 + 4x^2 - 5x + 7}{5x^4 - 12x^2 + 100}$

(c) $h(x) = \dfrac{2x^5 - 3x^2 + 15}{3x^2 + 4x - 5}$

19. Let $f(x) = \sqrt{x^2 + 1} - x$ and $g(x) = \sqrt{x^2 + x} - x$.

(a) Use a calculator or a computer to complete the following table.

x	100	1,000	10,000	100,000	1,000,000
$f(x)$					
$g(x)$					

(b) Use the results you recorded in the table above to estimate $\lim\limits_{x \to \infty} f(x)$ and $\lim\limits_{x \to \infty} g(x)$.

(c) What do your answers to part (b) imply about the graphs of f and g?

20. Give examples of polynomials $p(x)$ and $q(x)$ such that $\lim\limits_{x \to \infty} p(x) = \infty$, $\lim\limits_{x \to \infty} q(x) = \infty$, and

(a) $\lim\limits_{x \to \infty} \dfrac{p(x)}{q(x)} = \infty$. (b) $\lim\limits_{x \to \infty} \dfrac{p(x)}{q(x)} = 2$.

(c) $\lim\limits_{x \to \infty} \dfrac{p(x)}{q(x)} = 0$.

(d) $\lim\limits_{x \to \infty} \dfrac{p(x)}{q(x)} = 1$ and $\lim\limits_{x \to \infty} \left(p(x) - q(x) \right) = 3$.

21. For each part of the preceding problem, give a geometric explanation of how the graphs of p and q are related. [**HINT:** Graph your polynomials p and q on the same set of axes.]

22. Let $p(x) = ax^n + q(x)$ where $q(x)$ is a polynomial of degree $m < n$.

(a) What is $\lim\limits_{x \to \infty} \dfrac{q(x)}{x^n}$? (b) What is $\lim\limits_{x \to -\infty} \dfrac{q(x)}{x^n}$?

(c) If $a > 0$ and n is *even*, what is $\lim\limits_{x \to \infty} p(x)$?

(d) If $a > 0$ and n is *even*, what is $\lim\limits_{x \to -\infty} p(x)$?

(e) How do your answers to parts (c) and (d) change if n is *odd*?

23. Suppose that the average rate of change of a function f over the interval from $x = 3$ to $x = 3 + h$ is $4e^h - 5\cos(2h)$. What is $f'(3)$?

24. Let $f(x) = \dfrac{2x^2 + x + a}{x^2 - 2x - 3}$.

(a) Can the number a be chosen so that $\lim\limits_{x \to 3} f(x)$ exists? Explain your answer.

(b) Can the number a be chosen so that $\lim\limits_{x \to \infty} f(x)$ exists? Explain your answer.

25. Explain why $\lim\limits_{x \to \infty} \cos x$ does not exist.

26. Evaluate each of the following limits, if it exists. [**HINT:** Graphs may help.]

(a) $\lim\limits_{x \to \infty} \dfrac{\sin x}{x}$ (b) $\lim\limits_{x \to -\infty} \dfrac{x^3}{2 + \cos x}$

27. Suppose that neither $\lim\limits_{x \to a} f(x)$ nor $\lim\limits_{x \to a} g(x)$ exists. Find functions f and g so that the following limits exist or explain why it is impossible to do so.

(a) $\lim\limits_{x \to a} \left(f(x) + g(x) \right)$ (b) $\lim\limits_{x \to a} \left(f(x) \cdot g(x) \right)$

28. Let f be a function defined for all real numbers. Which of the following statements *must* be true about f? Which *might* be true? Which *must* be false? Justify your answers.

    (a) $\lim\limits_{x \to a} f(x) = f(a)$

    (b) If $\lim\limits_{x \to 0} \dfrac{f(x)}{x} = 2$, then $f(0) = 0$.

    (c) If $\lim\limits_{x \to 0} \dfrac{f(x)}{x} = 1$, then $\lim\limits_{x \to 0} f(x) = 0$.

    (d) If $\lim\limits_{x \to 1^-} f(x) = 1$ and $\lim\limits_{x \to 1^+} f(x) = 3$, then $\lim\limits_{x \to 1} f(x) = 2$.

    (e) If $\lim\limits_{x \to 2} f(x) = 3$, then 3 is in the range of f.

    (f) If $\lim\limits_{x \to 0} \dfrac{f(x) - f(0)}{x} = 3$, then $f'(0) = 3$.

2.8 Chapter Summary

This chapter is the heart of Calculus I. Its main idea, the **derivative**, is the most important idea in calculus.

Amount Functions, Rate Functions, Slope Functions.   Section 2.1 approached the derivative informally and intuitively, interpreting it (depending on the situation) either as the **rate function** or as the **slope function** derived from an original function, or **amount function**. *Any* function can be thought of as "original"; to any function we can associate a rate function that describes how fast the original function varies.

For example, suppose that $s(t)$ describes the **position** of a moving object at time t; then its rate function $s'(t)$ (usually written $v(t)$) describes the object's **velocity**. If v is considered original, then *its* rate function $v'(t)$ (usually written $a(t)$) describes the object's **acceleration**.

The **racetrack principle** relates the behavior of rate functions to that of amount functions. In its simplest form, it says that if g grows *faster* than f over an interval, then g grows *more* than f over that interval—not a very surprising fact, but one that will prove useful later.

For a linear function the associated rate function (or slope function) is constant—its value is the slope itself. For a nonlinear function the question of slope is stickier—what does it mean for a *curve* to have slope?

The answer—which we pursued in Section 2.2—is best understood graphically: The slope of a curve at a point is the slope of the **tangent line** at that point. Zooming helps us find this slope:

> *To find the slope of a curve at a point, zoom in until the curve looks straight, then find the slope.*

This trick really works, even for nonlinear functions. It works because almost all calculus functions are "locally straight."

Derivatives, Geometrically.   Sections 2.3 and 2.4 are as important as any in this book. They described in detail how a function and its derivatives (first and second) are related graphically. The key idea is worth repeating:

> *The derivative f' of a function f tells whether and how fast the graph of f is rising.*

Almost everything in Sections 2.3 and 2.4 derived from that fact. Where a function increases or decreases, where it's concave up or concave down, how many and what types of stationary points it has—derivatives carry all such information.

The Derivative as a Limit.   The formal definition of derivative is as a **limit of difference quotients**:

$$f'(x) = \lim_{h \to 0} \frac{f(x+h) - f(x)}{h}.$$

Section 2.5 explored this important limit and what it means in various settings. For a given function f, the difference quotient has two main interpretations: It can be thought of either as the **slope of the secant line from x to $x+h$** or as the **average rate of change** of f from x to $x+h$. From these interpretations it follows that the derivative itself also has two main interpretations. The derivative $f'(x)$ can be thought of either as the **slope of the tangent line** at x or as the **instantaneous rate of change** of f at x. Both viewpoints are important, and we'll use both throughout this book.

Limits.   The derivative is defined as a **limit**; Sections 2.6 and 2.7 treated the idea of limit in its various forms. The basic idea is simple: $\lim_{x \to a} f(x) = L$ means that $f(x)$ approaches L as x approaches a. Section 2.6 explored this definition and some of its variants in symbolic, graphical, and numerical forms.

Once understood, the limit is applied to define **continuity** of a function f at a point $x = a$ of its domain. In graphical terms, continuity of f means that the f-graph is unbroken at $x = a$; limits offer the right analytic language for describing this phenomenon.

Section 2.7 extended the idea of limits to **infinite limits** and **limits at infinity**—limits that involve the ∞ symbol. Such limits correspond graphically to **vertical and horizontal asymptotes**, respectively. Section 2.7 also described how to combine old limits algebraically to form new ones and, in the same spirit, how to combine old continuous functions to form new ones.

A Summary Problem

Here's a problem that calls on ideas from various parts of this chapter.

EXAMPLE 1   In each case, decide whether a function with the given properties exists. If so, sketch the graph of such a function. If not, explain why there cannot be such a function.

a. $\lim_{x \to \infty} f(x) = 5$ and $\lim_{x \to \infty} f'(x) = 1$.

b. $f(x) < 0$ and $f'(x) > 0$ for all x.

c. $f(x) < 0$ and $f'(x) < 0$ for all x.

d. $f''(x) > 0$ and $f(x) < 0$ for all x.

e. $f(x) > 0$, $f'(x) > 0$, and $f''(x) < 0$ for all x.

f. $f(x) > 0$, $f'(x) > 0$, and $f''(x) < 0$ for all $x > 0$.

Solution (Note: Read these solutions with graphs at hand. Use either some form of plotting technology or the graphs in Appendix I.)

a. No function with these properties can exist. If $\lim_{x \to \infty} f'(x) = 1$, then, for large x, $f'(x) \approx 1$. This means that, for large x, f increases at the rate of about one output unit for each input unit. Such a function cannot have a horizontal asymptote as $x \to \infty$, as the statement $\lim_{x \to \infty} f(x) = 5$ implies.

Plot it!

b. The conditions say that f should be *negative* and *increasing* for all x. The function $f(x) = -e^{-x}$ ◀ has these properties.

c. The conditions say that f should be *negative* and *decreasing* for all x. The function $f(x) = -e^x$ has these properties.

d. The conditions require that f be *concave up* and *negative* for all x. This is impossible. A graph that's concave up cannot stay below the x-axis forever—think about it.

e. The conditions require that f be *positive, increasing,* and *concave down* for all x. This is impossible, for the same reason as in part (d): A graph that's concave down cannot stay above the x-axis forever.

f. The conditions mean that for positive x, f is *positive, increasing,* and *concave down*. The function $f(x) = \ln(x + 1)$ has all these properties. ■

Derivatives of Elementary Functions

3.1 Derivatives of Power Functions and Polynomials

Introduction

In Chapter 2 we introduced the idea of the derivative (or rate) function and discussed geometric relationships between the functions f and f'. Next we looked closely at $f'(a)$, the value of the derivative function for a specific input $x = a$—how it's defined formally as a limit, how slopes of secant and tangent lines arise, and how they're connected to average and instantaneous rates of change. Now it's time to *calculate* some derivatives symbolically.

Derivatives: The Symbolic View

If a function f has a simple symbolic formula, then it's reasonable to hope that the derivative f' has one, too. This hope is well founded. Indeed, we'll learn in this chapter how to find a formula for the derivative of any **elementary function**, i.e., any function "built" from algebraic, exponential, logarithmic, and trigonometric ingredients. ▶ Even for a monster such as

$$f(x) = \ln\left(\frac{\sin(2x)}{1 + 3x^4}\right),$$

For much more information on elementary functions, see Section 1.5.

we'll soon have little trouble in writing down a derivative formula.

But first things first. Before we can differentiate ▶ a built-up function such as f, we'll need to know derivatives of its most basic ingredients: the natural logarithm functions, the sine function, the fourth power function, and so on. Once we know these basic derivatives, we'll see how to combine them appropriately to differentiate *any* elementary function.

Recall: "differentiate" means "find the derivative of."

Antiderivatives

Similarly, f is an antiderivative of f′.

The relation between functions and derivatives runs both ways. If f is the derivative of F, then F is an **antiderivative** of f. ◄ (*Antidifferentiation* means finding antiderivatives.) Throughout this chapter we'll keep an eye on both derivatives and antiderivatives.

Other Notations for the Derivative

The prime notation f' for the derivative of f is typographically convenient. It also reminds us of a derivative's "ancestry": Every f' comes from some original f.

Leibniz Notation. Alternatives to the prime notation exist. Depending on context, one may be more convenient than the others. To illustrate what's available, let $y = f(x) = x^2$. The following notations all say the same thing:

$$f'(x) = 2x; \qquad \frac{dy}{dx} = 2x; \qquad \frac{d}{dx}\left(x^2\right) = 2x; \qquad \frac{df}{dx} = 2x.$$

So do these:

$$f'(3) = 6; \qquad \left.\frac{dy}{dx}\right|_{x=3} = 6; \qquad \left.\frac{d}{dx}\left(x^2\right)\right|_{x=3} = 6; \qquad \left.\frac{df}{dx}\right|_{x=3} = 6.$$

(We'll soon see just below *why* these statements are true. Here the point is the notation.) The dy/dx form of the derivative is known as **Leibniz notation**, after Gottfried Wilhelm Leibniz (1646–1716), a contemporary of Newton. The "fractional" appearance of Leibniz notation recalls the definition of the derivative as a limit of difference *quotients*:

$$\frac{dy}{dx} = \lim_{\Delta x \to 0} \frac{\Delta y}{\Delta x}.$$

Notations for Higher Derivatives. Both prime and Leibniz notations exist for second, third, and higher derivatives, as well. The following notations all say the same thing about the function defined by $y = f(x) = x^2$.

$$f''(x) = 2; \qquad \frac{d^2 y}{dx^2} = 2; \qquad \frac{d^2}{dx^2}\left(x^2\right) = 2; \qquad \frac{d^2 f}{dx^2} = 2.$$

We'll use both the prime and the Leibniz notations in this book, as convenience dictates.

The Derivative as a Function: The Definition Revisited

In earlier work we formally defined $f'(a)$, the derivative of f at $x = a$. The following definition is essentially the same; using the generic x in place of the specific a emphasizes that f' is a *function*.

> **Definition** Let f be a function. The derivative of f, denoted by f', is the function defined for an input x by
>
> $$f'(x) = \lim_{h \to 0} \frac{f(x+h) - f(x)}{h}.$$

Observe:

Domain The domain of f' is the set of inputs x for which the limit exists. In the best cases (more often than not, for our purposes), the domain of f' is the full domain of f. ▶ For some functions f the domain of f' omits a few (usually obvious) points in the domain of f. ▶

Finding the Limit The limit in the definition can almost never be found simply by plugging in $h = 0$; some trick or algebraic manipulation is almost always needed to eradicate the troublesome denominator.

Beyond Algebra If f is given by a graph, by a table of values, in nonalgebraic form, or even as a complicated algebraic formula, the preceding limit may be hard or impossible to find exactly. All is not lost: In such cases, graphical or numerical estimates are usually possible.

The domain of f' can't possibly be bigger than the domain of f.

For $f(x) = |x|$, the domain of f' is smaller than the domain of f.

Derivatives of Power Functions

Our first order of business is to find derivatives of **power functions**—i.e., functions of the form $f(x) = x^k$, where k is any constant. ▶ Each of the following expressions defines a power function: ▶

The power k need not be an integer.

What's k in each case?

$$x, \quad x^2, \quad \sqrt{x}, \quad \frac{1}{\sqrt{x}}, \quad \frac{x^5}{\sqrt{x^3}}, \quad x^\pi.$$

We start with nonnegative integer powers.

EXAMPLE 1 Find derivative formulas for the linear, quadratic, and cubic power functions $l(x) = x$, $q(x) = x^2$, and $c(x) = x^3$. Interpret results graphically.

Solution The function l is linear, with constant slope 1. Its derivative is therefore the constant function $l'(x) = 1$. Reassuringly, the limit definition says the same thing:

$$l'(x) = \lim_{h \to 0} \frac{l(x+h) - l(x)}{h} = \lim_{h \to 0} \frac{x + h - x}{h} = 1.$$

Higher powers of x require a little more algebraic work. ▶ For q:

Follow each step; watch for a crucial cancellation.

$$q'(x) = \lim_{h \to 0} \frac{q(x+h) - q(x)}{h}$$

$$= \lim_{h \to 0} \frac{(x+h)^2 - x^2}{h}$$

$$= \lim_{h \to 0} \frac{x^2 + 2xh + h^2 - x^2}{h}$$

$$= \lim_{h \to 0} (2x + h)$$

$$= 2x.$$

The cubic function c requires more of the same:

$$c'(x) = \lim_{h \to 0} \frac{c(x+h) - c(x)}{h}$$

$$= \lim_{h \to 0} \frac{(x+h)^3 - x^3}{h}$$

$$= \lim_{h \to 0} \frac{x^3 + 3x^2h + 3xh^2 + h^3 - x^3}{h}$$

$$= \lim_{h \to 0} \left(3x^2 + 3xh + h^2\right)$$

$$= 3x^2.$$

Do they look right? Do the derivative graphs plausibly represent slopes on the original graphs?

In graphical form, our results look like this: ◄

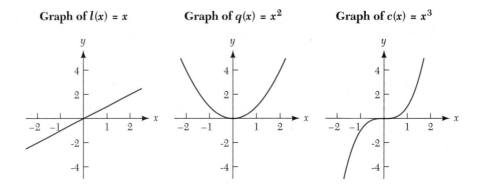

Graph of $l(x) = x$ **Graph of $q(x) = x^2$**   **Graph of $c(x) = x^3$**

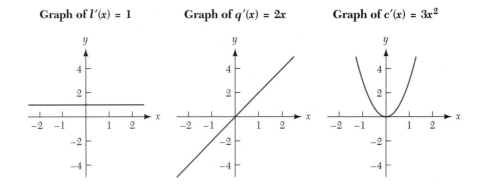

Graph of $l'(x) = 1$ **Graph of $q'(x) = 2x$**   **Graph of $c'(x) = 3x^2$**

The answers aren't surprising; we've drawn such graphs before. The limit calculations assure us, for the first time, that our derivative formulas are exact.   ■

What can be said about power functions x^k for which k is *not* a nonnegative integer?

EXAMPLE 2   Find derivative formulas for the power functions $s(x) = \sqrt{x} = x^{1/2}$ and $n(x) = 1/x = x^{-1}$. What is the domain of each derivative function? Interpret the results graphically.

Solution This example, like the last one, calls for limit computations. Here we'll need slightly more clever algebraic methods to eliminate the troublesome denominator h. For s,

$$s'(x) = \lim_{h \to 0} \frac{s(x+h) - s(x)}{h}$$

$$= \lim_{h \to 0} \frac{\sqrt{x+h} - \sqrt{x}}{h}$$

$$= \lim_{h \to 0} \frac{\sqrt{x+h} - \sqrt{x}}{h} \cdot \frac{\sqrt{x+h} + \sqrt{x}}{\sqrt{x+h} + \sqrt{x}}$$

$$= \lim_{h \to 0} \frac{1}{\sqrt{x+h} + \sqrt{x}}$$

$$= \frac{1}{2\sqrt{x}} = \frac{1}{2} x^{-1/2}.$$

(See the trick on the third line? Multiplying top and bottom by the conjugate expression $\sqrt{x+h} + \sqrt{x}$ helped us simplify the numerator.)
For n,

$$n'(x) = \lim_{h \to 0} \frac{n(x+h) - n(x)}{h}$$

$$= \lim_{h \to 0} \frac{\frac{1}{x+h} - \frac{1}{x}}{h}$$

$$= \lim_{h \to 0} \frac{\frac{1}{x+h} - \frac{1}{x}}{h} \cdot \frac{(x+h)(x)}{(x+h)(x)}$$

$$= \lim_{h \to 0} \frac{-h}{h\, x\, (x+h)}$$

$$= \lim_{h \to 0} \frac{-1}{x(x+h)}$$

$$= \frac{-1}{x^2} = -1x^{-2} = -x^{-2}.$$

Here are all the graphs:

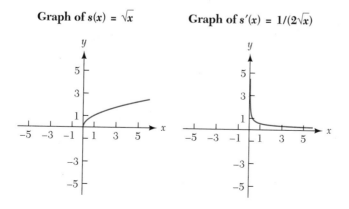

Graph of $s(x) = \sqrt{x}$ **Graph of $s'(x) = 1/(2\sqrt{x})$**

Graph of $n(x) = 1/x$

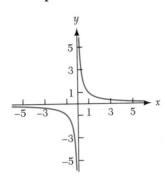

Graph of $n'(x) = -1/x^2$

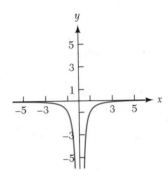

Observe the domains: The derivative s' is defined wherever s is defined—except at $x = 0$, where the graph of s has a vertical tangent line. The derivative n' has the same domain (all nonzero reals) as n itself.   ■

A General Power Rule

Each preceding example is an instance of a more general fact:

> **Theorem 1** (**Power Rule for Derivatives**) Let k be any real constant. If $f(x) = x^k$, then $f'(x) = kx^{k-1}$.

What does the rule say if $k = 0$?

(We've just seen that the rule holds for $k = 1$, $k = 2$, $k = 3$, $k = -1$, and $k = 1/2$.) ◄ We won't prove this theorem rigorously, but at the end of this section we indicate how the **binomial theorem** can be used to show that it holds for all nonnegative integers.

Combinations of Power Functions: The Sum and Constant Rules

Knowing the derivatives of "pure" power functions, it's easy to find derivatives of *combinations* of power functions (such as polynomials), piece by piece.

EXAMPLE 3   Let $p(x) = 3x^5 + 7x^4 - x^2/3 + 11$. Find $p'(x)$.

Solution   We break p apart, differentiate each piece separately, and then reassemble the results. Here are the details, in Leibniz notation:

$$\frac{dy}{dx} = \frac{d}{dx}\left(3x^5 + 7x^4 - \frac{1}{3}x^2 + 11\right)$$

$$= 3\frac{d}{dx}\left(x^5\right) + 7\frac{d}{dx}\left(x^4\right) - \frac{1}{3}\frac{d}{dx}\left(x^2\right) + \frac{d}{dx}(11)$$

$$= 3 \cdot 5x^4 + 7 \cdot 4x^3 - \frac{1}{3} \cdot 2x + 11 \cdot 0$$

$$= 15x^4 + 28x^3 - \frac{2}{3}x.$$

We just made two leaps of faith, having to do with sums and constants, which we'll justify in a moment. First, observe graphically that the result *looks* right:

Graphs of p and p'

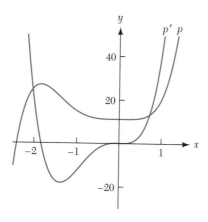

■

The Sum and Constant Rules for Derivatives

We made two reasonable-seeming assumptions in Example 3. Here they are explicitly, stated as theorems:

> **Theorem 2** (**The Sum Rule for Derivatives**) Let f and g be differentiable functions, and let $h = f + g$. Then
> $$h'(x) = (f + g)'(x) = f'(x) + g'(x).$$

> **Theorem 3** (**The Constant Rule for Derivatives**) If k is any constant, f is any differentiable function, and $g(x) = k \cdot f(x)$, then
> $$g'(x) = (k \cdot f)'(x) = k \cdot f'(x).$$

Observe these fine points.

In Words   In words, the theorems say:

> *The derivative of a sum of functions is the sum of their derivatives.*

> *The derivative of a constant times a function is the constant times the derivative of the function.*

Derivatives and Limits   The sum and constant rules for *derivatives* can be proved via the analogous rules for *limits*. The argument for the sum rule runs like this:

$$\frac{d}{dx}\big(f(x) + g(x)\big) = \lim_{h \to 0} \frac{\big(f(x+h) + g(x+h)\big) - \big(f(x) + g(x)\big)}{h}$$

$$= \lim_{h \to 0} \frac{f(x+h) - f(x)}{h} + \lim_{h \to 0} \frac{g(x+h) - g(x)}{h}$$

$$= f'(x) + g'(x).$$

Geometric Thinking   The constant rule can be interpreted geometrically. Recall that the graph of $kf(x)$ is obtained by stretching the graph of f vertically, with factor k. How does such a stretch affect a tangent line to the f-graph at $x = a$?

Every constant rule gives the answer: A vertical k-stretch multiplies slopes of everything—the graph as well as the tangent line—by the same factor k. ◀

Can you interpret the sum rule graphically?

Higher Derivatives and Graphical Analysis

Higher derivatives are found symbolically exactly as first derivatives are—we just repeat the process.

As we saw in Chapter 2, geometric properties of the graph of a function f are closely linked to the first and second derivative functions f' and f''. Having formulas for all these functions makes the job of analyzing such properties much easier.

EXAMPLE 4   Following are graphs of several functions of the form $f(x) = x^3 + ax$; a takes integer values from -3 to 3.

Graphs of $f(x) = x^3 + ax$,
$a = -3, -2, -1, 0, 1, 2, 3$

Which graph is which? Which graphs have stationary points? Which have inflection points? Where, exactly, *are* these points?

Solution Since each function has the form $f(x) = x^3 + ax$, the respective first and second derivative formulas are

$$f'(x) = 3x^2 + a \qquad \text{and} \qquad f''(x) = 6x.$$

These formulas tell us everything we need to know.

Observe first that, in each case, $f'(0) = a$. Thus, the constant a represents a given graph's *slope at* $x = 0$. This makes it easy to see which graph is which. ▶

Label each graph with its a-value.

Stationary points occur where $f'(x) = 0$, i.e., at values of x for which $3x^2 + a = 0$. If $a > 0$, this equation has *no* solutions. The picture agrees: Three of the graphs—those with $a > 0$—have no stationary points. If $a = 0$, the equation $f'(x) = 3x^2 = 0$ has just one solution, at $x = 0$. Similarly, if $a < 0$, then $f'(x) = 3x^2 + a = 0$ has two solutions, at $x = \pm\sqrt{-a/3}$. ▶

Again, the graphs agree.

The simple formula $f''(x) = 6x$ tells all about concavity. Regardless of the value of a, $f''(0) = 0$, $f''(x) < 0$ for $x < 0$, and $f''(x) > 0$ for $x > 0$. Graphically, this means (as the picture shows) that each of the preceding graphs has an inflection point at $x = 0$, where the direction of concavity changes. ■

Antiderivatives

We've just seen how to find the derivative function f' for certain functions f. Now we'll consider the opposite problem, called **antidifferentiation**:

Given a function f, find another function F such that $F' = f$.

Recall that the function F is called an antiderivative of f.

For polynomials and power functions, finding antiderivatives is not much harder than finding derivatives. (Of course, we need to stay clear about which problem we're tackling at any given time!)

EXAMPLE 5 Find antiderivative functions Q and S for $q(x) = x^2$ and $s(x) = \sqrt{x}$. Are your answers unique?

Solution Since $(x^3)' = 3x^2$, $Q(x) = x^3/3$ is a natural guess. It's easy to check this guess using the constant rule for derivatives:

$$Q'(x) = \left(\frac{1}{3}x^3\right)' = \frac{1}{3} \cdot 3x^2 = x^2.$$

Finding an antiderivative of s takes a little more thought. The power rule for derivatives suggests that we try a function of the form $S(x) = kx^{3/2}$—but for what constant k? Differentiating gives the answer: For the desired k,

$$S'(x) = k \cdot \frac{3}{2} \cdot x^{1/2} = x^{1/2}.$$

Therefore, $k = 2/3$; the function $S(x) = 2x^{3/2}/3$ works.

These answers aren't (quite) unique. For example, each of the functions

$$\frac{x^3}{3} + 1, \qquad \frac{x^3}{3} - 5, \qquad \text{and} \qquad \frac{x^3}{3} - \pi^2 + e - \ln\left(15 + \sin^2(3)\right)$$

is an antiderivative of q. So is *any* function of the form $x^3/3 + C$, for any constant C. ■

Antiderivatives Aren't (Quite) Unique

The previous example illustrates an important general principle:

> If F is an antiderivative for f, then so is $F + C$, for any constant C.

Interpret this graphically.

We'll return to this question in the next chapter.

The reason is simple: Any two functions that differ by an additive constant have the same derivative. ◄   The implication runs the other way, too: ◄

> If $F'(x) = G'(x)$ for all x, then $F(x) = G(x) + C$ for some constant C.

For the particular function $q(x) = x^2$, these two facts imply just what the preceding example suggests: (i) For any constant C, $Q(x) = x^3/3 + C$ is an antiderivative; and (ii) *every* antiderivative has this form.

Finding Antiderivative Formulas

The powers x^k need not be integers!

We've seen how to differentiate any polynomial in x: Use Theorem 1 to differentiate each power x^k; then combine results, using the sum and constant rules for derivatives. ◄

We can *antidifferentiate* combinations of powers in much the same way. The power rule for derivatives, read "backwards," gets us started.

T h e o r e m   4 (Power Rule for Antiderivatives)   For any constants C and k $(k \neq -1)$, $\dfrac{x^{k+1}}{k+1} + C$ is an antiderivative of x^k.

Proof.   With derivative rules, the proof is an easy computation:

$$\frac{d}{dx}\left(\frac{x^{k+1}}{k+1} + C\right) = \frac{(k+1)x^k}{k+1} + 0 = x^k,$$

just as we claimed.   ∎

The Exception: $k = -1$.   Notice that the preceding theorem can't possibly hold for $k = -1$, since division by zero isn't allowed. Nevertheless, the exceptional function $f(x) = 1/x$ *has* a well-behaved antiderivative, though not of the usual power form.

F a c t For $x > 0$, the natural logarithm function $\ln(x)$ is an antiderivative of $1/x$.

We'll revisit it soon.

We don't yet have the tools to prove this remarkable fact, ◄   but the following graphs make it plausible.

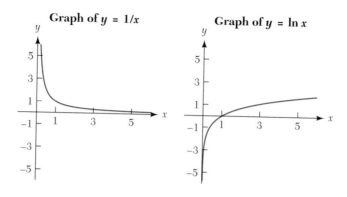

Graph of $y = 1/x$

Graph of $y = \ln x$

Antiderivatives of Combinations

The sum and constant rules for derivatives run in reverse: Sums and constant multiples behave just as well for antiderivatives as for derivatives.

EXAMPLE 6 Earlier we differentiated $p(x) = 3x^5 + 7x^4 - x^2/3 + 11$. Now let's antidifferentiate p.

Solution Again we rend p asunder, antidifferentiate the fragments separately, and reassemble. An antiderivative of $3x^5$ is $x^6/2$; an antiderivative of $7x^4$ is $7x^5/5$; an antiderivative of $x^2/3$ is $x^3/9$; an antiderivative of 11 is $11x$. Thus, for any constant C,

$$\frac{1}{2}x^6 + \frac{7}{5}x^5 - \frac{1}{9}x^3 + 11x + C$$

is an antiderivative of p. The result is easy to check for correctness: Differentiating the preceding expression really does give p. ▶

■ *See for yourself!*

> **Easier One Way than the Other.** The concepts of derivative and antiderivative, like those of parent and child, are nothing more than different views of the same functional relationship. Neither concept is any more (or any less) difficult than the other. Moreover, for power functions and polynomials, finding symbolic derivatives and antiderivatives is about equally easy.
>
> It may be surprising, then, that finding *derivative* formulas for elementary functions is, as a rule, much easier than finding *antiderivative* formulas. We'll polish off the derivatives-of-elementary-functions problem quite quickly; indeed, all the ideas are routine enough to fit comfortably into relatively modest computer software. Finding antiderivative formulas, on the other hand, is a greater challenge for both humans and computers. Indeed, most elementary functions have antiderivatives that aren't elementary, i.e., they can't be written as formulas in the usual sense.

Guess and Check. With the power rule, finding antiderivatives of polynomials and other power functions is easy, and checking answers is easier still. There's a moral:

When finding antiderivatives, guess freely—but check answers carefully.

EXAMPLE 7 Consider the function $f(x) = x^3 - x^2 - 2x$, which follows graphically at left. Four antiderivatives of f, labeled F, G, H, and K, follow at right. Find a formula for each antiderivative.

Graph of $f(x) = x^3 - x^2 - 2x$

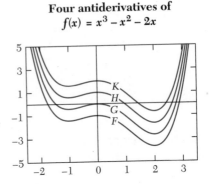

Four antiderivatives of $f(x) = x^3 - x^2 - 2x$

Solution First notice the resemblance among F, G, H, and K. Graphically, all four graphs are vertical translations of each other; symbolically, all four functions differ from each other by a constant. This is as it should be, since $F' = G' = H' = K' = f$. ◄

Functions that differ by a constant have the same derivative.

Now we find formulas. Theorem 4 (plus a bit of guessing and checking) shows that every antiderivative of f has the form

$$\frac{x^4}{4} - \frac{x^3}{3} - x^2 + C,$$

Check this by differentiation.

for some constant C. ◄ In particular, F, G, H, and K all have this symbolic form; they differ only in the value of C. For example, the graph shows that $F(0) = -1$; thus,

$$-1 = F(0) = C \implies F(x) = \frac{x^4}{4} - \frac{x^3}{3} - x^2 - 1.$$

For similar reasons, G, H, and K correspond to $C = 0$, $C = 1$, and $C = 2$, respectively. ∎

The Power Rule and the Binomial Theorem (Optional)

See any college algebra book for a more complete treatment.

The binomial theorem of elementary algebra ◄ is an explicit formula for an expression of the form $(a+b)^n$, where n is any positive integer. The theorem says, in brief,

$$(a + b)^n = a^n + na^{n-1}b + \frac{n(n - 1)}{2}a^{n-2}b^2 + \cdots + nab^{n-1} + b^n.$$

To an algebraist, what the ellipsis ($\ldots$) stands for is the heart of the theorem. For us, only the first few terms matter. Example 8 shows why and how.

EXAMPLE 8 Let $y = x^n$, where n is any positive integer. Show that $dy/dx = nx^{n-1}$.

Solution By definition,

$$\frac{dy}{dx} = \lim_{h \to 0} \frac{(x + h)^n - x^n}{h}.$$

Applying the binomial theorem in the numerator gives

$$\frac{dy}{dx} = \lim_{h \to 0} \frac{\left(x^n + nx^{n-1}h + \frac{n(n-1)}{2}x^{n-2}h^2 + \cdots\right) - x^n}{h}$$

$$= \lim_{h \to 0} \left(nx^{n-1} + h \cdot \text{stuff}\right)$$

$$= nx^{n-1},$$

as claimed. The key idea is that all terms after the first two involve at least the second power of h and so these terms vanish in the limit—even after we divide by the original denominator. ∎

BASICS

For each function f in Exercises 1–10, find an algebraic expression for f', then plot graphs of the expressions for f and f' on the same axes to check your answer for reasonableness.

1. $f(x) = 3x^2$
2. $f(x) = 7x^{-4}$
3. $f(x) = 4x^5 + 3x^2 - x$
4. $f(x) = 4\sqrt{x} + 1/x^3 \times^{-3}$
5. $f(x) = 3x^{-1/2} + 4x^{3/2}$

6. $f(x) = -7x^{1/4}$
7. $f(x) = x^3 + \sqrt[3]{x}$
8. $f(x) = 4\sqrt{x^3} - 5x^{-3/2}$
9. $f(x) = 1/\sqrt[3]{x^2} + 5$
10. $f(x) = 4\sqrt[5]{x^3} + 5/\sqrt[3]{x^4}$

11–20. Find the second derivative of each function in Exercises 1–10. Then sketch graphs of the expressions for f and f'' on the same axes to check your answer for reasonableness.

21–30. Find two antiderivatives of each function in Exercises 1–10.

31. When an oil tank is drained for cleaning, there are $V(t) = 100{,}000 - 4000t + 40t^2$ gallons of oil left in the tank t minutes after the drain valve is opened.
    (a) At what average rate does oil drain from the tank during the first 20 minutes?
    (b) At what rate does oil drain out of the tank 20 minutes after the drain valve is opened?
    (c) Explain what $V''(t)$ says about the rate at which oil is draining from the tank.

32. Let $f(x) = e^x$. Use graphs to explain why $f'(x) \neq xe^{x-1}$.

Graph of f

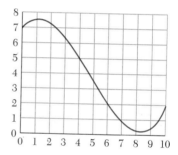

33. Let f be the function shown above, F be an antiderivative of f, and suppose that $F(0) = 10$. (The function f has a local maximum at $x = 1.14$, a point of inflection at $x = 4.67$, and a local minimum at $x = 8.19$.)
    (a) What is the slope of F at $x = 4$? Justify your answer.
    (b) Where is F increasing? Justify your answer.
    (c) Where in the interval $[0, 5]$ does F achieve its largest value? Its smallest value? Justify your answers.

34. Use the limit definition of the derivative function to find f' for each function f below.
    (a) $f(x) = (x + 1)^2$
    (b) $f(x) = 2x^3$
    (c) $f(x) = \sqrt{x + 3}$
    (d) $f(x) = \dfrac{1}{x + 5}$

FURTHER EXERCISES

Find the derivative of each function in Exercises 35–43.

35. $f(x) = (x + 2)^3$
36. $f(x) = 4(x - 1)^5$
37. $f(x) = (x + 4)^{-1}$
38. $f(x) = (x - 2)^{100}$
39. $f(x) = 3(x + 1)^{-2}$

40. $f(x) = \sqrt{x - 1}$
41. $f(x) = \sqrt[3]{x + 7}$
42. $f(x) = 1/\sqrt{x + 4}$
43. $f(x) = 6(x + 4)^{-5/3}$

44–57. Find the second derivative of each of the functions in Exercises 35–43.

58–71. Find an antiderivative of each of the functions in Exercises 35–43.

72. Show that $(1 + x)^r \geq 1 + rx$ when $x \geq 0$ and $r \geq 1$. [**HINT:** Use the racetrack principle.]

73. (a) Over which intervals is $f(x) = (1000 - x)^2 + x^2$ increasing? Decreasing? [**HINT:** $(1000 - x)^2 = (x - 1000)^2$.]

(b) Use part (a) to decide whether 1000^2 is bigger or smaller than $998^2 + 2^2$. Check your answer using a calculator.

(c) Generalize your result from part (a) to the function $f(x) = (c - x)^n + x^n$, where c is any positive number and n is an *even* positive integer. Use this result to decide whether $10,000^{100}$ is bigger or smaller than $9000^{100} + 1000^{100}$.

74. Let f and F be the functions defined in Exercise 33.

(a) At which values of x, if any, does F have points of inflection? Explain.

(b) Where is F concave down? Justify your answer.

(c) Estimate the value of $F''(5)$.

(d) Explain why F'' achieves its minimum value over the interval $[0, 10]$ at $x = 4.67$.

(e) Where in the interval $[0, 10]$ does F'' achieve its largest value? Justify your answer.

75. Let f and F be the functions defined in Exercise 33, and let L be the line tangent to F at $x = 0$.

(a) Find an equation for L.

(b) Does L lie above or below the curve $y = F(x)$ at $x = 1$? Justify your answer.

76. Let F be the function defined in Exercise 33.

(a) Is it possible that $F(5) = 5$? Explain.

(b) Is it possible that $F(5) = 15$? Explain.

(c) Is $F(10) - F(5) < 0$? Explain.

(d) Is $F(10) - F(7) > 6$? Explain.

77. Let f and F be the functions defined in Exercise 33.

(a) Let g be the line through the points $(2, 7)$ and $(6, 2)$. This line lies below f on the interval $[2, 5]$, so $g(x) < f(x)$ on this interval. Use this fact to compute a lower bound on the value of $F(5) - F(3)$. [**HINT:** Use an antiderivative of g.]

(b) Let h be the line through the points $(2, 8)$ and $(6, 3)$. This line lies above f on the interval $[2, 6]$, so $h(x) > f(x)$ on this interval. Use this fact to compute an upper bound on the value of $F(5) - F(3)$.

78. The function F graphed below is an *antiderivative* of a function f. F has a root at $x = 8.8$ and stationary points at $x = 3.2$, $x = 6$, and $x = 13.3$; F has points of inflection at $x = 4.5$ and $x = 10.5$. [**NOTE:** Although the questions that follow ask about f and f', graphs of f and f' are not shown.]

(a) Where is f positive? Justify your answer.

(b) How many roots does f have in the interval $[0, 15]$? Where are they located?

(c) Rank the five values -2, $f(3)$, $f(6)$, $f(9)$, and $f(14)$ in increasing order.

$3, 6, 7, 6, 3,$

(d) Where is f increasing? Justify your answer.

(e) Where does f achieve its maximum value in the interval $[0, 15]$? Justify your answer.

(f) Rank the five values 0, $f'(3)$, $f'(6)$, $f'(10)$, and $f'(13)$ in increasing order.

Graph of F

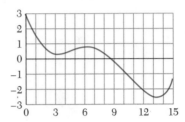

79. Find an equation of the line tangent to the curve $y = x^3 - 6x^2$ at its point of inflection.

80. Suppose that f is a function such that $f(0) = 0$ and f' is the function shown below. Let F be an antiderivative of f. [**NOTE:** The graph is of f', not f or F.]

Graph of f'

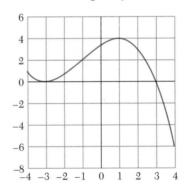

(a) Is F concave up or concave down at $x = -3$? At $x = 0$? At $x = 1$? At $x = 3$? Justify your answers.

(b) Is F increasing, decreasing, or stationary at $x = -3$? At $x = 0$? At $x = 1$? At $x = 3$? Justify your answers.

(c) Where in the interval $[-4, 3]$ does F achieve its minimum value? Justify your answer.

(d) How many stationary points does F have in the interval $[-4, 3]$? Where are they located? Justify your answer.

(e) How many points of inflection does F have in the interval $[-4, 4]$? Where are they located? Justify your answer.

81. (a) How are your answers to Exercise 80 affected if the condition $f(0) = 0$ is changed to $f(0) = 20$?

(b) How are your answers to Exercise 80 affected if the condition $f(0) = 0$ is changed to $f(0) = -15$?

82. Suppose that $y(x)$ is proportional to x^3. Show that $y'(x)$ is proportional to x^2.

83. How many points of inflection does the graph of $y = 2x^6 + 9x^5 + 10x^4 - x + 2$ have? Where are they located?

84. Let $f(x) = 3x^2 - 8x^{-2}$. Evaluate $\displaystyle\lim_{h \to 0} \frac{f(2+h) - f(2)}{h}$.

85. Find a symbolic expression for the nth derivative of each of the following functions.
    (a) $f(x) = 1/x$ (b) $f(x) = \sqrt{x}$

86. Let $p(x) = x^n + a_1 x^{n-1} + a_2 x^{n-2} + \cdots + a_{n-1}x + a_n$.
    (a) Evaluate $\dfrac{d^n}{dx^n} p(x)$.
    (b) Show that if k is an integer such that $1 \le k \le n$, $p^{(k)}(0) = k!a_k$.

87. Suppose that g is an nth degree polynomial. Explain why $\dfrac{d^{n+1}}{dx^{n+1}} g(x) = 0$.

88. Let $f(x) = x^n$ where $n \ge 2$ is an integer.
    (a) Show that f is concave up on the interval $(0, \infty)$.
    (b) What can be said about the concavity of f on the interval $(-\infty, 0)$. [**HINT:** Does it matter whether n is even?]

89. The function $f(x) = \dfrac{(x + 0.00001)^8 - x^8}{0.00001}$ is closely approximated by the function $g(x) = ax^b$ for appropriate values of a and b (i.e., for any x, $|f(x) - g(x)|$ is, in some sense, a "small" number). What are these values? Explain why g approximates f so well.

90. The figure below shows the graph of $f(x) = \sqrt{x}$ and its tangent line at $x = a$. Find the value of a.

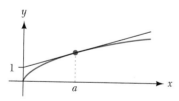

91. Find values of a and b so that the line $2x + 3y = a$ is tangent to the graph of $f(x) = bx^2$ at $x = 3$.

92. Find the value of k for which the graph of $f(x) = 2x^2 + k/x$ has a point of inflection at $x = -1$.

93. Let f be a function defined by $f(x) = Ax^2 + Bx + C$ with the following properties: (i) $f(0) = 2$; (ii) $f'(2) = 10$; (iii) $f''(10) = 4$. Find the value of $A + B + C$.

94. Find the value of b for which the function $y = x^4 + bx^2 + 8x + 1$ has both a horizontal tangent line and a point of inflection at the same value of x.

95. Find c so that the line $y = 4x + 3$ is tangent to the curve $y = x^2 + c$.

96. A fly is crawling from left to right along the edge of a curved barrier on the xy-plane whose shape is the graph

of $y = 7 - x^2$. A spider waits at the point $(4, 0)$. Find the distance between the two creatures when they first see each other. [**HINT:** Draw a picture.]

97. The arithmetic mean of n numbers $0 \le a_1 \le a_2 \le \cdots \le a_n$ is $A_n = (a_1 + a_2 + \cdots + a_n)/n$; their geometric mean is $G_n = (a_1 \cdot a_2 \cdots a_n)^{1/n}$.
    (a) Use algebra to show that $G_2 \le A_2$. [**HINT:** $a_1 \cdot a_2 = \frac{1}{4}\left((a_1 + a_2)^2 - (a_1 - a_2)^2\right)$.]
    (b) Show that $G_2 \le A_2$. [**HINT:** Let $f(a_2) = (a_1 + a_2)/2 - \sqrt{a_1 a_2}$. Use the racetrack principle to show that $f(a_2) \ge 0$.]
    (c) Show that $G_n \le A_n$ when $n \ge 2$. [**HINT:** Show that $f(a_n) = A_n - G_n \ge 0$.]

98. Let m, n be positive integers, and a, b be real numbers.
    (a) Show that $x^n - a^n = (x - a)\left(x^{n-1} + x^{n-2}a + x^{n-3}a^2 + \cdots + xa^{n-2} + a^{n-1}\right)$ by multiplying out the expression on the right.
    (b) Let $f(x) = x^n$. Use part (a) to show that $f'(a) = \displaystyle\lim_{x \to a} \frac{x^n - a^n}{x - a} = na^{n-1}$.
    (c) Let $g(x) = x^{-n}$. Use part (b) to show that $g'(a) = -na^{-n-1}$. [**HINT:** $\left(x^{-n} - a^{-n}\right)/(x - a) = (a^n - x^n)/\left(x^n a^n(x - a)\right)$.]
    (d) Let $h(x) = x^{m/n}$. Use part (a) to show that $h'(a) = \dfrac{m}{n} a^{(m-n)/n}$.
    $$\left[\textbf{HINT:}\ \frac{x^{m/n} - a^{m/n}}{x - a} = \frac{\left(x^{1/n}\right)^m - \left(a^{1/n}\right)^m}{\left(x^{1/n}\right)^n - \left(a^{1/n}\right)^n}.\right]$$
    (e) Let $k(x) = x^{-m/n}$. Show that $k'(a) = -\dfrac{m}{n} a^{-(m+n)/n}$.

99. In the previous exercise you showed that if r is a rational number, then $\dfrac{d}{dx}(x^r) = rx^{r-1}$. In this exercise we extend this result to the case where r is a real number.
    Let r be a real number and p, q be rational numbers such that $0 < p < r < q$.
    (a) Assume that $x > 0$. Explain why
    $$\frac{x^p - 1}{x - 1} < \frac{x^r - 1}{x - 1} < \frac{x^q - 1}{x - 1}.$$
    [**HINT:** Consider the cases $0 < x < 1$ and $x > 1$ separately.]
    (b) Use part (a) to show that $p \le \displaystyle\lim_{x \to 1} \frac{x^r - 1}{x - 1} \le q$.
    (c) Explain why part (b) implies that $\displaystyle\lim_{x \to 1} \frac{x^r - 1}{x - 1} = r$.
    (d) Suppose that c is an arbitrary real number and $a \ne 0$. Show that
    $$\frac{x^c - a^c}{x - a} = \left(\frac{w^c - 1}{w - 1}\right)a^{c-1} = -w^c \left(\frac{w^{-c} - 1}{w - 1}\right)a^{c-1}.$$
    (e) Use parts (c) and (d) to show that $(x^c)' = cx^{c-1}$ where c is any real number.

3.2 Using Derivative and Antiderivative Formulas

Explicit formulas for derivatives and antiderivatives can be immensely useful when applying calculus to real problems. This section illustrates just a few of many such applications.

Designer Graphs

Using derivative formulas, we can fit graph fragments together to produce given shapes. (Doing so is useful, for instance, in computer-assisted drawing applications.)

EXAMPLE 1   The figure below shows a line segment and a piece of a parabola.

Bits of a line segment and a parabola

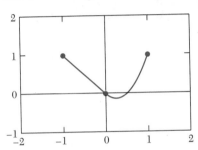

Find an equation for the parabola.

Solution   A parabola's equation has the form $p(x) = a + bx + cx^2$. We need to find numerical values for a, b, and c. The picture shows (at least) three features that we'll use to find the needed values:

i. The parabola passes through $(0, 0)$, so $p(0) = 0$. But $p(0) = a + b \cdot 0 + c \cdot 0^2 = a$, so $a = 0$. Thus, $p(x) = bx + cx^2$, with b and c still to be found.

Look closely to be sure you agree!

ii. At $x = 0$ the parabola and the line segment have the *same slope*. ◀   A look at the segment shows that this slope is -1. Therefore, $p'(0) = -1$. We'll use this information to find b. For any x, $p'(x) = b + 2cx$, so $p'(0) = b$. Therefore $b = -1$, and so $p(x) = -x + cx^2$. We still need to find c.

iii. The parabola passes through $(1, 1)$, so $p(1) = 1$. But $p(x) = -x + cx^2$, so $p(1) = -1 + c$; for $p(1) = 1$ to hold, we must have $c = 2$.

Try plotting both $p(x)$ and $y = -x$. Is the picture "right"?

Thus, $p(x) = -x + 2x^2$. ◀   ∎

Modeling Motion: Acceleration, Velocity, and Position

A falling object has, at any moment, an acceleration a, a velocity v, and a height h. All three quantities are natural functions of time: If t denotes time (measured in seconds after some reference time $t = 0$) and height is measured in feet above some reference point, then

$$h'(t) = v(t) \quad \text{and} \quad v'(t) = a(t).$$

Equivalently, h is an antiderivative of v and v is an antiderivative of a.

According to Newton's laws of motion, an object in free fall, i.e, affected only by gravity, ▶ has a *constant* (vertical) acceleration, about -32 feet per second per second. In symbols: $a(t) = -32$. ▶

Since v is an antiderivative of a,

$$v(t) = -32t + C$$

for some constant C. Which constant? It depends on the setting—because $v(0) = C$, ▶ C represents the **initial velocity** (i.e., the velocity at time 0). Therefore, we write

$$v(t) = -32t + v_0.$$

Antidifferentiating again gives a formula for $h(t)$:

$$h(t) = -16t^2 + v_0\, t + C.$$

Again $h(0) = C$, so we write

$$h(t) = -16t^2 + v_0\, t + h_0,$$

where h_0 represents **initial height**.

EXAMPLE 2   A cannonball is shot upward, with initial height 5 feet and initial velocity 288 feet per second. ▶ Assuming free-fall conditions, how high does the cannonball rise? How long does it remain airborne?

Solution   From the preceding information (with $v_0 = 288$ and $h_0 = 5$), the velocity and height functions have these formulas:

$$v(t) = -32t + 288 \qquad \text{and} \qquad h(t) = -16t^2 + 288t + 5.$$

The ball is highest when $v(t) = 0$; solving this equation gives $t = 9$ seconds. Thus, the cannonball rises to $h(9) = 1301$ feet; ▶ it stays airborne until $h(t) = 0$. Solving *this* equation gives $t \approx 18.02$ seconds. ∎

Maximum–Minimum Problems and the Derivative

Derivative formulas, if they're simple enough, can make locating maxima and minima of f an easy matter. The key idea is that, for a differentiable function f, *local* maxima and minima can occur only at stationary points, i.e., where $f'(x) = 0$. In searching for the largest and smallest values of f on an interval, therefore, the only natural candidates are stationary points and, in some cases, endpoints. To put it more precisely:

> Let f be differentiable on an interval I. Maximum and minimum values of f on I occur, if at all, only at stationary points of f or at endpoints (if any) of I.

EXAMPLE 3   Let $f(x) = x^2 - x$ and $I = [0, 2]$. Where on I does f take maximum and minimum values? ▶

Right margin notes:

See Section 2.1 for more on free fall.

The acceleration is negative because the object falls.

Plug $t = 0$ into the formula for t.

This translates to around 200 mph, which is possible for an old-fashioned cannon.

About 1/4 mile.

Although this problem can be solved without calculus, with calculus it's easier.

Solution   Since $f'(x) = 2x - 1$, there's just one stationary point, at $x = 1/2$. The graph of f is a parabola opening upward, with vertex at $x = 1/2$. Therefore, for x in $[0, 2]$, f takes its minimum value at $x = 1/2$ (the stationary point of f) and its maximum value at $x = 2$ (the right endpoint of I). ■

Real-World Maxima and Minima.   In real-life problems, the first—and often hardest—step is translation into mathematical language.

EXAMPLE 4   A gardener wants to enclose 1000 square feet of land, using as little fencing as possible. The garden is to be rectangular; one side, unfenced, lies along a river. How much fencing is needed? (**NOTE**: This problem and others like it appear in more detail in Appendix D.)

Solution   Here's the setup:

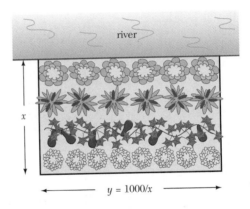

Let x and y denote the lengths shown. (By assumption, $xy = 1000$, so $y = 1000/x$.) If L denotes the total length of fence, then

$$L = 2x + y = 2x + \frac{1000}{x}.$$

We want to minimize L.

Plot your own.

A plot of L ◀ shows a minimum somewhere near $x = 20$. With the derivative we can locate this stationary point precisely:

$$L'(x) = 2 - \frac{1000}{x^2} = 0 \iff x = 10\sqrt{5} \approx 22.3607.$$

For this value of x, $y = 1000/x = 20\sqrt{5}$, and $L = 40\sqrt{5} \approx 89.44$. In agricultural terms: *The best dimensions for a garden, in feet, are $10\sqrt{5} \times 20\sqrt{5}$. The garden needs $40\sqrt{5}$ feet of fence.* (More realistically, the garden is about $22.5' \times 45'$ and needs about 90 feet of fence.) ■

BASICS

Where is each of the following functions increasing? Decreasing? Find *all* local maxima and minima. (Express your answers exactly—e.g., as $\sqrt{3}$—rather than in decimal form.)

1. $f(x) = x^3 - 3x^2 + 5$
2. $f(x) = -8x^9 + 9x^8 - 7$
3. $g(x) = 7x^9 - 18x^7 + 63$
4. $h(x) = 5x^6 + 6x^5 - 15x^4$

In Exercises 5–8, find the maximum and the minimum value of the function over the given interval.

5. $f(x) = x^3 - 3x + 5$; $[0, 2]$
6. $f(x) = 1 + 2x^3 - x^4$; $[-1, 2]$
7. $f(x) = \sqrt{x} - 3/x$; $[0.25, 1]$
8. $f(x) = x^3 - x^2$; $[-2, 0]$
9. Find the maximum and the minimum value of $f(x) = x^4 - 3x^2 + 6$ over each interval below.
   (a) $[-2, 2]$ (c) $[-1, 2]$
   (b) $[-4, -2]$ (d) $[-1, 1]$
10. A particle moves on the x-axis in such a way that its position at time $t > 0$ is given by $x(t) = 3t^5 - 25t^3 + 60t$. For which values of t is the particle moving to the left?
11. A particle moves along a straight line so that its velocity is given by $v(t) = t^2$. How far does the particle travel between $t = 1$ and $t = 3$?

FURTHER EXERCISES

12. Find the maximum and minimum values of the derivative of $f(x) = 6x^2 - x^4$ on the interval $[-2, 2]$.
13. For which value(s) of x does the slope of the line tangent to the curve $y = -x^3 + 3x^2 + 1$ take on its largest value? Justify your answer.
14. Suppose that $q(x)$ is a polynomial of degree $n \geq 2$. Explain why q has at most $n - 1$ local extrema. [**HINT:** $q'(x)$ is a polynomial of degree $n - 1$.]
15. Let $f(x) = ax^3 + bx^2 + cx + d$.
    (a) Show that f has either no local extrema or a local maximum and a local minimum.
    (b) Show that if they exist, the local extrema of f are located at $x = \dfrac{-b \pm \sqrt{b^2 - 3ac}}{3a}$.
16. Is there a real number x such that $4x^3 - x^4 = 30$? Prove your answer.
17. Let p be the parabola defined by $p(x) = ax^2 + bx + c$ where a, b, and c are constants. Show that the vertex of the parabola is at $x = -b/2a$.
18. (a) Find constants a and b so that the polynomial $p(x) = x^3 + ax + b$ has a local minimum at $(2, -9)$.
    (b) Draw a graph of p.
19. (a) Find constants a, b, and c so that the quadratic polynomial $p(x) = ax^2 + bx + c$ is zero at $x = 1$, decreasing when $x < 2$, and increasing when $x > 2$.
    (b) Draw a graph of p.
20. (a) Find a cubic polynomial $q(x) = ax^3 + bx^2 + cx + d$ that has a local maximum at $(0, 2)$ and a local minimum at $(5, 0)$.
    (b) Draw a graph of q.
21. For which values of the constant k does the graph of $y = x^3 - 3x^2 + k$ have just one x-intercept? Two x-intercepts? Three x-intercepts? Justify your answers.
22. For which values of k, if any, does the function $f(x) = (8x + k)/x^2$ have a local minimum at $x = 4$?
23. The altitude $h(t)$ of a rocket from its moment of launch until its fall to Earth is given by the formula $h(t) =$

$-t^4 + 8t^3 + 375$. (Altitude is measured in feet above sea level; time from launch is measured in seconds.)
   (a) From what altitude is the rocket launched?
   (b) When does the rocket fall into the ocean?
   (c) What is the rocket's maximum altitude?
   (d) What is the rocket's maximum upward velocity?
   (e) What is the rocket's maximum downward velocity?
   (f) What is the rocket's maximum upward acceleration?
   (g) What is the rocket's maximum downward acceleration?
24. An object moves along a straight line in such a way that it is slowing down with an acceleration of -8 meters per second per second. The object has an initial velocity of 14 meters per second. Find the distance the object travels during the first 3 seconds.
25. Suppose that the brakes are applied on a car traveling 50 mph and that this gives the car a constant negative acceleration of 20 ft/sec^2. How long will it take the car to come to a stop? How far will the car travel before stopping?
26. A car traveling at a constant speed of 80 mph along a straight highway fails to stop at a stop sign. Three seconds later, a highway patrol car starts from a point of rest at the stop sign and maintains a constant acceleration of 8 ft/sec^2. How long will it take the patrol car to overtake the speeding automobile? How far from the stop sign will this occur? What is the speed of the patrol car when it overtakes the automobile?
27. Suppose that a rectangle has its base on the x-axis and its two upper corners on the curve $y = 2(1 - x^2)$.
   (a) What is the maximum perimeter of such a rectangle?
   (b) Show that there is no minimum perimeter rectangle. [**HINT:** The sides of a rectangle must have nonzero length.]
28. Find the shortest distance from the curve $xy = 4$ to the origin. [**HINT:** Minimize the *square* of the distance function.]

29. What is the area of the largest rectangle whose base is on the x-axis and whose upper corners are on the curve $y = 12 - x^2$?

30. The cost of fuel for a luxury liner is proportional to the square of the speed v. At a speed of 10 km/hour, fuel costs $3000/hour. Other fixed costs (e.g., labor) amount to $12,000/hour. Assuming that the ship is to make a trip of total distance D, find the speed that minimizes the cost of the trip. Does this speed depend on D?

31. An open box is made by cutting squares of side w inches from the four corners of a sheet of cardboard that is 24 inches by 32 inches and then folding up the sides. What should w be to maximize the volume of the box?

32. An open box with a capacity of 36,000 cubic inches is to be twice as long as it is wide. The material for the box costs $0.10 per square foot. What are the dimensions of the least expensive box? How much does it cost?

33. Coffee mugs can be manufactured for 60¢ each. At a price of $1 each, 1000 can be sold. For each penny the price is lowered, 50 more mugs can be sold. What price maximizes profit? How many mugs are sold at this price?

3.3 Derivatives of Exponential and Logarithm Functions

Algebraic versus Transcendental Functions

Using the limit definition to find the derivative of a function with an algebraic formula can be messy, but it's usually routine.

EXAMPLE 1   If $f(x) = \sqrt{x+1}$, find $f'(x)$.

Solution   Here's the limit computation:

$$
\begin{aligned}
f'(x) &= \lim_{h \to 0} \frac{f(x+h) - f(x)}{h} \\
&= \lim_{h \to 0} \frac{\sqrt{x+h+1} - \sqrt{x+1}}{h} \\
&= \lim_{h \to 0} \frac{\sqrt{x+h+1} - \sqrt{x+1}}{h} \cdot \frac{\sqrt{x+h+1} + \sqrt{x+1}}{\sqrt{x+h+1} + \sqrt{x+1}} \\
&= \lim_{h \to 0} \frac{(x+h+1) - (x+1)}{h \cdot (\sqrt{x+h+1} + \sqrt{x+1})} \\
&= \lim_{h \to 0} \frac{1}{\sqrt{x+h+1} + \sqrt{x+1}} \\
&= \frac{1}{2\sqrt{x+1}}.
\end{aligned}
$$

Was this computation really necessary? Could we have used, instead, the (known) derivative of $\sqrt{x}$?

We used an algebraic trick in the third line; otherwise the calculation was straightforward. ◀   ∎

Nonalgebraic functions—those without algebraic formulas—are called **transcendental**. Exponential, logarithm, and trigonometric functions are all transcendental.

Finding derivatives of transcendental functions poses a special challenge. Without algebraic formulas, the limits that define derivatives usually don't succumb to the usual algebraic tricks. We'll need other methods. In another sense, transcendental functions are like all others: Given the graph of any function f, we can graphically estimate the behavior of f'. For each of the following functions, we begin with a graphical approach.

Assuming the Basics. Throughout this section we assume—usually without comment—basic properties of the exponential and logarithm functions. For quick reviews, see Appendices E and F.

Derivatives of Exponential Functions: Graphical Evidence

The Natural Exponential Function

In Example 6, page 112, we conjectured, from graphical evidence, this remarkable property of the natural exponential function: If $f(x) = e^x$, then $f'(x) = e^x$.

Does something similar hold for exponential functions with other bases? Here are graphs of several:

Graphs of three exponentials

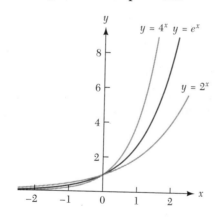

Observe these features.

Similar Slope Behavior Slopes of all three graphs behave similarly. As we move from left to right on any graph, slopes start small (but positive) and increase through larger and larger positive values.

Common Point, Different Slopes The three graphs cross—but have different slopes—at $x = 0$. The e^x-graph has slope 1 at $x = 0$; hence, at $x = 0$ the graphs of $y = 2^x$ and $y = 4^x$ have slopes respectively less than and greater than 1.

Derivatives at $x = 0$: A Closer Look. All graphs of the form $y = b^x$ pass through the point $(0, 1)$—but with different slopes. Zooming in on this point shows how these slopes (i.e., the respective derivatives at $x = 0$) differ from each other:

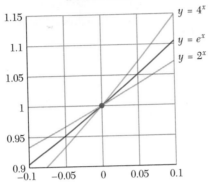

**Graphs of three exponentials:
a closer look**

Notice:

- At this magnification, each exponential curve looks like a straight line—namely, its tangent line at $x = 0$.
- Eyeball estimates of the slopes of these lines ◀ say:

Make these estimates!

$$\frac{d}{dx}2^x\bigg|_{x=0} \approx 0.7; \qquad \frac{d}{dx}e^x\bigg|_{x=0} \approx 1; \qquad \frac{d}{dx}4^x\bigg|_{x=0} \approx 1.4.$$

There are other ways to define e; we'll see one later.

The middle estimate is crucial. The property of having slope 1 at the origin is sometimes used to define the base e. ◀

> **Definition** The number e is the base for which the curve $y = b^x$ passes through $(0, 1)$ with slope 1.

Derivatives of Exponentials: Again, Analytically

With graphical evidence as intuition, we're ready to sharpen our knowledge of derivatives of exponential functions, using the limit definition of derivative. We'll show:

> **Theorem 5** If $f(x) = e^x$, then $f'(x) = e^x$. If $g(x) = b^x$, then $g'(x) = b^x \ln b$.

Before plunging into calculations, let's observe some facts.

- The first formula is a special case of the second. (Set $b = e$.)
- The second formula says that, for any positive base b, $g'(x) = kg(x)$, where $k = \ln b$ is a constant. Thus, every exponential function g has this important property:

  The instantaneous rate of change of g is proportional to the value of g.

  Many real-life varying quantities (populations, undisturbed bank balances, masses of radioactive minerals, comprehensive fees at St. Lena College, etc.) exhibit this property. That's why exponential functions are essential tools in mathematical modeling.
- If $g(x) = 2^x$, then $g'(x) = 2^x \ln 2 \approx 0.693 \cdot 2^x$. If $h(x) = 4^x$, then $h'(x) = 4^x \ln 4 \approx 1.386 \cdot 4^x$. This result agrees with what we saw graphically.

Calculating the Derivative of b^x

Let $g(x) = b^x$. We'll show that $g'(x) = b^x \ln b$.

Step 1: Why $g'(x) = kg(x)$. The computation starts simply:

$$g'(x) = \lim_{h \to 0} \frac{g(x+h) - g(x)}{h}$$

$$= \lim_{h \to 0} \frac{b^{x+h} - b^x}{h} = \lim_{h \to 0} b^x \cdot \frac{b^h - 1}{h}$$

$$= b^x \cdot \lim_{h \to 0} \frac{b^h - 1}{h} = b^x \cdot k.$$

Two remarks are needed:

- Pulling b^x outside the limit on the last line was legal because the limit depends on h, not on x.
- If the last limit—we called it k—exists, then we've shown what we claimed in Step 1: $(b^x)' = kb^x$.

Step 2: Why k Exists. Does the limit k exist? What's k's numerical value?

The key is to interpret the limit k as the slope of the line tangent to the g-graph at $x = 0$. Since $g(x) = b^x$,

$$k = \lim_{h \to 0} \frac{b^h - 1}{h} = \lim_{h \to 0} \frac{b^h - b^0}{h} = \lim_{h \to 0} \frac{g(h) - g(0)}{h}.$$

The last form is familiar by now—it's the *definition* of the slope of the tangent line at $x = 0$. We saw graphically that such tangent lines exist; their slopes depend on b.

Step 3: Why $k = \ln b$. We'll show in a moment that the numerical value of k (i.e., the slope of the line tangent to $y = b^x$ at $x = 0$) is $\ln b$. First, let's see some numerical evidence for the case $b = 10$.

Our claim, numerically, is that

$$\lim_{h \to 0} \frac{b^h - 1}{h} = \ln 10 \approx 2.3026.$$

A machine did the work.   Let's tabulate values of the limit expression for small inputs h: ◀

The slope at $x = 0$ of $y = 10^x$—numerical estimates						
h	-0.01	-0.001	-0.0001	0.0001	0.001	0.01
$(10^h - 1)/h$	2.2763	2.2999	2.3023	2.3029	2.3052	2.3293

The numbers for $b = 10$ look convincing. To prove that $k = \ln b$ for *any* positive base b, we'll reduce the situation to the best understood case, $b = e$. A key relationship holds ◀ between $f(x) = e^x$ and $g(x) = b^x$:

See Appendix E for more details.

$$g(x) = b^x = e^{x \ln b} = f\big((\ln b) \cdot x\big).$$

In graphical language, g is obtained from f by a *horizontal compression*, with factor $\ln b$. ◀ By assumption, the f-graph has slope 1 at $x = 0$. The g-graph, therefore, has slope $\ln b$ at $x = 0$. We're finished.

We studied horizontal squashing and stretching in Section 1.2.

Derivatives of Logarithm Functions

To find the derivative of the logarithm function $f(x) = \log_b x$, we'll use what we know about its inverse $f^{-1}(x) = b^x$. Here, in advance, are our conclusions. Notice again the "naturality" of e:

T h e o r e m 6 For any positive base $b \neq 1$,

$$\big(\log_b x \big)' = \frac{1}{x} \cdot \frac{1}{\ln b}.$$

If $b = e$, then

$$(\ln x)' = \frac{1}{x}.$$

We'll show these facts geometrically; a formal proof would amount only to translating everything into symbolic form.

A Picture "Proof"

The key idea is that, for any base $b \neq 1$, the functions b^x and $\log_b x$ are *inverses*. Geometrically, this relationship means ◀ that either graph can be obtained from the other by reflection through the line $y = x$. The following figure, which shows the case $b = e$, illustrates a crucial point: how such a reflection affects tangent lines.

As we saw, for example, in Section 1.5.

Inverse functions and their derivatives

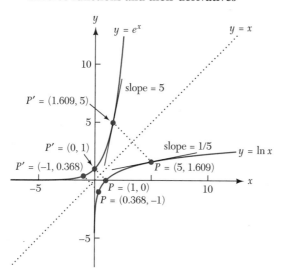

Notice the following features of the graph.

Symmetric Points A point $P = (x, y)$ lies on one graph if and only if the point $P' = (y, x)$ lies on the other. Several such pairs are shown.

Symmetric Tangent Lines Consider the lines tangent to the two graphs at corresponding points P and P'. (One such pair of tangent segments is shown.) Like the graphs themselves, these tangent lines are symmetric with respect to the line $y = x$. It follows ▶ that *the slopes of the tangent lines at P and P' are reciprocals.*

This is the key fact!

Reciprocal Slopes The slope of the tangent line at $P' = (\ln a, a)$ is the derivative of the function $y = e^x$ evaluated at $x = \ln a$. We saw earlier that $(e^x)' = e^x$. Therefore, the tangent line at P' has slope $e^{\ln a} = a$. It follows that the tangent line at $P = (a, \ln a)$ has slope $1/a$, just as the theorem predicts.

So far, we've explained that the theorem holds if $b = e$. For *any* base b, $\log_b x = \ln x / \ln b$. Applying the constant rule for derivatives gives the general derivative formula.

Antiderivatives of Exponential and Logarithm Functions

The derivative rules for exponential functions are readily reversed, so the following theorem closely resembles Theorem 5.

T h e o r e m 7 Let C be any constant and b any positive base. If $f(x) = e^x$, then $F(x) = e^x + C$ is an antiderivative of f. If $g(x) = b^x$, then $G(x) = \dfrac{b^x}{\ln b} + C$ is an antiderivative of g.

Notice:

Its Own Antiderivative, Naturally   The natural exponential function e^x is not only its own derivative but its own antiderivative. ◄

For exactly the same reason!

Other Bases   The second antiderivative formula is best checked by differentiation. Notice how we use the constant rule:

$$\left(\frac{b^x}{\ln b} + C \right)' = \frac{1}{\ln b} \cdot \left(b^x \right)' = \frac{1}{\ln b} \cdot b^x \ln b = b^x.$$

Antiderivatives of logarithm functions—even the natural logarithm—are a little more complicated. For the moment, we'll just *state* a formula.

> **Fact**   Let C be any constant. If $f(x) = \ln x$, then $F(x) = x \ln x - x + C$ is an antiderivative of f.

This claim needs some explanation.

Rabbit from a Hat?   The preceding antiderivative formula seems to have been plucked from thin air. Where did it come from? Why is it right? *Is it right?* Tools to be developed soon will let us show—by differentiation— that the formula is correct. ◄   Showing how one might *produce* such a formula will have to wait a little longer.

Also see the exercises at the end of this section.

Graphical Evidence   Formal proofs must wait, but graphical evidence for the antiderivative formula is available right now. Here's some:

Antiderivatives of $y = \ln x$

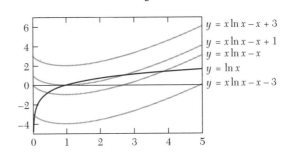

The antiderivatives differ from each other only by a vertical shift; all have the same derivative.

The picture shows graphs of $y = x \ln x - x + C$ for various values of C. ◄ Each such graph looks plausibly like an antiderivative of $y = \ln x$ (bold), as it should.

EXAMPLE 2   Find an antiderivative of $f(x) = 3e^x - 2 \ln x$. How many are there?

We're using the sum and constant rules for derivatives.

Solution   The theorems give antiderivatives of e^x and $\ln x$. Combining them appropriately ◄   tells us that $3e^x - 2 \cdot (x \ln x - x)$ is one suitable antiderivative. So, therefore, is any function of the form

$$F(x) = 3e^x - 2 \cdot (x \ln x - x) + C,$$

where C is any constant.

■

A Sample Problem

Here's a problem with ingredients from this section.

EXAMPLE 3   Let $f(x) = e^x = \exp(x)$. Find a linear function L and a quadratic function Q that closely approximate f near $x = 0$, as follows:

a. Find constants a and b so that the linear function $L(x) = a + bx$ has the same value and first derivative as f at $x = 0$—i.e., so that $L(0) = f(0)$ and $L'(0) = f'(0)$.

b. Find constants a, b, and c for which the quadratic function $Q(x) = a + bx + cx^2$ has the same value and first *two* derivatives as f at $x = 0$—i.e., so that $Q(0) = f(0)$, $Q'(0) = f'(0)$, and $Q''(0) = f''(0)$.

c. Plot f, L, and Q on the same axes. How closely do L and Q approximate f on the interval $[-0.5, 0.5]$?

Solution   Observe for later reference that $f(x) = f'(x) = f''(x)$; in particular, $f(0) = f'(0) = f''(0) = 1$.

a. Let $L(x) = a + bx$. We need values for a and b. Note first that $L(0) = a$. To have $L(0) = f(0) = 1$, we must have $a = 1$. Similarly, $L'(x) = b$. To have $L'(0) = b = f'(0) = 1$, we must have $b = 1$. Thus, $L(x) = 1 + x$.

b. Let $Q(x) = a + bx + cx^2$. We need values for a, b, and c. As in part (a),

$$Q(0) = f(0) = 1 \quad \Longrightarrow \quad a = 1;$$

$$Q'(0) = f(0) = 1 \quad \Longrightarrow \quad b = 1.$$

Thus, $Q(x) = 1 + x + cx^2$, and we need only find c. Since $Q''(x) = 2c$, ▶   *Check this!*
we have

$$Q''(0) = 2c = f''(0) = 1 \Longrightarrow c = 1/2.$$

Thus, $Q(x) = 1 + x + x^2/2$.

c. Here are graphs of all three functions (the exponential function is bold):

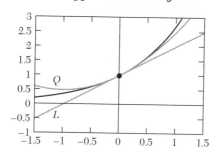

Two approximations to $y = e^x$

Use a calculator to check these numbers.

The picture shows that near $x = 0$, both L and Q approximate f well—Q even better than L. A close look reveals that, for x in the interval $[-0.5, 0.5]$, ◄

$$|f(x) - L(x)| \leq f(0.5) - L(0.5) \approx 0.149;$$

$$|f(x) - Q(x)| \leq f(0.5) - Q(0.5) \approx 0.024. \qquad \blacksquare$$

BASICS

For each function f in Exercises 1–15, find an algebraic expression for f'. Then plot graphs of the expressions for f and f' on the same axes to check your answer for reasonableness.

1. $f(x) = 2e^x + \pi$
2. $f(x) = e^5 - 4e^x$
3. $f(x) = e^x + x^e + e$
4. $f(x) = 2e^{x+1}$
5. $f(x) = 2^x + x^2 + 2$
6. $f(x) = 3^x$
7. $f(x) = 2 \cdot 3^{x-1}$
8. $f(x) = e^\pi + \pi^x + x^\pi$
9. $f(x) = (\ln 5) e^x$
10. $f(x) = (\ln 5)^x$
11. $f(x) = x^{\ln 5}$
12. $f(x) = (\ln 5) 5^x$
13. $f(x) = (\ln 5) x^5$
14. $f(x) = -2 \ln x$
15. $f(x) = 3 \log_2 x$

16–30. Find the second derivative of each function in Exercises 1–15. Then sketch graphs of the expressions for f and f'' on the same axes to check your answer for reasonableness.

31–45. Find an antiderivative of each function in Exercises 1–15.

46. What is the slope of the curve $y = 3^x$ at $x = 0$?

47. What is the slope of the curve $y = \log_3 x$ at $x = 1$?

48. On January 1, 1950, the population of Boomtown was 50,000. Since then the size of the population has been accurately modeled by the function $P(t) = 50,000(0.98)^t$,

where t is the number of years since January 1, 1950.
(a) What was the population of Boomtown on January 1, 1990?
(b) At what rate was the population of Boomtown changing on January 1, 1990? Was it increasing or decreasing then?

49. The position of a particle on the x-axis at time $t > 0$ seconds, is $x(t) = \ln t$ meters.
(a) Find the average velocity of the particle over the interval $1 \leq t \leq e$.
(b) Find the instantaneous velocity of the particle at time $t = e$.

50. Find an equation of the line tangent to the curve $y = e^x$ at $x = 0$.

51. Let $f(x) = \sqrt{x} - \ln x$.
(a) Show that f achieves its (global) minimum value at $x = 4$.
(b) Show that f has an inflection point at $x = 16$.

52. Let $f(x) = \ln x - e^{x-1}$.
(a) Show that f achieves its (global) maximum value when $x = 1$. [**HINT:** $e^{x-1} = e^x/e$.]
(b) Show that f is concave down everywhere on its domain.

FURTHER EXERCISES

53. Let f, L, and Q be as in Example 3.
(a) Find the value of the constant d for which the cubic function $C(x) = Q(x) + dx^3$ has the properties $C(0) = f(0)$, $C'(0) = f'(0)$, $C''(0) = f''(0)$, and $C'''(0) = f'''(0)$.
(b) Plot f, L, Q, and C on the same axes over the interval $[-3, 3]$.
(c) Find an interval over which $|f(x) - L(x)| < 0.1$.
(d) Find an interval over which $|f(x) - Q(x)| < 0.1$.
(e) Find an interval over which $|f(x) - C(x)| < 0.1$.

54. Let $f(x) = \ln x$.
(a) Find values of the constants a and b so that the linear function $L(x) = a + b(x - 1)$ has the properties $L(1) = f(1)$ and $L'(1) = f'(1)$.
(b) Find the value of the constant c for which the quadratic function $Q(x) = L(x) + c(x - 1)^2$ has the properties $Q(1) = f(1)$, $Q'(1) = f'(1)$, and $Q''(1) = f''(1)$.
(c) Find the value of the constant d for which the cubic function $C(x) = Q(x) + d(x - 1)^3$ has the properties

$C(1) = f(1)$, $C'(1) = f'(1)$, $C''(1) = f''(1)$, and $C'''(1) = f'''(1)$.
(d) Graph f, L, Q, and C on the same axes over the interval $[0.1, 3]$.
(e) Find an interval over which $|f(x) - L(x)| < 0.1$.
(f) Find an interval over which $|f(x) - Q(x)| < 0.1$.
(g) Find an interval over which $|f(x) - C(x)| < 0.1$.

55. Which of the following functions are solutions of the differential equation $f''(x) = f(x)$? Which are solutions of $f'''(x) = f'(x)$?

(i) $f(x) = \sqrt{3}$
(ii) $f(x) = x + 2$
(iii) $f(x) = e^x$
(iv) $f(x) = 2^x$
(v) $f(x) = \ln x$

56. Let n be a positive integer.
(a) Find an expression for $\dfrac{d^n}{dx^n} e^x$.
(b) Find an expression for $\dfrac{d^n}{dx^n} \ln x$.

57. Suppose that $b > 0$. Find an antiderivative of $\log_b x$. [**HINT:** Use the Fact on p. 204.]

58. (a) Use the Fact on p. 204 to show that $(x \ln x)' = 1 + \ln x$.
    (b) Suppose that $b > 0$. Evaluate $(x \log_b x)'$.

59. Let $f(x) = \ln(kx)$ where k is a positive constant.
    (a) Show that $f'(x) = x^{-1}$. [**HINT:** $\ln(kx) = \ln k + \ln x$.]
    (b) Find an antiderivative of f.

60. Let $f(x) = b^x$.
    (a) Find all values of $b > 0$ such that $f'(0) > 0$.
    (b) Find all values of $b > 0$ such that $f'(1) > f(1)$.

61. Let $g(x) = \log_b x$.
    (a) Find all values of $b > 0$ such that $g'(1) > 1$.
    (b) Find all values of $b > 0$ such that $g'(1) > g(1)$.

62. Let $f(x) = e^{-x} = \exp(-x)$. Show that $f'(x) = -e^{-x}$. [**HINT:** $e^{-x} = (1/e)^x$.]

63. Find the derivative of $f(x) = \ln(1/x)$. [**HINT:** $\ln(1/x) = -\ln x$.]

64. Let $f(x) = 2e^{x+3}$.
    (a) Explain how a graph of f can be obtained from a graph of the natural exponential function using *only* vertical stretching.
    (b) Use part (a) to find a symbolic expression for $f'(x)$.

65. Let $f(x) = 3e^{x+2}$.
    (a) Explain how a graph of f can be obtained from a graph of the natural exponential function using *only* horizontal translation.
    (b) Use part (a) to find a symbolic expression for $f'(x)$.

66. Find the x-intercept of the line tangent to the curve $y = 2^x$ at $x = 0$. [**HINT:** Draw a picture.]

67. Does there exist a real number a such that the line tangent to the curve $y = e^x$ at $x = a$ passes through the origin? If so, find it. If not, explain why there is no number with this property.

68. Does there exist a real number a such that the line tangent to the curve $y = x - \ln x$ at $x = a$ passes through the origin? If so, find it. If not, explain why there is no number with this property.

69. Give an example of a function f with the property that $f'(x) \neq f(x)$ but $f''(x) = f'(x)$.

70. Let $f(x) = Ae^x$ and $g(x) = e^{x+k}$ where A and k are constants.
    (a) Show (by differentiation) that $f'(x) = f(x)$.
    (b) Show (by differentiation) that $g'(x) = g(x)$.
    (c) Find the value of A that makes $f(0) = 3$.
    (d) Find the value of k that makes $g(0) = 3$.
    (e) Show that if $A > 0$, there is a value of k such that $f(x) = g(x)$ for all x.
    (f) Show that if $A < 0$, there is a value of k such that $f(x) = -g(x)$ for all x.
    (g) Find a function h such that $h'(x) = h(x)$ and $h(0) = -2$.

71. Let $f(x) = e^{kx} = \exp(kx)$ and $g(x) = Af(x)$ where A and k are constants.

    (a) Explain why $f(x) = b^x$ for some positive number b. What is b?
    (b) Use part (a) and Theorem 5 to show that $f'(x) = ke^{kx} = kf(x)$.
    (c) Use part (b) to show that $g'(x) = kg(x)$.
    (d) Find a function h such that $h'(x) = 3h(x)$ and $h(0) = 2$.
    (e) Find a function h such that $h'(x) = -2h(x)$ and $h(0) = 3$.

72. (a) Let A be a constant. Show that $y(x) = 5 + Ae^x$ is a solution of the differential equation $y' = y - 5$.
    (b) Find the solution of the differential equation $y' = y - 5$ with the property $y(0) = 6$.
    (c) Find the solution of the differential equation $y' = y - 5$ that passes through the point $(1, 0)$.

73. Let $f(x) = \ln |x|$, $g(x) = x^{-1}$, and
    $$h(x) = \begin{cases} -2 + \ln x & \text{if } x > 0 \\ 3 + \ln(-x) & \text{if } x < 0. \end{cases}$$
    (a) What is the domain of f? Of g?
    (b) Plot graphs of f, g, and h on the same set of axes.
    (c) Explain why f is an antiderivative of g. [**HINT:** f is an even function.]
    (d) Show that h is an antiderivative of g (i.e., show that $h'(x) = g(x)$ for all $x \neq 0$).

74. Let $f(x) = e^{kx}$ and $g(x) = e^{-kx}$; k is a constant.
    (a) Show that $f''(x) = k^2 f(x)$ and $g''(x) = k^2 g(x)$.
    (b) Let $h(x) = Af(x) + Bg(x)$ where A and B are any real numbers. Show that $h''(x) = k^2 h(x)$.
    (c) Let h be the function defined in part (b). Choose values of A and B so that $h''(x) = 4h(x)$, $h(0) = 2$, and $h'(0) = 8$.

75. Using the result of the previous exercise find a function h such that $h''(x) = 4h(x)$ for all x, $h'(0) = 4$, and $h(0) = 4$.

76. Suppose that A is a constant. Verify that $x(t) = 1 + t + Ae^t$ is a solution of the differential equation $x' = x - t$.

77. Suppose that A is a constant. Verify that $y(x) = Ae^x - x - 1$ is a solution of the differential equation $yy' = y^2 + xy$.

78. Suppose that A and B are constants. Verify that $y(x) = A \ln x + B + x$ is a solution of the differential equation $xy'' + y' = 1$.

79. Suppose that A and B are constants. Verify that $y(x) = Ax^{-1} + B + \ln x$ is a solution of the differential equation $x^2 y'' + 2xy' = 1$.

80. Let $f(x) = x - e \ln x$.
    (a) Prove that $f(\pi) > 0$. [**HINT:** Show that f has a global minimum at $x = e$.]
    (b) Use part (a) to show that $e^\pi > \pi^e$.

81. (a) Show that the distance between the point $(0, 0)$ and a point (a, b) on the curve $y = e^{x/2}$ is least when a is the solution of the equation $2a + e^a = 0$. [**HINT:** Minimize the square of the distance.]
    (b) Estimate the distance in part (a).

82. Let $f(x) = 2^x$ and $g(x) = 7x/3 + 1$. What is the slope of the line tangent to the curve $y = f(x)$ at the point in $[0, 3]$ where the vertical distance between f and g is the greatest?

83. Find the area of the smallest rectangle that has one side on the positive x-axis, one side on the negative y-axis, a corner at the origin, and a corner on the curve $y = \ln x$. [**HINT**: Use Exercise 58.]

84. (a) Find the minimum value of $f(x) = e^x - 1 - x$.
    (b) Explain why part (a) shows that $e^x \geq 1 + x$ for all x.
    (c) Use the racetrack principle to show that $e^x \geq 1 + x$ for all x.   [**HINT**: Consider the cases $x > 0$ and $x < 0$ separately.]
    (d) Show that $e^x \geq 1 + x + x^2/2$ for all $x \geq 0$.

85. (a) Show that $(1 + 1/n)^n \leq e$ for any integer $n \geq 1$. [**HINT**: Let $x = 1/n$ and use part (b) of the previous exercise.]
    (b) Show that $e \leq (1 + 1/n)^{n+1}$ for any integer $n \geq 1$. [**HINT**: Proceed as in part (a) but with $x = -1/(n + 1)$.]
    (c) Let $a_n = (1 + 1/n)^n$, $b_n = (1 + 1/n)^{n+1}$. Show that $\lim_{n \to \infty}(b_n - a_n) = 0$ [**NOTE**: It is a fact that $\lim_{n \to \infty} (1 + 1/n)^n = e$, but we don't yet have the tools necessary to prove this fact.]

86. (a) Show that $e^x > 2x$ for all $x \geq 1$.   [**HINT**: $e^x > 2$ for all $x \geq 1$.]
    (b) Use part (a) to show that $e^x > x^2$ for all $x \geq 1$.

87. Show that $1 - 1/x \leq \ln x \leq x - 1$ when $x > 0$. [**HINT**: Consider the cases $x \geq 1$ and $0 < x \leq 1$ separately.]

88. (a) Use the racetrack principle to show that $0 \leq \ln x \leq \sqrt{x} - 1/\sqrt{x}$ if $x \geq 1$.
    (b) Show that $\sqrt{x} - 1/\sqrt{x} \leq \ln x \leq 0$ if $0 < x \leq 1$.
    (c) Use part (b) to show that $\lim_{x \to 0^+} x \ln x = 0$.

89. (a) Prove that $\lim_{x \to \infty} \ln x/x = 0$.   [**HINT**: Use the previous exercise.]
    (b) Prove that $\lim_{x \to \infty} \ln x/\sqrt{x} = 0$.   [**HINT**: $\ln x/\sqrt{x} = 2\ln\left(\sqrt{x}\right)/\sqrt{x}$.]
    (c) Prove that if r is a positive real number, then $\lim_{x \to \infty} \ln x/x^r = 0$.

90. In this exercise we will prove that $\lim_{x \to \infty} x^r/e^x = 0$ for any real number r.
    (a) Prove that $2\ln x - x < 0$ for all $x > 0$.
    (b) Use part (a) to prove that $\lim_{x \to \infty}(r \ln x - x) = -\infty$ if $r \geq 1$.   [**HINT**: If $r \geq 1$, then $r^2 \geq r$.]
    (c) Use part (b) to prove that $\lim_{x \to \infty} x^r/e^x = 0$ if $r \geq 1$. [**HINT**: $x^r = e^{r \ln x}$.]
    (d) Use part (c) to prove that $\lim_{x \to \infty} x^r/e^x = 0$ if $r < 1$.

91. (a) Assume that $x \geq 1$. Use the inequality $\ln x \leq x - 1$ to show that $x \ln x - x \leq x^2/2 - x - 1/2$. [**HINT**: Use Exercise 58 and the racetrack principle.]
    (b) Prove that $(x - 1)/2 \leq \ln x$ if $1 \leq x \leq 3$. [**HINT**: Show that $f(x) = \ln x - (x - 1)/2$ is decreasing on the interval $[1, 2]$ and increasing on the interval $[2, 3]$.]

(c) Use the inequality in part (b) to find a quadratic polynomial p with the property that $p(x) \leq x \ln x - x$ if $1 \leq x \leq 3$.

92. (a) Use the fact that $1/x \leq 1$ if $x \geq 1$ to show that $\ln x \leq x - 1$ if $x \geq 1$.
    (b) Use the fact that $1/x^2 \geq 1/x$ if $0 < x \leq 1$ to show that $1 - 1/x \leq \ln x$ if $0 < x \leq 1$.

93. (a) Assume that $x \geq 1$. Explain how the relation $\ln x \leq 2(\sqrt{x} - 1)$ can be derived from the relation $1/x \leq 1/\sqrt{x}$.
    (b) Assume that $0 < x \leq 1$. Explain how the relation $\ln x \geq 2(1 - 1/\sqrt{x})$ can be derived from the relation $1/x \leq 1/x^{3/2}$.

94. The derivative of a logarithmic function $f(x) = \log_b x$ is proportional to $1/x$. In fact, the constant of proportionality is $f'(1)$—the slope of the line tangent to the graph of f at $x = 1$. Justify each step in the following derivation of this result.   [**HINT**: In line (iv), $w = h/x$.]

(i) $\dfrac{d}{dx} \log_b x = \lim_{h \to 0} \dfrac{\log_b (x + h) - \log_b x}{h}$

(ii) $= \lim_{h \to 0} \dfrac{\log_b \left(\dfrac{x + h}{x}\right)}{h}$

(iii) $= \lim_{h \to 0} \dfrac{\log_b \left(1 + \dfrac{h}{x}\right)}{h}$

(iv) $= \lim_{w \to 0} \dfrac{\log_b (1 + w)}{xw}$

(v) $= \dfrac{1}{x}\left(\lim_{w \to 0} \dfrac{\log_b (1 + w)}{w}\right)$

(vi) $= \dfrac{f'(1)}{x}$.

95. The equation $a^x = 1 + x$ has at least one solution in the interval $[0, \infty)$ for every $a > 0$. For which values of $a > 0$ does this equation have a nonzero solution? For these values of a, how many nonzero solutions are there?

96. Find all the solutions of the equation $2^x = 2x$. Prove that you have found all the solutions.   [**HINT**: Use the racetrack principle.]

97. The equation $b^x = bx$ has at least one solution in the interval $[0, \infty)$ for every $b > 0$. For which values of $b > 0$ does this equation have more than one solution?

Evaluate the limits in Exercises 98–101 by recognizing each as the definition of a derivative.

98. $\lim_{t \to 0} \dfrac{e^t - 1}{t}$

99. $\lim_{x \to 0} \dfrac{2^{3+x} - 2^3}{x}$

100. $\lim_{s \to 0} \dfrac{\log_5 (25 + s) - 2}{s}$

101. $\lim_{z \to e} \dfrac{\ln z - 1}{z - e}$

102. Evaluate $\lim_{x \to \pi} \dfrac{e^x - e^\pi}{x^e - \pi^e}$.

3.4 Derivatives of Trigonometric Functions

We've already studied the trigonometric functions and some of their derivatives informally, using graphical techniques. In this section we take a more systematic point of view. (To review trigonometric basics, see Appendix G.)

Derivatives of the Sine and Cosine Functions—Graphical Evidence

Graphs suggest almost irresistibly that $(\sin x)' = \cos x$. ▶

We've already succumbed to the temptation, e.g., in Section 2.3.

Graph of $f(x) = \sin x$

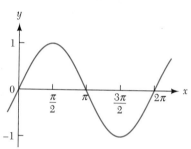

Graph of $f'(x) = \cos x$

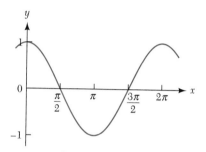

It's equally tempting to guess that $(\cos x)' = -\sin x$, as the following graphs show.

Graph of $f(x) = \cos x$

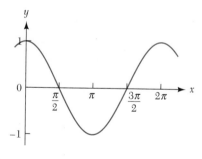

Graph of $f'(x) = -\sin x$

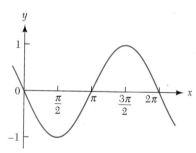

Convincing as they are, graphs aren't proofs. Rigor requires that we deal directly with the limit that defines the derivative.

Derivatives of the Sine and Cosine Functions—Analytic Proofs

Differentiating the Sine Function.   To show rigorously that $(\sin x)' = \cos x$, we need—somehow—to calculate the limit

$$\frac{dy}{dx} = \lim_{h \to 0} \frac{\sin(x + h) - \sin(x)}{h}.$$

It appears in Appendix G.

Algebra isn't enough. We need the addition formula for sines: ◄

$$\sin(x + h) = \sin(x)\cos(h) + \cos(x)\sin(h).$$

Check each step.

Now we can do some algebra: ◄

$$\lim_{h \to 0} \frac{\sin(x + h) - \sin(x)}{h} = \lim_{h \to 0} \frac{(\sin(x)\cos(h) + \cos(x)\sin(h)) - \sin(x)}{h}$$

$$= \lim_{h \to 0} \left((\sin x) \cdot \frac{\cos h - 1}{h} + (\cos x) \cdot \frac{\sin h}{h} \right)$$

$$= (\sin x) \cdot \lim_{h \to 0} \frac{\cos h - 1}{h} + (\cos x) \cdot \lim_{h \to 0} \frac{\sin h}{h}.$$

They depend on x, not h; that's why we could pull them outside the limits.

"Nontrivial" is the mathematician's euphemism for "hard."

(As far as the limit is concerned, $\sin x$ and $\cos x$ are constants. ◄)
To finish the computation we need two nontrivial ◄ facts:

$$\lim_{h \to 0} \frac{\cos h - 1}{h} = 0 \quad \text{and} \quad \lim_{h \to 0} \frac{\sin h}{h} = 1.$$

If we assume these facts for the moment, the conclusion follows easily:

$$\frac{d}{dx}\sin x = (\sin x) \cdot \lim_{h \to 0} \frac{\cos h - 1}{h} + (\cos x) \cdot \lim_{h \to 0} \frac{\sin h}{h}$$

$$= (\sin x) \cdot 0 + (\cos x) \cdot 1$$

$$= \cos x.$$

Two Challenging Limits.   Each of the limits in the preceding calculation can be interpreted as a certain derivative.

- The limit

$$\lim_{h \to 0} \frac{\cos h - 1}{h} = \lim_{h \to 0} \frac{\cos h - \cos 0}{h - 0}$$

defines the derivative of the cosine function at $x = 0$. The cosine graph is horizontal there, so the limit is zero.

- A similar argument applies to

$$\lim_{h \to 0} \frac{\sin h}{h} = \lim_{h \to 0} \frac{\sin h - \sin 0}{h - 0}.$$

The limit *defines* the derivative of the sine function at $x = 0$. In Example 4, page 152, we saw convincing graphical and numerical evidence that this derivative is 1. For a more rigorous symbolic proof, see Appendix H.

Differentiating the Cosine Function.   To show that $(\cos x)' = -\sin x$, we could mimic the argument just given for the sine function. Instead, we'll deduce the derivative from what we know already about the sine function.

As their graphs show, the sine and cosine functions differ by a horizontal translation of $\pi/2$ units. In symbols:

$$\cos x = \sin(x + \pi/2) \quad \text{and} \quad \cos(x + \pi/2) = -\sin x.$$

It follows from these facts ▶ that the cosine graph has the same slope at x as the sine graph has at $x + \pi/2$. Using the preceding identities and the fact—which we just showed—that $(\sin x)' = \cos x$, we can express all this symbolically:

Think about them!

$$\frac{d}{dx}\cos x = \frac{d}{dx}\sin(x + \pi/2) = \cos(x + \pi/2) = -\sin x.$$

Derivatives of Other Trigonometric Functions

Each of the remaining trigonometric functions (tangent, secant, cotangent and cosecant) is an algebraic combination of the sine and cosine functions. Not surprisingly, so is each derivative. With ingenious algebra we could calculate, from scratch, the derivative of each of these functions. It's more efficient, though, to wait ▶ until we develop explicit rules for derivatives of products and quotients. To allay unbearable suspense, here's a graphical peek at what's to come:

The wait will be short. In the meantime, see the exercises at the end of this section.

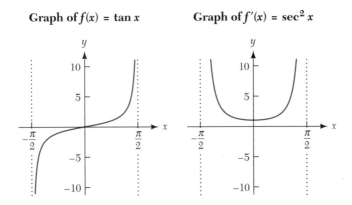

Graph of $f(x) = \tan x$ **Graph of $f'(x) = \sec^2 x$**

As it should, the second graph represents the slope behavior of the first. It amounts to circumstantial evidence—but only that—for the fact that $(\tan x)' = \sec^2 x$.

Antiderivatives of Sines and Cosines

With just a little care for signs, ▶ our derivative formulas can be reversed to give antiderivatives. Here are the details. They're easy to check by differentiation.

That's signs, not sines.

> **Theorem 8** Let C be any constant. If $f(x) = \sin x$, then $F(x) = -\cos x + C$ is an antiderivative of f. If $g(x) = \cos x$, then $G(x) = \sin x + C$ is an antiderivative of g.

EXAMPLE 1   Find an antiderivative of $f(x) = 3\sin x - 4\cos x$. How many antiderivatives are there?

We're using the sum and constant rules.

S o l u t i o n   Theorem 8 gives antiderivatives of $\sin x$ and $\cos x$. Combining them appropriately ◄   tells us that $-3\cos x - 4\sin x$ is one suitable antiderivative. So, therefore, is *any* function of the form

$$F(x) = -3\cos x - 4\sin x + C,$$

Plot both f and F. Does the picture show what it should?

where C is any constant. ◄   ■

A Sample Problem

Here's a problem with ingredients developed in this section.

E X A M P L E   2   Here's a graph of $f(x) = x + \cos x$. Two interesting points are bulleted.

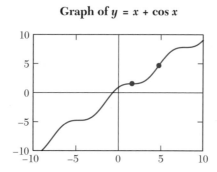

Graph of $y = x + \cos x$

Are there any local maxima or minima? Are there any inflection points?

S o l u t i o n   The graph appears to have flat spots, such as at the left bullet, but are they local maxima, local minima, or just flat spots? The picture alone doesn't tell us for sure. ◄

Would zooming help? If a plotting device is at hand, see for yourself.

With derivatives we can settle the question easily. Since $f(x) = x + \cos x$, $f'(x) = 1 - \sin x$ and $f''(x) = -\cos x$. Thus,

$$f'(x) = 0 \iff \sin x = 1 \iff x = \frac{\pi}{2} + 2k\pi;$$

$$f''(x) = 0 \iff \cos x = 0 \iff x = \frac{\pi}{2} + k\pi.$$

In particular, the first bulleted point is both a stationary point and an inflection point. In addition, f has stationary points at intervals of 2π and inflection points at intervals of π. (The right bullet is another inflection point.)

Notice, too, that $f'(x) = 1 - \sin x \geq 0$ for all x. Thus, the first derivative never changes sign, and the stationary points are nothing but that—they are neither local maxima nor local minima.   ■

Further Transcendental Meditations (Optional)

Hyperbolic Functions. Exponential, logarithmic, and trigonometric functions are not the only useful transcendental functions. Others include the **hyperbolic sine** and **hyperbolic cosine** functions, denoted by **sinh** and **cosh**, respectively, and defined by

$$\sinh(x) = \frac{e^x - e^{-x}}{2} \quad \text{and} \quad \cosh(x) = \frac{e^x + e^{-x}}{2}.$$

Hyperbolic functions arise naturally in various engineering applications. Under appropriate conditions, for instance, the cosh function models the shape of a hanging cable. ▶

Plot cosh *on your calculator to see what this means.*

As the definitions show, sinh and cosh are near relatives of the natural exponential function. Less obvious but equally important are links between hyperbolic and trigonometric functions. For example, sinh and cosh—just like sin and cos—can be combined to form such other hyperbolic functions as tanh and sech. More details on hyperbolic functions appear in the exercises at the end of this section.

Recognizing Transcendental Functions. Proving rigorously that a function is transcendental can be difficult. Consider the sine function, for example. We defined the sine function using the unit circle, not any algebraic formula. Nevertheless, the sine function might conceivably have some algebraic formula.

In fact, the sine function has *no* algebraic formula, and neither does any other trigonometric, logarithm, or exponential function. The following observations don't *prove* that the sine function is transcendental, but they give some idea of what's involved.

- Could $\sin x = p(x)$ for some polynomial $p(x)$? No. A polynomial of degree n has at most n roots; the sine function has infinitely many roots.

- Could the sine function be a rational function? That is, could $\sin x = p(x)/q(x)$ for some polynomials p and q? No again, for the same reason. A rational function has only finitely many roots.

- The sine function is 2π-periodic. Of the raw algebraic functions—powers, roots, and so on—none is periodic. Creating a periodic function from nonperiodic components would be difficult.

BASICS

For each function f in Exercises 1–4, find an algebraic expression for f', then plot graphs of the expressions for f and f' on the same axes to check your answer for reasonableness.

1. $f(x) = 2\sin x$
2. $f(x) = 3\cos x$
3. $f(x) = 3\sin x - 2\cos x$
4. $f(x) = 2\cos x - 3\sin x$

5–8. Find the second derivative of each function in Exercises 1–4. Then sketch graphs of the expressions for f and f'' on the same axes to check your answer for reasonableness.

9–12. Find an antiderivative of each function in Exercises 1–4.

13. Let $f(x) = x - \cos x$.
    (a) Does f have any stationary points in the interval $[0, 2\pi]$? If so, where?
    (b) Where in the interval $[0, 2\pi]$ is f increasing?
    (c) Find the maximum and minimum values of f over the interval $[0, \pi]$.
    (d) Where is f concave down?
    (e) Where is f increasing most rapidly? Evaluate f' there.

14. Repeat the previous exercise with $f(x) = x + \cos x$.

In Exercises 15–20, draw a graph of the function over the interval $[0, 2\pi]$. Label local extrema, global extrema, and points of inflection with both coordinates. (Use exact values rather than decimal approximations for the coordinates.)

$\frac{1}{4} - \cos^{u} 4x$

15. $f(x) = \dfrac{1}{2}x + \sin x$

16. $g(x) = x + \sin x$

17. $h(x) = \dfrac{3}{2}x + \sin x$

18. $f(x) = \sin x + \cos x$

19. $g(x) = 2\sin x - \sqrt{3}x$

20. $h(x) = \sqrt{3}\sin x + 3\cos x$

FURTHER EXERCISES

21. Let $f(x) = \sin x$.
    (a) Find values of the constants a and b so that the linear function $L(x) = a + bx$ has the properties $L(0) = f(0)$ and $L'(0) = f'(0)$.
    (b) Find the value of the constant c for which the quadratic function $Q(x) = L(x) + cx^2$ has the properties $Q(0) = f(0)$, $Q'(0) = f'(0)$, and $Q''(0) = f''(0)$.
    (c) Find the value of the constant d for which the cubic function $C(x) = Q(x) + dx^3$ has the properties $C(0) = f(0)$, $C'(0) = f'(0)$, $C''(0) = f''(0)$, and $C'''(0) = f'''(0)$.
    (d) Plot f, L, Q, and C on the same axes over the interval $[-4, 4]$.
    (e) Find an interval over which $|f(x) - L(x)| < 0.1$.
    (f) Find an interval over which $|f(x) - Q(x)| < 0.1$.
    (g) Find an interval over which $|f(x) - C(x)| < 0.1$.
22. Repeat the previous problem with $f(x) = \cos x$.
23. The range R of a projectile whose speed is v meters per second and whose angle of elevation is x radians is given by $R = \left(v^2/g\right)\sin(2x)$ where $g \approx 9.8$ meters per second per second is the acceleration due to gravity.
    (a) What angle of elevation gives the projectile maximum range? Justify your answer.
    (b) What is the maximum range of the projectile?
24. Let k be a real number. Find a trigonometric identity for $\cos(x + k)$ by differentiating both sides of the identity $\sin(x + k) = \sin x \cos k + \cos x \sin k$.
25. Let $f(x) = \cos x + \frac{1}{2}\cos(2x)$.
    (a) Determine where f has local maxima and minima over the interval $[0, 5]$.
    (b) Estimate where f has inflection points in the interval $[0, 5]$.
26. Repeat the previous problem using the function $f(x) = \sin x - \frac{1}{2}\cos(2x)$.
27. Find a function that satisfies each set of conditions below.
    (a) $f'(z) = \cos(2z)$, $f(\pi/2) = 1$
    (b) $f'(x) = 2x^3 + \sin(4x)$, $f(0) = 1$
    (c) $g'(w) = \sin(w + 2)$, $g(0) = 3$
28. Let $f(x) = \cos(kx)$ and $g(x) = \sin(kx)$; k is a constant.
    (a) Show that $f''(x) = -k^2 f(x)$ and $g''(x) = -k^2 g(x)$.
    (b) Let $h(x) = Af(x) + Bg(x)$, where A and B are any real numbers. Show that $h''(x) = -k^2 h(x)$.
    (c) Let h be the function defined in part (b). Choose values of A and B so that $h''(x) = -4h(x)$, $h(0) = 2$, and $h'(0) = 8$.

29. Let $f(x) = A\cos x + B\sin x$ where A and B are constants.
    (a) Show that $f''(x) = -f(x)$.
    (b) Show that $f^{(4)}(x) = f(x)$.
    (c) Find a function f such that $f''(x) = -f(x)$, $f'(\pi/3) = -1$, and $f(\pi/3) = \sqrt{3}$.
30. Let $f(x) = \sin(x^2)$. Show that $f'(x) \neq \cos(x^2)$ in a way that your classmates would find convincing. [**HINT:** Compare graphs.]
31. Let $f(x) = 2 - \sin x$ and $g(x) = 1 + \cos x$.
    (a) Find the maximum vertical distance between the curves $y = f(x)$ and $y = g(x)$ over the interval $[0, 2\pi]$.
    (b) Find the slopes of the lines tangent to f and g at the point in the interval $[0, 2\pi]$ where the vertical distance between f and g is greatest.
32. Let $f(x) = \cos x$ and $g(x) = \sqrt{3}\sin x$.
    (a) Find the maximum vertical distance between the curves $y = f(x)$ and $y = g(x)$ over the interval $[0, \pi]$.
    (b) Find the slopes of the lines tangent to f and g at the point in the interval $[0, \pi]$ where the the vertical distance between f and g is greatest.
33. Suppose that two sides of a rectangle are on the coordinate axes and that a corner of the rectangle lies on the curve $y = 2\cos x$, $0 \le x \le \pi/2$. Is it possible that the perimeter of the rectangle is $3\pi/2$? If so, find the coordinates of the corners of a rectangle with these properties. If not, find the coordinates of the corners of the rectangle with the greatest possible perimeter.
34. Suppose that one side of a rectangle is on the x-axis and that two corners of the rectangle lie on the curve $y = 2\sqrt{2}\sin x$, $0 \le x \le \pi$. Among all such rectangles, find the perimeter of the one with maximum perimeter.
35. Suppose that two sides of a rectangle are on the coordinate axes and that a corner of the rectangle lies on the curve $y = k\cos x$, where $\pi/3 \le k \le \pi/2$.
    (a) Show that k can be chosen so that the rectangle with maximum perimeter is a square.
    (b) Estimate the value of k in part (a).
36. (a) What is the 50th derivative of $f(x) = \cos x$?
    (b) What is the 51st derivative of $g(x) = \sin x$?
    (c) What is the 52nd derivative of $g(x) = \sin x$?

37. Show that $\dfrac{d^n}{dx^n} \sin x = \sin(x + n\pi/2)$ when n is an integer such that $n \geq 1$. [**HINTS:** $\sin(x + \pi/2) = \cos x$; $\sin(x + \pi) = -\sin x$.]

38. Suppose that the line tangent to the curve $y = \sin x$ at $x = z$ passes through the origin. Show that $\tan z = z$.

39. (a) Use the fact that $(\tan x)' = \sec^2 x$ to show that $x = 0$ is the only solution of the equation $x = \tan x$ in the interval $(-\pi/2, \pi/2)$.
    (b) Suppose that $k \geq 1$ is an integer. Explain why the equation $x = \tan x$ has exactly one solution between $k\pi$ and $(2k+1)\pi/2$.

40. Show that $\cos x + \sin x \leq 1 + x$ for all $x \geq 0$.

41. Show that $2 + \sin x - \cos x \leq e^x$ if $x \geq 0$.
    [**HINT:** Use the previous exercise and the fact that $1 + x \leq e^x$ for all $x \geq 0$.]

42. Let $f(x) = \sin x$.
    (a) Find an equation of the line tangent to f at $x = 0$.
    (b) Find an equation of the line tangent to f at $x = \pi/3$.
    (c) Which of the tangent lines in parts (a) and (b) yields a better approximation of $\sin(\pi/6)$? How could you have used f'' to predict which tangent line would yield the better approximation?

43. Let $f(x) = 2x/\pi$.
    (a) Prove that $f(x) \leq \sin x$ when $0 \leq x \leq \pi/2$.
    (b) Show that the function $g(x) = \sin x - f(x)$ achieves its maximum value over the interval $[0, \pi/2]$ at the point where the slope of the line tangent to the sine function is $2/\pi$.

44. (a) Use the inequality $\cos x \leq 1$ to prove that $\sin x \leq x$ for all $x \geq 0$.
    (b) Use part (a) to show that $1 - x^2/2 \leq \cos x$ for all $x \geq 0$.
    (c) Explain why the inequality in part (b) is true for all x. [**HINT:** $\cos x$ is an even function.]

45. (a) Use the result in part (b) of the previous exercise to prove that $x - x^3/6 \leq \sin x$ if $x \geq 0$.
    (b) Use part (a) to prove that $\cos x \leq 1 - x^2/2 + x^4/24$.
    (c) Prove that $\sin x \leq x - x^3/6 + x^5/120$ if $x \geq 0$.

Evaluate the limits in Exercises 46–49 by recognizing each as the definition of a derivative.

46. $\displaystyle\lim_{x \to \pi/2} \dfrac{\cos x}{x - \pi/2}$

47. $\displaystyle\lim_{x \to \pi} \dfrac{\sin x}{x - \pi}$

48. $\displaystyle\lim_{h \to 0} \dfrac{\sin(3\pi/2 + h) + 1}{h}$

49. $\displaystyle\lim_{h \to 0} \dfrac{2\cos(5\pi/4 + h) + \sqrt{2}}{h}$

50. Evaluate $\displaystyle\lim_{h \to 0} \dfrac{\cos(x - h) - \cos(x + h)}{h}$.

51. Let $f(x) = \sin(x + k)$ where k is a constant.
    (a) Use the limit definition of the derivative to show that $f'(x) = \cos(x + k)$.
    (b) Interpret the result in part (a) graphically.
    [**HINT:** How is the graph of f related to the graph of the sine function?]

(c) Let $g(x) = \cos x$. Use the result in part (a) to show that $g'(x) = -\sin x$. [**HINT:** $\cos x = \sin(x + \pi/2)$.]

(d) Let $g(x) = \cos(x + k)$. Use the result in part (a) to show that $g'(x) = -\sin(x+k)$. [**HINT:** $\cos(x+k) = \sin(x + (k + \pi/2))$.]

(e) Let $h(x) = 3\sin(x - 2)$. Use the results in parts (a) and (d) to find $h'(x)$, $h''(x)$, and an antiderivative of h.

52. (a) Let $f(x) = \sin(kx)$ where k is a constant. Provide a justification for each step in the following proof that $f'(x) = k\cos(kx)$. [**HINT:** In line (iii), $w = kh$.]

(i) $\dfrac{d}{dx} \sin(kx) = \displaystyle\lim_{h \to 0} \dfrac{\sin(k(x + h)) - \sin(kx)}{h}$

(ii) $\quad = \displaystyle\lim_{h \to 0} \dfrac{\sin(kx + kh) - \sin(kx)}{h}$

(iii) $\quad = \displaystyle\lim_{w \to 0} k \cdot \dfrac{\sin(kx + w) - \sin(kx)}{w}$

(iv) $\quad = \displaystyle\lim_{w \to 0} k \cdot \dfrac{\sin(kx + w) - \sin(kx)}{w}$

(v) $\quad = k \cdot \left(\displaystyle\lim_{w \to 0} \dfrac{\sin(kx + w) - \sin(kx)}{w} \right)$

(vi) $\quad = k\cos(kx)$.

(b) Use the result in part (a) to find the derivative of $g(x) = 3\sin(2x)$.

53. Let $f(x) = \cos(kx)$ where k is a constant.
    (a) Use the limit definition of the derivative to prove that $f'(x) = -k\sin(kx)$.
    (b) Use the result in part (a) to find the derivative of $g(x) = 2\cos(3x)$.

54. Let $f(x) = \tan x$. Justify each step in the following proof that $f'(x) = \sec^2 x$.

(i) $\dfrac{d}{dx} \tan x = \displaystyle\lim_{h \to 0} \dfrac{\tan(x + h) - \tan x}{h}$

(ii) $\quad = \displaystyle\lim_{h \to 0} \dfrac{\dfrac{\sin(x + h)}{\cos(x + h)} - \dfrac{\sin x}{\cos x}}{h}$

(iii) $\quad = \displaystyle\lim_{h \to 0} \dfrac{\cos x \sin(x + h) - \sin x \cos(x + h)}{h(\cos x \cos(x + h))}$

(iv) $\quad = \displaystyle\lim_{h \to 0} \dfrac{\cos(-x)\sin(x + h) + \sin(-x)\cos(x + h)}{h(\cos x \cos(x + h))}$

(v) $\quad = \displaystyle\lim_{h \to 0} \dfrac{\sin h}{h(\cos x \cos(x + h))}$

(vi) $\quad = \displaystyle\lim_{h \to 0} \left(\dfrac{\sin h}{h} \cdot \dfrac{1}{\cos x \cos(x + h)} \right)$

(vii) $\quad = \dfrac{1}{\cos^2 x}$

(viii) $\quad = \sec^2 x$.

55. Let $f(x) = \cot x$. Use the limit definition of the derivative to show that $f'(x) = -\csc^2 x$.

56. Let $f(x) = \sec x$. Use the limit definition of the derivative to show that $f'(x) = \sec x \tan x$.

57. Let $f(x) = \csc x$. Use the limit definition of the derivative to show that $f'(x) = -\csc x \cot x$.

58. (a) Show (algebraically) that the hyperbolic cosine (cosh) is an even function.
    (b) Is the hyperbolic sine function (sinh) even, odd, or neither? Justify your answer.

59. (a) Show that $(\sinh x)' = \cosh x$.
    (b) Show that $(\cosh x)' = \sinh x$.

60. (a) Plot the hyperbola defined by the equation $x^2 - y^2 = 1$.
    (b) Show that $\cosh^2 x - \sinh^2 x = 1$.

61. Show that $\sinh(x + y) = \sinh x \cosh y + \cosh x \sinh y$.

62. Show that $\cosh(x + y) = \cosh x \cosh y + \sinh x \sinh y$.

63. Show that $(\cosh x + \sinh x)^k = \cosh(kx) + \sinh(kx)$ for any number k.

64. Suppose that $\sinh x = 4/5$. Evaluate $\cosh x$.

65. (a) Show that $y = \sinh x$ satisfies the differential equation $y'' = y$, and the initial conditions $y(0) = 1$ and $y'(0) = 1$.
    (b) Show that $y = \cosh x$ satisfies the differential equation $y'' = y$, and the initial conditions $y(0) = 1$ and $y'(0) = 0$.

66. The hyperbolic tangent function *tanh* is defined by the equation $\tanh x = \sinh x / \cosh x$.
    (a) Plot $\tanh x$.
    (b) Suppose that $\tanh x = 3/4$. Evaluate $\cosh x$ and $\sinh x$.

3.5 New Derivatives from Old: The Product and Quotient Rules

The theme of producing new functions from old runs throughout calculus. In Section 1.6 we discussed various operations—algebraic operations, composition, inversion, and so on—that build new functions from old. In this section and the next, we'll see how *derivatives* of these new functions are related to derivatives of the old functions.

The derivative of a built-up function f reflects (not surprisingly) both the *materials* from which f is built and the *operations* used to combine them. We've already seen this principle in action in simple cases.

EXAMPLE 1   Let $f(x) = x^2 + 3 \sin x$. What's $f'(x)$? Why?

Solution   According to the sum and constant rules, ◀

We discussed them in Section 3.1.

$$f'(x) = (x^2 + 3 \cdot \sin x)' = (x^2)' + 3 \cdot (\sin x)' = 2x + 3\cos x.$$

If we let $g(x) = x^2$ and $h(x) = \sin x$, $f(x) = g(x) + 3 \cdot h(x)$, so the same result looks like this:

$$f'(x) = (g(x) + 3 \cdot h(x))' = g'(x) + 3 \cdot h'(x).$$

In this (especially simple) case, f' is built from g' and h' *exactly* as f is built from g and h. ∎

Not every situation is so simple. Derivatives of products and quotients of functions need more care than sums and constant multiples. Sadly, the "naïve" rules for products and quotients—$(fg)' = f'g'$ and $(f/g)' = f'/g'$—are false.

EXAMPLE 2 Show by examples that the naïve product and quotient rules are false.

Solution Almost any functions f and g will do. If, say, $f(x) = x$ and $g(x) = 1$, then $f(x)g(x) = x$ and $f(x)/g(x) = x$. But

$$\big(f(x) \cdot g(x) \big)' = \big(x \big)' = 1 \neq f'(x) \cdot g'(x) = 1 \cdot 0 = 0.$$

The "false quotient rule" also produces nonsense:

$$\left(\frac{f(x)}{g(x)} \right)' = (x)' = 1 \neq \frac{f'(x)}{g'(x)} = \frac{1}{0}. \qquad \blacksquare$$

The (True) Product and Quotient Rules

Following, stated as theorems, are the *true* product and quotient rules. Before considering *why* they hold, we'll illustrate how to use them.

In each theorem, u and v are differentiable functions. ▶

For some reason, it's traditional in calculus to use u and v (not f and g) in this context. We won't buck tradition.

> **Theorem 9** (**The Product Rule**) If $p(x) = u(x) \cdot v(x)$, then
>
> $$p'(x) = \big(u(x) \cdot v(x) \big)' = u'(x) \cdot v(x) + u(x) \cdot v'(x).$$
>
> Equivalently, $(uv)' = u'v + uv'$.

> **Theorem 10** (**The Quotient Rule**) If $q(x) = u(x)/v(x)$, then
>
> $$q'(x) = \left(\frac{u(x)}{v(x)} \right)' = \frac{v(x)u'(x) - u(x)v'(x)}{v(x)^2}.$$
>
> Equivalently, $(u/v)' = \dfrac{vu' - uv'}{v^2}$.

EXAMPLE 3 (**Antiderivatives of ln x**) In Section 3.3 we claimed, but didn't prove, ▶ that functions of the form $F(x) = x \ln x - x + C$ are antiderivatives of the natural logarithm function. Prove it now.

We gave graphical evidence. See page 204.

Solution With the product rule, nothing could be easier. Watch:

$$(x \ln x - x + C)' = 1 \cdot \ln x + x \frac{1}{x} - 1 = \ln x,$$

just as claimed. $\blacksquare$

EXAMPLE 4 Let $p(x) = x^2 \sin x$ and $q(x) = x^2 / \sin x$. Find p' and q'.

Solution In the preceding notation, $u(x) = x^2$ and $v(x) = \sin x$. Thus, $u' = 2x$ and $v' = \cos x$. The rest is "pattern-matching":

$$p'(x) = (uv)' = u'v + uv' = 2x \sin x + x^2 \cos x;$$

$$q'(x) = (u/v)' = \frac{vu' - uv'}{v^2} = \frac{(\sin x)2x - x^2 \cos x}{(\sin x)^2}.$$

(Some algebraic simplification of answers is possible, but since our main point concerns the derivatives, we'll leave these answers as they are.) ∎

EXAMPLE 5 The exponential function is its own derivative. What's the derivative of $r(x) = \exp(-x)$.

Solution Recall some basic exponential algebra:

$$r(x) = \exp(-x) = e^{-x} = \frac{1}{e^x}.$$

We leave some algebraic details to you.

Now use the quotient rule: ◀

$$r'(x) = \left(\frac{1}{\exp(x)} \right)' = \frac{-e^x}{(e^x)^2} = -\frac{1}{e^x} = -e^{-x} = -\exp(-x).$$

∎

EXAMPLE 6 Let f be a differentiable function; let $g = 1/f$. Find a formula for g' in terms of f and f'.

Solution We *could* apply the quotient rule to $1/f$, but we'll take another tack. First notice that $1 = g(x) f(x)$ for all sensible inputs x. Differentiating both sides of this equation ◀ gives

Use the product rule on the right.

$$0 = g'f + gf' = g'f + \frac{f'}{f}.$$

Solving for g' gives the answer:

$$\left(\frac{1}{f} \right)' = g' = -\frac{f'}{f^2}.$$

∎

Derivatives of the "Other" Trigonometric Functions

In Section 3.4 we showed carefully that $(\sin x)' = \cos x$ and $(\cos x)' = -\sin x$. The quotient rule makes easy work of finding derivatives of the remaining trigonometric functions. We'll do one such computation, leaving the rest as exercises.

EXAMPLE 7 Find and simplify the derivative of $\tan x$.

Solution Use the quotient rule:

$$(\tan x)' = \left(\frac{\sin x}{\cos x} \right)' = \frac{\cos^2 x + \sin^2 x}{\cos^2 x} = \frac{1}{\cos^2 x}.$$

To avoid fractions, the answer is usually written as $\sec^2 x$.

The full story of trigonometric derivatives fits snugly in a table:

Derivatives of Trigonometric Functions					
Function	$\sin x$	$\cos x$	$\tan x$	$\cot x$	$\sec x$	$\csc x$
Derivative	$\cos x$	$-\sin x$	$\sec^2 x$	$-\csc^2 x$	$\sec x \tan x$	$-\csc x \cot x$

Evidence for the Product Rule

A formal proof of the product rule ▶ involves difference quotients. Here we'll give two less formal plausibility arguments.

One appears in Appendix H.

Exhibit A: It Works for Linear Functions. Suppose that $f(x) = ax + b$ and $g(x) = cx + d$. We'll differentiate $f \cdot g$ twice—with and without the product rule.
First, notice that

$$f(x) \cdot g(x) = (ax + b) \cdot (cx + d) = acx^2 + (ad + bc)x + bd.$$

The last expression is a polynomial in x. Differentiating it directly gives

$$(fg)' = 2acx + (ad + bc).$$

Happily, the product rule agrees:

$$(fg)' = f'g + fg' = a \cdot (cx + d) + (ax + b) \cdot c = 2acx + ad + bc.$$

These computations show that the product rule gives the right answer for *linear* functions f and g. Most functions, of course, aren't linear. However, every differentiable function is "almost linear" in the sense that its graph looks straight if viewed at sufficiently small scale. ▶

A rigorous proof of the product rule can be constructed from this idea.

Exhibit B: It Works in Practice. Both U.S. population and U.S. per-capita income vary with time; we'll denote them by $p(t)$ and $i(t)$ respectively. The product function $w(t) = p(t)i(t)$ describes the *total annual U.S. income*—another function of t. Let's ask:

How fast was total annual income growing at $t = 1990$?

In 1990, the population of the United States was about 250 million and was increasing at the rate of about 3 million per year. Per-capita income was about $15,000 and was growing at about $1000 per year. In symbols:

$$p(1990) = 250 \text{ million}; \qquad p'(1990) = 3 \text{ million};$$
$$i(1990) = 15,000; \qquad i'(1990) = 1000.$$

To answer the question, observe first that growth in w comes from two sources: rising population and rising per-capita income. Over a year, rising population produces about 3 million new people, each earning about $15,000, or a total of about

$$p'(1990) \cdot i(1990) = 3 \text{ million} \cdot 15,000 = 45 \text{ billion}$$

dollars of new income. Similarly, rising personal income accounts for about

$$p(1990) \cdot i'(1990) = 250 \text{ million} \cdot 1000 = 250 \text{ billion}$$

dollars of new income. In all, therefore, income rises at the rate of

$$p'(1990) \cdot i(1990) + p(1990) \cdot i'(1990) = 45 \text{ billion} + 250 \text{ billion}$$
$$= 295 \text{ billion}$$

dollars per year—just as the product rule predicts.

Evidence for the Quotient Rule

Quotients are similar to products; in the same sense, the quotient rule follows directly from the product rule.

EXAMPLE 8 Deduce the quotient rule from the product rule. (Assume that the derivative of the quotient exists.) ◄

It does exist; showing that it does requires a proof similar to the one for the product rule in Appendix H.

The function q is defined wherever $v(x) \neq 0$.

Use the product rule.

Solution Let u and v be differentiable functions, and let $q(x) = u(x)/v(x)$ be the quotient function. ◄ We want $q'(x)$.

By definition, $qv = u$. Differentiating both sides ◄ of this equation gives

$$u' = (qv)' = q'v + qv' = q'v + \frac{u}{v}v' = u'.$$

Solving this last equation for q' gives the quotient rule:

$$q' = \frac{vu' - uv'}{v^2}$$

∎

Mixed Examples

Functions built from *several* products and quotients need several applications of the rules.

EXAMPLE 9 Let u, v, and w be differentiable functions. Find a formula for $(uvw)'$.

Solution Used carefully, the product rule is enough:

$$
\begin{aligned}
(uvw)' &= ((uv) \cdot w)' && \text{[by grouping]} \\
&= (uv)' \cdot w + (uv) \cdot w' && \text{[product rule]} \\
&= (u'v + uv') \cdot w + uvw' && \text{[product rule]} \\
&= u'vw + uv'w + uvw'.
\end{aligned}
$$

Notice the pleasing symmetry of the result: Each term involves one derivative. ∎

EXAMPLE 10 Differentiate $x e^x \sin x$ and $x/(e^x \sin x)$.

Solution By the rule just derived,

$$\left(x e^x \sin x \right)' = 1 \cdot e^x \cdot \sin x + x \cdot e^x \cdot \sin x + x \cdot e^x \cdot \cos x.$$

The second expression requires the quotient and product rules:

$$
\begin{aligned}
\left(\frac{x}{e^x \sin x} \right)' &= \frac{(e^x \sin x)(x)' - x(e^x \sin x)'}{(e^x \sin x)^2} \\
&= \frac{(e^x \sin x) - x(e^x \sin x + e^x \cos x)}{(e^x \sin x)^2} && \text{[product rule]} \\
&= \frac{(\sin x)(1 - x) - x \cos x}{e^x \sin^2 x} && \text{[simplified]}.
\end{aligned}
$$

∎

Sometimes other ingredients, such as constants, are involved.

EXAMPLE 11 Differentiate $e^{2x}\sin(3x)$.

Solution In the preceding section we saw how to handle constants:

$$\left(e^{2x}\right)' = 2e^{2x}; \qquad (\sin(3x))' = 3\cos(3x).$$

Now, by the product rule,

$$\left(e^{2x}\sin(3x)\right)' = 2e^{2x}\sin(3x) + 3\cos(3x)e^{2x}. \qquad \blacksquare$$

Guessing and Checking Antiderivatives: A Sample Problem

Finding derivatives is relatively easy. With techniques from this section, we can handle a great many functions. Finding *antiderivatives* is often harder. Making a guess—and then checking it—is often an effective technique.

EXAMPLE 12 Find antiderivatives for

$$f(x) = x^2\cos x + 2x\sin x \quad \text{and} \quad g(x) = \frac{x\cos x - \sin x}{x^2}.$$

Solution The basic forms of f and g suggest the results of applying the product and quotient rules, respectively. With this in mind, it's natural to guess

$$F(x) = x^2\sin x + C \quad \text{and} \quad G(x) = \frac{\sin x}{x} + C$$

as possible antiderivatives. ▶ With the product and quotient rules, it's easy to check our guesses. Just as we hoped,

In each, C stands for any constant.

$$F'(x) = (x^2\sin x)' = x^2\cos x + 2x\sin x;$$

$$G'(x) = \left(\frac{\sin x}{x}\right)' = \frac{x\cos x - \sin x}{x^2}. \qquad \blacksquare$$

BASICS

In Exercises 1–4 find the derivative of the function in two ways: (1) using the product rule, and (2) by multiplying out before differentiating.

1. $h(x) = x^3(x^2 - 4)$
2. $h(x) = (x + 4)^2$
3. $h(x) = e^x \cdot e^{2x}$
4. $h(x) = \tan x \cdot \csc x$

In Exercises 5–8, find the derivative of the function in two ways: (1) using the quotient rule, and (2) algebraically simplifying, then differentiating.

5. $h(x) = \dfrac{x^2}{x^3}$
6. $h(x) = \dfrac{1 - x^2}{1 + x}$

7. $h(x) = \dfrac{1}{e^x}$

8. $h(x) = \dfrac{\cot x}{\csc x}$

9. Let $h(x) = f(x) \cdot g(x)$, and $j(x) = f(x)/g(x)$. Fill in the missing entries in the table below using the information about f and g given in the following table and the definitions of h and j.

x	$f(x)$	$f'(x)$	$g(x)$	$g'(x)$	$h'(x)$	$j'(x)$
-2	1	-1	-3	4		$-1/9$
-1	0	-2	1	1	-2	
0	-1	2	-2	1		
1	2	-2	-1	2	6	
2	3	-1	2	-2		1

In Exercises 10–25, find an algebraic expression for f', then plot graphs of the expressions for f and f' on the same axes to check your answer for reasonableness.

10. $f(x) = x \sin x$

11. $f(x) = x^2 \cos x$

12. $f(x) = \sqrt{x} \ln x$

13. $f(x) = x^3 e^x$

14. $f(x) = \sin x \cos x$

15. $f(x) = (x^2 + 3)e^x$

16. $f(x) = x^{3/2} \sin x$

17. $f(x) = \dfrac{\sin x}{x^2}$

18. $f(x) = \dfrac{\cos x}{x^2 + e^{2x}}$

19. $f(x) = \dfrac{(x+1)^2}{x^2+1}$

20. $f(x) = \dfrac{\sin x}{\cos(2x)}$

21. $f(x) = \cot(3x)$

22. $f(x) = \sec(x+3)$

23. $f(x) = 3\csc(2x)$

24. $f(x) = 4\tan(x+5)$

25. $f(x) = x^3 e^{2x} \ln x$

In Exercises 26–37, find an antiderivative of the function.

26. $g(x) = \sec^2 x$

27. $g(x) = 3\sec x \tan x$

28. $g(x) = \sec^2(x+1)$

29. $g(x) = \csc^2(2x)$

30. $g(x) = 2x \cos x - x^2 \sin x$

31. $g(x) = (x+1)e^x$

32. $g(x) = e^{2x} \sin x - 2e^{2x} \cos x$

33. $g(x) = \dfrac{xe^x - e^x}{x^2}$

34. $g(x) = 2xe^x + x^2 e^x$

35. $g(x) = \dfrac{\sin x}{x} + \ln x \cos x$

36. $g(x) = e^x \sin x + e^x \cos x$

37. $g(x) = e^x \sin x - e^x \cos x$

38. Suppose that f and g are differentiable functions and that $h(x) = f(x)g(x)$. Find symbolic expressions involving only numerical constants and derivatives of f and g for each of the following.

(a) $h''(x)$ (b) $h'''(x)$

FURTHER EXERCISES

39. Let $f(x) = xe^x$.

(a) Find values of the constants a and b so that the linear function $L(x) = ax + b$ has the properties $L(0) = f(0)$ and $L'(0) = f'(0)$.

(b) Find the value of the constant c for which the quadratic function $Q(x) = L(x) + cx^2$ has the properties $Q(0) = f(0)$, $Q'(0) = f'(0)$, and $Q''(0) = f''(0)$.

(c) Plot f, L, and Q on the same axes over the interval $[-3, 3]$.

(d) Find an interval over which $|f(x) - L(x)| < 0.1$.

(e) Find an interval over which $|f(x) - Q(x)| < 0.1$.

40. Let $f(x) = \dfrac{x}{3 + x^2}$.

(a) Find values of the constants a and b so that the linear function $L(x) = a(x - 1) + b$ has the properties $L(1) = f(1)$ and $L'(1) = f'(1)$.

(b) Find the value of the constant c for which the quadratic function $Q(x) = L(x) + c(x - 1)^2$ has the properties $Q(1) = f(1)$, $Q'(1) = f'(1)$, and $Q''(1) = f''(1)$.

(c) Graph f, L, and Q on the same axes over the interval $[-3, 3]$.

(d) Find an interval over which $|f(x) - L(x)| < 0.1$.

(e) Find an interval over which $|f(x) - Q(x)| < 0.1$.

41. Suppose that f is the function shown below.

Graph of f

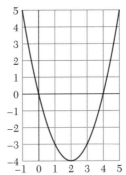

(a) Let $h(x) = x^2 f(x)$. Evaluate $h'(2)$ and $h'(4)$.

(b) Let $m(x) = f(x)/(x^2 + 1)$. Evaluate $m'(0)$.

42. Suppose that $f(1) = 2$ and f' is the function shown below. Let $m(x) = x^3 f(x)$.

Graph of f'

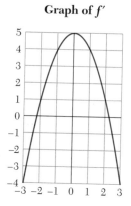

(a) Evaluate $m'(1)$.
(b) Show that m is increasing at 2.
(c) Estimate $m''(1)$.

43. Suppose that $f(0) = 3$ and f' is the function shown below. Let $g(x) = e^x f(x)$.

Graph of f'

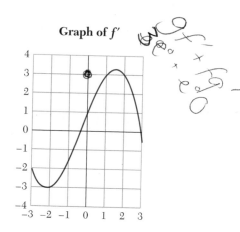

(a) Evaluate $g'(0)$.
(b) Is g increasing at $x = 1$? At $x = 2$? At $x = -2$? Justify your answers.
(c) Estimate $g''(0)$.
(d) Is g concave up at $x = 1$? At $x = 2$? Justify your answers.

44. Let f be the function described in the previous problem, and let $g(x) = \sin(x) f(x)$.
(a) Evaluate $g'(0)$.
(b) Is g increasing at $x = \pi/2$? At $x = \pi/2$? Justify your answers.
(c) Estimate $g''(0)$.
(d) Is g concave up at $x = \pi/2$? Justify your answer.

45. Let f be the function described in the previous problem, and let $g(x) = f(x) \cdot f(x) = f(x)^2$.
(a) Evaluate $g'(0)$.
(b) Is g increasing at $x = 1$? At $x = -1$? Justify your answers.
(c) Estimate $g''(0)$.
(d) Is g concave down at $x = 1$? Justify your answer.

46. A line is drawn that is tangent to the curve $y = x^2$ at a point in the first quadrant. This tangent line, along with the curve $y = x^2$ and the x-axis form a region whose area is 9. Find the coordinates of the point of tangency.

47. Let $f(x) = xe^x$. Find values of the parameters A and B so that $F(x) = (Ax + B)e^x$ is an antiderivative of f.

48. Let $g(x) = x^2 e^x$. Find values of the parameters A, B, and C so that $G(x) = (Ax^2 + Bx + C)e^x$ is an antiderivative of g.

49. Let $h(x) = e^{2x} \sin x$. Find values of the parameters A and B, so that $H(x) = e^{2x}(A \cos x + B \sin x)$ is an antiderivative of h.

50. Consider the function $f(x) = \dfrac{\ln x}{x}$ on the interval $(1, \infty)$.
(a) Where does f achieve its maximum value? What is this value?
(b) Find all inflection points in the graph of f.

51. Let $f(x) = \dfrac{x + \sin x}{\cos x}$.
(a) Find $f'(x)$.
(b) Find an equation of the line tangent to the graph of f at $x = 0$.

52. Let $f(x) = e^x \cos x$.
(a) At which value of x in the interval $[0, \pi]$ does f achieve its maximum value? Its minimum value?
(b) At which value(s) of x in the interval $[0, \pi]$ does f have an inflection point?

53. Find the maximum and minimum values of $g(x) = e^{-x} \sin x$ over the interval $[0, \infty)$.

54. Let $f(x) = 2x + x \cos x + \sin x$.
(a) Show that f has a local minimum at $x = (2k + 1)\pi$, where $k = 0, 1, 2, \ldots$.
(b) Show that f has a local maximum at $x = -(2k + 1)\pi$, where $k = 0, 1, 2, \ldots$.

55. Suppose that $N(t)$, the number of bacteria in a culture at time t, is proportional to $25 + te^{-t/20}$.
(a) At what time during the interval $0 \le t \le 100$ is the number of bacteria the smallest?
(b) At what time during the interval $0 \le t \le 100$ is the number of bacteria the greatest?
(c) At what time during the interval $0 \le t \le 100$ is the rate of change of the number of bacteria a minimum?

56. What is the area of the largest triangle that can be formed in the first quadrant by the x-axis, the y-axis, and a line tangent to the graph of $y = e^{-x}$?

57. (a) Show that $x + x^{-2} \geq 2$ if $0 < x \leq 1$.
    [**HINT:** Use the racetrack principle.]
    (b) Show that $\sin x + \tan x \geq 2x$ if $0 \leq x < \pi/2$.
    [**HINT:** Use part (a) to show that $\cos x + \sec^2 x \geq 2$ if $0 \leq x < \pi/2$.]

58. Show that $\dfrac{x}{1+x} \leq \ln(1+x) \leq x$ for all real numbers $x \geq 0$. [**HINT:** Use the racetrack principle.]

59. Suppose that f is a differentiable function, $g(x) = \big(f(x)\big)^2$, and $h(x) = \big(f(x)\big)^3$.
    (a) Use the product rule to show that $\big(g(x)\big)' = 2f(x)f'(x)$. [**HINT:** $\big(f(x)\big)^2 = f(x) \cdot f(x)$.]
    (b) Use part (a) and the product rule to find an expression for $\big(h(x)\big)'$.

60. Suppose that f is a differentiable function and $g(x) = \big(f(x)\big)^2$.
    (a) Show that every stationary point of f is a stationary point of g.
    (b) If x_0 is a local maximum point of f, is x_0 also a local maximum point of g? Explain.

61. (a) Show that $F(x) = x(\ln x - 1)$ is an antiderivative of $f(x) = \ln x$.
    (b) Use the result in part (a) to find an antiderivative of $g(x) = \log_b x$.

62. Let $f(x) = \dfrac{\ln x}{k} - \dfrac{kx}{x+1}$ where k is a constant.
    (a) Show that if $0 < k \leq 2$, f has no local extrema.
    (b) Show that f has local extrema if $k > 2$. Determine how many and where they are located.

63. Evaluate the following limits by recognizing each as the definition of a derivative.
    (a) $\displaystyle\lim_{x \to \pi} \dfrac{x \cos x + \pi}{x - \pi}$
    (b) $\displaystyle\lim_{t \to 0} \dfrac{1}{t} \cdot \left(\dfrac{t^2 - 3t + 4}{5t + 2} - 2 \right)$

64. Suppose that f, g, and h are differentiable functions and that $h(x) = f(x)g(x)$. Show that if $h'(1) \neq 0$, then $\big(f(1)\big)^2 + \big(g(1)\big)^2 > 0$.

65. Suppose that the functions f and g have domain $\mathbb{R}$ and the following properties:
    (i) $f(x + y) = f(x)f(y)$
    (ii) $g(x + y) = g(x)g(y)$
    (iii) $f(0) \neq 0$
    (iv) $g(0) \neq 0$
    (v) $f'(0) = 1$
    (vi) $g'(0) = 1$
    (a) Show that $f(0) = 1$. [**HINT:** Let $x = y = 0$ in (i) and use part (a).]
    (b) Show that $f(x) \neq 0$ for all x. [**HINT:** Let $y = -x$ in (i).]
    (c) Show that $f'(x) = f(x)$ for all x. [**HINT:** Use the definition of the derivative.]
    (d) Let $k(x) = f(x)/g(x)$. Show that k is defined for all real inputs and find an expression for $k'(x)$.
    (e) Use part (d) to show that $f(x) = g(x) = e^x$.

66. (An alternative proof of the product rule.) Suppose that f and g are differentiable functions and that $h(x) = f(x)g(x)$.
    (a) Use the definition of the derivative to show that $\left(\big(f(x)\big)^2\right)' = 2f(x)f'(x)$. [**HINT:** Use the algebraic identity $a^2 - b^2 = (a - b)(a + b)$.]
    (b) Show that
    $$h(x) = \frac{1}{2}\left(\big(f(x) + g(x)\big)^2 - \big(f(x)\big)^2 - \big(g(x)\big)^2 \right).$$
    (c) Use parts (a) and (b) to show that $h'(x) = f(x)g'(x) + f'(x)g(x)$.

3.6 New Derivatives from Old: The Chain Rule

In the last section we differentiated new functions constructed by *algebraic* operations on simpler functions. In this section we continue the same theme with respect to the operation of *composition*.

Derivatives of Compositions

With the sum, constant multiple, product, and quotient rules we can differentiate any *algebraic* combination of elementary functions—even monsters such as

$$\frac{x^2 \sin x}{\tan x} - \frac{\sin x \ln x + \cos x}{x^5 + 1} \cdot \left(2^x + \frac{3^x}{\ln x} \right).$$

The Question

All that remains before we can differentiate *all* elementary functions is to handle functions such as

$$\sin\left(x^2\right), \quad (\sin x)^2, \quad \text{and} \quad \sin^3\left(x^2\right)$$

constructed by *composing* simpler functions.

The simpler functions involved here are ▶

$$f(u) = \sin u, \qquad g(u) = u^2, \qquad \text{and} \qquad h(u) = u^3.$$

The "neutral" variable name u avoids conflict with x, which was used differently earlier.

Finding f', g', and h' is (by now) no problem; the real question is how to combine everything to find derivatives of composite functions. In simplest form, our question is:

How do we combine f, g, f', and g' to find $(f \circ g)'$?

The Answer

The **chain rule** ▶ is the missing link. In stating it we assume, as usual, that f and g are "nice enough" (i.e., that f, g, f', and g' have large enough domains to permit the desired operations). ▶

The name is apt: Composed functions are "chained" together in series.

More fine print after the theorem.

Theorem 11 (**The Chain Rule**) Let f and g be well-behaved functions. Then

$$(f \circ g)'(a) = f'(g(a)) \cdot g'(a).$$

This important theorem deserves several closer looks:

The Derivative Exists The theorem guarantees (among other things) that the derivative $(f \circ g)'(a)$ exists, provided that both $f'(g(a))$ and $g'(a)$ exist.

Why a Product? The theorem says that the derivative of the composition $f \circ g$ is a certain product of the derivatives f' and g'. Why is a *product* appropriate? Interpreting derivatives as rates of change gives some clue. Suppose, say, that $y = f(u)$, $u = g(x)$, $f'(u) = 3$, and $g'(x) = 2$. This means, in rate language, that y changes three times as fast as u, and u changes twice as fast as x. It sounds reasonable, then, that y should change six times as fast as x. (Additional evidence for the chain rule appears later in this section.)

Which Product? The chain rule involves a certain product of derivatives of f and g. But take care: The factors $f'(g(a))$ and $g'(a)$ show that f' and g' are evaluated for *different inputs*. ◄

In other words, the naïve chain rule $(f \circ g)' = f' \circ g'$ is false!

Other Notations In the theorem, we used a specific input, $x = a$, to emphasize the point just made. Since the theorem holds for all sensible inputs a, we can also think of it as a relation among functions. From this point of view, we might prefer to state the chain rule as

$$(f \circ g)'(x) = f'(g(x)) \cdot g'(x)$$

or, equivalently, as

$$(f \circ g)' = (f' \circ g) \cdot g'.$$

A fully rigorous proof of the chain rule runs quickly into delicate technical problems. We'll settle for plausible evidence. First, however, let's look at some examples.

EXAMPLE 1 Differentiate $\sin(x^2)$ and $\sin^2 x$. Are the answers the same? Should they be?

Solution No and no. The trick is to keep clear exactly what's meant, in context, by f, g, f', and g'. If we let $f(u) = \sin u$ and $g(u) = u^2$, then

$$\sin(x^2) = (f \circ g)(x) \qquad \text{and} \qquad \sin^2 x = (g \circ f)(x).$$

From the chain rule and the derivatives of f and g:

$$\left(\sin(x^2)\right)' = (f \circ g)'(x) = f'(g(x)) \cdot g'(x) = \cos\left(x^2\right) \cdot 2x.$$

Similarly,

$$\left(\sin^2 x\right)' = (g \circ f)'(x) = g'(f(x)) \cdot f'(x) = 2\sin x \cdot \cos x.$$

The two answers are different—as they should be. After all, composition isn't commutative. ∎

EXAMPLE 2 Differentiate $\sin(3x)$ and $\sin(x + 3)$, using the chain rule.

Solution Let $f(u) = \sin u$, $g(x) = 3x$, and $h(x) = x + 3$. Then

$$\left(\sin(3x)\right)' = (f \circ g)'(x) = f'(g(x)) \cdot g'(x) = \cos(3x) \cdot 3.$$

Similarly,

$$\left(\sin(x + 3)\right)' = (f \circ h)'(x) = f'(h(x)) \cdot h'(x) = \cos(x + 3) \cdot 1. \qquad ∎$$

EXAMPLE 3 Differentiate $\sin^3(x^2)$.

Solution We need to keep our wits about us here. Let $f(u) = \sin u$, $g(u) = u^2$, and $h(u) = u^3$; then

$$(h \circ f \circ g)(x) = \left(\sin(x^2)\right)^3 = \sin^3(x^2).$$

To handle the "three-deep" composition $h \circ f \circ g$, we group it in the form $h \circ (f \circ g)$ and use the chain rule *twice*:

$$(h \circ (f \circ g))' = (h' \circ (f \circ g)) \cdot (f \circ g)' = (h' \circ (f \circ g)) \cdot (f' \circ g \cdot g').$$

For our particular f, g, and h, we get

$$\left((\sin x^2)^3\right)' = 3\left(\sin x^2\right)^2 \cdot \left(\sin x^2\right)'$$

$$= 3\left(\sin x^2\right)^2 \cdot \left(\cos x^2 \cdot 2x\right)$$

$$= 6x\left(\sin x^2\right)^2 \cos x^2.$$

■

Working from Outside In: How to Use the Chain Rule. Composing two functions can be thought of as *nesting* one within the other. The use of nested parentheses, as in $f(g(x))$, reflects this point of view: Here f is the "outer" function and g the "inner." In this language, the chain rule says, in effect, to *work from outside in*, one level at a time. More precisely:

> To differentiate $f(g(x))$, first differentiate the outer function, leaving the inner function alone. Then multiply by the derivative of the inner function.

This rule works nicely for *any* level of nesting, as the next example illustrates.

EXAMPLE 4 Differentiate $\sin(\sin(\sin x))$.

Solution Carefully watch the de-nesting, step by step:

$$\left(\sin(\sin(\sin x))\right)' = \cos\left(\sin(\sin x)\right) \cdot \left(\sin(\sin x)\right)'$$

$$= \cos\left(\sin(\sin x)\right) \cdot \cos(\sin x) \cdot (\sin x)'$$

$$= \cos\left(\sin(\sin x)\right) \cdot \cos(\sin x) \cdot \cos x.$$

■

EXAMPLE 5 Let n be any real number. Differentiate $\sin^n x$; then differentiate $f(x)^n$, where f is any differentiable function.

Solution From the chain rule, ▶

$$\left((\sin x)^n\right)' = n\,(\sin x)^{n-1} \cdot \cos x.$$

Here the "outer" function is u^n.

The same idea works for *any* "inner" function f:

$$\left(f(x)^n\right)' = n\,f(x)^{n-1} \cdot f'(x).$$

(The last formula is sometimes called the **general power rule**.)

■

Evidence for the Chain Rule

As we did for the other rules, we'll argue informally.

Exhibit A: It Works for Linear Functions. Let $f(x) = ax + b$ and $g(x) = cx + d$. We'll differentiate $f \circ g$ twice—with and without the chain rule. Notice first that

$$(f \circ g)(x) = a(cx + d) + b = acx + ad + b.$$

The last form shows that $f \circ g$ is another linear function; its derivative is the constant ac.

The chain rule gives the same result. Because $f'(x) = a$ and $g'(x) = c$, it's easy to calculate that $f'(g(x)) \cdot g'(x) = ac$, just as claimed.

Exhibit B: Differentiable Functions Are Almost Linear.   Let f and g be any differentiable functions. Because f and g are differentiable, they're almost linear (as we have remarked before and have seen graphically): Near the point of tangency, f and g are well approximated by their tangent lines. The slope of the g-graph at $x = a$ is $g'(a)$, so near $x = a$, g is well approximated by a linear function l with slope $g'(a)$ (i.e., $g(x) \approx l(x) = g'(a)x + k$, for some constant k; the value of k doesn't matter here).

For convenience, let's write $g(a) = b$. For reasons like the earlier ones, f is well approximated near $x = b$ by a linear function m, this time with slope $f'(b)$. In other words, $f(x) \approx m(x) = f'(b)x + K$ for some other constant K.

From all the foregoing, $(f \circ g)(x) \approx (m \circ l)(x)$. Because l and m are linear, Exhibit A applies. It says that

$$(m \circ l)' = f'(b) \cdot g'(a) = f'(g(a)) \cdot g'(a),$$

just as the chain rule predicts.

The Chain Rule in Leibniz Notation

When we don't want to name functions, for example.

The chain rule is easy to remember, and sometimes convenient to state, ◄   in Leibniz form. Here's what it says, informally (we assume that y and u are well-behaved functions):

Let y be a function of u and u a function of x. Then y is a (composite) function of x; the derivative of y with respect to x is

$$\frac{dy}{dx} = \frac{dy}{du} \cdot \frac{du}{dx}.$$

E X A M P L E 6   Differentiate $y = \sin\left(x^2\right)$ yet again.

S o l u t i o n   We have $y = \sin u$ and $u = x^2$, so

$$\frac{dy}{dx} = \frac{dy}{du} \cdot \frac{du}{dx} = \cos u \cdot 2x = 2x \cdot \cos x^2.$$

(At the last step we replaced u with x^2.)   ■

Derivatives of Inverses and the Chain Rule

The chain rule can be used to relate the (known) derivative of a function to the (unknown) derivative of its *inverse*.

Recall that if f and g are inverse functions, the equation

$$(f \circ g)(x) = f(g(x)) = x$$

must hold for all x in the domain of g. Differentiating both sides—the left side needs the chain rule—gives

$$(f \circ g)'(x) = f'(g(x)) \cdot g'(x) = 1.$$

With a little algebra, our desired formula now follows:

Fact Let f and g be inverse functions. Then

$$g'(x) = \frac{1}{f'(g(x))}$$

for all x for which the right side is defined.

In the next section we'll use this formula to find derivatives of inverse trigonometric functions. For the moment we'll content ourselves with a simpler example.

EXAMPLE 7 Use the fact that e^x and $\ln x$ are inverses and the fact that $(e^x)' = e^x$ to find (again!) the derivative of $\ln x$.

Solution Let $f(x) = e^x$ (so that $f'(x) = e^x$) and $g(x) = \ln x$. Then by the preceding formula,

$$g'(x) = (\ln x)' = \frac{1}{f'(g(x))} = \frac{1}{e^{\ln x}} = \frac{1}{x}. \qquad \blacksquare$$

Guessing and Checking Antiderivatives: A Sample Problem

With the chain rule in our tool kit, we can now easily differentiate almost any elementary function. Finding *antiderivatives*, however, is another matter. Making a guess and then checking it is often the best strategy.

EXAMPLE 8 Find antiderivatives for

$$f(x) = x\cos(x^2) \qquad \text{and} \qquad g(x) = \frac{\cos x}{\sin x}.$$

Solution The basic form of f suggests the result of applying the chain rule, perhaps to the expression $\sin(x^2)$. A check shows that our guess is *almost* correct:

$$\left(\sin(x^2)\right)' = \cos(x^2) \cdot 2x = 2x\cos(x^2).$$

To dispose of that extra factor of 2, we simply divide it out. It follows ▶ that, for *Check for yourself.* any constant C,

$$F(x) = \frac{\sin(x^2)}{2} + C$$

is an antiderivative of f.

It's a little harder to recognize g as a derivative, but if we notice that the numerator is the derivative of the denominator, we may suspect that a logarithm function is involved. So it is—the guess $G(x) = \ln(\sin x) + C$ turns out to be correct:

$$G'(x) = \left(\ln(\sin x) + C\right)' = \frac{\cos x}{\sin x},$$

as desired. $\blacksquare$

BASICS

1. Let $h(x) = f(g(x))$. Use the information about f and g given in the table below to fill in the missing information about h and h'.

x	$f(x)$	$f'(x)$	$g(x)$	$g'(x)$	$h(x)$	$h'(x)$
1	1	2	4	3		
2	2	1	3	4		
3	4	3	1	2		
4	3	4	2	1		

In Exercises 2–11, assume that g is a function such that $g'(x)$ exists for all x. Find the derivative of the function f. (Answers will involve g and g'.)

2. $f(x) = (g(x))^n$

3. $f(x) = e^{g(x)}$

4. $f(x) = \ln(g(x))$

5. $f(x) = \sin(g(x))$

6. $f(x) = \tan(g(x))$

7. $f(x) = g(x^n)$

8. $f(x) = g(e^x)$

9. $f(x) = g(\ln x)$

10. $f(x) = g(\sin x)$

11. $f(x) = g(\tan x)$

In Exercises 12–41, find the derivative of the function f.

12. $f(x) = (x^2 + 3)^{29}$

13. $f(x) = (1 + e^x)^{33}$

14. $f(x) = \sqrt{2 + \sin x}$

15. $f(x) = \sin(e^{3x})$

16. $f(x) = \cos(x^2)$

17. $f(x) = \tan(\sqrt[3]{x})$

18. $f(x) = \ln(4x^2 + 3)$

19. $f(x) = \sqrt{x^2 + e^{-3x}}$

20. $f(x) = \cos(\sqrt{x} + e^{-x})$

21. $f(x) = (2x^{3/4} + 5x^{-1/6})^{12}$

22. $f(x) = (x^{3/2} - 4x^{2/3} + 17)^{421}$

23. $f(x) = (x + \ln x)^{14}$

24. $f(x) = \ln(x^2 + 3\sin x)$

25. $f(x) = \sin(e^{2x} + \tan(3x))$

26. $f(x) = \cos^3(x^4)$

27. $f(x) = \ln(e^{3x} + \cos^2 x)$

28. $f(x) = \tan(e^{x^2})$

29. $f(x) = \sin^5(3x^4)$

30. $f(x) = \sqrt{4 + \cos^2(3x)}$

31. $f(x) = \sqrt[3]{\cos(4x^2)}$

32. $f(x) = \cos^2(e^{3x})$

33. $f(x) = x^2(4x^3 + 5)^6$

34. $f(x) = x\ln(2 + \sin(3x))$

35. $f(x) = e^{x^3}\cos(4x^5 + 2)$

36. $f(x) = \sin^2(x\cos x)$

37. $f(x) = \sqrt{3 + x\cos x}$

38. $f(x) = e^{x\sin x}$

39. $f(x) = \tan(x^2\ln x)$

40. $f(x) = \cos(xe^{2x})$

41. $f(x) = x^2\sin(e^x\ln x)$

FURTHER EXERCISES

42. Let $h(x) = f\left(\dfrac{g(x)}{x^2 + 1}\right)$. If $f(1) = 3$, $f(2) = 5$, $f'(1) = 7$, $f'(2) = 11$, $g(1) = 2$, and $g'(1) = 4$, what is $h'(1)$?

43. Let $f(x) = \ln(2 + \cos x)$.
    (a) What is the domain of f?
    (b) At which values of x does f have local maxima? Local minima?
    (c) At which values of x does f have inflection points?

44. Suppose that f is the function shown to the right and that $g(x) = f(x^2)$.
    (a) For which values of x is $g'(x) = 0$?
    (b) Is g increasing or decreasing at -1?
    (c) Is g' positive or negative over the interval $(\sqrt{2}, \sqrt{5})$?

Graph of f

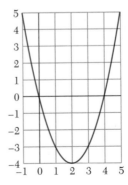

45. Suppose that $f(1) = 2$, that f' is the function shown below, and that $k(x) = f(x^3)$. Evaluate $k'(-1)$.

Graph of f'

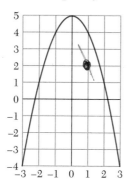

In Exercises 46–48, f and g are the functions defined by the graphs below.

Graph of f

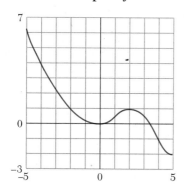

Graph of g

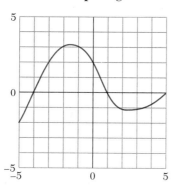

46. Let $h(x) = (f \circ g)(x)$.
    (a) Evaluate $h(-2)$, $h(1)$, and $h(3)$.
    (b) Is $h'(-1)$ positive, negative, or equal to zero? Justify your answer.
    (c) Estimate $h'(-2)$, $h'(1)$, and $h'(3)$.
    (d) What values of x correspond to stationary points of $h(x)$?
47. Let $j(x) = f(x^4)$. Is j increasing or decreasing at $x = -1$? Justify your answer.
48. Let $k(x) = f(e^{-x})$.
    (a) Determine whether k is increasing, decreasing, or stationary at $x = -1$, $x = 0$, and $x = 1$. Justify your answers.
    (b) Show that $x = -\ln 2$ is a stationary point of k and determine whether it corresponds to a local maxima or a local minima. Justify your answer.
49. Let $h(x) = (f \circ g)(x)$ and $j(x) = f(x) \cdot g(x)$. Fill in the missing entries in the table below.

x	$f(x)$	$f'(x)$	$g(x)$	$g'(x)$	$h(x)$	$h'(x)$	$j(x)$	$j'(x)$
-1	3	2	1		0	$-1/2$	3	
0	0	$1/2$	-1	1			0	$-1/2$
1		-5	0			2	0	

In Exercises 50–61, find an antiderivative of the function f.

50. $f(x) = xe^{x^2}$
51. $f(x) = 2x \cos\left(x^2\right)$
52. $f(x) = x^2 \sin\left(2x^3\right)$
53. $f(x) = x\left(x^2 + 3\right)^4$
54. $f(x) = x^3\left(2x^4 + 6\right)^5$
55. $f(x) = \dfrac{3x}{4 + x^2}$
56. $f(x) = \dfrac{x}{\sqrt{1 + x^2}}$
57. $f(x) = x \sin\left(1 - x^2\right)$
58. $f(x) = \dfrac{e^{\sqrt{x}}}{\sqrt{x}}$
59. $f(x) = e^{3x} \sec^2\left(e^{3x}\right)$
60. $f(x) = e^{\cos x} \sin x$
61. $f(x) = e^x \cos\left(e^x\right) \sin\left(e^x\right)$

In Exercises 62–65, find an antiderivative of the function f. [**HINT:** $\left(\ln(f(x))\right)' = f'(x)/f(x)$.]

62. $f(x) = \dfrac{3x^2 - 5}{x^3 - 5x}$
63. $f(x) = \dfrac{2x + 3}{x^2 + 3x + 5}$
64. $f(x) = \dfrac{e^x}{1 + e^x}$
65. $f(x) = \dfrac{\cos x}{2 + \sin x}$

66. Let f be an even function and g be an odd function.
    (a) Show that f' is an odd function. [**HINT:** $f(-x) = f(-1 \cdot x)$.]
    (b) Show that g' is an even function.
    (c) Show that f'' is an even function. [**HINT:** Use part (b).]
    (d) Is g'' even, odd, or neither? Justify your answer.
67. Suppose that f is a differentiable function. If f is periodic with period P, is f' also periodic? Explain.

68. Some values of the derivative of a function f are shown in the table below:

x	0	1	2	3	4	5	6
$f'(x)$	4	-3	2	0	-1	-6	1

[NOTE: No explicit formula for f is given!]

(a) Let $g(x) = f(x+3)$. For which values of x can you evaluate $g'(x)$? Evaluate $g'(x)$ for these values of x.

(b) Let $h(x) = 3f(x)$. For which values of x can you evaluate $h'(x)$? Evaluate $h'(x)$ for these values of x.

(c) Let $j(x) = f(3x)$. For which values of x can you evaluate $j'(x)$? Evaluate $j'(x)$ for these values of x.

(d) Let $k(x) = f(x) - 3$. For which values of x can you evaluate $k'(x)$? Evaluate $k'(x)$ for these values of x.

69. Suppose that f is a function such that $f'(x) = x \cos x$.

(a) Let $g(x) = f(x + \pi)$. Evaluate $g'(0)$.

(b) Let $h(x) = f(2x)$. Evaluate $h'(\pi/2)$.

(c) Let $j(x) = 2f(x) + 3$. Find an expression for $j'(x)$.

70. Suppose that F is an antiderivative of a function f.

(a) Find an antiderivative of $g(x) = f(x) + 3$.

(b) Find an antiderivative of $h(x) = f(x + 3)$.

(c) Find an antiderivative of $j(x) = 3f(x)$.

(d) Find an antiderivative of $k(x) = f(3x)$.

71. Let k be a constant and f a differentiable function. If $g(x) = f(x + k)$, then $g'(x) = f'(x + k)$.

(a) Justify (with a short sentence) each step in the following proof of this result:

(i) $\quad g'(x) = \lim\limits_{h \to 0} \dfrac{g(x+h) - g(x)}{h}$

(ii) $\quad = \lim\limits_{h \to 0} \dfrac{f(x+k+h) - f(x+k)}{h}$

(iii) $\quad = \lim\limits_{h \to 0} \dfrac{f(w+h) - f(w)}{h}$

(iii) $\quad$ where $w = x + k$

(iv) $\quad = f'(w)$

(v) $\quad = f'(x + k)$.

(b) Explain this result geometrically.

72. Let k be a constant and f a differentiable function. If $g(x) = f(kx)$, then $g'(x) = kf'(kx)$. Justify (with a short sentence) each step in the following proof of this result: [HINT: In line (iv), $w = kh$.]

(i) $\quad g'(x) = \lim\limits_{h \to 0} \dfrac{g(x+h) - g(x)}{h}$

(ii) $\quad = \lim\limits_{h \to 0} \dfrac{f(kx+kh) - f(kx)}{h}$

(iii) $\quad = \lim\limits_{w \to 0} k \cdot \dfrac{f(kx+w) - f(kx)}{w}$

(iv) $\quad = kf'(kx)$

73. Suppose that f is a function such that $f(\pi/2) = 17$ and $f'(x) = 2e^x \cos x$.

(a) Find an equation of the line tangent to the graph of f at $x = \pi/2$.

(b) Let $L(x)$ be the line tangent to the graph of f at $x = \pi/2$. Is $L(2) < f(2)$? Justify your answer.

74. (a) Suppose that f is a function such that $f'(x) > 0$ for all x. If $g(x) = f(f(x))$, must g be increasing for all x? Explain your answer.

(b) Suppose that f is a function such that $f'(x) < 0$ for all x. If $g(x) = f(f(x))$, must g be decreasing for all x? Explain your answer.

75. Use the identity $\dfrac{f(x)}{g(x)} = f(x) \cdot \left(g(x)\right)^{-1}$ together with the product and chain rules to derive the quotient rule.

76. Let $h = f \circ g$. Show that $h''(a) = f''\left(g(a)\right)\left(g'(a)\right)^2 + f'\left(g(a)\right)g''(a)$.

77. Let $f(x) = (1 + e^x)^{-1}$.

(a) Evaluate $\lim\limits_{x \to \infty} f(x)$.

(b) Evaluate $\lim\limits_{x \to -\infty} f(x)$.

(c) Explain why f is invertible.

(d) What are the domain and range of f^{-1}?

78. Let $f(x) = \sin(|x|)$.

(a) What is the domain of f? Of f'?

(b) Find an expression for $f'(x)$.

79. Suppose that $f'(x) = g(x)$ and that $h(x) = x^2$. Express $\left(f(h(x))\right)'$ in terms of g and x.

80. Suppose that f and g are functions such that

(i) $f(0) = 0$

(ii) $g(0) = 1$

(iii) $f'(x) = g(x)$ for all x

(iv) $g'(x) = -f(x)$ for all x

(a) Show that $h(x) = \left(f(x)\right)^2 + \left(g(x)\right)^2 = 1$ for all x. [HINT: Show that $h(0) = 1$ and $h'(x) = 0$ for all x.]

(b) Let $k(x) = \left(F(x) - f(x)\right)^2 + \left(G(x) - g(x)\right)^2$ where F and G are another pair of functions that satisfy the properties (i)–(iv). Show that $k'(x) = 0$ for all x.

(c) Use part (b) to show that $F(x) = f(x) = \sin x$ and that $G(x) = g(x) = \cos x$ for all x.

81. Let $f(x) = \sin x + \cos x$ on the interval $[-3\pi/4, \pi/4]$.

(a) Show that f has an inverse.

(b) Find an expression for $\left(f^{-1}\right)'(x)$.
[HINT: $(\cos x - \sin x)^2 = 2 - (\sin x + \cos x)^2$.]

82. Let $g(x) = f(x^2)$ where f is a function with the following properties:

(i) f, f', and f'' have domain $\mathbb{R}$

(ii) $f(0) = 1$

(iii) $f'(x) > 0$ for all $x > 0$

(iv) $f'(x) > 0$ for all $x < 0$

(v) f is concave down on $(-\infty, 0)$

(vi) f is concave up on $(0, \infty)$

(a) Where does g have local minima?

(b) Where is g concave up?

83. Evaluate $\lim\limits_{x \to \pi} \dfrac{5^{\sin x} - 1}{x - \pi}$ exactly (i.e., as a value of an elementary function).

3.7 Implicit Differentiation

Implicitly vs. Explicitly Defined Functions

The equation $y = -3x + 5$ defines y *explicitly* as a function (we'll call it f) of x; for an input x, the output is given by the explicit rule $f(x) = -3x + 5$.

Consider, by contrast, the equivalent equation $6x + 2y = 10$. Although it doesn't display y explicitly as a function of x, the new equation defines y *implicitly* as a function of x. Solving for y gives the same explicit formula we had before:

$$6x + 2y = 10 \iff 2y = -6x + 10 \iff y = -3x + 5.$$

For the linear equation $6x + 2y = 10$, it was easy to find an explicit formula for y. Things aren't always so simple. More complicated equations are much harder— or even impossible—to solve explicitly for y. Even simple equations, moreover, may not define y uniquely as a function of x, because the graph of an *equation* need not be the graph of a *function*.

> **EXAMPLE 1**   Does the equation $x - y^2 = 0$ define y as a function of x?

> **Solution**   No. Solving algebraically for y shows why:

$$x - y^2 = 0 \implies y^2 = x \implies y = \pm\sqrt{x}.$$

In effect, the equation defines *two* functions of x: $f_1(x) = \sqrt{x}$ and $f_2(x) = -\sqrt{x}$, each defined for nonnegative inputs x. The picture agrees:

Graph of $x - y^2 = 0$

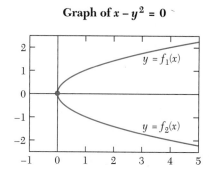

The "sideways" parabola is not the graph of a function, because each value of x corresponds to *two* values of y. ▶ The upper and lower halves of the graph do define functions of x—the same functions (f_1 and f_2) we wrote explicitly before. ■

It fails the vertical line test.

> **EXAMPLE 2**   Does the equation $5x^2 - 6xy + 5y^2 = 16$ implicitly define y as one function of x? As several functions of x?

> **Solution**   Here's the graph, an ellipse. The points $(2, 2)$ and $(1, -1)$ are bulleted.

Graph of $5x^2 - 6xy + 5y^2 = 16$

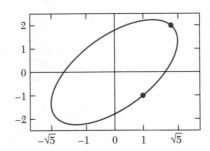

Like the preceding graph, this one defines two different functions y of x, corresponding to the "upper" and "lower" halves of the ellipse. (The upper half, passing through $(2, 2)$, defines a function g_1 for which $g_1(2) = 2$. The lower half defines another function g_2, for which $g_2(1) = -1$. Below we'll find derivatives at these points.) The domain of each function is the interval $[-\sqrt{5}, \sqrt{5}]$. Solving for y is harder this time; we won't bother. ■

Implicit Functions, Implicit Derivatives

All three of the equations $6x + 2y = 10$, $x = y^2$, and $5x^2 - 6xy + 5y^2 = 16$ implicitly define one or more functions of x. How can we differentiate such functions?

If we can solve *explicitly* for y, there's no problem. For $x = y^2$, for instance,

$$f_1(x) = \sqrt{x} \implies f_1'(x) = \frac{1}{2\sqrt{x}};$$

$$f_2(x) = -\sqrt{x} \implies f_2'(x) = -\frac{1}{2\sqrt{x}}.$$

What if we can't (or don't want to) solve explicitly for y? In that case, we use **implicit differentiation**, a technique for differentiating an *equation* without first solving it for one of the variables. Here's the idea for an equation in x and y:

> **Implicit differentiation.** *Differentiate both sides of the equation, treating y as an (implicitly defined) function of x; use the chain rule. The resulting equation involves x, y, and dy/dx; it can be solved, if necessary, for dy/dx.*

We'll illustrate with examples.

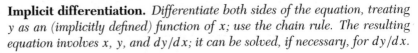

EXAMPLE 3   Let $x = y^2$. Find dy/dx by implicit differentiation.

Solution   We differentiate both sides of the equation, treating y as a function of x:

$$x = y^2 \implies \frac{d}{dx}(x) = \frac{d}{dx}(y^2) \implies 1 = 2y\frac{dy}{dx}.$$

Note the use of the chain rule: Since y is a function of x, $(y^2)' = 2yy'$. Solving for dy/dx is easy:

$$1 = 2y\frac{dy}{dx} \implies \frac{dy}{dx} = \frac{1}{2y}.$$

This agrees—as it should—with *both* of our earlier answers:

$$y = \sqrt{x} \implies y' = \frac{1}{2\sqrt{x}} = \frac{1}{2y};$$

$$y = -\sqrt{x} \implies y' = -\frac{1}{2\sqrt{x}} = \frac{1}{2y}. \qquad \blacksquare$$

EXAMPLE 4   As the graph in Example 2 shows, the equation $5x^2 - 6xy + 5y^2 = 16$ defines y implicitly as a function of x near the point $(2, 2)$. Find dy/dx in terms of x and y. What is its value at the point $(2, 2)$? At $(1, -1)$? What do these values mean?

Solution   First we differentiate our equation with respect to x: ▶

$$5x^2 - 6xy + 5y^2 = 16 \implies 10x - 6y - 6x\frac{dy}{dx} + 10y\frac{dy}{dx} = 0.$$

Don't forget: y is a function of x, so the chain rule applies.

(The middle term, $-6xy = -6x \cdot y$, is a product, so we used the product rule.)
   Now we solve for dy/dx:

$$10x - 6y - 6x\frac{dy}{dx} + 10y\frac{dy}{dx} = 0 \implies \frac{dy}{dx}(-6x + 10y) = 6y - 10x$$

$$\implies \frac{dy}{dx} = \frac{6y - 10x}{10y - 6x}.$$

At $(2, 2)$, therefore, $dy/dx = -1$; at $(1, -1)$, $dy/dx = 1$. ▶   Geometrically, these results mean that the curve shown has slopes -1 and 1, respectively, at the bulleted points $(2, 2)$ and $(1, -1)$. ▶ ■

Check these simple calculations.

Convince yourself that these results are graphically plausible.

Implicit Differentiation: Pros and Cons.   Implicit differentiation neatly avoids the hard problem of solving complicated equations explicitly for y. As usual, there's a price to pay for such convenience: Derivative formulas obtained via implicit differentiation usually involve both x and y.

A Sample Problem

Here's a problem that applies ideas of this section.

EXAMPLE 5   Use implicit differentiation to find points on the unit circle $x^2 + y^2 = 1$ at which the tangent line has slope 1.

Solution   Implicit differentiation with respect to x gives ▶

$$x^2 + y^2 = 1 \implies 2x + 2y\frac{dy}{dx} = 0 \implies \frac{dy}{dx} = \frac{-x}{y}.$$

Check the easy steps.

Thus, $dy/dx = 1$ if and only if $y = -x$, i.e., where the line $y = -x$ intersects the unit circle. From a picture ▶   it's easy to see that these intersections occur at the points $(-\sqrt{2}/2, \sqrt{2}/2)$ and $(\sqrt{2}/2, -\sqrt{2}/2)$. ■

Draw your own!

BASICS

1. This problem refers to the equation $5x^2 - 6xy + 5y^2 = 16$ and its graph—see Example 2 (page 233) and Example 4 (page 235).

   (a) Find the two points (x, y) on the ellipse at which the tangent line is horizontal.

   (b) Find the points (x, y) on the ellipse at which the tangent line is vertical. (At these points, the derivative dy/dx doesn't exist.)

2. This problem refers to the equation $5x^2 - 6xy + 5y^2 = 16$ and its graph—see Example 2 (page 233) and Example 4 (page 235).

   (a) The graph appears to be symmetric about the origin. To see that it really *is* symmetric, show that if (x_0, y_0) is on the graph, then so is the "opposite" point $(-x_0, -y_0)$.

   (b) Let (x_0, y_0) be any point on the graph. The previous part says that the "opposite" point is $(-x_0, -y_0)$. Show that the tangent lines at these two points have the same slope. (This is further evidence of the graph's symmetry.)

   (c) The line $y = -x$ intersects the ellipse at two points. Find these two points; then find the slope at each point.

3. All parts of this exercise are about the equation $y^2 - x^2 = 1$ and its graph, called a **hyperbola**.

   (a) The equation $y^2 - x^2 = 1$ defines two different functions $y = f_1(x)$ and $y = f_2(x)$. Find formulas for these functions.

   (b) Plot the equation $y^2 - x^2 = 1$. [**HINT:** One way is to plot both functions f_1 and f_2.]

   (c) Use implicit differentiation to find dy/dx.

   (d) Find the points on the curve $y^2 - x^2 = 1$ at which the tangent line is horizontal.

   (e) For any constant $c > 0$, the line $x = c$ intersects the hyperbola at two points. (Draw a picture to convince yourself.) Find these two points and the slope of the hyperbola at each.

   (f) What happens in the previous part as $c \to \infty$?

4. Repeat the previous problem, but use the hyperbola $y^2 - 2x^2 = 1$.

5. This problem refers to the curves with equations $x^2 + xy + y^2 = 1$ and $x^2 - xy + y^2 = 1$, shown.

   (a) Find dy/dx for each of the curves.

   (b) Decide which curve is which. [**HINT:** One way is to compare the derivatives at a common point, such as $(0, 1)$.)]

   (c) Draw tangent lines to the curves at the four intersection points. Find the slope of each one.

   (d) Find and label all points at which each curve is horizontal.

   (e) Find and label all points at which each curve is vertical.

   (f) Draw the line $y = x$ into the picture. Use derivatives to show that this line intersects both curves perpendicularly.

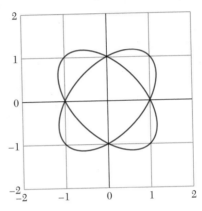

Graphs of $x^2 + xy + y^2 = 1$ and $x^2 - xy + y^2 = 1$

6. The equation $2x + 3y = 6$ implicitly defines y as a function of x.

   (a) By solving for y, find an *explicit* formula for y as a function of x. Use the result to find dy/dx.

   (b) Find dy/dx by implicit differentiation; compare your result with the previous part.

7. The equation $4x^2y - 3y = x^3$ implicitly defines y as a function of x.

   (a) Use implicit differentiation to find dy/dx.

   (b) Write y as an explicit function of x and compute dy/dx directly.

8. The equation $x^2 + y^2 = 1$ defines *two* different functions of x:

   $$x^2 + y^2 = 1 \implies y = \pm\sqrt{1 - x^2}.$$

   (a) Find dy/dx by implicit differentiation.

   (b) Find dy/dx (directly) for each of the functions $y = \pm\sqrt{1 - x^2}$. Compare your results with those found in part (a).

Each of the equations in Exercises 9–18 defines a curve in the xy plane. Use implicit differentiation to find an expression for dy/dx.

9. $x^2 + 2y^2 = 4$

10. $4x^2 - 9y^2 = 36$

11. $y^2 - x^2 = x^2y^2$

12. $y^2(2 - x) = x^3$

13. $x^3 + y^3 = 3xy$

14. $y^4 = y^2 - x^2$

15. $x^{2/3} + y^{2/3} = 1$
16. $\left(x^2 + y^2 - 2y\right)^2 = x^2 + y^2$
17. $\left(x^2 + y^2\right)^2 = x^2 - y^2$
18. $\left(x^2 + y^2\right)^2 = 2xy$

19. Plot the curve defined by the equation $y^2 = x^3$ and find equations of the *two* lines tangent to the curve at $x = 1$.
20. Plot the curve defined by the equation $y^4 = 4y^2 - x^2$ and find equations of the *four* lines tangent to the curve at $x = 1$.

3.8 Inverse Trigonometric Functions and Their Derivatives

The natural logarithm and exponential functions are inverses; each "undoes" the other. In symbols: ▶

$$\ln(e^x) = x \qquad \text{and} \qquad e^{\ln x} = x.$$

The first equation holds for all x, the second only for $x > 0$.

In this section we find and discuss inverses for several trigonometric functions.

Nomenclature: "Arc" or "Inverse"?

Two common forms are used to denote inverse trigonometric functions:

arcsin x, arccos x, *and* arctan x *are synonyms for* $\sin^{-1} x$, $\cos^{-1} x$, *and* $\tan^{-1} x$.

Each form has its advantages. The first makes explicit the connection with arcs on the unit circle; the second reminds us that function inversion is involved. We'll use both. ▶

Don't confuse $\sin^{-1} x$ with $1/\sin x$. They're entirely different.

A Technical Problem and Its Solution: Restricting Domains

To be invertible, a function must be one-to-one; graphically speaking, it must pass the "horizontal line test." Alas, *no* trigonometric function has this property. Far from it—trigonometric functions, being 2π-periodic, repeat themselves on intervals of length 2π, hitting each output value *infinitely often*. The next example illustrates the problem.

E X A M P L E 1   By any sensible definition, ▶ arcsin(0.5) should be a number whose sine is 0.5. Find such a number. How many are there?

We'll give a formal definition later.

S o l u t i o n   The following picture shows four inputs x for which $\sin x = 0.5$.

Values of x for which sin x = 0.5

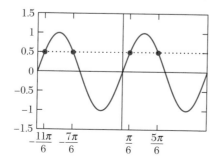

The bulleted points show that

$$0.5 = \sin\left(-\frac{11\pi}{6}\right) = \sin\left(-\frac{7\pi}{6}\right) = \sin\left(\frac{\pi}{6}\right) = \sin\left(\frac{5\pi}{6}\right).$$

Moreover, this list is far from exhaustive; infinitely many more solutions of the equation $\sin x = 0.5$ exist outside our viewing window.

Which of these numbers deserves to be called arcsin(0.5)? None especially, but if *forced* to choose, we might reasonably select $\pi/6$, the solution nearest to zero. Any scientific calculator would agree. According to a TI-82 calculator, for instance,

$$\sin^{-1}(0.5) = 0.5235987756 \approx \frac{\pi}{6}.$$ ∎

Surgery on Domains

The problem just seen—choosing from among reasonable candidates—arises because the trigonometric functions aren't one-to-one. Our solution is crude but effective: Amputate the offending part of each function's domain, and thus *make* the function one-to-one. Graphs will show us where to cut.

Inverting the Sine Function

The sine function is increasing, and therefore one-to-one, on the interval $[-\pi/2, \pi/2]$. (There are, of course, many other intervals on which the sine function is one-to-one. Among all such intervals, this one, centered at 0, seems most natural.) The following picture shows both the ordinary sine curve (drawn dotted) and the special piece (drawn solid) in question here:

Restricting the sine function

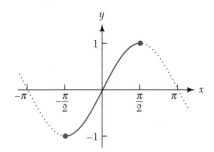

Restricted to this domain, the sine function is invertible. Its inverse is defined formally as follows:

> **Definition** For x in $[-1, 1]$, $y = \sin^{-1} x$ (or arcsin x) is defined by the conditions
>
> $$x = \sin y \qquad \text{and} \qquad -\frac{\pi}{2} \le y \le \frac{\pi}{2}.$$
>
> In words: arcsin x is the (unique) angle between $-\pi/2$ and $\pi/2$ whose sine is x.

Note the following features.

Graphical Symmetry The graphs of $\sin x$ (restricted!) and $\sin^{-1} x$ show the expected reflective symmetry:

Sine and arcsine

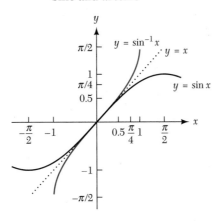

Domain and Range The sine function (as we've restricted it) has domain $[-\pi/2, \pi/2]$ and range $[-1, 1]$. The inverse sine function has it the other way around: Its domain is $[-1, 1]$, and its range is $[-\pi/2, \pi/2]$.

Collapsing Equations By the definition of inverse function, these equations hold:

$$\sin\left(\sin^{-1} x\right) = x \quad \text{if } -1 \le x \le 1;$$

$$\sin^{-1}\left(\sin x\right) = x \quad \text{if } -\pi/2 \le x \le \pi/2$$

Inverting the Cosine Function

The cosine function ▶ becomes one-to-one, and therefore invertible, when restricted to the interval ▶ $[0, \pi]$. This understood, the graphs and formal definition follow naturally.

The situation for $\cos x$ differs only in details from that for $\sin x$.

Restricting $\cos x$ to $[-\pi/2, \pi/2]$ doesn't work. Do you see why not?

Restricting the cosine function

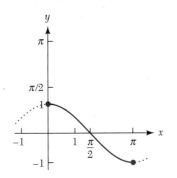

Graphs of cosine and arccosine

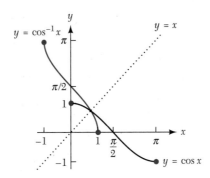

> **Definition** For x in $[-1, 1]$, $y = \cos^{-1} x$ (or arccos x) is defined by the conditions
>
> $$x = \cos y \quad \text{and} \quad 0 \le y \le \pi.$$
>
> In words: arccos x is the (unique) angle between 0 and π whose cosine is x.

Note the following features.

Graphical Symmetry The graphs of cos and $\cos^{-1}$ show the usual symmetry for inverse functions—they're symmetric with respect to the line $y = x$.

Domain and Range We restricted the cosine function to have domain $[0, \pi]$ and range $[-1, 1]$. The *inverse* cosine function therefore has domain $[-1, 1]$ and range $[0, \pi]$.

Collapsing Equations As with the arcsine, two equations express the inverse relationship between cos and arccos:

$$\cos\left(\cos^{-1} x\right) = x \qquad \text{if } -1 \le x \le 1;$$

$$\cos^{-1}\left(\cos x\right) = x \qquad \text{if } 0 \le x \le \pi.$$

Inverting the Tangent Function

The tangent function is one-to-one if restricted to any of its "branches." We'll use the middle branch, through the origin:

The tangent function: one-to-one on the middle branch

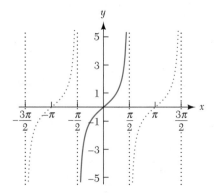

The graph shows this. Do you agree?

Unlike the sine and cosine functions, the tangent function hits *all* real numbers as outputs. ◄ The arctangent function therefore accepts *all* real inputs.

> **Definition** Let x be any real number. Then $y = \tan^{-1} x$ (or $y = \arctan x$) means that
>
> $$x = \tan y \quad \text{and} \quad -\pi/2 < y < \pi/2.$$

Note the following features.

Symmetry The domain of $\arctan x$ (i.e., the range of $\tan x$) is infinite, so any picture of its graph is incomplete. The following figure conveys this idea. Notice the usual symmetry between a function and its inverse.

The tangent function and its inverse

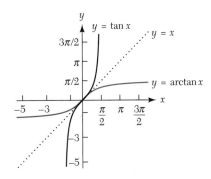

Infinite Domain The tangent function (as restricted) has domain $(-\pi/2, \pi/2)$ and range $(-\infty, \infty)$. The arctangent function turns things around: It has range $(-\pi/2, \pi/2)$ and domain $(-\infty, \infty)$.

Other Inverse Trigonometric Functions

All six trigonometric functions have inverses (as your calculator knows). As with $\arcsin x$, $\arccos x$, and $\arctan x$, defining $\operatorname{arcsec} x$, $\operatorname{arccsc} x$, and $\operatorname{arccot} x$ requires care with domains and ranges. Among these other functions, the arcsecant is most useful; it appears occasionally in antiderivatives.

For any legal ▶ input t, $\sec t = 1/\cos t$. The simplest definition of $\operatorname{arcsec} x$—based on $\arccos x$—reflects this fact:

There's trouble if $\cos t = 0$.

> **Definition** For any x with $|x| \geq 1$,
>
> $$\operatorname{arcsec}(x) = \sec^{-1}(x) = \cos^{-1}(1/x).$$

In particular,

$$y = \operatorname{arcsec}(x) = \arccos\left(\frac{1}{x}\right) \implies \cos y = \frac{1}{x} \implies \sec y = x,$$

as we'd expect. ▶

Plot your own copy of the arcsecant function; experiment with various domains. See the exercises at the end of this section for more on the arcsecant function.

Inverse Trigonometric Functions and the Unit Circle

We defined the sine, cosine, and tangent functions in terms of the unit circle. Their inverses can be interpreted in the same way. The following examples show how.

EXAMPLE 2   A unit circle follows; its arclengths—starting from the point $(1, 0)$—are in radians. Use the picture to estimate arctan 2 and arcsin(0.5).

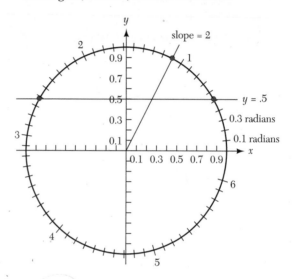

Arctangent, arcsine, and the unit circle

Don't recall? See
Appendix G.

S o l u t i o n   By definition, arctan 2 is an angle (or "arc") whose tangent is 2. Recall ◀ that the tangent of any angle is the *slope* of the line that angle determines. Therefore, we want an angle—the one between $-\pi/2$ and $\pi/2$—that corresponds to a slope of 2. To find this angle, we draw a line from the origin, with slope 2, and observe the radian measure of the angle it determines. The preceding picture shows that the line determines an arc of radian measure about 1.1; this is our estimate for arctan 2. ◀

According to a TI-82 calculator, arctan 2 ≈ 1.107148718.

Reading $\sin^{-1}(0.5)$ from the unit circle is similar. By definition, $\sin t$ is the y-coordinate of the point determined by an arc of length t on the unit circle. Therefore, arcsin(0.5) is the (one and only) angle between $-\pi/2$ and $\pi/2$ (i.e., to the right of the y-axis) whose corresponding y-coordinate is 0.5. To find this angle, we draw the line $y = 0.5$ (as shown in the preceding picture). It crosses the circle at a bulleted point, the end of an arc with radian measure about 0.52; therefore, arcsin(0.5) ≈ 0.52. ◀

This estimate is surprisingly good: The "right" answer is about 0.5235987756.

Combining Ordinary and Inverse Trigonometric Functions

In practice, inverse trigonometric functions are often combined with ordinary trigonometric functions. Diagrams like the following one can help demonstrate how the process works.

Relating inverse trigonometric functions: pictorial aids

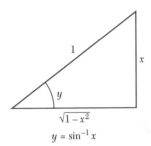

$y = \sin^{-1} x$

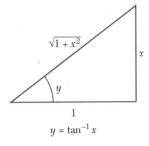

$y = \tan^{-1} x$

The first figure shows (among other things) that

$$\cos(\sin^{-1} x) = \frac{\text{adjacent}}{\text{hypotenuse}} = \sqrt{1 - x^2}.$$

From the second figure, similarly,

$$\sin(\tan^{-1} x) = \frac{\text{opposite}}{\text{hypotenuse}} = \frac{x}{\sqrt{1 + x^2}}.$$

Derivatives of Inverse Trigonometric Functions

We know derivatives of the ordinary trigonometric functions; with the chain rule, we can find derivatives of their inverses. After tabulating the derivatives, we'll derive them.

Derivatives of ordinary trigonometric functions are other trigonometric functions. Surprisingly, inverse trigonometric functions have *algebraic* derivatives:

Derivatives of Inverse Trigonometric Functions

Function	arcsin x	arccos x	arctan x	arcsec x
Derivative	$\dfrac{1}{\sqrt{1 - x^2}}$	$\dfrac{-1}{\sqrt{1 - x^2}}$	$\dfrac{1}{1 + x^2}$	$\dfrac{1}{|x|\sqrt{x^2 - 1}}$

Observe these features.

Restricted Domains The functions arcsin x, arccos x, and arcsec x have restricted domains. So, therefore, must their derivatives. ▶

For more detail, see the exercises at the end of this section.

Not Much Difference The derivatives of arcsin x and arccos x differ only in sign. Therefore, $(\arcsin x + \arccos x)' = 0$ for all legal inputs x. This is no accident, of course. For all x in $[-1, 1]$,

$$\arcsin x + \arccos x = \frac{\pi}{2}.$$

(The graphs of arcsin x and arccos x say the same thing.) ▶

Graph both on the same axes to see why.

Graphical Evidence We'll demonstrate by calculation *why* these formulas hold. Graphs show convincingly *that* they hold. Here, for instance, are graphs of the arctangent function and its claimed derivative:

Graph of $f(x) = \arctan x$ and
$f'(x) = 1/(1 + x^2)$

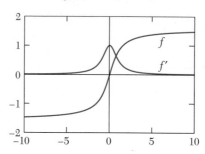

E X A M P L E 3   Explain why $(\arctan x)' = \dfrac{1}{1 + x^2}$.

S o l u t i o n   If $f(x) = \tan x$ and $g(x) = \arctan x$, then f and g are inverses, and $f'(x) = \sec^2 x$. Unraveling the derivative formula above gives

$$(\arctan x)' = \frac{1}{f'(g(x))} = \frac{1}{\sec^2(\arctan x)}.$$

Despite first appearances, the result is correct. We can simplify the last quantity using the trigonometric identity $\sec^2 t = 1 + \tan^2 t$:

$$\frac{1}{\sec^2(\arctan x)} = \frac{1}{1 + \tan^2(\arctan x)} = \frac{1}{1 + x^2}. \qquad \blacksquare$$

E X A M P L E 4   Explain why $(\arcsin x)' = \dfrac{1}{\sqrt{1 - x^2}}$.

S o l u t i o n   We play the same game as before, with slightly different twists.
Watch for the following combination of the cosine and arcsine functions. If $f(x) = \sin x$ and $g(x) = \arcsin x$, then $f'(x) = \cos x$, so ◄

$$(\arcsin x)' = \frac{1}{f'(g(x))}$$

$$= \frac{1}{\cos(\arcsin x)}$$

$$= \frac{1}{\sqrt{1 - \sin^2(\arcsin x)}}$$

$$= \frac{1}{\sqrt{1 - x^2}}. \qquad \blacksquare$$

The remaining derivative formulas can be found similarly.

BASICS

Find the numbers in Exercises 1–6 exactly (in radians).

1. $\arcsin(1)$
2. $\arcsin\left(\sqrt{3}/2\right)$
3. $\arccos(-1)$
4. $\arccos\left(-\sqrt{2}/2\right)$
5. $\arctan(1)$
6. $\arctan\left(\sqrt{3}\right)$

Use the unit circle on page 242 to estimate the numbers in Exercises 7–10.

7. $\arcsin(0.8)$
8. $\arccos(0.6)$
9. $\arctan(0.4)$
10. $\arctan(3)$

11. This table gives samples of what one calculator knows (and doesn't know) about inverse trigonometric functions:

Inverse Trigonometric Function Values

x	−2	−1	−0.5	−0.25
arcsin x		−1.57	−0.52	−0.25
arccos x		3.14	2.09	1.82
arctan x	−1.11	−0.79	−0.46	−0.24
x	0	0.25	0.5	1
arcsin x	0.00	0.25	0.52	1.57
arccos x	1.57	1.31	1.04	0.00
arctan x	0.00	0.24	0.46	0.79
x	2	5	100	
arcsin x				
arccos x				
arctan x	1.11	1.37	1.56	

(a) Can the gaps in the table above be filled in? Why or why not? Answer using the word "domain."

(b) In each column, the entries for arcsin x and arccos x add to 1.57. Surely this can't be coincidence! What's going on?

12. (a) The function $\operatorname{arcsec} x$ accepts as inputs only values of x for which $|x| \geq 1$. Why is this restriction necessary? [**HINT**: What's the domain of $\arccos x$?]

(b) What's the *range* of $\operatorname{arcsec} x$? [**HINT**: What's the range of $\arccos x$?]

Write each of the following as an algebraic expression. For example, $\cos(\arcsin x) = \sqrt{1 - x^2}$.

13. $\sin(\arccos x)$
14. $\tan(\arcsin x)$
15. $\sin(\arctan x)$
16. $\cos(\arctan x)$

In Exercises 17–28, find the derivative of the function f.

17. $f(x) = \arctan(2x)$
18. $f(x) = \arctan\left(x^2\right)$
19. $f(x) = \sqrt{\arcsin x}$
20. $f(x) = \arcsin(\sqrt{x})$
21. $f(x) = \arcsin\left(e^x\right)$
22. $f(x) = e^{\arctan x}$
23. $f(x) = \arctan(\ln x)$
24. $f(x) = \arccos(2x + 3)$
25. $f(x) = \operatorname{arcsec}(x/2)$
26. $f(x) = \arcsin x / \arccos x$
27. $f(x) = x^2 \arctan\left(\sqrt{x}\right)$
28. $f(x) = \ln(2 + \arcsin x)$

In Exercises 29–41, find an antiderivative of the function f. (Guess and check.)

29. $f(x) = 1/\left(1 + x^2\right)$
30. $f(x) = 1/\sqrt{1 - x^2}$
31. $f(x) = 2/\left(1 + 4x^2\right)$
32. $f(x) = 3/\left(9 + x^2\right)$
33. $f(x) = 1/\sqrt{9 - x^2}$
34. $f(x) = 1/\sqrt{1 - 4x^2}$
35. $f(x) = e^x/\left(1 + e^{2x}\right)$
36. $f(x) = x/\sqrt{1 - x^4}$
37. $f(x) = (2 \arctan x)/(1 + x^2)$
38. $f(x) = (\cos x)/\left(1 + \sin^2 x\right)$
39. $f(x) = e^{\arcsin x}/\sqrt{1 - x^2}$
40. $f(x) = 1/\left(x\left(1 + (\ln x)^2\right)\right)$
41. $f(x) = 1/\left(\left(1 + x^2\right)\arctan x\right)$

FURTHER EXERCISES

42. Find 4 values of x for which $\cos x = 0.5$.

43. Use a calculator to find 4 values of x (2 positive, 2 negative) for which $\tan x = 2$.

44. The graph of the tangent function has *vertical* asymptotes. Where are they? Does the graph of the *arctangent* function have asymptotes? If so, where? Why? [**HINT**: Look carefully at the graphs of the tangent function and its inverse. Explain what you see in terms of reflection across the line $y = x$.]

45. The equation arcsin (sin x) = x suggests that the graph of y = arcsin (sin x), over any interval, might be the straight line y = x. It isn't (see below).

Graphs of y = arcsin (sin x) and y = x

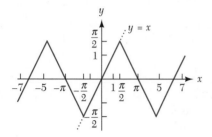

(a) Explain why arcsin(sin 5) ≠ 5.
(b) For which values of x *does* the equation arcsin(sin x) = x hold?
(c) Let $f(x) = \arcsin(\sin x)$. Show that $f'(x) = \dfrac{\cos x}{|\cos x|}$.
(d) How is the result in part (c) related to the graph shown above?

46. (a) What is the domain of the derivative of the arcsine function? Give a geometric explanation for why it is not the same as the domain of the arcsine function.
(b) What is the domain of the derivative of the arctangent function? Is it the same as the domain of the arctangent function?
(c) What is the domain of the derivative of the arcsecant function? Is it the same as the domain of the arcsecant function?

47. Let $f(x) = \arctan x$.
(a) Plot graphs of f and f' on the same axes.
(b) Evaluate $\lim_{x\to\infty} f(x)$, $\lim_{x\to-\infty} f(x)$, $\lim_{x\to\infty} f'(x)$, and $\lim_{x\to-\infty} f'(x)$. What do these results imply about the graph of the arctangent function?
(c) Is f even, odd, or neither?
(d) Show that f is increasing on $(-\infty, \infty)$. How does the graph of f' reflect this fact?
(e) Where is f concave up? Concave down? Does f have any inflection points? If so, where?

48. Let $f(x) = \arcsin x$.
(a) Plot graphs of f and f' on the same axes.
(b) Evaluate $\lim_{x\to1^-} f(x)$, $\lim_{x\to-1^+} f(x)$, $\lim_{x\to1^-} f'(x)$, and $\lim_{x\to-1^+} f'(x)$. What do these results imply about the graph of the arcsine function?
(c) Is f even, odd, or neither?
(d) Show that f is increasing on $(-1, 1)$. How does the graph of f' reflect this fact?
(e) Where is f concave up? Concave down? Does f have any inflection points? If so, where?
(f) How does the graph of f' reflect the information found in part (e)?

49. Let $f(x) = \arccos x$.
(a) Plot graphs of f and f' on the same axes.
(b) Evaluate $\lim_{x\to1^-} f(x)$, $\lim_{x\to-1^+} f(x)$, $\lim_{x\to1^-} f'(x)$, and $\lim_{x\to-1^+} f'(x)$. What do these results imply about the arccosine function?
(c) Is f even, odd, or neither?
(d) Show that f is decreasing on $(-1, 1)$. How does the graph of f' reflect this fact?
(e) Where is f concave up? Concave down? Does f have any inflection points? If so, where are they located.
(f) How does the graph of f' reflect the information found in part (e)?

50. Let $f(x) = \text{arcsec}\, x$.
(a) Plot graphs of f and f' using the plotting window $[-5, 5] \times [0, 5]$. [**NOTE:** If your calculator or computer doesn't recognize the arcsecant function, trick it by plotting $\arccos(1/x)$ instead.]
(b) Evaluate $\lim_{x\to1^+} f(x)$, $\lim_{x\to-1^-} f(x)$, $\lim_{x\to1^+} f'(x)$, and $\lim_{x\to-1^-} f'(x)$. What do these results imply about the graph of the arcsecant function?
(c) Evaluate $\lim_{x\to\infty} f(x)$, $\lim_{x\to-\infty} f(x)$, $\lim_{x\to\infty} f'(x)$, and $\lim_{x\to-\infty} f'(x)$. What do these results imply about the graph of the arcsecant function?
(d) Is f even, odd, or neither?
(e) Show that f is increasing everywhere in the interior of its domain. How does the graph of f' reflect this fact?
(f) Where is f concave up? Concave down? Does f have any inflection points? If so, where are they located?
(g) How does the graph of f' reflect the information found in part (f)?

51. Show that $x/(1 + x^2) \le \arctan x \le x$ for all $x \ge 0$.

52. (a) By drawing an appropriately labeled right triangle, show that $\arctan x + \arctan(1/x) = \pi/2$ if $x > 0$.
(b) Use calculus to show that $\arctan x + \arctan(1/x) = \pi/2$ for all $x > 0$. [**HINT:** Start by differentiating both sides of the equation.]
(c) Find an equation similar to the one in part (a) that is valid when $x < 0$.

Use calculus to prove the identities in Exercises 53–57. [**HINT:** See Exercise 52.]

53. $\arcsin(-x) = -\arcsin(x)$

54. $\arccos(-x) = \pi - \arccos(x)$

55. $\arccos x = \pi/2 - \arcsin x$

56. $\arctan\left(x/\sqrt{1 - x^2}\right) = \arcsin x$

57. $\arcsin\left(\dfrac{x - 1}{x + 1}\right) = 2\arctan(\sqrt{x}) - \pi/2$

58. (a) Show that $2\arcsin x = \arccos(1 - 2x^2)$ if $0 \le x \le 1$.
(b) Find an identity similar to the one in part (a) that is valid if $-1 \le x \le 0$.

59. Let $f(x) = \arctan\left(\dfrac{1+x}{1-x}\right)$.

    (a) Assume that $x < 1$. Show that there is a constant C such that $f(x) = C + \arctan x$.
    (b) Use the result in part (a) to evaluate $\lim_{x \to 1^-} f(x)$.
    (c) Evaluate $\lim_{x \to 1^+} f(x)$.

60. (a) Show that $\tan(\text{arcsec}\, x) = \sqrt{x^2 - 1}$ if $x \geq 1$.
    (b) Find an algebraic expression for $\tan(\text{arcsec}\, x)$ if $x \leq -1$.

61. (a) Derive the identity

$$\arctan x + \arctan y = \arctan\left(\frac{x+y}{1-xy}\right)$$

from the identity $\tan(x+y) = \dfrac{\tan x + \tan y}{1 - \tan x \tan y}$.

    (b) What conditions must be satisfied by x and y for the identity derived in part (a) to hold?

62. (a) Use the identity in part (a) of the previous problem to show that $\pi/4 = \arctan(1/2) + \arctan(1/3)$.
    (b) Show that $\pi/4 = 2\arctan(1/4) + \arctan(7/23)$. [**HINT:** Use the identity in part (a) of the previous problem twice.]

3.9 Chapter Summary

This chapter is about derivatives and antiderivatives of elementary functions—how to calculate and combine them, how to use them, and why they matter.

The Simplest Derivatives; First Uses. Section 3.1 introduced and used symbolic derivatives in the simplest possible setting: polynomials and power functions. Section 3.2 illustrated a few of the practical uses of derivative formulas.

Derivatives of Transcendental Functions. Sections 3.3 and 3.4 discussed derivatives of exponential, logarithmic, and trigonometric functions—the transcendental functions of elementary calculus. Finding derivatives of **algebraic functions** is—no surprise here—usually a matter of algebra. For **transcendental functions**, finding derivatives takes a little more ingenuity.

Antiderivatives. Differentiation is a two-way street: If f is the derivative of g, then g is an **antiderivative** of f. Throughout the chapter we kept an eye on the inverse problem of finding antiderivatives for given functions. In general, the problem is harder than the "forward" problem of finding derivatives; in Calculus II we'll return to finding antiderivatives.

New Derivatives from Old. Having found derivatives and antiderivatives of (almost) all the basic elementary functions, we turned in Sections 3.5 and 3.6 to combining those derivatives, in the appropriate ways, to find derivatives (and sometimes antiderivatives) of functions built in various ways from the basic elements. The key combinatorial rules are the **product and quotient rules** (for algebraic combinations) and the **chain rule**, for functions built by composition.

Derivatives of Functions Defined Implicitly. Sometimes functions are given implicitly by equations rather than explicitly by formulas. In such cases, the method of **implicit differentiation** applies; we discussed it in Section 3.7.

Inverse Trigonometric Functions. Inverse trigonometric functions ("arc" functions) are defined as inverses—in the sense that the natural log and exponential functions are inverses—of the trigonometric functions. In Section 3.8 we explored these functions and their derivatives, using tools (such as implicit differentiation) developed in earlier sections.

4

Applications of the Derivative

4.1 Differential Equations and Their Solutions

Such as Pythagoras's formula, $a^2 + b^2 = c^2$.

Algebraic equations ◄ describe relations among varying quantities. *Differential equations* go one step—a giant step—further: They describe relations among changing quantities and the rates at which they change. Only by solving differential equations can many real-life phenomena be usefully described.

> *Nature's voice is mathematics; its language is differential equations.*

The idea of a differential equation first appeared, informally, in Section 2.1. Here and in the next section we take a closer look.

Differential Equations: Basic Ideas

A **differential equation** (**DE**) is any equation that contains one or more derivatives. Each of the following is a DE: ◄

Observe the various derivative notations; all are used in practice.

$$f'(x) = 6x + 5; \qquad y' = y; \qquad \frac{dy}{dx} = 5;$$

$$g'(x) = kg(x); \qquad y''(t) = -32.$$

Or more than one.

(The last DE has **order two**, because it involves a second derivative. The others have **order one**.) To **solve** a differential equation means to find a *function* ◄ that satisfies the DE. Any such function is called a **solution** to the DE. The point bears repeating:

> *A solution to a DE is a function—not a number—that satisfies the DE.*

Simple examples will illustrate the ideas and terminology; watch too for some quirks of notation.

EXAMPLE 1 Solve the DE $f'(x) = 6x + 5$. How many solutions are there? How are they related to each other?

S o l u t i o n   The DE asks for an antiderivative f of the function $6x + 5$. It's easy to check ▶ that for any constant C, the function $f(x) = 3x^2 + 5x + C$ does the job. Thus, this DE has infinitely many solutions; they differ from each other by *additive* constants. ▶   ■

But do so now!

How do the solutions differ from each other graphically?

E X A M P L E 2   Solve the DE $y' = y$. How many solutions are there? How are they related to each other?

S o l u t i o n   Let's guess a solution. The DE $y' = y$ says that the desired function y is its own derivative. The natural exponential function has this property, so $y = e^t$ is a reasonable guess for a solution.

Once we've made it, our guess is easy to check:

$$y(t) = e^t \Longrightarrow y'(t) = e^t = y(t).$$

Thus $y' = y$, as we wanted, so $y(t) = e^t$ is indeed a solution of the DE.

Are there other solutions? Given the previous example we might try $y(t) = e^t + C$ for a nonzero constant C. A good try, but no dice:

$$y(t) = e^t + C \Longrightarrow y'(t) = e^t \neq y(t).$$

Other solutions *do* exist, however. For any real constant C, $y(t) = Ce^t$ wórks: ▶

$$y(t) = Ce^t \Longrightarrow y'(t) = Ce^t = y(t).$$

Is $C = 0$ OK? Why or why not?

Again, there are infinitely many solutions, but this time they differ from each other by *multiplicative* constants.   ■

Here are some morals from the examples:

No Arguments Here.   DEs are often stated without arguments. For example, all of the DEs

$$y' = y, \qquad y'(x) = y(x), \qquad \text{and} \qquad y'(t) = y(t)$$

say the same thing: *The desired function y is its own derivative.* The input variable name, or **argument**—either x or t in the preceding examples—can safely be assumed or is understood, so it's often omitted. Once gotten used to, the omission of arguments causes little confusion. Omitting inessential function arguments sometimes unclutters and simplifies a DE's appearance. For example, both

$$\frac{d^2 y}{dt^2} + 2t \frac{dy}{dt} + t = y(t) \qquad \text{and} \qquad y'' + 2ty' + t = y$$

say the same thing, but the latter form is more economical and (depending on the method) easier to handle.

Variable Names.   In writing DEs the symbol t, rather than x, often denotes the input variable. This choice is made because the letter t suggests *time*. A solution y (also known as $y(t)$), then, is naturally thought of as a quantity that varies with time.

Instead of x, u, v, or any other letter.

The use of t, ◄ however, is *only* an aid to intuition; mathematically speaking, one variable name is as good as another. DEs, moreover, can model physical situations that have nothing to do with time.

Checking Is Easy, Finding Is Hard.   Finding solutions to DEs, starting from scratch, can be very difficult. Many thick books treat the subject. *Checking* whether a given function satisfies a DE, by contrast, is usually much easier.

Initial Value Problems

A first-order DE involves only the first derivative.

A unique solution, in math-talk.

Examples 1 and 2 are typical of first-order DEs. ◄ Such DEs usually have infinite families of solutions, in which the family members differ from each other by additive or multiplicative constants. If we also stipulate that a solution function should have a particular output value for a particular input, then we expect only one solution. ◄ This extra stipulation is called an **initial condition**, and the combination of DE and initial condition is known as an **initial value problem** (**IVP**) .

EXAMPLE 3   Solve the IVP

$$f'(x) = 6x + 5; \qquad f(0) = 42.$$

Solution   As we saw, any function of the form $f(x) = 3x^2 + 5x + C$ is a solution of the DE. With the initial condition we can find C:

$$f(0) = 3 \cdot 0^2 + 5 \cdot 0 + C = 42 \Longrightarrow C = 42.$$

Thus $f(x) = 3x^2 + 5x + 42$ is the unique solution to the IVP.   ■

EXAMPLE 4   Solve the IVP

$$y' = y; \qquad y(0) = 42.$$

Solution   As we saw already, any function of the form $y(t) = Ce^t$ solves this DE. Again, the initial condition forces C to be 42, so $y(t) = 42e^t$ is the unique solution to the IVP.   ■

Solving DEs and IVPs: Guessing, Checking, and Parameters
Checking a Guess

Given a DE and a guess at a solution, it's straightforward to check whether the guess works. The next example shows what we mean.

EXAMPLE 5   Consider the DE $y' = k(y-T)$, where k and T are constants. Is $y(t) = T + Ae^{kt}$ a solution? Does the value of the constant A matter?

Solution   Deciding whether $y(t) = T + Ae^{kt}$ solves the DE is a straightforward calculation—we'll calculate both the left-hand side (LHS) and the right-hand side (RHS) of the DE, using this y, and compare our results:

$$\text{LHS} = y'(t) = k \cdot Ae^{kt};$$

$$\text{RHS} = k(y - T) = k(T + Ae^{kt} - T) = k \cdot Ae^{kt}.$$

The two sides are equal, so y is a solution.

So far the value of A hasn't mattered—y solves the DE for *any* constant A. Had we specified an initial condition, we would have needed a specific value of A. ▶   *See the next example.*   ∎

From One Solution, Many: Choosing Values of Parameters

The DE $y' = k(y - T)$ involves two constant parameters, k and T. The general solution, $y(t) = T + Ae^{kt}$, involves still another parameter, A.

Such parameters offer an important "leverage" advantage. Assigning them specific values as needed permits us, in effect, to solve infinitely many differential equations "for the price of one." We illustrate with a practical example.

EXAMPLE 6   Freshly poured coffee in a ceramic cup ▶ has a temperature of 190°F. As it cools, the coffee obeys **Newton's law of cooling**: ▶

In a better-insulated cup, say one made of foam, coffee might cool more slowly. (See the exercises.)

> *The rate at which an object cools is proportional to the temperature difference between the object and its environment.*

Assume that room temperature is a constant 65°F, and that after 2 minutes, the coffee has cooled to 160°F. Find a formula for the temperature at time t.

Cooling is actually more complicated than this simple—but still useful—model suggests.

Solution   Under the verbiage of this example lies the DE of Example 5. If $y(t)$ represents temperature at time t, then Newton's law says that for some constant k, ▶

We'll evaluate k soon.

$$y' = k(y - 65), \qquad y(0) = 190, \qquad \text{and} \qquad y(2) = 160.$$

As we saw in Example 5, a solution of this DE has the general form

$$y(t) = 65 + Ae^{kt}$$

for some constants A and k. All that remains is to give them numerical values.

From the initial condition $y(0) = 190$, it follows that

$$190 = y(0) = 65 + Ae^{k \cdot 0} = 65 + A \implies A = 125.$$

Similarly, $y(2) = 160$ means ▶ that

Check the steps.

$$160 = y(2) = 65 + 125e^{2k} \implies e^{2k} = \frac{95}{125}.$$

Take logs of both sides.

Solving the last equation ◄ gives $k = (\ln 95 - \ln 125)/2 \approx -0.13722$. Putting it all together gives (approximately)

$$y(t) = 65 + 125e^{-0.13722t}.$$

Does the answer make common sense? A graph of y over a 50-minute period suggests it does:

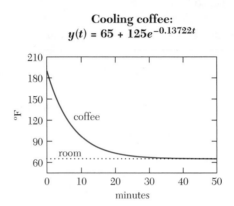

Cooling coffee:
$y(t) = 65 + 125e^{-0.13722t}$

As the temperature graph shows, the coffee cools quickly at the beginning, when it's hottest, and then more slowly as it nears room temperature. ■

Reading and Writing DEs and IVPs: The Language of Change

If $y = y(x)$ is *any* function of x, then the derivative function y' (that is, dy/dx) represents the instantaneous rate of change of y with respect to x. A differential equation, therefore, can ◄ be understood as a highly compressed statement about the *rate* at which some unknown function "grows." To *solve* a differential equation, therefore, means to find a function ◄ that "grows" in the way the DE stipulates. The DE $y' = y$, for instance, is a succinct way of saying:

And should!

Or family of functions.

> *The quantity y grows at a rate equal to y itself. Thought of as a function, y is its own derivative.*

As remarked above, functions of the form $y = Ce^x$ (but no others!) have this important property.

Translating to and from DE Language

That is, in both directions.

An ability to "translate" back and forth ◄ between the symbolic language of DEs and ordinary English statements about changing quantities and derivatives is important, both for understanding what DEs say and for applying them to solve worthwhile problems. The accompanying table illustrates several such translations. In each entry, y represents an unknown function $y(t)$ of a variable t; ◄ k denotes a *constant*.

We might have used x instead of t, but t naturally suggests time.

If the input variable t denotes time, then $y = y(t)$ represents a quantity that varies with time; $y' = y'(t)$ (that is, dy/dt) tells how fast ◄ y grows. (Think-ing of the input variable as time is intuitively convenient, but it's not necessary.

In the ordinary sense of the word "fast."

Sometimes the input variable denotes distance, position, or some other physical quantity.)

Translating between DE Language and English	
DE Version	**English Version**
$y' = 0$	The growth rate of y is zero (i.e., y remains constant).
$y' = 0.06y$	The growth rate of y is *proportional*, with proportionality constant 0.06, to y itself (i.e., y behaves like a bank account drawing 6% interest).
$y' = kt$	The growth rate of y at time t is *proportional* to t.
$y'' = k$	y has constant *second derivative* (i.e., y has constant "acceleration").
$y' = k(y - 65)$	y varies at a rate proportional to the *difference* between y and 65 .

Interpreting DEs in Real Contexts

Each DE in the table above describes how some (unknown) function y varies with time. Does any familiar, real-life quantity $y(t)$ vary in the manner described by each DE?

A DE chosen purely at random need not "model" any familiar real-world phenomenon. However, many simple DEs—including all those in the table above—do describe important, naturally occurring phenomena. The DE $y' = 0$ describes *any* quantity that remains unchanged over time—for instance the position of an immovable object or the value of a financial "deposit" under one's mattress.

In a similar vein, the DE $y' = 0.06y$ describes the value $y(t)$ at time t of a deposit that bears 6% interest. (We'll discuss financial interest in more detail in the next section.)

The DEs $y' = kt$ and $y'' = k$ are closely related; every solution to the first automatically satisfies the second. ▶ In physical language, both DEs describe quantities $y(t)$ with constant acceleration. For instance, $y(t)$ might represent the varying height of a free-falling object, because earth's gravity exerts a constant downward force, and so causes a constant downward acceleration. (Newton's second law of motion says so.)

Do you see why?

The DE $y' = k(y - 65)$, as we saw in Example 6, can be interpreted as an instance of Newton's law of cooling. If $y(t)$ is the temperature at time t of an object in 65°F room, then the DE says (à la Newton) that the object warms or cools at a rate *proportional to the difference between its own temperature and room temperature.*

Interpreting Initial Conditions

A DE such as $y' = y$ tells how fast a varying quantity $y(t)$ grows, but not how large y is at any specific time. An initial condition, such as $y(0) = 1$, adds a point of reference. It tells, in effect, the value from which y starts.

What initial conditions mean in real-world settings depends on the situation. In the various previously described scenarios an initial condition might specify, at some specific time, the value of a bank deposit, the vertical position of a falling object, or the temperature of a cooling object.

Notice that an "initial" condition need not refer specifically to time $t = 0$. In "DE-speak," *any* condition of the form $y(a) = b$ is called an initial condition. In modeling the growth of a bank account, for instance, the DE $y' = 0.06y$ describes a 6% interest rate. ◄   The initial condition $y(1995.753) = 100$ might describe an initial deposit of $100 on October 1, 1995.

With continuous compounding.

> **Hard Work, but Worth Doing.**   Solving DEs can be much more complicated than our simple examples suggest. A gigantic literature exists on the theory and methods of solving many sorts of differential equations. This body of literature attests to the depth and subtlety of the subject—but even more to its practical importance.

BASICS

In Exercises 1–10 decide whether the function is a solution of the differential equation given with it.

1. $y(t) = t^2/2; \quad y' = t$
2. $y(t) = t^2/2; \quad y' = y$
3. $y(t) = -e^t; \quad y' = -y$
4. $y(t) = e^{-t}; \quad y' = -y$
5. $y(t) = 42e^{-t}; \quad y' = -y$
6. $y(t) = Ce^{kt} + A; \quad y' = k(y - A)$
7. $y(t) = e^t + \frac{1}{2}t^2; \quad y' = y + t$
8. $y(t) = \frac{1}{2}e^{t^2}; \quad y' = ty$
9. $y(t) = e^{t^2/2}; \quad y' = ty$
10. $y(t) = \frac{1}{2}t^4 + \frac{3}{2}t^2 + \frac{1}{4}; \quad y' - \frac{2y}{t} = t^3$

FURTHER EXERCISES

11. This exercise relates to Example 6, which is about coffee cooling in a 65°F room. We found that for coffee in a ceramic cup, starting at 190°F, the solution function $y(t) = 65 + 125e^{-0.13722t}$ reasonably represents the (Fahrenheit) temperature after t minutes.
    (a) How long does the coffee in a ceramic cup remain above 100°F?
    (b) In a better-insulated cup, say one made of foam, coffee stays hot longer. Starting from 190°F, coffee in a foam cup is found to cool to 180°F in two minutes in a 65°F room. Let $z(t)$ represent the temperature of the coffee in a foam cup. Find a formula for $z(t)$.
    (c) How long does coffee in a foam cup stay above 100°F?
    (d) Plot $y(t)$ and $z(t)$ on the same axes.
12. (a) Check by differentiation that for any constant d,
    $$P(t) = \frac{1}{1 + de^{-t}}$$ is a solution of the DE $P' = P(1 - P)$. (This DE is a simple form of the so-called **logistic DE**; it's used to model certain populations.)
    (b) Solve the IVP $P' = P(1 - P); P(0) = 1/2$. [**HINT:** Use the initial condition to choose a value for d.]
    (c) Solve the IVP $P' = P(1 - P); P(0) = 2$.
    (d) Plot the solution functions for the previous two parts on the same axes; use the t-interval $0 \le t \le 5$.
13. Verify by direct calculation that if k, C, and d are con-

stants, then the function
$$P(t) = \frac{C}{1 + de^{-k \cdot C \cdot t}}$$
is a solution of the logistic DE $P' = kP(C - P)$. (We'll see the logistic DE in later applications.)

14. When the valve at the bottom of a cylindrical tank is opened, the depth of liquid in the tank drops at a rate proportional to the square root of the depth of liquid. (This fact, known as **Torricelli's law**, is named after Evangelista Torricelli [1608–1647], a contemporary of [and, briefly, secretary to] Galileo.) If $y(t)$ is the liquid's depth t minutes after the valve is opened, Torricelli's law can be expressed as the DE $y' = -k\sqrt{y}$, where k is some positive constant.
    (a) Does the water level drop faster when a tank is full or when it is half full? Explain your answer.
    (b) Verify by differentiation that, for any constants C and k, the function $y(t) = (C - kt)^2/4$ is a solution of Torricelli's DE $y' = -k\sqrt{y}$. (For physical reasons, it's OK to assume that $C - kt \ge 0$.)
    (c) Suppose that for a certain tank, $y(0) = 9$ feet and $y(20) = 4$ feet. Find an equation for $y(t)$. Show that the tank takes 60 minutes to empty entirely.
    (d) Plot $y(t)$ over the interval $0 \le t \le 60$.
    (e) At what time is the water level dropping at a rate of 0.1 feet per minute?

15. Let $h(t)$ represent the height, in meters above ground level, of an object (a helium balloon, for instance, or a cannonball, an airplane, or a toy rocket) at time t seconds. Let $v(t) = h'(t)$ represent the vertical velocity, in meters per second, at time t seconds. Note that $v(t) > 0$ means the object is rising; when $v(t) < 0$, the object falls.

    In this situation, the DE $v' = -9.8$ means the same thing as the English statement

    The object has constant downward acceleration of 9.8 meters per second per second.

    (The negative sign appears in the DE because the acceleration is downward.) Translate each of the following DEs into an English statement about the physical situation. (Specify units carefully.)
    (a) $v'(t) = 1$ (c) $v'(t) = -0.01(v(t))^2$.
    (b) $h'(t) = -3$

16. Let the situation and notation be as in the previous exercise.
    (a) Solve the IVP $v'(t) = 1$; $v(0) = 0$. Assuming these conditions, what is the object's upward velocity at $t = 10$ seconds? [**HINT:** If v' is a constant function, then v is a linear function.]
    (b) Solve the IVP $h'(t) = -3$; $h(0) = 100$. Find and interpret $h(10)$. [**HINT:** If h' is a constant function, then h is a linear function.]
    (c) Verify by differentiation that for any constant A, $v(t) = 100/(t + C)$ solves the DE $v' = -0.01v^2$.
    (d) Suppose that $v' = -0.01v^2$ and $v(0) = 5$. What's $v(30)$? What's $v(80)$? Interpret your answers in physical language. [**HINT:** See the previous part.]

17. An object moves along an east–west axis. Distance is measured in meters and time in seconds. Let $p(t)$, $v(t)$, and $a(t)$ denote, respectively, the object's position (in meters east of the starting point), eastward velocity (in meters per second), and eastward acceleration (in meters per second per second) at time t. Then $p' = v$ and $p'' = v' = a$.

    In this situation, the English statement

    The object has constant eastward acceleration 10 meters per second per second.

    says the same thing as the first-order DE $v' = 10$. Translate each of the following English statements into a first-order DE involving either v' or p'.
    (a) The object accelerates westward at a constant 5 meters per second per second.
    (b) The object travels eastward at a constant speed of 15 meters per second.
    (c) Because of friction, the object's eastward acceleration is proportional to the velocity. [**NOTE:** The proportionality constant will be negative; explain why.]

    (d) The object's eastward velocity is proportional to the square root of the object's position.

18. In learning theory, a "performance function" $P(t)$ may be used to measure someone's skill at a task (using units appropriate to the task) as a function of the training time t. The graph of a performance function is called a **learning curve**. For many tasks, performance improves quickly at first, but then tapers off (i.e., the rate of learning decreases) later, as the value of $P(t)$ approaches M, some maximal level of performance.
    (a) Explain how the DE $P' = k(M - P)$, where k is a positive constant, describes this situation. (Notice the similarity to Newton's law of cooling!)
    (b) Verify by differentiation that for any constant A, the function $P(t) = M - Ae^{-kt}$ solves the DE above.
    (c) For a certain learning activity, measurements show that $P(0) = 0.1M$ and $k = 0.05$; t is measured in hours. How long does it take to reach 90% of the maximum level of performance?

19. A flu epidemic spreads through a 3000-student college community at a rate proportional to the product of the number of members already infected and the number of those not yet infected. (This product measures the number of possible infectious contacts.)

    Let $P(t)$ represent the number of students infected after t days. Express the statement above as a differential equation.

20. It's a winter day in Frostbite Falls, Minnesota—10 degrees below zero Celsius. Boris and Natasha stop at a convenience store for hot (initially, $90°C$) coffee to warm them on the cold, windy walk home. Two types of cups are available: environmentally destructive foam and politically correct cardboard.

    Boris wants foam. "What do I care about the greenhouse effect?" he asks. "Besides, remember the proportionality constant k in Newton's law of cooling—$y'(t) = k \cdot (y(t) - C)$, where C is the environmental temperature, $y(t)$ is the Celsius temperature of the coffee at time t, and t is measured in minutes? Well, for foam, $k = -0.05$. For cardboard, k is a pathetic -0.08."

    "Do what you want, Boris," says Natasha. "I'd rather save the world. I'm having cardboard. And make mine decaf."
    (a) How long does Boris's coffee stay above 70 degrees? How about Natasha's? How hot is each cup after 5 minutes?
    (b) Redo part (a), assuming that Boris and Natasha drink their coffee in the overheated ($25°C$) convenience store.
    (c) What value of k is needed to assure that coffee, starting at $90°C$, is still at least $70°C$ after 5 minutes outdoors. [**HINT:** Use the solution function to set up an appropriate equation; then solve for k.]

4.2 More Differential Equations: Modeling Growth

Exponential Growth

Exponential functions have an important property, best expressed in terms of rates:

> *An exponential function grows at a rate proportional to its size, i.e., "with interest."*

We discussed this property in Section 3.3.

Thanks to this property, ◄ exponential functions model many real phenomena that also grow "with interest."

Check these claims carefully!

Let's interpret this property in the language of DEs and IVPs. For any constants A and k, let $y(t) = Ae^{kt}$. Two key properties of y matter here: ◄

$$y'(t) = Ake^{kt} = ky(t) \quad \text{and} \quad f(0) = Ae^0 = A.$$

In other words:

> **Theorem 1** For any constants A and k, the exponential function $y = Ae^{kt}$ solves the initial value problem
> $$y' = ky; \qquad y(0) = A.$$

This theorem offers considerable practical horsepower. Just by choosing k and A judiciously, we can solve a surprising number of useful and important DEs.

Modeling Things That Grow "with Interest"

Many real-world quantities grow or "decay" like exponential functions, at rates proportional to their size. The Consumer Price Index, the value of a financial deposit, college fees, ◄ radioactive decay, biological populations—all can be modeled, more or less accurately, by the IVP of Theorem 1.

Recall Uncle Eric, in Section 1.6.

The Federal Deficit: Keeping Rates Straight

Differential equations relate varying quantities to their derivatives, i.e., their instantaneous rates of change. Practical problems, on the other hand, often mention other sorts of rates; for instance, average rates, interest rates, and percentages. Teasing out the derivatives from all this loose talk may take some care.

EXAMPLE 1   At time $t = 0$ years, the federal deficit is \$300 billion and growing fast. "*How* fast?" worries Senator Smith. "What will the deficit be 6 years hence, ◄ when I face reelection?" In Senate hearings, Economist A ◄ predicts:

Politicians say "hence."

A is for "average."

> *The federal deficit will grow 15% each year.*

"Balderdash," says Economist I, ◄

I is for "instantaneous."

> *The federal deficit will grow at an instantaneous rate of 15% per year.*

Do the two economists disagree? How seriously? Does the difference matter?

Solution   Economists A and I agree on one thing: the deficit grows at a rate proportional to itself. In DE terms, if $y(t)$ represents the deficit in billions, t years out, A and I agree that for some constant k,

$$y'(t) = ky(t).$$

They *disagree*, however, on the value of k. Economist A's statement refers to an *average* rate: Over a 1-year interval, the deficit rises 15%. Economist I's statement means something else: At any *instant*, the deficit's rate of increase, if it remained unchanged for a year, would raise the deficit by 15%.

To understand the subtle difference and see its effects, we'll recast the predictions as IVPs. Combined with the initial value ($300 billion) of the deficit, I's claim comes to this:

$$y'(t) = 0.15 \cdot y(t); \qquad y(0) = 300.$$

A says something else:

$$y'(t) = k \cdot y(t); \qquad y(t+1) = 1.15 \cdot y(t); \qquad y(0) = 300.$$

We'll solve both IVPs and compare the results. I's claim fits perfectly into Theorem 1's template. Since $y'(t) = 0.15 \cdot y(t)$ and $y(0) = 300$, it follows that

$$y(t) = 300e^{0.15t}.$$

This is I's model for the deficit.

Solving A's IVP requires one more step. Since $y'(t) = k \cdot y(t)$ and $y(0) = 300$, it follows (from Theorem 1, again) that $y(t) = 300e^{kt}$.

Now we'll find a value for k. Our new formula $y(t) = 300e^{kt}$ gives $y(t+1) = 300e^{k(t+1)}$. Therefore, ▶   *Check the algebra carefully.*

$$y(t+1) = 1.15 \cdot y(t) \implies 300e^{k(t+1)} = 1.15 \cdot 300e^{kt}$$

$$\implies e^k = 1.15$$

$$\implies k = \ln 1.15 \approx 0.13976.$$

Economist A therefore models the deficit with the function $y(t) = 300e^{0.13976t}$.

What does each scenario predict for the next 6 years? Graphs answer best:

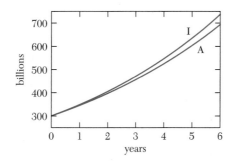

Dueling economists: growth of the deficit

A's and I's predictions diverge as time goes on. At $t = 6$, the difference is

$$300e^{0.15 \times 6} - 300e^{0.13976 \times 6} \approx 43.97 \text{ billion dollars—}$$

real money, even by government standards. ■

Money in the Bank

Bank interest serves as the most familiar example of growth with interest. Its crucial feature—shared by exponential growth in *all* forms—is that *the rate at which the account's value grows is proportional to the value itself.* The rich don't only get richer—they also get richer *faster*.

Bank interest may even be *too* familiar; some care is needed in translating banking language into mathematical terms. The next example illustrates this.

EXAMPLE 2   A savings account pays a **nominal** annual interest rate of 6%, compounded daily. Model the situation in DE language. How much does the account increase over 1 year?

Solution   If $b(t)$ represents the balance (in dollars) at time t (in years), the preceding terms mean that each day (in a 365-day year), the balance is multiplied by the same small factor:

$$b\left(t + \frac{1}{365}\right) = b(t) \cdot \left(1 + \frac{0.06}{365}\right). \tag{4.1}$$

With initial balance $b(0)$, for instance, here's what happens for the first 2 days:

$$b\left(\frac{1}{365}\right) = b(0) \cdot \left(1 + \frac{0.06}{365}\right); \qquad b\left(\frac{2}{365}\right) = b(0) \cdot \left(1 + \frac{0.06}{365}\right)^2.$$

Over a full year, therefore, the balance increases by the factor

$$\left(1 + \frac{0.06}{365}\right)^{365} \approx 1.0618.$$

The **effective annual interest rate**, therefore, is 6.18%; over a 1-year interval, the balance rises by this percentage.

How do interest rates, nominal or effective, translate into DE language?

Not continuously.

- In reality, the balance $b(t)$ varies discretely, ◄ at regular intervals and in jumps of at least 1 cent. Any DE model, being continuous, must commit *some* error. In practice, such errors are usually small enough to ignore.

- On a time scale measured in years, a single day is effectively an "instant." The *nominal* interest rate can therefore serve as k in the DE of Theorem 1. In other words, the balance $b(t)$ can be taken, with little error, to satisfy the DE

$$b'(t) = 0.06 \cdot b(t).$$

Theorem 1 solves the preceding DE. If the initial balance is b_0, then

$$b'(t) = 0.06 \cdot b(t) \quad \text{and} \quad b(0) = b_0 \Longrightarrow b(t) = b_0 \cdot e^{0.06t}.$$

After 1 year the balance is

$$b(1) = b_0 \cdot e^{0.06} \approx b_0 \cdot 1.0618.$$

Thus, the balance rises by 6.18%, the *effective* interest rate. ■

Interest Rates and Derivatives: Continuous Compounding.   The DE $b'(t) = 0.06 \cdot b(t)$ does a good—but not perfect—job of modeling 6% annual interest, compounded daily. The model would do better if interest were compounded hourly or even every minute. The DE represents **continuous compounding**, i.e., the limiting case as the number of compounding periods tends to infinity.

   Why does the DE $b'(t) = 0.06 \cdot b(t)$ approximate daily compounding as well as it does? Equation 4.1 and the following computation (for those who wish to follow it) explain. The result shows, too, how the frequency of compounding affects the "goodness" of approximation.

$$b'(t) \approx \frac{b(t + 1/365) - b(t)}{1/365}$$

$$= \left(b(t) \cdot \left(1 + \frac{0.06}{365} \right) - b(t) \right) \cdot 365$$

$$= b(t) \cdot \left(1 + \frac{0.06}{365} - 1 \right) \cdot 365$$

$$= 0.06 \cdot b(t).$$

Radioactive Decay

A sample of carbon-14 (a radioactive isotope of ordinary carbon) decays at time t (in years) at a rate proportional to its weight, with $k = -0.000121$. In DE language,

$$W'(t) = -0.000121 W(t).$$

(The constant k is negative because W shrinks. In banking terms, carbon-14 pays negative interest.)

EXAMPLE 3   If a sample of carbon-14 weighed W_0 pounds in 1991, how much will it weigh in 2991? What did it weigh in 991? How long does it take the sample to lose half its weight? (This period is called the sample's **half-life**.)

That is, t = 0 in 1991.

Solution   If $W(t)$ represents the sample's weight t years after 1991, ◄ then $W(t)$ solves the IVP

$$W'(t) = -0.000121W(t); \qquad W(0) = W_0.$$

According to Theorem 1, $W(t) = W_0 e^{-0.000121t}$. Therefore,

$$\text{weight in 2991} = W(1000) = W_0 e^{-0.121} \approx 0.88603 W_0;$$

$$\text{weight in 991} = W(-1000) = W_0 e^{0.121} \approx 1.12862 W_0.$$

The sample loses a bit more than 12% of its weight over 1000 years.

The sample's half-life is the value of t for which $W(t) = W_0/2$. To find it, we solve for t:

$$W(t) = \frac{W_0}{2} \iff W_0 e^{-0.000121t} = 0.5 W_0 \iff -0.000121t = \ln(0.5)$$

$$\iff t \approx 5728 \text{ years.} \qquad \blacksquare$$

Some Biological Populations Grow Exponentially

We'll soon see why such growth can't last forever.

Certain biological populations grow exponentially—at least for a while. ◄ A fruit fly population in laboratory conditions, for example, might reasonably grow at an instantaneous rate of 5% per day, i.e., at a rate proportional (with $k = 0.05$) to the population itself. In DE language, if the population $P(t)$ starts at P_0, then P satisfies the IVP

$$P'(t) = 0.05 P(t); \qquad P(0) = P_0.$$

Notice the analogy with ordinary interest: Each day's crop of new fruit flies can safely be thought of as "interest" added to the previous day's "balance."

EXAMPLE 4   A population of fruit flies grows at an instantaneous rate of 5% per day. How long does it take for the population to double?

Solution   Given an initial population (i.e., at $t = 0$) of P_0 flies, Theorem 1, on page 256, gives the formula

$$P(t) = P_0 e^{0.05t}.$$

A little algebra shows that the population doubles in less than 14 days:

$$P(t) = P_0 e^{0.05t} = 2 P_0 \iff 0.05t = \ln 2 \iff t \approx 13.8629. \qquad \blacksquare$$

Not All Populations Grow Exponentially: Logistic Growth

A logical, not just biological, impossibility.

Real biological populations can't grow exponentially forever. An initial "deposit" of 100 fruit flies, growing as just described, would number over 8 billion in a year. In a decade, the flies would outnumber Earth's molecules. ◄

Any exponentially-growing population sooner or later exceeds the limits (physical or biological) of its environment. This possibility applied to human population worried the 19th-century clergyman Thomas Malthus, who predicted an early and unpleasant end to the human race. ▶   Whether Malthus was right or wrong, exponentially-growing biological populations do eventually outgrow their environments. How real populations cope with this mathematical certainty (and with other vagaries of real life) varies: Some (e.g., lemmings and Dungeness crabs) have boom–bust cycles; others eat up their environments and die off completely.

Was Malthus right? It's too soon to tell. Some evidence may even suggest that Malthus was too optimistic.

A third (happier) possibility is **logistic growth**. As a population approaches an upper limit C (the environment's **carrying capacity**), growth slows. For our flies, logistic growth might lead to the lower of the following graphs.

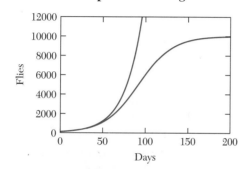

**Population growth:
exponential vs. logistic**

Toward the beginning the two growths are almost identical. This reflects the fact that when populations are small, the effects of crowding are scarcely felt, so logistic growth is essentially exponential.

Like exponential growth, logistic growth is characterized by a rate property, and therefore by a DE (the **logistic DE**):

The population's growth rate is proportional both to the population itself and to the difference between the carrying capacity and the population.

In symbols,

$$P' = kP(C - P),$$

where P represents the population, P' the population's growth rate, C the carrying capacity of the environment, and k a constant of proportionality. ▶

P and P' vary with time; k and C are parameters.

Wise Flies: A Case Study in Logistic Growth

A population of prudent fruit flies, eager to avoid crowding, chooses to reproduce logistically. This is their story.

On a certain day, the fly population is 1000 and growing at the (net) instantaneous rate of 50 flies per day. The environment can support 10,000 flies.

Questions.   Given what we know of growth *rates*, it's natural to wonder about population *amounts*:

- What is the population in 10 days? 100 days? 200 days?
- When is the population 5000? 9000? 9900?
- When does the population grow fastest? What's the population then?

Solving the Logistic DE

Modeling logistic population growth boils down to solving an IVP. If a population has P_0 members at $t = 0$ and grows logistically, then the population function $P(t)$ is a solution of the IVP

$$P' = kP(C - P); \qquad P(0) = P_0.$$

It gets easier with techniques from integral calculus.

Searching systematically for a solution, given our present tools, is a slightly messy process. ◄   Suffice it to say that, for any constants k, C, and d, the function

$$P(t) = \frac{C}{1 + d\,e^{-kCt}}$$

Try checking this; it's messy but not really hard.

solves the logistic DE $P' = kP(C - P)$. ◄   (The constants k and C in the solution are the same as those in the DE; as we'll see, d depends on the initial population.)

What Solutions Mean for Our Flies

To put things in context, let's define:

$$t = \text{time, in days, since the original measurement;}$$

Measuring population in thousands removes pesky zeros from later computations.

$$P(t) = \text{population of flies, in thousands, ◄ at time } t;$$

$$P'(t) = \text{growth rate, in thousands of flies per day, at time } t.$$

$P'(0) = 0.05$ corresponds to 50 flies/day.

Evaluating the Parameters.   We know three things: $P(0) = 1$, $P'(0) = 0.05$, ◄ and $C = 10$. We'll use what we know to evaluate k and d. Since

$$P'(0) = 0.05 = k \cdot P(0) \cdot (10 - P(0)) = k \cdot 1 \cdot 9,$$

it follows that $k = 0.05/9 \approx 0.00556$.

Now we can evaluate d. By assumption,

$$P(0) = 1 = \frac{10}{1 + d\,e^{-kC0}} = \frac{10}{1 + d},$$

so $d = 9$. Putting everything together gives, at last, an explicit formula for $P(t)$:

$$P(t) = \frac{C}{1 + d\,e^{-kCt}} = \frac{10}{1 + 9\,e^{-0.0556t}}.$$

Here's a graph of P:

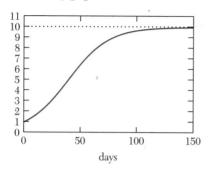

The fruit fly population, in thousands

days

Our Questions Answered: What the Graph and Formula Say.   With the formula and the graph we can answer—approximately at least—all the questions posed earlier.

- The graph shows that on day 100 the population was about 9700; room is quickly running out. Day 200 isn't shown, but the graph's asymptotic behavior shows that at $t = 200$ the population should still be just short of 10,000. The formula agrees:

$$P(200) = \frac{10}{1 + 9\,e^{-0.0556 \cdot 200}} \approx 9.99867.$$

  Thus, rounding to the nearest whole fly (the only reasonable way to round flies) the population is at full capacity.

- Both graph and formula show that the population reaches 5000 (half the upper limit) in about 40 days. Around day 80, the population reaches 9000—90% capacity. Around day 130, 99% capacity is reached. ▶

- Intuition says that under logistic conditions a population increases slowly at first (when there are few breeding members), faster as the population rises, and then slowly again it nears the upper limit. The graph agrees. Population grows fastest when the graph points most steeply upward, i.e., around day 40. ▶ At this time the population is about 5000—exactly half the carrying capacity. This isn't a coincidence. Under logistic conditions, it's a disadvantage for a population to be either too large or too small. Thus, intuitively, we'd expect growth conditions to be optimal at the halfway point.

All of these numbers can be checked with the formula.

Take a close look; do you agree?

Afterword: Discrete versus Continuous Growth

All mathematical models more or less simplify the phenomena they describe. The use of DEs and IVPs to model phenomena of growth is no exception to this rule.

One sense in which DEs and IVPs may describe reality imperfectly concerns the difference between discrete and continuous phenomena of change. Human population, for instance, grows discretely, in jumps of at least one unit. Hot coffee in a cool room, by contrast, cools continuously—its temperature takes all intermediate values on the way down.

DEs and IVPs are continuous models of growth. Solutions to DEs and IVPs are continuous functions, so they can't, by definition, describe discrete phenomena exactly. (Neither, for the same reason, can discrete models perfectly describe continuous phenomena.) Such philosophical problems, though undeniable, need not be fatal. In practice, differences between discrete and continuous growth models are often unimportant, especially when the "jumps" are small. It matters little, for instance, whether coffee temperature falls continuously or in tiny increments of 0.0001 degree. Perfection is impossible, but DEs and IVPs can be (and are) used to model growth—even discrete growth—easily, effectively, and accurately.

> **Discrete Models of Continuous Phenomena.**   DEs and IVPs are continuous entities that are often used to describe phenomena that vary discretely. Sometimes the relationship goes the other way: Continuous phenomena can be modeled discretely. Movies, digital watches, and electronic computing all depend, in various ways, on discrete approximations of continuous reality.

BASICS

Use Theorem 1, page 256, to solve the IVP's in Exercises 1–4. Check answers by differentiation.

1. $y' = 0.1y$; $y(0) = 100$
2. $y' = -0.0001y$; $y(0) = 1$
3. $y' = (\ln 2) \cdot y$; $y(2) = 4$
4. $y' = ky$; $y(10) = 2y(0)$ [**HINT:** Solve for k using the initial condition.]
5. Suppose that the "wise flies" described in this section had begun by breeding at the net rate of 100 (rather than 50) flies per day. How would things be different? More specifically, answer the following questions, and compare the results with those given on page 263.
   (a) What size will the population be in 10 days? 100 days?
   (b) When is the population 5000 flies? 9000? 9900?
   (c) When does the population grow fastest? What's the population then?

FURTHER EXERCISES

6. First National Bank bank advertises 8% interest, compounded continuously. Second National Bank advertises 10% interest, compounded continuously—but also charges depositors a $100 yearly administrative fee for the privilege of banking there. (For simplicity, assume that the $100 fee is deducted continuously over the full year.) Let $P(t)$ denote the value in dollars of a deposit after t years.
   (a) First National Bank's policy amounts to the DE $P'(t) = 0.08P(t)$ (or just $P' = 0.08P$). Explain why.
   (b) Second National Bank's policy amounts to the DE $P'(t) = 0.1P(t) - 100$ (or just $P' = 0.1P - 100$). Explain why.
   (c) Solve the IVP $P' = 0.08P$, $P(0) = 1000$. If $1000 is deposited now in First National Bank, how much will it be worth in 10 years?
   (d) Solve the IVP $P' = 0.1P - 100$, $P(0) = 1000$. [**HINT:** The DE can be rewritten like that for Newton's law of cooling.]
   (e) If $1000 is deposited now in Second National Bank, how much will it be worth 10 years?

7. A certain menacing biological culture (aka The Blob) grows at a rate proportional to its size. When it arrived unnoticed one Wednesday noon in Chicago's Loop, it weighed just 1 g. By the 4:00 p.m. rush hour it weighed 4 g.

   The Blob has its "eye" on the Sears Tower, a tasty morsel weighing around 3,000,000,000,000 g (i.e., 3×10^{12} g). As soon as it weighs 1000 times as much (i.e., 3×10^{15} g), The Blob intends to *eat* the Sears Tower. By what time must The Blob be stopped? Will Friday's rush-hour commuters be delayed?

8. A mold grows at a rate proportional to the amount present. Initially, its weight is 2 g; after two days, it weighs 5 g. How much does it weigh after eight days?

9. A bacterial culture is placed in a large glass bottle. Suppose that the volume of the culture doubles every hour, and the bottle is full after one day.
   (a) If the culture was placed in the bottle at time $t = 0$ hours, when was the bottle half full?
   (b) Assume that the bottle is "almost empty" when the culture occupies less than 1% of its volume. How long is the bottle "almost empty"?

10. The police guard gave Sara a cold look, but his voice was polite as he directed her to the room she sought. "Don't touch anything, please, Ms. Abrams. The Chief said I had to let you in, but he said to tell you to mind your fingers." "Thank you," Sara replied coolly. "The Chief knows he can trust me." The guard opened his mouth as if to speak, but he merely shook his head and withdrew.

Sara was standing in what appeared to be a combination bedroom and laboratory. A relative had found Dr. Howell's body on the floor of this room that morning. By 9:00 a.m., the coroner had completed his examination; he stated that death was due to a severe blow to the head, and that Dr. Howell had been dead for 36 to 40 hours. It seemed critical to Sara to know exactly when Dr. Howell had died so that she could eliminate certain suspects. But how could she possibly discover exactly when he was killed? Puzzled, she wandered around the small, cluttered room, being careful not to touch anything. The old doctor apparently had been conducting an experiment when he was killed. Sara absentmindedly read from the notebook that was lying open on the bench:

The fungus grows at a rate proportional to its current weight.

"Great," she thought, "I'm here to investigate a murder, and instead I'm getting a biology lesson." At a loss for what else to do, she continued reading.

To exemplify this biological truth, I place the fungus on a scale and record its weight at various times

10 g	5:30 p.m.
12 g	6:15 p.m.
13	

"Hmm," Sara mused, "the poor guy didn't even get to finish the last entry." Sara suddenly frowned in con-

centration. She searched her pockets until she found a pencil stub and an old receipt. When the guard entered the room a few minutes later, Sara had just finished scribbling on the receipt. She smiled as she shoved the receipt and the pencil stub back into her pocket.

"Don't worry," Sara said cheerfully, "I'm leaving. I now know exactly when Dr. Howell was killed." The guard looked sourly at Sara's back as she left the room.

When was Dr. Howell killed? How did Sara know?

11. Oil is pumped continuously from a well at a rate proportional to the amount of oil left in the well. Initially there were 1,000,000 barrels of oil in the well; 6 years later, 500,000 barrels remain.

(a) At what rate was the amount of oil in the well decreasing when there were 600,000 barrels of oil remaining?

(b) It will no longer be profitable to pump oil from the well when there are fewer than 50,000 barrels remaining. When should pumping stop?

12. Human skeletal fragments were brought to a laboratory for carbon dating. Analysis showed that the proportion of ^{14}C to ^{12}C is only 6.25% of the value in living tissue. How long ago did this person die?

13. A tank initially contains 100 gallons of water and 10 pounds of salt, thoroughly mixed. Pure water is added at the rate of 5 gallons per minute, and the mixture is drained off at the same rate. (Assume complete and instantaneous mixing.)

(a) Explain why $S(t)$, the amount of salt in the tank at time t, is the solution of the IVP

$$S' = -\frac{5}{100}S; \quad S(0) = 10.$$

(b) Using Theorem 1, page 256, find a solution of this IVP.

(c) How much salt is left in the tank after 1 hour?

4.3 Linear and Quadratic Approximation; Taylor Polynomials

In this section we consider how to find and use polynomial functions that approximate other (non-polynomial) functions. What does it mean for a function f to approximate another function g? How can we choose "good" polynomial approximations? How are derivatives involved?

We'll start with the simplest cases: linear and quadratic polynomials. Then we'll describe how to find polynomials of *any* degree, called **Taylor polynomials**, that approximate a given function.

Why Bother?   Often in mathematics and applied mathematics one wants to approximate one function—say, *f*—with another—say, *g*. Why bother? For several possible reasons: If *f* is complicated, poorly understood, or otherwise inconvenient, it's useful to replace *f* with a simpler, better behaved, more tractable, or better understood *g*.

Polynomial functions are simple, convenient, well understood, and easy to use, so it's natural to use them to approximate more complicated functions.

Tangent Lines and Linear Approximation

The line tangent to the graph of a function f at $x = x_0$ can be thought of in several ways. Geometrically, the tangent line is the straight line that best "fits" the graph of f at $x = x_0$. Analytically, the tangent line represents the linear function that best approximates f near $x = x_0$.

EXAMPLE 1   If $f(x) = \sqrt{x}$, then $f(64) = 8$. How closely does the tangent-line function approximate values of $f(x)$ for x near 64?

Verify this.

Solution   Since $f'(x) = 1/(2\sqrt{x})$, $f'(64) = 1/16$. Therefore, the line tangent to the f-graph at $(64, 8)$ has equation ◄

$$\ell(x) = f(64) + f'(64)(x - 64) = 8 + \frac{x - 64}{16} = 4 + \frac{x}{16}.$$

Here are graphs of both f and ℓ:

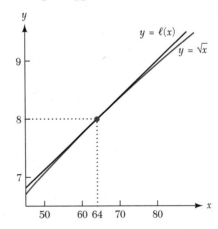

The tangent approximation at $x = 64$ to $\sqrt{x}$

Observe:

- The tangent line hugs the graph of f closely; ℓ should therefore approximate f well for x near $x = 64$—the nearer the better.

But only slightly if x is near 64.

- The graph of f is concave down, so it lies below its tangent line. Therefore, the tangent line function ℓ *overestimates* f ◄   for each $x \neq 64$.   ∎

Constant, Linear, and Quadratic Approximations

How might we approximate *any* function f near *any* convenient "base point" $x = x_0$? Example 1 showed that the tangent line at $x = x_0$ defines a good **linear approximation** to f at x_0. **Constant** and **quadratic approximations** are also possible.

E X A M P L E 2   Which constant function best approximates $f(x) = \sqrt{x}$ near $x = 64$? Which quadratic function?

S o l u t i o n   Since $f(64) = 8$, the constant function $c(x) = 8$ is the only sensible choice; c has the "right" value at $x = 64$.

How should we choose a good quadratic approximation q? For the linear approximation ℓ we required that

$$\ell(64) = f(64) = 8 \qquad \text{and} \qquad \ell'(64) = f'(64) = \frac{1}{16}.$$

It's reasonable, then, to ask that a good quadratic approximation q go one step further: ▶

Why is $f''(64) = -1/2048$? Check for yourself.

$$q(64) = 8; \qquad q'(64) = \frac{1}{16}; \qquad q''(64) = f''(64) = -\frac{1}{2048}. \qquad (4.2)$$

In other words, our desired quadratic function q should "agree" with f at $x = 64$, both in its value and in its first *two* derivatives.

The easiest way to construct such a q is to write it in the form ▶

Writing q in powers of $(x - 64)$, not x, greatly simplifies some of the algebra.

$$q(x) = a + b(x - 64) + c(x - 64)^2,$$

and then choose the coefficients a, b, and c appropriately. Since we'll soon need them, we first calculate several values and derivatives of q: ▶

Watch carefully—here's where the form of q pays off.

$$q(x) = a + b(x - 64) + c(x - 64)^2 \Longrightarrow q(64) = a;$$
$$q'(x) = b + 2c(x - 64) \Longrightarrow q'(64) = b;$$
$$q''(x) = 2c \Longrightarrow q''(64) = 2c.$$

Choosing a, b, and c is now easy:

Choosing a. Equation 4.2 requires that $q(64) = 8$. But $q(64) = a$, so $a = 8$.

Choosing b. Equation 4.2 requires that $q'(64) = 1/16$. But $q'(64) = b$, so $b = 1/16$.

Choosing c. Equation 4.2 requires that $q''(64) = -1/2048$. But $q''(64) = 2c$, so $c = -1/4096$. ▶

Where did the factor of 2 come from?

Putting everything together gives

$$q(x) = 8 + \frac{x - 64}{16} - \frac{(x - 64)^2}{4096}$$

as the best quadratic approximation to f at $x = 64$.

Graphs of c, ℓ, q, and f show what's happening:

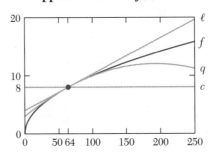

Constant, linear, and quadratic approximations to $f(x) = \sqrt{x}$

Observe:

- All three approximating functions (c, ℓ, and q) have the right value at $x = 64$.
- The graphs of ℓ and q (but not of c!) are tangent to the f-graph (i.e., have the same slope) at $x = 64$.

And much better than c.
- The function q fits f better than ℓ ◄ near $x = 64$. The fact that $q''(64) = f''(64)$ means that q and f have the same concavity at $x = 64$.

How well, numerically, do c, ℓ, and q approximate f? A table gives some idea:

Three Approximations to $f(x) = \sqrt{x}$							
x	50	63	63.9	64	64.1	65	80
$c(x)$	8.00000	8.00000	8.00000	8	8.00000	8.00000	8.00000
$\ell(x)$	7.12500	7.93750	7.99375	8	8.00625	8.06250	9.00000
$q(x)$	7.07715	7.93726	7.99375	8	8.00625	8.06226	8.93750
$f(x)$	7.07107	7.93725	7.99375	8	8.00625	8.06226	8.94427

The numbers agree with the graphs: q approximates f best, especially near $x = 64$. ∎

Definitions

Linear and quadratic approximations can be chosen for any well-behaved function f, at any base point x_0. The formal definition summarizes, in general language, what we did above for $f(x) = \sqrt{x}$ and $x_0 = 64$.

Definition Let f be any function for which $f'(x_0)$ and $f''(x_0)$ exist. The linear approximation to f, based at x_0, is the linear function

$$\ell(x) = f(x_0) + f'(x_0)(x - x_0).$$

The quadratic approximation to f, based at x_0, is the quadratic function

$$q(x) = f(x_0) + f'(x_0)(x - x_0) + \frac{f''(x_0)}{2}(x - x_0)^2.$$

EXAMPLE 3   Let $f(x) = 10^x$. Find ℓ and q, the linear and quadratic approximations to f, based at $x = 3$. Use each to estimate $f(3.1)$.

Solution   We need two derivatives and a value for f, all at $x = 3$. Calculation gives $f(x) = 10^x$, $f'(x) = 10^x \ln 10$, and $f''(x) = 10^x (\ln 10)^2$, so ▶

Check our work.

$$f(3) = 10^3 = 1000; \qquad f'(3) = 1000 \ln 10 \approx 2302.5851;$$
$$f''(3) = 1000(\ln 10)^2 \approx 5301.8981.$$

The definitions say:

$$\ell(x) = f(3) + f'(3) \cdot (x - 3) \approx 1000 + 2302.5851(x - 3);$$
$$q(x) = \ell(x) + \frac{f''(3)}{2} \cdot (x - 3)^2 \approx 1000 + 2302.5851(x - 3) + 2650.9491(x - 3)^2.$$

Numerical calculation shows:

$$\ell(3.1) \approx 1230.2585; \qquad q(3.1) \approx 1256.7680; \qquad f(3.1) \approx 1258.9254.$$

The following graphs, like the preceding numbers, show how much better q does than ℓ in fitting f:

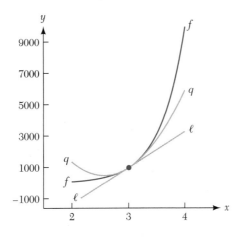

**Linear and quadratic approximation
to $f(x) = 10^x$**

The picture shows why ℓ fits f so poorly: f bends sharply away from ℓ, even quite near $x = 3$. The second derivative, f'', explains why. Any linear function ℓ has zero second derivative. Our target function f, by contrast, has a huge second derivative: $f''(3) \approx 5000$. Because f is so drastically nonlinear, it is badly approximated by any straight line—even the best-fitting one. ■

Taylor Polynomials

Linear and quadratic approximations ℓ and q are built to match f as well as possible at x_0; more precisely, ℓ and f have the same value and *first* derivative at x_0; q and f have the same value and first *two* derivatives at x_0.

Still better agreement is possible if polynomials of degree 3, 4, and higher are used to approximate f. An important example illustrates this idea.

> **Made to Order.**   The constant, linear, and quadratic approximations are sometimes called **zeroth-order**, **first-order**, and **second-order approximations** to f near x_0. In each case "order" refers to the number of derivatives of f with which a particular approximation agrees at x_0.

E X A M P L E 4   Find a fifth-degree polynomial p_5 that agrees with $f(x) = e^x$ to order 5 at $x_0 = 0$—i.e., such that

$$p_5(0) = f(0); \qquad p_5'(0) = f'(0); \qquad p_5''(0) = f''(0);$$
$$p_5^{(3)}(0) = f^{(3)}(0); \qquad p_5^{(4)}(0) = f^{(4)}(0); \qquad p_5^{(5)}(0) = f^{(5)}(0).$$

That's why we chose this f!

S o l u t i o n   We need values for several derivatives of f, all at $x = 0$. The fact that $f(x) = e^x$ makes calculating derivatives very, very easy: ◀

$$f(x) = e^x \implies f(0) = f'(0) = f''(0) = f^{(3)}(0) = f^{(4)}(0) = f^{(5)}(0) = 1.$$

We want a polynomial p_5 with the same derivatives at $x = 0$. To begin, we write $p_5(x)$ in standard form:

$$p_5(x) = a_0 + a_1 x + a_2 x^2 + a_3 x^3 + a_4 x^4 + a_5 x^5,$$

For each i, a_i is the coefficient of the ith power x^i.

The coefficients a_i are still to be found. ◀

Finding a_0 through a_5 is surprisingly easy. The key idea is that the a_i are closely related to derivatives of p_5 at $x = 0$. Specifically: ◀

Check these easy but important calculations.

$$p_5(x) \;\; = a_0 + a_1 x + a_2 x^2 + a_3 x^3 + a_4 x^4 + a_5 x^5 \implies p_5(0) = a_0;$$
$$p_5'(x) \;\; = a_1 + 2a_2 x + 3a_3 x^2 + 4a_4 x^3 + 5a_5 x^4 \implies p_5'(0) = a_1;$$
$$p_5''(x) \;\; = 2a_2 + 3\cdot 2a_3 x + 4\cdot 3a_4 x^2 + 5\cdot 4a_5 x^3 \implies p_5''(0) = 2a_2;$$
$$p_5'''(x) \;\; = 3\cdot 2a_3 + 4\cdot 3\cdot 2a_4 x + 5\cdot 4\cdot 3a_5 x^2 \implies p_5'''(0) = 6a_3;$$
$$p_5^{(4)}(x) = 4\cdot 3\cdot 2a_4 + 5\cdot 4\cdot 3\cdot 2a_5 x \implies p_5^{(4)}(0) = 24a_4;$$
$$p_5^{(5)}(x) = 5\cdot 4\cdot 3\cdot 2a_5 \implies p_5^{(5)}(0) = 120a_5.$$

The pattern of coefficients should now be clear: *For each i, $p_5^{(i)}(0) = i!\,a_i$.* ▶ In other words:

The symbol $i!$ denotes the factorial of i. We will discuss factorials soon.

$$a_0 = p_5(0); \qquad a_1 = p_5'(0); \qquad a_2 = \frac{p_5''(0)}{2!};$$

$$a_3 = \frac{p_5'''(0)}{3!}; \qquad a_4 = \frac{p_5^{(4)}(0)}{4!}; \qquad a_5 = \frac{p_5^{(5)}(0)}{5!}.$$

Now we can find p_5. By design,

$$p_5(0) = f(0) = 1; \qquad p_5'(0) = f'(0) = 1;$$

$$p_5''(0) = f''(0) = 1; \quad \ldots \quad p_5^{(5)}(0) = f^{(5)}(0) = 1.$$

Thus,

$$a_0 = 1; \qquad a_1 = 1; \qquad a_2 = \frac{1}{2!};$$

$$a_3 = \frac{1}{3!}; \qquad a_4 = \frac{1}{4!}; \qquad a_5 = \frac{1}{5!}.$$

Our search is over:

$$p_5(x) = \frac{1}{0!} + \frac{x}{1!} + \frac{x^2}{2!} + \frac{x^3}{3!} + \frac{x^4}{4!} + \frac{x^5}{5!} = 1 + x + \frac{x^2}{2} + \frac{x^3}{6} + \frac{x^4}{24} + \frac{x^5}{120}.$$

How well does p_5 approximate f near $x = 0$? Very well indeed:

A fifth-order polynomial approximation to $f(x) = e^x$

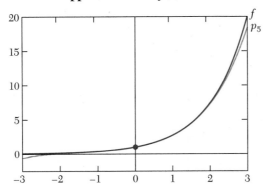

The graphs of f and p_5 are almost identical near $x = 0$. Compare, for instance, $f(1)$ and $p_5(1)$. Easy numerical calculations show that

$$p_5(1) = 1 + 1 + \frac{1}{2} + \frac{1}{6} + \frac{1}{24} + \frac{1}{120} = \frac{163}{60} \approx 2.717;$$

$$f(1) = e \approx 2.718.$$

∎

Factorials

The repeated product $i \cdot (i-1) \cdot (i-2) \cdots 2 \cdot 1$, denoted by $i!$, is called the **factorial** of i. (By convention, $0! = 1$.) Factorials appear often in mathematics, especially when (as we just saw) an operation is performed repeatedly. Factorials are important enough in practice that many calculators have a factorial key. ◄

Does yours? If so, try it. If not, calculate some factorials from scratch.

The Formal Definition

The polynomial p_5 just constructed is the fifth-order Taylor polynomial for $f(x) = e^x$ based at $x = 0$. Here is the general definition:

> **Definition** (Taylor polynomials) Let f be any function whose first n derivatives exist at $x = x_0$. The **Taylor polynomial** of order n, based at x_0, is defined by
>
> $$p_n(x) = f(x_0) + f'(x_0)(x - x_0) + \frac{f''(x_0)}{2!}(x - x_0)^2$$
> $$+ \frac{f^{(3)}(x_0)}{3!}(x - x_0)^3 + \cdots + \frac{f^{(n)}(x_0)}{n!}(x - x_0)^n.$$

> **Taylor's Idea?** The idea underlying Taylor polynomials predates Brook Taylor (1685–1731), for whom they are named. According to the author George Simmons, the idea appears in work of John Bernoulli published in 1694. The same versatile Bernoulli is also credited with having discovered—behind the scenes—the famous technical lemma known as l'Hôpital's Rule. (We'll see it later in this book.)

EXAMPLE 5   Let $f(x) = \sqrt{x}$. Find the first three Taylor polynomials p_1, p_2, and p_3, all based at $x_0 = 64$.

Check the last one; we found the others earlier.

Solution   We'll need the value and first three derivatives of f at $x = 64$: ◄

$$f(64) = 8; \qquad f'(64) = \frac{1}{16};$$
$$f''(64) = -\frac{1}{2048}; \qquad f'''(64) = \frac{3}{262144}.$$

The first two are familiar.

Plugging these values into the definition gives our results: ◄

$$p_1(x) = 8 + \frac{x - 64}{16};$$

$$p_2(x) = 8 + \frac{x - 64}{16} - \frac{(x - 64)^2}{2! \cdot 2048} = 8 + \frac{x - 64}{16} - \frac{(x - 64)^2}{4096};$$

$$p_3(x) = 8 + \frac{x - 64}{16} - \frac{(x - 64)^2}{4096} + \frac{3}{3! \cdot 262144}(x - 64)^3$$

$$= 8 + \frac{x - 64}{16} - \frac{(x - 64)^2}{4096} + \frac{(x - 64)^3}{524288}.$$

■

More on the Definition

The Taylor polynomial definition deserves a closer look.

Why Powers of $(x - x_0)$? Polynomials usually involve powers of x. ▶ Here, though, powers of $(x - x_0)$ ▶ are more convenient. For instance, the use of powers of $(x - 64)$ in Example 5 simplified our calculations, most of which involved $x = 64$.

x, x^2, x^3, etc.

$(x - x_0)$, $(x - x_0)^2$, $(x - x_0)^3$, etc.

Maclaurin Polynomials Taylor polynomials that happen to be based at $x = 0$ are known as **Maclaurin polynomials**. They look a little simpler than Taylor polynomials: ▶

$$P_n(x) = f(0) + f'(0)x + \frac{f''(0)}{2}x^2 + \cdots + \frac{f^{(n)}(0)}{n!}x^n.$$

They're simpler only in appearance. The idea is exactly the same.

In Example 4 we found the fifth-order Maclaurin polynomial for $f(x) = e^x$.

Order, Degree A Taylor polynomial P_n of order n is chosen so that its value and first n derivatives at $x = x_0$ agree with those of f. The definition shows that P_n also has **degree** n; i.e., it involves powers of $(x - x_0)$ up through the nth power. (If, by chance, $a_n = 0$, then P_n has smaller degree.)

Many Symbols, Just One Variable The definition involves many symbols: a, x, i, x_0, f, and n. While in this thicket, it is helpful to remember that x *is the only variable.* In specific cases, all the other symbols take *numerical* values.

From P_n to P_{n+1} Let f be a function and P_n the Taylor polynomial of order n based at x_0. The next Taylor polynomial, P_{n+1}, has just one more term:

$$P_{n+1}(x) = P_n(x) + \frac{f^{(n+1)}(x_0)}{(n+1)!}(x - x_0)^{n+1}.$$

Approximating Functions with Taylor Polynomials

Taylor polynomials are useful in the approximating of other, more complicated functions. For example, the sine function—which has no algebraic formula—is a good candidate for Taylor approximation.

EXAMPLE 6 Find P_1, P_3, P_5, and P_7, the Taylor (or Maclaurin) polynomials for $f(x) = \sin x$ based at $x = 0$. Plot everything on one set of axes. How closely does each polynomial approximate $\sin 1$?

Solution To find P_7 we need the value and the first through seventh derivatives of $f(x) = \sin x$ at $x = 0$. In order, they are $0, 1, 0, -1, 0, 1, 0, -1$. ▶ It follows that

Check this easy calculation.

$$P_7(x) = 0 + 1x + 0x^2 - \frac{1}{3!}x^3 + 0x^4 + \frac{1}{5!}x^5 + 0x^6 - \frac{1}{7!}x^7$$

and, therefore, that

$$P_1(x) = x; \qquad P_3(x) = x - \frac{x^3}{6};$$

$$P_5(x) = x - \frac{x^3}{6} + \frac{x^5}{120}; \qquad P_7(x) = x - \frac{x^3}{6} + \frac{x^5}{120} - \frac{x^7}{5040}.$$

Here are all five graphs:

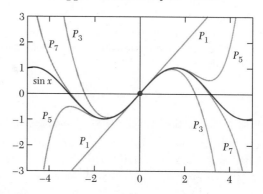

Several Taylor polynomial approximations to $f(x) = \sin x$

The picture shows:

Snug Fits near $x = 0$ All four Taylor polynomials fit the sine curve well near the base point $x = 0$. Differences appear as one moves away (in either direction) from $x = 0$.

Better and Better   Higher-order Taylor polynomials approximate the sine function better than lower-order ones. P_7, for instance, follows the sine curve much farther than P_1 or P_3.

Standard Shapes   The graphs of P_1 and P_3 have standard shapes for linear and cubic functions. ◄   The wigglier appearance ◄   of P_5 and P_7 reflects their higher degree.

Trace P_3 carefully to see this.

Wiggly is good in sine-curve approximating!

Intuition suggests that P_1, P_3, P_5, and P_7 should give successively closer estimates for $\sin 1 \approx 0.84147$. So they do:

$$P_1(1) = 1; \qquad\qquad P_3(1) = \frac{5}{6} \approx 0.83333;$$

$$P_5(1) = \frac{101}{120} \approx 0.84167; \qquad\qquad P_7(1) = \frac{4241}{5040} \approx 0.84147. \qquad\blacksquare$$

Bounding the Error of Linear Approximation (Optional)

An Error Bound Formula

How large, numerically, are the errors committed when Taylor polynomials approximate a function f? We'll address the question only for linear approximations. For higher-order approximations the ideas are similar but the calculations are more complicated.

If we compare the results of Example 2 and Example 3, we find that the error of a linear approximations depends on two factors:

1. the distance from x to x_0;

2. the size of the second derivative f'' near x_0; (Large values of $|f''|$ mean that the f-graph bends sharply away from its tangent line; small (i.e., near 0) values of $|f''|$ mean that the f-graph is nearly straight near x_0 and therefore closely approximated by its tangent line.)

The following theorem restates these ideas quantitatively. Throughout, f is a differentiable function, ℓ is the linear approximation based at $x = x_0$, and I is an interval containing x_0.

T h e o r e m 2 (**An Error Bound for Linear Approximation**) Suppose that the inequality

$$|f''(x)| \le K$$

holds for all x in I. Then, for all x in I,

$$|f(x) - \ell(x)| \le \frac{K}{2}(x - x_0)^2.$$

Observe:

The Error Function The function $f(x) - \ell(x)$ on the left is called an **error function**; we'll call it $e(x)$. For a given x, $e(x)$ is the **error** committed by ℓ in approximating f. Ideally, $e(x) \approx 0$.

An Error Bound Formula The inequality

$$|f(x) - \ell(x)| \le \frac{K}{2}(x - x_0)^2$$

is (appropriately) called an **error bound formula**; it guarantees that the error function is no larger than the *computable* quantity on the right.

It's OK to Overestimate K Any upper bound for $|f''(x)|$ on I can serve as K in the inequality. Choosing K as small as possible improves the error bound, but the inequality holds even for larger values of K. ▶

It's safe to overestimate K but not to underestimate.

Worst-Case Scenario The function $f(x) = K(x-x_0)^2/2$ represents the worst case that the hypothesis allows. For this f,

$$f(x_0) = 0, \qquad f'(x_0) = 0; \qquad \text{and} \qquad f''(x) = K.$$

In this case, the linear approximation is simply $\ell(x) = 0$, and the error ℓ commits is as large as the theorem permits:

$$|f(x) - \ell(x)| = f(x) = \frac{K}{2}(x - x_0)^2.$$

Before proving the error bound theorem, we'll show it in action.

Satisfaction Guaranteed

The practical value of an error bound formula is its guarantee of the accuracy of an approximation.

EXAMPLE 7 We showed that $\ell(x) = 4 + x/16$ is the linear approximation to $f(x) = \sqrt{x}$ at $x = 64$. What accuracy does the error bound formula guarantee if ℓ approximates f on the interval $[50, 80]$? What does the formula guarantee for the interval $[63, 65]$?

Check ours.

Solution   First we need a suitable K. Calculation ◄ shows that

$$f''(x) = \frac{d^2}{dx^2}\left(\sqrt{x}\right) = -\frac{1}{4x^{3/2}} \implies |f''(x)| = \frac{1}{4x^{3/2}}.$$

A moment's thought—or a look at the graph of $|f''|$—shows that $|f''(x)|$ decreases as x increases. For the interval $[50, 80]$, therefore, we may use $K = |f''(50)| \le$ 0.00071. ◄ On $[63, 65]$, $K = |f''(63)| \le 0.0005$ works.

$|f''(x)|$ is largest at the left endpoint.

Watch all inequalities carefully.

All ingredients are ready. The theorem says: ◄

• For x in $[50, 80]$, the largest possible error is no more than

$$|\sqrt{x} - \ell(x)| \le \frac{0.00071}{2}(x - 64)^2 \le \frac{0.00071}{2}(80 - 64)^2 \approx 0.09.$$

(We used $x = 80$, not $x = 50$, to make $(x - 64)^2$ as large as possible.) In other words, on $[50, 80]$, ℓ approximates f to about *one-decimal-place accuracy.*

• For x in $[63, 65]$, the largest possible error is even less:

$$|\sqrt{x} - \ell(x)| \le \frac{0.0005}{2}(x - 64)^2 \le \frac{0.0005}{2}(65 - 64)^2 \approx 0.00025.$$

In other words, on $[63, 65]$, ℓ approximates f to about *three-decimal-place accuracy.*

The table on page 268 shows that these error estimates are conservative: ℓ and q behave a little better than the theorem predicts. ∎

A Proof of the Error Bound Formula

We use f, ℓ, and K as in the theorem. From these data we'll construct two new functions—an error function e and a **bounding function** b, defined as follows:

$$e(x) = f(x) - \ell(x); \qquad b(x) = K\frac{(x - x_0)^2}{2}.$$

The same proof, with minor variations, shows that the inequality holds for $x < x_0$ and in absolute value.

The theorem claims—and we need to show—that for x in I, $|e(x)| \le b(x)$. To simplify things slightly, we'll show only that $e(x) \le b(x)$ for $x \ge x_0$. ◄

To our aid come three key properties of e and b:

$$e(x_0) = 0 = b(x_0); \qquad e'(x_0) = 0 = b'(x_0); \qquad |e''(x)| \le K = b''(x). \quad (4.3)$$

Make them!

(All three properties of e follow from the construction of ℓ: By design, $\ell(x_0) = f(x_0)$, $\ell'(x_0) = f'(x_0)$, and $e''(x) = f''(x) - l''(x) = f''(x)$. All three properties of b come from direct calculations.) ◄

We met it first in Section 2.1.

To finish the proof, we use the racetrack principle. ◄ Recall what it says about *any* functions g and h:

If $g(x_0) = h(x_0)$ and $g'(x) \le h'(x)$ for $x \ge x_0$, then $g(x) \le h(x)$ for $x \ge x_0$.

Let's apply the racetrack principle to e and b. As we saw earlier, $e(x_0) = 0 = b(x_0)$. We'll be finished, says the racetrack principle, if we can somehow show that $e'(x) \le b'(x)$ for $x \ge x_0$. To do so, we apply the racetrack principle *again*, but this time to e' and b' and their derivatives e'' and b''. Equations 4.3 contain everything we need: $e'(0) = b'(0)$ and $e''(x) \le b''(x)$. Thus, the racetrack principle says that $e'(x) \le b'(x)$ for $x \ge x_0$—exactly what we need to finish the proof. ■

BASICS

For each function in Exercises 1–5, first find the Taylor polynomial P_n of order n for the function f with base point x_0. Then plot both f and P_n on the same axes. Choose your plotting window to show clearly the relationship between f and P_n.

1. $f(x) = \dfrac{1}{1-x}$, $n = 3$, $x_0 = 0$

2. $f(x) = \sin x + \cos x$, $n = 4$, $x_0 = 0$

3. $f(x) = \ln x$, $n = 3$, $x_0 = 1$

4. $f(x) = \tan x$, $n = 2$, $x_0 = 0$

5. $f(x) = \sqrt{x}$, $n = 3$, $x_0 = 4$

In Exercises 6–11, find linear and quadratic approximations of each function at $x_0 = 0$. Then, find an interval over which each approximating polynomial makes an error no greater than 0.01.

6. $f(x) = \sin x$

7. $f(x) = \cos x$

8. $f(x) = \tan x$

9. $f(x) = e^x$

10. $f(x) = \arctan x$

11. $f(x) = \arcsin x$

12. The line tangent to the curve $y = g(x)$ at the point $(3, 5)$ intersects the y-axis at the point $(0, 10)$.
    (a) What is $g'(3)$?
    (b) Estimate $g(2.95)$ using a linear approximation.

FURTHER EXERCISES

13. In Example 2 we worked with the quadratic function $q(x) = a + b(x - 64) + c(x - 64)^2$. Here are more details:
    (a) We claimed that $q'(x) = b + 2c(x - 64)$ and that $q''(x) = 2c$. Verify these claims by differentiation.
    (b) Rewrite q in powers of x, i.e., in the form $q(x) = A + Bx + Cx^2$.
    (c) Rewrite $q'(x) = b + 2c(x - 64)$ in powers of x. Does your answer agree, as it should, with the previous part?

14. This exercise refers to Example 6 (p. 273). Throughout this problem, $f(x) = \sin x$, and $P_n(x)$ is the Taylor (Maclaurin) polynomial of order n based at $x = 0$.
    (a) Every function mentioned in Example 6—f, P_1, P_3, P_5, and P_7—is odd. How is this fact "reflected" in their graphs?
    (b) Find the *even-order* Maclaurin polynomials P_2, P_4, P_6, and P_8 for $f(x) = \sin x$. How are they related to P_1, P_3, P_5, and P_7?

15. Throughout this exercise, $g(x) = \cos x$ and $P_n(x)$ is the Taylor (Maclaurin) polynomial of order n based at $x = 0$. For use below, here are graphs of g, P_0, P_2, P_4, and P_6.

Several Taylor polynomial approximations to $f(x) = \cos x$

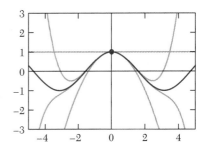

(a) Find formulas for P_0 through P_6, the Maclaurin polynomials through order 6 for g based at $x = 0$. Use your results to label the graphs shown above.
(b) Is g odd, even, or neither? What about the Maclaurin polynomials for g found in part (a)? How do the graphs "reflect" the situation?
(c) In Example 6 we found the Maclaurin polynomials P_1, P_3, P_5, and P_7 for $f(x) = \sin x$. Find the derivatives $P_1'(x)$, $P_3'(x)$, $P_5'(x)$, $P_7'(x)$, and $f'(x)$ of these functions. How are the results related to the rest of this problem?

16. Suppose that f is a function such that $f'(x) = \sin\left(x^2\right)$ and $f(1) = 0$. [**NOTE:** No explicit formula for f is given.]
    (a) Estimate the value of $f(0.5)$ using a linear approximation.
    (b) Is the estimate you computed in part (a) greater than or less than the exact value of $f(0.5)$? Explain.
    (c) Estimate the value of $f(0.5)$ using a quadratic approximation.

17. Let g be a well-behaved function defined on the interval $[0, 10]$ such that the graph of g passes through the point $(5, 2)$ and the derivative of g is the function sketched below.

Graph of g′

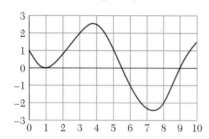

    (a) Estimate $g(0)$ using a linear approximation.
    (b) Estimate $g(8)$ using a quadratic approximation.

18. Suppose that h is a well-behaved function such that $h(2) = 3$, $h'(2) = -2$, and $-2 \le h''(x) \le 1$ for all x in $(0, 4)$.
    (a) Find a linear approximation to h.
    (b) Show that $0 \le h(3) \le 2$.

19. At time $t = 0$ seconds a car passed a reference point, heading eastward at 25 meters per second, with eastward acceleration 2 meters per second per second. Let $p(t)$ be the car's eastward position at time t seconds, in meters east of the reference point. (In particular, $p(0) = 0$.)
    (a) Find the linear approximation function ℓ_p to p based at $t_0 = 0$. Use it to predict where the car will be at time $t = 1$, and where the car was at time $t = -1$.
    (b) Find the quadratic approximation function q_p to p based at $t_0 = 0$. Use it to predict where the car will be at time $t = 1$, and where the car was at time $t = -1$.
    (c) Let $v(t)$ be the car's eastward velocity, in meters per second, at time t. Find the linear approximation ℓ_v to v based at $t_0 = 0$. What does it predict at time $t = 1$ second?
    (d) (Requires the error-bound theorem.) The car's motor and brakes are powerful enough to produce acceleration anywhere from -3 to 3 meters per second per second. Use this fact and Theorem 2, page 275,

to decide how much much error the linear approximation ℓ_p can commit, at worst, over the interval $0 \le t \le 1$.

20. Repeat the previous exercise, but assume that at $t = 0$ the car was heading eastward at 25 meters per second, and had eastward acceleration -2 meters per second per second.

21. At time $t = 0$ seconds an object was dropped from a 100-meter cliff. Its downward acceleration, due to gravity, was 9.8 meters per second per second. Let $h(t)$ be the height in meters of the object above ground level at time t seconds.
    (a) Find the linear approximation ℓ to h, based at $t_0 = 0$. What does it predict for $t = 1$?
    (b) Find the quadratic approximation q to h, based at $t_0 = 0$. What does it predict for $t = 1$?
    (c) Plot the functions ℓ and q on the same axes; use $0 \le t \le 6$. Which model—ℓ or q—seems more realistic? Does either model describe the real situation perfectly? Why or why not?

22. Theorem 2, page 275, gives a bound on the error committed by a linear approximation. Here, without proof, is a similar result for quadratic approximation:

    An error bound for quadratic approximation. *Suppose that the inequality $|f'''(x)| \le K$ holds for all x in an interval I. Then for all x in I,*

    $$|f(x) - q(x)| \le \frac{K}{6}(x - x_0)^3.$$

    Let $f(x) = \cos(x)$.
    (a) Find the quadratic approximation to f based at $x_0 = 0$.
    (b) Use the result above to bound the error committed by q in approximating f on the interval $[-1, 1]$.
    (c) Plot both f and q on the same axes; use $-1 \le x \le q$. How is the picture consistent with what you found in the previous part?

23. Let $f(x) = \ln(x)$.
    (a) Find the quadratic approximation q to f, based at $x_0 = 1$.
    (b) Use the error-bound result from the previous problem to bound the error committed by q in approximating f on the interval $[1/2, 3/2]$.
    (c) Plot both f and q on the same axes; use $1/2 \le x \le 3/2$. How are the graphs consistent with what you found in the previous part?

24. Estimate the value of each of the following expressions using a linear approximation. Then compute the difference between your estimate and the value given by a scientific calculator.
    (a) $\sqrt{103}$ (c) $\tan 31°$
    (b) $\sqrt[3]{29}$ (d) 0.8^{10}

25. For each part of the previous exercise, compare the "actual" approximation error with the theoretical error bound.

26. Let f be a function, and let ℓ and q be the linear and quadratic approximations to f based at x_0. This exercise is about the two functions E_1 and E_2 defined by

$$E_1(x) = f(x) - \ell(x) \quad \text{and} \quad E_2(x) = f(x) - q(x).$$

Observe that E_1 and E_2 measure the *difference* between $f(x)$ and either $\ell(x)$ or $q(x)$—i.e., E_1 and E_2 measure the *errors* ℓ and q commit in approximating f. (Ideally, $E_1(x) \approx 0$ and $E_2(x) \approx 0$ when $x \approx 0$.)

(a) Let $f(x) = e^x$. Find formulas for the functions ℓ, q, E_1, and E_2. (Use the base point $x_0 = 0$.)

(b) Plot f, ℓ, and q on one set of axes; use $-1 \le x \le 1$.

How does the picture show that q is a better approximation to f than ℓ?

(c) Plot E_1 and E_2 on one set of axes; use $-1 \le x \le 1$. How does *this* picture show that q is a better approximation to f than ℓ?

(d) The two graphs in the previous part should resemble a quadratic and a cubic curve, respectively. Do they? Which resembles which?

(e) Are the curves in part (d) *really* quadratic and cubic, or do they just look that way?

27. Repeat the previous exercise, but use $f(x) = \ln(x)$, based at $x = 1$.

28. Repeat Exercise 26, but use $f(x) = x^5$, based at $x = 1$.

4.4 Newton's Method: Finding Roots

Consider these problems:

- Let $f(x) = 2x^3 - x - 2$. Find a number r for which $f(r) = 0$.
- Estimate $\sqrt{2}$ with high numerical accuracy.
- Find a number r such that

$$150r = (1 + r)^{31} - 1.$$

Each of these problems involves finding a **root** of some function f, i.e., a number r such that $f(r) = 0$. None can be solved easily or directly by algebraic methods. Newton's method offers a simple and efficient alternative. We'll use it to solve all three problems.

Roots, Graphically

A number r is a root of a function f if $f(r) = 0$. Graphically speaking, a root of f is an x-intercept of the f-graph. Given a graph of f, it's easy to find roots—approximately—just by looking. Following, for instance, are two views of the graph of $f(x) = 2x^3 - x - 2$.

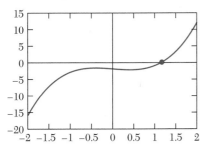
Graph of $f(x) = 2x^3 - x - 2$

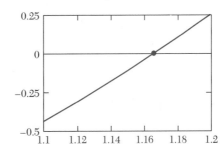
A closer view: $f(x) = 2x^3 - x - 2$

The first graph shows a root somewhere near $x = 1$; the second locates it more precisely, near $x = 1.17$. Further zooming might reveal additional digits. Newton's method offers a more efficient strategy, so we won't bother.

EXAMPLE 1   Graphically estimate a root of $f(r) = (1 + r)^{31} - 1 - 150r$. (Such a root solves the equation $150r = (1 + r)^{31} - 1$, already mentioned.)

Solution   The graph

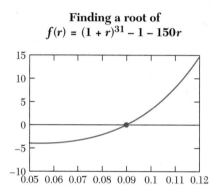

Finding a root of
$$f(r) = (1 + r)^{31} - 1 - 150r$$

shows a root somewhere near $r = 0.09$. A numerical check shows that $f(0.09) \approx -0.0382$, so our guess isn't exact. We'll refine it with Newton's method.   ∎

Roots, Algebraically

Roots can sometimes be found algebraically, by solving equations. Unfortunately, such methods often fail; unless a function has an algebraic formula—a simple one—solving algebraically for roots may be hard or impossible. For linear functions, roots are easy to find. ◄

Don't skip the next example; we'll refer to it later.

EXAMPLE 2   A line passes through the point (x_0, y_0) slope m. Find its x-intercept.

Solution   The line has point–slope equation $y - y_0 = m(x - x_0)$. To find the x-intercept, we set $y = 0$ and solve for x:

$$- y_0 = m(x - x_0) \iff x = x_0 - \frac{y_0}{m}. \tag{4.4}$$

Unless $y_0 = 0$, in which case the line is the x-axis itself.

(This formula fails, of course, if $m = 0$. The line is horizontal and so has no x-intercept.) ◄   ∎

Newton's Idea: Ride the Tangent Line

Most functions aren't linear and so don't lend themselves easily to algebraic root finding. When algebra fails, Newton's method often succeeds.

Newton's method starts with one approximation to a root and produces another, better approximation. The following picture illustrates one **Newton step**.

Zeroing in on a root: one Newton step

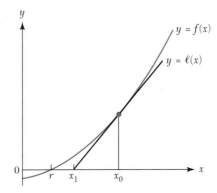

Observe:

- In the picture, r is the sought-after root, x_0 is an **initial guess** for r, and x_1 is an *improved* approximation to r—the result of one Newton step.
- If f happens to be linear, then for any initial guess x_0, the next estimate, x_1, is an *exact* root of f. In other words, for linear functions Newton's method succeeds exactly in only one step.

Repeating the preceding process produces a sequence $x_0, x_1, x_2, x_3, \ldots$ of successive approximations to r. As the picture suggests, successive approximations converge quickly toward the root, r.

Finding a root: several Newton steps

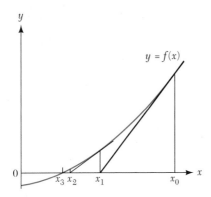

Each approximation is obtained from its predecessor in the same way:

   x_{n+1} *is the x-intercept of the tangent line at* $x = x_n$.

Equation 4.4, with $m = f'(x_0)$, gives an **iteration rule**, i.e., an explicit recipe for moving from one estimate to the next.

> **Fact** (Newton's Iteration Formula) Let f be a function and x_0 an initial guess at a root. For $n \geq 0$,
>
> $$x_{n+1} = x_n - \frac{f(x_n)}{f'(x_n)}. \tag{4.5}$$

Notice:

Iteration The Fact defines x_1, x_2, x_2, x_3, ... **iteratively**; that is, each term is defined by its predecessor. Each successive use of the formula is called an **iteration**.

Found a Root? Then Stop If x_n happens to *be* a root of f, then $x_{n+1} = x_n$. Here's why:

$$x_{n+1} = x_n - \frac{f(x_n)}{f'(x_n)} = x_n - \frac{0}{f'(x_n)} = x_n.$$

N is for Newton.

The Iteration Function Consider the new function (we'll call it the **iteration function**) N ◀ defined by $N(x) = x - f(x)/f'(x)$. Then

$$x_1 = N(x_0), \qquad x_2 = N(x_1), \qquad x_3 = N(x_2), \qquad \ldots, \qquad x_{n+1} = N(x_n).$$

Newton's method, then, amounts to applying N *repeatedly* to x_0; r is a root of f if and only if $N(r) = r$. Such an r is called a **fixed point** of the function N. (The modern theory of dynamical systems studies how inputs behave under repeated applications of a function.)

> **Aren't Graphs Enough?**   Isaac Newton didn't have computer graphics. We do, so why should we bother with Newton's method? Why not just zoom in, as often as necessary, to find roots to any desired accuracy?
>
> One reason is that the process can be tedious, especially if many decimal places of accuracy are required. Moreover, some graphics hardware and software display no more than 8 or 10 digits.
>
> A more important reason concerns efficiency of effort. Whether performed by human or machine, plotting any function often requires dozens, if not hundreds, of function evaluations— an "expensive" proposition. Newton's method is much more efficient. For most functions, successive Newton estimates converge to a root with remarkable speed: Each estimate has about twice as many correct digits as its predecessor.

EXAMPLE 3   We estimated graphically that $f(x) = 2x^3 - x - 2$ has a root near $x = 1.17$. Use Newton's method to do better.

Solution   Since $f(x) = 2x^3 - x - 2$, $f'(x) = 6x^2 - 1$. By the preceding Fact,

$$x_{n+1} = x_n - \frac{f(x_n)}{f'(x_n)} = x_n - \frac{2x_n^3 - x_n - 2}{6x_n^2 - 1}.$$

Check some of these numbers.

If, say, $x_0 = 1$, then ◀

$$x_1 = 1.2, \qquad x_2 \approx 1.16649, \qquad x_3 \approx 1.16537, \quad \text{and} \quad x_4 \approx 1.16537.$$

Our guess, then, is $r \approx 1.16537$. It checks out numerically (graphically, too): $f(1.16537) \approx -0.00002.$ ∎

Newton's Method in Action

Successful applications of calculus ideas to real problems often depend, sooner or later, on finding roots. Newton's method can simplify the process.

EXAMPLE 4   Which point P on the parabola $y = x^2$ is closest to $(2, 1)$?

Solution   Here's the picture:

On $y = x^2$, P is closest to $(2, 1)$

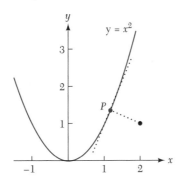

From the picture alone we can only estimate P. However, the picture does suggest the crucial idea: ▶

We'll return to this idea more formally in Section 4.6.

   The line segment from $(2, 1)$ to P is perpendicular to the tangent line at P.

We'll translate this condition into an equation, then solve it to find P.

   Since P lies on the parabola, its coordinates have the form (r, r^2). Thus, the line tangent to $y = x^2$ at P has slope $2r$, ▶ and the line segment from $(2, 1)$ to P has slope $(r^2 - 1)/(r - 2)$. The perpendicularity condition, therefore, reads symbolically as follows : ▶

*Because
$y = x^2 \implies dy/dx = 2x$.
Perpendicular lines have negative reciprocal slopes.*

$$\frac{r^2 - 1}{r - 2} = -\frac{1}{2r}.$$

Equivalently, $2r^3 - r - 2 = 0$. ▶
   This last equation should look familiar—we solved it in Example 3, with Newton's method, and got $r \approx 1.16537$. (The result looks graphically plausible, too.)

Convince yourself of the equivalence.

∎

EXAMPLE 5   Calculators "know" that $\sqrt{2} \approx 1.4142136$. Use Newton's method to derive such an estimate.

Solution   Our target, $\sqrt{2}$, is a root of $f(x) = x^2 - 2$. Applied to f, Newton's formula is especially simple: ▶

Check the last algebraic step.

$$x_{n+1} = x_n - \frac{f(x_n)}{f'(x_n)} = x_n - \frac{x_n^2 - 2}{2x_n} = \frac{x_n}{2} + \frac{1}{x_n}.$$

Check some with a calculator.

Starting with $x_0 = 1$, successive estimates ◄ quickly zero in on $\sqrt{2}$:

$$x_1 = 1.5, \quad x_2 \approx 1.4166667, \quad x_3 \approx 1.4142157, \quad x_4 \approx 1.4142136. \quad \blacksquare$$

Other Root-Finding Methods. Newton's method is only one of several common root-finding techniques. The **bisection, or interval-halving, method** (discussed further in Section 4.10) works by isolating a root in smaller and smaller intervals. The **secant method** resembles Newton's tangent method, except that secant lines replace the tangent lines of Newton's method.

Root-Grubbing for Money: IRAs and Newton's Method

U.S. tax law allows some taxpayers to deposit $2000 per year in tax-sheltered individual retirement accounts (IRAs). Doing so raises interesting questions: How much will such an account be worth after 30 years of faithful contributions? What interest rate will produce $300,000 after 30 years?

Expressed as a decimal number, not a percent.

The **compound amount formula** helps answer such questions. If p dollars are deposited yearly at effective annual interest rate r, ◄ then the total value $V(n)$ after n years is given by

$$V(n) = \frac{p}{r}\left((1+r)^{n+1} - 1\right).$$

How nice? It depends, for example, on inflation and tax rates 30 years hence.

For an IRA paying 6% over 30 years, we'd use $p = 2000$, $r = 0.06$, and $n = 30$. The result is a nice ◄ little nest egg:

$$V(30) = \frac{2000}{0.06}\left((1.06)^{31} - 1\right) \approx \$169{,}603.35.$$

EXAMPLE 6   Riskier investments might yield higher interest rates for one's IRA. What interest rate would ensure a value of $300,000 after 30 years of $2000 contributions?

Solution   We need a value of r for which

$$V(30) = \frac{2000}{r}\left((1+r)^{31} - 1\right) = 300{,}000.$$

Solving this equation algebraically for r is hard or impossible, so we resort to Newton's method. A little preliminary algebra simplifies the work:

$$\frac{2000}{r}\left((1+r)^{31} - 1\right) = 300{,}000 \iff \left((1+r)^{31} - 1\right) - 150r = 0.$$

In Example 1, page 280, we estimated $r \approx 0.09$ as a root of the last equation. With

$$f(r) = \left((1+r)^{31} - 1\right) - 150r, \qquad f'(r) = 31(1+r)^{30} - 150,$$

By hand the computations are messy. With a calculator they're easy.

and initial guess $x_0 = 0.09$, Newton's method gives ◄

$$x_1 \approx 0.09014631, \qquad x_2 \approx 0.09014585, \qquad x_3 \approx 0.09014585.$$

After just three Newton steps, the answer is clear: An annual interest rate of 9.015% will suffice. $\blacksquare$

When Things Go Wrong

Newton's method, though impressive when it works, sometimes fails. Several problems can arise.

Horizontal Tangent Lines.   If $f'(x_n) = 0$, then the recursion formula

$$x_{n+1} = x_n - \frac{f(x_n)}{f'(x_n)}$$

makes no sense, and the process fails. In graphical terms, the problem is that the tangent line at x_n is horizontal and so never meets the x-axis.

Near-Horizontal Tangent Lines.   The preceding problem occurs only rarely in practice; hitting a point of horizontal tangency *exactly* is unlikely. Trouble may still occur, however, if x_n lands too near a root of f'. The next example shows how this can happen.

EXAMPLE 7   Let $f(x) = \sin x$ and $x_0 = 1.3$. Among the roots of f, $r = 0$ lies closest to x_0. What does Newton's method say?

Solution   Applied to f, with initial guess $x_0 = 1.3$, Newton's method gives

$$x_1 \approx -2.3021; \qquad x_2 \approx -3.4166; \qquad x_3 \approx -3.1344; \qquad x_4 \approx -3.1416.$$

We found a root but it's the wrong one. The picture shows what happened:

A shallow tangent: finding the wrong root

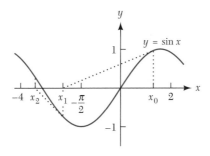

At x_0 the tangent line was too nearly horizontal; riding it to the x-axis took us far away from the expected root. ∎

Cycling.   In rare (but interesting) circumstances, successive Newton's approximations cycle, repeating endlessly without converging to a root.

EXAMPLE 8   It's easy to check (by factoring the formula) that the cubic polynomial $f(x) = x^3 - 6x^2 + 7x + 2$ has three roots: $x = 2$, $x = 2 - \sqrt{5} \approx -0.24$, and $x = 2 + \sqrt{5} \approx 4.24$. But what happens if Newton's method is applied to f, with initial guess $x_0 = 1$? *Why* does it happen?

S o l u t i o n   Successive estimates x_0, x_1, x_2, x_3, ... follow the repeating pattern 1, 3, 1, 3, ... The following picture shows why:

Cycling around with Newton's method

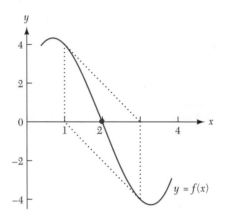

One that won't, however, is $x_0 = 3$. Do you see why not?

By careful planning, we chose the worst possible initial guess x_0. Almost any other initial estimate x_0 would lead successfully to a root. ◄   If $x_0 = 0.999$, for example, Newton's method finds the negative root $2 - \sqrt{5}$. If $x_0 = 1.001$, Newton's method finds the middle root, at $x = 2$. ∎

> **The Butterfly Effect.**   Tiny changes in the initial guess x_0 for Newton's method may cause dramatically different results. In Example 8, for instance, the initial inputs $x_0 = 1.000$, $x_0 = 0.999$, and $x_0 = 1.001$, although almost identical, produced completely different outputs.
>
> This phenomenon—small causes producing large effects—occurs often in mathematics and in the real world. It's sometimes called the **butterfly effect**: A butterfly's wing flap in Kansas may eventually trigger a typhoon in Taiwan. Mathematicians use a more prosaic name: **sensitive dependence on initial conditions**.

BASICS

1. By hand, find the first three Newton estimates for $\sqrt{5}$, starting with $x_0 = 2$. Write answers as fractions.
   [**HINT**: First use the Fact on page 282 to write a simple formula for x_{n+1} in terms of x_n.]
2. Use Newton's method with the value of x_0 indicated to compute approximate values of each of the following expressions. (Iterate until successive approximations agree to three decimal places.)
   (a) $\sqrt[3]{100}$; $x_0 = 4.5$ (b) $(13)^{-2}$; $x_0 = 0.01$

3. (a) Use Newton's method to find a solution of the equation $x^5 + 4x - 3 = 0$ in the interval $[0, 1]$ to three-decimal-place accuracy.
   (b) Does the equation in part (a) have any other solutions? Justify your answer.
4. Find a solution of the equation $x \tan x = 1$ in the interval $[0, \pi/2]$ to three-decimal accuracy.

FURTHER EXERCISES

5. For the sample problems on the first page of this section, try to find roots by solving algebraically. (The point is to see that this can't always be done.)

6. Showing your work, derive Equation 4.4, page 280. Use the result to find the x-intercepts of each of the following lines:
   (a) Through $(1, 1)$, with slope 1.
   (b) Through $(1, 2)$, with slope 3.
   (c) The line with equation $y = mx + b$.
   (d) The line tangent to the graph $y = x^2$ at $x = 1$.
   (e) The line tangent to the graph $y = x^2 - 3$ at $x = 1$.

7. Use Equation 4.4 to derive Equation 4.5. Show all work.

8. Here's another way to find the point P on the parabola $y = x^2$ that's closest to $(2, 1)$:
   (a) Let $d(r)$ be the distance from $(2, 1)$ to $P = (r, r^2)$. Write a formula for $d(r)$.
   (b) At the desired value of r, d has a local minimum, so $d'(r) = 0$. Write out and simplify the equation $d'(r) = 0$. Does the result look familiar?

9. Let $f(x) = x^2 - a$.
   (a) Explain why the two numbers x and a/x bracket $\sqrt{a}$.
   (b) Show that Newton's method applied to f gives the iteration function $N(x) = (x + a/x)/2$—i.e., the average of x and a/x.
   (c) What happens if $x_n = \sqrt{a}$?

10. Let $f(x) = x^3 - 2x^2 - 5x + 7 = (x-3)(x-1)(x+2)+1$.
    (a) Explain why it is natural to guess that f has roots near $x = 3$, $x = 1$, and $x = -2$.
    (b) Use Newton's method to estimate the values of the roots of f to four decimal places.

11. Let $f(x) = x^3 - 10x^2 + 22x + 6$.
    (a) Find each of the three roots of f to three decimal places.
    (b) What happens when Newton's method is used with $x_0 = 2$? Explain why this happens.
    (c) Newton's method behaves oddly if $x_0 = 1.39$. How? Why?

12. Let $h(x) = 179.5x^6 - 436.8x^4 + 257.3x^2 - 1$.
    (a) $h(x)$ has three positive roots. Use Newton's method to find each one to three decimal places.
    (b) $h(x)$ has three negative roots. How are they related to the three positive roots you found in part (a)?

13. Find (approximately) the minimum value of $g(x) = 10x^{-2} + 3x^2 + x - 4$ when $1 \le x \le 10$.

14. Let $f(x) = x^2 \sin x^2$ on the interval $[0, 3]$.
    (a) Find (approximately) each of the following numbers to two decimal places using Newton's method:
        (i) The minimum value of f.
        (ii) The maximum value of f.
        (iii) The inflection points of f.
    (b) Sketch a graph of f and label both coordinates of each of the points you found above.

15. Apply Newton's method to $f(x) = x^{1/3}$, starting from $x_0 = 1$. Explain what happens.

16. Apply Newton's method to $f(x) = e^{-x}$, starting from $x_0 = 0$. Explain what happens.

4.5 Splines: Connecting the Dots

How does one draw a curve through several—perhaps many—points in a plane? (**NOTE:** In this section, "curve" means any unbroken line, with or without corners.) The answer depends partly on who does the drawing. For a human artist, it's easy to draw freehand curves that pass through, say, five given points. Indeed, there are many ways to do so. Here are four possibilities:

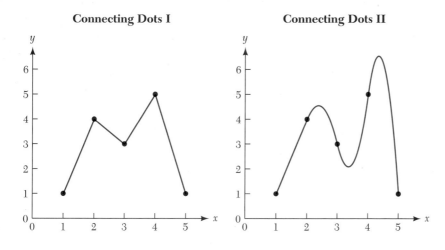

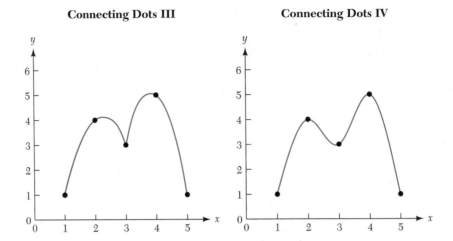

For humans, such ambiguity is no problem. We can easily choose (on aesthetic, physical, or other grounds) one curve through the given points that meets our purposes. Computers, on the other hand, need precise instructions in their own language—mathematics. Calculus provides the tools both to *find* curves that fit data points and to *describe* those curves in terms computers can understand.

Splines: Basic Ideas and Key Words

The idea behind splines is to "tie" small curve segments, called **elements**, together to form a single larger curve, called a **spline**. The points at which elements are tied together (the dots in the preceding pictures) are called **knots**.

Our connect-the-dots pictures illustrate these ideas. Picture I, formed from line segments, shows a **linear spline**. Picture II shows a **quadratic spline**, formed by joining pieces of four parabolas. Picture III shows another (perhaps less pleasing) possible way of joining parabolic arcs. Picture IV shows a **cubic spline**, formed by joining pieces of four cubic arcs. (A cubic arc is defined by a cubic polynomial, $a + bx + cx^2 + dx^3$.)

Linear, quadratic, and cubic splines are formed by joining pieces of lines, parabolas, or cubic curves end to end. Individual lines, parabolas, and cubic arcs are smooth and unbroken, so any new curve built from such pieces is also smooth and unbroken, except perhaps at the knots. With help from derivatives, we'll fit such elements together properly, even at the knots.

Linear Splines: Continuous but Crooked

The simplest splines are formed from straight lines.

EXAMPLE 1   Write a formula for the linear spline in Picture I.

S o l u t i o n   The curve in question is **piecewise-linear**, i.e., built from several straight-line pieces. The formula for a piecewise-linear function has several
parts, one for each piece. The first piece in Picture I, for instance, is part of the
line joining the points $(1, 1)$ and $(2, 4)$. This line has slope 3, so its equation in
point–slope form ▶ is $y - 1 = 3(x - 1)$, or, equivalently, $y = 3x - 2$. This equation describes only the first piece, where $1 \le x \le 2$. Similar reasoning applied to
the other spline elements gives the following multiline formula for the spline function S:

Convince yourself of this.

$$S(x) = \begin{cases} 3x - 2 & \text{if } 1 \le x \le 2; \\ -x + 6 & \text{if } 2 \le x \le 3; \\ 2x - 3 & \text{if } 3 \le x \le 4; \\ -4x + 21 & \text{if } 4 \le x \le 5. \end{cases}$$

The formula, like the graph, is piecewise-linear: On each piece of its domain the
function S is defined by a linear polynomial.   ◼

A linear spline, by definition, is a piecewise-linear function that's also continuous. For piecewise-linear functions, continuity holds as long as the linear pieces
join properly at their ends.

Quadratic Splines: Smoother

Linear splines, although simple, have an ugly kink at each knot (unless, by pure
chance, successive slopes match up). One way to avoid such kinks is to use a
quadratic spline, formed by linking parabolic pieces. ▶

*Each piece is defined by a
quadratic equation.*

> **When Is a Parabola Degenerate?**   A quadratic polynomial q has the form $q(x) = a + bx + cx^2$, where a, b, and c are constants. The graph of a quadratic polynomial is almost always a parabola. There's one exception: If $c = 0$, then $q(x) = a + bx + 0x^2 = a + bx$ and q is actually a *linear* polynomial; its graph is a straight line.
>
> To allow for this possibility, it's often best to interpret the word "quadratic" as meaning "having powers no greater than two." Similarly, cubic polynomials involve powers no greater than three. A polynomial that happens to have less than its "advertised" power is sometimes called **degenerate**. (Mathematicians attach no stigma to degeneracy.) From this point of view, a line can be thought of as a degenerate parabola.

Linear splines are crooked because only one possible line segment can join one
knot to the next. Quadratic segments are more flexible—literally! Given any two
successive knots, we can find a parabola that joins them, with any desired slope at
the left knot. Here, for instance, are several parabolas joining $(0, 0)$, and $(1, 1)$,

See the exercises at the end of this section for more on these parabolas.

each labeled with its slope at $(0, 0)$: ◄

Parabolas joining $(0, 0)$ and $(1, 1)$:
m = slope at $(0, 0)$

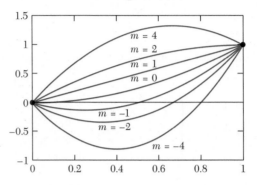

(One "parabola" is actually a line—which one?) By choosing slopes judiciously at each knot, we can iron out the kinks. The next example shows exactly how.

E X A M P L E 2   Use quadratic pieces to form a smooth spline through the same knots as in Example 1:

x	1	2	3	4	5
y	1	4	3	5	1

From among the infinite possibilities.

Smoothness is an issue only where two curves join.

Remember: A straight line is a degenerate quadratic.

We saw earlier that $S_1'(2) = 3$.

S o l u t i o n   To join each knot to the next, we choose ◄ a parabolic curve. We produce, in other words, a **piecewise-quadratic function**. To avoid kinks, we make sure that the slopes of "incoming" and "outgoing" elements match at each knot. The first spline element, S_1 (from $x = 1$ to $x = 2$), is simplest. Smoothness is no problem at the first knot, ◄ so any quadratic curve joining $(1, 1)$ to $(2, 4)$ will do. The simplest choice, perhaps, is a straight line; ◄ we start with that. We saw in Example 1 that the linear function $S_1(x) = 3x - 2$ works. In particular, $S_1'(2) = 3$.

The second spline element (from $x = 2$ to $x = 3$) is more interesting. We need a quadratic polynomial $S_2(x)$ that joins the second and third knots and also agrees in slope with S_1 at $x = 2$. ◄ In other words, our quadratic polynomial S_2 should satisfy three conditions:

$$S_2(2) = 4; \qquad S_2'(2) = 3; \qquad S_2(3) = 3.$$

How can we find such a polynomial? The following strategy simplifies the algebra considerably. *Write S_2 in powers of $(x - 2)$*, like this:

$$S_2(x) = a + b(x - 2) + c(x - 2)^2.$$

Using powers of $(x - 2)$, rather than powers of x or powers of $(x - 3)$, conveniently "fixes" attention at the current starting point, and makes it easy to choose the constants a, b, and c properly.

Setting $x = 2$ in the preceding expression gives $S_2(2) = a$. Thus, to arrange that $S_2(2) = 4$, we need $a = 4$. Choosing b is similar but starts with a derivative:

$$S_2(x) = a + b(x - 2) + c(x - 2)^2 \implies S_2'(x) = b + 2c(x - 2)$$
$$\implies S_2'(2) = b \implies b = 3.$$

Now we know that $S_2(x) = 4 + 3(x - 2) + c(x - 2)^2$; only c is still to be found. To find it, we use the remaining condition, ▶ $S_2(3) = 3$, as follows:

$$3 = S_2(3) = 4 + 3(3 - 2) + c(3 - 2)^2 = 7 + c \implies c = -4.$$

We've already used the other two conditions, $S_2(2) = 4$ and $S_2'(2) = 3$.

Everything is now in place: ▶

$$\begin{aligned} S_2(x) &= a + b(x - 2) + c(x - 2)^2 \\ &= 4 + 3(x - 2) - 4(x - 2)^2 \\ &= -18 + 19x - 4x^2. \end{aligned}$$

Check for yourself that S_2 does indeed satisfy all three of the preceding requirements.

(The last expression above can be found by multiplying out the second-last expression. The last version, although shorter typographically, is not necessarily simpler—the preceding version shows better how the coefficients were found.)

We find the remaining spline elements in just the same way. The next one, S_3, must pass through the third and fourth knots and, to ensure smoothness, must agree in slope with S_2 at $x = 3$. The preceding formula shows that $S_2'(x) = 19 - 8x$, so $S_2'(3) = -5$. Thus, S_3 must satisfy these three conditions:

$$S_3(3) = 3; \qquad S_3'(3) = -5; \qquad S_3(4) = 5.$$

It follows, as before, ▶ that

$$S_3(x) = a + b(x - 3) + c(x - 3)^2 = 3 - 5(x - 3) + 7(x - 3)^2 = 81 - 47x + 7x^2.$$

Convince yourself; only the last coefficient takes much effort.

Similarly, S_4 is chosen to satisfy ▶

$$S_4(4) = 5; \qquad S_4'(4) = S_3'(4) = 9; \qquad S_4(5) = 1.$$

Verify that $S_3'(4) = 3$.

The result ▶ is

$$\begin{aligned} S_4(x) &= a + b(x - 4) + c(x - 4)^2 \\ &= 5 + 9(x - 4) - 13(x - 4)^2 \\ &= -239 + 113x - 13x^2. \end{aligned}$$

The details are left to you; see the exercises at the end of this section.

Joining the elements produces this quadratic spline:

$$S(x) = \begin{cases} S_1(x) = 1 + 3(x - 1) & \text{if } 1 \le x \le 2; \\ S_2(x) = 4 + 3(x - 2) - 4(x - 2)^2 & \text{if } 2 \le x \le 3; \\ S_3(x) = 3 - 5(x - 3) + 7(x - 3)^2 & \text{if } 3 \le x \le 4; \\ S_4(x) = 5 + 9(x - 4) - 13(x - 4)^2 & \text{if } 4 \le x \le 5. \end{cases}$$

The picture is worth a careful look.

Following is the graphical result. ◄   The spline S is the bold curve; the quadratic elements S_1 through S_4 are extended beyond the knots to show how they fit together *at* the knots.

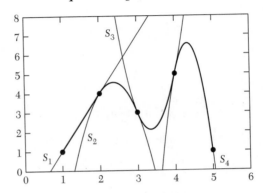

**Smoothing the knots:
quadratic spline elements**

Example 2 illustrates the formal definition.

> **Definition** A quadratic spline is a continuous, piecewise-quadratic function S whose first derivative, S', is also continuous.

This definition may seem forbiddingly technical. Here it means simply that (as we arranged) incoming and outgoing quadratic elements agree at each knot, both in their values and in their first derivatives.

Quadratic Splining: The Process in General

The technique just illustrated works for any collection (x_0, y_0), (x_1, y_1), . . . , (x_n, y_n) of knots, ordered from left to right. Here's the drill:

Start   Choose any quadratic curve S_1 joining (x_0, y_0) to (x_1, y_1). (A line segment is one possibility; other choices are possible.) For later use, calculate $m_1 = S_1'(x_1)$.

Fit the Next Parabola   Find a quadratic curve S_2 joining (x_1, y_1) to (x_2, y_2), such that $S_2'(x_1) = m_1$. ◄   To do so, find coefficients a, b, and c so that $S_2(x) = a + b(x - x_1) + c(x - x_1)^2$. Then, for later use, calculate $m_2 = S_2'(x_2)$.

m_1 is the "incoming slope" at (x_1, y_1).

Continue   Proceed as before. For each successive pair of knots (x_i, y_i) and (x_{i+1}, y_{i+1}), choose a quadratic curve S_{i+1} joining the knots, with $S_{i+1}'(x_{i-1}) = m_i$. ◄

m_i is the "outgoing slope" at (x_i, y_i).

How Smooth are Quadratic Splines?

A quadratic spline S has a continuous first derivative. The graph of S therefore has no kinks or corners. The second derivative S'', on the other hand, is *discontinuous* at each knot. ▶ This means, graphically, that the concavity of the S-graph can change abruptly at each knot. A close look at the preceding graph shows this happening. ▶ With cubic splines, we'll smooth things even further.

Unless, by sheer coincidence, successive elements have equal second derivatives at a knot.

Imagine riding a roller coaster along a quadratic spline. What would you feel at each knot?

Counting the Ways: Degrees of Freedom. There is just one line through two given points, or through one point with given slope. Either way, a straight line is completely determined by two conditions.

The reason, algebraically, is that the generic linear equation $y = a + bx$ involves *two* parameters, a and b; different choices of a and b give different lines. In mathematical language, we have **two degrees of freedom** in choosing a line; specifying two points (or one point and a slope) "uses up" both degrees of freedom.

By contrast, a quadratic polynomial $y = a + bx + cx^2$ involves *three* parameters. Thus, three degrees of freedom are available when we choose quadratics. The practical result is that a quadratic spline element—a parabola—can be chosen to satisfy three conditions. Two degrees of freedom are used to hit the left and right knots, but the third degree of freedom remains available. Our method "spends" the extra degree of freedom in specifying a derivative at the left knot, thus matching incoming and outgoing slopes.

Cubic spline elements, of the form $y = a + bx + cx^2 + dx^3$, offer four degrees of freedom. As with linear and cubic elements, hitting the knots costs two degrees of freedom. We'll spend the remaining two on matching two derivatives of incoming and outgoing elements at the left knot.

Cubic Splines: Smoother Still

A **cubic spline**, by definition, is a piecewise-cubic function S for which the first two derivatives, S' and S'', are continuous. Like linear and quadratic splines, cubic splines are constructed by piecing elements together, with each element spanning two knots. With cubic equations ▶ we can ensure that, at each knot, incoming and outgoing elements agree in their values and in their first two derivatives. ▶ By avoiding abrupt changes in concavity, cubic elements produce smoother, more pleasing splines than quadratic elements can achieve.

Cubic equations offer four degrees of freedom, one for each coefficient.

With quadratic splines we could manage just one derivative.

EXAMPLE 3 Construct a cubic spline through the same knots:

x	1	2	3	4	5
y	1	4	3	5	1

Solution Smoothness is no problem at the first knot. In theory, therefore, we can start by choosing *any* cubic curve S_1 that joins $(1, 1)$ to $(2, 4)$. Infinitely many possibilities exist—any values for $S_1'(1)$ and $S_1''(1)$ are possible. In practice, some choices give better results than others. Moreover, the resulting spline depends strongly on how the first element is chosen. Here, for instance, are two different cubic splines through the same knots:

Two cubic splines through the same knots

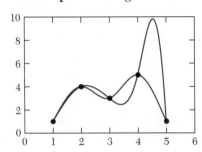

The two splines differ only in how the first element is chosen. As the picture shows, a tiny difference in the first element produces huge differences later on.

Faced with such exquisite sensitivity to initial choices, mathematicians have found systems for choosing the first cubic spline element. ◀ The **natural spline method**, for instance, arranges things so that $S''(x_0) = 0 = S''(x_n)$, i.e., so that S has zero concavity at the first and last knots. (The "tamer" curve in the preceding picture is a natural spline.) Full details on how natural splining works are beyond the scope of this discussion. ◀ We'll simply remark that the unlikely-looking cubic

Or for choosing all the elements simultaneously.

Numerical analysis courses discuss natural and other splining methods.

$$S_1(x) = 1 + \frac{123}{28}(x - 1) - \frac{39}{28}(x - 1)^3$$

Check that $S_1(1) = 1$ and $S_1(2) = 4$.

can be used to start things off. ◀ For impending use, we notice that $S_1'(2) = 3/14$ and $S_1''(2) = -117/14$.

For the next spline element, from $x = 2$ to $x = 3$, we need a cubic polynomial $S_2(x)$ that satisfies four conditions: ◀

The arithmetic will soon get messy. Bear with us.

$$S_2(2) = 4; \qquad\qquad S_2'(2) = \frac{3}{14} = S_1'(2);$$

$$S_2''(2) = -\frac{117}{14} = S_1''(2); \qquad S_2(3) = 3.$$

As before, writing S_2 in powers of $(x - 2)$, as in

$$S_2(x) = a + b(x - 2) + c(x - 2)^2 + d(x - 2)^3,$$

simplifies our search.

We choose the coefficients a and b exactly as in Example 2. The preceding expression shows ◀ that $S_2(2) = a$, so $a = 4$. Similarly,

Just plug in $x = 2$.

$$S_2'(x) = b + 2c(x - 2) + 3d(x - 2)^2 \implies S_2'(2) = b \implies b = \frac{3}{14}.$$

To find c, we differentiate again:

$$S_2''(x) = 2c + 6d(x - 2) \implies S_2''(2) = 2c \implies c = -\frac{117}{28}.$$

Now we know that $S_2(x) = 4 + \frac{3}{14}(x - 2) - \frac{117}{28}(x - 2)^2 + d(x - 2)^3$; we need only find d. To do so, we use the last remaining condition, $S_2(3) = 3$: ▶

Check the arithmetic if you like, but be sure to check the logic.

$$3 = S_2(3) = 4 + \frac{3}{14}(3 - 2) - \frac{117}{28}(3 - 2)^2 + d(3 - 2)^3 \implies d = \frac{83}{28}.$$

We're finished at last:

$$S_2(x) = 4 + \frac{3}{14}(x - 2) - \frac{117}{28}(x - 2)^2 + \frac{83}{28}(x - 2)^3.$$

For later reference, ▶ straightforward calculation shows that $S_2'(3) = 3/4$ and $S_2''(3) = 66/7$.

We'll need these numbers to fit the next element.

We find the remaining spline elements in the same way. The next one, S_3, must satisfy four conditions:

$$S_3(3) = 3; \qquad\qquad S_3'(3) = \frac{3}{4} = S_2'(3);$$

$$S_3''(3) = \frac{66}{7} = S_2''(3); \qquad S_3(4) = 5.$$

When all the dust settles, calculations like these reveal that our desired spline has the following piecewise-cubic form: ▶

It isn't necessary to check all calculations; the idea's the thing.

$$S(x) = \begin{cases} S_1(x) = 1 + \dfrac{123}{28}(x - 1) - \dfrac{39}{28}(x - 1)^3 & \text{if } 1 \leq x \leq 2; \\[2mm] S_2(x) = 4 + \dfrac{3}{11}(x - 2) - \dfrac{117}{28}(x - 2)^2 + \dfrac{83}{28}(x - 2)^3 & \text{if } 2 \leq x \leq 3; \\[2mm] S_3(x) = 3 + \dfrac{3}{4}(x - 3) + \dfrac{33}{7}(x - 3)^2 - \dfrac{97}{28}(x - 3)^3 & \text{if } 3 \leq x \leq 4; \\[2mm] S_4(x) = 5 - \dfrac{3}{14}(x - 4) - \dfrac{159}{28}(x - 4)^2 + \dfrac{53}{28}(x - 4)^3 & \text{if } 4 \leq x \leq 5. \end{cases}$$

∎

Postscript and Confession

The general idea of linear, quadratic, and cubic splines—our main interest in this section—is simple. Elements of the desired type are linked end to end; as many derivatives as possible are made to agree at knots.

The necessary calculations, although routine, can be tedious and repetitive, especially when dozens or hundreds of knots are involved. In real-life practice, such calculations are left to machines. Specialized methods that extend the ideas mentioned here are used to speed up and simplify the machines' work.

BASICS

1. What does it mean for a function to be "piecewise-linear?" Explain briefly in words.
2. Describe (i.e., give a formula for) a piecewise-linear function whose graph passes through the points $(0, 0)$, $(1, 1)$, and $(2, 4)$.
3. A linear spline is given by the multiline formula—one element is missing:

$$S(x) = \begin{cases} -2x + 7, & \text{if } 1 \leq x \leq 2 \\ \vdots & \vdots \\ -3x + 13, & \text{if } 3 \leq x \leq 4 \\ x - 3, & \text{if } 4 \leq x \leq 5 \end{cases}$$

Draw the spline; then fill in the missing line of the formula.

FURTHER EXERCISES

4. (a) Which upper-case letters of the alphabet could be graphs of piecewise-linear functions?
   (b) For two of the letters you chose in the previous part, describe (i.e., give a formula for) a piecewise-linear function whose graph has the right shape. (There are several ways to do this.)

5. Verify the last three lines of the multiline formula for the linear spline S in Example 1. [**HINT:** For each piece, start with the point–slope form of the line.]

6. In Example 2 we claimed that

   $$S_4(x) = 5 + 9(x-4) - 13(x-4)^2 = -239 + 113x - 13x^2$$

   satisfies the three conditions (i) $S_4(4) = 5$; (ii) $S_4'(4) = 9$; (iii) $S_4(5) = 1$. Verify that all three conditions really hold.

In Exercises 7–10, find a quadratic polynomial that satisfies the three given conditions. Then plot the polynomial on an appropriate interval to *show* that it has the desired properties.

7. $Q(0) = 0$; $Q'(0) = 3$; $Q(1) = 5$.
8. $Q(0) = 0$; $Q'(0) = 3$; $Q(10) = 5$.
9. $Q(1) = 0$; $Q'(1) = 3$; $Q(3) = 0$.
10. $Q(-17.3) = 11.1$; $Q'(-17.3) = -1.7$; $Q(-17.1) = 11.5$.

11. The picture on page 290 shows several quadratic curves, each joining $(0, 0)$ and $(1, 1)$, each labeled with its slope at $x = 0$. Find a quadratic equation for each curve.

12. Let (x_0, y_0) and (x_1, y_1) be any two points with $x_0 \neq x_1$, and let m_0 be any real number. Consider the quadratic function q defined by

    $$q(x) = y_0 + m_0(x - x_0) + \frac{y_1 - y_0 - m_0(x_1 - x_0)}{(x_1 - x_0)^2}(x - x_0)^2.$$

    (a) Find $q(x)$ in the case $(x_0, y_0) = (2, 4)$, $(x_1, y_1) = (3, 3)$, $m_0 = 3$. Then check that $q(2) = 4$, $q(3) = 3$, and $q'(2) = 3$.
    (b) Plot your function q from the previous part. How does the graph show that $q(2) = 4$, $q(3) = 3$, and $q'(2) = 3$?
    (c) Show (work with the general formula above) that $q(x_0) = y_0$, $q(x_1) = y_1$, and $q'(x_0) = m_0$. What do these properties mean about the graph of q?
    (d) Give an example in which $q(x)$ turns out to be a linear function. Interpret your example in graphical terms.

13. In each part below, find a quadratic spline S through the three knots

x	0	1	2
y	1	-2	3

    (a) Choose the first element by assuming that $S'(0) = -3$.

(b) Choose the first element by assuming that $S'(0) = 0$.
(c) The graphs of the two splines described in (a) and (b) follow. Which is which? Why?

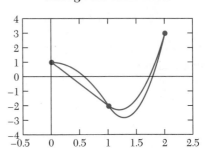

Two quadratic splines through the same knots

14. (For algebra buffs.) Let (x_0, y_0) and (x_1, y_1) be any two points with $x_0 \neq x_1$, and let d_1 and d_2 be any real numbers. Consider the cubic function c defined by

    $$c(x) = y_0 + d_1(x - x_0) + \frac{1}{2}d_2(x - x_0)^2 +$$
    $$\frac{y_1 - y_0 - d_1(x_1 - x_0) - \frac{1}{2}d_2(x_1 - x_0)^2}{(x_1 - x_0)^3}(x - x_0)^3$$

    Show by (careful!) calculation that c satisfies the four conditions (i) $c(x_0) = y_0$; (ii) $c'(x_0) = d_1$; (iii) $c''(x_0) = d_2$; (iv) $c(x_1) = y_1$. (These four conditions mean that c is the "general" spline element joining (x_0, y_0) to (x_1, y_1) with first two derivatives d_1 and d_2 at (x_0, y_0)).

In Exercises 15–18, use the formula in the previous exercise (even if you didn't do the previous exercise) to find a cubic function with the given properties. Plot each result.

15. $c(0) = 0$; $c'(0) = 0$; $c''(0) = 0$; $c(1) = 0$.
16. $c(0) = 0$; $c'(0) = 0$; $c''(0) = 0$; $c(1) = 1$.
17. $c(0) = 1$; $c'(0) = 0$; $c''(0) = -1$; $c(1) = 1/2$.
18. $c(0) = 1$; $c'(0) = 1$; $c''(0) = 1$; $c(1) = 8/3$.

19. In each part below, find a cubic spline S through the three knots

x	0	1	2
y	1	-2	3

    (a) Choose the first element by assuming that $S'(0) = -3$, $S''(0) = 0$.
    (b) Choose the first element by assuming that $S'(0) = 0$, $S''(0) = 0$.

(c) Below are graphs of the two splines you just found. Which is which? Why?

**Two cubic splines
through the same knots**

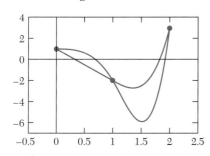

20. By definition, a quadratic spline is a piecewise-linear function whose derivative is also continuous. In fact, the derivative of a quadratic spline is a linear spline. Verify this in the case of Example 2 by writing a (multiline) formula for S'. Then plot both S' and S on the same axes. [**HINT:** To find S', differentiate S line by line.]

4.6 Optimization

Optimization and the Derivative

Optimization is a two-dollar word for the general problem of finding maximum or minimum values of functions. By now it should be clear that the derivative is a useful optimization tool: Anywhere f' is zero is a natural candidate for a maximum or minimum of f. Several subtleties can arise, however. A root of f' might, for instance, correspond to a local maximum, a local minimum, or neither. A maximum, in turn, could be global or merely local.

To help sort things out, we'll collect and summarize the language, the techniques, and a few of the subtleties involved in optimization problems.

Stationary Points and Critical Points. Let f be a function and f' its derivative. As we've seen, f increases where $f' > 0$ and decreases where $f' < 0$. Therefore, any point x_0 at which f' changes sign is naturally of interest.

Barring bizarre circumstances, ▶ f' can change sign at x_0 in only two ways: *We do bar them.*

Either $f'(x_0) = 0$ or $f'(x_0)$ doesn't exist.

If the first condition holds, x_0 is a **stationary point** of f. If *either* condition holds, then x_0 is a **critical point** of f.

EXAMPLE 1   The function $f(x) = |x|$ ▶ has no stationary points, because $f'(x)$ is never zero. The critical point $x_0 = 0$ (where the derivative fails to exist) is a local minimum point of f. ■   *Draw (or imagine) the graph of f.*

Local vs. Global Extreme Values. Local maximum and minimum values of a function may or may not be global. ▶ The function $f(x) = x/2 + \sin(x)$, for example, has many local maxima and minima on $(-\infty, \infty)$, but none is global—this   *Formal definitions of these words appear in Section 2.3.*

f is unbounded above and below on $(-\infty, \infty)$. The graph agrees:

Local but not global maxima and minima:
$$f(x) = x/2 + \sin x$$

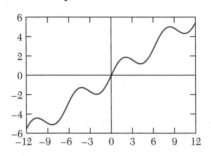

Which Functions Have Global Maxima and Minima? Continuous Functions on Closed Intervals. As we just saw for $f(x) = x/2 + \sin(x)$ on $(-\infty, \infty)$, a function may have neither a maximum nor a minimum on a given interval. All is not lost. The **extreme value theorem** ◄ guarantees that a continuous function f on a closed interval $[a, b]$ must assume both a global maximum and a global minimum.

We'll study it carefully in Section 4.10.

Another look at the preceding graph shows that this property holds for $f(x) = x/2 + \sin(x)$ ◄ on the closed interval $[-12, 12]$. In this case, f takes its largest and smallest values (approximately ± 5.5) at the interval's endpoints, $x = \pm 12$.

Evidently a continuous function.

Finding Maxima and Minima of f on $[a, b]$. If f is continuous on a *closed* interval $[a, b]$, then global maxima and minima of f on $[a, b]$ ◄ must exist. The following Fact is a key to finding them.

The extreme value theorem guarantees that they exist.

> **Fact** A continuous function f on a closed interval $[a, b]$ can assume its maximum and minimum values only at critical points in (a, b) or at endpoints of $[a, b]$.

(We will state and prove an equivalent fact in the first lemma on page 335.) The Fact says that the only possible candidates for maximum and minimum points of f are (1) the endpoints a and b, (2) roots of f', and (3) points at which f' doesn't exist. In practice, this slate of candidates is usually short. Choosing among them is usually easy, with or without computer graphics.

EXAMPLE 2 Find the maximum and minimum values of $f(x) = x/2 + \sin(x)$ on $[0, 3]$.

Solution First we find critical points. Since f' exists everywhere, it's enough to find roots of f'. An easy computation does the trick:

$$f'(x) = \frac{1}{2} + \cos x = 0 \iff \cos x = -\frac{1}{2}.$$

Right? See for yourself.

The only such x in $[0, 3]$ is $x = 2\pi/3 \approx 2.094$. ◄

According to the preceding Fact, the maximum and minimum values of f on $[0, 3]$ can occur only at 0, 3, or $2\pi/3$. All that remains is to check f at these values:

$$f(0) = 0;$$

$$f\left(\frac{2\pi}{3}\right) = \frac{\pi}{3} + \frac{\sqrt{3}}{2} \approx 1.91;$$

$$f(3) = \frac{3}{2} + \sin 3 \approx 1.64.$$

The conclusion for f on $[0, 3]$ is now clear: f assumes its minimum value, 0, at $x = 0$ and its maximum value, 1.91, at $x = 2\pi/3$. ∎

Finding Maxima and Minima on Any Interval. The preceding Fact applies only to *closed* intervals. On an arbitrary interval, f may or may not assume global maximum and minimum values. In practice, however, it's usually clear either from context or from graphics whether extreme values exist and how to find them. Critical points are still key.

EXAMPLE 3 Which point on the line $y = -2x + 2$ is closest to the origin?

Solution Theory temporarily aside, it's intuitively clear that *some* point on the line minimizes the distance in question. To find it, let's define the function

$$f(x) = \text{distance from } (0, 0) \text{ to } (x, -2x + 2)$$

as follows; $f(x)$ is the dotted distance.

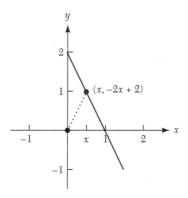

Which point on $y = -2x + 2$ is closest to $(0, 0)$?

An explicit formula for f is easy ▶ to find:

$$f(x) = \sqrt{x^2 + (-2x + 2)^2} = \sqrt{5x^2 - 8x + 4}.$$

Use the distance formula in the plane.

Intuition (or a graph) says that f has just one local (and therefore global) minimum; theory tells us to look for a stationary point. Finding one involves symbolic calculus ▶ and algebra:

Use the chain rule.

$$f'(x) = \frac{10x - 8}{2\sqrt{5x^2 - 8x + 4}} = 0 \iff 10x - 8 = 0 \iff x = \frac{4}{5}.$$

Convince yourself that the answer is reasonable.

Thus, the minimum distance occurs at the point $(4/5, 2/5)$; ◄   the distance itself is $\sqrt{(4/5)^2 + (2/5)^2} \approx 0.8944$. (A slightly different approach is to minimize the *square* of the distance; this eliminates square roots and simplifies the algebra.)   ■

Objective Functions, Constraint Equations, and Implicit Derivatives

Many optimization problems have the same general form. Some convenient jargon will help us describe it.

Objective Function: Describes the quantity to be maximized or minimized. In Example 3, for instance, we minimized the distance D from the origin to a point (x, y) on a certain line. This distance—the objective function—has the form $D(x, y) = \sqrt{x^2 + y^2}$. (As written here, the objective function depends on *two* variables, x and y. In Example 3 we rewrote the objective function using just one variable, x.)

Constraint Equation: Describes a condition that must be satisfied by the variables in an optimization problem. In Example 3, we required (i.e., "constrained") (x, y) to lie on the line $y = -2x + 2$; this is our constraint equation, or simply the **constraint**.

Constrained Optimization Problem: Any optimization problem that (like Example 3) involves maximizing or minimizing some objective function subject to one or more constraint equations.

To solve Example 3, we used the constraint equation to rewrite the objective function to depend on only one variable: $f(x) = \sqrt{5x^2 - 8x + 4}$. Then we used ordinary differentiation to minimize f. Implicit differentiation offers a slightly different (and sometimes simpler) approach to optimizing an objective function with a constraint. Two easy examples will illustrate. ◄

Watch the use of technical language, too.

   EXAMPLE 4   Two nonnegative numbers sum to 10. How large can their *product* be?

   Solution   Let x and y be the two numbers. Then our **objective function** is $P = xy$. Not just any x and y will do; the **constraint equation** says that $x + y = 10$. Legal values of x are in the interval $[0, 10]$. Note that if either $x = 0$ or $x = 10$, then $P = 0$. Thus, P must be largest at some critical value of x, where $dP/dx = 0$. We find it by differentiation.

   We *could* use the constraint $x + y = 10$ to solve for y and then rewrite P as a function of x alone. Instead, we differentiate both the objective function and the constraint equation

$$P = xy; \qquad x + y = 10$$

implicitly with respect to x (i.e., treating y as an implicit function of x). Here are the results: ◄

$$\frac{dP}{dx} = x\frac{dy}{dx} + y; \qquad 1 + \frac{dy}{dx} = 0.$$

Check the result for yourself.

At a critical point x, $dP/dx = 0$. This makes, in all, three equations (including the constraint equation) that must hold at a critical point:

$$0 = x\frac{dy}{dx} + y; \qquad 1 + \frac{dy}{dx} = 0; \qquad x + y = 10.$$

Substituting the second equation into the first gives

$$\frac{dy}{dx} = -1 \Longrightarrow 0 = -x + y \Longrightarrow x = y.$$

Thus, the objective function is greatest when the two factors x and y are equal. Since $x + y = 10$, $x = y = 5$ and $P = 25$. ■

The answer isn't too surprising. Neither is the answer to the following (closely related) problem.

EXAMPLE 5   Two nonnegative numbers multiply to 25. How small can their *sum* be? ▶

First make a guess. (The preceding example should help.)

Solution   The new objective function and constraint equation are, respectively,

$$S = x + y \qquad \text{and} \qquad xy = 25.$$

Differentiating both of these equations implicitly with respect to x and setting $dS/dx = 0$ gives a familiar result:

$$\frac{dS}{dx} = 1 + \frac{dy}{dx} = 0; \qquad y + x\frac{dy}{dx} = 0.$$

It follows—by the same calculation as in the preceding example—that at a critical point, $x = y$. Thus (just as before), $x = y = 5$ at the critical point, and so $S = 10$ is minimum. ■

The Best Pop Cans: An Extended Example

Real-world calculus optimization problems usually involve translating English prose into—and out of—mathematical language. Once a problem is translated into mathematical terms, more than one solution avenue may be open. Mathematical solutions, once found, require interpretation in the original context.

The next example, chosen to illustrate the preceding points, has to do with the design of ordinary 12-ounce (355-ml) aluminum soft-drink containers. Do such cans have the right dimensions? What does "right" mean? Why do other products come packed in cans of other shapes?

EXAMPLE 6   An ordinary pop can has a volume of 355 cm³. Find the dimensions (radius r and height h) that minimize the surface area of such a can.

Solution   We'll offer two solutions, the first using ordinary derivatives and the second using *implicit* derivatives.

Solution 1:   The can's volume V and surface area A are given by

$$V = 355 = \pi r^2 h; \qquad A = 2\pi rh + 2\pi r^2.$$

From the first equation, $h = 355/(\pi r^2)$, so

$$A = 2\pi rh + 2\pi r^2 = 2\pi r \cdot \frac{355}{\pi r^2} + 2\pi r^2 = \frac{710}{r} + 2\pi r^2.$$

Therefore,

$$\frac{dA}{dr} = -\frac{710}{r^2} + 4\pi r = 0 \implies 710 = 4\pi r^3$$

$$\implies r = \left(\frac{710}{4\pi}\right)^{1/3} \approx 3.837.$$

Not a local maximum or just a flat spot.

It's easy to check (either graphically or by taking another derivative) that A has a local minimum ◄ here. Thus, the can's optimal dimensions are $r \approx 3.84\,\text{cm}$ and $h = 355/\left(\pi r^2\right) \approx 7.67\,\text{cm} = 2r$. In particular, the optimal can has a *square* profile!

Solution 2: We start with the same equations: $A = 2\pi rh + 2\pi r^2$ and $V = \pi r^2 h = 355$. Differentiating both equations implicitly with respect to r gives

$$\frac{dA}{dr} = 2\pi h + 2\pi r \frac{dh}{dr} + 4\pi r;$$

$$\frac{dV}{dr} = 2\pi rh + \pi r^2 \frac{dh}{dr} = 0 \implies \frac{dh}{dr} = -\frac{2h}{r}.$$

Now we substitute the last result into the expression for dA/dr and simplify:

$$\frac{dA}{dr} = 2\pi h + 2\pi r \left(-\frac{2h}{r}\right) + 4\pi r = 0 \implies h = 2r.$$

Thus, the optimal can has a square profile. ∎

EXAMPLE 7   Example 6 showed that, to minimize surface area, a pop can should have a square profile. Why don't real pop cans have this shape?

Solution   Minimizing surface area does not necessarily minimize the *cost* of a can. Surface area ignores, for instance, the extra cost of the can's top, which is thicker than the sides. ◄ ∎

Look carefully at an aluminum pop can. Is the top obviously thicker?

EXAMPLE 8   It's reasonable to estimate that the top of a can is three times as thick as the sides and bottoms. Choose h and r to minimize the amount of aluminum in the can.

Solution   For simplicity, let's pretend that the top is made of three layers of aluminum and still minimize total area A. The argument runs as before, except that now

$$A = \text{sides} + \text{bottom} + \text{top} = 2\pi rh + \pi r^2 + 3\pi r^2 \implies A = 710/r + 4\pi r^2.$$

Now

$$\frac{dA}{dr} = \frac{-710}{r^2} + 8\pi r = 0 \implies r = \left(\frac{710}{8\pi}\right)^{1/3} \approx 3.046.$$

Therefore, the optimal dimensions are now $r \approx 3.046$ cm and $h = 12.182$ cm. ∎

EXAMPLE 9   What are the dimensions of a real pop can? Are they "right"?

Solution   Measuring a real can gives $h \approx 12.5$ cm and $r \approx 3.1$ cm. These values correspond well to the numbers computed before—they're a bit greater, since pop cans have some room to spare. Moreover, the tops of pop cans do seem to be about three times as thick as the sides.   ■

On Beyond Soda Pop

Derivatives are good for more than designing aluminum cans for carbonated beverages. They work with noncarbonated beverages, too.

EXAMPLE 10   Fruit juice is often sold in 6-ounce ($\approx$ 168-cm^3) *steel* cans. Assuming that the top of such a can is no thicker than the sides, find the dimensions that minimize surface area. Are the results realistic?

Solution   Mimicking the foregoing work shows, as in Example 6, that $h = 2r$ for the optimal can. From this it follows that $r \approx 2.99$ cm and $h \approx 5.98$ cm.

Actual juice cans have a more elongated profile than these numbers suggest, perhaps in order to fit the hand.   ■

BASICS

1. Which point on the line $y = -2x + 2$ is closest to $(0, 1)$? [**HINT:** See Example 3 (page 299).]

2. A triangle has legs on the positive x- and y-axes, and its hypotenuse passes through the point $(2, 1)$. Which such triangle has the smallest area?

3. This exercise offers another approach to Example 3 (page 299). This time, let $g(x) =$ square of the distance from $(0, 0)$ to $(x, -2x + 2) = x^2 + (2 - 2x)^2$. (Instead of minimizing the distance from the origin to the line, we'll minimize the square of the distance.)
   (a) Find the minimum value of $g(x)$ and the value of x for which it occurs.
   (b) Is the answer to the previous part consistent with what was found in Example 3 (page 299)? Why or why not?

4. This exercise has to do with finding a point P on the right half of the parabola $y = 1 - x^2$ that's closest to the origin.
   (a) Plot the parabola carefully in the rectangle $[0, 1] \times [0, 1]$. (Use equal size units on both axes.) Then use your plot to estimate the coordinates of P.
   (b) Use calculus to find the coordinates of P. [**HINT:** Mimic either the previous exercise or Example 3 (page 299).]
   (c) Show that the line from the origin to P is perpendicular to the line that's tangent to the parabola at P. How does this property appear on your graph?

5. Redo the previous exercise, but use the parabola $y = 1/4 - x^2$. What accounts for the difference between this exercise and the previous one?

FURTHER EXERCISES

6. Do Example 4 (page 300) without using implicit differentiation. Instead, use the equations $P = xy$ and $x + y = 10$ to rewrite P as a function of only one variable, x. Then maximize P over an appropriate interval.

7. Do Example 5 (page 301) without using implicit differentiation. Instead, use the equations $S = x + y$ and $xy = 25$ to rewrite S as a function of only one variable, x. Then minimize S.

8. Suppose that a projectile (e.g., a thrown baseball or a kicked football) leaves the origin with initial velocity v_0, is aimed along the line $y = mx$, and is influenced only by gravity. Then the projectile's path, or "trajectory,"

satisfies the equation

$$y = mx - \frac{g}{2v_0^2}(1 + m^2)x^2,$$

where g is the constant (downward) acceleration due to gravity. (Numerically, $g \approx 9.8$ meters/second$^2 \approx 32$ feet/second^2.)
   (a) A football is kicked with initial velocity $v_0 = 30$ meters per second and initial slope $m = 1$. (Use $g = 9.8$ meters/second^2.) How high does the football rise? Where does it land?
   (b) Plot the football's trajectory, labeling important points found in the previous part.

9. Use the trajectory formula of Exercise 8 as needed.
   (a) A baseball is thrown with initial velocity $v_0 = 128$ feet per second, with initial slope $m = 1/2$. (Use $g = 32$ feet per second².) How high does the baseball rise? Where does it land?
   (b) Another baseball is thrown with initial velocity $v_0 = 128$ feet per second, but with initial slope $m = 2$. How high does this baseball rise? Where does it land?
   (c) Plot the trajectories of both baseballs, labeling the important points found in the previous parts.

10. In the situation of Exercise 8, the projectile has range
$$R = \frac{2v_0^2}{g} \frac{m}{1+m^2}.$$
(The range is the horizontal distance covered before landing.)
   (a) Use the formula of Exercise 8 to show why the range formula above holds.
   (b) Suppose that v_0 is fixed, but m is variable. What value of m maximizes the range?

11. To find the point on the line $x + y = 1$ that's closest to the origin, we can minimize the objective function $D = x^2 + y^2$ (the square of the distance from (x, y) to $(0, 0)$) subject to the constraint $x + y = 1$. Differentiating gives two equations that must hold at a critical point:

$$\frac{dD}{dx} = 2x + 2y\frac{dy}{dx} = 0 \quad \text{and} \quad 1 + \frac{dy}{dx} = 0.$$

   Plugging the second equation into the first gives $x = y$. Thus the minimum distance occurs where both $x = y$ and $x + y = 1$, i.e., at the point $(1/2, 1/2)$.

   In each part that follows, find the point on the given line that's closest to the origin. In all parts, a, b, and c are nonzero constants.
   (a) $x + y = c$
   (b) $ax + y = c$
   (c) $ax + by = c$

12. In each part, a farmer wants to fence a rectangular area as inexpensively as possible. Assume that fencing materials cost $1 per foot.
   (a) Suppose that $40 is available for the project. How much area can be enclosed? [**HINT:** Let x and y be the two dimensions of the fence, so that the area A is xy and the perimeter P is $2x + 2y$. Imitate the method of Example 4 (page 300) to maximize the objective $A = xy$ subject to the constraint $2x + 2y = 40$.]
   (b) Suppose that 100 square feet must be enclosed. What is the least possible cost? [**HINT:** Minimize $P = 2x + 2y$ subject to the constraint $xy = 100$.]
   (c) Discuss the relation between the two parts above. Did the two approaches give the same result?

13. In each part that follows, a farmer wants to fence a rectangular area as inexpensively as possible. Fences must run east–west and north–south. Assume that fencing ma-

terials cost $h per foot in the east–west direction and $v per foot in the north-south direction.
   (a) Suppose that a total budget $b is available for the project. How much area can be enclosed? What are the best dimensions for the fence? [**HINT:** Let x and y be the east–west and north–south dimensions of the fence, respectively. Then the area is $A = xy$ and the total cost is $C = 2hx + 2vy$. Maximize $A = xy$ subject to $C = 2hx + 2vy = b$.]
   (b) Suppose that a total area of a square feet must be enclosed. What is the least possible cost? What are the best dimensions for the fence? [**HINT:** Minimize $C = 2hx + 2vy$ subject to $A = xy = a$.]
   (c) Are the two parts above consistent with each other? In each case, how much is spent for east–west fence? How much is spent for north–south fence?

14. A rectangle has its base on the x-axis, a vertex on the y-axis, and a vertex on the curve $y = e^{-x^2}$.
   (a) What choice of vertices gives the largest area?
   (b) Show that one of the vertices found in part (a) is at an inflection point of the curve.

15. Show that the optimal fruit juice can has the dimensions claimed in Example 10.

16. Soup cans are to be manufactured to contain a given volume V. No waste is involved in cutting the material for the vertical sides, but the top and bottom (circles of radius r) are cut from squares that measure $2r$ units on a side. Thus the area A of material consumed for a can of height h is $A = 2\pi rh + 8r^2$. Find the ratio of height to diameter for the most economical can (i.e., the can requiring the least material). [**HINT:** Use the fact that $V = \pi r^2 h$ to express A as a function of r.]

17. A piece of wire 100 cm long is going to be cut into several pieces and used to construct the skeleton of a rectangular box with a square base.
   (a) What are the dimensions of the box with the largest volume?
   (b) What are the dimensions of the box with the largest surface area?

18. Federalist Express limits the size of a mailable parcel: The longest side plus the girth (the perimeter of a cross-section perpendicular to the longest side) may not exceed 108 inches.
   (a) Which mailable cube has largest volume?
   (b) Find the dimensions of the mailable rectangular parcel with a square base that has the largest volume. [**HINT:** Consider two cases, depending on which side is longest.]
   (c) Among mailable right circular cylinders, which has the largest volume? [**HINT:** Again, consider two cases.]
   (d) Give an example of a mailable parcel with larger volume than a nonmailable parcel.

19. It's 10:30 p.m. Your snowmobile is out of gas and you are 3 miles due south of a major highway. The nearest service station is on the highway 6 miles east; it closes at midnight. You can walk 4 miles per hour on roads, but only 3 miles per hour through snowy fields. Can you make it to the service station? What route is best?

20. The water depth in Moon River x miles downstream from Hard Rock is $D(x) = 20x + 10$ feet; the width of the river is $W(x) = 10(x^2 - 8x + 22)$ feet. To create Moon Lake, a dam is to be built downstream from Hard Rock. For engineering reasons, the dam cannot be more than 130 feet high.

    (a) For which values of x is $0 \le D(x) \le 130$?
    (b) How far downstream can the dam be built? If the dam were constructed at this point, how wide would it be? How high?

(c) What are the dimensions of the widest dam that could be constructed?
(d) What are the dimensions of the narrowest dam that could be constructed?
(e) If the cost of building the dam is proportional to the product of the length and the depth of the dam, where should the cheapest dam be located?

21. A truck driving over a flat interstate at a constant speed of 50 mph gets 4 miles to the gallon. Fuel costs $1.19 per gallon. The truck loses a tenth of a mile per gallon in fuel efficiency for each mile per hour increase in speed. Drivers are paid $27.50 per hour in wages and benefits. Fixed costs for running the truck are $11.33 per hour. A trip of 300 miles is planned. What speed minimizes operating expenses?

4.7 Calculus for Money: Derivatives in Economics

Mathematics has been applied to finance for a very long time. A 3000-year-old Babylonian problem (cited in *A History of Mathematics*, by Carl Boyer) goes as follows.

> *Ten brothers received 1;40 minas of silver, and brother over brother received a constant difference. If the eighth brother received 6 shekels, find out how much each earned.*

(A mina is 60 shekels. The Babylonians used a base-60 number system, so the notation 1;40 means 1 mina plus 40 shekels, or 100 shekels.)

Much has changed since Old Babylon. Calculus and economics were invented, and base 60 disappeared from common use. ▶ Financial applications of mathematics, however, remain very much alive, and calculus—especially the derivative—is among modern economists' first tools of choice. In this section we hint briefly at one application of calculus to economics.

Ten-brother families are rare nowadays, too.

Derivatives in Economics

Economics is full of *rates*—rates of interest, tax, inflation, unemployment, return on investment, and many more. Most rates can somehow be thought of as derivatives. Naturally, therefore, many economic calculations involve calculus with derivatives.

Economic Functions: Functions of What?

To apply calculus, we need functions with which to work. Economics offers plenty. Here are just a few examples, informally described: ▶

See the exercises at the end of this section for more details.

$HW(t)$ = average U.S. hourly manufacturing wage at time t;

$U(t)$ = number of unemployed U.S. workers at time t;

$ND(t)$ = cumulative U.S. national debt at time t;

$C(q)$ = seller's total cost of q bushels of soybeans;

$p(q)$ = price per bushel at which q bushels of soybeans can be sold.

The first three of these functions vary with *time*, so their derivatives measure rates with respect to time. To say that $U'(1993.5) = -50,000$, for instance, could mean that halfway through 1993 the unemployment population was dropping at an instantaneous rate of 50,000 people per year. ◄

Notice the units of U': people/year.

The functions C and p are a little different: For them the independent variable represents *quantity* (here, the quantity of soybeans), not time. Thus, their derivatives measure rates of change with respect to quantity, not time. To say, for instance, that $C'(2000) = 5$ might mean that at 2000 bushels the seller's total cost is increasing at the rate of $5 per bushel. ◄   To put it another way, the 2001st bushel costs the seller $5. ◄

Notice the units of C': dollars/bushel.

Read further for more on comparing these two interpretations of the derivative.

Discrete versus Continuous Phenomena

Technically speaking, the cost and price functions C and p accept only *whole-number* inputs q: Grain traders sell only whole bushels. A variable that takes only isolated values (e.g., integers) is called a **discrete variable**; a **continuous variable** can take *any* real value. Although many financial functions involve discrete variables, it's common to approximate them—as we will later in this section—with functions of continuous variables. Doing so makes the powerful tools of calculus available.

Modeling Reality Mathematically.   To model any phenomenon (e.g., soybean sales on the Chicago Board, snowshoe hare and fox populations in the Arctic, heating requirements in Minnesota, demand for hard Montana wheat, or the spread of AIDS) mathematically means to *describe* or *represent* the phenomenon quantitatively, in mathematical language.

Every mathematical model simplifies, to some extent, the phenomenon it describes. Economic models, for example, often involve "pretending" that a discrete variable is continuous. A useful model is simple enough to permit calculations but detailed enough to capture essential qualities of the phenomenon studied.

Mathematical modeling involves three steps: (i) *describing* the phenomenon in mathematical terms, using functions, variables, equations, and other mathematical ingredients; (ii) *deducing* mathematical consequences of the model; and (iii) *interpreting* mathematical results in terms of the phenomenon studied.

Mathematical economists use all sorts of functions. For simplicity, in this section we treat only functions that (like C and p) vary with quantity. Our first example uses several such functions in context.

EXAMPLE 1   Lotta Mays, an experienced grain trader, wonders how many bushels of soybeans to buy—and then sell—on the Chicago Board of Trade. Lotta knows that the per-bushel price p (in dollars) she can get for her beans depends on the quantity q (in bushels) she wants to sell: Too many beans on the market drives prices down. Here's her best guess for a **price function**:

$$p(q) = 8 - \frac{q}{1000}.$$

If Lotta sells q bushels at $p(q)$ dollars each, she'll raise a total of $q \cdot p(q)$ dollars. Thus, her **total revenue function** has the formula ◄

The variable q is in bushels, and R is in dollars.

$$R(q) = q \cdot p(q) = q \left(8 - \frac{q}{1000} \right) = 8q - \frac{q^2}{1000}.$$

Before selling any beans, Lotta must buy some. As a large wholesale customer, she can buy any amount of beans for $5 a bushel—$q$ bushels cost $5q$ dollars. In addition, Lotta has fixed expenses of $1000 for buying and selling her beans (handling fees, taxes, commissions, etc.). Lotta's **total cost function** therefore has the formula

$$C(q) = 5q + 1000.$$

How many bushels should Lotta buy and sell? At what price will she sell them? How much will she make?

Solution   Lotta wants to maximize her **profit**—the difference between revenue and cost—on the transaction. Like R and C, the total profit P depends on q:

$$P(q) = R(q) - C(q) = 8q - \frac{q^2}{1000} - (5q + 1000) = 3q - \frac{q^2}{1000} - 1000.$$

Lotta first plots P, R, and C together:

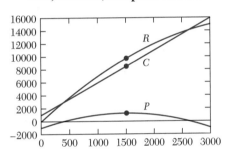

Cost, revenue, and profit functions

The bulleted point on the P-graph is clearly of interest: It suggests that the maximum possible profit—a little more than $1000—occurs for $q = 1500$. The other two bulleted points are also important. They mark the spot at which the R-graph is *farthest above* the C-graph, i.e., where revenue most exceeds cost.

Symbolic calculations agree with the graphs. Differentiating our formulas for C, R, and P gives

$$C'(q) = 5; \qquad R'(q) = 8 - \frac{2q}{1000}; \qquad P'(q) = 3 - \frac{2q}{1000}.$$

Setting $P'(q) = 0$ ▶ gives $q = 1500$—as the graph shows. At this optimal quantity, *Check our work.*

$$C(1500) = 8500; \qquad R(1500) = 9750;$$

$$P(1500) = 1250; \qquad p(1500) = 6.5.$$

All is now revealed: Lotta earns the maximum profit, $1250, by selling 1500 bushels at $6.50 per bushel. ∎

On the Margin: Derivatives of Revenue, Cost, and Profit

The term "marginal" will be explained soon.

The derivatives R', C', and P' from Example 1, though simple, deserve a look: ◄

Marginal cost, marginal revenue, and marginal profit functions

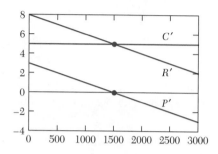

Again, $q = 1500$ is the quantity of interest. As the bullets show,

$$C'(1500) = R'(1500) = 5; \qquad P'(1500) = R'(1500) - C'(1500) = 0.$$

In other words, at the critical value $q = 1500$, C and R increase at the same rate, and P is stationary.

Mathematicians usually interpret derivatives as rates of change. The statement $P'(1000) = 1$ means, in rate language, that after selling 1000 bushels, the trader is profiting at the rate of \$1 per bushel.

Economists often interpret derivatives such as R', C', and P' as **marginal** quantities. In this language, the statement $P'(1000) = 2$ means that, after selling 1000 bushels, the trader makes \$2 profit on the *next bushel*. (The 1001st bushel is "on the margin.") For any q, **marginal profit at** q is defined as the additional profit to be made on the *next* (i.e., the $(q + 1)$st item. The fact that (as we found earlier) $P'(1500) = 0$ means, in the marginal sense, that zero profit will be made on the 1501st item. ◄

This would be a good time to quit the business.

Marginal cost and **marginal revenue** have similar meanings. In Example 1 we saw that $C'(1500) = 5 = R'(1500)$. In marginal language, this means that the 1501st bushel is a wash: It costs Lotta \$5 and sells for the same amount.

Here's the general marginal idea for *any* function f that depends on quantity q:

Marginal f is the increase in f due to the $(q + 1)$st item.

Marginals and Instantaneous Rates: Economists versus Mathematicians

How different are the marginal and rate interpretations of the derivative? Are mathematicians and economists really at odds?

Not seriously. In practice, rates and marginals seldom differ by much. A little symbolic algebra shows why. Suppose that $P(q)$ describes the profit to be made on q items. Then the marginal profit function $MP(q)$ is defined by

$$MP(q) = \text{profit on } (q + 1)\text{st item} = P(q + 1) - P(q).$$

The derivative function $P'(q)$, by contrast, is defined by the limit

$$P'(q) = \lim_{h \to 0} \frac{P(q + h) - P(q)}{h}.$$

Setting $h = 1$ in the last expression gives

$$P'(q) \approx \frac{P(q + 1) - P(q)}{1} = MP(q).$$

Thus, $P'(q)$ and $MP(q)$ are at least approximately equal; mathematicians and economists are on the same wavelength.

In practice, the difference between MP and P' is seldom worth fretting over. One reason is that P, P', and MP are all mere models of reality and so are inherently inexact (although *not* inherently useless). Another reason is that values of q are usually large, so setting $h = 1$ in the difference quotient gives a good approximation. If, say, P is the profit function in the example, and $q = 2000$, then (as easy calculations show)

$$P'(2000) = -1; \qquad MP(2000) = P(2001) - P(2000) = -1.001.$$

The tiny difference isn't worth a fight.

Varying Parameters

Economic strategies often depend sensitively on various parameters—the wholesale price of soybeans, the prime interest rate, and the rate of inflation—that change unpredictably from day to day. As parameters change, so do the results of economic calculations. The *form* of the calculations, however, may remain essentially the same.

EXAMPLE 2   Lotta Mays (see Example 1) finds a new wholesale supplier of soybeans at $4.50 a bushel. What should she do?

Solution   Lotta's price and revenue functions p and R are unchanged, but her cost and profit functions C and P are new. Here are full details:

$$p(q) = 8 - \frac{q}{1000}; \qquad R(q) = q \cdot p(q) = 8q - \frac{q^2}{1000};$$

$$C(q) = 4.5q + 1000; \qquad P(q) = R(q) - C(q) = 3.5q - \frac{q^2}{1000} - 1000.$$

We maximize P just as before:

$$P'(q) = 3.5 - \frac{2q}{1000} = 0 \implies q = 1750.$$

Since $P(1750) = 2062.5$ and $p(1750) = 6.25$, we have this result: Lotta should now sell 1750 bushels at $6.25 a bushel; she'll make $2062.50. (Notice the bottom line: Lotta passed *half* her 50-cent-per-bushel savings on to her customers.)   ■

Treating the wholesale price as a parameter—without a specific value—can solve a whole lotta problems at once.

EXAMPLE 3   The wholesale price of beans seems to change every day; Lotta's sick of working the same problem again and again. How many beans should she sell if the price to her is b dollars a bushel? At what price? How much will she make or lose?

Solution   For any value of b, the functions look like this:

$$p(q) = 8 - \frac{q}{1000}; \qquad R(q) = q \cdot p(q) = 8q - \frac{q^2}{1000};$$

$$C(q) = b \cdot q + 1000; \qquad P(q) = R(q) - C(q) = (8 - b)q - \frac{q^2}{1000} - 1000.$$

We'll maximize P in the usual way:

$$P'(q) = 8 - b - \frac{2q}{1000} = 0 \Longrightarrow q = 4000 - 500b.$$

For this quantity the price formula gives

$$p(4000 - 500b) = 8 - \frac{4000 - 500b}{1000} = 4 + \frac{b}{2}.$$

A little algebra shows, too, that

$$P(4000 - 500b) = 15{,}000 - 4000b + 250b^2.$$

Lotta has her general result: If beans wholesale for b dollars a bushel, she should sell at $4 + b/2$ dollars per bushel. Doing so, she'll sell $4000 - 500b$ bushels, for a total profit of $15{,}000 - 4000b + 250b^2$ dollars.   ∎

BASICS

1. This exercise is about several functions described informally in this section. In each part, first specify reasonable units for the input and output variables. Then give a plausible domain and range for the function, and draw a graph that could represent the function in question. (No special knowledge of economics is required—if in doubt, guess!)

   (a) $HW(t) = $ average U.S. hourly manufacturing wage at time t

   (b) $U(t) = $ number of unemployed U.S. workers at time t

   (c) $ND(t) = $ cumulative U.S. national debt at time t

2. What does "marginal tax rate" mean? Is this use of "marginal" consistent with those in marginal cost, marginal revenue, and marginal profit?

FURTHER EXERCISES

3. In Example 1 Lotta Mays guessed that the formula

   $$p(q) = 8 - \frac{q}{1000}$$

   might reasonably describe the selling price per bushel of q bushels of soybeans.

   (a) Plot p; choose *xrange* and *yrange* to make sense in context. What are the units on each axis?

   (b) State a sensible domain for the function p.

   (c) Using your answer from the previous part, state a plausible range for the function p.

   (d) The function p above is decreasing. Why would Lotta choose a decreasing function? Explain briefly in words.

   (e) The function p above is linear. Lotta chose a linear function purely for simplicity, but is that a good choice? Would you expect a real price function to be linear? If so, why? If not, sketch a more convincing price function and explain why it is a more realistic choice.

4. This exercise refers to Example 2.

   (a) On one set of axes, plot the functions C, R, and P in Example 2. (A similar picture appears in Example 1.) What do the results show graphically about Lotta's maximum possible profit?

   (b) Calculate $C'(q)$, $R'(q)$, and $P'(q)$. Use the results to find the marginal cost, marginal revenue, and marginal profit at $q = 1500$. What does it mean, in context, that the marginal profit at $q = 1500$ is positive?

   (c) Find a value of q for which the marginal profit at q is negative. What are the marginal revenue and marginal cost at this q?

5. Consider again the situation in Example 1; use the functions C, p, R, and P given there.

   (a) Suppose that Lotta wants only to break even. What's the smallest quantity of soybeans she can sell? What's the greatest quantity? Explain your answers. [**HINT**: Look first at the P-graph on page 307.]

   (b) Suppose that Lotta wants to maximize her total revenue—regardless of profit. How many bushels should she sell? At what price? What is her bottom line?

6. Redo the previous exercise in the situation of Example 2.

7. In Example 3 we concluded: If beans sell wholesale for b dollars per bushel, Lotta should sell $4000 - 500b$ bushels, at $4 + b/2$ dollars per bushel, for an optimal profit of $15,000 - 4000b + 250b^2$ dollars.

   (a) What does our conclusion mean if $b = 5$? If $b = 4.5$? Do the results agree with earlier work?

   (b) What's Lotta's best strategy if $b = 7$?

   (c) Not all values for b make realistic sense. In this context, the sensible values for b lie between $b = 0$ and $b = 8$. (Prices can't be negative, and our price function says that nobody buys above \$8.) Among these possible values of b, what's the largest one for which Lotta can at least break even?

8. A lightning rod manufacturer has monthly overhead costs of \$6000. Materials cost \$1.00 per unit. If no more than 4500 lightning rods are made in a month, labor cost is \$0.40 per unit; for each unit over 4500, the manufacturer must pay time-and-a-half for labor. The manufacturer can sell 4000 units per month at \$7.00 per unit. Monthly sales are expected to rise by 100 for each \$0.10 reduction in price. How many lightning rods should be manufactured each month to maximize profit?

9. A successful retail store must control its inventory. Too much inventory results in excessive interest costs, extra warehouse fees, and the danger of obsolescence. Too small an inventory requires more paperwork for reordering, extra delivery charges, and increases the likelihood of shortages.

   If microwave ovens are ordered in batches of size n, on average $n/2$ ovens will be in stock at any time. An appliance store estimates that it costs \$25 to hold an oven in stock for a year.

   The cost of ordering n ovens is \$250 plus \$2.50 for each oven.

   Assume that the store sells 1000 microwave ovens each year and that it always orders the ovens in "lots" of size n. What lot size will result in the smallest total yearly cost? Is your answer reasonable? Why or why not?

4.8 Related Rates

Related Amounts, Related Rates.   Calculus was invented to describe and predict phenomena of change: planetary motion, objects in free fall, varying populations, etc. In many practical applications, several related quantities vary *together*; naturally, the rates at which they vary are also related to each other. With calculus we can describe and calculate such **related rates.**

EXAMPLE 1   A spacecraft travels directly toward Mars. Describe several amounts that vary with time. Interpret their derivatives as rates of change.

Solution   Many quantities vary with time, including the spacecraft's distance from Earth, the spacecraft's distance from Mars, the distance and direction from Earth to Mars, the amount of fuel consumed, and so on. These quantities are related to each other in various ways; at any time t, for instance, the fourth quantity depends on the first two.

Each such quantity is a function of time; its derivative, therefore, is the rate of change of the given quantity with respect to time. For the preceding quantities, the derivatives represent, respectively, the rocket's velocity away from Earth, its velocity toward Mars, Earth's velocity relative to Mars, and the rate of fuel consumption. ▶ These rates, like their corresponding amounts, are related to each other. ■

Each rate's units depend on corresponding units of amount.

The following (more mundane) example illustrates the technique (as well as the idea) of related rates.

*The shadow cast by P on
the x-axis moves to the right
at 3 units per second. Can
you picture the situation?*

EXAMPLE 2   A point P moves from left to right along the curve $y = x^2$, at a constant *horizontal* speed of 3 units/second. ◄   How fast does the y-coordinate of P increase at the moment when P passes through $(1, 1)$? When P passes $(2, 4)$?

Solution   As P moves, many different quantities vary with time. The two of interest here are x and y, the coordinates of P. Written as functions, they are

$$x(t) = x\text{-coordinate of } P \text{ at time } t;$$

$$y(t) = y\text{-coordinate of } P \text{ at time } t.$$

Because P moves along the parabola $y = x^2$, the functions x and y satisfy the equation

$$y(t) = \big(x(t)\big)^2. \tag{4.6}$$

Differentiating both sides of Equation 4.6 with respect to t gives

$$\frac{dy}{dt} = 2x(t)\frac{dx}{dt}, \tag{4.7}$$

or, more compactly, $y' = 2xx'$. With Equation 4.7 we can answer our questions. By assumption, $dx/dt = 3$, for all t. At the point $(1, 1)$, $x = 1$, so

$$\frac{dy}{dt} = 2x(t)\frac{dx}{dt} = 2 \cdot 1 \cdot 3 = 6 \text{ units per second.}$$

When $(x, y) = (2, 4)$,

$$\frac{dy}{dt} = 2x(t)\frac{dx}{dt} = 2 \cdot 2 \cdot 3 = 12 \text{ units per second.}$$

Notice, finally, that although x increases at the constant rate of 3 units per second, y increases faster and faster. ◄   ∎

*Think again about the
situation. Is it obvious that
y should grow faster and
faster?*

What We Just Did: Related Rates in a Nutshell.   The generic related rates problem asks for one rate of change, given information about one or more others. *Solving* related rates problems can be hard, but the underlying idea is simple:

> *Start with an equation relating two or more quantities that vary with time. Differentiate both sides with respect to time, using the chain rule. The resulting equation relates the rates at which the quantities vary. Use it to solve for the desired rate.*

As usual with story problems, the first step—identifying the quantities in question and setting up the original equation—is often the hardest.

Textbook authors love related-rates problems; the supply is almost inexhaustible. Many old favorites involve right triangles and the Pythagorean rule (sometimes thinly disguised).

EXAMPLE 3   **(Pythagoras, Newton, and the State Patrol)**   It's spring break at Huxley College; the annual migration to Fort Lauderdale has begun.

Hovering 1000 feet directly above the southbound interstate is the helicopter state patrol, with Officer Ingkvist on the radar gun.

A car with Huxley decals passes underneath and continues south. Jason and Medea, fresh from their calculus midterm, aren't watching the speedometer. Officer Ingkvist fires. In moments, the results are in: At the moment of firing the car was 2000 feet from the helicopter and moving away at 85 feet per second. "Got 'em!" mutters Ingkvist under her breath.

What happens next?

Solution In traffic court, Jason and Medea deny all wrongdoing. "The speed limit is 65 mph," they tell the judge. "We were clocked at 85 feet per second. But a simple calculation shows that 65 mph is 95 feet per second. Hence, clearly, ▶ we couldn't have been speeding." "Watch your language," replies the judge, "but what you say makes sense. Explain yourself, Ingkvist. Use that blackboard. Be clear. And cite your sources."

Under duress, even students say "hence" and "clearly."

Officer Ingkvist rises calmly; she's been through this before. Do judges *never* learn? She draws this diagram:

Officer Ingkvist's diagram

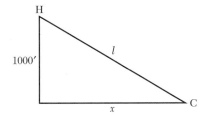

"In the diagram, your honor, H is the helicopter, C is the car, l is the straight-line distance from H to C, x is the horizontal distance from H to C, and 1000 feet is the constant height of the helicopter. The amounts x and l vary with time, of course. Pythagoras says, however, that at *any* time t, x and l satisfy the equation $l^2 = x^2 + 1000^2$. Newton says that differentiating this relation between l and x with respect to t relates the rates of change of l and x, as follows:

$$l^2 = x^2 + 1000^2 \implies 2l \cdot \frac{dl}{dt} = 2x \cdot \frac{dx}{dt} \implies \frac{dx}{dt} = \frac{l}{x} \cdot \frac{dl}{dt}.$$

At the time in question, dl/dt was 85 and l was 2000, so ▶ x was $\sqrt{2000^2 - 1000^2} = 1000\sqrt{3}$. At that moment, therefore, Jason and Medea's southward speed was

Here officer Ingkvist used the Pythagorean equation.

$$\frac{dx}{dt} = \frac{l}{x} \cdot \frac{dl}{dt} = \frac{2000}{1000\sqrt{3}} \cdot 85 \approx 98.1495457622 \text{ feet per second.}$$

Any questions?"

"Just one," says the judge. "Who are Pythagoras and Newton, and why won't they testify in person?"

"Never mind," say Jason and Medea. "We'll pay the ticket." ∎

BASICS

1. The width of a rectangle is increasing at a rate of 2 cm/sec, and its length is increasing at a rate of 3 cm/sec. At what rate is the area of the rectangle increasing when its width is 4 cm and its length is 5 cm?

2. Digging in your backyard, you accidentally break a town water line. Water, bubbling up at a rate of 1 in³/sec, is forming a circular pond of depth $\frac{1}{2}$ inch in your backyard.

How quickly is this pond covering your lawn? (That is, how quickly is the surface area of this pond increasing?)

3. Two cars leave an intersection simultaneously. One travels north at 30 mph, the other travels east at 40 mph. How fast is the distance between them changing after 5 minutes?

FURTHER EXERCISES

4. Boyle's Law states that if the temperature of a gas remains constant, then the pressure P and the volume V of the gas satisfy the equation $PV = c$, where c is a constant. If the volume is decreasing at the rate of 10 in³ per second, how fast is the pressure increasing when the pressure is 100 lbs/in² and the volume is 20 in³?

5. Two aircraft are in the vicinity of the Erehwon air traffic control center in Neverland. Pachyderm flight 1003 from Celesteville to Paris is approaching from the south, and Peterpan flight 366 from London to Wendytown is approaching from the east. Both are at an altitude of 33,000 feet on paths that will take them directly over the air traffic control center. The Pachyderm flight is 36 nautical miles from the center and approaching it at a rate of 410 knots. The Peterpan flight is 41 nautical miles from the center and approaching it at a rate of 455 knots. [NOTE: One **nautical mile** is equal to 1852 meters. One **knot** is one nautical mile per hour.]
   (a) How close will the planes come to each other? Will they violate the FAA's minimum separation requirement of 5 nautical miles?
   (b) How many minutes before the time of closest approach?

6. Sand is being dumped on a pile in such a way that it always forms a cone whose radius equals its height. If the sand is being dumped at a rate of 10 cubic feet per minute, at what rate is the height of the pile increasing when there is 1000 cubic feet of sand in the pile?

7. A tanker accident has spilled oil in Pristine Bay. Oil-eating bacteria are gobbling 5 ft³/hour. The oil slick has the form of a circular cylinder. When the radius of the cylinder is 500 ft, the thickness of the slick is 0.01 ft and decreasing at a rate of 0.001 ft/hr. At what rate is the area of the slick changing at this time? Is the area of the slick increasing or decreasing?

8. Oil is leaking from an ocean tanker at the rate of 5000 liters per minute. The leakage results in a circular oil slick. The depth of the oil varies linearly from a maximum of 5 cm at the point of leakage to a minimum of 0.5 cm at the outside edge of the slick. How fast is the radius

of the slick increasing 4 hours after the tanker started leaking? [NOTE: 1 liter = 1000 cm³]

9. A spherical iron ball 8 in. in diameter is coated with a layer of ice of uniform thickness.
   (a) If the ice melts at the rate of 10 in³/min, how fast is the thickness of the ice decreasing when it is 2 in thick?
   (b) How fast is the outer surface area of the ice decreasing at this time?

10. Bazooka Joe is blowing a spherical bubble gum bubble. Let V be the volume of gum in the bubble, R the inside radius of the bubble, and T the thickness of the bubble. V, T, and R are functions of time t.
    (a) Write a formula for V in terms of T and R. [HINT: Draw a picture.]
    (b) Assume that the amount of bubble gum in the bubble is not changing. What is $V'(t)$?
    (c) After 5 minutes of blowing, the bubble is 3 feet in diameter and 0.01 feet thick. If the inside radius of the bubble is expanding at a rate of 0.5 feet per minute, how fast is the thickness changing? [HINT: Remember that the volume of gum in the bubble does not change over time.]

11. At a certain moment, one bicyclist is 4 miles east of an intersection, traveling toward the intersection at the rate of 9 miles/hour. At the same time a second bicyclist is 3 miles south of the intersection, traveling away from the intersection at the rate of 10 miles/hour. Is the distance between the bicyclists increasing or decreasing at that moment? At what rate?

12. A radio transmitter is located 3 miles from a straight section of interstate highway. A truck is traveling away from the transmitter along the highway at a speed of 65 miles per hour. How fast is the distance between the truck and the transmitter increasing when they are 5 miles apart?

13. A baseball diamond is a square 90 ft on a side. A runner travels from home plate to first base at 20 ft/sec. How fast is the runner's distance from second base changing when the runner is halfway to first base?

14. A baseball diamond is a square 90 ft on a side. A batter hits a ball along the third base line at 100 ft/sec and runs to first base at 25 ft/sec.
    (a) At what rate is the distance between the ball and first base changing when the ball is halfway to third base?
    (b) At what rate is the distance between the ball and the player changing when the ball is halfway to third base?

15. A highway patrol plane flies one mile above a straight section of rural interstate highway at a steady ground speed of 120 miles per hour. The pilot sees an oncoming car and, with radar, determines that the line-of-sight distance from the plane to the car is 1.5 miles and that this distance is decreasing at a rate of 136 miles per hour. Should the driver of the car be given a ticket for speeding? Tell it to the judge. (Assume the judge knows calculus.)

16. A coffee filter has the shape of an inverted cone. Water drains out of the filter at a rate of 10 cm³/min. When the depth of water in the cone is 8 cm, the depth is decreasing at 2 cm/min. What is the ratio of the height of the cone to its radius?

17. An inverted cone has height 10 cm and radius 2 cm. It is partially filled with liquid, which is oozing through the sides at a rate proportional to the area of the cone in contact with the liquid. Liquid is also being poured in the top of the cone at a rate of 1 cm³/min. When the depth of the liquid in the cone is 4 cm, the depth is decreasing at a rate of 0.1 cm/min. At what rate must liquid be poured into the top of the cone to keep the liquid at a constant depth of 4 cm?

18. The minute and hour hands on the face of a town clock are 7 feet and 5 feet long, respectively. How fast is the distance between the tips of the hands increasing when the clock reads 9:00?

19. A swimming pool is 15 feet wide, 40 feet long, 3 feet deep at one end, and 10 feet deep at the other end (see the sketch). Water is being added to it at the rate of 25 cubic feet per minute.
    (a) How fast is the water level rising when there is 2000 cubic feet of water in the pool?

(b) How fast is the water level rising when there is 3000 cubic feet of water in the pool?

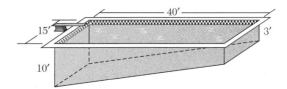

20. A rocket rises vertically from a point on the ground that is 100 m from an observer at ground level. The observer notes that the angle of elevation is increasing at a rate of 12° per second when the angle of elevation is 60°. Find the speed of the rocket at that instant.

21. A Ferris wheel, 50 feet in diameter, revolves at the rate of 10 radians per minute. How fast is a passenger rising when the passenger is 15 feet higher than the center of the Ferris wheel and is rising?

22. The ends of a horizontal trough 10 ft long are isosceles trapezoids with lower base 3 ft, upper base 5 ft, and altitude 2 ft. If the water level is rising at a rate of 0.25 inches per min. when the depth is 1 ft, how fast is water entering the trough?

23. A railroad bridge is 20 ft above, and at right angles to, a river. A person in a train traveling at 60 mph passes over the center of the bridge at the same instant that a person in a motor boat traveling 20 mph passes under the center of the bridge. How fast are the two people separating 10 seconds later?

24. A ladder, 12 feet long, is leaning against a wall. If the foot of the ladder slides away from the wall along level ground, what is the rate of change of the top of the ladder with respect to the distance of the foot of the ladder from the wall when the foot is 6 ft from the wall?

25. An airplane flying east at 400 mph goes over a certain town at 11:30 a.m., and a second plane, flying northeast at 500 mph goes over the same town at noon. How fast are they separating at 1:00 p.m.?

4.9 Parametric Equations, Parametric Curves

Here the parameter is a variable. In other situations, parameters are often constants.

A point P wanders about the xy-plane, tracing its path as it goes. As P travels, its coordinates $x = f(t)$ and $y = g(t)$ are functions of time. In this situation t is called a **parameter**, ▶ f and g are **coordinate functions**, and the figure traced out by P is a **parametric curve.** ▶
   Our first example illustrates the idea and some specialized vocabulary.

Most plotting programs and graphing calculators handle parametric curves. Try yours.

EXAMPLE 1   At any time t with $0 \leq t \leq 10$, the coordinates of P are given by the **parametric equations**

$$x = t - 2\sin t; \qquad y = 2 - 2\cos t.$$

What curve does P trace out? Where is P at $t = 1$? In which direction is P moving?

Check several entries.

S o l u t i o n   The simplest (but most tedious) way to draw a curve is to calculate many points (x, y), plot each one, and "connect the dots." The first step works as usual: For many inputs t, calculate the corresponding x and y. Here's a sampler of results, ◄ rounded to two decimals:

Parametric Plot Points for $x = t - 2 \sin t$, $y = 2 - 2 \cos t$											
t	0	0.1	0.2	0.3	0.7	0.8	0.9	1.0	...	9.8	9.9	10
x	0	−0.10	−0.20	−0.29	−0.59	−0.63	−0.67	−0.68	...	10.53	10.82	11.09
y	0	0.01	0.04	0.09	0.47	0.61	0.76	0.92	...	3.86	3.78	3.68

Here's the result, a curve C:

The parametric curve
$x = t - 2\sin t,\, y = 2 - 2\cos t,\, 0 \leq t \leq 10$

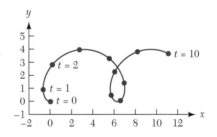

Notice:

- Points corresponding to *integer* values of t are bulleted. At $t = 1$, P has coordinates $(-0.68, 0.92)$; at this instant, P is heading almost due north.

x = 6, for instance.

- The full curve C is not the graph of a function $y = f(x)$; some x-values correspond to more than one y-value. ◄ Certain *pieces* of C, however, do define such functions. (The part of C from $t = 2$ to $t = 5$ does so.)

Most graphing calculators don't show t-values graphically, but some calculate t-values numerically.

- The picture shows the x- and y-axes *but no t-axis*: t-values are indicated only by the bulleted points. In most parametric curves t-values don't appear at all. ◄

In this example, t represents time.

- The bullets on the graph appear at equal *time* ◄ intervals but not at equal *distances* from each other, because P speeds up and slows down as it moves. ◄ We'll soon see how to calculate the speed of a parametric curve at a point.

When is P moving fastest? Slowest?

- If t measures time, we can visualize C dynamically, as a curve traced by a moving point. Curves defined by ordinary equations in x and y are *static* objects, lying passively on the page. ∎

A Sampler of Parametric Curves

Parametric curves come in mind-boggling variety. Any choice of two equations $x = f(t)$ and $y = g(t)$ and a t-interval produces a parametric curve. Surprisingly often, the result is beautiful, useful, interesting, or all three. The next several examples hint at some of the possibilities and at connections between parametric curves and ordinary function graphs.

EXAMPLE 2 **(Every Ordinary Function Graph Can Be Written in Parametric Form)** To produce a sine curve, for example, we can use

$$x = t; \qquad y = \sin t; \qquad -2\pi \le t \le 2\pi.$$

Here's the graph. Integer multiples of $\pi/2$ are bulleted.

The parametric curve
$x = t, y = \sin t, -2\pi \le t \le 2\pi$

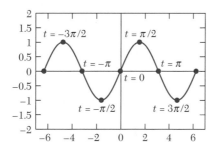

■

EXAMPLE 3 **(Parametric Curves May Have Loops, Cusps, Vertical Tangents, and Other Peculiar Features)** Consider the curve

$$x = 2\cos(t) + 2\cos(4t); \qquad y = \sin t + \sin(4t); \qquad 0 \le t \le 5.$$

Here it is. Bullets show the starting and ending points.

The parametric curve $x = 2\cos(t) + 2\cos(4t)$,
$y = \sin(t) + \sin(4t), 0 \le t \le 5$

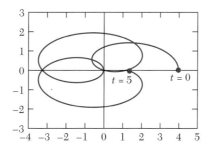

■

The curve in Example 3 has this property.

EXAMPLE 4   **(The Unit Circle)**   If a parametric curve's coordinate functions are periodic (i.e., repeat themselves), then the curve's shape reflects this fact. ◄ The simplest and most important such curve is the **unit circle**, often written in xy form as $x^2 + y^2 = 1$. The simplest *parametric* description of the unit circle is as follows:

$$x = \cos t; \qquad y = \sin t; \qquad 0 \le t \le 2\pi.$$

Here's the picture. Integer multiples of $\pi/2$ are bulleted.

The parametric curve
$x = \cos t, y = \sin t, 0 \le t \le 2\pi$

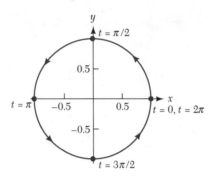

Notice the following facts:

- As usual, the parameter t *can* be thought of as time. Alternatively, t can be thought of as the radian measure of the angle determined by the x-axis and the line from the origin to P. Then, for any *angle* t, $(x, y) = (\cos t, \sin t)$ is the point on the unit circle lying t radians counterclockwise from $(1, 0)$.

- Because $x = \cos t$ and $y = \sin t$,

$$x^2 + y^2 = (\cos t)^2 + (\sin t)^2 = 1.$$

This shows that every point on the parametric curve satisfies the Cartesian equation $x^2 + y^2 = 1$ and so—as we intended—lies on the unit circle.
    "Reducing" two parametric equations to xy-form, as we just did, is called **eliminating the parameter**. (Doing so is sometimes possible, sometimes not.)

- The circle is traced *once* as t runs from $t = 0$ to $t = 2\pi$. With a larger t-interval, the same circle would be traced repeatedly; with a smaller t-interval, the circle would be traced only partially.   ■

EXAMPLE 5   **(Other Circles)**   The idea in Example 4 extends to *all* circles. If (a, b) is any point in the plane and $r > 0$ is any radius, then the parametric equations

$$x = a + r \cos t; \qquad y = b + r \sin t; \qquad \text{and} \qquad 0 \le t \le 2\pi$$

produce the circle with center (a, b) and radius r. (See the exercises at the end of this section for more details.)   ■

EXAMPLE 6   **(Other Curves with Periodic Coordinate Functions)**
Many curves defined by periodic (i.e., repeating) coordinate functions have striking, beautiful shapes. A typical **Lissajou curve**, for instance, is defined by

$$x = \sin(5t); \qquad y = \sin(6t); \qquad 0 \le t \le 2\pi.$$

(Replacing 5 and 6 with other integers would produce other Lissajou figures.) Our curve looks like this:

The Lissajou curve
$$x = \sin(5t), y = \sin(6t), 0 \le t \le 2\pi$$

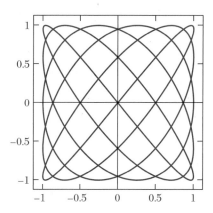

EXAMPLE 7   **(Same Curve, Different Parametric Equations)**   Different pairs of parametric equations may produce exactly the same curve in the xy-plane. In such cases, labeling t-values can make differences appear. For example, setting

$$x = t; \qquad y = t^2; \qquad \text{and} \qquad -1 \le t \le 1$$

produces a parabolic arc. So does

$$x = t^3; \qquad y = t^6; \qquad -1 \le t \le 1.$$

Following are both arcs. The bullets mark 0.25-second time intervals.

The curve $x = t, y = t^2, -1 \le t \le 1$ The curve $x = t^3, y = t^6, -1 \le t \le 1$

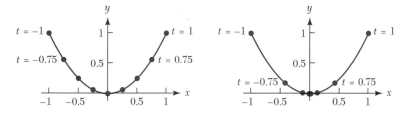

The curves are geometrically identical; they differ only in the ways in which they are traced out.

Calculus with Parametric Curves: Speed and Slope

Suppose that a parametric curve C has differentiable coordinate functions $x = f(t)$ and $y = g(t)$. What do the derivatives f' and g' say about the situation?

Speed

As the bullets on some of the foregoing graphs show, speed varies with time.

If t tells time, then $\big(f(t), g(t)\big)$ gives P's *position* in the xy-plane at time t, and it makes sense to consider the speed of P at time t. ◄ Since the derivatives f' and g' tell, respectively, how fast x and y increase, their appearance is no surprise:

> **Definition** Suppose that the position of a point P at time t, $a \le t \le b$, is given by differentiable coordinate functions $x = f(t)$ and $y = g(t)$. Then the speed at time t is given by
>
> $$\text{speed of } P \text{ at time } t = \sqrt{f'(t)^2 + g'(t)^2}.$$

Making Sense.   Is the definition "right"? Definitions aren't subject to proof, but they *should* appeal to common sense. Does this one?

Speed, properly defined, should tell how fast the distance traveled by P increases with respect to time. Showing carefully that the preceding definition actually does so involves a rather subtle definition of **arclength**, i.e., distance measured along a curve. We won't define arclength here, but the next example shows that the definition makes good sense for linear coordinate functions.

EXAMPLE 8   At time t seconds, a point P has coordinates $x = at + b$ and $y = ct + d$; a, b, c, and d are constants. How fast is P moving at time t_0?

Solution   The definition says that the speed of P at any time t is constant:

$$\sqrt{f'(t)^2 + g'(t)^2} = \sqrt{a^2 + c^2}.$$

Let's see why this result makes sense.

See the exercises at the end of this section for more on this point.

Check our algebra.

Notice first that because P has *linear* coordinate functions, P *moves along a straight line.* ◄ When $t = t_0$, P has coordinates $(x_0, y_0) = (at_0 + b, ct_0 + d)$; by time t_1, P has moved to $(at_1 + b, ct_1 + d)$—a distance of ◄

$$\sqrt{(at_1 + b - at_0 - b)^2 + (ct_1 + d - ct_0 - d)^2} = (t_1 - t_0)\sqrt{a^2 + c^2}$$

units. Therefore, from $t = t_0$ to $t = t_1$, P has

$$\text{average speed} = \frac{\text{distance}}{\text{time}} = \frac{(t_1 - t_0)\sqrt{a^2 + c^2}}{t_1 - t_0} = \sqrt{a^2 + c^2}.$$

This shows that P has the same average speed—$\sqrt{a^2 + c^2}$—over any time interval. Thus, the instantaneous speed of P at any time t_0 is also $\sqrt{a^2 + c^2}$, as the definition says. ∎

Look first at the picture. Try to guess reasonable answers.

EXAMPLE 9   Consider again the parametric curve

$$x = f(t) = t - 2\sin t; \qquad y = g(t) = 2 - 2\cos t$$

of Example 1 on page 316. Find the speed at $t = 3$. When does P move fastest? ◄

Solution   The definition says that at any time t,

$$\text{speed} = \sqrt{(f'(t))^2 + (g'(t))^2} = \sqrt{(1 - 2\cos t)^2 + (2\sin t)^2} = \sqrt{5 - 4\cos t}.$$

(The last step uses a trigonometric identity.) ▶   *Check for yourself.*

The rest is easy. At $t = 3$ the speed is $\sqrt{5 - 4\cos 3} \approx 2.99$. The speed formula also shows—and the picture agrees—that the speed $\sqrt{5 - 4\cos t}$ is greatest when $\cos t = -1$, e.g., when $t = \pi$.   ∎

Slopes on Parametric Curves

On an ordinary, $y = f(x)$-style curve, $f'(x)$ gives the slope at (x, y). Not surprisingly, slopes on parametric curves also involve derivatives.

Not all functions have derivatives; not all parametric curves t have slopes. To avoid needless complications, we'll work only with smooth parametric curves, defined as follows:

Definition The parametric curve C defined by

$$x = f(t); \qquad y = g(t); \qquad a \le t \le b$$

is **smooth** if f' and g' are continuous functions of t, and f' and g' are not simultaneously zero.

(The second requirement says that the moving point P *never stops.*) ▶   For smooth curves, the next theorem tells how to find the slope at a point.

See the exercises at the end of this section for more on this idea.

Theorem 3 Let the smooth parametric curve C be defined as before. If $f'(t) \ne 0$, then the slope dy/dx at the point $(x, y) = (f(t), g(t))$ is given by

$$\frac{dy}{dx} = \frac{g'(t)}{f'(t)} = \frac{dy/dt}{dx/dt}.$$

Before proving the theorem, let's use it.

EXAMPLE 10   Consider the parametric curve

$$x = f(t) = t - 2\sin t; \qquad y = g(t) = 2 - 2\cos t$$

of Example 1, on page 316. Find the slope at $t = 1$. Where is the curve horizontal? Where is it vertical? ▶

Guess first, by looking at the picture.

Solution   The theorem says that

$$\text{slope at } t = \frac{g'(t)}{f'(t)} = \frac{2\sin t}{1 - 2\cos t},$$

unless $f'(t) = 0$ (where the denominator is zero). Setting $t = 1$ gives

$$\text{slope at time } 1 = \frac{g'(1)}{f'(1)} = \frac{2 \sin 1}{1 - 2 \cos 1} \approx -20.88.$$

The picture agrees: At $(f(1), g(1)) \approx (-0.68, 0.92)$, the curve has large negative slope.

The general slope formula

$$\frac{dy}{dx} = \frac{2 \sin t}{1 - 2 \cos t}$$

shows that the curve is *horizontal* only when the numerator is zero, i.e., when t is an integer multiple of π. Again the picture agrees. At $t = \pi$, for instance, $(x, y) = (\pi, 4)$. At this point the curve appears to be horizontal.

Otherwise, the slope of C is finite.

The curve C can be *vertical* only where the preceding denominator is zero, ◄ i.e., when $\cos t = 1/2$. One such value is $t = \pi/3 \approx 1.05$; again the picture agrees. ■

Proof of the Theorem

Remember—we're assuming that f is smooth.

Suppose that $f'(t_0) \neq 0$; then either $f'(t_0) > 0$ or $f'(t_0) < 0$. If $f'(t_0) > 0$, then continuity of f' ◄ implies that $f'(t) > 0$ for t near t_0 and therefore that $x = f(t)$ is *increasing* near $t = t_0$. If $f'(t) < 0$, then (by the same reasoning) $x = f(t)$ is *decreasing* near $t = t_0$. In either case, it follows that if t is near t_0 but $t \neq t_0$, then $f(t) \neq f(t_0)$. This ensures, in turn, that all of the following limit computations

Denominators don't vanish unexpectedly.

make sense. ◄ Notice first that

$$\text{slope} = \left.\frac{dy}{dx}\right|_{t=t_0} = \lim_{t \to t_0} \frac{y - y_0}{x - x_0} = \lim_{t \to t_0} \frac{g(t) - g(t_0)}{f(t) - f(t_0)}.$$

A common sort of trick in proofs of calculus theorems.

Now for a little trick. ◄ We'll divide the preceding numerator and denominator by $(t - t_0)$, take limits of everything in sight, and see what happens. Here goes:

$$\left.\frac{dy}{dx}\right|_{t=t_0} = \lim_{t \to t_0} \frac{\frac{g(t) - g(t_0)}{t - t_0}}{\frac{f(t) - f(t_0)}{t - t_0}} = \frac{\lim_{t \to t_0} \frac{g(t) - g(t_0)}{t - t_0}}{\lim_{t \to t_0} \frac{f(t) - f(t_0)}{t - t_0}} = \frac{g'(t_0)}{f'(t_0)},$$

which is what we wanted to show. (The second-last step is OK because, by assumption, the limit in the denominator is not zero.) ■

Modeling with Parametric Equations

Parametric equations often help model physical processes that involve quantities that vary in both time and space. Descriptions of **trajectories**—paths taken by projectiles—are one such setting. ◄

Projectiles could be baseballs, BBs, ballistic missiles, etc.

EXAMPLE 11   A major-league fastball leaves the pitcher's hand traveling horizontally, with initial speed 150 ft/sec and initial height 7 feet. Ignoring wind resistance, ◄ describe the ball's trajectory. When, and at what speed, does the ball cross the plate, 60.5 feet away? Is the pitch a strike? If the batter, catcher, and umpire all miss the ball, where does it land?

Major-league batters don't ignore wind resistance. See the exercises at the end of this section for a more realistic example.

Solution   We use an xy-coordinate system with its origin at the pitcher's feet and with home plate at $(60.5, 0)$; the ball starts at $(0, 7)$.

Since we're ignoring wind resistance, the ball, once released, is influenced only by gravity. ▶ Thus, the ball's horizontal acceleration ▶ is zero; its vertical acceleration due to gravity is −32 ft/sec/sec. ▶

If $x = f(t)$, $y = g(t)$, and t measures time in seconds since the ball's release, our information boils down to this:

$$f''(t) = 0; \qquad f'(0) = 150; \qquad f(0) = 0;$$
$$g''(t) = -32; \qquad g'(0) = 0; \qquad g(0) = 7.$$

For more on falling objects, with and without air resistance, see pp. 102–105.

Recall that acceleration is the second derivative of position.

Why is the vertical acceleration negative?

These data lead to simple formulas for f and g. Here is the argument for f:

$$f''(t) = 0 \implies f'(t) = a \qquad \text{and} \qquad f(t) = at + b,$$

where a and b are constants, as yet unknown. Finding them is easy—a combination of the facts that $f(t) = at + b$, $f(0) = 0$, and $f'(0) = 150$ gives $a = 150$ and $b = 0$, so $f(t) = 150t$. A similar argument applies to g, with the result that

$$x = f(t) = 150t; \qquad y = g(t) = 7 - 16t^2.$$

These equations make sense, of course, only while the ball remains airborne, i.e., until $y = 7 - 16t^2 = 0$, or $t = \sqrt{7}/4 \approx 0.661$ seconds. Here's the parametric curve for $0 \le t \le 0.661$:

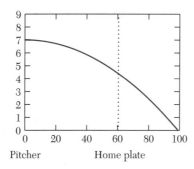

A baseball's trajectory:
$$x = 150t, y = 7 - 16t^2, 0 \le t \le 0.661$$

Pitcher   Home plate

Notice the parabolic shape. In fact, every free-falling object either moves vertically or follows a parabolic trajectory. ▶

Using the picture and the formulas, we can answer our original questions. The ball crosses the plate when $x = f(t) = 150t = 60.5$, i.e., when $t = 121/300 \approx 0.4033$ seconds. At this time, ▶

$$\text{speed} = \sqrt{f'(0.4033)^2 + g'(0.4033)^2} \approx 150.55$$

See the exercises for more on parabolic trajectories.

Check the arithmetic.

feet per second. At the same time, the ball's height is $y = g(0.4033) \approx 4.4$ feet—high but in the strike zone for an average batter. Left untouched, the ball hits the ground at $t = 0.661$; at this time, $x = f(t) = 150 \cdot 0.661 \approx 99.2$ feet—almost 40 feet behind home plate. ∎

BASICS

Plot each parametric curve below. (Using a machine is fine, but then draw or copy your own curve on paper.) In each case, mark the direction of travel and label the points corresponding to $t = -1$, $t = 0$, and $t = 1$.

1. $x = t$, $y = \sqrt{1 - t^2}$, $-1 \leq t \leq 1$
2. $x = t$, $y = -\sqrt{1 - t^2}$, $-1 \leq t \leq 1$
3. $x = \sqrt{1 - t^2}$, $y = t$, $-1 \leq t \leq 1$
4. $x = -\sqrt{1 - t^2}$, $y = t$, $-1 \leq t \leq 1$
5. $x = \sin(\pi t)$, $y = \cos(\pi t)$, $-1 \leq t \leq 1$

FURTHER EXERCISES

6. Each parametric "curve" below is actually a line segment. In each part, state the beginning point ($t = 0$) and ending point ($t = 1$) of the segment. Then state an equation in x and y for the line each segment determines.
    (a) $x = 2 + 3t$, $y = 1 + 2t$, $0 \leq t \leq 1$
    (b) $x = 2 + 3(1 - t)$, $y = 1 + 2(1 - t)$, $0 \leq t \leq 1$
    (c) $x = t$, $y = mt + b$, $0 \leq t \leq 1$
    (d) $x = a + bt$, $y = c + dt$, $0 \leq t \leq 1$
    (e) $x = x_0 + (x_1 - x_0)t$, $y = y_0 + (y_1 - y_0)t$, $0 \leq t \leq 1$
7. A parametric curve $x = f(t)$, $y = g(t)$, $a \leq t \leq b$ has **constant speed** if—what else?—its speed is constant in t. Which of the curves in Exercises 1–5 have constant speed?
8. Show that if a curve C has linear coordinate functions $x = at + b$ and $y = ct + d$, then C has constant speed. (See the previous exercise.)
9. Consider the curve C shown in Example 1; suppose that t tells time in seconds.
    (a) At which bulleted points would you expect P to be moving quickly? Slowly? Why?
    (b) Use the curve to estimate the speed of P at $t = 3$. (One approach: Estimate how far P travels—i.e., the length of the curve—over the one-second interval from $t = 2.5$ to $t = 3.5$. If d is this distance, then d distance units per second is a reasonable speed estimate at $t = 3$.)
    (c) Use the curve to estimate the speed of P at $t = 6$.
10. Plot the parametric curve
    $$x = t^3; \qquad y = \sin t^3; \qquad -2 \leq t \leq 2.$$
    What familiar curve is produced? Why does the result happen?
11. Let (a, b) be any point in the plane, and $r > 0$ any positive number. Consider the parametric equations
    $$x = a + r \cos t; \qquad y = b + r \sin t; \qquad 0 \leq t \leq 2\pi.$$
    (a) Plot the parametric curve defined above for $(a, b) = (2, 1)$ and $r = 2$. Describe your result in words.
    (b) Show by calculation that if x and y are as above, then $(x - a)^2 + (y - b)^2 = r^2$. Conclude that the curve defined above is the circle with center (a, b) and radius r.
    (c) Write parametric equations for the circle of radius $\sqrt{13}$, centered at $(2, 3)$.

(d) What "curve" results from the equations above if $r = 0$?
12. Let a and b be any positive numbers, and let a parametric curve C be defined by
    $$x = a \cos t; \qquad y = b \sin t; \qquad 0 \leq t \leq 2\pi.$$
    The resulting curve is called an **ellipse**.
    (a) Plot the curve defined above for $a = 2$ and $b = 1$. Describe C in words. Where is the "center" of C? Why do you think the quantities $2a$ and $2b$ are called the **major and minor axes** of C.
    (b) What curve results if $0 \leq t \leq 4\pi$? Why?
    (c) Write parametric equations for an ellipse with major axis 10 and minor axis 6.
    (d) Write parametric equations for another ellipse with major axis 10 and minor axis 6.
    (e) Show that for all t, $\dfrac{x^2}{a^2} + \dfrac{y^2}{b^2} = 1$.
    (f) How does the "ellipse" look if $a = b$? How does its xy-equation look?
    (g) How does an ellipse look if $a = 1000$ and $b = 1$?
    (h) How does an ellipse look if $a = 1$ and $b = 1000$?
13. Consider the Lissajou curve in Example 6.
    (a) Label the points corresponding to $t = 0$, $t = 0.1$, and $t = \pi/2$. Add some arrows to the curve to indicate direction.
    (b) How often in the interval $0 \leq t \leq 2\pi$ does the tracing point P return to $(0, 0)$?
    (c) How would the picture be different if the t-interval $0 \leq t \leq 4\pi$ were used?
14. We stated in this section that if a parametric curve C has linear coefficient functions $x = f(t) = at + b$ and $y = g(t) = ct + b$, then C is a straight line (or part of a line). This exercise explores that fact.
    (a) Plot the parametric curve $x = 2t$, $y = 3t + 4$, $0 \leq t \leq 1$. Where does the curve start? Where does it end? What is its shape?
    (b) Eliminate the variable t in the two equations above to find a single equation in x and y for the line of the previous part.
    (c) Find the slope at $t = t_0$ of the parametric curve $x = 2t$, $y = 3t + 4$, $0 \leq t \leq 1$. Why doesn't the answer depend on t?

15. This exercise pursues the idea that if a parametric curve C has linear coefficient functions $x = f(t) = at + b$ and $y = g(t) = ct + d$, then C is a straight line (or part of a line).

    (a) Consider the parametric curve $x = at + b$, $y = ct + d$, $t_0 \le t \le t_1$. Where does the curve start? Where does it end?

    (b) Assume that $a \ne 0$. (Don't assume that $c \ne 0$.) Eliminate the variable t in the two equations above to find one equation in x and y. What line does it describe?

    (c) Assume that $c \ne 0$. (Don't assume that $a \ne 0$.) Eliminate the variable t in the two equations above to find one equation in x and y. What line does it describe?

    (d) What happens if a and c are both zero?

16. This exercise explores the technical assumptions in the definition of smooth curves.

    (a) Consider the "curve" defined for all t by $x = 0$, $y = 0$. Plot C. Does C "deserve" to have a slope at $t = 0$? If so, what slope? If not, why not? Is C smooth in the sense of the definition?

    (b) Consider the curve defined for $-1 \le t \le 1$ by $x = t^3$ and $y = t^3$. Plot C. Does C "deserve" to have a slope at $t = 0$? What does Theorem 3 say in this case?

    (c) Consider the curve defined for $-1 \le t \le 1$ by $x = t$ and $y = t$. Plot C. Does this C "deserve" to have a slope at $t = 0$? What does Theorem 3 say about slope this time?

17. This exercise concerns Example 11 (p. 322).

    (a) How could the model be made more realistic? What additional information would be needed?

    (b) Use the conditions $g''(t) = -32$, $g'(0) = 0$, and $g(0) = 7$ to show that $g(t) = 7 - 16t^2$, as claimed in the Example.

    (c) Find a formula for $s(t)$, the ball's speed at time t. Plot $s(t)$ over an appropriate interval. How does the graph's shape reflect the physical situation?

18. Use parametric equations to describe the trajectory of a baseball thrown exactly as in Example 11 (p. 322), except that this time the initial velocity is 100 feet per second. Plot the result over an appropriate interval. Is the pitch a strike? At what speed does it cross the plate? (The strike zone is roughly from 1.5 to 4.5 feet above the ground at home plate.)

19. Consider a baseball thrown horizontally, starting from height 7 feet (as in Example 11), but with initial speed s_0.

    (a) Explain briefly why the parametric equations

$$x = s_0 t, \quad y = 7 - 16t^2$$

    describe the ball's trajectory (ignoring wind resistance).

    (b) When does the ball reach home plate?

    (c) Is the trajectory parabolic? Why?

20. This exercise is again about the situation in Example 11, but takes wind resistance into account. In practice, wind resistance *does* affect a baseball's trajectory. One possible model of wind resistance (we omit the details) leads to the parametric equations

$$x = \frac{\ln(150 \cdot k \cdot t + 1)}{k}; \quad y = 7 - 16t^2,$$

where the constant k can be thought of as the ball's "drag coefficient"—the smoother the ball, the lower the value of k. For a typical baseball, $k = 0.003$ is reasonable. With this value of k we get

$$x = \frac{1000 \ln(0.45t + 1)}{3}; \quad y = 7 - 16t^2.$$

The resulting trajectory is plotted below; the curve from Example 11 is also shown:

Baseball trajectories with and without air drag

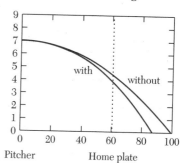

In each part below, use the graphs to give approximate answers; then use appropriate formulas to improve your results.

    (a) At what time does the air-dragged ball cross the plate? (**NOTE:** The graphs don't show time.) How much longer does it take to reach the plate than the drag-less ball?

    (b) At what height does the air-dragged ball cross the plate?

    (c) At what speed does the air-dragged ball cross the plate?

    (d) Where does the air-dragged ball land?

21. The situation is as in the previous exercise, except that this time the ball is scuffed, so that its "drag coefficient" is 0.005, not 0.003. Plot the new trajectory. Then answer the same questions as in the previous exercise.

4.10 Why Continuity Matters

In Section 2.6 we defined continuity in terms of limits: f is **continuous** at $x = a$ if

$$f(x) \to f(a) \quad \text{as} \quad x \to a.$$

The standard elementary functions—algebraic, exponential, logarithm, and trigonometric functions—are continuous wherever they're defined. Combinations—sums, products, quotients, compositions, etc.—of elementary functions are also continuous wherever they're defined, thanks to the analogous properties of limits. (See, for example, Theorem 2, page 169.)

Indeed, continuity is so much the rule in calculus that its importance may be hard to appreciate. This section aims to remedy that oversight.

The Importance of Being Continuous: The Intermediate Value Theorem and the Extreme Value Theorem

Continuity is clearly a desirable property of a function. The **intermediate value theorem (IVT)** and the **extreme value theorem (EVT)** give two good reasons why. We won't prove either, but we'll indicate why the property each guarantees is practically useful.

Each theorem pertains to a continuous function f on a *closed and bounded* interval. ◄

See the exercises at the end of this section for more on why $[a, b]$ must be closed.

> **Theorem 4** (Intermediate Value Theorem) Let f be continuous on the closed and bounded interval $[a, b]$, and let y be any number between $f(a)$ and $f(b)$. Then, for some input c between a and b, $f(c) = y$.

> **Theorem 5** (Extreme Value Theorem) Let f be continuous on the closed, bounded interval $[a, b]$. Then f assumes both a maximum value and a minimum value somewhere on $[a, b]$.

Existential Philosophy

The EVT and the IVT are **existence theorems**. Each guarantees the existence of (at least) one point in the domain with some desirable property. Neither theorem says *where* in $[a, b]$ these points may fall or *how many* there may be. Such high-toned philosophy may suggest that these theorems, whatever their theoretical importance, have little practical use. That's wrong, as we'll see.

The IVT, Pictorially

The IVT says that if f is continuous on $[a, b]$, then the range of f contains not just $f(a)$ and $f(b)$ but *everything in between*. This means that the graph of a continuous function f is unbroken in the following sense.

Enroute from $(a, f(a))$ to $(b, f(b))$, the graph of f crosses every horizontal line between $y = f(a)$ and $y = f(b)$.

Here's the picture:

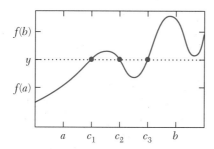

The IVT:
f "hits" every intermediate value

The EVT, Pictorially

The EVT guarantees that if f is continuous on the closed interval $[a, b]$, then f attains both a maximum and a minimum somewhere therein. In the case pictured as follows, these extreme values are attained at $x = x_{\max}$ and $x = x_{\min}$, respectively.

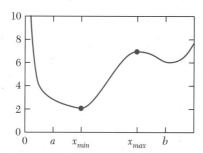

The EVT:
On $[a, b]$, f assumes extreme values

Both hypotheses of the EVT—that f be *continuous* and that the interval be closed— are necessary. If either fails, f need not assume a maximum or a minimum. In the case just pictured, for instance, f does not achieve a maximum on the *open* interval $(0, b)$ even though f is continuous there.

From Abstract Theory, a Concrete Payoff

The IVT is a theoretical result, but it has useful, concrete consequences. One of them, the bisection method, or interval-halving method, is a simple but effective technique for solving equations. First, let's take care of some preliminaries.

A number r is a root (or zero) of a function f if $f(r) = 0$.

EXAMPLE 1   The function $f(x) = x^2 - 1 = (x - 1)(x + 1)$ has *two* roots, at $x = \pm 1$. The function $g(x) = x^2$ has *one* root, at $x = 0$. Since $h(x) = x^2 + 1 \geq 1$ for all x, h has *no* roots. How many roots can any quadratic polynomial have?

Solution   The **quadratic formula**

$$r = \frac{-b \pm \sqrt{b^2 - 4ac}}{2a}$$

finds all the roots—there may be 0, 1, or 2—of any quadratic polynomial $p(x) = ax^2 + bx + c$. ∎

Digging Systematically for Roots

Finding roots by algebraic solution techniques is often difficult or impossible. Even when algebraic methods (such as the quadratic formula) do succeed, finding decimal equivalents may remain a problem. It's clear, for example, that $x = \sqrt{2}$ is a root of $x^2 - 2$, but how can we find an accurate decimal value?

The bisection method will help. It's based on this key fact:

> **Fact** If f is continuous on $[a, b]$, and $f(a)$ and $f(b)$ have opposite signs, then $[a, b]$ contains a root of f. Hence, the midpoint $(a+b)/2$ of $[a, b]$ differs from a root of f by no more than $(b-a)/2$—the "radius" of $[a, b]$.

This fact follows directly from the IVT. Because f is negative at one end of $[a, b]$ and positive at the other, the IVT guarantees that $f(r) = 0$ somewhere in between—i.e., that a root r lies somewhere in $[a, b]$. Lacking further information, the midpoint of $[a, b]$ is as good an estimate of r as any. By applying this fact repeatedly, we can approximate a root as closely as we wish.

EXAMPLE 2   We use the bisection algorithm to estimate $\sqrt{2}$ accurately. Since $\sqrt{2}$ is a root of $f(x) = x^2 - 2$ and certainly lies somewhere in $[1, 2]$, we have all we need to get started.

Solution   We'll converge on an answer, step by step:

Step 1: Since $f(1) = -1$ and $f(2) = 2$, our Fact says that $m_1 = 1.5$—the midpoint of $[1, 2]$—differs from a root of f by less than 0.5.

Step 2: Since $f(1.5) = 0.25 > 0$, the Fact applies to the new—and shorter—interval $[1, 1.5]$. Conclusion: $m_2 = 1.25$, the new midpoint, differs from a root by less than 0.25.

Step 3: Since $f(1.25) = -0.4375 < 0$, our *latest* interval is $[1.25, 1.50]$. Conclusion: $m_3 = 1.375$ differs from a root by less than 0.125.

We can continue these steps as long as we like; in practice, we would continue until we were sure of some predetermined accuracy. (The next few estimates are 1.4375, 1.40625, 1.421875, 1.4140625, 1.41796875, and 1.416015625, not far from $\sqrt{2} \approx 1.41421356237$.) ∎

Satisfaction Guaranteed—Up to a Point

Each step *doubles* the guaranteed precision of our estimate. In other words, the nth midpoint m_n can't possibly miss a root c by more than the radius of the nth interval. More precisely,

$$|m_n - c| < \frac{b - a}{2^n}.$$

On the other hand, there's no guarantee of *ever* finding a root exactly. ▶   In real life, excellent approximations are almost always good enough.

If a and b are rational numbers, then so is each midpoint m_n. Thus, if a root r is irrational, the method cannot ever locate r exactly.

A Bisection Algorithm

The bisection method is an **algorithm**—a finite sequence of explicit instructions which, starting from appropriate inputs, reliably produces a desired output. ("Explicit" means, roughly, "executable by a machine." Computer programs are prime examples of algorithms.)

To emphasize the algorithmic nature of the bisection method, we'll state it explicitly in **pseudocode**—a sort of generic computer language. (Computerese is printed in the "typewriter" typeface.) Notice that the error tolerance is stated up front, as an input to the program. Once a root is found to within that tolerance, the procedure stops. (Without a stopping rule, a program might run forever.)

Input: A continuous function f, an interval $[a, b]$, and a tolerance $t > 0$. (For simplicity, we'll assume that $f(a) < 0$ and $f(b) > 0$.)

Output: An approximate root of f, i.e., a number m that differs from a root by less than the tolerance t.

Step 1: (Find midpoint) `m:=(b+a)/2`

Step 2: (Done?) `if b-a<2t then output m; QUIT`

Step 3: (New interval) `if f(m)<0 then a:=m else b:=m`

Step 4: (Start again) `return to step 1`

A Sample Problem

Here's a problem that exploits ideas discussed in this section.

EXAMPLE 3   Let $f(x) = \sqrt{x}$.

a. What does the EVT say about f on the interval $[0, 1]$?

b. Although f is continuous on $[0, \infty)$, it has no maximum value on this interval. Doesn't this contradict the EVT?

c. Although f is continuous on $(0, 1]$, it assumes no minimum value on this interval. Doesn't *this* contradict the EVT?

d. Show that the IVT holds for f on $[0, 4]$ and $y = 1$.

Solution

a. The EVT says that f must achieve a maximum and a minimum somewhere on $[0, 1]$. It does: $f(0) = 0$ is the minimum value and $f(1) = 1$ is the maximum value.

b, c. There's no contradiction in either part. The EVT applies only for closed and bounded intervals. Neither $[1, \infty)$ nor $(0, 1]$ is closed and bounded.

d. The IVT says that f must assume all values between $f(0) = 0$ and $f(4) = 2$. In particular, $y = 1$ is one such value. Sure enough, $f(1) = 1$, as the IVT requires.

■

BASICS

1. Let f and g be the functions shown graphically below. Use the graphs to answer the following questions; no formulas are given (or needed). Notice that f is continuous except at $x = -3$, $x = -1$, $x = 1$, $x = 2$, and $x = 4$.

Graph of f

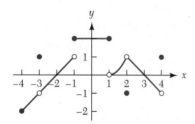

Graph of g

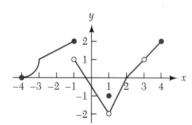

(a) Over what intervals is g continuous?

(b) Which of the following functions are continuous at $x = 2$? (Justify your answers.)

(i) $f(x) + g(x)$ (iii) $\dfrac{f(x)}{g(x)}$

(ii) $f(x) \cdot g(x)$ (iv) $\dfrac{g(x)}{f(x)}$

2. Let $f(x) = 1/x$.
   (a) What does the EVT say about f on the interval $[0.1, 1]$?
   (b) Although f is continuous on $[1, \infty)$, it has no minimum value on this interval. Why doesn't this contradict the EVT?
   (c) Although f is continuous on $(0, 1]$, it assumes no maximum value on this interval. Why doesn't this contradict the EVT?
   (d) What does the EVT say about f on the interval $[0, 1]$?
   (e) Although $f(-1) = -1$ and $f(1) = 1$, there is no value of c for which $f(c) = 0$. Why doesn't this contradict the IVT?
   (f) Show that the IVT holds for f on $[0.1, 1]$ and $y = 5$.

3. Use the IVT to show that $f(x) = x^5 + x^3 - 5x + 2$ has a root in $[0, 1]$.

4. Use paper and pencil alone to find the first four midpoint approximations to $\sqrt{3}$, starting with the interval $[1, 2]$. Express your answers as fractions.

5. Use a calculator to find the first six midpoint approximations to a root of $\cos x - x$, starting on $[0, 1]$. Then use graphics or any other method to find the root to within 0.001.

FURTHER EXERCISES

6. Let f and g be the functions defined in Exercise 1. Neither f nor g is continuous on the entire interval $[-4, 4]$, so neither satisfies the hypothesis of the EVT or the IVT. Still, it's possible that either theorem's conclusion might hold.
   (a) Does the conclusion of the IVT hold for f on $[-4, 4]$? Explain.
   (b) Does the conclusion of the IVT hold for g on $[-4, 4]$? Explain.
   (c) Does the conclusion of the EVT hold for f on $[-4, 4]$? Explain.
   (d) Does the conclusion of the EVT hold for g on $[-4, 4]$? Explain.
   (e) The function g is continuous on $[-4, -1]$, so the

EVT holds. What are the maximum and minimum values of g on $[-4, -1]$? Where are they attained?

7. Suppose that f is a continuous function defined for all real numbers and that f achieves a maximum value of 5 and a minimum value of -7. Determine which of the following statements *must* be true, which *might* be true, and which *cannot* be true.
   (a) The maximum value of $f(|x|)$ is 7.
   (b) The minimum value of $f(|x|)$ is 0.
   (c) The maximum value of $|f(x)|$ is 7.
   (d) The minimum value of $|f(x)|$ is 5.

8. Suppose that f is a continuous function and $f(3) = -1$. Evaluate $\lim\limits_{x \to 3} f(x)$.

9. Determine which of the following statements *must* be true, which *might* be true, and which *cannot* be true.

    (a) If $f(1) < 0$ and $f(5) > 0$, then there must be a number c in $(1, 5)$ such that $f(c) = 0$.

    (b) If f is continuous on $[1, 5]$, $f(1) > 0$, and $f(5) < 0$, then there must be a number c in $(1, 5)$ such that $f(c) = 0$.

    (c) If f is continuous on $[1, 5]$ and there is a number c in $[1, 5]$ such that $f(c) = 0$, then $f(1) \cdot f(5) \le 0$.

    (d) If f is continuous and has no roots in $[1, 5]$, then $f(1) \cdot f(5) > 0$.

10. Use the IVT to find an interval in which $f(x) = x^5 + x^3 + 5x + 2$ has a root.

11. Let

$$f(x) = \begin{cases} ax^2 + 3, & \text{if } x < 2 \\ 3x - 5, & \text{if } x \ge 2 . \end{cases}$$

    Find the value of a that makes f continuous at $x = 2$.

12. Suppose that f is defined on $[-5, 5]$, that $f(-5) = 2$ and $f(5) = -2$. If f assumes every value in $[-2, 2]$, must f be continuous? Explain. [**HINT:** The answer is *no*. Draw a graphical example that illustrates this.]

13. Use the IVT to prove this fixed-point theorem: If f is continuous on $[0, 1]$ and has range contained in $[0, 1]$, then f has a fixed point—i.e., an input x for which $f(x) = x$. [**HINT:** Consider $g(x) = f(x) - x$.]

14. Let

$$f(x) = \begin{cases} x, & \text{if } x \text{ is rational} \\ 0, & \text{otherwise.} \end{cases}$$

    Show that f is continuous at 0.

4.11 Why Differentiability Matters; The Mean Value Theorem

Recall the limit definition of derivative, from Section 2.5:

> **Definition** Let f be a function defined near and at $x = a$. Then
>
> $$f'(a) = \lim_{h \to 0} \frac{f(a + h) - f(a)}{h}.$$

For a given function f and $x = a$, the limit in the definition may or may not exist. If it does exist, we say that f is **differentiable** at $x = a$.

The standard elementary functions—algebraic, exponential, logarithm, and trigonometric functions—are differentiable wherever they're defined. Combinations—sums, products, quotients, compositions, and so on—of elementary functions are also differentiable where they're defined, thanks to the constant, sum, product, quotient, and chain rules.

Like continuity, differentiability is the rule in calculus—so much so, perhaps, that its importance might be overlooked. This section aims to avoid that pitfall.

Differentiable Functions Are Continuous

The absolute value function $g(x) = |x|$ is *continuous* at $x = 0$, but not *differentiable* at $x = 0$. On the other hand, a differentiable function must be continuous.

> **Theorem 6** If f is differentiable at $x = a$, then f is also continuous at $x = a$.

Proof.   The theorem is both graphically plausible—it's hard to imagine a discontinuous graph having a tangent line—and easy to prove. To prove the theorem, we need only show that

$$f(x) \to f(a) \quad \text{as} \quad x \to a.$$

(This is the formal definition of continuity of f at $x = a$.) The definition of differentiability at $x = a$ means that, as $x \to a$,

$$\frac{f(x) - f(a)}{x - a} \to f'(a).$$

From this fact it follows easily that as $x \to a$,

$$f(x) - f(a) = \frac{f(x) - f(a)}{x - a} \cdot (x - a) \to f'(a) \cdot 0 = 0,$$

as desired. ∎

Theory and Practice: A Case Study

In later mathematics courses, such as elementary real analysis, the theory of calculus is the main subject.

Calculus is more than a set of powerful methods for solving important problems. It is also a mathematical *theory*—a coherent, rigorously developed body of definitions and theorems. We don't aim here (or elsewhere) to present the theory of calculus in full detail or with full rigor. ◄ This section is, instead, a sort of case study—a close look at one important theorem in its mathematical context.

> **The Guts (and the Beauty) of the Machine.**   Using calculus, like driving a car, can be done effectively without perfect or constant attention to the machine's inner workings. Why bother with theory?
>
>   One good reason is that ignorance of underlying principles leaves one vulnerable in unexpected situations—no owner's manual can cover everything. A better reason concerns the difference between driver's education and calculus. Driving is a useful skill, but mathematics is a science.
>
>   Another good reason is that the theory of calculus is a human creation of unparalleled power and beauty, putting even the internal combustion engine to shame.

The Mean Value Theorem in Context

We encountered them in the preceding section.

The intermediate value theorem and the extreme value theorem ◄ guarantee that a continuous function assumes certain desirable values on a closed interval. The **mean value theorem**, similar in spirit, guarantees that the derivative of a well-behaved function assumes a certain desirable mean value.

> **T h e o r e m 7** (Mean Value Theorem) Suppose that f is continuous on the closed interval $[a, b]$ and differentiable on the open interval (a, b). Then, for some c between a and b,
>
> $$f'(c) = \frac{f(b) - f(a)}{b - a}.$$

Taken out of context, the mean value theorem (MVT) is difficult to understand, ▶ let alone appreciate for what it is: one of the two or three "big" theorems of calculus. Most of what follows aims to put the theorem in context. What do its words mean? Why is the name appropriate? What does it say? Are the hypotheses really necessary? What questions does the theorem answer? Why do the answers matter? What makes the MVT a big theorem?

Few statements of any kind make much sense out of context. Ask a politician.

What the MVT Says

The MVT says that, under appropriate hypotheses on f, the derivative f' must assume a certain mean value. Interpreting f' either as an instantaneous rate or as a slope makes everything plausible.

The MVT—A Motorist's View

If a car averages 60 mph over an interval of time, then, at some instant, the speedometer must have read exactly 60. The MVT says the same thing, a little more generally. If $f(t)$ represents a car's position in miles at time t in hours, then (as we saw in Section 2.1) $f'(t)$ represents instantaneous velocity at time t, and $(f(b) - f(a))/(b - a)$ represents average velocity over the interval $a \le t \le b$. In this language, the MVT equation

$$f'(c) = \frac{f(b) - f(a)}{b - a}$$

reads as follows:

> *At some time c between a and b, the instantaneous velocity is equal to the average (i.e., mean) velocity over the entire interval.*

The MVT—A Graphical View

On the graph of f, each side of the MVT equation represents a *slope*—of the tangent line at $x = c$ and of the secant line from a to b, respectively. The theorem asserts that, for at least one value of c between a and b, these two slopes are equal. In the case shown graphically as follows, *two* such c's exist.

The mean value theorem

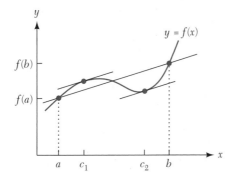

Mathematicians like to state theorems as generally as possible.

The Hypothesis of Differentiability; Why It Matters

The MVT requires that $f'(x)$ should exist for every x in (a, b). (Most often, $f'(x)$ exists at the endpoints, too, but that isn't strictly necessary.) ◄

This differentiability hypothesis can't be diluted. If f isn't differentiable at even *one* point in (a, b), the MVT can fail. The function $f(x) = |x|$ on the interval $[-1, 1]$ illustrates this problem.

Two Natural Questions, Their Answers, and How the MVT Helps

The MVT has important theoretical uses; many amount to putting intuitively reasonable results on a sound logical footing. Here we illustrate two such uses; we'll see others later.

Question 1: When Is a Function Constant?

It's clear either from the definition of derivative or from graphical arguments that the derivative of a constant function is the zero function. In other words,

If $f(x) = k$ for all inputs x, then $f'(x) = 0$.

Is the converse true? In other words, does our supposed theorem really hold?

> **Theorem 8** If $f'(x) = 0$ for all x in an interval I, then f is constant on I.

Geometric intuition supports the theorem; if the graph of f has a horizontal tangent line everywhere, then it's natural to expect the graph itself to be horizontal. Not surprisingly, the theorem is true. What *is* surprising is that proving this "obvious" fact requires a gun of the MVT's caliber.

We don't care which constant.

Proof. To show that f is constant, ◄ it's enough to show that for any two inputs $x = a$ and $x = b$, $f(a) = f(b)$. According to the MVT, there's some input c between a and b for which

$$f'(c) = \frac{f(b) - f(a)}{b - a}.$$

We don't know which c works, but here it doesn't matter. By hypothesis, $f'(c) = 0$ for *any* input c. Thus,

$$0 = f'(c) = \frac{f(b) - f(a)}{b - a} \implies f(b) - f(a) = 0 \implies f(b) = f(a).$$

A simple but important consequence follows:

Any two antiderivatives of the same function differ by a constant.

More succinctly stated:

> **Theorem 9** If $F'(x) = G'(x)$ for all x in an interval I, then $F(x) = G(x) + C$ for some constant C.

One would prove this theorem applying Theorem 8 to the difference function $F(x) - G(x)$.

Question 2: When Is a Function Increasing?

Up to now we've often stressed—but never proved—the general principle that the sign of f' determines whether f increases or decreases over an interval. The MVT offers a simple but rigorous proof. First let's make a concise statement:

> **Theorem 10** Suppose that $f'(x) > 0$ for all x in an interval I. Then f is increasing on I.

Proof. It's enough to show that, for any a and b in I with $a < b$, $f(a) < f(b)$. By the MVT, there's a c in (a, b) ▶ for which

$$f(b) - f(a) = (b - a) \cdot f'(c).$$

As before, the location of c doesn't matter.

Since both factors on the right must be positive, ▶ so must $f(b) - f(a)$, so $f(b) > f(a)$.

Why?

The MVT and the Racetrack Principle

One version of the racetrack principle ▶ says, in functional language:

We introduced it in Section 2.1, using canine language.

If $f(a) = g(a)$ and $f'(x) \leq g'(x)$ for $x \geq a$, then $f(x) \leq g(x)$ for $x \geq a$.

The racetrack principle (like so much else) can be proved by applying the MVT to the "right" function—in this case, to the function $h(x) = g(x) - f(x)$. Let's do so.
    The hypotheses say that $h(a) = 0$ and $h'(x) \geq 0$ if $x \geq a$. The MVT assures that for any $x \geq a$, there's a number $c \geq a$ such that

$$h'(c) = \frac{h(x) - h(a)}{x - a} = \frac{h(x)}{x - a}.$$

By hypothesis, moreover, $h'(c) \geq 0$. The rest is easy. ▶ If $x \geq a$, then

But check each step.

$$g(x) - f(x) = h(x) = h(x) - h(a) = h'(c) \cdot (x - a) \geq 0 \cdot (x - a) = 0. \quad \blacksquare$$

Proving the MVT, Step by Step (Optional)

Understood graphically, the MVT may seem so natural as to require no proof at all. ▶ That impression, although understandable, is wrong. A rigorous proof of the MVT involves several steps and appeals to several other results.

For excellent reasons, the same applies to other powerful, important theorems of mathematics.

Two Lemmas

We'll sketch the outline of a proof, based on two preliminary lemmas. ▶

*The term **lemma** denotes an auxiliary result, used to help prove another.*

> **Lemma 1** Suppose that f assumes either its maximum or its minimum on $[a, b]$ at a point c between a and b. If $f'(c)$ exists, then $f'(c) = 0$.

Proof. We'll show what happens if $f(c)$ is a maximum (the minimum case is similar). By the definition of derivative,

$$f'(c) = \lim_{h \to 0} \frac{f(c+h) - f(c)}{h} = \lim_{h \to 0^-} \frac{f(c+h) - f(c)}{h} = \lim_{h \to 0^+} \frac{f(c+h) - f(c)}{h}.$$

(Because the ordinary limit exists, so must the left-hand and right-hand limits, and all three are equal.) Notice first that since f assumes a maximum at c, the numerator $f(c+h) - f(c)$ in all three limit expressions must be negative. The denominator h, on the other hand, is negative if $h \to 0^-$ and positive if $h \to 0^+$. Therefore,

$$f'(c) = \lim_{h \to 0^-} \frac{f(c+h) - f(c)}{h} = \lim_{h \to 0^-} [\text{positive stuff}] \geq 0.$$

Similarly,

$$f'(c) = \lim_{h \to 0^+} \frac{f(c+h) - f(c)}{h} = \lim_{h \to 0^+} [\text{negative stuff}] \leq 0.$$

The only possible conclusion is that $f'(c) = 0$, as desired.   ■

Lemma 2 (Rolle's Theorem) Suppose that f is continuous on $[a, b]$ and differentiable on (a, b), and that $f(a) = f(b)$. Then, for some c between a and b, $f'(c) = 0$.

Remark. This lemma, known as **Rolle's theorem**, is nothing more (or less) than the MVT in the special case that $f(a) = f(b)$. The picture, too, is similar:

Rolle's theorem

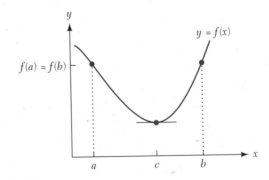

Proof of Rolle's Theorem. Because f is continuous on $[a, b]$, the extreme value theorem guarantees that f assumes both a maximum and a minimum somewhere in $[a, b]$. If both maximum and minimum occur at the endpoints a and b, then, since $f(a) = f(b)$, f must be a constant function. In this case, there's nothing more to do—any c works.

The alternative is that at least one of the maximum and the minimum lies at some c *between* a and b. But, by Lemma 1, $f'(c) = 0$. ■

All Together Now: Proving the MVT

Using Rolle's theorem to prove the MVT requires a technical step. Using f, a, and b, we create a new function to which Rolle's theorem applies.

Proof of the Mean Value Theorem.   Given f as in the mean value theorem, consider the linear function l whose graph is the secant line to f from a to b, as shown:

Proving the mean value theorem: f and l

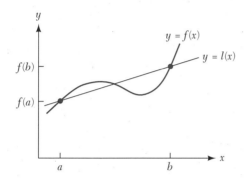

The precise formula for l doesn't matter. What does matter is that

$$l(a) = f(a); \qquad l(b) = f(b); \qquad \text{and} \qquad l'(x) = \frac{f(b) - f(a)}{b - a}.$$

Finally, we consider the difference function $g(x) = f(x) - l(x)$. By design, $g(a) = 0 = g(b)$, so Rolle's theorem applies to g. The rest is bookkeeping: For some c in (a, b),

$$0 = g'(c) = f'(c) - l'(c) = f'(c) - \frac{f(b) - f(a)}{b - a}.$$

We're finished. ■

Sample Problems

Here are two sample problems that show the mean value theorem in action.

EXAMPLE 1   Show, using derivatives, that if $x \geq 0$, then $\sin x \leq x$.

Solution   One approach uses the racetrack principle. If $f(x) = \sin x$ and $g(x) = x$, then $f'(x) = \cos x$ and $g'(x) = 1$, so $f'(x) \leq g'(x)$ for all x. By the racetrack principle, $f(x) \leq g(x)$, as desired. ■

EXAMPLE 2   Suppose that the function f is differentiable for all x and that f has n roots. Show that f has *at least* $n - 1$ stationary points.

Solution   We'll show that between any two roots, there's at least one stationary point. Suppose, then, that x_0 and x_1 are roots, so $f(x_0) = 0 = f(x_1)$. By the mean value theorem, there's a c between x_0 and x_1 such that

$$\frac{f(x_1) - f(x_0)}{x_1 - x_0} = f'(c).$$

But since x_0 and x_1 are roots, the left side is zero, so c is a stationary point.   ■

BASICS

1. Throughout this exercise, $f(x) = |x|$.
   (a) Does f satisfy all hypotheses of the MVT on $[a, b] = [-1, 1]$? Explain.
   (b) Does f satisfy the conclusion of the MVT on $[a, b] = [-1, 1]$? Explain.
   (c) Does f satisfy all hypotheses of the MVT on $[a, b] = [1, 2]$? Explain.

   (d) Does f satisfy the conclusion of the MVT on $[a, b] = [1, 2]$? If so, find at least one suitable value of c. How many such c's exist?

2. Let $f(x) = x^7 - x^5 - x^4 + 2x + 1$. Explain why there must exist a point c in the interval $(-1, 1)$ such that the line tangent to the graph of f at the point $(c, f(c))$ has slope 2.

FURTHER EXERCISES

3. Throughout this exercise, $f(x) = x^2$. Because f is differentiable everywhere, it satisfies all hypotheses of the MVT on any interval $[a, b]$.
   (a) What does the MVT say about f on $[a, b] = [-1, 1]$? Find all suitable values of c. How many are there?
   (b) What does the MVT say about f on $[a, b] = [1, 2]$? Find all suitable values of c. How many are there?
   (c) (A little harder) Show that for $f(x) = x^2$ and any interval $[a, b]$, the MVT's number c is the midpoint of $[a, b]$.

4. Let $g(x) = x^{2/3} - 1$. Notice that although $g(-1) = g(1) = 0$ there is no point c in the interval $(-1, 1)$ where $g'(c) = 0$. Why doesn't this contradict Rolle's Theorem?

5. The sketch below shows portions of the graph of a continuous function f.

Graph of f

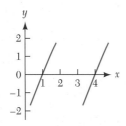

(a) Explain why f must have a root in the interval $(1, 4)$.

(b) Assume that f' is a continuous function. Explain why f' must have a root in the interval $(1, 4)$.

(c) Assume that f' is a continuous function. Explain why the graph of f must have a horizontal tangent line somewhere between $x = 1$ and $x = 4$.

(d) Why is it necessary to assume that f' is continuous in parts (b) and (c)? Sketch a function f that is continuous on the interval $(1, 4)$ for which $f'(x) \neq 0$ anywhere in this interval.

6. Let f be a continuous function that is positive on the interval $[0, 4]$ and differentiable on $(0, 4)$. Answer "yes" or "no" to each of the following questions about f. If you answer "yes" explain how you know that the statement is true. (Your answer should include references to appropriate theorems.) If you answer "no" give an example of a function that shows that the statement is not always true.

   (a) Is it always possible to find a c in $[0, 4]$ such that $f(x) \leq f(c)$ for all x in $[0, 4]$?

   (b) Is it always possible to find a c in $[0, 4]$ such that the instantaneous rate of change of f at c equals the average rate of change of f over the interval $[0, 4]$?

   (c) Is it always possible to find a c in $[0, 4]$ such that the line tangent to the graph of f at $(c, f(c))$ is horizontal?

7. Let f be a function such that $f(0) = 0$ and $\frac{1}{2} \leq f'(x) \leq 1$ for all values of x.
   (a) Use the Mean Value Theorem to explain why $f(2)$ cannot be equal to 3.
   (b) Find upper and lower bounds for the value of $f(3)$.
   (c) Find upper and lower bounds for the value of $f(-1)$.

8. Use the Mean Value Theorem to show that $|\sin b - \sin a| \leq |b - a|$. [**HINT:** $|\cos x| \leq 1$ for all x.]

9. Let f be a function that is differentiable everywhere. Furthermore, assume that the line $y = 3x - 1$ is tangent to f at $x = 1$ and that the line tangent to f at $x = 3$ passes through the points $(0, 5)$ and $(4, 1)$.
   (a) Evaluate $f'(1)$ and $f'(3)$.
   (b) Evaluate $f(1)$ and $f(3)$.
   (c) Explain why there must be a number c in the interval $[1, 3]$ such that $f'(c) = 0$.

(d) Suppose that it is known that $f''(x) < 0$ for all x in $[1, 3]$. Explain why this implies that f has a unique maximum value over the interval $[1, 3]$.

10. Suppose that f is differentiable for all values of x, $f(-3) = -3$, $f(3) = 3$, and $|f'(x)| \leq 1$ for all x in $[-3, 3]$. Show that $f(0) = 0$.

11. At 1:00 p.m. a truck driver picked up a fare card at the entrance of a tollway. At 2:15 p.m., the trucker pulled up to a toll booth 100 miles down the road. After computing the trucker's fare, the toll booth operator summoned a highway patrol officer who issued a speeding ticket to the trucker. (The speed limit on the tollway is 65 mph.)
    (a) The trucker claimed that he hadn't been speeding. Is this possible? Explain.
    (b) The fine for speeding is $35 plus $2 for each mph by which the speed limit is exceeded. What is the trucker's minimum fine?

4.12 Chapter Summary

This chapter offered a potpourri of applications of the derivative andantiderivative. (The sampling is a skimpy one, unfortunately—these ideas have far more uses than any textbook could possibly cover).

Differential Equations. A **differential equation** (**DE**) is any equation that involves derivatives. To **solve** a differential equation means to find a function ▶ that satisfies the DE.

Yes, a function; not a number!

Typical differential equations have many solutions. Every function of the form

$$y = Ce^x,$$

for example, solves the differential equation ▶

$$y' = y.$$

The world's most important DE.

Adding an **initial condition** to a DE creates an **initial value problem**, or **IVP**. An IVP typically has only one solution. The only solution of the IVP ▶

$$y' = y; \qquad y(0) = 1,$$

The world's most important IVP, perhaps.

for instance, is the function

$$y = e^x.$$

Sections 4.1 and 4.2 treated DEs, first briefly in theory and then in more detail, in applications related to population growth.

Linear and Quadratic Approximation. The graph of any well-behaved function is approximated—more or less closely, and near the point of tangency—by its tangent line. This means, in functional language, that f is approximated well by the linear function, called the **linear approximation** to f at x_0, whose graph is the tangent line. Section 4.3 explored the meaning and ramifications of this fact and took matters one step further, to quadratic approximation. In graphical terms, the **quadratic approximation** to f at x_0 is the function whose graph is the parabola that best fits f

at x_0. Going further still, Section 4.3 introduced **Taylor polynomials**, polynomials of any degree that approximate nonpolynomial functions.

Newton's Method. Finding roots of any but the simplest functions—e.g., linear functions—is difficult or impossible algebraically. **Newton's method** (like the bisection method of Section 4.10) offers a numerical alternative. To estimate a root of f near an initial guess x_0, Newton's method "rides the tangent line" from the f-graph at x_0 to the x-axis. The result is seldom *exactly* a root, but repeating Newton's method usually produces—remarkably quickly—a highly accurate approximation to a root. Section 4.4 treated the idea and typical applications of Newton's method—and suggested some ideas for "breaking" it.

Splines. Section 4.5 is about **splining**—fitting together pieces of simple curves (linear, quadratic, or cubic) to formlarger curves of desired shapes. Derivatives are the basic tool: Bymatching one or more derivatives of "incoming" and "outgoing" functions at each knot, curves of any desired smoothness can be constructed.

Optimization. Among the derivative's most important and useful applications is in finding **extreme values**—i.e., maximum and minimum values—of functions. As Section 4.6 illustrated, solving optimization problems usually means identifying an appropriate function and then using the derivative to look for and classify its **critical points**. We stressed problems posed in words. Translation of English descriptions into mathematically useful forms is important wherever mathematics is applied.

Economic Applications. Calculus ideas apply to many areas, including economic measurement and prediction. In Section 4.7 we applied derivative ideas (using the **marginal** language favored by economists) to maximize a grain trader's profit on various transactions.

Related Rates. In practical situations, several quantities vary together in time, and the rates at which these quantities vary are themselves related. The method of **related rates** is a technique for discovering such relationships.

A typical problem begins with an equation relating several quantities, all varying with time. Differentiating everything in sight—usually with respect to time—produces a new equation, one that relates the rates at which the quantities vary.

Parametric Equations and Parametric Curves. Many interesting plane curves—e.g., circles—are not graphs of functions. Parametrically defined curves offer much more flexibility. Defining x and y separately as **coordinate functions** of another variable—say t—lets us draw all sorts of interesting,beautiful, and complicated curves. If the coordinate functions are differentiable in t, then calculus ideas apply; we use them to define and calculate slopes and speeds.

Continuity and Differentiability. Sections 4.10 and 4.11 are brief excursions into the theory of calculus. They explored, respectively, the mathematical significance of continuity and differentiability. The most important results are three "value theorems": the **intermediate value theorem**, the **extreme value theorem**, and, most important of all, the **mean value theorem**. All three results are implicit in much of the machinery of calculus.

5

The Integral

5.1 Areas and Integrals

The **tangent-line problem** and the **area problem** are the two main geometric problems of calculus. ▶ As we've seen, the idea of derivative, together with the rules for computing derivatives, solves the tangent-line problem. For an impressive variety of functions, even complicated ones, it's now a routine matter to describe the tangent line at any point on the graph.

These problems have been "big" for many centuries: The ancient Greeks worked hard and ingeniously at both. Had Greek mathematicians had our algebraic advantages, they would have gone farther than they did.

The Area Problem and the Integral

The general area problem asks how to measure the area of a plane region. For special regions—rectangles, triangles, squares, circles, trapezoids, and on on—well-known ▶ formulas do the job.

And long known—the Greeks knew them.

Area problems are more challenging, and far more interesting, for regions bounded by less familiar curves, such as graphs of functions. The most important area problem of this type is to measure an area like the shaded one in the following picture.

What's the shaded area?

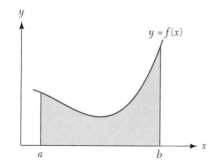

Stated carefully, the problem reads as follows.

Find the area of the region bounded above by the graph of f, below by the x-axis, on the left by the vertical line $x = a$, and on the right by the vertical line $x = b$.

The shorthand phrase is less tedious but also less precise. If in doubt, use the full form.

That's a mouthful; we usually say just "the area under the graph of f from a to b." ◀

Signed Area

In this chapter it is often convenient to consider **signed area** as opposed to area in the everyday sense. The adjective "signed" means that *all area below the x-axis counts as negative.* (In the preceding picture, all the shaded area is above the x-axis, so the question of sign doesn't arise.) We'll give more details and examples later.

A compact notation for signed area is even more convenient:

Definition (**The Integral as Signed Area**) Let f be a function defined for $a \le x \le b$. Either of the equivalent expressions

$$\int_a^b f \quad \text{or} \quad \int_a^b f(x)\,dx$$

denotes the signed area bounded by $x = a$, $x = b$, $y = f(x)$, and the x-axis.

Observe:

In Words For either $\int_a^b f$ or $\int_a^b f(x)\,dx$, read "the **integral** of f from a to b." The function f is the **integrand**.

Which Is Better? Both $\int_a^b f$ and $\int_a^b f(x)\,dx$ denote the same area. The first form looks simpler; for now, we'll usually choose it. The other (dx) form has advantages that will sometimes matter later on. ◀

Differences between the two forms are akin to those between the dy/dx and f′ notations for derivative.

How to Find It? So far, all we've done is define the symbolic expression $\int_a^b f$ to *stand for* area. The problem of *calculating* areas remains.

EXAMPLE 1 Several areas follow, labeled as integrals. Use familiar area formulas to evaluate each integral.

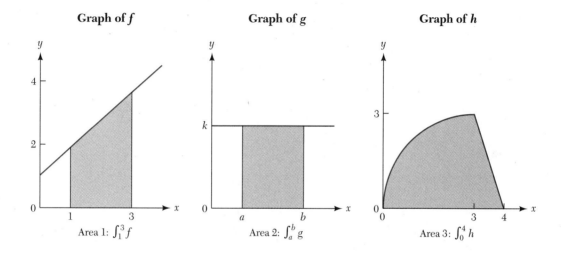

Graph of f

Graph of g

Graph of h

Area 1: $\int_1^3 f$ Area 2: $\int_a^b g$ Area 3: $\int_0^4 h$

Solution All three areas are easy to find. Area 1 is a trapezoid, with base 2 and vertical sides 2 and 4; hence, the area ▶ is 6. Area 2 is a rectangle; by the usual formula (area = base × height), its area is $(b - a) \cdot k$. Area 3 is a quarter circle plus a triangle; a close look at these figures reveals a total area of $9\pi/4 + 3/2$, or about 8.569. In integral notation:

A trapezoid with base b and heights h_1 and h_2 has area $b(h_1 + h_2)/2$.

$$\int_1^3 f = 6; \qquad \int_a^b g = k(b - a); \qquad \int_0^4 h = \frac{9\pi}{4} + \frac{3}{2}. \qquad \blacksquare$$

Signed Area: Positive and Negative Contributions

Integrals measure *signed area*. In calculating integrals, therefore, area above the graph of f and beneath the x-axis counts as negative. Thus if an integrand f takes negative values in an interval $[a, b]$, then so may the integral $\int_a^b f$.

EXAMPLE 2   Let $f(x) = 1 - x^2$. Find (or estimate) values for the integrals $I_1 = \int_0^2 f$ and $I_2 = \int_{-2}^2 f$.

Solution   The areas in question look like this:

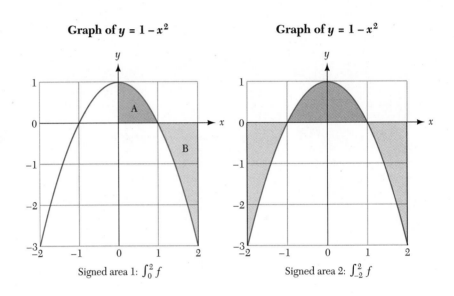

Graph of $y = 1 - x^2$ · Signed area 1: $\int_0^2 f$

Graph of $y = 1 - x^2$ · Signed area 2: $\int_{-2}^2 f$

Soon it will be easy.

Note that each square has area 1.

For now, computing I_1 and I_2 *exactly* is difficult. ◄ *Estimating* the integrals, on the other hand, is easy. Eyeballing general sizes ◄ suggests that A has area around 2/3; similarly, B has area around 3/2. From these guesses, we estimate (counting B as negative!) that

$$I_1 = \int_0^2 f \approx \frac{2}{3} - \frac{3}{2} = -\frac{5}{6}.$$

It is in the ballpark; the exact answer turns out to be −2/3.

Convince yourself; how is evenness of f involved?

How close is this estimate? We can't yet tell for sure, but it seems to be in the ballpark. ◄ In any event, symmetry shows ◄ that

$$I_2 = 2 \cdot I_1 \approx -\frac{5}{3}. \qquad \blacksquare$$

EXAMPLE 3   Let $g(x) = x^3$. Find or estimate $\int_{-1}^1 g(x)\, dx$ and $\int_0^1 g(x)\, dx$.

Solution   Here's the graph:

Graph of $y = x^3$

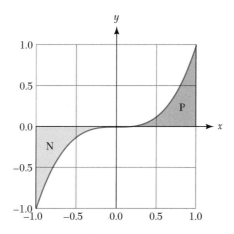

Because of the graph's symmetry ▶ the positive and negative areas P and N cancel. Hence

Or, equivalently, because g is odd.

$$\int_{-1}^{1} g(x)\, dx = 0.$$

(*Any* odd integrand, over *any* interval that's symmetric about $x = 0$, behaves the same way.)

Symmetry is no help with $\int_0^1 g$. A close look suggests that P's area is about that of one small square, or 0.25. In other words,

$$\int_0^1 g(x)\, dx \approx 0.25.$$

(We'll see soon that—and why—this estimate is actually exact.) ■

Properties of the Integral

Viewing the integral as area makes many of the integral's important properties simple and natural. For example, this picture

Adding areas: $\int_a^b f = \int_a^c f + \int_c^b f$

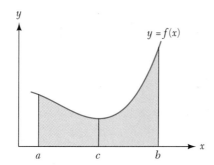

is a convincing illustration of the following simple, but often useful, property of the integral:

$$\text{If } a < c < b, \text{ then } \int_a^b f = \int_a^c f + \int_c^b f.$$

Requiring f and g to be continuous ensures that all the integrals in question exist.

The next theorem collects several similar properties of the integral. ◄

Theorem 1 (New Integrals From Old) Let f and g be continuous functions on $[a, b]$; let k denote a real constant. Then

1. $\int_a^b \big(f(x) \pm g(x)\big)\, dx = \int_a^b f(x)\, dx \pm \int_a^b g(x)\, dx.$

2. $\int_a^b k f(x)\, dx = k \int_a^b f(x)\, dx.$

3. If $f(x) \le g(x)$ for all x in $[a, b]$, then
   $$\int_a^b f(x)\, dx \le \int_a^b g(x)\, dx.$$

4. If $a < c < b$, then $\int_a^b f(x)\, dx = \int_a^c f(x)\, dx + \int_c^b f(x)\, dx.$

Here are some remarks on the theorem:

Like the Derivative The first two of the four properties are akin to properties of the derivative: Like the derivative, the integral behaves well with respect to sums and constant multiples. The following picture illustrates the **constant multiple rule** (the second identity in the theorem) for integrals.

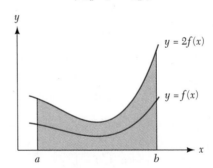

Constant multiple rule for integrals:
$$\text{why } \int_a^b 2f = 2\int_a^b f$$

A Useful Case The third property is often used in the case where either f or g is a constant. Here's one useful version:

Fact (Bounding an Integral) Suppose that, for some numbers m and M, the inequality $m \le f(x) \le M$ holds for all x in $[a, b]$. Then

$$m \cdot (b - a) = \int_a^b m\, dx \le \int_a^b f(x)\, dx \le \int_a^b M\, dx = M \cdot (b - a).$$

A picture makes the point best. ▶

The numbers m and M are lower and upper bounds for f on [a, b].

Bounding areas: why $m(b - a) \leq \int_a^b f \leq M(b - a)$

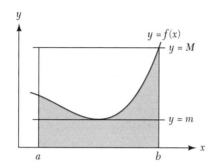

As the picture illustrates, the value of the integral (i.e., the area of the shaded region) lies between the areas of the smaller and the larger rectangles.

The next three examples illustrate uses of the theorem.

EXAMPLE 4   The integral $I = \int_1^3 f(x)\,dx$, where $f(x) = 1/x$, measures the shaded area in the following picture. Use the linear functions g_1, g_2, and g_3 to estimate I.

Three ways to estimate $\int_1^3 f(x)\,dx$

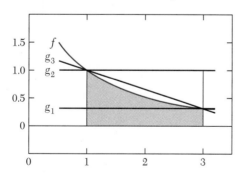

Solution   The shaded area isn't a rectangle or other simple shape, so we can't find I exactly from simple area formulas. We can, however, easily estimate I, using the third part of Theorem 1. As the picture shows, $f(x) \geq g_1(x) = 1/3$ for all x in $[1, 3]$. Therefore, by Theorem 1, ▶

The right-hand integral is easy—it corresponds to a rectangle.

$$\int_1^3 f(x)\,dx \geq \int_1^3 g_1(x)\,dx = \frac{2}{3}.$$

Similarly, $f(x) \le g_2(x) = 1$, for all x in $[1, 3]$, so

$$\int_1^3 f(x)\,dx \le \int_1^3 g_2(x)\,dx = 2.$$

This is an upper bound for I—but not a very good one, as the picture shows. Using g_3 instead of g_2 gives a better upper bound:

$$\int_1^3 f(x)\,dx \le \int_1^3 g_3(x)\,dx = \frac{4}{3}.$$

(The right-hand integral represents a trapezoidal area, so an elementary area formula applies.) Combining these results gives both lower and upper bounds for I: $2/3 \le I \le 4/3$. (By choosing the upper and lower bounding functions more cleverly, we could improve these estimates.) ■

It is a fact—a remarkable one.

EXAMPLE 5   Using the fact ◄   that $\int_0^\pi \sin x\,dx = 2$, find $\int_0^{\pi/2} \sin x\,dx$ and $\int_0^\pi (3\sin x + 2\cos x)\,dx$.

Draw them to convince yourself!

Solution   Symmetry of the sine and cosine graphs ◄   shows that

$$\int_0^{\pi/2} \sin x\,dx = \frac{1}{2} \int_0^\pi \sin x\,dx = 1; \qquad \int_0^\pi \cos x\,dx = 0.$$

Therefore, by the sum and constant multiple rules,

$$\int_0^\pi (3\sin x + 2\cos x)\,dx = 3 \int_0^\pi \sin x\,dx + 2 \int_0^\pi \cos x\,dx = 6 + 0 = 6.$$

Does it?

A look at the graph should make the general size of the answer plausible. ◄

Why $\int_0^\pi (3\sin x + 2\cos x)\,dx = 6$

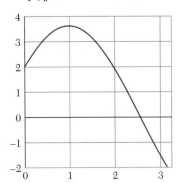

The "net area" does seem to be about 6. ■

EXAMPLE 6   For any positive constant b,

$$\int_0^b 1\,dx = b \qquad \text{and} \qquad \int_0^b x\,dx = \frac{b^2}{2}.$$

Explain why. Then use the theorem to find $\int_0^b (Cx + D)\, dx$ for any constants C and D.

Solution Graphs ▶ of the functions $f(x) = 1$ and $g(x) = x$ over the *Draw them!* interval $[0, b]$ show that the first two integrals have the claimed values. To finish, we use the sum and constant multiple rules for integrals:

$$\int_0^b (Cx + D)\, dx = \int_0^b (Cx)\, dx + \int_0^b D\, dx$$

$$= C \int_0^b x\, dx + \int_0^b D\, dx$$

$$= C \frac{b^2}{2} + Db. \qquad \blacksquare$$

Average Value and the Integral

In the following picture, the rectangle is chosen to have the same area as the shaded region:

Equal areas: average value of a function

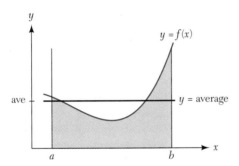

The height of that rectangle is, in a natural way, the **average value** of the function f over the interval $[a, b]$. The following definition states in analytic language what the picture shows geometrically.

Definition Let f be defined on an interval $[a, b]$. The quantity

$$\frac{\int_a^b f(x)\, dx}{b - a}$$

is the average value of f over $[a, b]$.

The ideas of integral and average value are close cousins; understanding either helps with the other. The preceding picture summarizes this two-way relationship. The average value is the *height* of a certain rectangle, the integral is the *area* of the same rectangle.

EXAMPLE 7   The following graphs show speed functions s_A and s_B for two cars A and B.

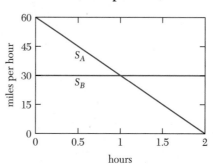

**Speeds of cars A and B
(miles per hour)**

What's the average value of each function over the interval $[0, 2]$? What do the results mean in "car talk?"

Convince yourself.

Solution   A look at areas shows readily ◄ that

$$\int_0^2 s_A(t)\,dt = 60 = \int_0^2 s_B(t)\,dt.$$

Thus, using the preceding definition, *both* speed functions have average value 30 over the interval $[0, 2]$.

We'll explain exactly why this is true later in this chapter.

The answers make excellent automotive sense. The area under each graph represents the total distance—60 miles—that each car covers in the 2-hour interval. ◄ Dividing the distance by the time elapsed gives an average speed of 30 miles per hour (mph) for each car.   ■

Interpreting the Integral

So far, we've interpreted the integral $\int_a^b f$ mainly in geometric terms. The integral has many other natural and important interpretations, several of which we'll see in this book. Physical motion is one important example.

Speed and Distance

If a function $f(t)$ represents the speed of a moving object at time t, then $\int_a^b f(t)\,dt$ represents the distance traveled by the object over the interval $a \le t \le b$. Why is this true?

As it is for car B in Example 7.

If the speed function $f(t)$ happens to be constant, ◄ say, k—then

$$\int_a^b f(t)\,dt = k \cdot (b - a) = \text{speed} \times \text{time} = \text{distance},$$

as claimed. Remarkably, integrating the speed function gives distance traveled even if $f(t)$ *isn't* constant. Example 7 should make this result believable; in the following sections we'll see exactly why it's true.

Velocity and Net Distance

Speed, by definition, is always positive. Velocity can be either positive or negative, depending on the direction of motion. As a result, integration of a velocity function $v(t)$ gives slightly different information from integration of a speed function. If $v(t)$ tells the velocity of an object moving along a line at time t, then $\int_a^b f(t)\,dt$ represents the *net* distance traveled over the interval $a \le t \le b$. In particular, net distance, like velocity, can be either positive or negative.

Integrating from Right to Left: A Technicality

The English description of $\int_a^b f(x)\,dx$—"the integral of f from $x = a$ to $x = b$"— implies a *direction*: x starts at a and ends at b. ▶ It's usually natural to think of x as moving from left to right. Up to now, we've done just that: In every integral $\int_a^b f$ we've treated so far, we've assumed or stated that $a \le b$. For example, we've discussed

The variable name x isn't sacred; t is another popular choice.

$$\int_0^\pi \sin x\,dx, \quad \text{but not} \quad \int_\pi^0 \sin x\,dx \quad \text{or} \quad \int_\pi^0 (3\sin x + 2\cos x)\,dx.$$

Sometimes the latter sort of "right-to-left" integrals *do* arise. ▶ The following convention, based on signs, handles all the possibilities with a minimum of fuss:

We'll see some in the next section.

$$\int_a^b f(x)\,dx = -\int_b^a f(x)\,dx. \tag{5.1}$$

The convention says (for instance) that, since $\int_0^1 x^3\,dx = 0.25$, ▶

See Example 3.

$$\int_1^0 x^3\,dx = -\int_0^1 x^3\,dx = -0.25.$$

Why does Equation 5.1 make sense? The speed–distance context offers one answer:

> If $f(t)$ *represents speed at time t, and $a < b$, then $\int_a^b f(t)\,dt$ gives the (positive) distance traveled from time a to time b.*

It's physically reasonable, therefore, to consider as negative the distance covered over the reverse interval. ▶

Even if Equation 5.1 seems arbitrary now, it will prove useful in later work.

Integrals and Areas: The Story So Far and a Look Ahead

In this section we defined the integral $\int_a^b f$ as a certain area. Then we observed some properties and uses of the integral, arguing mainly from geometric intuition. What's missing so far is any really practical or effective method of *measuring* areas. In the simplest cases elementary area formulas (for squares, rectangles, and so on) are some help. For all *but* the simplest functions, however, such formulas are useless. In the next three sections we will offer two different practical methods of calculating integrals. In Sections 5.2 and 5.3, we'll relate the new idea of the integral to the older (and better understood) idea of the derivative. This key connection between derivative and integral, known as the **fundamental theorem of calculus**, is among the most important results of mathematics. In Section 5.4 we'll give another definition of the integral, this time as a limit of approximating sums. Logically, the situation resembles that for the derivative, which we first defined informally as the slope of a line and later, more formally, as a limit of differences.

Must Area Exist?   There seems to be little question that areas under simple graphs—lines, parabolas, sinusoids, and so on—exist. The situation is less clear for "not-nice" (e.g., discontinuous) functions, whose graphs may be extremely ragged. Consider, for instance, this graph:

The graph of $y = \sin(1/x)$

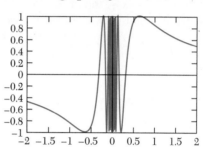

Can such a nasty curve sensibly be said to bound area?

The answer turns out (for this curve, anyway) to be yes. Defining area (let alone computing it) in such unpromising circumstances, however, requires special care. In Section 5.4 we'll redefine the integral, using limits, partly in order to handle cases like this one. In practice, there's no need to worry unduly; the vast majority of standard calculus functions behave nicely enough to accord well with our everyday intuition about area.

---

1. Let g be the function shown graphically below. When asked to estimate $\int_1^2 g(x)\,dx$, a group of calculus students submitted the following answers: -4, 4, 45, and 450. Only one of these responses is reasonable; the others are "obviously" incorrect. Which is the reasonable one? Why?

Graph of g

2. Let f be the function shown graphically below.

Graph of f

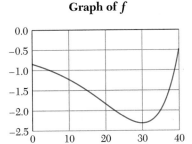

(a) Which of the following is the best estimate of $\int_0^{40} f(x)\,dx$: -65, -30, 0, 35, 60? Justify your answer.

(b) Show that the average value of f over the interval $[10, 30]$ is between -2 and -1.5.

3. The graph of a function f is shown below.

Graph of f

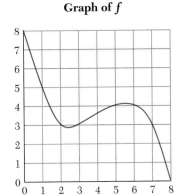

(a) Which of the following is the best estimate of $\int_1^6 f(x)\,dx$: -24, 9, 20, 38? Justify your answer.

(b) Find positive integers A and B such that $A \le \int_3^7 f(x)\,dx \le B$. Explain how you know that the values of A and B you chose have the desired properties.

(c) $\int_6^8 f(x)\,dx \approx 4$. Does this approximation overestimate or underestimate the exact value of the integral? Justify your answer.

(d) Estimate the average value of f over the interval $[0, 2]$.

(e) Is your estimate in part (d) too large or too small? Justify your answer.

4. For each function below, use a graph of f to evaluate
$$\int_0^2 f(x)\,dx, \quad \int_1^4 f(x)\,dx, \quad \int_{-5}^{-1} f(x)\,dx, \text{ and } \int_{-2}^3 f(x)\,dx.$$

(a) $f(x) = 3x$ (c) $f(x) = 5 - 2x$

(b) $f(x) = 2x + 5$

5. The graph of a function f (shown below) consists of two straight lines and two one-quarter circles.

Graph of f

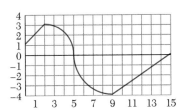

Evaluate each of the following integrals.

(a) $\displaystyle\int_0^2 f(x)\,dx$ (f) $\displaystyle\int_0^{15} f(x)\,dx$

(b) $\displaystyle\int_2^5 f(x)\,dx$ (g) $\displaystyle\int_0^{15} |f(x)|\,dx$

(c) $\displaystyle\int_0^5 f(x)\,dx$ (h) $\displaystyle\int_{15}^9 f(x)\,dx$

(d) $\displaystyle\int_5^9 f(x)\,dx$ (i) $\displaystyle\int_{12}^{15} f(x)\,dx$

(e) $\displaystyle\int_4^4 f(x)\,dx$ (j) $\displaystyle\int_9^{12} f(x)\,dx$

6. Suppose that $\int_{-2}^5 f(x)\,dx = 18$, $\int_{-2}^5 g(x)\,dx = 5$, and $\int_{-2}^5 h(x)\,dx = -11$. Evaluate each of the following integrals.

(a) $\displaystyle\int_{-2}^5 \big(f(x) + g(x)\big)\,dx$

(b) $\displaystyle\int_{-2}^5 \big(f(x) + h(x)\big)\,dx$

(c) $\displaystyle\int_{-2}^5 \big(f(x) + g(x) - h(x)\big)\,dx$

(d) $\displaystyle\int_{-2}^5 3f(x)\,dx$

(e) $\displaystyle\int_{-2}^5 -4g(x)\,dx$

(f) $\displaystyle\int_5^{-2} f(x)\,dx$

(g) $\displaystyle\int_{-2}^5 \big(h(x) + 1\big)\,dx$

(h) $\displaystyle\int_0^7 g(x - 2)\,dx$

FURTHER EXERCISES

7. Four students disagree on the value of the integral $\int_0^{\pi/2} \cos^8 x\, dx$. Jack argues for $\pi \approx 3.14$, Joan for $35\pi/256 \approx 0.43$, Ed for $2\pi/9 - 1 \approx -0.30$, and Lesley for $\pi/4 \approx 0.79$. Use the graph below to determine who is right. (One *is* right!) Explain how you know that the other values are incorrect. [**HINT**: $\big(\cos(1/2)\big)^8 < 0.4$.]

Graph of $y = \cos^8 x$

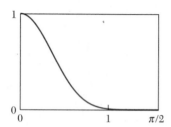

8. The graph of a function h is shown below.

Graph of h

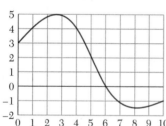

List, from smallest to largest: (i) $h'(5)$; (ii) the average value of h over the interval $[0, 10]$; (iii) the average rate of change of h over the interval $[0, 10]$; (iv) $\int_0^{10} h(x)\, dx$; (v) $\int_0^5 h(x)\, dx$; (vi) $\int_6^{10} h(x)\, dx$; (vii) $\int_5^5 h(x)\, dx$.

9. Estimate each of the quantities (i)–(vi) listed in Exercise 8.

10. Let h be the function shown graphically in Exercise 8 and suppose that $h(t)$ represents the *eastward* velocity (in meters/second) of an object moving along an east-west axis at time t seconds. Estimate each of the following quantities (use appropriate units!) and interpret in the language of velocity, distance, etc.

   (a) $\int_0^{10} h(t)\, dt$ (d) $|h(8)|$

   (b) $\dfrac{1}{10}\int_0^{10} h(t)\, dt$ (e) $\int_0^{10} |h(t)|\, dt$

   (c) $\dfrac{1}{10}\big(h(10) - h(0)\big)$ (f) $\dfrac{1}{10}\int_0^{10} |h(t)|\, dt$

11. Suppose that a car travels on an east-west road with eastward velocity $v(t) = 60 - 20t$ mph at time t hours.

   (a) Evaluate $\int_0^4 v(t)\, dt$. Interpret the answer in car talk.
   (b) Find the average value of $v(t)$ on the interval $[0, 4]$. Show your result graphically.
   (c) Let $s(t)$ be the car's speed at time t. Evaluate $\int_0^4 s(t)\, dt$ and interpret the answer in car talk.
   (d) What is the car's average speed between $t = 0$ and $t = 4$?

12. Sketch the graph of a continuous function f such that $\int_{-2}^3 f(t)\, dt = -10$ and

   (a) $\int_{-2}^3 |f(t)|\, dt = 10$.   (b) $\int_{-2}^3 |f(t)|\, dt \neq 10$.

13. For each function below, use a graph of f to evaluate $\int_0^2 f(x)\, dx$, $\int_1^4 f(x)\, dx$, $\int_{-5}^{-1} f(x)\, dx$, and $\int_{-2}^3 f(x)\, dx$. (a is a positive constant.)

   (a) $f(x) = 1$ (c) $f(x) = -a$
   (b) $f(x) = a$

14. Let f be the function shown graphically in Exercise 5.

   (a) Show that $\int_2^4 f(x)\, dx > 5$. [**HINT**: Draw the trapezoid with vertices at $(2, 0)$, $\big(2, f(2)\big)$, $(4, 0)$, and $\big(4, f(4)\big)$.]

   (b) Show that $\int_6^9 f(x)\, dx < -9$.

   (c) Is $\int_0^6 f(x)\, dx < 11$?

   (d) Is $\int_2^7 f(x)\, dx$ positive or negative?

15. (a) Evaluate $\int_{-3}^3 (x + 2)\, dx$.

   (b) Evaluate $\int_{-3}^3 |x + 2|\, dx$.

   (c) Evaluate $\int_{-3}^3 \big(|x| + 2\big)\, dx$.

   (d) Evaluate $\int_{-3}^3 \big(2 - |x|\big)\, dx$.

16. Evaluate $\int_0^2 f(x)\, dx$, where
$$f(x) = \begin{cases} 1 + x, & \text{if } 0 \leq x \leq 1 \\ 2 - x, & \text{if } 1 < x \leq 2. \end{cases}$$
[**HINT**: Sketch a graph of the integrand.]

17. Evaluate $\int_0^1 \sqrt{1 - (x - 1)^2}\, dx$. [**HINT**: Sketch a graph of the integrand.]

18. Evaluate $\int_1^3 \left(6 - \sqrt{4 - (x - 3)^2}\, dx\right) dx$.

19. Evaluate $\int_0^3 f(x)\,dx$, where

$$f(x) = \begin{cases} 1 - \sqrt{1 - x^2}, & \text{if } 0 \leq x \leq 1 \\ 1 + \sqrt{4 - (x - 3)^2}, & \text{if } 1 < x \leq 3. \end{cases}$$

20. Explain why

$$\int_1^2 x^3\,dx = \int_3^4 (x - 2)^3\,dx = \int_{-3}^{-2} (x + 4)^3\,dx.$$

21. Suppose that (i) $\int_0^2 f(x)\,dx = 2$; (ii) $\int_1^2 f(x)\,dx = -1$; (iii) $\int_2^4 f(x)\,dx = 7$.

    (a) Evaluate $\displaystyle\int_1^4 f(x)\,dx$.

    (b) Evaluate $\displaystyle\int_0^4 3f(x)\,dx$.

    (c) Evaluate $\displaystyle\int_0^1 f(x)\,dx$.

    (d) Evaluate $\displaystyle\int_0^1 f(x + 1)\,dx$.

    (e) Evaluate $\displaystyle\int_0^1 \left(f(x) + 1\right)dx$.

    (f) Evaluate $\displaystyle\int_2^4 f(x - 2)\,dx$.

    (g) Evaluate $\displaystyle\int_2^4 \left(f(x) - 2\right)dx$.

    (h) Explain why f must be negative somewhere in the interval $[1, 2]$.

    (i) Explain why $f(x) \geq 3$ somewhere in the interval $[2, 4]$.

    (j) Draw the graph of a function f with properties (i)–(iii).

22. Suppose that f is a function such that $\int_0^3 f(x)\,dx = -1$.
    (a) Suppose that f is an *even* function. Explain why $\int_{-3}^3 f(x)\,dx = -2$.
    (b) If f is an *odd* function, what is the value of $\int_{-3}^3 f(x)\,dx$? Explain.

23. Evaluate each of the following integrals using graphical arguments.   [**HINT:** See the previous exercise.]

    (a) $\displaystyle\int_0^\pi \cos x\,dx$

    (c) $\displaystyle\int_{-2}^2 (7x^5 + 3)\,dx$

    (b) $\displaystyle\int_{\pi/2}^{3\pi/2} \sin x\,dx$

    (d) $\displaystyle\int_{-1}^1 (4x^3 - 2x)\,dx$

24. Evaluate each of the following integrals using graphical arguments and the fact that $\int_0^\pi \sin x\,dx = 2$.

    (a) $\displaystyle\int_0^{2\pi} \sin x\,dx$

    (d) $\displaystyle\int_0^{\pi/2} \sin x\,dx$

    (b) $\displaystyle\int_0^{2\pi} |\sin x|\,dx$

    (e) $\displaystyle\int_{-\pi/2}^{\pi/2} \cos x\,dx$

    (c) $\displaystyle\int_0^\pi (1 + \sin x)\,dx$

    (f) $\displaystyle\int_0^{\pi/2} (x + \cos x)\,dx$

(g) $\displaystyle\int_0^{100\pi} \sin x\,dx$

(i) $\displaystyle\int_0^{100\pi} \cos x\,dx$

(h) $\displaystyle\int_0^{100\pi} |\sin x|\,dx$

25. Explain why $\displaystyle\int_1^3 \frac{1 - x}{x^2}\,dx < \int_1^2 \frac{1 - x}{x^2}\,dx$.

26. Is $\int_0^2 (2x^3 - x^2)\,dx$ greater than $\int_0^2 (3x^2 + x)\,dx$? Justify your answer.

27. (a) Explain why $\displaystyle\int_0^{2\pi} (1 + \cos x)\,dx \geq 0$.

    (b) Show that $\displaystyle\int_0^{2\pi} (1 + \cos x)\,dx > 0$.   [**HINT:** Consider the intervals $[0, \pi]$ and $[\pi, 2\pi]$ separately.]

28. Let $f(x) = \frac{1}{2}x + \cos x$. Show that $1.3 \leq \int_0^3 f(x)\,dx \leq 3.5$.   [**HINT:** Find the maximum and minimum values of f over the interval $[0, 3]$.]

29. Show that $4.5 \leq \displaystyle\int_1^3 e^x \sin x\,dx \leq 15$.

30. Show that $0 \leq \displaystyle\int_0^\pi x \sin x\,dx \leq \pi^2/2$.   [**HINT:** $x \sin x \leq x$ when $0 \leq x \leq \pi$.]

31. Show that $\pi/6 \leq \displaystyle\int_{\pi/6}^{\pi/2} \sin x\,dx \leq \pi/3$.

32. Show that $-\pi/3 \leq \displaystyle\int_{2\pi/3}^\pi \cos x\,dx \leq -\pi/6$.

33. (a) Show that $\pi(1 + 1/e) \leq \displaystyle\int_0^{2\pi} e^{\sin x}\,dx \leq \pi(1 + e)$.
    [**HINT:** When $0 \leq x \leq \pi$, $0 \leq \sin x \leq 1$. When $\pi \leq x \leq 2\pi$, $-1 \leq \sin x \leq 0$.]
    (b) Use the result in part (a) to find upper and lower bounds for $\int_0^{50\pi} e^{\sin x}\,dx$.

34. (a) Show that $\pi/6 \leq \displaystyle\int_0^{\pi/2} \cos x\,dx \leq 5\pi/12$.
    [**HINT:** $1/2 \leq \cos x \leq 1$ when $0 \leq x \leq \pi/3$.]
    (b) Using ideas similar to those in part (a), find upper and lower bounds on the value of $\int_0^{\sqrt{\pi/2}} \cos(x^2)\,dx$.

35. Show that $0.4 \leq \displaystyle\int_0^1 \sin(e^x)\,dx \leq 1$.

36. Sketch the graph of a function f with the property that

$$\left| \int_1^5 f(x)\,dx \right| < \int_1^5 |f(x)|\,dx.$$

Explain why the function f you sketched has this property.

37. Show that $\dfrac{\pi}{2} < \displaystyle\int_0^\pi \cos(\sin x)\,dx \leq \pi$.   [**HINT:** $\cos 1 > \cos(\pi/3) = 1/2$.]

38. (a) Let $y = f(x)$ be the line through the points $(0, \arctan 0)$ and $(1, \arctan 1)$. Explain why $f(x) \le \arctan x$ when $0 \le x \le 1$. [**HINT:** Consider the concavity of the arctangent function.]

   (b) Find an equation of the line tangent to the curve $y = \arctan x$ at $x = 0$.

   (c) Show that $\dfrac{\pi}{8} \le \displaystyle\int_0^1 \arctan x\, dx \le \dfrac{1}{2}$.

39. (a) Prove that $2x/\pi \le \sin x \le x$ when $0 \le x \le 1$.

   (b) Use part (a) to show that $\pi/16 \le \displaystyle\int_0^{\pi/4} \sin x\, dx \le \pi^2/32$.

40. (a) Show that $1 + x \le e^x \le 1 + 3x$ when $0 \le x \le 1$.

   (b) Use part (a) to find upper and lower bounds on $\displaystyle\int_0^1 e^x\, dx$.

41. (a) Suppose that $f(x) \ge 0$ when $0 \le x \le 1$. Show that $\displaystyle\int_0^1 x^k f(x)\, dx \le \int_0^1 f(x)\, dx$ for all integers $k \ge 0$.

   (b) Does the result in part (a) remain valid if the assumption "$f(x) \ge 0$ when $0 \le x \le 1$" is not made? Explain.

42. Show that $\displaystyle\int_0^{\pi/4} \cos(2x)\, dx \ne \int_0^{\pi/2} \cos x\, dx$. Which integral is larger? Justify your answer.

43. (a) Show that $\displaystyle\int_0^{\pi/2} \sin^2 x\, dx = \int_0^{\pi/2} \cos^2 x\, dx$.

   [**HINT:** $\cos x = -\sin(x - \pi/2) = \sin(\pi/2 - x)$.]

   (b) Show that $\displaystyle\int_0^{\pi/2} \sin^2 x\, dx + \int_0^{\pi/2} \cos^2 x\, dx = \pi/2$.

   (c) Show that $\displaystyle\int_0^{\pi/2} \sin^2 x\, dx = \pi/4$.

44. (a) Show that $\displaystyle\int_0^{\pi/2} \sqrt{1 + \cos(2x)}\, dx = \sqrt{2}\int_0^{\pi/2} \cos x\, dx$.

   [**HINT:** $2\cos^2 x = 1 + \cos(2x)$ for all x.]

   (b) Does $\displaystyle\int_{\pi/2}^{\pi} \sqrt{1 + \cos(2x)}\, dx = \sqrt{2}\int_{\pi/2}^{\pi} \cos x\, dx$? Justify your answer.

45. Suppose that f is an odd function and continuous everywhere. Explain why the average value of f over any interval $[-a, a]$ is zero.

46. Suppose that f is an even function and continuous everywhere. Explain why the average value of f over the interval $[-a, a]$ is equal to the average value of f over the interval $[0, a]$.

47. Suppose that f is a continuous function and that $-2 \le f(x) \le 5$ if $1 \le x \le 3$. Show that the average value of f over $[1, 3]$ is between -2 and 5.

48. Suppose that f is a continuous function. If the average value of f over the interval $[0, 1]$ is 2 and the average value of f over the interval $[1, 3]$ is 4, what is the average value of f over the interval $[0, 3]$?

49. Suppose that f is a continuous function. If the average value of f over the interval $[-3, 1]$ is 2 and the average value of f over the interval $[-3, 7]$ is 5, what is the average value of f over the interval $[1, 7]$?

50. Show that the expression $\displaystyle\int_a^b \big(f(x) - c\big)^2\, dx$ assumes its minimum value when c is the average value of f over the interval $[a, b]$.

51. Suppose that f is continuous on $[0, 1]$. Explain why $\displaystyle\int_0^1 f(x)\, dx = \int_0^1 f(1 - x)\, dx$. [**HINT:** How is the graph of $g(x) = f(1 - x)$ related to the graph of f?]

52. Suppose that f is continuous on $[a, b]$. Show that $\left| \displaystyle\int_a^b f(x)\, dx \right| \le \int_a^b |f(x)|\, dx$. [**HINT:** $-|f(x)| \le f(x) \le |f(x)|$.]

53. Suppose that f and g are continuous on $[a, b]$ and $\displaystyle\int_a^b f(x)\, dx \le \int_a^b g(x)\, dx$.

   (a) Must $f(x) \le g(x)$ for every x in $[a, b]$? If so, explain why. If not, give a counterexample.

   (b) Must there be a number c such that $a \le c \le b$ and $f(c) \le g(c)$? If so, explain why. If not, give a counterexample.

54. Archimedes (ca. 250 B.C.E.) proved that the area under a parabolic arch is $2bh/3$, where b is the width of the base of the arch and h is the height. [**NOTE:** Archimedes actually proved a more general theorem about the area of any region cut off from a parabola by a line.]

   (a) Use Archimedes's result to show that $\displaystyle\int_{-a}^a x^2\, dx = 2a^3/3$. [**HINT:** Draw the curve $y = x^2$ and the line $y = a^2$.]

   (b) Use part (a) and the fact that x^2 is an even function to show that $\displaystyle\int_0^a x^2\, dx = a^3/3$. [**HINT:** $\displaystyle\int_{-a}^a x^2\, dx = \int_{-a}^0 x^2\, dx + \int_0^a x^2\, dx$.]

   (c) Use part (b) to show that $\displaystyle\int_a^b x^2\, dx = (b^3 - a^3)/3$. [**HINT:** $\displaystyle\int_a^b x^2\, dx = \int_0^b x^2\, dx - \int_0^a x^2\, dx$.]

55. Let $f(x) = 2x + 3$.

   (a) Sketch a graph of f in the viewing window $[1, 2] \times [5, 9]$.

   (b) Evaluate $\displaystyle\int_1^2 f(x)\, dx$. Shade the region of the part (a) graph represented by this integral.

   (c) Show that $f^{-1}(x) = (x - 3)/2$.

   (d) Evaluate $\displaystyle\int_5^9 f^{-1}(x)\, dx$ and shade the region of the part (a) graph represented by this integral.

   (e) Verify that
   $$\int_1^2 f(x)\, dx = 2f(2) - f(1) - \int_{f(1)}^{f(2)} f^{-1}(x)\, dx.$$

56. Suppose that f is a continuous function and invertible on $[a, b]$. Then it can be shown that
   $$\int_a^b f(x)\, dx = bf(b) - af(a) - \int_{f(a)}^{f(b)} f^{-1}(x)\, dx.$$

(a) Give a graphical proof of this result when $0 < a < b$, and $0 < f(a) < f(b)$. [**HINT:** Draw a sketch of f in the viewing window $[0, b] \times [0, f(b)]$.]

(b) Show that $\displaystyle\int_1^e \ln x\, dx = e - \int_0^1 e^x\, dx$.

(c) Show that $\displaystyle\int_a^b \sqrt{x}\, dx = b^{3/2} - a^{3/2} - \int_{\sqrt{a}}^{\sqrt{b}} x^2\, dx$.

57. Use the results stated in Exercises 54 and 56 to show that $\int_0^a \sqrt{x}\, dx = 2a^{3/2}/3$.

5.2 The Area Function

In Section 5.1 we defined the integral $\int_a^b f$ to be the signed area of the region from $x = a$ to $x = b$, bounded above or below by the graph of f. For given f, a, and b, this area is a certain fixed *number* (positive, negative, or zero!). In this section we define and study a *function* based on the signed area bounded by the graph of f. This area function, which we'll denote by A_f, is "built" from f, using the integral.

We've played the new-function-from-old game before. In the most important case, we built the derivative function f' from an "original" function f, and then studied what each tells about the other. Here we repeat the process, this time with f and A_f. The relationship between f and A_f will prove remarkably similar to that between f' and f.

Defining the Area Function

First comes the formal definition:

> **Definition** Let f be a function and a any point of its domain. For any input x, the **area function** A_f is defined by the rule
> $$A_f(x) = \int_a^x f(t)\, dt = \text{signed area defined by } f, \text{ from } a \text{ to } x.$$

A generic picture illustrates the idea:

The area function: $A_f(x)$ = the shaded area

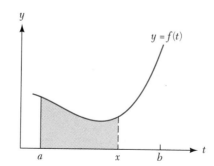

The idea is best understood through pictorial examples. First, three brief remarks.

Variable Names: Why So Many?   Why did we use another variable, t, in the defining expression $A_f(x) = \int_a^x f(t)\,dt$? We did it because the letter x was already taken; we used it to denote the right-hand boundary of the defining region. We introduced t to avoid using x in two different ways.

The Domain of A_f   For which inputs x does the rule $A_f(x) = \int_a^x f$ make sense? Barring bad behavior (e.g., discontinuities) of f, A_f has the same domain as f itself. In particular, $A_f(x)$ makes sense even if x is to the left of a; in this case we use the sign convention ◀ $\int_a^x f = -\int_x^a f$.

We discussed this convention in the preceding section.

The Choice of a   The role of a is to fix one edge (the left edge, in the preceding picture) of the region whose area gives $A_f(x)$. The other edge varies freely. As the definition says, any a in the domain of f is a legal choice. We'll soon see that different choices of a give different (but only slightly different) functions A_f.

EXAMPLE 1   Let $f(x) = 3$ and $a = 0$. Describe the area function $A_f(x) = \int_0^x f(t)\,dt$ for positive inputs x. Does A_f have a simple formula?

Solution   Because $f(x) = 3$, $A_f(x) = 3x$. The first of the following pictures shows why; ◀ in the second, graphs of f and A_f appear together. ◀

A rectangle with height 3 and base x has area $3x$.

For now we ignore negative inputs x; we'll return to them in the next example.

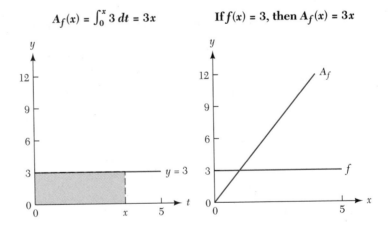

$$A_f(x) = \int_0^x 3\,dt = 3x \qquad\qquad \text{If } f(x) = 3, \text{ then } A_f(x) = 3x$$

Notice too that $A_f(x) = 3x$ is an *antiderivative* of $f(x) = 3$. This result is no accident, as we'll see. ◀

We'll return to this important fact in the next example.

EXAMPLE 2   As before, let $f(x) = 3$ and $a = 0$. Discuss the area function $A_f(x) = \int_0^x f(t)\,dt$ for *negative* values of x. Does the formula $A_f(x) = 3x$ still hold?

Solution   Yes—$A_f(x) = 3x$ for *all* values of x. The key step is a careful look at signs. Here are the pictures:

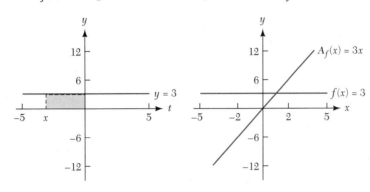

$A_f(x)$ for negative x   **If $f(x) = 3$, then $A_f(x) = 3x$**

To make sense of them, recall from the preceding section how we handle right-to-left integrals. Since $x < 0$,

$$\int_0^x f = -\int_x^0 f = -\text{(shaded area)}.$$

Since $x < 0$ and the rectangle has height 3, $3x$ is, just as claimed, *minus the shaded area*. Combining this example with the preceding one leads to this conclusion:

If $f(x) = 3$ and $a = 0$, then $A_f(x) = \int_0^x f = 3x$ for all x.   ■

A Happy Moral.   What $\int_a^x f$ means geometrically is clear if $x \geq a$. ▶   If $x < a$, the situation looks trickier; the preceding example was all about this possibility. Working Example 2 took some care with plus and minus signs. When the dust settled, though, the happiest possible outcome was revealed:

See the first picture in this section if you've forgotten.

> *The same formula for $A_f(x)$ works for all x, regardless of whether $x < a$ or $x > a$.*

This is evidence that our signed area conventions are not just convenient, but "right."

The same principle holds for any well behaved function f. Consequently, we'll seldom worry about (or even mention) the issue of whether $x < a$ or $x \geq a$.

EXAMPLE 3   Consider the linear function $f(x) = x$ and the area function $A_f(x) = \int_0^x t\, dt$. What's A_f now?

Solution   As the left-hand graph that follows shows, the region defining $A_f(x)$ for a positive input x is a triangle, with base x and height x. ▶   The formula for A_f is therefore

We skip the case of negative x, for reasons outlined earlier.

$$A_f(x) = \frac{\text{base} \times \text{height}}{2} = \frac{x^2}{2}.$$

Graphs of $f(x)$ and $A_f(x)$ appear together at the right.

$$A_f(x) = \int_0^x t\, dt = x^2/2 \qquad\qquad \textbf{If } f(x) = x,\textbf{ then } A_f(x) = x^2/2$$

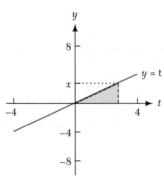

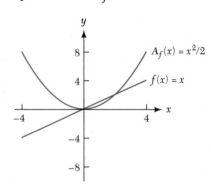

This is still no accident. Stay tuned.

Again, A_f turns out to be an antiderivative of f. ◄   ■

EXAMPLE 4   (Another f, same a)   Consider the new linear function $f(x) = 2 - x$. Find the new area function $A_f(x) = \int_0^x f$. Is A_f again an antiderivative of f?

Notice the different shadings above and below the t-axis.

Solution   The usual area picture applies; area below the x-axis counts as negative. ◄

$$A_f(x) = \int_0^x (2 - t)\, dt \qquad\qquad \begin{array}{c}\textbf{If } f(x) = 2 - x,\\ \textbf{then } A_f(x) = 2x - x^2/2\end{array}$$

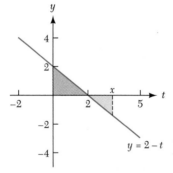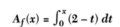

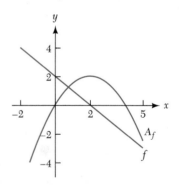

With a little care, one can use triangles (or trapezoids, if $x < 2$) to find the net shaded area. On the left-hand of the two preceding graphs, the upper triangle has area 2 and the lower triangle has area $(x - 2)^2/2$. Thus, the net area for given x is ◄

Check the algebra.

$$A_f(x) = 2 - \frac{(x - 2)^2}{2} = 2x - \frac{x^2}{2}.$$

The exercises at the end of this section pursue this point further.

[**NOTE:** The picture shows a value of x greater than 2. If $0 < x < 2$, the picture is different but the resulting formula is the same.] ◄   The graphs of both f and A_f appear at the right.

Another, perhaps easier, approach to the problem is to use additive properties of the integral ▶ and ideas from earlier examples:

As outlined in Section 5.1.

$$A_f(x) = \int_0^x (2-t)\, dt = \int_0^x 2\, dt - \int_0^x t\, dt = 2x - \frac{x^2}{2}.$$

Either way, the point is the same:

In all examples so far, the area function A_f is an antiderivative of f.   ■

EXAMPLE 5   (Same *f*, another *a*)   Consider the same linear function, $f(x) = 2 - x$, this time setting $a = 1$ so that $A_f(x) = \int_1^x f$. What's the area function now?

Solution   The only difference from the previous example concerns the lower endpoint of integration. Thus, the new function A_f differs from the old one only by the *area of the trapezoidal region* from $x = 0$ to $x = 1$. ▶ In symbols,

This area is 3/2—convince yourself.

$$A_f(x) = \int_1^x f(t)\, dt = \int_0^x f(t)\, dt - \int_0^1 f(t)\, dt$$

$$= 2x - \frac{x^2}{2} - \frac{3}{2}.$$

The graphs support this result:

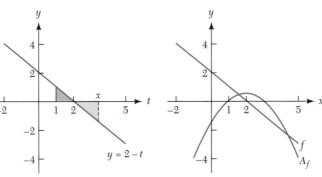

$$A_f(x) = \int_1^x (2-t)\, dt$$

If $f(x) = 2 - x$ and $a = 1$,
then $A_f(x) = 2x - x^2/2 - 3/2$

Yet again, A_f is an antiderivative of f; ▶ this time, it's the one for which $A_f(1) = 0$.   ■

Two antiderivatives of the same f can differ only by a constant, as they do here.

In each of the preceding examples we used an elementary area formula to find an explicit algebraic expression for the area function. In the next example, no such simple formula is available.

EXAMPLE 6   Let $f(x) = 1/x$ and $A_f(x) = \int_1^x f(t)dt$. Discuss the area function A_f. Does it look familiar?

Solution   The graph that follows on the left shows $f(x)$; the shaded area represents $A_f(3)$.

Graph of $A_f(x)$

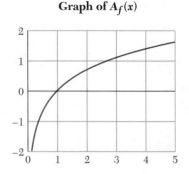

Graph of $f(x) = 1/x$

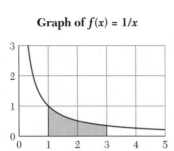

Note that $A_f(x) < 0$ for $x < 1$. Do you see why?

Values of A_f can be estimated by use of the grid. The shaded area, for example, seems to be a little more than one unit, so $A_f(3)$ is a little more than 1. By plotting such estimates we can produce a rough graph of A_f, such as the one at upper right. ◄

The graph of A_f should look familiar; in fact, it shows the natural logarithm function. We'll explain this striking result in the next section. ∎

Properties of A_f : What the Examples Show

In each of the preceding examples, direct calculation showed the area function A_f to be an antiderivative of f. This was no accident:

> **Theorem 2** (The Fundamental Theorem of Calculus, Informal Version) For any well-behaved function f and any base point a, A_f is an antiderivative of f.

In the next section we'll discuss and prove this theorem, which is one of the most important in mathematics. In the rest of this section, we continue to amass evidence for it.

To do so, we list various properties of A_f, each following directly from the definition. Added to the earlier concrete examples, these properties build a strong case for the fundamental theorem of calculus.

Properties of A_f

Let f be a continuous function, a a point of its domain, and $A_f(x) = \int_a^x f(t)\,dt$. Then:

- $A_f(a) = 0$.
- Where f is positive, A_f is increasing.
- Where f is negative, A_f is decreasing.
- Where f is zero, A_f has a stationary point.
- Where f is increasing, A_f is concave up.
- Where f is decreasing, A_f is concave down.

Use, for example, the graphs in Example 5 to convince yourself.

Examples—especially in graphical form—are the best evidence for each claim. ◄ Each claim restates or sharpens the general principle of the fundamental theorem.

BASICS

1. Let $f(x) = x$ and $A_f(x) = \int_0^x f(t)\,dt$. It was shown in the text that $A_f(x) = x^2/2$ if $x > 0$. Show that this formula is correct even if $x < 0$.

2. Let $F(x) = \int_0^x f(t)\,dt$, $G(x) = \int_1^x f(t)\,dt$, and $H(x) = \int_{-2}^x f(t)\,dt$.

   (a) Suppose that $f(x) = 1$. Find symbolic expressions for the functions F, G, and H. Are F, G, and H antiderivatives of f?

   (b) Suppose that $f(x) = 3x$. Find symbolic expressions for the functions F, G, and H. Are F, G, and H antiderivatives of f?

   (c) Suppose that $f(x) = 3x + 1$. Find symbolic expressions for the functions F, G, and H. Are F, G, and H antiderivatives of f?   [**HINT:** Use your answers to parts (a) and (b).]

3. Let $F(x) = \int_0^x f(t)\,dt$, $G(x) = \int_2^x f(t)\,dt$, and $H(x) = \int_{-1}^x f(t)\,dt$.

   (a) Suppose that $f(x) = a$, a positive constant. Find symbolic expressions for the functions F, G, and H. Are F, G, and H antiderivatives of f?

   (b) Suppose that $f(x) = bx$, where b is a positive constant. Find symbolic expressions for the functions F, G, and H. Are F, G, and H antiderivatives of f?

   (c) Suppose that $f(x) = a + bx$, where a and b are positive constants. Find symbolic expressions for the functions F, G, and H. Are F, G, and H antiderivatives of f?

4. Let a be a constant, $f(x) = 2 - x$, and $A_f(x) = \int_a^x f(t)\,dt$.
   (a) Evaluate $A_f(a)$.
   (b) Find a symbolic expression for $A_f(x)$.
      [**HINT:** $\int_a^x (2 - t)\,dt = \int_0^x (2 - t)\,dt - \int_0^a (2 - t)\,dt$.]
   (c) Show that A_f is an antiderivative of f.
   (d) On the same axes, plot f and A_f for $a = 0$, $a = 2$, and $a = -1$.

5. Let $F(x) = \int_a^x f(t)\,dt$ and $G(x) = \int_b^x f(t)\,dt$, where a and b are constants, and f is a continuous function. Show that $G(x) = F(x) + C$, where C is a constant.

6. Suppose that F is an antiderivative of a differentiable function f.
   (a) If F is increasing on $[a, b]$, what is true about f?
   (b) If f is negative on $[a, b]$, what is true about F?
   (c) If f' is positive on $[a, b]$, what is true about F?
   (d) If F is concave down on $[a, b]$, what is true about f'?
   (e) Suppose that G is another antiderivative of f. What relationship exists between F and G?

FURTHER EXERCISES

7. Let $f(x) = \sin x$ and $A_f(x) = \int_0^x f(t)\,dt$.
   (a) It can be shown that $A_f(\pi/2) = 1$. Use this fact (and symmetry) to evaluate $A_f(0)$, $A_f(\pi)$, $A_f(3\pi/2)$, $A_f(2\pi)$, $A_f(-\pi/2)$, $A_f(-\pi)$, $A_f(-3\pi/2)$, and $A_f(-2\pi)$.
   (b) Explain why A_f is a periodic function.   [**HINT:** Use the fact that f is 2π periodic.]
   (c) Show that $0 \le A_f(x) \le 2$ for all x.
   (d) Sketch graphs of f and $A_f(x)$ on the same axes over the interval $[-2\pi, 2\pi]$.

8. Use the fact that $\displaystyle\int_0^{\pi/2} \cos x\,dx = 1$ to show that $-1 \le \displaystyle\int_0^x \cos t\,dt \le 1$.

9. Suppose that f is a continuous function and that $\int_0^x f(t)\,dt = \sin(x^2)$.

   (a) Show that $\displaystyle\int_{\sqrt{\pi/2}}^x f(t)\,dt = \sin(x^2) - 1$.

   (b) Find an expression for $\displaystyle\int_{-\sqrt{3\pi/2}}^x f(t)\,dt$.

10. Let $f(x) = 2 - |x|$ and $A_f(x) = \int_0^x f(t)\,dt$.
    (a) Find a symbolic expression for A_f.   [**HINT:** Write A_f as a piecewise-defined function.]
    (b) Plot f and A_f on the same axes.
    (c) Where is A_f increasing? What is true about f on these intervals?
    (d) Where is A_f decreasing? What is true about f on these intervals?
    (e) Where does A_f have local extrema? What is true about f at these points?
    (f) Where is A_f concave up? What is true about f on these intervals?
    (g) Where is A_f concave down? What is true about f on these intervals?
    (h) Does A_f have any inflection points? If so, where are they located? What is true about f at these points?
    (i) How would your answers to parts (c)–(h) change if the definition of A_f were changed to $A_f(x) = \int_3^x f(t)\,dt$?

11. Let $A_f(x) = \int_0^x f(t)\,dt$ where f is the function graphed below.

Graph of f

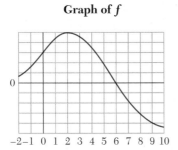

−2−1 0 1 2 3 4 5 6 7 8 9 10

(a) Which is larger: $A_f(1)$ or $A_f(5)$? Justify your answer.

(b) Which is larger: $A_f(7)$ or $A_f(10)$? Justify your answer.

(c) Which is larger: $A_f(-2)$ or $A_f(-1)$? Justify your answer.

(d) Where is A_f increasing?

(e) Explain why A_f has a stationary point at $x = 6$. Is this a local maximum or a local minimum?

(f) Let $F(x) = \int_{-2}^x f(t)\,dt$. Explain why $A_f(x) = F(x) + C$ where C is a negative constant.

(g) Rank the five numbers 0, $A_f(-1)-A_f(-2)$, $A_f(0)-A_f(-1)$, $A_f(1)-A_f(0)$, and $A_f(2)-A_f(1)$ in increasing order.

(h) The five numbers $A_f(4) - A_f(3)$, $A_f(5) - A_f(4)$, $A_f(6) - A_f(5)$, $A_f(7) - A_f(6)$, and $A_f(8) - A_f(7)$ can each be interpreted as the slope of a secant line of A_f. What do these values suggest about the concavity of A_f on the interval $[3, 8]$?

12. Let $A_f(x) = \int_c^x f(t)\,dt$, where c is a constant and f is a continuous function.

(a) Prove that if f is positive on $[a, b]$, then A_f is increasing on $[a, b]$. [**HINT:** Let y and z be numbers such that $a \le y < z \le b$. Compare $A_f(y)$ and $A_f(z)$.]

(b) Use part (a) to show that if f is negative on $[a, b]$, then A_f is decreasing on $[a, b]$. [**HINT:** The function g defined by $g(x) = -f(x)$ is positive wherever f is negative.]

(c) Use parts (a) and (b) to show that A_f has a local maximum or a local minium wherever f changes sign.

13. Suppose that $G(x) = \int_1^x g(t)\,dt$. Evaluate each of the following expressions in terms of G.

(a) $\displaystyle\int_1^3 g(u)\,du$

(b) $\displaystyle\int_0^1 g(x)\,dx$

(c) $\displaystyle\int_{-2}^2 g(t)\,dt$

14. Suppose that $F(t) = \int_{-1}^t \sqrt{1 + x^4}\,dx$. Evaluate each of the following in terms of F.

(a) $\displaystyle\int_{-1}^4 \sqrt{1 + u^4}\,du$

(b) $\displaystyle\int_0^{-1} \sqrt{1 + z^4}\,dz$

(c) $\displaystyle\int_{-2}^3 \sqrt{1 + t^4}\,dt$

15. Let $g(x) = \int_1^x f(t)\,dt$ and $h(x) = \int_3^x f(t)\,dt$ where f is a function with the following properties:

(i) $f'(x) \ge 0$ when $0 \le x \le 3$ and $6 \le x \le 7$; $f'(x) \le 0$ when $3 \le x \le 6$.

(ii) $f''(x) \le 0$ when $0 \le x \le 5$; $f''(x) \ge 0$ when $5 \le x \le 7$.

(iii) f has the values shown below

x	0	1	2	3	4	5	6	7
$f(x)$	0	3	5	6	4	0	−2	0

(a) Sketch a graph of f.

(b) Evaluate $g(1)$.

(c) Explain why $h(x) < g(x)$ when $0 \le x \le 7$.

(d) Rank the four values $h(0)$, $g(3)$, $h(4)$, and $g(7)$ in increasing order.

(e) How many roots does g have in the interval $[0, 7]$? Justify your answer.

(f) How many roots does h have in the interval $[0, 7]$? Justify your answer.

16. Let $A_f(x) = \int_c^x f(t)\,dt$, where c is a constant and f is a continuous function.

(a) Prove that if f is increasing on $[a, b]$, then A_f is concave up on $[a, b]$. [**HINT:** A function g is concave up on an interval if and only if $\big(g(x) - g(a)\big)/(x - a) < \big(g(b) - g(a)\big)/(b - a)$ for every a, b, and x in the interval such that $a < x < b$.]

(b) Use part (a) to show that if f is decreasing on $[a, b]$, then A_f is concave down on $[a, b]$.

(c) Suppose that f has a local maximum at a. Explain why A_f has an inflection point at a.

17. Let $G(x) = \int_{-3}^x f(t)\,dt$ and $H(x) = \int_2^x f(t)\,dt$ where f is the function graphed on the next page. [**NOTE:** The graph of f is made up of straight lines and a semicircle.]

(a) How are the values of $G(x)$ and $H(x)$ related? Give a geometric explanation of this relationship.

(b) On which subintervals of $[-5, 5]$, if any, is H increasing?

(c) Explain why G has a local minimum at $x = 1$.

(d) Where in the interval $[-5, -2]$ does G achieve its minimum value? Its maximum value? What are these values?

(e) Where in the interval $[-5, 5]$ does H achieve its minimum value? Its maximum value? What are these values?

(f) On which subintervals of $[-5, 5]$, if any, is G concave down?

(g) Where does H have points of inflection?

(c) Let $f(x) = \sqrt{1 - x^2}$ and let $A_f(x) = \int_0^x f(t)\,dt$. Show that A_f is an antiderivative of f.

Graph of f

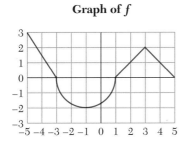

Graph of $y = \sqrt{1 - t^2}$

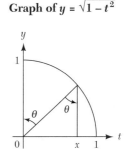

18. Let $F(x) = \int_a^x f(t)\,dt$ and $G(x) = \int_a^x g(t)\,dt$, where f and g are continuous functions. Suppose that $F(a) = G(a)$ and $f(x) \le g(x)$ when $x \ge a$. Show that $F(x) \le G(x)$ when $x \ge a$.

19. Let $F(x) = \int_a^x f(t)\,dt$ and $G(x) = \int_b^x g(t)\,dt$, where f and g are continuous functions. Suppose that $F(c) = G(c)$ and $f(x) \le g(x)$ when $x \ge c$. Show that $F(x) \le G(x)$ when $x \ge c$.

20. (a) Use the figure to show that $\int_0^x \sqrt{1 - t^2}\,dt = \frac{1}{2}x\sqrt{1 - x^2} + \frac{1}{2}\arcsin x$ if $0 \le x \le 1$.

   [**HINT:** The area of a circular sector of radius r and angle θ is $r^2\theta/2$.]

   (b) Show that $\int_0^x \sqrt{1 - t^2}\,dt = \frac{1}{2}x\sqrt{1 - x^2} + \frac{1}{2}\arcsin x$ if $-1 \le x \le 0$.

21. Let $f(x) = \sqrt{1 - x^2}$ and let $A_f(x) = \int_{-1/2}^x f(t)\,dt$.

   (a) Use the result stated in part (a) of the previous exercise to find a symbolic expression for A_f.

   (b) Is A_f an antiderivative of f? Justify your answer.

22. Let $a > 0$ be a constant. Use a sketch similar to that given in Exercise 20 to find an expression for $\int_0^x \sqrt{a^2 - t^2}\,dt$ when $-a \le x \le a$. [**HINT:** Check that your answer produces the correct value when $x = a$.]

23. Let $f(x) = x^2$ and $A_f(x) = \int_0^x f(t)\,dt$.

   (a) Use the result of Exercise 54 in Section 5.1 to find a symbolic expression for $A_f(x)$.

   (b) Is A_f an antiderivative of f? Justify your answer.

24. Let $f(x) = x^2$ and $A_f(x) = \int_{-3}^x f(t)\,dt$.

   (a) Use the result of Exercise 54 in Section 5.1 to find a symbolic expression for $A_f(x)$.

   (b) Is A_f an antiderivative of f? Justify your answer.

5.3 The Fundamental Theorem of Calculus

What the Theorem Says

In the last section we defined and studied the area function A_f, built from an original function f and a base point a. All the evidence so far points to this conclusion:

   A_f is an antiderivative of f.

The formal statement of this fact is the most important theorem of our subject.

Here it is, complete with mathematical fine print:

Theorem 3 (The Fundamental Theorem of Calculus, Formal Version)   Let f be a continuous function, defined on an open interval I containing a. The function A_f with rule

$$A_f(x) = \int_a^x f(t)\, dt$$

is defined for every x in I, and

$$\frac{d}{dx}\left(A_f(x)\right) = f(x).$$

We'll say much more about the FTC (our shorthand) and provide a proof at the end of this section. First, some remarks.

Why Continuous?   Requiring f to be continuous assures that the integral $\int_a^x f$ exists. If f is discontinuous (e.g., blows up somewhere), A_f may have problems, too.

Why Is I Open?   Working on an open interval I is a technical convenience. Doing so avoids troubles that might occur if the domain of f had gaps and endpoints.

In Other Symbols   The last equation of the FTC is sometimes written as follows:

$$\frac{d}{dx}\left(\int_a^x f(t)\, dt\right) = f(x).$$

Which form is clearer?

This form avoids the symbol A_f but requires more careful unpacking. ◄

EXAMPLE 1   Let $f(x) = 2x \sin(x^2)$ and let $A_f(x) = \int_0^x f(t)\, dt$. Find a symbolic formula for A_f; interpret results graphically.

Solution   For the simple functions of the last section, we used elementary area formulas (for rectangles, triangles, trapezoids, and so on) to find explicit formulas for A_f. Here, simple formulas aren't enough. What, for instance, is the exact value of $A_f(2)$, shown shaded in the following figure?

Graph of $f(x) = 2x \sin(x^2)$; $A_f(2)$ is net shaded area

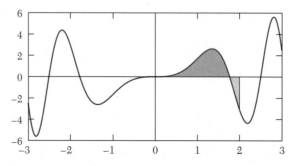

The FTC solves our problem. It says that A_f is *some* antiderivative of f; all we need do is find the right one.

Notice first that for any constant C, $F(x) = -\cos(x^2) + C$ is an antiderivative of $f(x) = 2x\sin(x^2)$. ▶ *Every* antiderivative of f—including A_f—has this form. ▶ Therefore, for some constant C,

Differentiate to see for yourself.

See the Fact that follows this example.

$$A_f(x) = -\cos(x^2) + C.$$

To choose the right value of C, we use the fact that $A_f(0) = 0$. ▶ In other words,

Why? Because $\int_0^0 f = 0$.

$$0 = A_f(0) = -\cos 0 + C \implies C = 1.$$

We've found our formula: $A_f(x) = -\cos(x^2) + 1$.

The following graphs agree; as it should, the A_f graph seems to describe the growth of signed area based at 0.

**If $f(x) = 2x\sin(x^2)$,
then $A_f(x) = -\cos(x^2) + 1$**

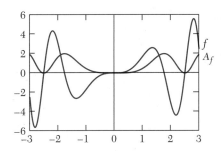

Notice, finally, that $A_f(2) = -\cos(4) + 1 \approx 1.65364$. ▶   ■

Is this result reasonable, given the general size of units?

Finding All Antiderivatives of a Function

In Example 1 we claimed that any two antiderivatives of the function $f(x) = 2x\sin(x^2)$ differ only by an additive constant. This important principle deserves specific mention:

> **Fact** Suppose that $F'(x) = G'(x)$ for all x in an interval I. Then, for some constant C, $F(x) = G(x) + C$ for all x in I.

The Fact means, in practice, that finding *one* antiderivative for a function f on an interval I is tantamount to finding *all* antiderivatives: If $F(x)$ is *any* antiderivative, then every other antiderivative is of the form $F(x) + C$. (We gave more details on this Fact in Section 4.11.)

Continuous Functions Have Antiderivatives.   The fundamental theorem guarantees (among other things) that every continuous function f on I *has* an antiderivative on I—namely, the area function A_f.

For ordinary, well-behaved functions of elementary calculus, this is no great surprise. For naughtier functions, such as the one that follows, the situation is murkier.

**Does $f(x) = x \sin(1/x)$
have an antiderivative?**

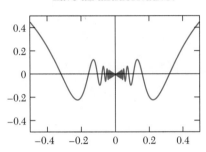

The fundamental theorem says that, naughty or not, f *has* an antiderivative. (*Finding* one in symbolic form may be difficult or impossible; the FTC offers no help there.)

Why the FTC Is Fundamental

The FTC deserves its high-sounding name for both theoretical and practical reasons. It's fundamental *theoretically* because it connects the two main concepts of the calculus: the derivative and the integral. Each is a sort of inverse of the other.

The FTC's *practical* consequences are at least as important; implicitly or explicitly, we'll use them again and again. Most important, the FTC leads to a practical method of calculating specific integrals. ◄

So far, we've relied entirely on elementary area formulas.

Using the FTC

Before proving the FTC, let's use it to prove another theorem, closely related to the FTC, but handier in computations.

> **T h e o r e m 4** (Fundamental Theorem, Second Version)   Let f be continuous on $[a, b]$, and let F be *any* antiderivative of f. Then
>
> $$\int_a^b f(x)\,dx = F(b) - F(a).$$

Proof.   This result follows readily from the original FTC. If F is any antiderivative of f, then $F(x)$ can differ from $A_f(x)$—another antiderivative of f, according to the FTC—only by some constant C. In other words, $F(x) = A_f(x) + C$. But then

$$F(b) - F(a) = \big(A_f(b) + C\big) - \big(A_f(a) + C\big) = A_f(b) - A_f(a)$$

$$= \int_a^b f - \int_a^a f = \int_a^b f,$$

just as claimed.   ■

Calculating Areas with Antiderivatives

EXAMPLE 2   In Section 5.1 we claimed—but didn't show—that the shaded regions N and P have signed areas -0.25 and 0.25, respectively. Show it now.

Graph of $y = x^3$

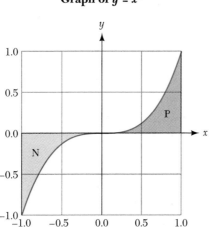

Solution   Let $f(x) = x^3$. The signed areas N and P are given by integrals:

$$\text{Signed area } N = \int_{-1}^{0} f(x)\,dx; \qquad \text{Signed area } P = \int_{0}^{1} f(x)\,dx.$$

Because $F(x) = x^4/4$ is an antiderivative of $f(x)$, Theorem 4 applies:

$$\int_{-1}^{0} f(x)\,dx = F(0) - F(-1) = -\frac{1}{4}; \qquad \text{and} \qquad \int_{0}^{1} f(x)\,dx = F(1) - F(0) = \frac{1}{4}$$

■

Bracket Notation.   The **bracket notation** offers a convenient shorthand for calculating integrals. For any function F,

$$F(x)\Big]_a^b \qquad \text{means} \qquad F(b) - F(a).$$

Watch for brackets in the next example.

EXAMPLE 3   Evaluate $\int_0^1 x^2\,dx$ and $\int_0^1 x^{10}\,dx$. Interpret each integral as an area.

S o l u t i o n   Theorem 4 says that it's enough to find *any* antiderivative, plug in endpoints, and subtract. We do just that:

$$\int_0^1 x^2\,dx = \frac{x^3}{3}\Bigg]_0^1 = \frac{1}{3}; \qquad \int_0^1 x^{10}\,dx = \frac{x^{11}}{11}\Bigg]_0^1 = \frac{1}{11}.$$

Each integral measures an area:

Areas under two curves: $y = x^{10}$ and $y = x^2$

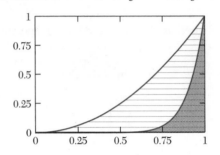

The lightly and darkly shaded areas represent, respectively, the integrals $\int_0^1 x^2\,dx = 1/3$ and $\int_0^1 x^{10}\,dx = 1/11$. The picture shows that the answers are numerically reasonable.   ∎

E X A M P L E 4   Calculate the shaded area in the following picture.

Area under the graph of
$y = 1/x$ from 1 to b

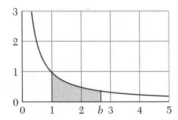

Which value of b makes the area 1?

S o l u t i o n   The shaded area is the value of the integral $\int_1^b 1/x\,dx$. The function $\ln x$ is an antiderivative for $1/x$, so the FTC says that

$$\int_1^b \frac{1}{x}\,dx = \ln x\Big]_1^b = \ln b - \ln 1 = \ln b.$$

If $b = e$ (as in the picture) then the area is $\ln e = 1$.   ∎

From Rate to Amount

Restated slightly, the second version of the fundamental theorem reads like this:

Fact Let f be a well-behaved function on $[a, b]$, with derivative f'. Then

$$\int_a^b f'(t) \, dt = f(b) - f(a).$$

In words,

> *Integrating f' (the rate function) over $[a, b]$ gives the change in f (the amount function) over the same interval.*

This fact has important practical uses; with it, we can deduce *amount* information from *rate* information.

EXAMPLE 5 Demand for electric power varies more or less predictably over the course of a day. Drawing on experience, engineers in a mythical small town (a bedroom suburb, perhaps) use the formula

$$r(t) = 4 + \sin(0.263t + 4.7) + \cos(0.526t + 9.4)$$

to model the rate of power consumption, in megawatts, t hours after midnight. A graph of r follows. Is the model reasonable? What does the shaded area represent?

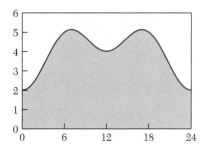

Rate of power use (megawatts) over 24 hours

Solution The graph looks believable (though the model is certainly oversimplified). Residential power demand peaks in the morning, declines during working hours, and rises again in the evening. The formula for r is complicated, but the ingredients—periodic functions and various constants—look right. ▶

The shaded area is the integral $\int_0^{24} r(t) \, dt$. To understand it in the present context, let $a(t)$ be the total *amount* of power consumed up to time t, starting at

Remember how additive and multiplicative constants affect periodic functions?

By definition, therefore,
a = 0 when t = 0.

midnight. ◄   Then $a'(t) = r(t)$; by the preceding Fact,

$$\int_0^{24} r(t)\, dt = a(24) - a(0) = a(24) = \text{total consumption, all day.}$$

With the "classic" FTC (and, from earlier information, a little help in finding an antiderivative) we can calculate this amount. It's easy to check (but harder to guess) that

$$4t - \frac{\cos(0.263t + 4.7)}{0.263} + \frac{\sin(0.526t + 9.4)}{0.526}$$

is an antiderivative for $r(t)$. The rest is straight calculation:

$$\int_0^{24} r(t)\, dt = 4t - \frac{\cos(0.263t + 4.7)}{0.263} + \frac{\sin(0.526t + 9.4)}{0.526} \Bigg]_0^{24}$$

$$\approx 95.781 \text{ megawatt hours.} \qquad\blacksquare$$

When *Not* to Use the FTC

Beautiful, too.

The FTC, useful ◄ as it is, won't solve all our integral problems. Evaluating integrals by antidifferentiation requires first finding a usable antiderivative. For a surprising number of integrands, even apparently simple ones, finding an antiderivative formula is hard or even impossible. None of the following functions, for instance, has an elementary antiderivative:

$$\sin(x^2); \qquad \frac{x}{\ln x}; \qquad 3 + \sin x + 0.3\arcsin(\sin(7x)).$$

Nevertheless, all have perfectly sensible integrals over finite intervals. The following picture shows what $\int_0^{10} f$ means for the last of the preceding functions.

An FTC-resistant integral:

$$\int_0^{10}(3 + \sin x + 0.3\arcsin(\sin(7x)))\, dx$$

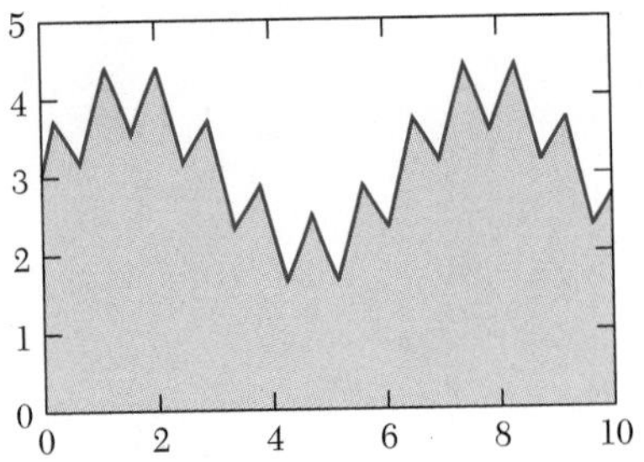

Can you estimate it? We get
about 30, looking at the
graph.

With or without an antiderivative formula, the integral certainly exists. ◄   Soon we'll study techniques (that *don't* involve antiderivatives) for estimating integrals like the preceding one.

A Useful Notation for Antiderivatives—and a Caution

The **indefinite integral notation** for antiderivatives, as in

$$\int x^2\,dx = \frac{x^3}{3} + C \qquad \text{and} \qquad \int \cos x\,dx = \sin x + C$$

can be a convenient shorthand. It reflects, too, the fundamental connection between antiderivatives and integrals. A little care is necessary, however.

The Whole Family $\int f(x)\,dx$ denotes a family of functions, one for each value of C.

Similar Signs, Different Meanings The **definite integral** $\int_a^b f(x)\,dx$ and the **indefinite integral** $\int f(x)\,dx$ have very different meanings: The first is a *number*, and the second is a family of *functions*.

Proving the FTC

The FTC is clearly useful. By now, we hope, it's plausible. But why is it *true*?

The Idea of the Proof

Recall the setup: f is a continuous function, defined on an interval I containing $x = a$, and

$$A_f(x) = \int_a^x f(t)\,dt.$$

The theorem asserts that, for x in I,

$$A_f'(x) = \lim_{h\to 0} \frac{A_f(x+h) - A_f(x)}{h} = f(x).$$

We'll work directly with the difference quotient to prove this.

Notice first that $A_f(x+h) - A_f(x) = \int_a^{x+h} f - \int_a^x f = \int_x^{x+h} f$. The following pictures show why.

$A_f(x + h) - A_f(x)$ = the shaded area

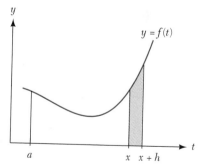

$A_f(x + h) - A_f(x)$ — a closer look

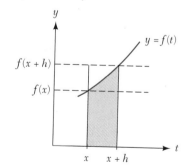

This second picture shows that, for small positive values of h,

$$A_f(x+h) - A_f(x) = \int_x^{x+h} f \approx f(x) \cdot h.$$

(The last quantity is the area of the shorter rectangle.) Thus, for h near 0,

$$\frac{A_f(x+h) - A_f(x)}{h} \approx \frac{f(x) \cdot h}{h} = f(x).$$

Therefore, as $h \to 0$, the difference quotient tends to $f(x)$, as desired. ■

More on the Proof: Fine Print

A rigorous proof can be constructed from the preceding informal argument. The missing ingredient is a precise approach to the approximation $\int_x^{x+h} f \approx f(x) \cdot h$. It helps if f happens to be increasing near x—as it is in the preceding figure. In that case, ◄

Check each step!

$$f(x) \cdot h \le A_f(x+h) - A_f(x) = \text{shaded area} \le f(x+h) \cdot h,$$

so

$$f(x) \le \frac{A_f(x+h) - A_f(x)}{h} \le f(x+h).$$

As $h \to 0$, the quantities on both left and right tend to $f(x)$. So, therefore, must the difference quotient. (A similar idea works if f is not increasing, although some technical details differ.)

BASICS

1. Let $F(x) = \int_0^x f(t)\,dt$ where f is the function whose graph is shown below.

(a) Which of the following is the graph of F? Justify your answer.

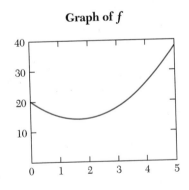

Graph of f

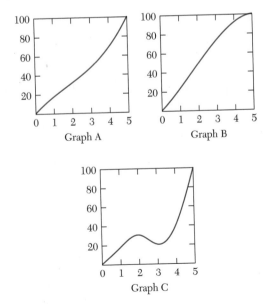

Graph A

Graph B

Graph C

(b) Suppose that $g(0) = 2$ and $g'(x) = f(x)$ for all x. Explain how the graphs of F and g are related.

2. Let $F(x) = \int_0^x f(t)\,dt$ where f is the function graphed below. (The graph of f is made up of straight lines and a semicircle.)

Graph of f

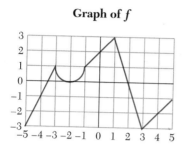

(a) Evaluate $F(-1)$, $F(0)$, and $F(2)$.
(b) Identify all the critical points of F in the interval $[-5, 5]$.
(c) Identify all the inflection points of F in the interval $[-5, 5]$.

(d) What is the average value of f over the interval $[-5, 5]$?
(e) Let $G(x) = \int_0^x F(t)\,dt$. On which subintervals of $[-5, 5]$, if any, is G concave upward?

3. Evaluate each of the following definite integrals using the Fundamental Theorem of Calculus.

(a) $\int_1^4 (x + x^{3/2})\,dx$

(b) $\int_0^\pi \cos x\,dx$

(c) $\int_{-2}^5 \frac{dx}{x + 3}$

(d) $\int_0^b x^2\,dx$

(e) $\int_1^b x^n\,dx \quad [n \neq -1]$

(f) $\int_2^{2.001} \frac{x^5}{1000}\,dx$

(g) $\int_0^{0.001} \frac{\cos x}{1000}\,dx$

(h) $\int_0^{\sqrt{\pi}} x \sin(x^2)\,dx$

FURTHER EXERCISES

4. Let $F(x) = \int_0^x f(t)\,dt$ where $f(t)$ is the function shown below:

Graph of f

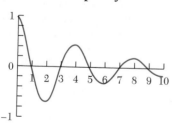

(a) Does $F(x)$ have any local maxima within the interval $[0, 10]$? If so, where are they located?
(b) At what value of x does $F(x)$ attain its minimum value on the interval $[1, 10]$?
(c) On which subinterval(s) of $[0, 10]$, if any, is the graph of $F(x)$ concave up? Justify your answer.

5. Show that if $f(a) \neq 0$, then $\dfrac{d}{dx}\left[\int_a^x f(t)\,dt\right] \neq \int_a^x \left[\dfrac{d}{dt} f(t)\right] dt$.

6. Let $g(x) = \int_1^x f(t)\,dt$ and $h(x) = \int_3^x f(t)\,dt$ where f is a function with the following properties:

(i) $f'(x) \geq 0$ when $0 \leq x \leq 3$ and $6 \leq x \leq 7$; $f'(x) \leq 0$ when $3 \leq x \leq 6$.
(ii) $f''(x) \leq 0$ when $0 \leq x \leq 5$; $f''(x) \geq 0$ when $5 \leq x \leq 7$.
(iii) f has the values shown below

x	0	1	2	3	4	5	6	7
$f(x)$	0	2	4	5	4	0	-2	0

(a) Sketch a graph of f.
(b) Evaluate $g'(4)$.
(c) Where is the graph of g concave down? Justify your answer.
(d) Find the value of x where $g(x)$ is maximum on the interval $[0, 7]$.
(e) Find the value of x where $h(x)$ is minimum on the interval $[0, 7]$.
(f) What is the average value of f' over the interval $[2, 5]$?

7. Let $g(u) = \int_1^u f(x)\,dx$ where f is the function graphed below.

Graph of f

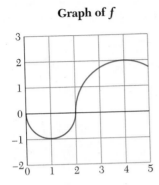

(a) Evaluate $g'(4)$.

(b) Where in the interval $[0, 5]$ is g concave up? Justify your answer.

(c) How many roots does g have in the interval $[0, 5]$? Justify your answer.

(d) Rank the six numbers $-1, 0, 1, g(0), g(2),$ and $g(4)$.

(e) Is the average value of g' over the interval $[0, 3]$ greater than 1? Justify your answer.

8. Suppose that g is a differentiable function, that the graph of g passes through the point $(5, 1)$, and that the derivative of g is the function sketched below. Let $G(x) = \int_0^x g'(t)\,dt$.

Graph of g'

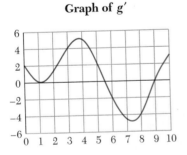

(a) How are the graphs of g and G related? Explain.

(b) Explain why $g(x) = 1 + \int_5^x g'(t)\,dt$.

(c) How many solutions of the equation $g(x) = 0$ exist in the interval $[0, 10]$? Explain.

(d) How many solutions of the equation $G(x) = 0$ exist in the interval $[0, 10]$? Explain.

9. Suppose that $\int_0^x f(t)\,dt = 3x^2 + e^x - \cos x$. Find $f(2)$.

10. Find an equation of the line tangent to the graph of
$$F(x) = \int_1^x \sqrt[3]{t^2 + 7}\,dt \text{ at } x = 1.$$

11. Suppose that $f'(x) = \sqrt[3]{1 + x^2}$ and $f(1) = 0$.

(a) Explain why $f(x) = \int_1^x \sqrt[3]{1 + t^2}\,dt$.

(b) Express $\int_0^3 \sqrt[3]{1 + x^2}\,dx$ in terms of f.

12. Let $F(x) = \int_0^x \sqrt[3]{1 + t^2}\,dt$. Evaluate $\frac{d}{dx}\left[F(x^2)\right]$.

13. Let f be a function with the following properties:

  (i) $f'(x) = ax^2 + bx$ (iii) $f''(1) = 18$

  (ii) $f'(1) = 6$ (iv) $\int_1^2 f(x)\,dx = 18$

Find an algebraic expression for $f(x)$.

14. Let $f(x) = \begin{cases} -2, & \text{if } x < 0 \\ 1, & \text{if } x \geq 0 \end{cases}$ and

let $F(x) = \int_{-2}^x f(t)\,dt$.

(a) Evaluate $\int_{-1}^1 f(x)\,dx$.

(b) Sketch a graph of $F(x)$ and $f(x)$ on the same axes for x in $[-2, 2]$.

(c) Does $F'(1)$ exist? If so, evaluate it using the definition of the derivative. If not, explain why it doesn't exist.

(d) Does $F'(-1)$ exist? If so, evaluate it using the definition of the derivative. If not, explain why it doesn't exist.

(e) Explain why $F'(0)$ doesn't exist. Why doesn't this contradict the Fundamental Theorem of Calculus?

15. A company is planning to phase out a product because demand for it is declining. Demand for the product is currently 800 units/month and dropping by 10 units/month each month. To maintain customer and employee relations, the company has announced that it will continue to produce the product for one more year. At the present time, the company is producing the product at a rate of 900 units/month and 1680 units of the product are in inventory.

(a) Give expressions for the demand and production rates as functions of time t (take $t = 0$ as the present).

(b) Let $D(t)$ be the demand rate for the product at time t. Explain why $\int_0^t D(s)\,ds$ is the total demand for the product between now and time t.

(c) Give an expression for the inventory at time t. [**HINT:** The amount of the product in inventory is the difference between supply and demand at the end of t months.]

(d) The company would like to reduce production at a constant rate of R units/month, with R chosen so the product inventory is zero at the end of 12 months. Find a suitable R.

16. The function $\ln x$ is sometimes *defined* for $x > 0$ by integration:

$$\ln x = \int_1^x \frac{dt}{t}.$$

Assume $a > 0$ and $x > 0$. Derive the identity $\ln(ax) = \ln a + \ln x$ by carrying out and justifying each of the following steps:

(a) Show that $\left(\int_1^{ax} \frac{dt}{t} \right)' = \frac{1}{x}.$ [**HINT:** Use the Chain Rule.]

(b) Explain why the result in part (a) implies that that $\ln(ax) = \ln x + C$.

(c) By choosing an appropriate x, show that the value of the constant C in part (b) is $\ln a$.

5.4 Approximating Sums: The Integral as a Limit

The integral $\int_a^b f$, as defined so far, is the signed ▶ area of the region bounded by the graph of f from $x = a$ to $x = b$. In the last section, we saw in the fundamental theorem of calculus the crucial inverse relationship between integrals and derivatives. Using the FTC, we can readily calculate any integral for which we can find an antiderivative.

Remember what "signed" means? It's important.

Despite their close relationship, integration and antidifferentiation are not the same thing. For many integrals (even simple-looking ones such as $\int_0^1 \sin(x^2)\,dx$), antidifferentiation is useless, because no usable antiderivative formula exists. Nevertheless, such integrals clearly exist, as graphs show.

In this section we take a new approach to the integral, ▶ defining it this time as a certain limit of "approximating sums." The limit definition offers new and useful perspectives on the integral and what it means. Equally important, it lends itself well—better than the FTC—to the problem of estimating integrals numerically.

The approach is new; the integral is the same!

This section resembles Section 2.5, where we first met the limit definition of the derivative. There, as here, the general problem was to translate an intuitively reasonable geometric concept ▶ into precise analytic language. For both derivative and integral, the idea of limit proves crucial; it links approximations to exact values.

There, the slope of the tangent line; here, the area under a graph.

Estimating Integrals by Approximating Sums

The formal definition of integral as a limit involves some formidable-seeming terminology and notation. The underlying ideas, however, are natural and straightforward, especially when expressed graphically.

To back up this claim, we'll approach the definition via a leisurely example, introducing terms and ideas as we go.

EXAMPLE 1   Part of the graph of $f(x) = x^3 - 3x^2 + 8$ follows; the shaded area represents the integral $\int_{0.5}^{3.5} f(x)\,dx$. ▶ Use the FTC to evaluate the integral *exactly*. Then use approximating sums to *estimate* the same integral.

First, estimate the area roughly, by eye alone. Write your answer in the margin.

What's the area?

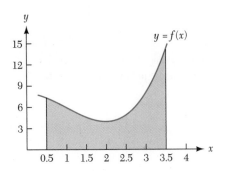

Check the steps; they're routine.

> Solution   With the FTC, an exact answer is easy to find by antidifferentiation. ◀

$$\int_{0.5}^{3.5} \left(x^3 - 3x^2 + 8\right) dx = \left(\frac{x^4}{4} - x^3 + 8x\right)\Bigg]_{0.5}^{3.5} = \frac{75}{4} = 18.75.$$

How might we estimate this area? The following figures suggest four plausible strategies.

Left sum, 10 subdivisions

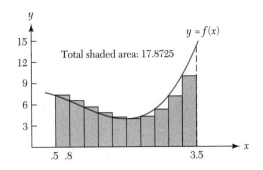

Right sum, 10 subdivisions

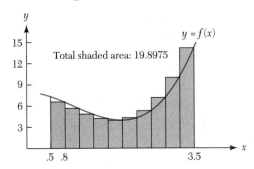

Midpoint sum, 10 subdivisions

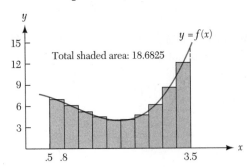

Trapezoid sum, 6 subdivisions

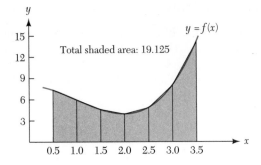

In each case we *approximated* the desired area with a sum of simpler areas—rectangles or trapezoids. (For trapezoids we used six, not ten, subdivisions in order to show more clearly the difference between the integral and the trapezoid estimate.) Adding up the areas of these simpler figures gives, in each case, a natural estimate to the desired area. ▶

Study the pictures; which estimate looks best? Which is best?

Calculating Approximate Areas

How, exactly, did we calculate each of the preceding area estimates?

A close look at each picture gives the answer. Consider first L_{10}, the **left approximating sum with ten subdivisions**. The shaded area consists of ten rectangles; each has base 0.3 (i.e., $(3.5 - 0.5)/10$). The rectangles' *heights* vary: The height of each is the value of the function f at the left edge of the base—hence the name "left sum."

The second rectangle, for example, has the x-interval $[0.8, 1.1]$ for its base; its height is $f(0.8)$, i.e., the value of f at the *left* endpoint. Since $f(x) = x^3 - 3x^2 + 8$, $f(0.8) = 6.592$. Therefore, for the second rectangle,

$$\text{Area} = 0.3 \cdot 6.592 = 1.9776.$$

The *total* area of all ten left rectangles is therefore the sum

$$
\begin{aligned}
L_{10} &= f(0.5) \cdot 0.3 + f(0.8) \cdot 0.3 + f(1.1) \cdot 0.3 + \cdots + f(3.2) \cdot 0.3 \\
&= 7.375 \cdot 0.3 + 6.592 \cdot 0.3 + 5.701 \cdot 0.3 + \cdots + 7.159 \cdot 0.3 + 10.048 \cdot 0.3 \\
&= 17.8725.
\end{aligned}
$$

The **right approximating sum with ten subdivisions**, R_{10}, differs only slightly—the height of each rectangle is the value of f at the *right* endpoint of the base interval. ▶ Hence, the total area of all ten "right" rectangles is the sum

Think this sentence through carefully. Do you see how the picture says the same thing?

$$
\begin{aligned}
R_{10} &= f(0.8) \cdot 0.3 + f(1.1) \cdot 0.3 + f(1.4) \cdot 0.3 + \cdots + f(3.5) \cdot 0.3 \\
&= 6.592 \cdot 0.3 + 5.701 \cdot 0.3 + \cdots + 10.048 \cdot 0.3 + 14.125 \cdot 0.3 \\
&= 19.8975.
\end{aligned}
$$

For M_{10}, the **midpoint approximating sum with ten subdivisions**, heights are evaluated at the *midpoint* of each subinterval. Therefore,

$$
\begin{aligned}
M_{10} &= f(0.65) \cdot 0.3 + f(0.95) \cdot 0.3 + f(1.25) \cdot 0.3 + \cdots + f(3.35) \cdot 0.3 \\
&= 7.007 \cdot 0.3 + 6.150 \cdot 0.3 + 5.226 \cdot 0.3 + \cdots + 11.928 \cdot 0.3 \\
&= 18.6825.
\end{aligned}
$$

The use of a **trapezoid approximating sum** looks, geometrically, like a good idea. Calculating it, moreover, is almost as easy as calculating the rectangular sums. A key idea ▶ simplifies the work: The area of a trapezoid is the average of the areas of the two rectangles determined by its shorter and longer sides. From this

A picture makes the idea clear; draw one.

fact a useful observation follows:

> *The trapezoid approximation with n subdivisions is the average of the corresponding left and right approximations.*

The numerical details are left to you.

For *our* trapezoid sum (we'll call it T_6): ◄

$$T_6 = \frac{1}{2}(L_6 + R_6)$$

$$= \frac{1}{2}(17.4375 + 20.8125) = 19.125.$$

Sigma Notation; Partitions

Sums with many terms (the preceding sums, for instance) are tedious to write. **Sigma (Σ) notation** is more efficient than brute force. Even better, it shows clearly the similarities and the differences among summands.

EXAMPLE 2   Discuss and evaluate each expression:

$$\sum_{k=1}^{4} k; \qquad \sum_{j=1}^{4} j; \qquad \sum_{i=1}^{10} 3i.$$

Solution   By definition, $\sum_{k=1}^{4} k = 1 + 2 + 3 + 4$, or simply 10. Although the second sum is typographically slightly different, it means exactly the same thing: $\sum_{j=1}^{4} j = 1 + 2 + 3 + 4 = 10$. The moral: *The name of the **index variable** ◄ doesn't matter.*

First it was k, then j; i through n are all popular choices.

The third expression is more complicated, but the ideas are the same:

$$\sum_{i=1}^{10} 3i = 3 \cdot 1 + 3 \cdot 2 + 3 \cdot 3 + \cdots + 3 \cdot 9 + 3 \cdot 10$$

$$= 3 \cdot (1 + 2 + \cdots + 10) = 3 \cdot 55 = 165.$$

(Notice that the common factor can be pulled outside the sum: $\sum_{i=1}^{10} 3i = 3 \sum_{i=1}^{10} i$. This is the Σ-version of the distributive law.)   ■

EXAMPLE 3   Use sigma notation to rewrite the left, right, and midpoint approximating sums (L_{10}, R_{10}, and M_{10}) we already calculated.

Solution   All three approximating sums are based (literally!) on subdividing the interval $[0.5, 3.5]$ into ten equal subintervals, each of length $\Delta x = 0.3$. Here are their endpoints, listed in order:

$$0.5 < 0.8 < 1.1 < 1.4 < \cdots < 3.2 < 3.5.$$

(This ordered set of 11 endpoints is called a **partition** of the x-interval $[0.5, 3.5]$.) For convenience, let's name these endpoints, in the same order, $x_0 = 0.5$, $x_1 = 0.8, \ldots, x_{10} = 3.5$.

Now sigma notation is convenient:

$$L_{10} = \sum_{i=0}^{9} f(x_i)\,\Delta x = f(x_0)\Delta x + f(x_1)\Delta x + f(x_2)\Delta x + \cdots + f(x_9)\Delta x$$

$$= f(0.5) \cdot 0.3 + f(0.8) \cdot 0.3 + \cdots + f(3.2) \cdot 0.3;$$

$$R_{10} = \sum_{i=1}^{10} f(x_i)\,\Delta x = f(0.8) \cdot 0.3 + f(1.1) \cdot 0.3 + \cdots + f(3.5) \cdot 0.3;$$

$$M_{10} = \sum_{i=0}^{9} f\left(\frac{x_i + x_{i+1}}{2}\right)\Delta x = f(0.65) \cdot 0.3 + f(0.95) \cdot 0.3 + \cdots + f(3.35) \cdot 0.3.$$

The Σ form makes the approximating sums appear simple and compact. ▶

If we'd rather see numbers than symbols, we can always use the fact ▶ that $x_i = 0.5 + 0.3i$ to rewrite our sums. Here are two of them:

Why does the R_{10} sum run from $i = 1$ to $i = 10$, not from $i = 0$ to $i = 9$?

It is a fact. Do you see why?

$$L_{10} = \sum_{i=0}^{9} f(x_i)\,\Delta x = 0.3 \cdot \sum_{i=0}^{9} f\left(0.5 + 0.3i\right);$$

$$M_{10} = \sum_{i=0}^{9} f\left(\frac{x_i + x_{i+1}}{2}\right)\Delta x = 0.3 \cdot \sum_{i=0}^{9} f\left(0.65 + 0.3i\right). \qquad \blacksquare$$

Getting Better All the Time

It stands to reason that an approximating sum of any type should better approximate an integral as the number of subdivisions increases. The following pictures (still pertaining to $\int_{0.5}^{3.5} f(x)\,dx$) agree.

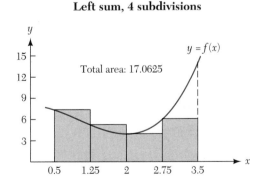

Left sum, 4 subdivisions

Total area: 17.0625

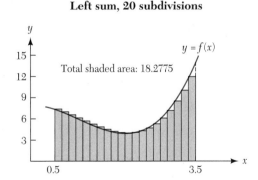

Left sum, 20 subdivisions

Total shaded area: 18.2775

Neither left approximation appears perfect, but the error committed with 20 subdivisions seems much less. Numerical values of various sums (all computed by machine, of course) show the same thing:

		Approximating sums for the integral $\int_{0.5}^{3.5} f(x)\,dx$		
n	Left Sum	Right Sum	Midpoint Sum	Trapezoid Sum
2	17.0625	27.1875	17.0625	22.125
4	17.0625	22.125	18.3281	19.5938
8	17.6953	20.2266	18.6445	18.9609
16	18.1699	19.4355	18.7236	18.8027
32	18.4468	19.0796	18.7434	18.7632
64	18.5951	18.9115	18.7484	18.7533
128	18.6717	18.8299	18.7496	18.7508
256	18.7107	18.7898	18.7499	18.7502

Riemann Sums and the Limit Definition of Integral

Riemann Sums and Their Ingredients

After the German mathematician G. F. B. Riemann (1826–66); he gave the first modern definition of the integral.

Left, right, and midpoint approximating sums are all special types of **Riemann sums**. ◄ Roughly speaking, any "rectangular" approximating sum is a Riemann sum. Riemann's idea of an approximating sum to an integral $\int_a^b f(x)\,dx$ is very liberal. He allows rectangles with unequal bases, and their heights can be determined anywhere—even randomly—within their respective subintervals.

Following, for instance, is the graphical version of a somewhat irregular Riemann sum for our by-now-familiar integral:

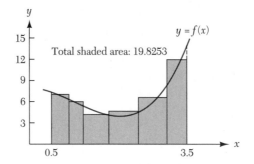

A Riemann sum, 6 (unequal) subdivisions

It may not be a *good* approximation, but it's a legitimate one.

We don't require such generality, but the formal definition of Riemann sum follows, for the record. The necessary ingredients are a function f and an interval $I = [a, b]$.

Definition Let I be partitioned into n subintervals by any $n + 1$ points

$$a = x_0 < x_1 < x_2 < \cdots < x_{n-1} < x_n = b;$$

let $\Delta x_i = x_i - x_{i-1}$ denote the width of the ith subinterval. Within each subinterval $[x_{i-1}, x_i]$, choose any point c_i. The sum

$$\sum_{i=1}^{n} f(c_i)\Delta x_i = f(c_1)\Delta x_1 + f(c_2)\Delta x_2 + \cdots + f(c_n)\Delta x_n$$

is a Riemann sum with n subdivisions for f on $[a, b]$.

Our left, right, and midpoint sums fit this definition with room to spare. Each is built from a **regular partition** of $[a, b]$ (i.e., one with subintervals of equal length) and some simple, consistent, scheme for choosing the **evaluation points** (or "sampling points") c_i. A *trapezoid* approximating sum, on the other hand, isn't quite a Riemann sum, since its approximating figures aren't rectangles. Nevertheless, trapezoidal sums do an excellent job of approximating the integral.

The Integral Defined as a Limit

Graphical intuition suggests that all of the approximating sums—L_n, R_n, M_n, and T_n—should approach the true signed area $\int_{0.5}^{3.5} f(x)\,dx$ as n increases. Numerical evidence ▶ argues irresistibly for a value somewhere near 18.75.

The limit definition of integral makes these ideas precise:

See the preceding table to dispel any lingering doubts.

Definition Let the function f be defined on the interval $I = [a, b]$. The **integral of f over I**, denoted $\int_a^b f(x)\,dx$, is the number to which all Riemann sums S_n tend as n tends to infinity and as the widths of all subdivisions tend to zero. In symbols:

$$\int_a^b f(x)\,dx = \lim_{n \to \infty} S_n = \lim_{n \to \infty} \sum_{i=1}^{n} f(c_i)\,\Delta x_i,$$

if the limit exists.

Here are some comments on the definition.

When Is a Function Integrable? A function for which the limit in the definition exists is called **integrable** on $[a, b]$. In practice, being integrable on $[a, b]$ is easy: Almost every calculus-style function that's defined and *bounded* (i.e., doesn't blow up or down) on $[a, b]$ is integrable. The surprise is the difficulty of finding a *non*integrable function. ▶

Well-Behaved Functions The limit in the definition, taken at face value, is a slippery customer. Understanding every ramification of permitting arbitrary partitions and sampling points, for example, can be tricky. Luckily, the matter is much simpler for the well-behaved functions (e.g., continuous functions) we typically encounter in calculus. For such functions,

Here's one: $f(x) = 1$ if x is rational; $f(x) = 0$ if x is irrational.

almost any respectable sort of approximating sum does what we'd expect—approaches the true value of the integral as n tends to infinity.

Notes on the dx Notation

The limit definition helps explain the otherwise mysterious dx in the notation $\int_a^b f(x)\,dx$. Consider, for example, the case of a *right* approximating sum with n equal subdivisions, each of length $(b-a)/n = \Delta x$. Then, by definition,

$$\int_a^b f(x)\,dx = \lim_{n\to\infty} \sum_{i=1}^n f(x_i)\,\Delta x.$$

Now the left side looks much like the right; dx on the left corresponds in a natural way to Δx on the right. ◄

We've seen this before:
$dy/dx = \lim_{\Delta x\to 0}\Delta y/\Delta x.$

Approximating Sums and Computers

Approximating sums offer simple, effective, and accurate approximations to almost any integral (simple for a computer or calculator, anyway). Calculation of approximating sums *by hand* is impractical except for simple integrands and few subdivisions. Calculation of approximating sums is the sort of "bureaucratic" task computers do best—it's simple, repetitive, and tedious.

BASICS

1. Let g be the function graphed below. Estimate the value of $\int_{-5}^5 g(x)\,dx$ by evaluating left, right, and midpoint sums each with 5 equal subintervals. Then draw sketches that illustrate these sums graphically.

Graph of g

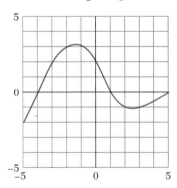

2. Let f be the function graphed to the right.
   (a) Show that $\int_0^8 f(x)\,dx < 45$.
   (b) Is $20 < \int_1^7 f(x)\,dx$? Justify your answer using a picture.

(c) Estimate the value of $\int_1^7 f(x)\,dx$ using a right sum with three equal subintervals.
(d) Estimate the value of $\int_1^7 f(x)\,dx$ using a left sum with three equal subintervals.
(e) Estimate the value of $\int_0^8 f(x)\,dx$ using a right sum with four equal subintervals.
(f) Estimate the value of $\int_0^8 f(x)\,dx$ using a left sum with four equal subintervals.

Graph of f

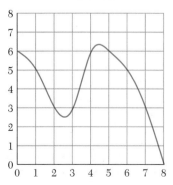

3. Let $I = \int_0^{20} x^2\, dx$. Use sigma notation to write an expression for R_{10}, the right Riemann sum approximation to I with 10 equal subintervals.

4. At time t (measured in hours), $0 \le t \le 24$, a firm uses electricity at the rate of $E(t)$ kilowatts. The power company's rate schedule indicates that the cost per kilowatt-hour at time t is $c(t)$ dollars. Assume that both E and c are continuous functions.

(a) Set up an N-term left sum that approximates the cost of the electricity consumed in a 24-hour period.

(b) What definite integral is approximated by the sum you found in part (a)?

FURTHER EXERCISES

5. Let $I = \int_0^5 \sqrt[3]{2x}\, dx$. Use sigma notation to write the left, right, and midpoint approximations to I. (Assume that $N = 10$ equal subintervals are used.)

6. Let $I = \int_0^5 \sqrt{3x}\, dx$. Use sigma notation to write the left, right, and midpoint approximations to I. (Assume that N equal subintervals are used.)

7. Find a definite integral that is approximated by the right sum $\dfrac{2}{100} \displaystyle\sum_{k=1}^{100} \sin\left(\dfrac{2k}{100}\right)$. What is the value of this integral?

8. Find a definite integral that is approximated by the midpoint sum

$$\frac{2}{10} \sum_{k=1}^{40} \cos\left(\frac{2k-1}{10}\right).$$

What is the value of this integral? [**HINT:** Write out the first few terms of the sum.]

9. The following table contains values of a continuous function f at several values of x.

x	1	2	3	4	5	6
$f(x)$	0.14	0.21	0.28	0.36	0.44	0.54

(a) Estimate $\int_2^5 f(x)\, dx$ using a left sum with 3 equal subintervals.

(b) Repeat part (a) using the trapezoid rule.

(c) Draw sketches that illustrate the computations you did in parts (a) and (b).

(d) Show that $f(2) \cdot 2 + f(4) \cdot 3$ is a Riemann sum approximation to $\int_1^6 f(x)\, dx$. Justify your answer. [**HINT:** Draw a picture.]

10. The rate at which the world's oil is being consumed is increasing. Suppose that the rate (measured in billions

of barrels per year) is given by the function $r(t)$, where t is measured in years and $t = 0$ is January 1, 1990.

(a) Write a definite integral that represents the total quantity of oil used between the start of 1990 and the start of 1995.

(b) Suppose that $r(t) = 32e^{0.05t}$. Find an approximate value for the definite integral from part (a) using a right sum with $n = 5$ equal subintervals.

(c) Interpret each of the five terms in the sum from part (b) in terms of oil consumption.

(d) Evaluate the definite integral from part (a) exactly using the FTC.

11. Evaluate $\lim_{n \to \infty} \frac{1}{n} \sum_{j=1}^{n} \left(\frac{j}{n}\right)^3$ by expressing it as a definite integral and then evaluating this integral using the Fundamental Theorem of Calculus.

12. Evaluate $\lim_{n \to \infty} \frac{2}{n} \sum_{j=1}^{n} \left(\frac{2j}{n}\right)^3$ by expressing it as a definite integral and then evaluating this integral using the Fundamental Theorem of Calculus.

13. (a) Write the sum

$$\frac{5}{2}(2.3)^2 + \frac{5}{6}(2.8)^2 + \frac{5}{12}(3.3)^2 + \frac{5}{20}(3.8)^2$$

using sigma notation. [**HINT:** $2 = 1 \cdot 2$, $6 = 2 \cdot 3$, $12 = 3 \cdot 4$, and $20 = 4 \cdot 5$.]

(b) Is the sum in part (a) a Riemann sum approximation to $\int_0^4 x^2\, dx$? Explain.

14. For each of the following definite integrals, draw a sketch illustrating the approximations L_4, R_4, M_4, and T_4. Then compute each approximation and compare it with the exact value of the integral computed using the FTC. How could you have used the sketches to predict the results of these comparisons?

(a) $\displaystyle\int_{-2}^{2} x^2\, dx$

(b) $\displaystyle\int_{-2}^{2} x^3\, dx$

(c) $\displaystyle\int_{-2}^{2} (x+1)\, dx$

(d) $\displaystyle\int_{-\pi/2}^{\pi/2} \sin x\, dx$

15. The following table gives speedometer readings at various times over a one-hour interval.

Speed Readings Over One Hour							
Time (min)	0	10	20	30	40	50	60
Speed (mph)	42	38	36	57	0	55	51

(a) Draw a plausible speed graph for the one-hour period.

(b) Estimate the total distance traveled using
   (i) A trapezoid approximating sum, 6 subdivisions.
   (ii) A left approximating sum, 6 subdivisions.
   (iii) A midpoint approximating sum, 3 subdivisions.
   Which answer is most convincing? Why?

(c) Draw a plausible distance graph for the one-hour period.

16. Find a definite integral for which $5 \sum_{k=1}^{3} f(5k)$ is
   (a) A left Riemann sum with 3 equal subintervals.
   (b) A right Riemann sum with 3 equal subintervals.
   (c) A midpoint Riemann sum with 3 equal subintervals.

17. Find a definite integral for which $S = 2\big(f(2) + f(4) + f(6) + f(8)\big)$ is
   (a) A left Riemann sum with 4 equal subintervals.
   (b) A right Riemann sum with 4 equal subintervals.
   (c) A midpoint Riemann sum with 4 equal subintervals.

5.5 Approximating Sums: Interpretations and Applications

In the preceding section we defined the integral as a limit of approximating sums. In symbols,

$$\int_{a}^{b} f(x)\,dx = \lim_{n \to \infty} \sum_{i=1}^{n} f(c_i)\,\Delta x_i.$$

The limit involves some subtleties, but the basic idea of sums (left sums, right sums, midpoint sums, and so on) approximating an integral is natural and straightforward, especially if approached graphically.

In this section we examine approximating sums more closely. In the process we'll both see the integral in another light and apply the idea of approximating sums to several problems of measurement.

Different Views of Approximating Sums

Approximating sums are *discrete* (i.e., stepwise) approximations to integrals of *continuously* varying functions. This important relationship can be interpreted in several useful ways. We discuss three.

Approximating Sums and Simpler Areas

The simplest way to view approximating sums is geometric. Geometrically, approximating sums represent simpler versions (usually polygons) of curvy regions. ▶ The following shaded region, for example, might be called a **polygonal approximation** to the area from a to b under the graph of f.

Most of the pictures in this section and the last reflect this point of view.

A polygonal approximation to a curved area

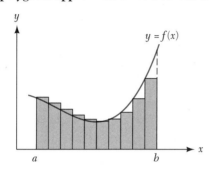

The total area of the polygonal region is the value of a left sum; it's evidently a reasonable approximation to $\int_a^b f$, the true area under the graph.

Approximating Sums and Simpler Functions

Another view of approximating sums for an integral $\int_a^b f$ focuses on the function f rather than on the area under graph. From this point of view, an approximating sum amounts to *replacement of the original function f with a new, simpler function*, one that's linear on each subinterval of the partition in question. In the following graph, for example, the function f of the preceding picture was replaced with a six-step linear approximation.

Replacing f with a piecewise-linear function

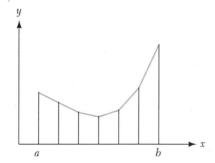

The area beneath is the value of the trapezoid approximating sum T_6.

Weighted Sums

A Riemann sum (left, right, midpoint, or any other kind) for $\int_a^b f$ has the form

$$\sum_{i=1}^n f(c_i)\,\Delta x_i = f(c_1)\Delta x_1 + f(c_2)\Delta x_2 + \cdots + f(c_n)\Delta x_n.$$

An ordinary average is another type of weighted sum.

In other words, an approximating sum is a **weighted sum** ◄ of values of f at various points c_i (the sampling points) in the interval $[a, b]$; the weights are the Δx_i. The exact choice of sampling points and the weights attached depend on the type of approximating sum chosen.

EXAMPLE 1   Let $I = \int_0^1 \sin x \, dx$. Let R_4 and M_4 be the midpoint and right approximating sums for I, each with four subdivisions. What are the sampling points? What are the weights? What are the weighted sums?

Convince yourself that the forms are correct.

Solution   Calculation of M_4 and R_4 is routine: ◄

$$R_4 = \sin(0.25) \cdot 0.25 + \sin(0.5) \cdot 0.25 +$$
$$\sin(0.75) \cdot 0.25 + \sin(1) \cdot 0.25 \approx 0.56248;$$
$$M_4 = \sin(0.125) \cdot 0.25 + \sin(0.375) \cdot 0.25 +$$
$$\sin(0.625) \cdot 0.25 + \sin(0.875) \cdot 0.25 \approx 0.460897.$$

The sampling points for R_4 are 0.25, 0.5, 0.75, and 1—the *right* endpoints of four equal subintervals. For M_4, the sampling points are 0.125, 0.375, 0.625, and 0.875, the midpoints of the same subintervals.

The weights in an approximating sum are the coefficients of the sample values $f(c_i)$. In the preceding sums, every sample value has weight 1/4, the length of the subinterval from which it comes. ◄ This result seems just; there's no obvious reason to favor one value over another. The weighted sums are the numerical values of R_4 and M_4, shown earlier.   ■

Unequal subintervals would give unequal weights.

Applications of Approximating Sums

Approximating sums are good for more than formally defining the integral. Indeed, compared to symbolic integrals, approximating sums are often simpler conceptually, handier in applications, and easier to calculate—especially for machines. For integrals that involve complicated formulas or no formulas at all, ◄ approximating sums may be the only practical resort.

Such integrals are common in practice.

Beyond Formulas—Integration with Tabular Data

Integration is important in the real world. Alas, real-world functions seldom come with handy formulas attached. The FTC, for all its other glories, is not much good without symbolic formulas on which to act. Approximating sums are the answer.

EXAMPLE 2   If $v(t)$ is the speed (in miles per hour) of a car at time t (in hours), then $\int_a^b v(t)\,dt$ = miles traveled from $t = a$ to $t = b$. To measure distance traveled from $t = 0$ to $t = 1$, all we need is the integral $\int_0^1 v(t)\,dt$.

Real cars don't adhere to explicit speed formulas. What we *can* observe, practically speaking, are numerical speed data such as the following ones. (Notice the irregular time intervals; real data sets often have gaps.)

Speed Readings over One Hour										
Time (min)	0	5	10	15	20	30	35	40	45	55	60
Speed (mph)	35	38	36	57	71	55	51	23	10	27	35

How far did the car travel in one hour?

Solution   Without more information (e.g., about speeds *between* the times measured), we'll never know exactly. But we can use approximating sums to estimate $\int_0^1 v(t)\,dt$ as well as possible.

These discrete speed data are mathematically compatible with many possible speed functions. The simplest possibility is drawn by "connecting the dots" (lower left): ▶

We converted time from minutes to hours.

A possible speed function (hrs vs. mph)

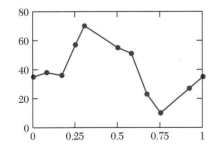

Estimating distance: total area = 40.375

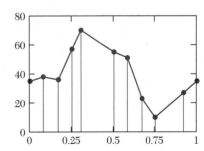

The total bounded area (upper right) is easily computed using ten trapezoids. ▶ We get 40.375—a reasonable estimate for distance traveled over the hour.

The piecewise-linear speed function just plotted isn't the only possibility. *Any* curve through the known data points makes mathematical sense. ▶ The important

We'll spare you the arithmetic.

Some curves are physically more likely than others.

point is that the trapezoid estimate depends only on the 11 observed data, not on the particular curve drawn through them. The following picture illustrates this idea; it shows a new speed function ◄ but the same trapezoid estimate (40.375 miles) for distance.

Better or worse than the old? What do you think?

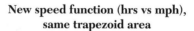

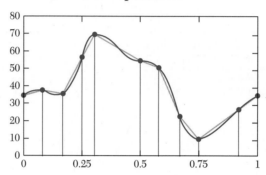

New speed function (hrs vs mph), same trapezoid area

■

Areas in the Plane; Measuring by Slicing

We first defined the integral in terms of area bounded by the graph of one function. Not surprisingly, integrals can also be used to measure areas of more general plane areas, e.g., areas bounded by two or more graphs. Approximating sums show how and why.

EXAMPLE 3   Using an integral, measure the shaded area in the following picture. (It's bounded by the graphs of $y = \sin x$ and $y = \cos x$.)

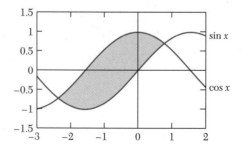

Area between two curves:
$y = \sin x$ and $y = \cos x$

Solution   The following picture suggests an approximate solution. We slice the area into vertical strips, approximate each strip with a rectangle, then add up the areas of the rectangles. Here's the picture for ten rectangles: ▶

The first rectangle has height 0.

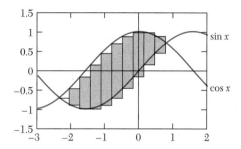

Approximating area with 10 "left-rule" rectangles

Look carefully:

- The curves intersect at $x = -3\pi/4$ and $x = \pi/4$. ▶ (Because $\cos x - \sin x \geq 0$ over the interval $[-3\pi/4, \pi/4]$, the integral gives the true shaded area desired, even though part of the area lies below the x-axis.)

  Do you see why?

- The height of each rectangle is determined at its left edge, $x = x_i$; the height is $\cos(x_i) - \sin(x_i)$.

- The total area of all ten rectangles is L_{10}, the left sum with ten equal subdivisions for the function $\cos x - \sin x$ on the interval $[-3\pi/4, \pi/4]$. (It isn't necessary to solve the problem at hand, but calculating L_{10} is easy by machine; we get 2.80513.)

Geometric intuition says that as $n \to \infty$, the area L_n tends to the exact area we want. By the limit definition of integral, L_n also tends to the integral in question. We conclude:

The area betwen the curves is the integral $\displaystyle\int_{-3\pi/4}^{\pi/4} (\cos x - \sin x)\, dx.$

Finding an exact answer is now easy, thanks to the FTC: ▶

But check the details.

$$\text{Area} = \int_{-3\pi/4}^{\pi/4} (\cos x - \sin x)\, dx$$

$$= \sin x + \cos x \big]_{-3\pi/4}^{\pi/4}$$

$$= 2\sqrt{2} \approx 2.82843.$$

∎

Two Basic Area Formulas

The simplest plane regions for area calculations are bounded above by the graph of one function, below by the x-axis, and on the left and right by vertical lines, as follows:

Area at its simplest

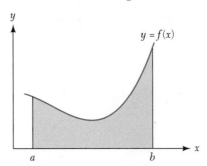

In this simplest case, the integral $\int_a^b f(x)\,dx$ measures the shaded area.

More interesting, but slightly more complicated, regions may be bounded by various types of curves—not necessarily all graphs of functions. Two common types follow. ◀

Some plane regions are of neither type shown. However, many regions, even quite complicated ones, are combinations of the two following types.

Region 1: top and bottom curves   **Region 2: left and right curves**

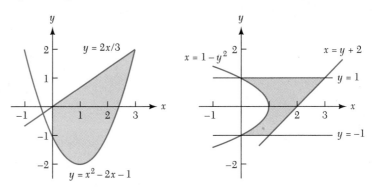

The following rules apply to finding areas such as the preceding ones.

Fact Let f and g be continuous functions.

Integrating in x Let R be the region bounded above by $y = g(x)$, below by $y = f(x)$, on the left by $x = a$, and on the right by $x = b$. Then R has area

$$\int_a^b \big(g(x) - f(x) \big)\,dx.$$

Integrating in y Let R be the region bounded on the right by $x = g(y)$, on the left by $x = f(y)$, below by $y = c$, and above by $y = d$. Then R has area

$$\int_c^d \big(g(y) - f(y) \big)\,dy.$$

For Region 2 in the preceding pictures, and other similar regions, integrating in y may be much simpler.

Using the Rules

Using the rules effectively may require a combination of graphical, algebraic, and symbolic operations. We illustrate by example.

E X A M P L E 4   Find the area of Region 1 in the preceding picture.

S o l u t i o n   The graphs *appear* to intersect at $x = 3$. Indeed they do: By the quadratic formula, ▶

Convince yourself of this!

$$\frac{2}{3}x = x^2 - 2x - 1 \iff x = 3 \text{ or } x = -\frac{1}{3}.$$

(This shows, in fact, that the two graphs intersect also at $x = -1/3$, to the left of the shaded region.) Now, by the first rule (and a close look at the picture):

$$\text{Area} = \int_0^3 \left(\tfrac{2}{3}x - (x^2 - 2x - 1)\right) dx.$$

An easy antiderivative calculation shows the result to be 6.   ■

E X A M P L E 5   Find the area of Region 2 in the preceding picture.

S o l u t i o n   Integration in x seems impractical, since defining upper and lower curves is troublesome. However, a close look at the figure shows that the second form of the preceding Fact applies, as follows:

$$\text{Area} = \int_{-1}^1 \left(y + 2 - (1 - y^2)\right) dy.$$

Another easy integration—this time in the variable y—shows that the area is 8/3.   ■

Any Way You Slice It

Some areas can be calculated by integration either in x or in y.

E X A M P L E 6   Find the area of the region R bounded by the curves $x = 0$, $y = 2$, and $y = e^x$.

Solution   Here's the picture:

Graphs of $y = e^x$, $y = 2$

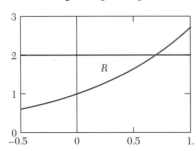

The two curves intersect where $e^x = 2$, i.e., at $x = \ln 2$. Therefore

$$\text{Area of } R = \int_0^{\ln 2} \left(2 - e^x\right) dx.$$

Why is this legitimate?

To find the same area in terms of y, we rewrite the equation $y = e^x$ in the equivalent form $x = \ln y$; ◄ the associated curve becomes the right boundary. Then

$$\text{Area of } R = \int_1^2 \ln y \, dy.$$

Do you recall this? If not, check it by differentiation.

Either integral is easy to calculate. For the second one (recalling that $y \ln y - y$ is an antiderivative of $\ln y$), ◄ we get

$$\text{Area} = y \ln y - y\big]_1^2 = 2 \ln 2 - 1 \approx 0.3863. \quad ■$$

BASICS

1. Let A be the region between the curves $y = x$ and $y = x^2$.
   (a) Draw a sketch of the region A.
   (b) Use 5 left rule rectangles to approximate the area of A.
       (Give a numerical answer.)
   (c) Draw a picture (similar to the one on p. 391) that illustrates the sum you computed in part (b). Use this picture to decide whether your answer overestimates or underestimates the area of A.
   (d) Find the area of A *exactly* by integration.

2. Find the area of the region bounded by the curves $y = \sin x$ and $y = \cos x$, and the lines $x = 1$ and $x = 3$ *exactly* (i.e., express your answer in terms of values of elementary functions rather than as a decimal number).

3. In this exercise, integration in x is used to find the area of Region 2 shown on page 392.
   (a) Find the area of the region bounded below by the x-axis, above by the line $y = 1$, on the right by the line $x = y + 2$, and on the left by the curve $x = 1 - y^2$.
   (b) Find the area of the region bounded above by the x-axis, below by the line $y = -1$, on the right by the line $x = y + 2$, and on the left by the curve $x = 1 - y^2$.
   (c) Find the area of the region bounded by the lines $x = 1$, $x = y + 2$, and $y = 1$.
   (d) Use the results of parts (a), (b), and (c) to find the area of Region 2.

4. This exercise provides yet another approach to finding the area of Region 2 shown on page 392.
   (a) Find the area of the region bounded by the line $x = 0$ and the curve $x = 1 - y^2$.
   (b) Find the area of the region bounded by the lines $x = y + 2$, $y = -1$, and $x = 3$.
   (c) Find the area of the region bounded by the lines $x = 0$, $x = 3$, $y = 1$, and $y = -1$.
   (d) Use the results of parts (a), (b), and (c) to find the area of Region 2.

In Exercises 5–16, sketch the region bounded by the given curves and find the area of the region *exactly*.

5. $y = x^4, \quad y = 1$

6. $y = x, \quad y = x^3$

7. $y = x^2, \quad y = x^3$

8. $y = x^2 - 1, \quad y = x + 1$

9. $x = y^2 - 4, \quad y = 2 - x$

10. $y = \sqrt{x}, \quad y = 0, \quad x = 4$

11. $y = \sqrt{x}, \quad y = x^2$

12. $y = 9(4x + 5)^{-1}, \quad y = 2 - x$

13. $y = 9(4x^2 + 5)^{-1}, \quad y = 2 - x^2$

14. $y = 2 + \cos x, \quad y = \sec^2 x, \quad x = -\pi/4, \quad x = \pi/6$

15. $y = e^x, \quad y = 0, \quad x = 0, \quad x = 1$

16. $y = 2^x, \quad y = 5^x, \quad x = -1, \quad x = 1$

5.6 Chapter Summary

This chapter introduced the **integral**. The derivative and the integral are the two most important concepts of calculus. Remarkably, they're closely related; the **fundamental theorem of calculus** describes how and why.

The Integral as Signed Area.   Section 5.1 introduced the integral graphically, as measuring **signed area**. ("Signed" means that area below the x-axis counts negatively.) The graphical viewpoint, though insufficient for *exact* calculations, explains clearly what the integral is and why some of its elementary properties hold.

The Area Function.   Given a function f and a starting point $x = a$, we defined the related **area function** A_f:

$$A_f(x) = \int_a^x f = \text{signed area bounded by } f \text{ from } a \text{ to } x.$$

Simple examples suggest that the area function is an **antiderivative** for f. This idea, suitably embellished, is the simplest form of the **fundamental theorem of calculus**.

The Fundamental Theorem.   Put more formally, the fundamental theorem says that if A_f is defined as we just stated, then

$$\frac{d}{dx}\left(A_f(x)\right) = f(x).$$

This result connects derivatives and integrals, the two main calculus ideas. For that reason, it deserves to be called *fundamental*.

The same idea, restated in another form, is called the **second version of the fundamental theorem**. If F is *any* antiderivative of f, then

$$\int_a^b f(x)\,dx = F(b) - F(a).$$

In other words,

> *To find $\int_a^b f$, find any antiderivative, plug in endpoints, and subtract.*

The Integral as a Limit of Approximating Sums.   Like the derivative, the integral is defined rigorously as a limit of approximating sums. Section 5.4 described the procedure. Although symbolically messy, approximating sums make good graphical sense. Several varieties were introduced, compared, and calculated.

Approximating Sums and Applications of the Integral.   Thinking of the integral as a limit of sums suggests a variety of applications. Several were presented. The main point is that the integral is capable of various interpretations.

One application of the integral concerns rates and amounts:

If $f(t)$ tells the rate at which a quantity varies at time t, then $\int_a^b f$ tells the amount by which the same quantity changes over the interval from $t = a$ to $t = b$.

Another application of the integral is to calculate areas of plane regions defined by graphs of functions. Such regions come in many shapes and forms. Choosing the best integral to calculate a given area can be a challenging problem. We illustrated various techniques for doing so.

6

Finding Antiderivatives

6.1 Antiderivatives: The Idea

Integrals, Antiderivatives, and the FTC

This chapter is about antidifferentiation—finding, for a given function f, a new function F for which $F' = f$. The fundamental theorem of calculus ▶ (FTC) offers ample motivation for doing so. It says that if f is a continuous function and F is any antiderivative of f, then

See Section 5.3.

$$\int_a^b f(x)\,dx = F(b) - F(a).$$

Finding a nice ▶ antiderivative function F, in effect, translates an integral problem into something easier: plugging a and b into F. A simple example illustrates this principle and its importance.

Here, "nice" means "easy to evaluate."

EXAMPLE 1 Calculate $\displaystyle\int_0^\pi \sin x\,dx$. What does the answer mean graphically?

Solution Any function of the form $F(x) = -\cos x + C$ is an antiderivative of $f(x) = \sin x$. For simplicity, let's use $C = 0$. ▶ By the FTC,

Other values of C give the same answer.

$$\int_0^\pi \sin x\,dx = -\cos x\Big]_0^\pi = -\cos\pi + \cos 0 = 2.$$

Geometrically, the integral represents the area under one arch (from $x = 0$ to $x = \pi$) of the sine function. Surprisingly, this area measures exactly 2 square units. ∎

Finding Antiderivatives

To use the FTC, we need first to find an antiderivative function F. Must there *be* an antiderivative? If so, how many antiderivatives are there?

It's reasonable to do so. If f is very badly behaved, $\int_a^b f$ may not even exist.

Existence.   If f is continuous (we'll assume so throughout this chapter) ◄ , the short answer—but not the whole answer—is yes: f does have an antiderivative. In fact, the other version of the FTC says that the area function

$$A_f(x) = \int_a^x f(t)\,dt$$

is an antiderivative of f.

It's nice to know in the abstract that every continuous function f has an antiderivative F. A better question for us, however, is whether f has an antiderivative function that we can use. To be useful for evaluating integrals, an antiderivative F must have a concrete *formula*, one that accepts numerical inputs and produces numerical outputs.

Most recently in Example 1. See also Section 5.3.

See the Fact on page 367.

How Many Antiderivatives?   As we've seen repeatedly, ◄ antiderivatives are not unique. Far from it: If $F(x)$ is an antiderivative of $f(x)$, then so is $F(x)+C$, for any constant C. Thus, a given function f has infinitely many different antiderivatives.

Not *very* different, however: We showed in Section 5.3 ◄ that any two antiderivatives can differ *only* by an additive constant on any domain interval. Finding *one* antiderivative is essentially tantamount to finding *all* antiderivatives.

Elementary Functions, Derivatives, and Antiderivatives

Functions with formulas are called **elementary**; they're built from the standard function **elements**—polynomials, trigonometric and exponential functions, and their inverses. ◄

Roughly speaking, elementary functions are those an inexpensive scientific calculator can handle.

Every elementary function has an elementary derivative. (This is just a pompous way of saying that differentiating a formula gives another formula.) The situation for antiderivatives is different: An antiderivative of an elementary function may or may not be elementary. Even deciding whether a function *has* an elementary antiderivative—let alone finding one—can be very difficult indeed. Thus, the generic problem of this chapter is as follows:

> *Given an elementary function f, find an elementary function F such that $F' = f$.*

Either by human or by machine. Maple's `int` command doesn't always succeed, for example.

It cannot always be solved. ◄   This problem is sometimes called **integration in closed form**; an elementary solution (if one exists) is called a **closed-form solution**.

Antidifferentiation in Closed Form: Art, Science, or Parlor Trick?

The uncertainty just discussed adds an element of intrigue to our problem, elevating it somewhat above the mundane level of routine calculation. (Differentiation, by contrast, *is* routine calculation.) Finding antiderivatives can be challenging, but the rewards are proportional to the effort.

We'll attack the antiderivative problem with a modest arsenal of antidifferentiation techniques. Like differentiation rules, antidifferentiation techniques can reduce a complicated problem to one or more simpler ones. But unlike differentiation rules, antidifferentiation rules don't always succeed; choosing the "right" antidifferentiation rule or rules combines art, science, intuition, and (sometimes) luck.

Notation and Terminology

We'll use several standard notations and technical terms in our tour of the antiderivative problem

Indefinite Integrals The expression

$$\int f(x)\,dx$$

is called an **indefinite integral**; f is the **integrand** and x is the **variable of integration.** The indefinite integral is a symbolic shorthand for the antiderivative problem. ▶ For example,

$$\int \cos x\,dx = \sin x + C$$

means that the antiderivatives of $\cos x$ are precisely the functions $\sin x + C$, where C, the **constant of integration**, may take any value.

The name and notation aren't perfect; integrals and antiderivatives aren't really the same thing.

One caveat is as follows. Interpreted scrupulously, the indefinite integral $\int f(x)\,dx$ denotes not just one function but many—the "family" of all possible antiderivatives of f. ▶ The distinction isn't always important in practice. Writing, say, $\int \cos x\,dx = \sin x$ seldom causes confusion.

For nice functions f, antiderivatives differ only by additive constants.

Definite Integrals The expression

$$\int_a^b f(x)\,dx,$$

with endpoints definitely specified, is called a **definite integral**. ▶

A definite integral has one value, not many.

The FTC links definite and indefinite integrals. In words,

The definite integral $\int_a^b f(x)\,dx$ equals the change $F(b) - F(a)$ from $x = a$ to $x = b$ in any indefinite integral.

> **Antidifferentiation by Machine.** Many high-powered mathematical computer programs attack the (hard) problem of elementary antidifferentiation. *Maple*'s `int` command, for example, tries to find antiderivatives in closed form. For the good reasons already cited, any antiderivative program sometimes fails; a given problem may have no closed-form solution.
>
> Until recent times there was no reliable method of deciding, once and for all, whether a function has an elementary antiderivative. In the late 1960s, R. H. Risch discovered an algorithm that determines whether a function has an elementary antiderivative and, if it does, finds one. Although the Risch algorithm and its more recent variants can be implemented by computer, most modern software uses faster, less cumbersome methods.

Elementary Antiderivatives: Why Bother Looking?

Many useful functions do have elementary antiderivatives. We'll have great fun finding them. But why bother? The question is a serious one. As we'll soon see (starting in Chapter 7 of Volume 2), numerical methods let us compute many definite integrals to high accuracy without fussing over antiderivatives. Following are three good reasons for our trouble.

Formulas for Success Elementary functions—those with formulas—are in many ways the simplest and most convenient possible functions. In applied mathematics, finding a formula that usefully models an interesting phenomenon represents the best imaginable outcome.

As a simple but important example, consider how an object moves along a straight line, starting from rest at time $t = 0$. If the object's acceleration a is constant (as in free fall), then the velocity function v is linear and the distance-traveled function s is quadratic. More precisely,

If $a(t) = a$, then $v(t) = at$, and $s(t) = \dfrac{at^2}{2}$.

Thus, a simple *formula* for acceleration leads in this case to almost equally simple *formulas* for velocity and position, its first two antiderivatives.

Not all applications work so smoothly, of course. Complex combinations of forces may produce complicated acceleration functions, perhaps without elementary antiderivatives. Nevertheless, the point remains: The practical and theoretical value of *having* antiderivative formulas amply justifies some effort in finding them.

Predicting Planetary Motion. Planetary motion, the subject of **celestial mechanics**, has fascinated observers throughout history. The search for formulas to describe and predict planetary motion has continued for many thousands of years.

Nicolaus Copernicus, Johannes Kepler, and Galileo Galilei, working in the 16th and early 17th centuries, began the task of describing planetary and gravitational phenomena in modern mathematical terms. Isaac Newton, in his *Principia* (1687), used the tools of calculus to extend and unify earlier work. Newton showed, for instance, that his inverse-square law for gravitation (gravity varies inversely with the square of the distance) implies Kepler's first law—that the orbit of a planet is an ellipse with the sun as one focus.

Exact Answers Evaluation of integrals in closed form gives exact answers. Numerical results, no matter how accurate, are only approximate. The difference can be important. For example, the fact that

$$\int_{-1}^{1} \frac{2}{1 + x^2}\, dx \approx 3.141592614$$

is easy to show numerically. It's not very exciting, however; one number is much like another. But the fact that

$$\int_{-1}^{1} \frac{2}{1 + x^2}\, dx = 2\arctan x \bigg]_{-1}^{1} = \pi,$$

which depends on antidifferentiation, is an interesting surprise—how in the world does π arise?

Integrals with Parameters: Many for the Price of One Many useful integrals involve parameters—constants whose values we don't want to specify in advance. Consider, for example, the integral

$$\int_{0}^{\pi} \sin(kx)\, dx,$$

for k a nonzero integer. ▶ Without a specific value for k, numerical methods are helpless. ▶ Antidifferentiation solves the problem for all values of k at one blow: ▶

If $k = 0$, the problem is trivial.

Think about it. Do you see why?

Check details.

$$\int_0^\pi \sin(kx)\,dx = -\frac{\cos(kx)}{k}\Bigg]_0^\pi = \frac{1 - \cos(k\pi)}{k}.$$

Thus, the parity of k determines the value of the integral: It's zero if k is even, $2/k$ if k is odd.

Derivatives and Antiderivatives: Basic Strategies

Antidifferentiation is differentiation in reverse. The expert antidifferentiator, therefore, has a ready mastery of the easier skill of differentiation and remembers this rule:

> *Antidifferentiation is hard; differentiation is easy. Therefore, check antiderivatives by differentiation.*

EXAMPLE 2   Is $\int e^x \sin x\,dx = e^x \cos x + C$?

Solution   No. Here's why: $(e^x \cos x)' = e^x \cos x - e^x \sin x \neq e^x \sin x$. ■

Caveat.   Some care is necessary. When checking antiderivatives, one should realize that some functions (especially trigonometric functions) can be written in more than one way.

EXAMPLE 3   Is $\int \sin(2x)\,dx = \sin^2(x) + C$?

Solution   Let's check. By the chain rule,

$$\left(\sin^2(x)\right)' = 2 \sin x \cos x.$$

We hoped for $\sin(2x)$, so we're tempted to answer no. The right answer, however, is yes. One of the many trigonometric identities says that

$$2 \sin x \cos x = \sin(2x).$$

We stand corrected. ■

Basic Antiderivative Formulas

Differentiation techniques (e.g., the product and quotient rules) start with a few known derivatives; combining them properly produces new derivatives.

Antidifferentiation, too, requires a basis of known formulas to which others are reduced. We'll use the following ones.

Basic Antiderivative Formulas		
$\displaystyle\int x^k\, dx$	$=$	$\displaystyle\frac{x^{k+1}}{k+1} + C$ (if $k \neq -1$)
$\displaystyle\int \frac{1}{x}\, dx$	$=$	$\ln|x| + C$
$\displaystyle\int e^x\, dx$	$=$	$e^x + C$
$\displaystyle\int a^x\, dx$	$=$	$\displaystyle\frac{a^x}{\ln a} + C$ (if $a \neq 1$)
$\displaystyle\int \sin x\, dx$	$=$	$-\cos x + C$
$\displaystyle\int \cos x\, dx$	$=$	$\sin x + C$
$\displaystyle\int \sec^2 x\, dx$	$=$	$\tan x + C$
$\displaystyle\int \frac{dx}{\sqrt{1-x^2}}$	$=$	$\arcsin x + C$
$\displaystyle\int \frac{dx}{1+x^2}$	$=$	$\arctan x + C$

Notice:

Why They're True Antiderivative rules are nothing more than derivative rules in reverse. The rule $\int \cos x\, dx = \sin x + C$, for instance, means simply that

$$(\sin x + C)' = \cos x + 0 = \cos x.$$

A Subtlety: Logs and Absolute Values The formula $\int \frac{1}{x}\, dx = \ln|x| + C$ has an unexpected ingredient: the absolute value. After all, $\ln x$ itself is an antiderivative for $1/x$—why bother with an absolute value?

The answer has to do with domains. The function $f(x) = 1/x$ is defined for all $x \neq 0$; ideally, an antiderivative should have the same domain. Unfortunately, the unadorned logarithm function $F(x) = \ln x$ accepts only *positive* inputs x. Use of $G(x) = \ln|x|$ solves this problem neatly. This G has both the right domain (positive and negative numbers) and the right derivative. The following graphs give convincing evidence. ◄

For more formal arguments, see the exercises at the end of this section.

Graph of $y = 1/x$

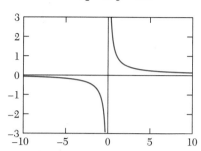

Graph of $y = \ln |x|$

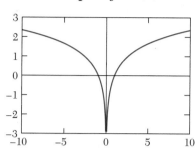

Combining Antiderivatives: Sums and Constant Multiples

Every derivative rule, read backwards, says something about antiderivatives. For sums and constant multiples, the rules are especially simple. ▶ If f and g are continuous functions and k is any constant, then

We'll read the chain rule backwards in the next section.

$$\int \big(f(x) + g(x)\big)\, dx = \int f(x)\, dx + \int g(x)\, dx;$$

$$\int k \cdot f(x)\, dx = k \cdot \int f(x)\, dx.$$

EXAMPLE 4   Find $\displaystyle\int (3\cos x - 2e^x)\, dx$.

Solution   Use the rules and the basic formulas:

$$\int (3\cos x - 2e^x)\, dx = \int 3\cos x\, dx - \int 2e^x\, dx$$

$$= 3\int \cos x\, dx - 2\int e^x\, dx$$

$$= 3\sin x - 2e^x + C.$$

The answer, as usual, is easy to check by differentiation.   ■

Guess and Check

Educated guessing is a perfectly respectable strategy for antidifferentiation. Because differentiation is simple, most guesses are easily checked. Incorrect guesses, moreover, are often easy to repair. The next two examples show what we mean.

EXAMPLE 5   Find $\int \cos(2x)\, dx$ by guessing.

Solution   Since $\int \cos x\, dx = \sin x + C$, we might—just guessing—try $\sin(2x)$ as an antiderivative. Differentiation shows that we missed (but by only a factor of 2).

$$\big(\sin(2x)\big)' = \cos(2x) \cdot 2 \neq \cos(2x).$$

A minor alteration is all that's needed:

$$\left(\frac{\sin(2x)}{2}\right)' = \cos(2x) \implies \int \cos(2x)\, dx = \frac{\sin(2x)}{2} + C.$$ ■

EXAMPLE 6   Find $\int 6/(1 + 4x^2)\,dx$ by creative guessing.

Check this; use the chain rule carefully.

Solution   The integrand's form suggests an arctangent, perhaps with some constants attached, as in $F(x) = A\arctan(Bx)$. Differentiation gives ◄

$$F'(x) = \big(A\arctan(Bx)\big)' = \frac{A}{1 + (Bx)^2}\cdot B = \frac{AB}{1 + B^2x^2}.$$

Our desired antiderivative therefore has $AB = 6$ and $B^2 = 4$. Setting $A = 3$ and $B = 2$ does nicely; therefore, $F(x) = 3\arctan(2x)$ is a suitable antiderivative.   ■

BASICS

In Exercises 1–12, evaluate the antiderivative. Check your answer by differentiation.

1. $\int (3x^5 + 4x^{-2})\,dx$

2. $\int \dfrac{dx}{3x}$

3. $\int \dfrac{dx}{4\sqrt{1 - x^2}}$

4. $\int \dfrac{3}{x^2 + 1}\,dx$

5. $\int 3e^{4x}\,dx$

6. $\int \big(2\sin(3x) - 4\cos(5x)\big)\,dx$

7. $\int 4\sec^2(3x)\,dx$

8. $\int \big(1 + \sqrt{x}\big)^2\,dx$

9. $\int (x + 1)^2\,\sqrt[3]{x}\,dx$

10. $\int \dfrac{(3 - x)^2}{x}\,dx$

11. $\int \dfrac{(1 - x)^3}{\sqrt{x}}\,dx$

12. $\int e^x(1 - e^x)\,dx$

In Exercises 13–15, determine whether the antiderivative formula is true or false by differentiating the right side.

13. $\int e^{\sin x}\,dx = e^{-\cos x} + C$

14. $\int \dfrac{dx}{1 + 4x^2} = \arctan(2x) + C$

15. $\int (1 + \tan^2 x)\,dx = \tan x + C$

FURTHER EXERCISES

Some antiderivatives are difficult to *find* from scratch, but may be *checked* easily by differentiation. In Exercises 16–22, verify the antiderivative formula by differentiation.

16. $\int \ln x\,dx = x\ln x - x + C$

17. $\int \arctan x\,dx = x\arctan x - \dfrac{1}{2}\ln(1 + x^2) + C$

18. $\int \dfrac{dx}{x^2 + a^2} = \dfrac{1}{a}\arctan\left(\dfrac{x}{a}\right) + C$

19. $\int \tan x\,dx = \ln|\sec x| + C$

   [**HINT**: Recall that $\big(\ln|x|\big)' = 1/x$ for all $x \neq 0$.]

20. $\int \sec x\,dx = \ln|\sec x + \tan x| + C$

21. $\int \dfrac{dx}{1 - x^2} = \dfrac{1}{2}\ln\left|\dfrac{1 + x}{1 - x}\right| + C$

   [**HINT**: $\ln\left|\dfrac{1 + x}{1 - x}\right| = \ln|1 + x| - \ln|1 - x|.$]

22. $\int \arcsin x\,dx = x\arcsin x + \sqrt{1 - x^2} + C$

23. (a) Use the trigonometric identity $2\sin^2 x = 1 - \cos(2x)$ to evaluate $\int \sin^2 x\,dx$.

   (b) Explain why $\displaystyle\int_0^{2\pi} \sqrt{1 - \cos(2x)}\,dx \neq \sqrt{2}\int_0^{2\pi} \sin x\,dx$.

Some antiderivatives—trigonometric ones, especially—can be written in more than one way. In Exercises 24–26, verify the formula using the trigonometric identities $\sin^2 x + \cos^2 x = 1$, $\sin(2x) = 2 \cos x \sin x$, and $\cos(2x) = 2 \cos^2 x - 1$.

24. $\displaystyle\int 2 \sin x \cos x \, dx = \sin^2 x + C = -\cos^2 x + C$

25. $\displaystyle\int \cos(2x) \, dx = \frac{1}{2} \sin(2x) + C = \cos x \sin x + C$

26. $\displaystyle\int \cos^2 x \, dx = \frac{x}{2} + \frac{1}{2} \cos x \sin x + C$

$\qquad = \dfrac{x}{2} + \dfrac{1}{4} \sin(2x) + C$

In Exercises 27–36 evaluate the antiderivative. Check your answers by differentiation.

27. $\displaystyle\int \frac{x^2}{1+x^2} \, dx \quad \left[\textbf{HINT: } \frac{x^2}{1+x^2} = 1 - \frac{1}{1+x^2}.\right]$

28. $\displaystyle\int \frac{4x^2 + 3x + 2}{x+1} \, dx$

$\left[\textbf{HINT: } \dfrac{4x^2 + 3x + 2}{x+1} = 4x - 1 + \dfrac{3}{x+1}.\right]$

29. $\displaystyle\int \frac{x-1}{x+1} \, dx \quad [\textbf{HINT: } x - 1 = (x+1) - 2.]$

30. $\displaystyle\int \frac{dx}{\sqrt{x-1} + \sqrt{x+1}}$

$\left[\textbf{HINT: } \left(\sqrt{x-1} + \sqrt{x+1}\right)\left(\sqrt{x-1} - \sqrt{x+1}\right) = -2.\right]$

31. $\displaystyle\int \tan^2 x \, dx$

$\left[\textbf{HINT: } \tan^2 x = \dfrac{\sin^2 x}{\cos^2 x} = \dfrac{1 - \cos^2 x}{\cos^2 x} = \sec^2 x - 1.\right]$

32. $\displaystyle\int \frac{dx}{1 + \sin x} \, dx$

$[\textbf{HINT: } (1 + \sin x)(1 - \sin x) = 1 - \sin^2 x = \cos^2 x.]$

33. $\displaystyle\int \frac{dx}{1 + 9x^2} \quad [\textbf{HINT: } \text{See Example 6.}]$

34. $\displaystyle\int \frac{6}{9 + x^2} \, dx \quad [\textbf{HINT: } 9 + x^2 = 9(1 + (x/3)^2).]$

35. $\displaystyle\int \frac{2}{\sqrt{1 - 9x^2}} \, dx$

36. $\displaystyle\int \frac{8}{\sqrt{4 - x^2}} \, dx$

6.2 Antidifferentiation by Substitution

The simplest and most important antidifferentiation technique is called, variously, **direct substitution, *u*-substitution, ▶** or just plain **substitution**.

Antidifferentiation by any method is nothing more than differentiation in reverse. Substitution, in particular, can be understood as the chain rule running backward. We'll explain the connection more precisely later in this section. In the meantime, watch the chain rule appear again and again as we check our answers.

For obscure reasons, the letter u is almost invariably used in this setting. We'll follow convention.

Substitution: How it Works

We'll use examples ▶ to see both *how* substitution works and *that* it works.

Read them all carefully; they're the heart of this section.

EXAMPLE 1 Find $\int 2x \cos(x^2) \, dx$. (i.e., antidifferentiate $2x \cos(x^2)$).

Solution Let $u = u(x) = x^2$; then $du/dx = 2x$, so $du = 2x \, dx$. (Whoa—can du and dx legally be separated this way? The short answer is yes, but we'll return to such questions later.) Substituting everything into the original integral gives

$$\int \cos(x^2) \, 2x \, dx = \int \cos u \, du.$$

The last integral is simpler than the first. The rest is routine. We first antidifferentiate, then substitute back for u:

$$\int \cos u \, du = \sin u + C = \sin(x^2) + C.$$

Watch the chain rule in action.

Did it work? An easy differentiation shows that it did: ◀

$$\left(\sin(x^2) + C\right)' = \cos(x^2) \cdot 2x,$$

as we wanted.

∎

Separating du and dx might seem suspicious, for example.

That was easy, quick, and possibly dirty. ◀   Notice, though, that the method— dirty or not—did work. It produced an answer we could easily check.

EXAMPLE 2   Find $\int \dfrac{x}{1+x^2}\, dx$.

Solution   Let $u(x) = 1 + x^2$; then $du/dx = 2x$ and $du = 2x\, dx$. Now we substitute:

$$\int \frac{x}{1+x^2}\, dx = \frac{1}{2} \int \frac{2x}{1+x^2}\, dx = \frac{1}{2} \int \frac{du}{u}.$$

The first step was for convenience; multiplying by 2 inside the integral sign "made" the desired du. (We compensated for this by dividing by 2 outside.)

In its new, simpler form, the antiderivative problem is easier:

$$\frac{1}{2} \int \frac{du}{u} = \frac{1}{2} \ln|u| + C = \frac{1}{2}\ln(1+x^2) + C.$$

Checking the answer is another chain-rule exercise:

$$\left(\frac{1}{2}\ln(1+x^2) + C \right)' = \frac{1}{2} \cdot \frac{2x}{1+x^2} = \frac{x}{1+x^2}.$$

∎

The Idea of u-Substitution: Trading One Integral for Another

As the preceding examples show, a successful u-substitution transforms one indefinite integral into another, the second being simpler than the first. The process has three steps:

Substitute   Judiciously choose a function $u = u(x)$, and write $du = u'(x)\, dx$. Then substitute both u and du into the original integral, $\int f(x)\, dx$, to produce a new one of the form $\int g(u)\, du$.

Antidifferentiate   Solve $\int g(u)\, du$ as an antidifferentiation problem in u; i.e., find a function $G(u)$ for which $G'(u) = g(u)$.

Resubstitute   Substitute *back* to eliminate u. The result, $F(x) = G\big(u(x)\big)$, is an antiderivative of f; so is any function of the form $F(x) + C$.

Examples: A Substitution Sampler

Substitution is simple enough in theory. In practice, however, finding the right substitution is as much art as science. ◀   In both realms, practice (false starts and all) makes perfect. The following examples illustrate some of the possibilities and some useful tricks of the trade.

This makes success in substitution especially satisfying.

Choosing u: Wise and Foolish Choices.   The first step—choosing u properly—is the tricky part. Substitution works best when (as in the preceding examples) both

u and du appear conveniently in the original integrand. Morever, the new integral should be simpler or more familiar than the old.

EXAMPLE 3   Find $\int \sin^3 x \cos x \, dx$.

Solution   If $u = \sin x$, then $du = \cos x \, dx$. With u and du staring at us, the situation looks promising. The rest is routine:

$$\int \sin^3 x \cos x \, dx = \int u^3 \, du = \frac{u^4}{4} + C = \frac{\sin^4 x}{4} + C. \qquad \blacksquare$$

EXAMPLE 4   Find $\int \dfrac{x}{1 + x^4} \, dx$.

Solution   No obvious substitution suggests itself. We might try $u = 1 + x^4$, but then $du = 4x^3 \, dx$, and nothing similar appears in the given integrand. A better choice, it turns out, is $u = x^2$. Then $du = 2x \, dx$, and ▶

Watch the 2's carefully.

$$\int \frac{x}{1 + x^4} \, dx = \frac{1}{2} \int \frac{2x}{1 + x^4} \, dx = \frac{1}{2} \int \frac{du}{1 + u^2}.$$

We've made progress at last. The last antiderivative is standard:

$$\frac{1}{2} \int \frac{du}{1 + u^2} = \frac{1}{2} \arctan u + C = \frac{1}{2} \arctan(x^2) + C.$$

Checking the answer involves, as usual, the chain rule. ▶   ■   *The details are left to you.*

Traveling Constants.   Multiplicative constants can move freely (and legally) in and out of integrals. We exploit this freedom in the next example, as we did in Example 2.

EXAMPLE 5   Find $\int \dfrac{3x}{5x^2 + 7} \, dx$.

Solution   The numerator is almost (except for multiplicative constants) the derivative of the denominator; this brings $\int du/u$ to mind. If we set $u = 5x^2 + 7$, then $du = 10x \, dx$, and ▶

Watch the constants carefully.

$$\int \frac{3x}{5x^2 + 7} \, dx = \frac{3}{10} \int \frac{10x}{5x^2 + 7} \, dx$$

$$= \frac{3}{10} \int \frac{du}{u}$$

$$= \frac{3}{10} \ln |u| + C$$

$$= \frac{3}{10} \ln (5x^2 + 7) + C.$$

Differentiation shows that we're right:

$$\left(\frac{3}{10} \ln (5x^2 + 7) + C \right)' = \frac{3}{10} \frac{10x}{5x^2 + 7} = \frac{3x}{5x^2 + 7}. \qquad \blacksquare$$

Inverse Substitutions: Writing x and dx in Terms of u and du

In the examples so far, we've written u and du in terms of x and dx. Sometimes it's simpler to write x and dx in terms of u and du. The next two examples illustrate this strategy.

EXAMPLE 6   Find $\displaystyle\int \frac{x}{\sqrt{2x+3}}\, dx$.

Convince yourself.

Solution   We start as usual. If $u = \sqrt{2x+3}$, then ◀ $\quad du = dx/\sqrt{2x+3}$. Substitution gives

$$\int \frac{x}{\sqrt{2x+3}}\, dx = \int x\, \frac{dx}{\sqrt{2x+3}} = \int x\, du.$$

So far, so good, but what about that remaining x?

The trick is to write x in terms of u, as follows:

$$u = \sqrt{2x+3} \implies u^2 = 2x + 3 \implies \frac{u^2 - 3}{2} = x.$$

Now we're on our way:

$$\int x\, du = \int \frac{u^2 - 3}{2}\, du$$

$$= \frac{u^3}{6} - \frac{3u}{2} + C$$

$$= \frac{(2x+3)^{3/2}}{6} - \frac{3\sqrt{2x+3}}{2} + C.$$

For variety, let's redo the problem a bit differently, still using $u = \sqrt{2x+3}$. Notice first that

$$\frac{u^2 - 3}{2} = x \implies \frac{2u}{2}\, du = dx \implies u\, du = dx.$$

Finally, substitution for x, dx, and $\sqrt{2x+3}$ gives

$$\int \frac{x}{\sqrt{2x+3}}\, dx = \int \frac{u^2 - 3}{2u}\, u\, du = \int \frac{u^2 - 3}{2}\, du,$$

just as before.   ■

EXAMPLE 7   Find $\displaystyle\int \frac{dx}{1 + \sqrt{x}}$.

This is not the only possibility: $u = 1 + \sqrt{x}$ also works.

Solution   Let $u = \sqrt{x}$. ◀   Then easy calculations give

$$x = u^2 \quad \text{and} \quad dx = 2u\, du.$$

Substituting for x and dx gives

$$\int \frac{dx}{1 + \sqrt{x}} = \int \frac{2u}{1 + u}\, du.$$

The last integral is certainly simpler than the first. Another substitution makes it simpler still. If $v = 1 + u$, then $dv = du$ and $u = v - 1$, so ▶ *Check the algebra.*

$$\int \frac{2u}{1+u}\, du = \int \frac{2(v-1)}{v}\, dv = 2 \int \left(1 - \frac{1}{v}\right) dv = 2(v - \ln|v|) + C.$$

Finally, we substitute back for x:

$$2(v - \ln|v|) + C = 2(1 + u - \ln|1 + u|) + C = 2 + 2\sqrt{x} - 2\ln|1 + \sqrt{x}| + C. \quad \blacksquare$$

Substitution in Definite Integrals

The next examples show two ways of evaluating *definite* integrals by substitution.

EXAMPLE 8   Find $\int_0^{\sqrt{\pi/2}} 2x \, \cos(x^2)\, dx$.

Solution   (**Antidifferentiate in x; Plug in Endpoints**)   We write $f(x) = 2x \, \cos(x^2)$. If $F(x)$ is any antiderivative of $f(x)$, then (by the FTC) $F(\sqrt{\pi/2}) - F(0)$ is our answer.

We already found a suitable F: Using the substitution $u = x^2$, $du = 2x\, dx$, we saw that $F(x) = \sin(x^2)$ is an antiderivative of f. ▶   Thus,

We need only one antiderivative, so we set $C = 0$.

$$\int_0^{\sqrt{\pi/2}} 2x \, \cos(x^2)\, dx = \sin(x^2) \Big]_{x=0}^{x=\sqrt{\pi/2}} = \sin(\pi/2) - \sin 0 = 1. \quad \blacksquare$$

EXAMPLE 9   Find $\int_0^{\sqrt{\pi/2}} 2x \, \cos(x^2)\, dx$ again.

Solution   (**Substitute; Create a New Definite Integral**)   As before, let $u = x^2$ and $du = 2x\, dx$. At the endpoints $x = 0$ and $x = \sqrt{\pi/2}$, $u = 0$ and $u = \pi/2$, respectively. Therefore, substitution for u, du, *and the endpoints* gives

$$\int_0^{\sqrt{\pi/2}} 2x \, \cos(x^2)\, dx = \int_0^{\pi/2} \cos u \, du.$$

The last integral can be calculated just as it stands: ▶

No need to substitute x back in.

$$\int_0^{\pi/2} \cos u \, du = \sin u \Big]_0^{\pi/2} = \sin(\pi/2) - \sin 0 = 1,$$

as before.   ■

Look Again.   Notice carefully how we used substitution differently in the last two examples. In Example 8 we used substitution only as a temporary aid to find an antiderivative of the given integrand, with respect to the original variable x. In Example 9 we used substitution to transform the original definite integral into an entirely new definite integral, with new limits and a new integrand. The final result was, the same in each case.

Substitution in General: Why It Works

In each preceding example, substitution worked—it led successfully to a correct antiderivative. To see *why* substitution works, we'll describe it in general language.

For any indefinite integral $\int f(x)\,dx$, substitution means finding a function $u = u(x)$ such that $f(x)$ can be rewritten in the form

$$f(x) = g(u(x))\,u'(x)$$

for some new function g. To say that substitution works is to say that

$$\int f(x)\,dx = \int g(u)\,du,$$

i.e., that if G is a function for which $G'(u) = g(u)$, then $G(u(x))$ is an antiderivative of f.

The chain rule guarantees that $G(u(x))$ has the desired property:

$$\frac{d}{dx}\left(G(u(x)) \right) = G'(u(x)) \cdot u'(x) = g(u(x)) \cdot u'(x) = f(x),$$

so $G(u(x))$ is indeed the antiderivative we seek.

Substitution in *definite* integrals works for essentially the same reason. The fact that it does is sometimes known as the **change-of-variables theorem:**

Theorem 1 Let f, u, and g be continuous functions such that, for all x in $[a, b]$,

$$f(x) = g(u(x)) \cdot u'(x).$$

Then

$$\int_a^b f(x)\,dx = \int_{u(a)}^{u(b)} g(u)\,du.$$

Proof.   Let G be an antiderivative of g. By the FTC,

$$\int_{u(a)}^{u(b)} g(u)\,du = G(u(b)) - G(u(a)).$$

As we showed before (using the chain rule), $G(u(x))$ is an antiderivative for $f(x)$. Therefore,

$$\int_a^b f(x)\,dx = G(u(b)) - G(u(a)),$$

so the two integrals are equal.

Differentials: du, dx, and All That

The symbols du and dx are called the **differentials** of u and x, respectively. If $u = u(x)$, then du and dx are related in the expected way: $du = u'(x)\,dx$.

A rigorous (and rather subtle) theory of differentials exists in higher mathematics; we won't need it in this book. For us, the differential in an integral expression (dx in $\int f(x)\,dx$, for instance) is mainly an aid to memory: It reminds us of the variable of integration. In substitution problems, trading dx's for du's (or vice versa) is a useful bookkeeping device; it helps keep us honest with respect to Theorem 1.

BASICS

In Exercises 1–10, evaluate the antiderivative using the indicated substitution, then check your answer by differentiation.

1. $\displaystyle\int (4x+3)^{-3}\,dx; \quad u = 4x+3$

2. $\displaystyle\int x\sqrt{1+x^2}\,dx; \quad u = 1+x^2$

3. $\displaystyle\int e^{\sin x}\cos x\,dx; \quad u = \sin x$

4. $\displaystyle\int \frac{(\ln x)^3}{x}\,dx; \quad u = \ln x$

5. $\displaystyle\int \frac{\arctan x}{1+x^2}\,dx; \quad u = \arctan x$

6. $\displaystyle\int \frac{\sin\left(\sqrt{x}\right)}{\sqrt{x}}\,dx; \quad u = \sqrt{x}$

7. $\displaystyle\int \frac{e^{1/x}}{x^2}\,dx; \quad u = 1/x$

8. $\displaystyle\int x^3\sqrt{9-x^2}\,dx; \quad u = 9-x^2$

9. $\displaystyle\int \frac{e^x}{1+e^{2x}}\,dx; \quad u = e^x$

10. $\displaystyle\int \frac{x}{1+x}\,dx; \quad u = 1+x$

In Exercises 11–14, find real numbers a and b so that equality holds, then evaluate the definite integral.

11. $\displaystyle\int_{-2}^{1} \frac{x}{1+x^4}\,dx = \frac{1}{2}\int_{a}^{b} \frac{1}{1+u^2}\,du$

12. $\displaystyle\int_{-\sqrt{\pi/2}}^{\sqrt{\pi}} x\cos(3x^2)\,dx = \frac{1}{6}\int_{a}^{b} \cos u\,du$

13. $\displaystyle\int_{0}^{3} \frac{x}{(2x^2+1)^3}\,dx = \frac{1}{4}\int_{a}^{b} u^{-3}\,du$

14. $\displaystyle\int_{1}^{2} x^2 e^{x^3/4}\,dx = \frac{4}{3}\int_{a}^{b} e^u\,du$

FURTHER EXERCISES

15. Example 4 says that $\displaystyle\int \frac{x}{1+x^4}\,dx = \frac{1}{2}\arctan(x^2) + C.$ Verify this by differentiation.

16. Example 6 says that $\displaystyle\int \frac{x}{\sqrt{2x+3}}\,dx = \frac{(2x+3)^{3/2}}{6} - \frac{3\sqrt{2x+3}}{2} + C.$ Verify this by differentiation.

17. (a) Use the substitution $u = 1+\sqrt{x}$ to find the antiderivative $\displaystyle\int \frac{dx}{1+\sqrt{x}}$.

    (b) The expression obtained in part (a) and the expression for this antiderivative given in Example 7 are not identical. Show that both are correct.

18. Let $a > 0$ be a constant. Derive the antiderivative formula $\displaystyle\int \frac{dx}{\sqrt{a^2-x^2}} = \arcsin(x/a) + C$ from the antiderivative formula $\displaystyle\int \frac{dx}{\sqrt{1-x^2}} = \arcsin x + C.$

19. Let a and $b > 0$ be constants. Use the substitution $u = a + b/x$ to find the antiderivative $\displaystyle\int \frac{dx}{ax^2+bx}$.

20. Show that $\displaystyle\int \frac{dx}{1+e^x} = x - \ln(1+e^x) + C.$

    [**HINT:** $\dfrac{1}{u(1+u)} = \dfrac{1}{u} - \dfrac{1}{1+u}.$]

21. Suppose that $\int_0^{12} g(x)\,dx = \pi$. Evaluate $\int_0^4 g(3x)\,dx$.

22. Let $I = \int \sec^2 x \tan x\,dx$.
    (a) Use the substitution $u = \sec x$ to show that $I = \frac{1}{2}\sec^2 x + C$.
    (b) Use the substitution $u = \tan x$ to show that $I = \frac{1}{2}\tan^2 x + C$.
    (c) The expressions for I in parts (a) and (b) look different, but both are correct. Explain this apparent paradox.

In Exercises 23–69, evaluate the antiderivative, then check your answer by differentiation.

23. $\displaystyle\int \cos(2x+3)\,dx$

24. $\displaystyle\int \sin(2-3x)\,dx$

25. $\displaystyle\int (3x-2)^4\,dx$

26. $\displaystyle\int x\cos(1-x^2)\,dx$

27. $\displaystyle\int \frac{2x^3}{1+x^4}\,dx$

28. $\displaystyle\int x(3x+2)^4\,dx$

29. $\displaystyle\int \frac{dx}{1-2x}$

30. $\displaystyle\int \sqrt{3x-2}\,dx$

31. $\displaystyle\int x\sqrt{3-2x}\,dx$

32. $\displaystyle\int \frac{\ln x}{x}\,dx$

33. $\displaystyle\int \frac{\sqrt{1+1/x}}{x^2}\,dx$

34. $\displaystyle\int xe^{x^2}\,dx$

35. $\displaystyle\int x^3(x^4-1)^2\,dx$

36. $\displaystyle\int \frac{x^3}{1+x^2}\,dx$

37. $\displaystyle\int \frac{x^2}{1+x^6}\,dx$

38. $\displaystyle\int \frac{x}{\sqrt{1-x^4}}\,dx$

39. $\displaystyle\int x^2\sqrt{4x^3+5}\,dx$

40. $\displaystyle\int \frac{x}{\sqrt{1+x^2}}\,dx$

41. $\displaystyle\int \frac{x+4}{x^2+1}\,dx$

42. $\displaystyle\int x(1-x^2)^{15}\,dx$

43. $\displaystyle\int \frac{2x+3}{\left(x^2+3x+5\right)^4}\,dx$

44. $\displaystyle\int (x+2)(x^2+4x+5)^6\,dx$

45. $\displaystyle\int \frac{x+1}{\sqrt[3]{3x^2+6x+5}}\,dx$

46. $\displaystyle\int \frac{e^{2x}}{\left(1+e^{2x}\right)^3}\,dx$

47. $\displaystyle\int \frac{e^x}{(2e^x+3)^2}\,dx$

48. $\displaystyle\int \frac{e^{\sqrt{x}}}{\sqrt{x}}\,dx$

49. $\displaystyle\int \frac{2x+3}{(x+1)^2}\,dx$

50. $\displaystyle\int \frac{x^2}{x-3}\,dx$

51. $\displaystyle\int \tan x\,dx$

52. $\displaystyle\int \sec x\tan x\,dx$

53. $\displaystyle\int \frac{\arcsin x}{\sqrt{1-x^2}}\,dx$

54. $\displaystyle\int \sec x\tan x\sqrt{1+\sec x}\,dx$

55. $\displaystyle\int \frac{5x}{3x^2+4}\,dx$

56. $\displaystyle\int \frac{\cos x}{1+\sin^2 x}\,dx$

57. $\displaystyle\int \frac{\sec^2 x}{\sqrt{1-\tan^2 x}}\,dx$

58. $\displaystyle\int \tan^2 x\csc x\,dx$

59. $\displaystyle\int x\tan(x^2)\,dx$

60. $\displaystyle\int x^2\sec^2(x^3)\,dx$

61. $\displaystyle\int \frac{\cos x}{\sin^4 x}\,dx$

62. $\displaystyle\int x^4\sqrt[3]{x^5+6}\,dx$

63. $\displaystyle\int \ln(\cos x)\tan x\,dx$

64. $\displaystyle\int \frac{\left(1+\sqrt{x}\right)^3}{\sqrt{x}}\,dx$

65. $\displaystyle\int \frac{dx}{\sqrt{x}\left(\sqrt{x}+2\right)^3}$

66. $\displaystyle\int x^3\sqrt{x^2+2}\,dx$

67. $\displaystyle\int \sqrt{1+\sqrt{x}}\,dx$

68. $\displaystyle\int \frac{e^x}{e^{2x}+2e^x+1}\,dx$

69. $\displaystyle\int \frac{e^{\tan x}}{1-\sin^2 x}\,dx$

In Exercises 70–75, evaluate the definite integral.

70. $\displaystyle\int_0^2 \frac{x}{\left(1+x^2\right)^3}\,dx$

71. $\displaystyle\int_{-19}^8 \sqrt[3]{8-x}\,dx$

72. $\displaystyle\int_1^e \frac{\sin\left(\ln x\right)}{x}\,dx$

73. $\displaystyle\int_e^{4e} \frac{dx}{x\sqrt{\ln x}}$

74. $\displaystyle\int_0^\pi \sin^3 x\cos x\,dx$

75. $\displaystyle\int_{-\pi/2}^\pi e^{\cos x}\sin x\,dx$

76. Evaluate $\displaystyle\int_0^1 x\sqrt{1-x^4}\,dx$.   [**HINT**: Use a substitution to relate the given integral to one that can be evaluated geometrically.]

77. Suppose that the function f is continuous on the interval $[-1,1]$. Use the substitution $u=\pi-x$ to show that

$$\int_0^\pi xf(\sin x)\,dx = \frac{\pi}{2}\int_0^\pi f(\sin x)\,dx.$$

78. Suppose that the function f is continuous on the interval $[a,b]$. Show that

$$\int_a^b f(x)\,dx = \int_a^b f(a+b-x)\,dx.$$

79. Use the substitution $u=1-x$ to show that

$$\int_0^1 x^n(1-x)^m\,dx = \int_0^1 x^m(1-x)^n\,dx.$$

The substitution $u^n = ax + b$ can sometimes be used to find antiderivatives of expressions involving the form $\sqrt[n]{ax + b}$. Use this substitution to evaluate the following antiderivatives in Exercises 80–81.

80. $\displaystyle \int x\sqrt{2x + 1}\, dx$

81. $\displaystyle \int \frac{dx}{\sqrt{x} + \sqrt[3]{x}}$

82. Find the flaw in the following "proof" that
$$I = \int_{-1}^{1} \frac{dx}{x^2 + 1} = 0:$$

$$I = \int_{-1}^{1} \frac{dx}{x^2 + 1} = \int_{-1}^{1} \frac{x^{-2}}{1 + x^{-2}}\, dx = -\int_{-1}^{1} \frac{du}{1 + u^2} = -I$$

so $I = 0$.

83. Find the flaw in the following "proof" that $\int_{0}^{\pi} \sqrt{1 - \sin x}\, dx = 0$:

$$\int \sqrt{1 - \sin x}\, dx = \int \frac{\sqrt{1 - \sin^2 x}}{\sqrt{1 + \sin x}}\, dx$$

$$= \int \frac{\cos x}{\sqrt{1 + \sin x}}\, dx$$

$$\rightarrow \int \frac{du}{\sqrt{u}}$$

$$= 2\sqrt{u} + C \rightarrow 2\sqrt{1 + \sin x} + C.$$

Thus,

$$\int_{0}^{\pi} \sqrt{1 - \sin x}\, dx = 2\sqrt{1 + \sin \pi} - 2\sqrt{1 + \sin 0} = 0.$$

6.3 Integral Aids: Tables and Computers

Although finding symbolic antiderivatives from scratch can be difficult, there are excellent educational (and recreational) reasons for doing so: Finding the right substitution or other method may reveal satisfying order and pattern in apparent symbolic chaos.

Whatever its educational and aesthetic value, the process of symbolic antidifferentiation is difficult, tricky, and laborious enough that, in practice, calculus users freely resort to integral tables, computer software, or anything else that comes to hand.

Like all power tools, tables and mathematical software require some care. In this section we illustrate some of their possibilities and pitfalls.

Integral Tables

Standard scientific reference works contain extensive integral tables. ▶ The *CRC Handbook of Chemistry and Physics*, 48th edition, for instance, lists nearly 600 different antiderivative formulas on more than 40 pages. One of them reads as follows: ▶

"Antiderivative table" would be a better name.

We added the absolute value signs.

Fact (Formula 403)
$$\int \frac{dx}{a + be^{px}} = \frac{x}{a} - \frac{1}{ap} \ln \left| a + be^{px} \right|$$

Notice the missing (but understood) constant of integration; supplying one every time would be typographically tiresome.

Parameters and Pattern Matching

Integral tables use parameters—letters other than the variable of integration—to stand for numerical constants that can vary from problem to problem. With this device, a single integral formula can solve an entire family of similar problems. The next few examples concern the parameters in the preceding formula.

EXAMPLE 1   Verify Formula 403 by differentiation.

Watch the algebra carefully.

Solution   We differentiate the right side with respect to x. All other letters are constants. ◄

$$\left(\frac{x}{a} - \frac{1}{ap} \ln \left| a + be^{px} \right| \right)' = \frac{1}{a} - \frac{1}{ap} \frac{bpe^{px}}{a + be^{px}}$$

$$= \frac{1}{a} - \frac{1}{a} \frac{be^{px} + a - a}{a + be^{px}}$$

$$= \frac{1}{a} - \frac{1}{a} \left(1 - \frac{a}{a + be^{px}} \right)$$

$$= \frac{1}{a + be^{px}}.$$

That's what we wanted; the *CRC Handbook* got it right.   ∎

Getting It Right.   Not every equation in published integral tables is correct. Integral tables change over time as new antiderivative formulas come into favor and old ones fall into disuse. With each new version of an integral table, new typographical errors may creep in; some errors have lain undiscovered for years.

   Standard integral tables are sometimes used as benchmarks to check mathematical software programs for accuracy. This process sometimes uncovers errors—not in the software but in the tables. Some integral tables have been found to have error rates in the double-digit percentage range.

EXAMPLE 2   Use Formula 403 to find $\int \dfrac{5}{3 - 2e^{-x}} \, dx$.

Solution   Writing

$$\int \frac{5}{3 - 2e^{-x}} \, dx = 5 \int \frac{dx}{3 - 2e^{-x}}$$

Don't forget the 5 out front.

makes the parameter values directly apparent: $a = 3$, $b = -2$, $p = -1$. ◄   The result is

$$\int \frac{5}{3 - 2e^{-x}} \, dx = 5 \left(\frac{x}{3} + \frac{1}{3} \ln \left| 3 - 2e^{-x} \right| \right) + C.$$ ∎

Making the Shoe Fit

Antiderivatives that don't appear to fit an integral template can sometimes be made to do so. A u-substitution, some symbolic algebra, or a combination of both often effects this Cinderella-like transformation. ▶

Sometimes the shoe just won't fit. Remember the Cinderella story?

EXAMPLE 3 Find $I = \int \dfrac{\cos x}{5 + 2e^{3\sin x}} \, dx$.

Solution As it stands the integral resembles Formula 403 only vaguely. However, substituting $u = \sin x$ and $du = \cos x \, dx$ reveals the familiar pattern:

$$\int \frac{\cos x}{5 + 2e^{3\sin x}} \, dx = \int \frac{du}{5 + 2e^{3u}}.$$

Setting $a = 5$, $b = 2$, and $p = 3$ completes the problem:

$$I = \frac{u}{5} - \frac{\ln\left|5 + 2e^{3u}\right|}{15} + C = \frac{\sin x}{5} - \frac{\ln\left|5 + 2e^{3\sin x}\right|}{15} + C. \qquad ■$$

EXAMPLE 4 Find $I = \int \dfrac{e^x}{3e^{-x} + 2e^x} \, dx$.

Solution A touch of exponential algebra ▶ puts the integrand into the necessary form:

Multiplying top and bottom by e^{-x}.

$$\frac{e^x}{3e^{-x} + 2e^x} = \frac{e^x e^{-x}}{e^{-x}\left(3e^{-x} + 2e^x\right)} = \frac{1}{2 + 3e^{-2x}}.$$

Now we're ready for Formula 403, with $a = 2$, $b = 3$, $p = -2$:

$$I = \int \frac{dx}{2 + 3e^{-2x}} = \frac{x}{2} + \frac{1}{4}\ln\left|2 + 3e^{-2x}\right| + C. \qquad ■$$

Completing the Square

Quadratic expressions (of the form $ax^2 + bx + c$) appear often in antiderivative problems. Completing the square sometimes helps fit such integrands into an integral table's template. We illustrate with two examples.

EXAMPLE 5 Find $I = \int \dfrac{dx}{x^2 + 4x + 5}$.

Solution Rummaging through an integral table (such as the one at the end of this book) turns up two likely possibilities:

$$\int \frac{dx}{x^2 + a^2} = \frac{1}{a}\arctan\left(\frac{x}{a}\right); \qquad \int \frac{dx}{x^2 - a^2} = \frac{1}{2a}\ln\left|\frac{x - a}{x + a}\right|.$$

Which one, if either, applies?

To decide, we'll complete the square in the denominator:

$$x^2 + 4x + 5 = (x^2 + 4x + 4) + (5 - 4) = (x + 2)^2 + 1.$$

This, in turn, suggests the substitution $u = x + 2$, $du = dx$. Here's the result, it's easily checked by differentiation.

$$\int \frac{dx}{x^2 + 4x + 5} = \int \frac{du}{u^2 + 1} = \arctan u + C = \arctan(x + 2) + C. \qquad \blacksquare$$

EXAMPLE 6 Find $I = \int \dfrac{dx}{x^2 + 4x + 3}$.

Solution The same two possibilities suggest themselves as in Example 5. Again we complete the square:

$$x^2 + 4x + 3 = (x^2 + 4x + 4) + (3 - 4) = (x + 2)^2 - 1.$$

The rest follows easily. Substitution of $u = x + 2$ and $du = dx$ gives

$$\int \frac{dx}{x^2 + 4x + 3} = \int \frac{du}{u^2 - 1} = \frac{1}{2} \ln \left| \frac{u - 1}{u + 1} \right| + C = \frac{1}{2} \ln \left| \frac{x + 1}{x + 3} \right| + C. \qquad \blacksquare$$

A Reality Check: Do the Graphs Look Right?

Good sense dictates that results obtained by one method be tested by another, from time to time. Symbolic antiderivatives, especially, deserve such attention. Functions, after all, are more than purely formal expressions; they're graphical and numerical objects as well. If symbolic operations really make sense, they should stand up to graphical scrutiny. In other words, symbolically produced antiderivatives should "look right." Now and then, we'll check to make sure that they do.

Examples 5 and 6 invite such a look. Their symbolic results,

$$\int \frac{dx}{x^2 + 4x + 5} = \arctan(x + 2); \qquad \int \frac{dx}{x^2 + 4x + 3} = \frac{1}{2} \ln \left| \frac{x + 1}{x + 3} \right|,$$

are a bit mysterious. Although the integrands are almost identical, the antiderivatives look very different. Is there some mistake?

Let's approach this graphically. Here, first, are graphs of the two integrands:

Graphs of $f_1(x) = 1/(x^2 + 4x + 5)$ and $f_2(x) = 1/(x^2 + 4x + 3)$

Now the mystery begins to dissolve. Despite their typographical similarity, the two integrands are quite different: f_1 behaves tamely, while f_2 has vertical asymptotes at $x = -3$ and $x = -1$. As $x \to \pm\infty$, on the other hand, both f_1 and f_2 tend to zero.

Plots of the antiderivatives F_1 and F_2 show that they, too, behave as they should:

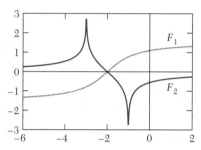

Graphs of $F_1(x) = \arctan(x + 2)$
and $F_2(x) = 0.5 \ln|(x + 1)/(x + 3)|$

Notice the following features.

Asymptotes   Like f_1, its antiderivative F_1 is tamely behaved. Like f_2, the antiderivative F_2 has vertical asymptotes at $x = -3$ and $x = -1$, but with different directions.

Long-Run Behavior   Both F_1 and F_2 appear to tend toward the horizontal as $x \to \pm\infty$. This is as it should be: Their respective derivatives both tend to zero in the long run.

Reduction Formulas: One Antiderivative in Terms of Another

Some formulas in integral tables have antiderivatives on *both* sides of an equation. Here's another entry from the *CRC Handbook*:

Fact (Formula 311)   If $n \neq 1$, then

$$\int \tan^n x \, dx = \frac{\tan^{n-1} x}{n - 1} - \int \tan^{n-2} x \, dx.$$

Applying such a **reduction formula**—repeatedly, perhaps—transforms one integral problem into another, with the second being easier than the first.

EXAMPLE 7   Find $\displaystyle\int \tan^4 x \, dx$.

Solution   Formula 311, with $n = 4$, says that

$$\int \tan^4 x \, dx = \frac{\tan^3 x}{3} - \int \tan^2 x \, dx.$$

The last integral may itself appear in an integral table. If so, we're finished. If not, we can apply Formula 311 again (with $n = 2$, this time).

$$\int \tan^2 x \, dx = \frac{\tan x}{1} - \int \tan^0 x \, dx = \tan x - x.$$

Combining the results,

$$\int \tan^4 x \, dx = \frac{\tan^3 x}{3} - \tan x + x + C. \qquad \blacksquare$$

Antidifferentiation by Machine

Old mathematical handbooks contain vast tables of trigonometric, exponential, and logarithm function values. Practically speaking, scientific calculators have rendered such tables obsolete.

To a lesser (but significant) degree, sophisticated mathematical software effectively replaces integral tables. *Derive, Maple, Mathematica,* and other programs "know" not just integral formulas but—much more challenging—how and when to apply them.

Symbolic Camouflage: Different Forms of the Same Result

Equivalent mathematical expressions can appear in radically different symbolic forms. Trigonometric expressions are especially tricky; the many "trigonometric identities" offer almost unlimited possibilities for disguise.

Computer symbolic operations often produce variant or unexpected forms of symbolic results. (What's convenient for a computer may be baffling to you and me—and vice versa.)

EXAMPLE 8   According to one integral table,

$$\int \sin^2 x \, dx = \frac{x}{2} - \frac{\sin(2x)}{4}.$$

Maple says:

```
> int( sin(x)^2, x );

                    - 1/2 cos(x) sin(x) + 1/2 x
```

Which is right?

Solution   Both are right (we forgive *Maple* the omitted constant of integration). This time the reason is simple. ◄   Stirring the trigonometric identity $\sin(2x) = 2 \sin x \cos x$ into the first formula readily produces the second. ◄   ■

Sometimes it's much harder. See for yourself.

When Nothing Works

Sometimes nothing seems to work. Symbolic methods, integral tables, algebraic tricks, and software may all fail to find an antiderivative in elementary form. Two reasons are possible: Either we haven't (or a computer hasn't) searched cleverly enough, or there *is* no elementary antiderivative. As we remarked at the beginning of this chapter, many elementary functions, even simple ones, don't have elementary antiderivatives, (i.e., antiderivatives with formulas). None of these does, for instance:

$$e^{x^2}, \qquad \sin(x^2), \qquad \frac{\sin x}{x}, \qquad \frac{x}{\ln x}.$$

Like all continuous functions, these have antiderivatives, but not elementary ones. Notice how *Maple* handles one of them:

```
> int( sin(x)/x, x );
```

$$Si(x)$$

Unable to find an elementary answer, *Maple* returned the function *Si*, which is *defined* to be an antiderivative of $\sin(x)/x$. In effect, *Maple* punted.

The Bright Side. Even when symbolic methods fail, all is not lost. Numerical methods, such as those we'll study in the next chapter, often succeed. For example, numerical methods will nicely handle *definite* integrals of all the preceding "problematic" functions.

BASICS

Evaluate the integrals in Exercises 1–31 using a table of integrals or a software package. Making simple u-substitutions or completing the square *first* may help.

1. $\displaystyle\int \frac{dx}{3 + 2e^{5x}}$

2. $\displaystyle\int \frac{dx}{x(2x + 3)}$

3. $\displaystyle\int \frac{dx}{x^2(3 - x)}$

4. $\displaystyle\int_{e}^{4e} x^2 \ln x \, dx$

5. $\displaystyle\int \tan^3(5x) \, dx$

6. $\displaystyle\int \frac{dx}{x^2\sqrt{2x + 1}}$

7. $\displaystyle\int x \sin(2x) \, dx$

8. $\displaystyle\int x^2 e^{3x} \, dx$

9. $\displaystyle\int e^{2x} \cos(3x) \, dx$

10. $\displaystyle\int \frac{2x + 3}{4x + 5} \, dx$

11. $\displaystyle\int \frac{dx}{4 - x^2}$

12. $\displaystyle\int x^2\sqrt{1 - 3x} \, dx$

13. $\displaystyle\int \frac{4x + 5}{(2x + 3)^2} \, dx$

14. $\displaystyle\int \frac{dx}{x\sqrt{3x - 2}}$

15. $\displaystyle\int \frac{dx}{4x^2 - 1}$

16. $\displaystyle\int \frac{dx}{(4x^2 - 9)^2}$

17. $\displaystyle\int \frac{x + 2}{2 + x^2} \, dx$

18. $\displaystyle\int \frac{3}{\sqrt{6x + x^2}} \, dx$

19. $\displaystyle\int \frac{5}{4x^2 + 20x + 16} \, dx$

20. $\displaystyle\int \frac{dx}{\sqrt{x^2 + 2x + 26}}$

21. $\displaystyle\int x^3 \cos(x^2) \, dx$

22. $\displaystyle\int \frac{x}{\sqrt{x^2 + 4x + 3}}\, dx$

23. $\displaystyle\int \frac{dx}{(x^2 + 3x + 2)^2}$

24. $\displaystyle\int x^2(\cos x + 3\sin x)\, dx$

25. $\displaystyle\int \frac{e^x}{e^{2x} - 2e^x + 5}\, dx$

26. $\displaystyle\int \cos x \sin x \sin^2(2\cos^2 x + 1)\, dx$

27. $\displaystyle\int \sqrt{x^2 + 4x + 1}\, dx$

28. $\displaystyle\int \frac{\cos x}{3\sin^2 x - 11\sin x - 4}\, dx$

29. $\displaystyle\int \frac{\cos x \sin x}{(\cos x - 4)(3\cos x + 1)}\, dx$

30. $\displaystyle\int \frac{e^{2x}}{\sqrt{e^{2x} - e^x + 1}}\, dx$

31. $\displaystyle\int x \sin(3x + 4)\, dx$

7

Numerical Integration

7.1 The Idea of Approximation

The definite integral $\int_a^b f(x)\,dx$ is defined formally as a certain limit of approximating sums. This chapter is about different types of approximating sums—left sums, right sums, midpoint sums, and others. We'll study both how such sums approximate integrals and how much error they commit in the process.

Calculating Approximating Sums: Help from Technology

Calculating approximating sums is easy—in theory. ▶ A right sum with 100 subdivisions for $\int_0^1 \sin x\,dx$, for instance, starts like this:

In practice, too, if n is small.

$$R_{100} = \sum_{i=1}^{100} \sin\left(\frac{i}{100}\right) \cdot \frac{1}{100}$$

$$= 0.01 \cdot [\sin(0.01) + \sin(0.02) + \sin(0.03) + \cdots + \sin(0.99) + \sin(1.00)].$$

Evaluating such monsters numerically is no fit job for a human. For computers and calculators, it's simple. ▶ How one uses a calculator or computer program to calculate approximating sums varies with the machine at hand. For the sake of

Computers exist to do such things.

Other software programs, programmable calculators, and so on.

These commands don't work in all Maple versions—the idea is the point.

example, we illustrate how one program—*Maple*—does the job. We use *Maple* only as an illustration; the ideas, not the details of syntax, are our main concern. Those ideas apply regardless of the form of technology ◄ at hand.

With *Maple* it's easy to calculate left, right, midpoint, and other approximating sums, even for large values of *n*. One version ◄ of *Maple* "knows" about the following sums, expressed in *Maple* syntax:

```
left(sin(x),x=0..1,10);          nleft(sin(x),x=0..1,10);
right(sin(x),x=0..1,10);         nright(sin(x),x=0..1,10);
midpoint(sin(x),x=0..1,10);      nmidpoint(sin(x),x=0..1,10);
trapezoid(sin(x),x=0..1,10);     ntrapezoid(sin(x),x=0..1,10);
simpson(sin(x),x=0..1,10);       nsimpson(sin(x),x=0..1,10);
```

(Commands beginning with n give decimal results.)

Maple Tips. To simplify reading *Maple* talk, know these conventions (which are illustrated in this section):

- Inputs (what the user types) follow an "input prompt"—here, the "greater than" symbol, >. Outputs (what *Maple* returns) are indented.

- To avoid roundoff errors, *Maple* doesn't write answers in decimal form unless explicitly told to do so.

EXAMPLE 1   Use *Maple* to calculate various approximating sums for $\int_0^1 \sin x \, dx$. Compare results with the "exact" answer.

Solution   The fundamental theorem of calculus (FTC) gives the exact answer:

$$\int_0^1 \sin x \, dx = -\cos x \Big]_0^1 = -\cos 1 + \cos 0 \approx 0.4596976941.$$

Now, for comparison:

```
> right(sin(x),x=0..1,4);
     1/4 sin(1/4) + 1/4 sin(1/2) + 1/4 sin(3/4) + 1/4 sin(1)
```

The pattern looks correct. Let's see some numbers:

```
> nright(sin(x),x=0..1,10);
                    0.5013880981
```

```
> nleft(sin(x),x=0..1,10);
                    0.4172409996
```

```
> nmidpoint(sin(x),x=0..1,10);
                    0.4598892908
```

The results look reasonable; notice M_{10}'s impressive accuracy. ∎

Tapping *Maple*. *Maple* is a package of powerful, useful, and convenient mathematical operations. We use it occasionally in this book—but not for just anything. Typical uses of *Maple* include:

- Simple but laborious *numerical* operations.

  Example: Calculate L_{100} for the integral $\int_0^1 \sin x\, dx$.

- Simple but laborious *symbolic* operations.

  Example: Check by differentiation that

  $$\frac{x\sqrt{1+4x^2}}{2} + \frac{\ln\left(2x + \sqrt{1+4x^2}\right)}{4}$$

  is an antiderivative for $\sqrt{1+4x^2}$.

- *Graphical* operations. Simple graphs are easily and quickly drawn by hand. Complicated graphs are no harder conceptually but take more time and effort to draw accurately. We'll use *Maple* or some other technology.

  Example: Let $f(x) = \sin(x^2)$. Estimate the maximum value of $|f''|$ for $0 \le x \le 1$, using a graph of f.

Notice the common theme:

We use technology if the result, rather than the process, of calculation is of interest.

Error: A Philosophical Aside

How *well* do approximating sums—L_n, R_n, M_n, and so on—approximate an integral I? If n (the number of subdivisions) is large, we expect such approximations to be close to the true value of I. But how close? Within 0.01? Within 0.00001? Within 0.000000000001? How can we be certain? ▶

To answer such questions, we'll look carefully at approximating sums and the sources of the errors they commit.

In Example 1 we computed I exactly. In many cases, I isn't known exactly.

Why Approximate?

Approximations and the errors they commit are important themes of this book. Following are three good reasons why.

No Closed-Form Solutions Many calculus problems, even simple-looking ones, cannot be solved in "closed form," i.e., using "exact" methods such as antidifferentiation and algebraic solution of equations. Consider, for example, this harmless-looking problem:

Maximize $f(x) = x \sin x$ on the interval $[0, 2]$.

Symbolic methods are of little use with this problem. (Try it yourself ▶ to see where symbolic methods fail.)

Differentiate, find roots, and so on.

Accurate Results Approximate methods (graphing, numerical integration, and so on) often "work" even if exact methods fail. To be sure, approximate methods yield only approximate results. If due care is taken, however, such estimates can be made highly accurate. The maximum problem in the preceding paragraph, for instance, is easy to solve approximately, by graphing.

Estimating Error   An approximation to an unknown quantity is of little value by itself. A useful guess includes a margin of error—i.e., a guarantee that the error committed is no worse than some computable amount. ◄   Finding and using such "accuracy guarantees" for approximations to integrals are the main subjects of this chapter.

Respectable opinion polls do this.

Exact Answers, Estimates, and Accuracy Guarantees

Approximating *known* quantities, such as $\int_0^1 \sin x \, dx$, is useful only as an exercise. Real approximation problems concern quantities we *don't know exactly*. ◄ To "solve" real approximation problems, we need two things:

Such as $\int_0^1 \sin(x^2) \, dx$; see Example 2.

1. An **estimate** for the unknown quantity

2. An **error bound**—i.e., an upper bound on how much error the approximation can commit

Error Bounds: What To Expect

We cannot expect to compute *exactly* the error a given approximation commits—that would be tantamount to solving the original problem exactly. Error bounds, therefore, are usually inequalities. The error estimate

$$|I - L_n| \leq \frac{K_1(b-a)^2}{2n}$$

is a typical example. (L_n is the left sum with n subdivisions for $I = \int_a^b f(x) \, dx$; K_1 is a certain constant we'll discuss later.) Notice:

- The left side of the inequality—the absolute value of the error—is what we want to estimate. We hope it's small!

- Bounding the error's absolute value means that we care (or know) more about the *size* than about the *sign* of the error. ◄

For the same reason, pollsters usually report error as, say, plus or minus 5%.

- The $\leq$ inequality guarantees that the error is *no worse than* the (computable) quantity on the right. Error estimates represent conservative, worst-case scenarios; the *actual* error might well be less.

The Bright Side

Working out approximations and error estimates by hand is tedious. With computing, it's easy.

Still More Errors: Roundoff.   Any approximation, such as L_{1000} for the integral $I = \int_0^1 x^2 \, dx$, can commit several types of error. One source of error is the method itself. We know, for instance, that because the integrand is increasing, *every* left sum underestimates I. Decimal rounding is another source of error. Even small roundoff errors may accumulate dangerously during delicate or repetitive calculations.

A careful treatment of roundoff errors is beyond the scope of this book. As a rule, we control—though never completely avoid—roundoff errors by avoiding very large-scale computations.

The Simplest Error Bounds; Monotone Functions

If the integrand f happens to be **monotone** (i.e., either increasing or decreasing, but not both) on $[a, b]$, then left and right sums always bracket the integral $\int_a^b f$. This simple idea leads to a simple but useful error bound. Let's see how, by example.

EXAMPLE 2   Let $I = \int_0^1 \sin(x^2)\, dx$. ▶   Also, let L_{100} and R_{100} denote left- and right-sum approximations of I, each with 100 subdivisions:

$$L_{100} \approx 0.30607; \qquad R_{100} \approx 0.31449.$$

Make your best guess at I. How far off could your guess be?

The integral I cannot be evaluated using the FTC. Why not? Because (for deep reasons) the antiderivative of $\sin(x^2)$ is not an elementary function.

Solution   The integrand, $\sin(x^2)$, *increases* everywhere on $[0, 1]$. Therefore:

For any n, L_n underestimates I; R_n overestimates I.

The following picture shows the situation for left sums.

Left sums underestimate
$$\int_0^1 \sin(x^2)\, dx$$

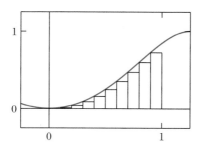

Thus, I lies somewhere between L_{100} and R_{100}, i.e., in the interval $[0.30607, 0.31449]$. Lacking further information, we use the *average*, 0.31028, to estimate I.

How much error could our guess possibly commit? Because the interval $[0.30607, 0.31449]$ has length 0.00842, the distance from I to the midpoint cannot exceed the "radius," 0.00421. In symbols,

$$|I - 0.31028| < 0.00421.$$

We conclude:

The estimate $I \approx 0.31028$ holds with margin of error ± 0.00421. ■

A General Error Bound

The example illustrates a principle ▶   that applies to every monotone function f and integral $I = \int_a^b f$.

See the next theorem.

All the Ingredients.   Let's assemble all our ingredients:

- The integrand f is monotone on $[a, b]$.
- The equally spaced points $a = x_0 < x_1 < x_2 < \cdots < x_n = b$ partition $[a, b]$ into n equal subintervals, ▶   each of width $\Delta x = (b - a)/n$.

Equal-sized subintervals simplify computations.

Check these carefully to review the notation,

- L_n and R_n are the left and right sums built from f, a, b, and n. In sigma notation, ◄

$$L_n = \sum_{i=1}^{n} f(x_{i-1})\Delta x; \qquad R_n = \sum_{i=1}^{n} f(x_i)\Delta x.$$

See Section 5.4.

- L_n and R_n are both estimates for I. Because I lies between L_n and R_n, their average, $(L_n + R_n)/2 = T_n$, ◄ is another natural estimate for I.

> **Theorem 1** (**Error Bounds for Left and Right Sums**) Suppose that f is monotone on $[a, b]$, and let $I = \int_a^b f(x)\, dx$. Then:
>
> $$|I - R_n| \le |f(b) - f(a)|\frac{(b-a)}{n};$$
>
> $$|I - L_n| \le |f(b) - f(a)|\frac{(b-a)}{n};$$
>
> $$|I - T_n| \le |f(b) - f(a)|\frac{(b-a)}{2n}.$$

Proof. That I lies between L_n and R_n is the key idea. The crucial question is how far apart they can be; if they're close together, then L_n, R_n, and T_n are all close to I. Here's the answer:

$$|R_n - L_n| = |f(b) - f(a)|\, \Delta x = |f(b) - f(a)|\frac{(b-a)}{n}. \tag{7.1}$$

From this equation, all inequalities in the theorem follow directly. ◄ The first inequality, for instance, holds since $|I - R_n| \le |L_n - R_n|$.

Convince yourself. Remember that I lies between L_n and R_n.

The proof for a decreasing f is almost identical.

Why does Equation 7.1 hold? A "picture proof" provides the simplest answer. For convenience, we'll use an *increasing* function f. ◄

The difference between R_n and L_n

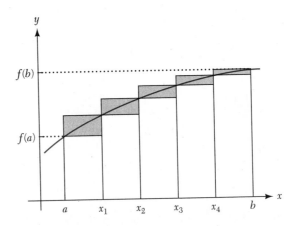

The shaded area represents the difference $|R_n - L_n|$—just what we want to measure. Notice:

- Each shaded box has *width* Δx.

- The *total height* of all n boxes is $|f(b) - f(a)|$. ▶

Therefore, the total shaded area is what we claimed:

$$|R_n - L_n| = |f(b) - f(a)|\Delta x = |f(b) - f(a)|\frac{(b-a)}{n}. \qquad ∎$$

Stack the shaded boxes vertically to see why.

Fine Print: From Pictures to Equations

Pictures are not proofs. For completeness, here's an analytic version (i.e., a translation into symbolic language) of the preceding discussion. As before, we assume that f is increasing on $[a, b]$.

$$|R_n - L_n| = \sum_{i=1}^{n} f(x_i)\,\Delta x - \sum_{i=1}^{n} f(x_{i-1})\,\Delta x$$

$$= \Delta x \sum_{i=1}^{n} (f(x_i) - f(x_{i-1}))$$

$$= \Delta x\,[f(x_1) - f(x_0) + f(x_2) - f(x_1) + \cdots + f(x_n) - f(x_{n-1})]$$

$$= \Delta x\,[f(b) - f(a)] = \Delta x\,|f(b) - f(a)|.$$

Using the Error-Bound Formula: Choosing n in Advance

With an error-bound formula, we can tell *in advance* how many subdivisions are needed for stipulated accuracy.

EXAMPLE 3   How large must n be to ensure that L_n estimates $I = \int_0^1 \sin x\,dx$ with error less than 0.01? Calculate such an L_n. Is it as good as claimed?

Solution   The error-bound formula says that, for this I,

$$|L_n - I| \le |\sin 1 - \sin 0| \cdot \frac{1}{n} \approx \frac{0.8415}{n}.$$

To ensure that $|L_n - I| \le 0.01$, it's enough to insist that

$$\frac{0.8415}{n} < 0.01.$$

We're OK, therefore, if $n > 84.15$; any integer greater than 84 will do. Let's check with *Maple*:

```
> nleft( sin(x), x=0..1, 85 );
                    0.4547425626
```

The exact value of I, by comparison, is

$$\int_0^1 \sin x\,dx = -\cos x\,\Big]_0^1 = -\cos 1 + \cos 0 \approx 0.4596976941.$$

Thus, L_{85} *under*estimates I by about 0.005—well within the error bound's prediction. ∎

BASICS

1. Let $f(x) = 1/\sqrt{x}$ and $I = \int_1^4 f(x)\,dx$.
   (a) Use the FTC to evaluate I exactly.
   (b) Compute L_3 and R_3 by hand. (You may use a scientific calculator.)
   (c) Use your results from parts (a) and (b) to compute the approximation errors $|I - L_3|$ and $|I - R_3|$.
   (d) Are the values in part (c) consistent with the bounds given by Theorem 1? Justify your answer.
   (e) What is the approximation error made by T_3? Is this consistent with Theorem 1?
   (f) Use the error bounds in Theorem 1 to find a value of n that guarantees that $|I - R_n| \le 0.005$ (i.e., that R_n approximates I to two-decimal-place accuracy).

2. Let $f(x) = x^3$ and $I = \int_0^1 f(x)\,dx$.
   (a) Evaluate I exactly, using the FTC.
   (b) Compute L_4 and R_4 by hand. (You may use a scientific calculator.)
   (c) Use your results from parts (a) and (b) to compute the approximation errors $|I - L_4|$ and $|I - R_4|$.
   (d) Are the values in part (c) consistent with the bounds given by Theorem 1? Justify your answer.
   (e) What is the error made by the approximation $I \approx T_4 = (L_4 + R_4)/2$? Is this consistent with Theorem 1?
   (f) Use the error bounds in Theorem 1 to find a value of n that guarantees that $|I - L_n| \le 0.005$ (i.e., that L_n approximates I to two-decimal-place accuracy).

3. Let $f(x) = e^{-x^2}$ and $I = \int_0^3 f(x)\,dx$. [**NOTE:** The function f has no elementary antiderivative, so the FTC cannot be used to compute I exactly.]
   (a) Sketch a graph of f over the interval $[0, 3]$ and use it to estimate I.
   (b) Compute L_{20} and R_{20}. Does either of these values overestimate I? Which one? Why?
   (c) Use the error bounds in Theorem 1 to find a value of n that guarantees that $|I - R_n| \le 0.005$ (i.e., that R_n approximates I to two-decimal-place accuracy).

For each of the integrals in Exercises 4–8, find a value of n for which Theorem 1 guarantees that $|I - L_n| \le 0.000005$. Justify your answers.

4. $\int_1^3 x\,dx$

5. $\int_1^2 x^2\,dx$

6. $\int_1^4 \sqrt{x}\,dx$

7. $\int_1^2 x^{-1}\,dx$

8. $\int_2^3 \sin x\,dx$

9–13. For each of the integrals in Exercises 4–8, find a value of n for which Theorem 1 guarantees that $|I - R_n| \le 0.000005$. Justify your answers.

14–18. For each of the integrals in Exercises 4–8, find a value of n for which Theorem 1 guarantees that $|I - T_n| \le 0.000005$. Justify your answers.

FURTHER EXERCISES

19. Let $I = \int_0^1 f(x)\,dx$. Suppose that f is a decreasing function on $[0, 1]$ such that $f(0) = 7$ and $f(1) = 4$, and that $L_{16} = 5.3172$.
    (a) Does L_{16} underestimate I? Justify your answer.
    (b) Find an upper bound on the value of $|I - L_{16}|$.
    (c) Evaluate R_{16}.
    (d) Evaluate T_{16} and find an upper bound on the value of $|I - T_{16}|$.

20. Let
$$I = \int_1^6 \frac{dx}{1 + f(x)}.$$
    Suppose that f is an increasing function on $[1, 6]$ such that $f(1) = 2$ and $f(6) = 9$, and that $R_{10} = 1.08536$.
    (a) Does R_{10} overestimate I? Justify your answer.
    (b) Find an upper bound on the value of $|I - R_{10}|$.
    (c) Evaluate L_{10}.

21. Let $I = \int_a^b f(x)\,dx$, and suppose that f is increasing on $[a, b]$. Explain why $L_n \le M_n \le R_n$ for any n. [**HINT:** Draw a picture of such an f.]

22. Let $I = \int_a^b f(x)\,dx$, and suppose that f is decreasing on $[a, b]$. Rank the values L_n, T_n, and R_n in increasing order.

23. Let $f(x) = \cos(1/x)$. Estimate the average value of f over the interval $[1, 5]$ within 0.01. Justify the accuracy of your estimate.

24. Let $I = \int_0^\pi \sin x\,dx$.
    (a) Use the FTC to evaluate I exactly.
    (b) Show that $|I - L_4| > 0.1$.
    (c) Explain why the result in part (b) doesn't contradict Theorem 1.

25. Let $I = \int_0^2 \sin(x^2)\,dx$. Using a left, right, or trapezoid sum, estimate I with an error no greater than ± 0.01. Justify your answer. [**HINT:** The integrand is not monotone over the interval $[0, 2]$.]

26. Let $I = \int_2^5 e^{\cos x}\,dx$. Using a left, right, or trapezoid sum, estimate I with an error no greater than ± 0.05. Justify your answer. [**HINT:** The integrand is not monotone over the interval of integration.]

Let $I = \int_3^8 f(x)\,dx$. In Exercises 27–33, indicate whether the statement must be true, may be true, or cannot be true. Justify your answers.

27. If $f'(x) > 0$ for all x in $[3, 8]$, then $L_{20} < I$.

28. If f is monotone on $[3, 8]$, $f(3) = 2$, and $f(8) = -4$, then $L_{1000} < I$.

29. If f is monotone on $[3, 8]$, $f(3) = 2$, and $f(8) = -4$, then $|I - L_{1000}| < 0.005$.

30. If f is monotone on $[3, 8]$, $f(3) = 5$, and $f(8) = 1$, then $I > T_{1000}$.

31. If f is monotone on $[3, 8]$, $f(3) = 5$, and $f(8) = 1$, then $|I - T_{1000}| \leq 0.05$.

32. If f is monotone on $[3, 8]$, $f(3) = 5$, and $f(8) = 1$, then $|I - L_n| > 0.1$ unless $n > 200$.

33. If $|R_{10} - I| < 0.05$, then $|R_{20} - I| < 0.05$.

34. Show that $R_n = L_n + \left(f(b) - f(a)\right) \cdot \dfrac{(b-a)}{n}$.

35. Show that $T_n = L_n + \dfrac{1}{2}\left(f(b) - f(a)\right) \cdot \dfrac{(b-a)}{n}$.

36. Show that if $f(a) = f(b)$, $L_n = R_n = T_n$. [**HINT:** See Exercises 34 and 35.]

37. Let f be an increasing function over the interval $[-2, 5]$, and define $I = \int_{-2}^5 f(x)\,dx$. Suppose that $L_{10} = 9.4132$ and $R_{10} = 9.5768$.

   (a) Evaluate T_{10}, the trapezoid approximation to I, and find an upper bound on $|I - T_{10}|$.

   (b) Suppose that $R_{50} = 9.5294$. Explain why $\frac{1}{2}(L_{10} + R_{50}) = 9.4713$ is an estimate of I that is guaranteed to be correct within $\pm\frac{1}{2}(R_{50} - L_{10}) = \pm\frac{1}{2}(9.5294 - 9.4132) = \pm 0.0581$.

38. Let f be a linear function. Show that T_n computes $\int_a^b f(x)\,dx$ exactly (i.e., with no approximation error). [**HINT:** Draw a picture.]

39. Let $I = \int_a^b f(x)\,dx$, and suppose that f is monotone on the interval $[a, b]$. Show that

$$|I - T_n| \leq |f(b) - f(a)| \cdot \frac{(b-a)}{2n}.$$

(This is the third inequality in Theorem 1.)

40. Let $I = \int_a^b f(x)\,dx$, and suppose that f is monotone on the interval $[a, b]$. Show that

$$|I - M_n| \leq |f(b) - f(a)| \cdot \frac{(b-a)}{n}.$$

41. Let $I = \int_1^{12} f(x)\,dx$, where f is a decreasing function such that $f(1) = 10$, $f(2) = 2$, and $f(12) = 1$.

   (a) If the error bounds in Theorem 1 are used, what is the smallest value of n that guarantees that L_n approximates I to two-decimal-place accuracy (i.e., that $|I - L_n| \leq 0.005$)?

   (b) If the error bounds in Theorem 1 are used, what is the smallest value of n that guarantees that $|L_n - \int_1^2 f(x)\,dx| \leq 0.004$?

   (c) If the error bounds in Theorem 1 are used, what is the smallest value of n that guarantees that $|L_n - \int_2^{12} f(x)\,dx| \leq 0.001$?

   (d) Show that the estimates computed in parts (b) and (c) can be combined to produce an estimate of I that is guaranteed to have two-decimal-place accuracy.

   (e) The values in parts (a) and (d) are both estimates of I with two-decimal-place accuracy. How does the amount of computational effort necessary to compute the estimate in part (d) compare with that needed to compute the estimate in part (a)? (The number of values of f used is a reasonable measure of computational effort.)

42. Let $I = \int_0^5 f(x)\,dx$, where $f(x) = e^{-x^2}$.

   (a) If the error bounds in Theorem 1 are used, what is the smallest value of n that guarantees that $|I - R_n| \leq 0.01$?

   (b) If the error bounds in Theorem 1 are used, what is the smallest value of n that guarantees that $|R_n - \int_0^2 f(x)\,dx| \leq 0.009$?

   (c) If the error bounds in Theorem 1 are used, what is the smallest value of n that guarantees that $|R_n - \int_2^5 f(x)\,dx| \leq 0.001$?

   (d) Show that the estimates computed in parts (b) and (c) can be combined to produce an estimate of I that is guaranteed to approximate I within 0.01.

   (e) The estimates in parts (a) and (d) both estimate I within 0.01. How does the amount of computational effort necessary to compute the estimate in part (d) compare with that needed to compute the estimate in part (a)? (The number of values of f used is a reasonable measure of computational effort.)

43. (a) For each of the following integrals, tabulate L_n and $I - L_n$ for $n = 2$, 8, 32, and 128. (Round answers to five decimal places.)

   (i) $\int_1^2 x^2\,dx$ (iii) $\int_1^2 x^{-1}\,dx$

   (ii) $\int_1^4 \sqrt{x}\,dx$ (iv) $\int_2^3 \sin x\,dx$

   (b) How do the *actual* approximation errors computed in part (a) compare with the bounds given by Theorem 1?

44. Repeat Exercise 43 using T_n rather than L_n.

45. (a) For each of integrals in Exercise 43, compute $(I - L_{2n})/(I - L_n)$ for $n = 8, 16, 32$, and 64.

   (b) Use the results you computed in part (a) to predict the approximation error made by L_{256}.

   (c) Explain why your results in part (a) lead to the conjecture that the magnitude of the approximation error made by L_n is proportional to $1/n$.

46. (a) For each of integrals in Exercise 43, compute $(I - T_{2n})/(I - T_n)$ for $n = 8, 16, 32$, and 64.

   (b) What do the results from part (a) suggest about how the magnitude of the approximation error made by T_n depends on n?

7.2 More on Error: Left and Right Sums and the First Derivative

The simplicity of the error-bound formulas in the preceding section was purchased at a price: The formulas apply to $\int_a^b f$ only if f is monotone.

Most integrands are not monotone. It's important, therefore, to have more forgiving error-bound formulas—ones that don't demand that f be monotone. In this section we'll derive such an error bound for left and right sums. ◄ As usual in mathematics, greater generality doesn't come for free: The new error-bound formula is a little subtler.

For simplicity, we'll mainly discuss left sums; right sums behave almost identically.

Error Bounds and Derivatives: A Continuing Theme

In the rest of this chapter we derive several error-bound formulas for approximations to $\int_a^b f$. Although these formulas differ from each other, each one somehow depends on a *derivative* ◄ of f. The connection between derivatives of f and the errors committed by various rules is an important theme of this chapter. Error bounds for left and right sums are the simplest in this sense: They can be expressed in terms of the *first* derivative, f'.

First-, second-, or higher-order.

Left-Rule Errors and the Size of $|f'|$

The Best Case: $f' = 0$. The left rule "pretends," in effect, that f is constant on each subinterval. If f happens to *be* constant on $[a, b]$, then the left rule commits zero error in estimating $\int_a^b f$. ◄ The first derivative tells whether this happy situation applies: f is a *constant* function if and only if f' is the *zero* function.

Do you believe all this? Draw a picture.

The Worst Case: $|f'|$ Is Large. Alas, most functions are not constant. The more f differs from a constant function, the worse the left rule behaves. Graphically speaking, the steeper the graph of f, the more error L_n commits. (The same principle holds whether f rises or falls; the only difference is whether L_n underestimates or overestimates.)

The quantity $|f'(x)|$ measures how steeply the f-graph rises or falls. If $|f'|$ is large on $[a, b]$, then f is far from constant, and we expect L_n to behave poorly. ◄

If, say, $f'(x) = 100$, then $f(x) = 100x + C$—very far from a constant function.

The next example illustrates how $|f'|$ affects left rule errors for three linear functions.

> **EXAMPLE 1**   How well does L_4 approximate each of the three integrals $\int_0^1 x \, dx$, $\int_0^1 2x \, dx$, and $\int_0^1 (4 - 4x) \, dx$? Relate the results to first derivatives.

Solution   Exact values of the integrals are easy to find: ▶

Use the FTC, or just look at graphs.

$$\int_0^1 x \, dx = \frac{1}{2}; \qquad \int_0^1 2x \, dx = 1; \qquad \int_0^1 (4 - 4x) \, dx = 2.$$

Pictures show best how well or poorly L_4 works for each integral. Shaded areas represent L_4 *errors*.

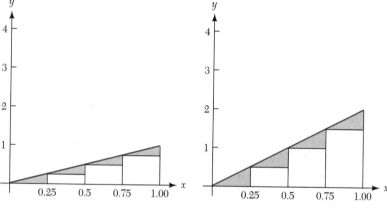

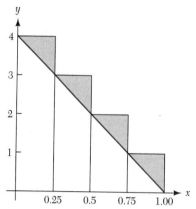

$$I = \int_0^1 x \, dx; \quad |L_4 - I| = .125 \qquad I = \int_0^1 2x \, dx; \quad |L_4 - I| = .25 \qquad I = \int_0^1 (4 - 4x) \, dx; \quad |L_4 - I| = .5$$

The pictures show that errors (whether overestimates or underestimates) increase as graphs get steeper, i.e., as f becomes farther from constant. Derivatives tell the same story: For the three given integrands, $|f'| = 1$, $|f'| = 2$, and $|f'| = |-4| = 4$, respectively.

Maple's numbers agree with the pictures:

```
> nleft(x,x=0..1,4), nleft(2*x,x=0..1,4), nleft(4-4*x,x=0..1,4);
        0.3750000000, 0.7500000000, 2.500000000
```

The following table shows all our numerical results.

L_4 **Errors: How Steepness Matters**			
Integral	$\int_0^1 x \, dx$	$\int_0^1 2x \, dx$	$\int_0^1 (4 - 4x) \, dx$
Exact Value	0.500	1.000	2.000
L_4	0.375	0.750	2.500
$I - L_4$	0.125	0.250	-0.500
$|I - L_4|$	0.125	0.250	0.500

The relative sizes of the errors are the main point:

For linear functions f, the L_4 error is proportional to $|f'|$.   ■

Left-Rule Errors and the Step Size

With a smaller step size Δx, ▶ L_n presumably commits less error. How much less?

I.e., with a larger n.

EXAMPLE 2   Compare the errors committed by L_4 and L_8 in estimating $I = \int_0^1 x^2 \, dx$.

Convince yourself.

Solution   The exact value of I is $1/3$. ◄   For comparison, here are values for L_4 and L_8:

$$L_4 = 0.2187500000, \qquad L_8 = 0.2734375000;$$

Use the fact that $I = 1/3$.   the errors they commit are ◄

$$|I - L_4| \approx 0.114583; \qquad |I - L_8| \approx 0.0598958.$$

The first error is about twice the second. The following graphs agree; errors are shown shaded.

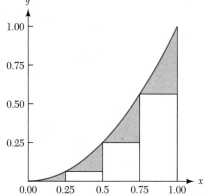

$I = \int_0^1 x^2\, dx;\ |I - L_4| \approx 0.115$

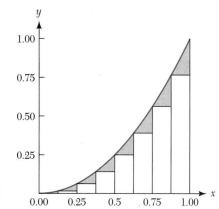

$I = \int_0^1 x^2\, dx;\ |I - L_8| \approx 0.06$

Take one!

A careful look ◄ at the geometry shows again that halving the step size (approximately) halves the error committed. Here's the main point:

I.e., inversely proportional to n.

> *The error L_n commits is (approximately) proportional to the step size Δx.* ◄   ■

Left-Rule Error Over One Subinterval: A Worst-Case Scenario

In the best cases, such errors tend to cancel each other out; in the worst cases, they add up.

As the preceding pictures show, the error L_n commits in approximating an integral I is the sum of small errors committed over each subinterval. ◄   How large, at worst, can the error committed over *one* subinterval be? We answer this question precisely in the next example. Note, however, that Examples 1 and 2 suggest that the answer depends both on Δx and on $|f'|$.

The subscript reminds us that we're bounding the first derivative.

EXAMPLE 3   Let $I = \int_a^b f$, and let K_1 ◄ be any upper bound for $|f'|$ on $[a, b]$. (In other words, $|f'(x)| < K_1$ for all x in $[a, b]$.) How much error, at worst, can L_n commit over the single subinterval $[x_{i-1}, x_i]$? (One subinterval has width $\Delta x = (b - a)/n$.)

Solution   Study the following picture.

Left rule error over one subdivision

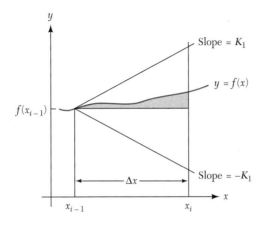

The picture shows:

What the Left Rule Pretends The left rule "pretends" that f is constant on $[x_{i-1}, x_i]$. The shaded area represents the error that results from this pretense.

What the Inequality Means The derivative inequality $|f'(x)| < K_1$ means ▶ that the f-graph over $[x_{i-1}, x_i]$ must stay in the wedge between the two lines shown; they have slopes $\pm K_1$. (To "escape" this wedge, f would have to increase faster than K_1 allows.) ▶

Think this through carefully.

When the Worst Happens The worst possible error—whether positive or negative—occurs if the f-graph is either the top line or the bottom line of the wedge. A careful look at the picture shows that these errors correspond to the areas of the two right triangles, one above and one below the horizontal line. Each of these triangles has area ▶

A rigorous proof requires the mean value theorem. We studied it in Section 4.11 of Volume 1.

Check details.

$$\frac{\text{base} \cdot \text{height}}{2} = \frac{\Delta x \cdot (K_1 \, \Delta x)}{2} = K_1 \frac{\Delta x^2}{2} = K_1 \frac{(b-a)^2}{2n^2}.$$

The last expression answers our question.   ■

A General Error-Bound Theorem

Example 3 shows how much error L_n can commit over *one* subinterval of $[a, b]$. Over all n subintervals, therefore, L_n commits no more than n times as much error. We've just proved a useful theorem:

Theorem 2 (Left- and Right-Rule Error Bounds)
Let $I = \int_a^b f(x)\,dx$, and let L_n be the left approximating sum for I, with n equal subdivisions. Suppose that for all x in $[a, b]$, $|f'(x)| \leq K_1$. Then

$$|I - L_n| \leq \frac{K_1(b-a)^2}{2n}.$$

(The same error bound applies for right sums.)

EXAMPLE 4   We found the actual errors L_4 and L_8 commit in estimating $I = \int_0^1 x^2 \, dx$:

$$|I - L_4| \approx 0.114583; \qquad |I - L_8| \approx 0.0598958.$$

What does the theorem predict?

Solution   First we need a value for K_1. Because $f'(x) = 2x$, the inequality

$$|f'(x)| = |2x| \le 2$$

holds for every x in $[0, 1]$; we take $K_1 = 2$. Now the theorem says:

$$|I - L_4| \le 2\frac{1}{2 \cdot 4} = 0.25; \qquad |I - L_8| \le 2\frac{1}{2 \cdot 8} = 0.125.$$

We're OK—the actual errors are considerably less than the theorem's guarantees.

■

EXAMPLE 5   Use L_{20} and L_{200} to estimate $I = \int_0^3 \sin(x^2) \, dx$. What does the theorem guarantee about the error committed by each?

Solution   First we compute L_{20} and L_{200}:

```
> nleft(sin(x^2), x=0..3, 20);
                0.7322735967

> nleft(sin(x^2), x=0..3, 200);
                0.7703692456
```

By definition, the number K_1 can be *any* upper bound for $|f'|$ on $[0, 3]$. How can we *find* such a K_1? Computing f', either by hand or by machine, is easy:

```
> diff(sin(x^2),x);
                       2
              2 x cos(x )
```

How large can $|2x \cos(x^2)|$ be if $0 \le x \le 3$? The simplest approach is graphical:

Graph of f': a value for K_1

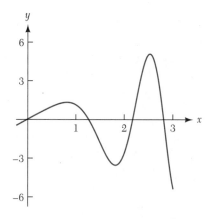

The graph shows that for all x in $[0, 3]$, $|f'(x)| < 6$, so $K_1 = 6$ will do.

Now we have all our ingredients. By the theorem,

$$|I - L_{20}| \le \frac{6(3-0)^2}{40} = 1.35;$$

$$|I - L_{200}| \le \frac{6(3-0)^2}{400} = 0.135.$$

The second inequality means that I lies no farther than 0.135 from $L_{200} \approx 0.770$, i.e., I lies somewhere in the interval $[0.635, 0.905]$.

Both of the preceding error bounds are admittedly unimpressive, given the amount of computation involved. The main point, however, is the comparison: A tenfold *increase* in n gives a tenfold *decrease*—i.e., an improvement—in the error bound. ■

EXAMPLE 6   How much error do L_4 and L_8 commit in estimating $I = \int_0^1 10x \, dx$? ▶   What does the error-bound theorem predict?

By the FTC, $I = 5$.

Solution   First, the numbers: ▶

Courtesy of Maple.

```
> int(10*x,x=0..1), nleft(10*x,x=0..1,4), nleft(10*x,x=0..1,8);
                    5, 3.75000000, 4.37500000
```

Thus, $I = 5$, $L_4 = 3.75$, and $L_8 = 4.375$. The *actual* errors are therefore

$$|I - L_4| = |5 - 3.75| = 1.25; \qquad |I - L_8| = |5 - 4.375| = 0.625.$$

What does the theorem say? Because f is linear, the situation is unusually simple: $f' = 10 = K_1$. ▶   Therefore, the theorem says that

Usually we'd have to bound $|f'|$.

$$|I - L_4| \le 10 \frac{1}{2 \cdot 4} = 1.25; \qquad |I - L_8| \le 10 \frac{1}{2 \cdot 8} = 0.625.$$

In short, our worst fears are realized: L_4 and L_8 behave as badly as the theorem allows. The following pictures illustrate this melancholy state of affairs (errors are shaded).

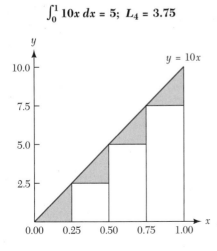

$\int_0^1 10x \, dx = 5; \; L_4 = 3.75$ $\int_0^1 10x \, dx = 5; \; L_8 = 4.375$

Morals. The theorem and examples suggest several general principles.

The Size, Not the Sign, of the Error The theorem bounds the *absolute value* of the error committed by L_n or R_n; it doesn't distinguish over-estimates from underestimates. That distinction may be clear, in particular examples, from pictures.

Finding K_1: Are Estimates OK? The constant K_1 in the theorem can be *any* upper bound for $|f'|$ on $[a, b]$. Finding such a bound can be hard, especially by hand. In practice, we usually resort (as in Example 6) to graphical estimates. Using such estimates may seem illicit, but it isn't. Overestimating K_1 is harmless—the error bound "works" even if K_1 is much larger than necessary. ◀ The only price paid for such sloppiness is a weaker error bound.

But don't underestimate K_1.

More and More Subdivisions? In theory, we can improve our L_n estimate by using thousands or millions of subdivisions. In practice, this is a bad idea. For one thing, roundoff errors would accumulate. More important, as we'll soon see, other approximating sums (e.g., midpoint sums) do much better with far fewer computations.

Worst-Case Scenarios Error-bound formulas such as the ones in this section represent worst-case scenarios: They guarantee that the error a method commits is *no worse than* the error bound. In practice, the *actual error* is often much less than the estimated error, as predicted by the error-bound formula. For $I = \int_0^3 \sin(x^2)\, dx$ and $L_{200} \approx 0.770$, for example, the estimated error is about 0.135. It can be shown that $I = \int_0^3 \sin x^2\, dx \approx 0.774$. The actual error, therefore, is only around 0.004.

BASICS

In Exercises 1–6, compute L_{10} and R_{10}, then compare these estimates with the exact value of the integral and check that the bound on the magnitude of the approximation error given in Theorem 2 is satisfied.

1. $\int_2^3 1\, dx$

2. $\int_1^3 x\, dx$

3. $\int_1^2 x^2\, dx$

4. $\int_1^4 \sqrt{x}\, dx$

5. $\int_1^2 x^{-1}\, dx$

6. $\int_2^3 \sin x\, dx$

7. Suppose that $-4 \le f'(x) \le 3$ if $1 \le x \le 2$. Explain why Theorem 2 does *not* guarantee that $|I - L_n| \le 3/(2n)$.

In Exercises 8–11, find a value of n for which Theorem 2 guarantees that L_n approximates the value of the integral within ± 0.005. Justify your answers.

8. $\int_0^3 e^{-x^2}\, dx$

9. $\int_0^2 \sin(x^2)\, dx$

10. $\int_0^1 (1 + x^2)^{-1}\, dx$

11. $\int_1^{10} \sin(1/x)\, dx$

FURTHER EXERCISES

12. Let $f(x) = e^x/x$, and let $I = \int_1^4 f(x)\, dx$.
    (a) Show that $0 \le f'(x) \le 3e^4/16$ if $1 \le x \le 4$.
    (b) Show that the bound on $|I - L_n|$ given by Theorem 2 is larger than the bound given by Theorem 1.

13. Let $I = \int_0^2 \sqrt{4 - x^2}\, dx$.
    (a) Explain why $I = \pi$. [**HINT**: Draw a picture.]

(b) Compute $|I - L_{10}|$.

(c) Why doesn't Theorem 2 provide a useful bound on the magnitude of the approximation error?

(d) Compare the magnitude of the actual approximation error made by L_{10} with the bound given in Theorem 1.

14. Let $I = \int_a^b f(x)\,dx$, and suppose that the function shown is f'.

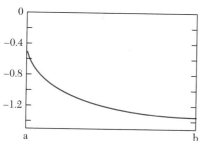

(a) Rank the values I, L_n, and R_n in increasing order. Justify your answer.
(b) Explain why $|I - R_n| \le (b-a)^2/n$.
(c) Does L_n overestimate $\int_a^b e^{-f(x)}\,dx$? Justify your answer.

15. Suppose that $f'(x) > 0$ and $f''(x) < 0$ if $a \le x \le b$. Use a picture to show that T_n is a more accurate estimate of $I = \int_a^b f(x)\,dx$ than L_n (i.e., that $|I - T_n| < |I - L_n|$).

16. Suppose that $f'(x) > 0$ and $f''(x) > 0$ if $a \le x \le b$. Is R_n a more accurate estimate of $I = \int_a^b f(x)\,dx$ than T_n? Justify your answer.

17. Let $I = \int_0^4 f(x)\,dx$, and suppose that f' is the function shown

Graph of f'

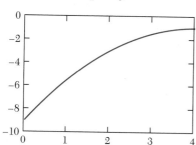

(a) Find a value of n that guarantees that L_n, the left-sum approximation to I with n equal subintervals, satisfies $|I - L_n| \le 0.0001$.
(b) Is $I < R_n$? Explain.
(c) Suppose that F is an antiderivative of f and that $f(2) = 0$. Find a value of n such that

$$\left| L_n - \int_2^4 F(x)\,dx \right| \le 0.01.$$

18. Give an example of a function f such that R_n always underestimates the value of $\int_0^5 f(x)\,dx$ by the maximum amount allowed by Theorem 2.

19. Adapt the argument given in Example 3 to prove that

$$|I - R_n| \le \frac{K_1\,(b-a)^2}{2n}.$$

20. Compute T_{10} and M_{10} for each of the integrals in Exercises 1–6.
    (a) How do the approximation errors made by T_{10} and M_{10} compare with those made by L_{10}?
    (b) How do the approximation errors made by T_{10} compare with those made by M_{10}?
    (c) For which of the integrals does T_{10} make no approximation error? What is true the first derivative of the integrand in each of these cases?
    (d) Which of the integrals does T_{10} underestimate? What is true about the second derivative of the integrand in each of these cases? [**HINT:** Sketch each integrand over the interval of integration.]
    (e) Which of the integrals does T_{10} overestimate? What is true about the second derivative of the integrand in each of these cases? [**HINT:** Sketch each integrand over the interval of integration.]
    (f) For which of the integrals does M_{10} make no approximation error? What is true about the first derivative of the integrand in each of these cases?
    (g) Which of the integrals does M_{10} underestimate? What is true about the second derivative of the integrand in each of these cases?
    (h) Which of the integrals does M_{10} overestimate? What is true about the second derivative of the integrand in each of these cases?

21. Let $I = \int_a^b f(x)\,dx$, and suppose that $0 \le f'(x) \le K_1$ for all x in $[a, b]$.
    (a) Show that $0 \le I - L_n \le \dfrac{K_1\,(b-a)^2}{2n}$.
    (b) Show that $-\dfrac{K_1\,(b-a)^2}{2n} \le I - R_n \le 0$.
    (c) Show that if $f(x) = x$, the magnitude of $I - L_n$ is as large as part (a) allows.
    (d) Parts (a) and (b) show that if f is nondecreasing on the interval $[a, b]$, L_n underestimates I, and R_n overestimates I. Note that the error bounds have the same magnitude but are opposite in sign. This suggests that the estimates L_n and R_n can be combined to produce a better estimate of I. Show how this can be done. Where have you seen this approximation to I before?

22. Adapt the argument given in Example 3 to prove that $|I - M_n| \le \dfrac{K_1\,(b-a)^2}{4n}$.

23. Prove that $|I - T_n| \le \dfrac{K_1\,(b-a)^2}{2n}$. [**HINT:** Use the identity $T_n = (L_n + R_n)/2$.]

24. Consider the following assertion: *If $0 < f'(x) < g'(x)$ for every x in $[a, b]$, then $|L_{10} - \int_a^b f(x)\,dx| < |L_{10} - \int_a^b g(x)\,dx|$.* Is this assertion true or false? Justify your answer.

7.3 Trapezoid Sums, Midpoint Sums, and the Second Derivative

Left and right sums, although conceptually simple, commit large errors. For most integrals, the trapezoid and midpoint rules do better. In this section we see how *much* better, and why.

The Rules and Their Properties

First come formal definitions. Throughout this discussion, $I = \int_a^b f(x)\,dx$, and the partition $a = x_0 < x_1 < x_2 < \cdots < x_n = b$ divides $[a, b]$ into n subintervals, each of length $\Delta x = (b - a)/n$.

> **Definition** (Midpoint Rule) A **midpoint sum** has the form
> $$M_n = \sum_{i=1}^{n} f\left(\frac{x_{i-1} + x_i}{2}\right) \cdot \Delta x.$$
> (In the ith summand, f is evaluated at the *midpoint* of the ith subinterval $[x_{i-1}, x_i]$.)
> (Trapezoid Rule) A **trapezoid sum** has the form
> $$T_n = \sum_{i=1}^{n} \frac{f(x_{i-1}) + f(x_i)}{2} \cdot \Delta x.$$
> (In the ith summand, the values $f(x_{i-1})$ and $f(x_i)$ are averaged.)

(Pictures illustrating trapezoid and midpoint sums appear in the following example and in Section 5.4.)

By the FTC, $I = 1/3$.

EXAMPLE 1   Use M_4 and T_4 to approximate $I = \int_0^1 x^2\,dx$. ◀ How much error does each commit?

We spare you the numerical computations, but check that the forms are right.

Solution   Using the FTC and the preceding definitions, ◀

$$M_4 = [f(1/8) + f(3/8) + f(5/8) + f(7/8)] \cdot \frac{1}{4}$$
$$= \left[(1/8)^2 + (3/8)^2 + (5/8)^2 + (7/8)^2\right] \cdot \frac{1}{4}$$
$$= \frac{63}{192};$$

$$T_4 = \left[\frac{f(0) + f(1/4)}{2} + \frac{f(1/4) + f(1/2)}{2} + \frac{f(1/2) + f(3/4)}{2} + \frac{f(3/4) + f(1)}{2}\right] \cdot \frac{1}{4}$$
$$= [f(0) + 2f(1/4) + 2f(1/2) + 2f(3/4) + f(1)] \cdot \frac{1}{8}$$
$$= \frac{66}{192};$$

$$I = \frac{1}{3} = \frac{64}{192}.$$

The following pictures illustrate the situation.

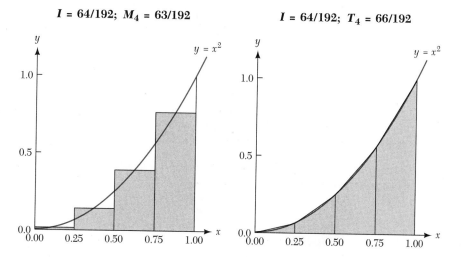

$I = 64/192$; $M_4 = 63/192$ $I = 64/192$; $T_4 = 66/192$

Notice, in particular, that M_4 *under*estimates I by 1/192; T_4 *over*estimates I by exactly twice as much. ■

M_n and Tangent Lines: Another View

It's natural to think of M_n as a sum of *rectangular* areas. As in the preceding picture, we replace the graph of f over the ith subinterval (with midpoint m_i) with a horizontal line at height $y = f(m_i)$.

An alternative but equivalent view of M_n often helps us assess the *sign* of the error:

Replace the graph of f over the ith subinterval with the tangent line at
$x = m_i$.

The following picture illustrates this idea in a simple case.

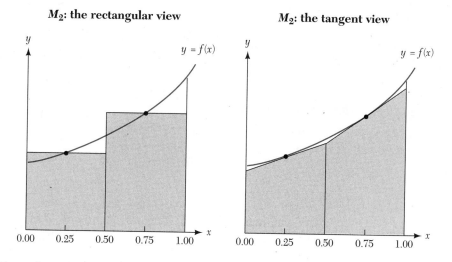

M_2: the rectangular view M_2: the tangent view

From this new viewpoint, M_n is a sum of *trapezoidal* areas; the top sides are determined by tangent lines.

Both approaches to M_n produce the same answer; the reason has nothing especially to do with tangent lines. The fortunate fact is that the area of *any* trapezoid (with parallel vertical sides, as shown) is its base times its *height at the midpoint*.

Concavity of f and the Direction of Error

It's graphically obvious that if f is concave *up* on $[a, b]$, every trapezoid sum T_n necessarily *overestimates* $I = \int_a^b f$. The "tangent interpretation" of M_n shows that the opposite principle applies to midpoint sums. In short, the concavity of f determines the direction of T_n and M_n errors: ◄

It's easier to learn this principle than to memorize it.

> If f is concave up on $[a, b]$, then T_n overestimates I and M_n underestimates.
> If f is concave down, the reverse holds.

Error Bounds for Midpoint and Trapezoid Sums

The evidence so far suggests that trapezoid and midpoint sums usually do better than left and right sums at approximating integrals. How much better, and why?

What Causes M_n and T_n Errors? The Role of f″

Left- and right-sum approximations to $\int_a^b f$ commit *zero* error only if f is constant, i.e., if $f' = 0$. Thus, left- and right-rule errors derive ◄ from the *first* derivative: Because $|f'|$ measures how far f differs from being constant, an upper bound K_1 for $|f'|$ plays a natural role in any worst-case error formula for L_n and R_n.

Pun intended.

Midpoint and trapezoid sums do better than left and right sums for a simple but important reason:

> M_n and T_n commit no error if f is a linear function.

Therefore, midpoint and trapezoid errors result not from the steepness of the f-graph but from its *concavity*:

> The more concave (upward or downward) the graph of f, the worse M_n and T_n behave.

An Error-Bound Theorem for M_n and T_n

Since $|f''|$ measures the concavity of f, its appearance in the following theorem should be no surprise. The usual ground rules apply: $I = \int_a^b f$, M_n and T_n are midpoint and trapezoid sums with n equal subdivisions, and f is "nice." ◄

"Nice" means that f has well-behaved derivatives f' and f'' on $[a, b]$.

> **Theorem 3** (Midpoint and Trapezoid Rule Error Bounds) Let $I = \int_a^b f(x)\, dx$, and let K_2 be an upper bound for $|f''|$ on $[a, b]$. Then
>
> $$|I - M_n| \le \frac{K_2(b-a)^3}{24n^2}; \qquad |I - T_n| \le \frac{K_2(b-a)^3}{12n^2}.$$

Notice:

Any Upper Bound Works The number K_2 ► can be *any* upper bound for $|f''|$ on $[a, b]$, i.e., any number for which the inequality

The subscript reminds us of the second derivative.

$$|f''(x)| \le K_2$$

holds for every x in $[a, b]$. Like K_1, ► K_2 can be estimated graphically. Ballpark estimates suffice; in fact, *over*estimating K_2 is the conservative strategy.

From the theorem on left- and right-rule error bounds.

The Ingredients The precise ingredients of the error bounds—the constants 12 and 24, the power n^2, and so on—are, for the moment, mysterious. We'll justify them soon.

Inequalities Both formulas are inequalities. The actual errors (on the left sides) may be less ► than the right sides predict.

But not more!

Dependence on n For a particular integral $I = \int_a^b f(x)\, dx$, the right sides of all the error formulas depend only on n; everything else is constant.

$$|I - M_n| \le \frac{C_1}{n^2}; \qquad |I - T_n| \le \frac{C_2}{n^2}; \qquad |I - L_n| \le \frac{C_3}{n}.$$

The first two estimates are "stronger" than the third because of the higher powers of n.

EXAMPLE 2   Use M_{10}, T_{10}, and L_{10} to approximate $I = \int_0^1 \sin(x^2)\, dx$. What do the error-bound theorems predict?

Solution   Here, first, are the numbers: ►

With technology they're easy to find.

$$M_{10} = 0.3098162947; \qquad T_{10} = 0.3111708112; \qquad L_{10} = 0.2690972619.$$

To find values for K_1 and K_2, we find derivatives and plot the results: ►

We could do all this by hand, but Maple simplifies the process. The second input computes the second derivative.

```
> f1 := diff(sin(x^2),x);
                    2
          f1 := 2 cos(x ) x

> f2 := diff(sin(x^2),x,x);
                  2   2            2
       f2 := - 4 sin(x ) x  + 2 cos(x )

> plot( {f1,f2}, x=0..1);
```

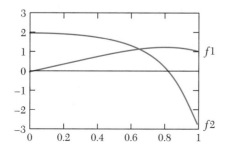

The graphs suggest reasonable values: $K_1 = 1.5$ and $K_2 = 2.5$ will do.   Only arithmetic remains:

$$|I - M_{10}| \le \frac{K_2(b-a)^3}{24n^2} = \frac{2.5 \cdot 1^3}{2400} = 0.00104167;$$

$$|I - T_{10}| \le \frac{K_2(b-a)^3}{12n^2} = \frac{2.5 \cdot 1^3}{1200} = 0.00208333;$$

$$|I - L_{10}| \le \frac{K_1(b-a)^2}{2n} = \frac{1.5 \cdot 1^2}{20} = 0.075.$$

The maximum possible errors committed by M_n, T_n, and L_n are about 0.001, 0.002, and 0.08, respectively.   ■

The same integral as before.

EXAMPLE 3   How many subdivisions does each method require to approximate $I = \int_0^1 \sin(x^2)\, dx$ ◄ with assured error less than 0.0001?

Check details.

Solution   Set the right sides equal to 0.0001 and solve for n. For M_n, T_n, and L_n, respectively, we get: ◄

$$\frac{2.5}{24n^2} \le \frac{1}{10,000} \iff n \ge 32.2749;$$

$$\frac{2.5}{12n^2} \le \frac{1}{10,000} \iff n \ge 45.6435;$$

$$\frac{1.5}{2n} \le \frac{1}{10,000} \iff n \ge 7500.$$

To be sure of approximating I to within 0.0001, the midpoint, trapezoid, and left rules need about 33, 46, and 7500 subdivisions, respectively. Notice the striking difference in efficiency: The left rule requires more than 200 times as much work as the midpoint rule!   ■

The Error Bounds Revisited: A Closer Look

Formal proofs use the mean value theorem.

That the preceding error bounds should *somehow* involve K_2 isn't surprising. Their particular forms, on the other hand, are hardly obvious from intuition. To end this section, we show informally ◄ how the various constants (12, 24, and so on) arise.

A Quadratic Function: The Worst Offender.   The preceding error bounds apply to any function f on $[a, b]$ for which $|f''(x)| \le K_2$. Among *all* such functions, the polynomial $q(x) = K_2 x^2/2$ represents a sort of extreme, because $q''(x) = K_2$ for all x. ◄   In other words, $|q''|$ is always as large as the theorem permits.

Convince yourself.

Given the "extreme" nature of q, we expect the maximum possible errors.

Errors on One Subinterval.   How much error, at worst, can M_n and T_n commit over one subinterval? We answer by computing directly with the function q. ◄   To simplify the algebra, we take $[0, h]$ as our subinterval.

The first step is to compare I_h (the exact integral of q over $[0, h]$) with M_h and T_h (the midpoint and trapezoid estimates over the same interval): ▶

Check the details—we omitted some simple ones.

$$I_h = \int_0^h q = \int_0^h \frac{K_2 x^2}{2} \, dx = \frac{K_2 x^3}{6} \Big]_0^h = \frac{K_2 h^3}{6};$$

$$M_h = q(h/2) \cdot h = \frac{K_2 (h/2)^2}{2} \cdot h = \frac{K_2 h^3}{8};$$

$$T_h = \frac{q(0) + q(h)}{2} \cdot h = \frac{K_2 h^3}{4}.$$

The errors committed over $[0, h]$, therefore, are

$$I_h - M_h = \frac{K_2 h^3}{6} - \frac{K_2 h^3}{8} = \frac{K_2 h^3}{24};$$

$$I_h - T_h = \frac{K_2 h^3}{6} - \frac{K_2 h^3}{4} = -\frac{K_2 h^3}{12}.$$

The mysterious constants 12 and 24 have appeared at last. ▶

As they should, the trapezoid and midpoint errors have opposite signs.

Error Over n Subintervals. Multiplying the preceding quantities by n gives the worst-case error over n subintervals. Replacing h with $(b - a)/n$ gives the error-bound formulas of the theorem.

BASICS

1. Let $I = \int_0^1 e^{x^2} \, dx$.
   (a) Compute M_2 and T_2 by hand. (You may use a scientific calculator.)
   (b) Compute L_{10}, R_{10}, M_{10}, and T_{10}. Which of these approximations underestimates the exact value of I? Justify your answer.
   (c) Using the error bounds in Theorem 3, what is the smallest value of n that guarantees that $|I - M_n| \le 0.0005$ (i.e., that M_n approximates I to three-decimal-place accuracy)?

2. (a) How large must n be to guarantee (using Theorem 3) that T_n estimates $\int_0^1 \sin x \, dx$ to within 10^{-10}?
   (b) Does T_n underestimate the exact answer? Explain.

3. Suppose that $f''(x) = \dfrac{e^x \cos x}{1 + x^2}$. Find an integer n such that T_n approximates $\int_0^5 f(x) \, dx$ within 0.001.

In Exercises 4–9, compute M_{10} and T_{10}, then compare these estimates with the exact value of the integral and check that the bound on the magnitude of the approximation error given in Theorem 3 is satisfied.

4. $\int_2^3 1 \, dx$

5. $\int_1^3 x \, dx$

6. $\int_1^2 x^2 \, dx$

7. $\int_1^4 \sqrt{x} \, dx$

8. $\int_1^2 x^{-1} \, dx$

9. $\int_2^3 \sin x \, dx$

In Exercises 10–13, find a value of n for which Theorem 3 guarantees that M_n approximates the value of the definite integral within ± 0.005. Justify your answers.

10. $\int_0^3 e^{-x^2} \, dx$

11. $\int_0^2 \sin(x^2) \, dx$

12. $\int_0^1 (1 + x^2)^{-1} \, dx$

13. $\int_1^{10} \sin(1/x) \, dx$

FURTHER EXERCISES

14. Let f be the function shown.

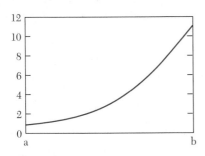

Estimates of $\int_a^b f(x)\,dx$ were computed using the left, right, midpoint, and trapezoid rules, each with the same number of subintervals. The answers obtained were 8.52974, 9.71090, 9.74890, and 11.04407. Which rule produced which estimate? Justify your answer, and explain why it does not depend on knowledge of the value of n used to compute the estimates.

15. Let $I = \int_a^b f(x)\,dx$, where f is the function shown.

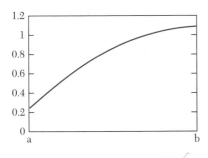

Rank the values I, L_{30}, R_{30}, T_{30}, and M_{30} in increasing order. Justify your answer.

16. Let $I = \int_a^b F(x)\,dx$, where F is an antiderivative of the function f in Exercise 15. Does T_{100} underestimate or overestimate the value of I? Explain.

17. Let $I = \int_a^b f(x)\,dx$, and suppose that the graph shows f''.

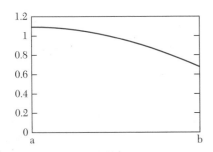

Indicate whether each of the following inequalities must be true, may be true, or cannot be true. Justify your answers.

(a) $L_{100} \le R_{100}$ (d) $I \le M_{100}$

(b) $T_{200} \le M_{200}$ (e) $I \le L_{200}$

(c) $M_{50} \le L_{50}$

18. Let $I = \int_0^4 f(x)\,dx$, and suppose that f' is the function shown.

Graph of f'

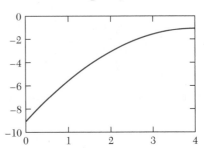

(a) Is $I < M_{10}$? Justify your answer.

(b) Let F be an antiderivative of f. Find a value of n such that $|M_n - \int_1^2 F(x)\,dx| \le 0.01$.

19. Let $I = \int_0^5 f(x)\,dx$, and suppose that the following inequalities are true for $0 \le x \le 5$:

   (i) $-3 \le f'(x) \le -1$;

   (ii) $2 \le f''(x) \le 6$.

Indicate whether each of the following statements must be true, may be true, or cannot be true. Justify your answers.

(a) $I - T_{10} < 0.000005$ (c) $|I - L_{10}| < |I - T_{10}|$

(b) $M_{10} < R_{10}$

20. Let $I = \int_{-1}^2 f(x)\,dx$, where f is a function with the following properties:

   (i) f is increasing on the interval $[-1, 2]$;

   (ii) f'' is the function shown;

Graph of f''

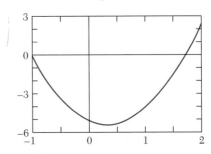

(iii) f has the values shown in the table.

x	-1.00	-0.25	0.50	1.25	2.00
$f(x)$	2.0000	26.522	48.755	68.328	86.790

(a) Compute R_4 and an upper bound on the value of $|I - R_4|$.

(b) Compute T_4 and an upper bound on the value of $|I - T_4|$.

Let $I = \int_0^{10} f(x)\,dx$. In Exercises 21–24, give an example of a function f with the given property.

21. M_{200} underestimates I, and the magnitude of the approximation error made is as large as Theorem 3 allows.

22. M_{200} overestimates I, and the magnitude of the approximation error made is as large as Theorem 3 allows.

23. T_{200} underestimates I, and the magnitude of the approximation error made is as large as Theorem 3 allows.

24. T_{200} overestimates I, and the magnitude of the approximation error made is as large as Theorem 3 allows.

25. (a) Compare the error bound for M_n to the error bound for M_{10n}. Approximately how many additional decimal places of accuracy are gained by using ten times as many subintervals?

(b) Repeat part (a) for L_n and L_{10n}.

(c) What do parts (a) and (b) suggest about the relative performance of these two numerical integration methods?

26. Let $I = \int_a^b f(x)\,dx$, and suppose that $0 \le f''(x) \le K_2$ for all x in $[a, b]$.

(a) Show that $0 \le I - M_n \le \dfrac{K_2\,(b-a)^3}{24n^2}$.

(b) Show that $-\dfrac{K_2\,(b-a)^3}{12n^2} \le I - T_n \le 0$.

(c) Parts (a) and (b) show that if f is concave up on the interval $[a, b]$, M_n underestimates I, and T_n overestimates I. This suggests that the estimates M_n and T_n can be combined to produce a better estimate of I. Show how this can be done. [**HINT:** Combine the estimates in such a way that the worst-case errors cancel.]

27. (a) Compute $(I - M_{10})/(I - T_{10})$ for each of the integrals in Exercises 4–9. Are the results what you anticipated? Explain.

(b) Compute the M_{10} and T_{10} approximations to $\int_0^1 \sqrt{x}\,dx$, then compute $(I - M_{10})/(I - T_{10})$. Does the result surprise you? What's going on here?

7.4 Simpson's Rule

The Story So Far: Integration Rules in Perspective

Simpson's rule, also known as the **parabolic rule**, completes our menu of methods for approximating the value of an integral $I = \int_a^b f$. As the trapezoid and midpoint rules improve the simpler left and right rules, so Simpson's rule improves the midpoint and trapezoid rules.

The following hierarchy of rules shows how this improvement proceeds, step by step.

1. The **left and right rules** find I exactly if f is any *constant* function. For arbitrary functions, therefore, L_n and R_n errors reflect the size of $|f'|$ (which measures how far f differs from being constant). So does the error-bound formula: ▶

$$|I - L_n| \le \frac{K_1(b-a)^2}{2n}.$$

The error bound for R_n is identical. K_1 is an upper bound for $|f'|$.

2. The **trapezoid and midpoint rules** find I exactly if f is any *linear* function. For arbitrary functions, therefore, T_n and M_n errors reflect the size of $|f''|$ (which measures how far f differs from linearity). So do the error-bound formulas: ▶

$$|I - T_n| \le \frac{K_2(b-a)^3}{12n^2}; \qquad |I - M_n| \le \frac{K_2(b-a)^3}{24n^2}.$$

K_2 is an upper bound for $|f''|$.

3. **Simpson's rule** finds I exactly if f is any *quadratic* function. For arbitrary functions, therefore, we'd expect S_n errors to depend on the *third* derivative $|f'''|$ (which measures how far f differs from being quadratic).

Why is this a bonus? We'll soon see.

By a lucky surprise, S_n does even better than expected. Although designed for *quadratic* integrands, Simpson's rule turns out to handle even *cubic* integrands exactly. The resulting bonus ◄ is that the S_n error-bound formula involves an even higher derivative—the fourth—and an even higher power of n: ◄

K_4 is an upper bound for $|f^{(iv)}|$; the details will come later.

$$|I - S_n| \le \frac{K_4(b-a)^5}{180n^4}.$$

The last formula needs explanation, of course; we include it here for comparison with the others. In the rest of this section we explain what its ingredients mean and show the rule in action.

Simpson's Rule: Definition and Interpretations

Simpson's method of approximating $I = \int_a^b f$ starts with a partition $a = x_0 < x_1 < x_2 < \cdots < x_n = b$ of $[a, b]$ into an even number, $n = 2m$, of subdivisions, each of length $\Delta x = (b-a)/n$. ◄ We can now state the formal definition:

The other rules don't "care" whether n is even or odd.

> **Definition**   Simpson's approximation to I is the sum
> $$S_n = \big(f(x_0) + 4f(x_1) + 2f(x_2) + 4f(x_3) + 2f(x_4)$$
> $$+ \cdots + 4f(x_{n-1}) + f(x_n)\big)\frac{\Delta x}{3}.$$
> In sigma notation,
> $$S_n = S_{2m} = \frac{\Delta x}{3}\sum_{i=1}^{m}\big(f(x_{2i-2}) + 4f(x_{2i-1}) + f(x_{2i})\big).$$

The definition raises obvious questions: Why must n be even? What accounts for the coefficients 2, 3, and 4 in the approximating sum? What does any of this have to do with quadratic functions? We address these questions by interpreting Simpson's rule in two different but related ways.

Simpson Sums as Weighted Averages of Trapezoid and Midpoint Sums

We already observed that for any n, the trapezoid sum T_n is the (ordinary) average $(L_n + R_n)/2$ of corresponding left and right sums. For a typical function over a small interval, the left- and right-rule estimates *straddle* the exact integral, so "splitting the difference" makes good sense.

In a similar way, Simpson's rule is built from the trapezoid and midpoint rules. Two ideas are key:

Which rule overestimates? Which underestimates? Concavity decides. Section 7.3 has more details.

- For a typical function over a small interval, the trapezoid and midpoint methods *commit errors with opposite sign*. ◄
- The trapezoid rule error bound is exactly *twice* the midpoint-rule error bound.

Put together, these facts say:

> *Over a small interval, the trapezoid error, is about twice as large as the midpoint error and opposite in sign.*

This principle, in turn, suggests a natural strategy: ▶ Estimate I using the value one-third of the way from M_n to T_n. The result is Simpson's estimate. In symbols, ▶

$$S_{2n} = \frac{T_n + 2M_n}{3}.$$

The errors made by T_n and M_n cancel each other out.

S_{2n} deserves its subscript because it involves evaluating f at the endpoints of $2n$ subintervals.

EXAMPLE 1 Make sense of the preceding formula for $n = 3$ and $I = \int_0^1 \sin x \, dx$.

Solution Is $S_6 = (T_3 + 2M_3)/3$? Let's check: ▶

Check the algebra.

$$T_3 = \frac{1}{6}\left[\sin(0) + 2\sin\frac{1}{3} + 2\sin\frac{2}{3} + \sin(1)\right];$$

$$M_3 = \frac{1}{3}\left[\sin\frac{1}{6} + \sin\frac{3}{6} + \sin\frac{5}{6}\right];$$

$$\frac{T_3 + 2M_3}{3} = \frac{1}{18}\left[\sin(0) + 2\sin\frac{1}{3} + 2\sin\frac{2}{3} + \sin(1)\right] + \frac{2}{9}\left[\sin\frac{1}{6} + \sin\frac{3}{6} + \sin\frac{5}{6}\right]$$

$$= \frac{1}{18}\left[\sin(0) + 4\sin\frac{1}{6} + 2\sin\frac{2}{6} + 4\sin\frac{3}{6} + 2\sin\frac{4}{6} + 4\sin\frac{5}{6} + \sin(1)\right]$$

$$= S_6.$$

The results are as they should be. ▶ The numbers work out nicely, too:

Did you see the coefficients 2, 3, and 4 pop up in the computation?

$$T_3 = 0.455433; \qquad M_3 = 0.461833; \qquad \frac{T_3 + 2M_3}{3} \approx 0.4597 = S_6. \qquad ■$$

Simpson's Rule and Approximation by Parabolic Arcs

A second, more geometric, approach to Simpson's rule concerns **parabolic approximation**. The other rules "pretend" either that f is constant (for left and right sums) or that it is linear (for trapezoid and midpoint sums) on subintervals. Simpson's rule pretends that f is *quadratic on successive pairs of subintervals*. The process, more precisely, is in three steps:

1. Partition $[a, b]$ into an even number n of equal subdivisions; consider successive *pairs* of subdivisions. ▶

We can do this because n is even.

2. Each such pair of subdivisions involves *three* equally-spaced partition points: x_{2i-2}, x_{2i-1}, and x_{2i}. Over each pair of subdivisions, replace the given f with a quadratic function q—the one whose graph (a parabola) passes through the three points $(x_{2i-2}, f(x_{2i-2}))$, $(x_{2i-1}, f(x_{2i-1}))$, and $(x_{2i}, f(x_{2i}))$. ▶ For algebraic reasons, *one* such function q exists. (If the three points happen to be collinear, then the coefficient of x^2 is zero, so q is actually linear.)

In other words, q "agrees" with f at three consecutive partition points.

3. Find the (signed) area under ▶ each small parabolic arc; the sum of all $n/2$ areas is Simpson's estimate S_n.

Or over, if $f < 0$.

The following picture illustrates the idea of Simpson's rule; the shaded area represents S_6. ◄

Notice that S_6 involves three parabolas, not six.

Simpson's rule: the idea

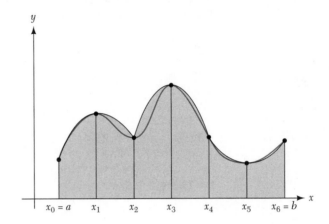

The third step looks daunting. Finding a new quadratic function for each of $n/2$ pairs of subintervals, for example, threatens lots of tedious algebra. It's not that bad, however. The area under the parabolic arc joining *any* three points $\left(x_{2i-2}, f(x_{2i-2})\right)$, $\left(x_{2i-1}, f(x_{2i-1})\right)$, and $\left(x_{2i}, f(x_{2i})\right)$ reduces, after the computational dust settles, to

$$\frac{\Delta x}{3}\left(f(x_{2i-2}) + 4f(x_{2i-1}) + f(x_{2i})\right);$$

that's precisely the ith summand in the preceding definition.

The "parabolic" approach to Simpson's rule has an immediate consequence: S_n commits zero error *for quadratic integrands*.

An Error Bound for Simpson's Rule

By design, Simpson's rule is exact for constant, linear, and quadratic polynomials—i.e., all functions for which $f''' = 0$. By good luck, Simpson commits no error on cubic integrands, either. The next example shows why.

EXAMPLE 2   Use S_2 to estimate $I = \int_a^b x^3\,dx$. How much error is committed?

Solution   Let's calculate: ◄

Here $x_0 = a$, $x_1 = (a + b)/2$, and $x_2 = b$.

$$I = \int_a^b x^3\,dx = \frac{x^4}{4}\Bigg]_a^b = \frac{b^4 - a^4}{4};$$

$$S_2 = \frac{b - a}{6}\left(a^3 + 4\left(\frac{a + b}{2}\right)^3 + b^3\right).$$

Straightforward algebra shows that the two quantities are equal; hence S_2 commits no error. ■

Added to all the preceding, Example 2 shows:

S_n commits zero error for constant, linear, quadratic, and cubic integrands, i.e., for any function f with $f^{(4)} = 0$. ▶

Both $f^{(iv)}$ and $f^{(4)}$ denote the fourth derivative of f.

It's no surprise, then, that the error bound for Simpson's rule involves K_4, an upper bound for $f^{(4)}$ on $[a, b]$.

Theorem 4 (**Error Bound for Simpson's Rule**)
Let $I = \int_a^b f(x)\, dx$, and let K_4 be an upper bound for $|f^{(4)}|$ on $[a, b]$. Then

$$|I - S_n| \leq \frac{K_4(b - a)^5}{180 n^4}.$$

Notice:

- As with K_1 and K_2 in earlier theorems, K_4 may be *any* upper bound for $|f^{(4)}|$. As before, rough estimates are best found graphically.

- The parameter n appears to the fourth power in the preceding denominator; this accounts for Simpson's remarkable efficiency.

EXAMPLE 3   Earlier we calculated that if $I = \int_0^1 \sin x\, dx$, then $S_6 = 0.45969967$. What does the error formula predict? How much actual error does S_6 commit?

Solution   Because $f(x) = \sin x$, an easy computation ▶ shows that   *Do it!* $|f^{(4)}(x)| = |\sin x| \leq 1$; $K_4 = 1$ will do. Thus, by the theorem,

$$|I - S_n| \leq \frac{K_4(b - a)^5}{180 n^4} = \frac{1 \cdot (1 - 0)^5}{180 \cdot 6^4} \approx 0.0000043.$$

The predicted error therefore occurs in the *sixth* decimal place!
    The actual error is even less:

$$I = \int_0^1 \sin x\, dx = -\cos x \Big]_0^1 = 1 - \cos 1 \approx 0.45969769,$$

so

$$|I - S_6| = |0.45969769 - 0.45969967| \approx 0.000002,$$

only about half what the theorem predicts. ■

In the final example, we estimate an unknown quantity.

EXAMPLE 4   Use S_{10} to estimate $I = \int_0^1 \sin(x^2)\, dx$. How much error, at worst, might S_{10} commit? How many subdivisions would be needed to approximate I with error less than 10^{-8}?

Thankfully—imagine doing this by hand.

Solution   We use *Maple*: ◄

```
> f := sin(x^2);
```
$$f := \sin(x^2)$$

```
> nsimpson( f, x=0..1, 10);
```
$$0.3102602344$$

```
> f4 := diff(f,x,x,x,x);
```
$$f4 := 16\sin(x^2)\,x^4 - 48\cos(x^2)\,x^2 - 12\sin(x^2)$$

(The last command calculates the fourth derivative.)

It would be hard by symbolic methods.

Bounding the fourth derivative is best done graphically: ◄

```
> plot(f4, x=0..1);
```

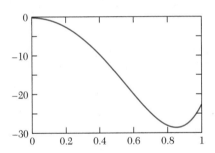

Given the graph, $K_4 = 30$ looks safe. The rest is arithmetic. By the theorem,

$$|I - S_{10}| \le \frac{30 \cdot (1-0)^5}{180 \cdot 10^4} \approx 0.000017.$$

The error, if any, enters around the fifth decimal place, so 0.3103 is a meaningful estimate for I.

To ensure error less than 10^{-8}, we need n so that

$$|I - S_n| \le \frac{30 \cdot (1-0)^5}{180 \cdot n^4} < 10^{-8}.$$

Solving the last inequality for n gives

$$\frac{1}{6 \cdot n^4} < 10^{-8} \iff n^4 > \frac{10^8}{6} \iff n > \frac{100}{\sqrt[4]{6}} \approx 63.9.$$

Why does n have to be even?

Any even n over 63 will do. ◄   ∎

BASICS

In Exercises 1–4, show (by hand computation) that S_2 evaluates the integral exactly.

1. $\int_a^b 1\,dx$

2. $\int_a^b x\,dx$

3. $\int_a^b x^2\,dx$

4. $\int_a^b x^3\,dx$

5. Show (by hand computation) that if S_2 is used to estimate $I = \int_a^b x^4\,dx$, the magnitude of the approximation error is equal to the upper bound given in Theorem 4.

6. Let $I = \int_0^2 e^{x^2}\,dx$.
   (a) Compute S_4 using a scientific calculator.
   (b) Bound the approximation error made by S_4.
   (c) Use the error-bound formula for Simpson's rule given in Theorem 4 to find a value of n that guarantees that S_n approximates I within 10^{-5}.

7. Let $I = \int_0^1 \cos(100x)\,dx$.
   (a) Compute the value of I exactly, using the FTC.
   (b) Compute S_{10}. Show that the *actual* approximation error is less than 0.05.
   (c) Find an upper bound on the magnitude of the approximation error. Why is this upper bound so enormous?

8. Let $I = \int_5^7 \cos x\,dx$.
   (a) Compute S_4, the Simpson's rule estimate of I with four equal subintervals, by hand. Use a calculator (*set in radian mode!*) to find the necessary values of the cosine function.
   (b) What does the error-bound formula say about the error made if I is approximated by S_4?
   (c) Does S_4 underestimate or overestimate I? By how much?

FURTHER EXERCISES

9. Suppose that f is positive, increasing, and concave down on the interval $[1, 7]$. Rank the following estimates of $\int_1^7 f(x)\,dx$ in increasing order: L_{100}, M_{100}, R_{100}, S_{100}, T_{100}.

10. Suppose S_{10} is the first Simpson sum guaranteed to approximate a certain definite integral I within 10^{-2}.
    (a) How much approximation error can S_{100} make?
    (b) What is the smallest value of n for which S_n must approximate I with an error no greater than $\pm 10^{-10}$?

11. Let $I = \int_{-1}^2 f(x)\,dx$, where f is a function such that $1 \le f^{(4)}(x) \le 8$ if $-1 \le x \le 2$, and f has the values shown in the table.

x	−1.00	−0.25	0.50	1.25	2.00
$f(x)$	2.0000	26.522	48.755	68.328	86.790

Compute S_4 and an upper bound on the value of $|I - S_4|$.

12. Compute an estimate of $\int_0^2 xe^{-x^3}\,dx$ that is guaranteed to be correct within ± 0.001. Be sure to explain carefully why your estimate has the desired accuracy.

13. Let $I = \int_0^1 \sin(\sin(x))\,dx$.
    (a) Compute bounds on the error that occurs if M_n approximates I for $n = 4$, 8, and 16, respectively.
    (b) Compute bounds on the error that occurs if S_n approximates I for $n = 4$, 8, and 16, respectively.

14. Let $I = \int_0^1 f(x)\,dx$, where the function f has the values shown in the table:

x	0.00	0.25	0.50	0.75	1.00
$f(x)$	1.307	1.096	1.018	1.173	1.435

(a) Compute estimates of I using left, right, midpoint, trapezoid, and Simpson's rules.
(b) A plot of the data makes it seem reasonable to assume that the graph of f is concave up on the interval $[0, 1]$. Use this assumption and your results from part (a) to find upper and lower bounds on I. Justify your choices.

15. Let $f(x) = e^{\sin x}$ and $I = \int_{-50\pi}^{150\pi} f(x)\,dx$.
    (a) It is straightforward to calculate that
    $$f^{(4)}(x) = (\cos^4 x - 6\cos^2 x \sin x - 4\cos^2 x + 3\sin^2 x + \sin x)e^{\sin x}.$$
    Use this to show that $\left|f^{(4)}(x)\right| \le 15e \approx 41$ for any x. [**HINT:** If x and y are real numbers, $|x + y| \le |x| + |y|$.]
    (b) Use a graph of $f^{(4)}(x)$ to show that $\left|f^{(4)}(x)\right| < 11$ for any x.

(c) What is the smallest value of n for which Theorem 4 guarantees that S_n approximates I within 0.001? Justify your answer.

(d) Explain why $I = 100 \int_0^{2\pi} f(x)\,dx$.

(e) What is the smallest value of n for which Theorem 4 guarantees that S_n approximates $\int_0^{2\pi} f(x)\,dx$ within 0.00001? Justify your answer.

(f) Explain how the results in parts (d) and (e) can be combined to produce an estimate of I with error guaranteed to be less than 0.001.

7.5 Chapter Summary

This chapter surveyed and compared several numerical methods for approximating definite integrals: the left and right rules, the trapezoid and midpoint rules, and Simpson's rule.

Numerical methods, although they give approximate answers, are generally easier to apply than symbolic methods of antidifferentiation. (The latter are surveyed in Chapters 6 and 9.)

Left and Right Rules. The left and right rules for approximating $\int_a^b f(x)\,dx$ amount to pretending that the integrand function remains constant over short subintervals. The error committed by these simple rules, therefore, is relatively large. Moreover, it depends on the size of the first derivative f', which can be thought of as a measure of how much f differs from being constant on its domain.

For typical functions, the errors committed by the left and right rules have opposite sign—if one overestimates, the other underestimates. It's natural, therefore, to average the results of the left and right rules; doing so gives the trapezoid rule.

Trapezoid and Midpoint Rules. The trapezoid and midpoint rules amount to pretending that the integrand f is *linear* over each subinterval; in this case, both rules give exact answers. The errors committed by these rules, therefore, are relatively smaller than those committed by the left and right rules. Moreover, the errors depend on the size of the second derivative f'', which can be thought of as a measure of how much f differs from being linear on its domain.

For typical functions, the errors committed by the trapezoid and midpoint rules have opposite sign—if one overestimates, the other underestimates. Moreover, the midpoint rule generally commits error about one-half that of the trapezoid rule. With that in mind, it's natural to use a weighted average of the trapezoid and midpoint approximations, with the midpoint approximation given double weight. Calculating this average produces Simpson's rule.

Simpson's Rule. Simpson's rule amounts to pretending that the integrand f is *quadratic* over each pair of subintervals; thus, Simpson's rule calculates integrals of quadratic functions exactly, with zero error.

The errors committed by Simpson's rule are often much smaller than those committed by the other rules. Moreover, those errors depend on the size of the fourth derivative $f^{(4)}$. Estimating this size may be difficult by hand, but it's far easier with computing.

8

Using the Definite Integral

8.1 Introduction

Measurement and the Definite Integral

Many important applications of calculus involve measuring something: the *area* of a plane region, the *volume* of a solid object, the *net distance* a moving object travels over an interval, the *length* of a curve from one point to another, the *work* ▶ done against gravity in raising a satellite into orbit, the *present value* ▶ of an income stream (allowing for interest and inflation), and so on. In a few simple cases, such quantities can be found by common sense alone. Measuring the area of a rectangular region, the length of a straight line segment, or the distance covered by an object moving at constant speed requires no big machinery from calculus. In practice, though, most regions aren't rectangular, most curves aren't straight, and most speeds aren't constant. In these more usual and more interesting situations, calculus tools—the definite integral, in particular—are indispensable.

In the physicist's sense—see Section 8.4.

In the economist's sense—see Section 8.5.

Technical Note: Assuming Good Behavior. Throughout this section we assume, to avoid unhelpful distractions, that all the integrals $\int_a^b f(x)\,dx$ we meet make good mathematical sense. To guarantee this, it's enough to assume that every integrand f is *continuous* on $[a, b]$, as we do from now on. In fact, discontinuous integrands *do* sometimes arise in practical applications. Even in such cases, however, the basic ideas of this section often apply, although perhaps in slightly different forms.

Definite Integrals and Area: Reprise. The relation between the definite integral and area isn't new. From the very beginning ▶ we've understood the definite integral geometrically, in terms of area. For any continuous function f on $[a, b]$,

See the definition and pictures in Section 5.1.

$$\int_a^b f(x)\,dx = \text{signed area bounded by } f\text{-graph for } a \leq x \leq b.$$

Remember that *signed area* means that any area under the *x*-axis counts as negative. Keeping track of positive and negative areas takes a little care, but the basic link between integrals and areas is by now familiar. Here's an easy example to bring the issue back to mind.

Check for yourself.

EXAMPLE 1   Easy calculations ◄ show that

$$I_1 = \int_0^\pi \sin x \, dx = 2; \qquad I_2 = \int_0^{2\pi} \sin x \, dx = 0.$$

Interpret these results in area language.

Draw one, or see Appendix I.

Solution   As its graph ◄ shows, $\sin x \geq 0$ on $[0, \pi]$, so I_1 measures the ordinary area—2 square units—under one arch of the sine curve.

The value of I_2 means that the net, or signed, area over the interval $[0, 2\pi]$ is zero. This makes good geometric sense: $\sin x \leq 0$ on the interval $[\pi, 2\pi]$, and the symmetry of the graph guarantees that the areas above and below the *x*-axis exactly cancel each other out.   ∎

Definite Integrals: Not Just for Area Anymore. Every definite integral $\int_a^b f(x) \, dx$ *can* be interpreted as a signed area, as just illustrated. But if definite integrals measured *only* area, they wouldn't deserve the fuss we make over them. In fact, definite integrals can be used to measure or model many quantities other than area. Volume, arclength, distance, work, mass, fluid pressure, accumulated financial value, and other quantities can all be calculated as definite integrals. Choosing the *right* integral and interpreting the result appropriately depend on the problem at hand.

EXAMPLE 2   We'll show in Section 8.3 that the length of the curve $y = f(x)$ from $x = a$ to $x = b$ is given by the integral

$$\int_a^b \sqrt{1 + f'(x)^2} \, dx.$$

Use this fact to interpret the integral

$$I = \int_0^1 \sqrt{1 + 4x^2} \, dx$$

in two different ways: (i) as the length of an appropriate curve and (ii) as an appropriate area. Evaluate these quantities numerically.

Solution   The integral I fits the arclength "template" if $f'(x) = 2x$. Clearly, $f(x) = x^2$ has this property. Therefore, I gives the length of the curve $y = x^2$ from $x = 0$ to $x = 1$. ◄

Important note: The curve we're measuring is not the integrand.

Any definite integral measures the signed area bounded by its integrand. In this case, therefore, I measures the area under the curve $y = \sqrt{1 + 4x^2}$ from $x = 0$ to $x = 1$.

The following pictures illustrate both interpretations of I.

$\int_0^1 \sqrt{1 + 4x^2}\, dx \approx 1.479$: **length of a curve**

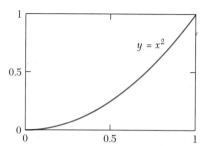

$\int_0^1 \sqrt{1 + 4x^2}\, dx \approx 1.479$: **area under a curve**

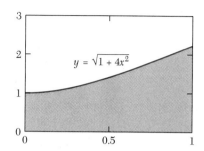

How large, numerically, is I? Antidifferentiation is algebraically messy, so we use M_{20}, the midpoint rule with 20 subdivisions. Here's the result, to three decimal places: ▶

$$I \approx M_{20} = 1.479.$$

See the exercises at the end of this section for more on this calculation.

Thus, both quantities (length and area) have the same approximate numerical value: 1.479 units of length and 1.479 units of area, respectively. ▶ ∎

The respective units might be inches and square inches.

Two Views of the Definite Integral

As the preceding example suggests, definite integrals can be used to measure many disparate quantities. Usually the key questions are which *function* to integrate, and over which *interval*.

In applying the integral in varied settings, it's useful to remember two different but closely related interpretations of a definite integral $\int_a^b f(x)\, dx$.

A Limit of Approximating Sums The integral is defined formally as a limit of approximating sums. Chapter 7 discusses and compares several kinds of approximating sums: left, right, trapezoid, midpoint, and so on. Using right sums, for instance, we can write

$$\int_a^b f(x)\, dx = \lim_{n \to \infty} \sum_{i=1}^{n} f(x_i) \Delta x,$$

where the inputs x_i are the right endpoints of n equal-length subintervals of $[a, b]$. From this point of view, the integral "adds up" small contributions, each of the form $f(x_i)\Delta x$.

Accumulated Change in an Antiderivative The fundamental theorem of calculus says that

$$\int_a^b f(x)\, dx = F(b) - F(a),$$

where, on the right, the function F can be any antiderivative of f on $[a, b]$. ▶ The difference $F(b) - F(a)$ represents, in a natural way, the **accumulated change** (or net change) in F over the interval $[a, b]$. In other words, to find the accumulated change in F over $[a, b]$, integrate f—the *rate function* associated with F—over $[a, b]$.

Remember: Antiderivatives are not unique. If F is one antiderivative of f, then so is $F + k$, where k is any constant.

Mathematically speaking, these two approaches to the integral are equivalent. The fundamental theorem of calculus says so; it guarantees that both methods give the same "answer." Having two different ways to think about the integral makes it more versatile in applications. Which viewpoint is better depends on the situation. The next example illustrates both viewpoints.

EXAMPLE 3   The function $v(t) = 10 + 20t - 10t^2$ gives a car's eastward velocity, in miles per hour at time t hours, for $0 \le t \le 3$. Calculate $I = \int_0^3 v(t)\,dt$; interpret I both as a limit of sums and as accumulated change in an antiderivative.

Solution   Calculating I symbolically is easy:

$$I = \int_0^3 v(t)\,dt = 10t + 10t^2 - \frac{10t^3}{3}\Bigg]_0^3 = 30.$$

What does the answer mean? The following pictures give two different views of I:

Graph of v: eastward velocity at time t

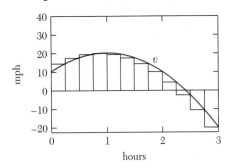

Velocity v and an antiderivative p

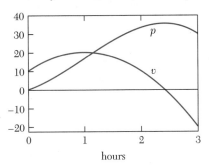

The left-hand picture suggests the limit-of-sums view of I. It shows R_{12}, the right approximating sum for I with 12 subdivisions. The signed area of each rectangle approximates the *eastward* distance covered by the car over that subinterval. (The last three rectangles contribute negative eastward distance, i.e., *westward* distance.) Therefore, the right sum R_{12} approximates the net accumulated eastward distance covered over the entire interval $0 \le t \le 3$. As the limit of such approximating sums, I represents the *exact* net eastward distance covered over $[0, 3]$—precisely 30 miles. Numerical evidence ◀ supports this idea:

We computed the right sums by machine, of course.

$$R_{12} \approx 25.938; \qquad R_{50} \approx 29.082; \qquad R_{200} \approx 29.774; \quad \dots \quad I = 30.$$

The right-hand picture illustrates the accumulated-change approach to I. Because the function v describes the car's eastward *velocity*, any antiderivative function p must describe the car's east-west *position*, given appropriate units and points of reference. Given our formula $v(t) = 10 + 20t - 10t^2$, we may as well choose the antiderivative function p defined by $p(t) = 10t + 10t^2 - 10t^3/3$. Since $p(0) = 0$, the car starts at position 0, and $p(t)$ represents the car's position at time t, measured in miles east of the starting point. Graphs of both v and p appear above. They show that

$$I = \int_0^3 v(t)\,dt = p(3) - p(0) = 30.$$

In words: Integrating the car's velocity over [0, 3] gives the accumulated change in position over the same interval. ∎

Which View: Sum or Antiderivative?

Which view of the integral is right for a given application? Should we calculate an approximating sum or look for an antiderivative?

The answer depends on the type and amount of information at hand. When an integrand presents itself in simple symbolic form, antidifferentiation is the obvious next step. But for data given graphically or in tabular form, using approximating sums is a natural strategy. The next example illustrates both possibilities.

EXAMPLE 4 It's harvest time in the Corn Belt, and Farmer Brown is about to put her new Deere 12-row combine harvester through its paces. This machine is "loaded": Farmer Brown has AC, AM, FM, CD, and 200 HP—but so does every other farmer in the county. What's really special on Brown's Deere is something completely new: a continuous graphical readout, in real time, of the machine's instantaneous rate of harvesting, in bushels of corn per minute.

Farmer Jones, one farm south, can only dream of such conveniences. His two-year-old combine can take occasional instantaneous rate-of-harvest readings, but plotting them in real time is out of the question.

Brown and Jones, neighborly rivals, agree to compare their machines' harvesting performance over a 60-minute period, starting at $t = 0$. An hour later they stop and compare their results—a plot and a table:

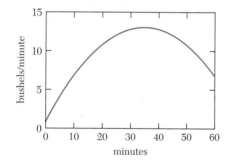

Farmer Brown's rate of harvest

Farmer Jones's Rate of Harvest					
t (min)	5	15	25	35	45	55
bu/min	4	9	12	13	12	9

Who harvests more?

Solution "Looks parabolic to me," says Farmer Brown. "That curve has a vertex at (35, 13), and it goes through (60, 27/4). Only one quadratic function has both of those properties, so my rate-of-harvest function r_B *must* have the formula

$$r_B(t) = 13 - \frac{(t - 35)^2}{100}. \blacktriangleright$$

Integrating that will tell me how many bushels of corn I accumulated over the hour." Here's her result:

$$\int_0^{60} r_B(t)\, dt = \int_0^{60} \left(13 - \frac{(t - 35)^2}{100} \right) dt = 585.$$

Why is this the only possible quadratic formula? See the exercises at the end of this section.

Farmer Jones wants to integrate *his* rate-of-harvest function, r_J. But without an explicit formula for r_j, there's nothing to antidifferentiate. "No problem," thinks Jones. "Approximating sums are always available, with or without an explicit formula for the integrand. Since I know values of r_J at the midpoints of six 10-minute intervals, I'll use M_6, the midpoint rule with six subdivisions." Here's his result:

$$\int_0^{60} r_J(t)\, dt \approx M_6 = \left(r_J(5) + r_J(15) + r_J(25) + r_J(35) + r_J(45) + r_J(55) \right) \cdot 10$$

$$= 590.$$

"Pretty close, but I'm up five bushels," says Jones.
    "Not so fast," says Brown. Why doesn't Brown concede? See the following exercises. ∎

BASICS

1. In Example 4, Farmer Brown claimed that $r_B(t) = 13 - (t - 35)^2/100$ is the only quadratic function whose graph has a vertex at (35, 13) and passes through (60, 27/4). Show this. [**HINT**: Any quadratic function can be written in the form $A + B(t - 35) + C(t - 35)^2$ for some constants A, B, and C. Assuming this, use the conditions given to find the appropriate values of A, B, and C.]

2. Why doesn't Farmer Brown concede in Example 4? To give one possible reason, find a quadratic function that "fits" Farmer Jones's data. (There is one, and only one.) Integrate this function to get another estimate for Farmer Jones's total harvest over the hour.

3. In the situation of Example 4, suppose we ignore the first and last 5 minutes and consider only the interval $5 \le t \le 55$.
   (a) Use an integral to calculate how much Farmer Brown harvests from $t = 5$ to $t = 55$.
   (b) Use the trapezoid rule to estimate how much Farmer Jones harvests over the same period.

4. This problem concerns Example 2, page 454, and the integral $I = \int_0^1 \sqrt{1 + 4x^2}\, dx$.
   (a) The left-hand picture in Example 2 illustrates the arclength interpretation of I. Use the picture to *estimate* the length of the curve $y = x^2$ from $x = 0$ to $x = 1$.
   (b) The right-hand picture in Example 2 illustrates the area interpretation of I. Use the picture to *estimate* the shaded area. Is your answer consistent with part (a)?
   (c) In Example 2 we used the midpoint approximation

$I \approx M_{20} = 1.4788$. How much error, at worst, can M_{20} commit? (Use techniques from Chapter 7.)
   (d) The integral I can be calculated symbolically, using Formula 36 from the Table of Integrals. Do so.

5. Let f be any function for which f' is continuous on $[a, b]$. (This technical requirement guarantees that the arclength integral makes sense.) Let C denote the graph of $y = f(x)$ from $x = a$ to $x = b$. We said in Example 2 that

$$\text{length of } C = \int_a^b \sqrt{1 + f'(x)^2}\, dx.$$

In Section 8.3 we'll explain why this formula holds. Here we assume that it holds and explore what it says.
   (a) Use the integral formula to find the length of the straight line segment joining (0, 0) and (a, b). Assume that $a > 0$. [**HINT**: Start by finding a function f whose graph over $[0, a]$ is the straight line segment in question. Could you have found the same answer in an easier way?]
   (b) Write an integral I that gives the length of the curve $y = x^2 + 1$ from $x = 0$ to $x = 1$. Estimate the value of I using M_{20}, the midpoint rule with 20 subdivisions.
   (c) Repeat part (b) for the two curves $y = \sin x$ and $y = 2 + \sin x$, in each case from $x = 0$ to $x = \pi$.
   (d) Let K be any real number, and let f be a well-behaved function. Use the integral formula to explain why the length of the graph $y = f(x) + K$ from $x = a$ to $x = b$ does not depend on K. Then give a geometric explanation for the same fact.

6. A moral of Example 2 is that a given integral can be interpreted in various ways. In each of the following parts, first calculate the given integral, then interpret the numerical result in the stated context using appropriate units of measurement.

   (a) $\int_0^3 f(t)\,dt$, where the function $f(t) = 5t^2 - 20t + 50$ tells a certain car's speed, in miles per hour, t hours after midnight

   (b) $\int_0^3 f(t)\,dt$, where the function $f(t) = 5t^2 - 20t + 50$ tells a certain car's speed, in feet per second, t seconds after midnight

   (c) $\int_0^3 f(t)\,dt$, where the function $f(t) = 5t^2 - 20t + 50$ tells a certain car's acceleration, in feet per minute per minute, t minutes after midnight

7. All parts of this problem concern the integral $I = \int_0^1 \sqrt{1+x}\,dx$.

   (a) Calculate I exactly by antidifferentiation. Then give a decimal equivalent, rounded to three decimal places.

   (b) I can be thought of as the area under some curve. Decide which curve, and draw the region in question. Does its area look about right, numerically?

   (c) I can also be thought of as the length of a graph $y = f(x)$, from $x = 0$ to $x = 1$. Find and plot such a function f. Does your graph appear to have about the right length?

   (d) Is more than one function f possible in part (c)? If so, find and plot another possible function.

8.2 Finding Volumes by Integration

Slicing

Imagine a solid object—a long loaf of French bread, say—lying along the x-axis in the xy-plane, between $x = a$ and $x = b$. The usual way to slice such a loaf is with cuts perpendicular to the x-axis. Mathematically speaking, the "knives" that cut such slices are vertical planes, each of the form $x = k$ for some constant k. One such cut, viewed from overhead, appears in the following picture.

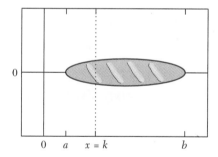

Slicing a loaf: an overhead view

The loaf's total volume is the sum of the volumes of all the slices. This obvious fact is not of much use in the kitchen. Mathematically, it lets us compute volumes by integration for a variety of solids.

Cross-Sectional Area and the Volume of One Slice

If the loaf is irregular, different slices have different volumes even if all have the same thickness. This difference arises from the fact that slices from the middle of the loaf have larger **cross-sectional area** (where the butter goes), and hence more volume, than slices from the narrow ends.

For any value of x between a and b, let $A(x)$ denote the cross-sectional area revealed by a knife cut at x (i.e., the area of the intersection of the loaf with the

For a loaf of white sandwich bread, A is essentially constant!

Mathematically, if not nutritionally.

vertical plane at x, perpendicular to the x-axis). Clearly, $A(x)$ varies with x, rising and falling with the loaf's thickness. ◄

The *exact* volume of a slice of bread is hard to compute, because the cross section $A(x)$ varies with x. In the best possible case, ◄ $A(x)$ is constant. In that event, the following simple but important fact applies:

> *If a slice with thickness Δx has constant cross-sectional area A, then the slice has volume $A \cdot \Delta x$.*

In a small x-interval (i.e., for a very thin slice), the cross-sectional area $A(x)$ is nearly constant. We can therefore approximate the volume of a slice of thickness Δx with

$$\text{Volume of one slice} \approx A(x)\Delta x,$$

where $A(x)$ is the cross-sectional area at any convenient value of x within the slice.

Reassembling Riemann's Loaf: Estimating Volume

Crumbless cuts, in this ideal world.

Cut on the dotted lines.

To estimate the volume of the loaf, let's slice it into n slabs of equal thickness $\Delta x = (b - a)/n$, with cuts ◄   at $x = x_0 = a$, $x = x_1$, $x = x_2, \ldots, x = x_n = b$. Here's an overhead view for $n = 4$: ◄

Four slices from the loaf

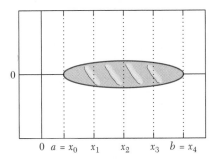

Using the cross-sectional areas $A(x_1)$, $A(x_2), \ldots, A(x_n)$ for the slabs leads to this estimate for total volume:

$$\text{Total volume} \approx A(x_1)\Delta x + A(x_2)\Delta x + \cdots + A(x_n)\Delta x.$$

Here is the crucial observation:

A right sum, to be precise.

> *This volume estimate is an approximating sum* ◄ *for the integral*
> $$\int_a^b A(x)\,dx$$
> *of the area function.*

Now we play the standard limit game: As n tends to infinity, the slab thickness Δx tends to zero, so the volume estimate tends to the true volume of the loaf. Meanwhile, the approximating sums tend to the true value of the integral.

We state our conclusion as a general theorem, applicable to *all* well-behaved solids:

> **Theorem 1** Suppose that a solid lies with its base on the xy-plane, between the vertical planes $x = a$ and $x = b$. For all x in $[a, b]$, let $A(x)$ denote the area of the cross section at x, perpendicular to the x-axis. If $A(x)$ is a continuous function, then
>
> $$\text{Volume} = \int_a^b A(x)\, dx.$$

Note:

Why the Integral Exists Requiring that $A(x)$ vary continuously with x ensures that the integral exists. For ordinary smooth, solid objects, A *is* continuous. More exotic objects can often be handled by mentally breaking them up into simpler objects to which the preceding formula applies.

Looking for a Formula Using the theorem requires an explicit formula for A. Finding one may be difficult. For the simple solids studied here, however, it's easy.

The Simplest Case For a solid with a *constant* cross section (e.g., a stick of butter or a glass of milk), the theorem merely restates elementary volume formulas. ▶

Convince yourself of this—say, for an ordinary circular cylinder.

Using the Theorem: Solids of Revolution

Among all solids, calculus users consistently prefer **solids of revolution**, i.e., solids formed by revolving some region in the xy-plane about an axis, often the x-axis. The following figure shows the solid formed by revolving the region under the curve $y = x^2$, from $x = 0$ to $x = 1$, about the x-axis:

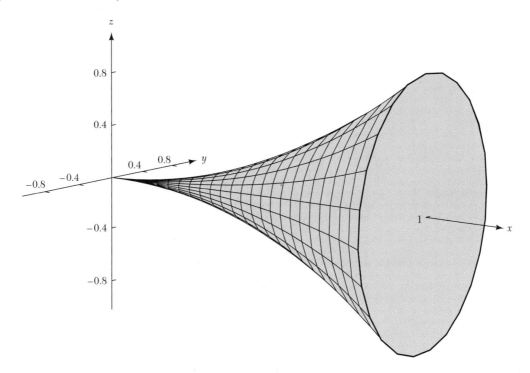

The preceding picture illustrates this.

Solids of revolution share one great advantage: Their cross sections perpendicular to the axis of rotation are *circular*. ◄ Slices of such solids, in other words, are **disks**. This property leads to simple, convenient volume computations.

EXAMPLE 1   Find the volume of the solid formed by revolving the following shaded region about the x-axis. ◄

It's the same solid that was just shown. Does either picture suggest a numerical estimate?

Rotate the shaded region about the x-axis

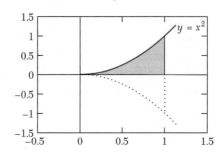

S o l u t i o n   For any x in $[0, 1]$, the cross section at x is a circle with radius $y = x^2$. Thus,

$$A(x) = \pi(x^2)^2 = \pi x^4.$$

By Theorem 14.6,

$$\text{Volume} = \int_0^1 A(x)\, dx = \int_0^1 \pi x^4\, dx = \frac{\pi}{5} \approx 0.628. \qquad \blacksquare$$

A similar computation applies whenever the region under the graph of a positive function $y = f(x)$, from $x = a$ to $x = b$, is revolved around the x-axis.

EXAMPLE 2   Find the volume of the sphere of radius r.

Draw your own picture!

S o l u t i o n   The sphere is the result of revolving the area under the curve $f(x) = \sqrt{r^2 - x^2}$, from $x = -r$ to $x = r$, about the x-axis. ◄ The cross section at x, therefore, is a circle with radius $f(x)$. Hence,

$$A(x) = \pi\big(f(x)\big)^2 = \pi(r^2 - x^2),$$

Check algebra.

and so ◄

$$\text{Volume} = \pi \int_{-r}^{r} (r^2 - x^2)\, dx = \pi \left[r^2 x - \frac{x^3}{3} \right]_{-r}^{r} = \frac{4\pi r^3}{3}. \qquad \blacksquare$$

Other Axes.   Rotation around the x-axis is not the only possibility. Whatever the axis, the principle of Theorem 1 applies: Volume is found by integrating the cross-sectional area.

EXAMPLE 3   Find the volume of the solid formed by rotating the region of Example 1 about the axis $y = -1$.

Solution   The solid formed in this way is a hollow "horn." For any x in $[0, 1]$ the cross section at x is a **washer**, ▶ with inner radius 1 and outer radius $1 + x^2$. The area $A(x)$ of such a washer is

A nuts-and-bolts-type washer.

$$A(x) = \pi \text{ (outer radius)}^2 - \pi \text{ (inner radius)}^2$$
$$= \pi (1 + x^2)^2 - \pi 1^2 = \pi (2x^2 + x^4).$$

By the theorem,

$$\text{Volume} = \int_0^1 A(x) \, dx = \pi \int_0^1 (2x^2 + x^4) \, dx = \frac{13\pi}{15} \approx 2.723. \quad ■$$

Noncircular Cross Sections

For solids with noncircular cross sections, $A(x)$ may be harder to find; otherwise, the idea is exactly the same.

EXAMPLE 4   A pyramid of height 1 has square horizontal cross sections; at the bottom, the edge length is 1. Find the volume. ▶

Draw your own picture.

Solution   Let the x-axis run vertically from $x = 0$ to $x = 1$. Since the edge length decreases *linearly* from 1 (at the base) to 0 (at the apex), the edge length $l(x)$ at height x is $1 - x$. Thus, the cross-sectional area at height x is

$$A(x) = l(x)^2 = (1 - x)^2 = 1 - 2x + x^2,$$

and the volume is

$$\text{Volume} = \int_0^1 A(x) \, dx = \int_0^1 (1 - 2x + x^2) \, dx = \frac{1}{3}. \quad ■$$

BASICS

In Exercises 1–4, sketch the region in the first quadrant bounded by the given curves, and find the volume of the solid that is formed when this region is revolved around the x-axis.

1. $y = x^3$, $y = 0$, $x = 8$

2. $y = x^4$, $y = 1$, $x = 0$

3. $y = x + 6$, $y = x^3$, $x = 0$

4. $y = x^2$, $y = x^3$

In Exercises 5–8, sketch the region bounded by the given curves, and find the volume of the solid of revolution that is formed when the region is revolved around the y-axis.

5. $y = \sqrt{x}$, $y = 0$, $x = 4$

6. $y = \sqrt{x}$, $y = x^2$

7. $y = x^2$, $y = 2^x$, $x = 0$

8. $y = e^x$, $y = 0$, $x = 0$, $x = 1$

9. Find a formula for the volume of a cone with radius r and height h. [**HINT:** Cross-sections parallel to the base are circles.]

10. Find a formula for the volume of a pyramid with height h and a square base, each edge of which has length ℓ. [**HINT:** Cross-sections parallel to the base are squares.]

FURTHER EXERCISES

11. Let $\mathcal{R}$ denote the region bounded by the curves $y = x^2$, $y = 6 - x$, and $y = 0$. Write an integral for the volume of the solid that is obtained when $\mathcal{R}$ is rotated about the line
    (a) $x = -2$ (b) $y = -1$

In Exercises 12–15, sketch the region bounded by the given curves, and find the volume of the solid of revolution that is formed when the region is revolved around the line $y = a$.

12. $y = \sqrt{x}$, $y = 0$, $x = 1$, $a = 1$

13. $y = x^2 - 1$, $y = x + 1$, $a = -1$

14. $x = y^2 - 4$, $y = 2 - x$, $a = 2$

15. $y = \sqrt{x}$, $y = x^2$, $a = -2$

16. The base of a certain solid is the region enclosed by $y = 1/x$, $y = 0$, $x = 1$, and $x = 4$. Every cross section of the solid taken perpendicular to the x-axis is an isosceles right triangle with its hypotenuse across the base. Find the volume of the solid.

17. The following table gives the circumference (in inches) of a pole at several heights (in feet). Use numerical integration (e.g., the trapezoid rule or Simpson's rule) to estimate the volume of the pole. (Assume that cross sections of the pole taken parallel to the ground are circles.)

Height	0	10	20	30	40	50	60
Circumference	16	14	10	5	3	2	1

18. Find the volume of the solid that is formed when the region bounded by the curves $y = \arctan x$, $y = 0$, and $x = 1$ is revolved around the y-axis.

19. A drinking glass has circular cross-sections. The glass has height 5 inches, bottom diameter 2 inches, and top diameter 3 inches. How much liquid can the glass hold?

20. Assume that the Earth is a sphere with circumference $C \approx 24,900$ miles.
    (a) Find the volume of the Earth north of latitude $45°$.
    (b) Find the volume of the Earth between the equator and latitude $45°$.

21. A spherical balloon with radius 3 inches is partially filled with water. If the water in the balloon is 4 inches deep at the deepest point, how much water is released when the balloon hits the wall and breaks?

22. A cylindrical gasoline tank with radius 4 feet and length 25 feet is buried on its side under a service station (i.e., its flat ends are perpendicular to the ground surface). If the gasoline in the tank is 6 feet deep, what is the volume of gasoline in the tank?

23. Let V be the volume of the solid that is formed when the region bounded by the curves $y = \arctan x$, $y = 0$, $x = -3$, and $x = -1$ is revolved around the x-axis.
    (a) Express V as a definite integral.
    (b) If the integral in part (a) is estimated using a left sum with $n = 10$ subintervals, does the result underestimate or overestimate V? Justify your answer.

24. Suppose that the volume of water required to fill a hollow object to a depth of h inches is $V(h) = 1.5h + \sin h$ cubic inches. What is the cross-sectional area of the object 1 inch above its base?

25. A fuel-oil tank is 10 feet long and has flat ends that are perpendicular to the ground surface. Cross-sections parallel to the flat ends have the shape of the ellipse $x^2/9 + y^2/36 = 1$. If the fuel oil in the tank is 9 feet deep, what is the volume of the fuel oil in the tank?

26. Suppose that two circular cylinders of radius R intersect in such a way that their axes meet at right angles. Find the volume of the space that is inside both cylinders. [**HINT**: Draw a sketch. The space that is inside both cylinders has an axis perpendicular to the axes of the cylinders. Furthermore, any cross-section of the space perpendicular to this axis is a square.]

Exercises 27–33 provide an introduction to the method of cylindrical shells.

27. Let $R > 0$ be a constant, $\Delta r = R/n$, and $r_k = k\,\Delta r$.
    (a) Show that the area of an annulus (i.e., a ring) with inner radius r and outer radius $r + \Delta r$ is $\pi(2r + \Delta r)\,\Delta r$.
    (b) Explain geometrically why

    $$\sum_{k=0}^{n-1} \pi(2r_k + \Delta r)\,\Delta r = \pi R^2.$$

    [**HINT**: The interior of a circle of radius R can be divided into a circle of radius Δr and $n-1$ concentric annuli, each with thickness Δr.]
    (c) Explain why

    $$\lim_{n \to \infty} \sum_{k=0}^{n-1} \pi(2r_k + \Delta r)\,\Delta r = \int_0^R 2\pi r \, dr.$$

    [**HINT**: First show that $\lim_{n \to \infty} \sum_{k=0}^{n-1} (\Delta r)^2 = 0$.]
    (d) Explain geometrically why

    $$\sum_{k=1}^{n} \pi(2r_k - \Delta r)\,\Delta r = \pi R^2.$$

28. Suppose that f is continuous and that $f(x) \geq 0$ if $a \leq x \leq b$. Let V be the volume of the solid obtained when the region bounded by the curve $y = f(x)$ and the lines $x = a$, $x = b$, and $y = 0$ is rotated about the y-axis.
    (a) Let $n \geq 1$ be an integer, $\Delta x = (b-a)/n$, and $x_k = a + k\,\Delta x$. Explain why

    $$V \approx \sum_{k=0}^{n-1} \pi (2x_k + \Delta x) f(x_k)\,\Delta x.$$

    [**HINT:** Think of the expression as the sum of the volume of a solid cylinder and the volumes of $n-1$ concentric hollow cylinders.]
    (b) Show that

    $$\sum_{k=0}^{n-1} 2\pi x_k f(x_k)\Delta x$$

    is a left Riemann sum approximation to the integral $\int_a^b 2\pi x f(x)\,dx$.
    (c) Show that

    $$\lim_{n\to\infty} \sum_{k=0}^{n-1} \pi (2x_k + \Delta x) f(x_k)\,\Delta x = \int_a^b 2\pi x f(x)\,dx.$$

    [**HINT:** Show that $\lim_{n\to\infty} \sum_{k=0}^{n-1} f(x_k)(\Delta x)^2 = 0$.]
    (d) Explain why it is reasonable to believe that $V = \int_a^b 2\pi x f(x)\,dx$.

29. Suppose that f is continuous and that $f(x) \geq 0$ if $a \leq x \leq b$. Let V be the volume of the solid obtained when the region bounded by the curve $y = f(x)$ and the lines $x = a$, $x = b$, and $y = 0$ is rotated about the y-axis. Also, let $n \geq 1$ be an integer, $\Delta x = (b-a)/n$, and $x_k = a + k\,\Delta x$.
    (a) Suppose that f is decreasing on the interval $[a, b]$. Explain why

    $$\sum_{k=1}^{n} \pi (2x_k - \Delta x) f(x_k)\Delta x \leq V \leq$$

    $$\sum_{k=0}^{n-1} \pi (2x_k + \Delta x) f(x_k)\Delta x.$$

    (b) Suppose that f is decreasing on the interval $[a, b]$. Show that $V = \int_a^b 2\pi x f(x)\,dx$.
    (c) Show that if f is increasing over the interval $[a, b]$, then $V = \int_a^b 2\pi x f(x)\,dx$.

30. Suppose that f is continuous and that $f(x) \geq 0$ if $a \leq x \leq b$. Let V be the volume of the solid obtained when the region bounded by the curve $y = f(x)$ and the lines $x = a$, $x = b$, and $y = 0$ is rotated about the y-axis.
    (a) Let r and Δx be numbers. Show that $\pi(r + \Delta x)^2 - \pi r^2 = 2\pi \Delta x(r + \frac{1}{2}\Delta x)$.
    [**HINT:** $x^2 - y^2 = (x - y)(x + y)$.]

(b) Let $\Delta x = (b-a)/n$ and $x_k = a + k\,\Delta x$. Explain why

$$V \approx \sum_{k=0}^{n-1} 2\pi \left(x_k + \frac{\Delta x}{2} \right) f\left(x_k + \frac{\Delta x}{2} \right) \Delta x.$$

(c) Explain why

$$\lim_{n\to\infty} \sum_{k=0}^{n-1} 2\pi \left(x_k + \frac{\Delta x}{2} \right) f\left(x_k + \frac{\Delta x}{2} \right) \Delta x$$

$$= \int_a^b 2\pi x f(x)\,dx.$$

31. Suppose that f is continuous and that $f(x) \geq 0$ if $a \leq x \leq b$. Let V be the volume of the solid obtained when the region bounded by the curve $y = f(x)$ and the lines $x = a$, $x = b$, and $y = 0$ is rotated about the y-axis.
    (a) Suppose that $|f'(x)| \leq K$ when $a \leq x \leq b$ and that $a \leq c < d \leq b$. Explain why $f(c) - K(d-c) \leq f(z) \leq f(c) + K(d-c)$ when $c \leq z \leq d$.
    (b) Let $\Delta x = (b-a)/n$ and $x_k = a + k\,\Delta x$. Use part (a) to explain why

    $$\sum_{k=0}^{n-1} \pi (2x_k - \Delta x)\big(f(x_k) - K\,\Delta x\big)\Delta x \leq V$$

    $$\leq \sum_{k=0}^{n-1} \pi (2x_k + \Delta x)\big(f(x_k) + K\,\Delta x\big)\Delta x.$$

    (c) Show that $V = \int_a^b 2\pi x f(x)\,dx$.

32. Let V be the volume of the solid of revolution that is formed when the region bounded by the curve $y = x^{1/3}$ and the lines $y = 0$, $x = 0$, and $x = 8$ is revolved around the y-axis.
    (a) Compute V using the method of disks and washers.
    (b) Compute V using the method of cylindrical shells.

33. Let V be the volume of the solid of revolution that is formed when the region bounded by the curve $y = x\sqrt{1 - x^2}$ and the lines $y = 0$, $x = 0$, and $x = 1$ is revolved around the y-axis.
    (a) Use the method of disks and washers to show that

    $$V = \int_0^{1/2} \frac{\pi}{2} \left(1 + \sqrt{1 - 4y^2} \right) dy$$

    $$- \int_0^{1/2} \frac{\pi}{2} \left(1 - \sqrt{1 - 4y^2} \right) dy.$$

    (b) Use the method of cylindrical shells to show that

    $$V = \int_0^1 2\pi x^2 \sqrt{1 - x^2}\,dx.$$

    (c) Show that $V = \pi^2/8$.

34. The rate at which a viscous fluid (e.g., water or blood) moves through a tube is not uniform. The fluid moves fastest near the center and slowest along the wall.

    Suppose that a fluid is flowing in a circular pipe of radius R centimeters and that the velocity of the fluid r centimeters from the center of the pipe is $v(r)$ centimeters per second. Explain why the rate (in cubic centimeters per second) at which the liquid is flowing through the pipe is the value of the integral

    $$2\pi \int_0^R r v(r)\, dr.$$

35. Suppose that a city is circular and that the population density decreases linearly to zero from the center to the rim.

    (a) Let K be the population density (in people/mi²) at the center of the city, and let R be the radius of the city. Explain why the population density at a distance r from the center of the city is $p(r) = K(1 - r/R)$ people/mi².

    (b) What is the population of the city? [**HINT**: Use an integral.]

8.3 Arclength

Integrals can be used to measure geometric quantities other than area and volume. **Arclength** offers another good example, which shows clearly the role of approximating sums.

EXAMPLE 1   How long is the curve $y = \sin x$ from $x = 0$ to $x = \pi$?

Part of the sine curve: how long is it?

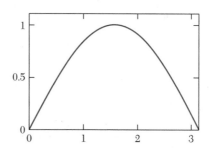

(Before continuing, try to guess the length.)

Use the distance formula.   **Solution**   Straight line segments are easy to measure. ◄ Curves pose more of a problem. In the spirit of this chapter, a natural strategy suggests itself:

Approximate the curve with straight line segments and add up their lengths.

Following is one such piecewise-linear approximation to the sine curve, built from four segments.

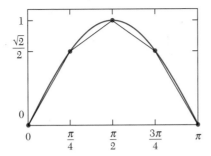

From the coordinates of the endpoints ▶ it's a routine (though tedious) matter to calculate the total length; here, it's about 3.79. ▶

They're shown as bullets.

It's OK to skip the numerical computation, but is the answer reasonable?

With more and shorter line segments, we'd expect the piecewise-linear "curves" to approach the sine curve itself. In particular, the polygonal length should approach L, the length of the sine curve. In other words, L should be the *limit* of the lengths of polygonal approximations to the curve.

To show this principle in action numerically, we tabulate the lengths of several piecewise-linear approximations to the elusive L. Each corresponds to subdividing the x-interval $[0, \pi]$ into n equal subintervals.

Approximating the Length of a Curve					
Number of segments	4	8	16	31	100
Polygonal length	3.7901	3.8125	3.8183	3.8197	3.8201

For now, 3.82 is our best guess. ■

Generic Problem, Generic Solution. The generic arclength problem reads like this:

Find the length of the f-graph from x = a to x = b.

The preceding example suggests a general strategy:

Approximate with piecewise-linear arcs, and take a limit.

Great, but how does it all work in practice? How is the integral involved? *Which* integral is involved?

To describe the situation more precisely, we need a picture.

**Approximating a curve C with
a polygonal arc Cₙ**

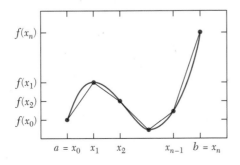

The picture has these ingredients:

1. C, the graph of f from $x = a$ to $x = b$
2. A partition of $[a, b]$ into n equal subintervals, with endpoints $x_0, x_1, \ldots, x_n$.
3. A polygonal arc C_n, made from n line segments joining the points $x_0, x_1, \ldots, x_n$ on C.

Now we can put the idea succinctly: Length of $C = \lim_{n \to \infty} (\text{length of } C_n)$.

How Long Is C_n? A Useful Estimate.   To measure C_n, we add the lengths of its n segments. The ith segment joins $(x_{i-1}, f(x_{i-1}))$ to $(x_i, f(x_i))$, so its length is

$$\sqrt{(f(x_i) - f(x_{i-1}))^2 + (x_i - x_{i-1})^2}.$$

The total length of C_n is therefore given by

$$\text{Length of } C_n = \sum_{i=1}^{n} \sqrt{(f(x_i) - f(x_{i-1}))^2 + (x_i - x_{i-1})^2}. \tag{8.1}$$

We rewrite Equation 8.1 as an approximating sum for an integral. The first step is pure algebra:

$$\text{Length of } C_n = \sum_{i=1}^{n} \sqrt{(f(x_i) - f(x_{i-1}))^2 + (x_i - x_{i-1})^2}$$

$$= \sum_{i=1}^{n} \sqrt{\left(\frac{f(x_i) - f(x_{i-1})}{x_i - x_{i-1}}\right)^2 + 1} \cdot (x_i - x_{i-1}).$$

This clumsy expression can be simplified. By the definition of derivative, the approximation

$$\frac{f(x_i) - f(x_{i-1})}{x_i - x_{i-1}} \approx f'(x_i)$$

holds if $x_i - x_{i-1}$ is small; it *is* small here if n is large.

Putting the pieces together and writing Δx_i for $(x_i - x_{i-1})$, we get

$$\text{Length of } C_n = \sum_{i=1}^{n} \sqrt{\left(\frac{f(x_i) - f(x_{i-1})}{x_i - x_{i-1}}\right)^2 + 1} \cdot (x_i - x_{i-1})$$

$$\approx \sum_{i=1}^{n} \sqrt{(f'(x_i))^2 + 1} \cdot \Delta x_i.$$

The last line is the payoff: The sum just obtained is an approximating sum for the integral

$$\int_a^b \sqrt{(f'(x))^2 + 1} \, dx.$$

The conclusion follows:

Fact (Arclength by Integration) The integral

$$\int_a^b \sqrt{(f'(x))^2 + 1} \, dx$$

gives the length of the f-graph from $x = a$ to $x = b$.

EXAMPLE 2 Write the length of the sine curve from $x = 0$ to $x = \pi$ as an integral. Estimate its value.

Solution By the preceding Fact, the length is the value of the integral

$$I = \int_0^\pi \sqrt{(\cos x)^2 + 1} \; dx.$$

The FTC won't help us evaluate I. Like many integrands that involve square roots, this one has no convenient antiderivative, so we estimate I by using the midpoint rule with 100 subdivisions. According to our computer, $I \approx 3.8202$—a reassuring result, given earlier work. ∎

Do Curves Have Length? Do piecewise-linear approximations successfully approximate a curve's arclength, as we claimed?

With modern computer graphics, the answer seems obvious. Almost every "curve" in this book, for instance, is really a collection of short line segments—shorter than the eye can perceive individually—strung end to end. In this case, appearances don't deceive: All but the very worst-behaved curves *can* be approximated in the sense just discussed.

Curves do have length.

BASICS

1. Compute the length of the line $y = x$ from $x = 0$ to $x = 1$. Does the answer agree with the usual distance formula?

2. Compute the length of the line $y = mx + b$ from $x = 0$ to $x = 1$. Does the answer agree with the usual distance formula?

3. Let C be the curve described by the equation $y = (1/3)x^3 + (1/4)x^{-1}$ over the interval $1 \le x \le 3$.
   (a) Sketch the curve C, and estimate its length.
   (b) Compute the length of C exactly. [**HINT:** If you do your algebra correctly, the square root disappears.]

4. Let C be the curve $y = (e^x + e^{-x})/2$ from $x = 1$ to $x = 4$. Find the length of C exactly.
   [**HINT:** $\left(\frac{1}{2}(e^x + e^{-x})\right)^2 = 1 + \left(\frac{1}{2}(e^x - e^{-x})\right)^2.$]

FURTHER EXERCISES

5. Explain in words why the length approximation *increases* as the number of linear segments increases.

6. Find the length of the curve $y = x^2$ from $x = 1$ to $x = 2$ within ± 0.005.

7. Find the length of the curve $y = e^x$ from $x = 0$ to $x = 1$ within ± 0.005.

8. Find a number α such that the length of the curve $y = x^2$ between $x = 1$ and $x = \alpha$ is greater than 8 and less than 9.

9. Let I be the length of the curve $y = f(x)$ from $x = a$ to $x = b$.
   (a) Suppose that the function f is increasing and concave down on the interval $[a, b]$. Show that L_n, the left-sum approximation computed with n equal subintervals, overestimates the value of the arclength integral.
   (b) Suppose that $4 \le g'(x) \le 7$ and $-3 \le g''(x) \le -2$ when $0 \le x \le 1$. Find an upper bound on the approximation error made when the arclength of the curve $y = g(x)$ from $x = 0$ to $x = 1$ is approximated by L_{10}.

10. Let J be the length of the curve $y = f(x)$ from $x = 0$ to $x = 4$ and let f' be the function shown. [**NOTE:** The graph is of f', not f.]

Graph of f'

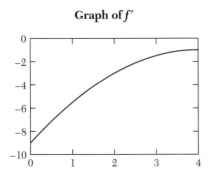

(a) Use the information in the following table to compute L_4, the left-sum approximation to J, using four equal subintervals.

x	0	1	2	3	4
$f'(x)$	-9	-5.5	-3	-1.5	-1

(b) Is $L_4 < J$? Justify your answer.

11. At time $t = 0$, a bug starts crawling along the path $y = \frac{1}{3}(x^2 + 2)^{3/2}$ with constant speed $v = 6$ feet per minute. If the bug starts its journey at the point $(1, \sqrt{3})$, where (approximately) is it 5 minutes later?

8.4 Work

How much work does a rocket do in lifting a satellite into orbit? How much work does a car do on the 934-mile drive from Perham, Minnesota, to Durham, North Carolina? How much work does a horse do in plowing the back 40? Is it more work to raise a 60-pound bucket from a 60-foot well or a 50-pound bucket from a 70-foot well?

Those are all good questions, with thoroughly practical consequences. Any kind of work requires fuel (hydrogen, gasoline, hay, cheeseburgers, ...). It's important to know how much is needed "on board." **Work** is the common ingredient (and the common word) in each situation mentioned. We need a mathematically useful definition.

The Meaning of Work

To a physicist, digging ditches is a lot of work; filing tax returns is almost none.

The everyday meaning of "work" is all too familiar; the word encompasses everything from completing tax returns to digging ditches. Physicists have a narrower, more precise definition. ◄ Work is done when a *force acts through a distance*. Without movement, *no* work is done; for instance, a house does no work by holding up its roof. Even Atlas, the Titan forced to bear the world on his shoulders, accomplished no work (except, perhaps, in lifting the globe)—fatiguing though his labors must have been.

Simplifying Assumptions: Are They Reasonable?

Throughout this section, we assume:

> *A force (either constant or variable) acts along a straight line; movement occurs along the same line.*

These assumptions are convenient, but are they reasonable? After all, few real-life physical phenomena are that simple. Forces may act at *angles* to the direction of motion, and motion may occur along *curves*. Moving cars, for instance, encounter many forces—gravity, air drag, rolling resistance, crosswinds, and so on—all of them varying over the course of a trip. Given the complex nature of physical reality, one might expect that nothing useful could survive our simplifying assumptions.

In fact, much does survive. Some interesting physical phenomena conform to our "straight-line" assumptions. Complicated combinations of forces, ▶ moreover, can often be understood as sums of simpler forces, each acting as we have assumed.

On moving cars, for example.

Work Done by a Constant Force

In the simplest instance, the force is constant along its line of action, and the "obvious" definition applies:

$$\text{Work} = \text{force} \times \text{distance}.$$

The units of work reflect this definition. In the United States, lifting a 30-pound toddler 3 feet to hip height, for instance, takes

$$30 \text{ pounds} \times 3 \text{ feet} = 90 \text{ foot-pounds}$$

of work. In Canada, ▶ lifting the same child to the same hip takes

$$133.44 \text{ newtons} \times 0.914 \text{ m} \approx 121.964 \text{ newton-meters}$$

of work. Other work units, such as inch-ounces and mile-pounds, are possible, but all have the same "dimension": units of *distance* times units of *force*.

In the metric system the basic unit of force is the newton, which is approximately 0.225 pound.

Work Done by a Variable Force

In typical physical situations, work is performed by a force that *varies* along its line of action. Even hoisting a toddler can be construed this way: On the way up, the child moves away from the center of the earth and so becomes "lighter." (For practical purposes, this effect is small enough to ignore. ▶) In more interesting situations, the force varies more dramatically:

But see Example 4 later in this section.

Springs   The farther a spring is stretched from its "natural length," the greater the force with which it pulls back. According to **Hooke's law**, an ideal spring's restoring force $F(x)$ is proportional to x, the distance the spring is stretched. Thus,

$$F(x) = kx$$

for some constant k, called the **spring constant**. ▶

Gravity   An object's weight decreases as it moves away from the earth. **Newton's law of universal gravitation** quantifies this commonplace observation: ▶

The force of earth's gravity on an object is inversely proportional to the square of the distance from the object to the center of the earth.

Thus, the force F required to lift an object (i.e., to counteract gravity) is

$$F(x) = \frac{k}{x^2}$$

if the object is x miles from the center of the earth. The numerical value of k ▶ depends on the object and on the units of measurement.

The numerical value of k depends on the units of measurement and on the "stiffness" of the spring.

Commonplace now, but not in Newton's time.

k is the proportionality constant.

Buckets and Ropes   The force required to draw water from a well decreases as the bucket rises—the higher the bucket, the less rope to be pulled up. We'll pursue this observation in a later example.

Pumping Iron   One way to "press" a heavy barbell is to exert a *constant* upward force (equal to the barbell's weight) all the way from shoulder level to the top of one's reach. Although physically simple, this strategy is not physiologically ideal. The mechanics of the human body make it easier to exert more upward force at some levels of arm extension than at others. Therefore, to the extent that the rules permit, a human weight lifter exerts a *varying* upward force through a barbell's travel. (But note: Whatever the weight lifter's strategy, the work of lifting is the same.)

Such examples argue for a definition that permits *variable* forces.

Definition (Work)   Let F be a continuous function. If the force $F(x)$ acts along an axis from $x = a$ to $x = b$, then the work done by the force F is

$$\int_a^b F(x)\,dx.$$

After a simple example, we'll discuss why this definition is reasonable.

EXAMPLE 1   Stretching a spring 1 foot beyond its natural length requires 10 pounds of force. How much work is done in stretching the spring from rest to 1 foot? From rest to a feet? How does the answer depend on a?

Solution   By Hooke's law, $F(x) = kx$. The given conditions say that $F(1) = 10$, so $k = 10$. By the preceding definition,

$$\text{Work to stretch 1 ft} = \int_0^1 10x\,dx = 10\frac{x^2}{2}\bigg]_0^1 = 5 \text{ foot-pounds};$$

$$\text{Work to stretch } a \text{ ft} = \int_0^a 10x\,dx = 10\frac{x^2}{2}\bigg]_0^a = 5a^2 \text{ foot-pounds}.$$

The last answer reveals an interesting relationship between force and work:

The force required to stretch a spring a feet from rest is proportional to a; the work done in the process is proportional to a^2.   ■

Work as an Integral: Why the Definition Makes Sense

Why is it "right" to define work as an integral? Does the integral really represent a physically meaningful quantity, one that deserves the name "work"?

Any sensible definition of work done by a variable force should at least agree with the simpler definition of work done by a constant force. Ours does: If $F(x) = k$, then

$$\text{Work} = \int_a^b F(x)\,dx = \int_a^b k\,dx = k(b - a) = \text{force} \times \text{distance}.$$

So far, so good; but why is this definition right for variable forces? We'll give two arguments.

Adding Small Contributions: The Integral as a Sum

The work integral $\int_a^b F(x)\,dx$ can be thought of as a limit of approximating sums. Specifically, let's slice the interval $[a, b]$ into n equal subintervals, each of length Δx, and let x_i be the right ▶ endpoint of the ith subinterval. Then

Left endpoints or midpoints would have done as well.

$$\int_a^b F(x)\,dx = \lim_{n \to \infty} \sum_{i=1}^n F(x_i)\,\Delta x.$$

The following picture should look familiar: ▶

Similar pictures appear in Section 5.4 and in Chapter 7.

Approximating work: a right sum

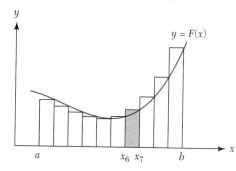

The sum $\sum_{i=1}^n F(x_i)\,\Delta x$ is the key quantity. Each summand $F(x_i)\,\Delta x$ represents the work done by the *constant* force $F(x_i)$ over the subinterval $[x_{i-1}, x_i]$. (The shaded area represents the work done over the subinterval $[x_6, x_7]$.) The sum adds up these small contributions. (The entire polygonal area represents this total.) Here's the punchline:

Each approximating sum is an estimate to the total work done over the full interval $[a, b]$.

As $n \to \infty$, therefore, the sums tend (by definition of the integral) to $\int_a^b F(x)\,dx$. Therefore, the integral plausibly measures the "true" work done.

Increments and Elements: A Mnemonic Argument

As further evidence for the integral definition of work, we offer the following mnemonic argument, using increments and elements.

Let x be any point in the interval $[a, b]$. Consider the tiny interval $[x, \ x+\Delta x]$, with **increment of distance** Δx. Because F remains essentially constant over any tiny interval, ▶ the work ΔW done over the interval $[x, \ x + \Delta x]$ satisfies

Every continuous function has this property.

$$\Delta W \approx F(x)\,\Delta x.$$

Letting Δx tend to zero now gives $dW/dx = F(x)$, or

$$dW = F(x)\,dx;$$

dW is called the **element of work.** "Adding up" these elements by integration gives the total work done from $x = a$ to $x = b$:

$$\text{Total work} = W = \int dW = \int_a^b F(x)\,dx.$$

(The equation $W = \int dW$ should be understood informally; it means simply that the integral adds up small contributions.)

Calculating Work: Miscellaneous Examples

The integral definition of work is simple enough. Deciding what to integrate and over what interval is usually the hardest part of applying the formula.

Working Against Gravity

Earth's gravity affects even very distant objects; it never dies out completely. How, then, is it possible to put objects into orbit? Calculus gives the answer.

EXAMPLE 2   Communications satellites weigh around 1000 pounds and orbit the earth at altitudes of about 15,000 miles. How much work is done against earth's gravity in lifting such a satellite into orbit?

Solution   By Newton's law of gravitation, discussed earlier in this section, the (variable) force F needed to counteract earth's gravity is

$$F(x) = \frac{k}{x^2},$$

where k is a constant and x denotes the distance from Earth's center. The fact that the satellite weighs 1000 pounds on earth's surface—about 4000 miles from earth's center—means that (allowing for approximation)

$$1000 = \frac{k}{4000^2} \implies k = 1.6 \times 10^{10} \implies F(x) = \frac{1.6 \times 10^{10}}{x^2}.$$

(Here, force is measured in pounds.) Therefore, the work done (in mile-pounds) in moving from $x = 4000$ to $x = 19,000$ ◄ is found by an easy integral calculation:

$x = 19,000$ *corresponds to an "altitude" of 15,000 miles.*

$$\int_{4000}^{19000} \frac{1.6 \times 10^{10}}{x^2} = -\frac{1.6 \times 10^{10}}{x}\Bigg]_{4000}^{19000} = 3,157,895 \text{ mile-pounds.}$$

That's heavy lifting, but not impossibly heavy—a loaded Boeing 747 accomplishes about the same work against gravity to reach cruising altitude.   ■

EXAMPLE 3   How much *additional* work is required to lift the satellite another 85,000 miles, to an altitude of 100,000 miles?

Solution   The next 85,000 miles of altitude come almost free. The extra work done in raising the satellite from $x = 19,000$ to $x = 104,000$ miles is a relatively trifling

$$\int_{19000}^{104000} \frac{1.6 \times 10^{10}}{x^2}\,dx = -\frac{1.6 \times 10^{10}}{x}\Bigg]_{19000}^{104000} \approx 688,259 \text{ mile-pounds,}$$

less than 22% of the work required to gain the first 15,000 miles of altitude.

A similar computation gives the work $W(a)$ done in lifting the satellite from the earth to *any* altitude a:

$$W(a) = \int_{4000}^{a} \frac{1.6 \times 10^{10}}{x^2} = -\frac{1.6 \times 10^{10}}{x}\bigg]_{4000}^{a} = 10^6\left(4 - \frac{16{,}000}{a}\right).$$

A close look at the last expression reveals a surprising conclusion, with important consequences for space flight:

> *Lifting the satellite to any altitude a, no matter how great, requires no more than 4×10^6 mile-pounds of work.*

Plotting W against a shows that W is bounded above—and suggests why space flight is possible.

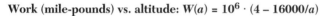
Work (mile-pounds) vs. altitude: $W(a) = 10^6 \cdot (4 - 16000/a)$

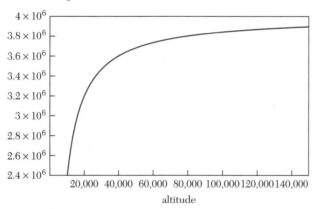

EXAMPLE 4   Gravity's force diminishes, however slightly, when a 30-pound toddler is lifted 3 feet to hip height. We argued earlier that this effect is negligible. Is it?

S o l u t i o n   Ignoring Newton's law, we calculated 90 foot-pounds of work. Now let's take Newton's law into account. All the computations are similar to the earlier ones; only the numbers are different. We need these facts: ▶   *Check details.*

- At $x = 4000$, $F(x) = k/x^2 = 30$, so $k = 4.8 \times 10^8$.
- Converting feet to miles gives 3 feet ≈ 0.0005681 miles. Thus, the toddler's vertical travel is from $x = 4000$ to $x = 4000.0005681$ miles.
- One mile-pound equals 5280 foot-pounds.

All the ingredients are now in place. The total work done, in foot-pounds, is

$$5280\int_{4000}^{4000.0005681} \frac{4.8 \times 10^8}{x^2}\, dx = -5280\,\frac{4.8 \times 10^8}{x}\bigg]_{4000}^{4000.0005681} \approx 89.99998719.$$

The "savings" from diminishing gravity is therefore only 0.00001281 foot-pounds—about the work done when a medium-sized raindrop falls 1 foot.

Guess before computing anything!

EXAMPLE 5   Is it more work to raise a 70-pound bucket of water from a 50-foot well or a 50-pound bucket from a 70-foot well? Rope weighs 0.25 pounds per foot. ◀

Draw a diagram illustrating the quantities involved.

Solution   Let x denote height, i.e., the distance (in feet) from either bucket to the bottom of its well. As the first bucket travels from $x = 0$ to $x = 50$, the force required to raise it varies with x. At height x, this force must counteract both the weight of the bucket and the weight of $50 - x$ feet of rope. ◀   Thus, for the first bucket, the force function is

$$F(x) = 70 + 0.25 \cdot (50 - x) = 82.5 - 0.25x;$$

the work done is

$$\int_0^{50} F(x)\, dx = \int_0^{50} (82.5 - 0.25x)\, dx = 3812.5 \text{ foot-pounds.}$$

For the second bucket, the computation is almost identical. This time

$$F(x) = 50 + 0.25 \cdot (70 - x) = 67.5 - 0.25x;$$

the work done is

$$\int_0^{70} F(x)\, dx = \int_0^{70} (67.5 - 0.25x)\, dx = 4112.5 \text{ foot-pounds.}$$

Because work = force × distance.

The answer stands to reason—the same amount of work is done on each bucket, ◀ but drawing up the longer rope takes more work. ■

BASICS

1. Suppose that the spring in a bathroom scale is stretched by 0.06 inches when a 115-pound person stands on the scale.

   (a) How much work is done on the spring when a 115-pound person steps on the scale?

   (b) How much work is done on the spring when a 175-pound person steps on the scale?

2. The cable attached to an elevator car weighs 5 pounds per foot. If 200 feet of cable must be wound onto a pulley to lift the car from the bottom floor to the top floor of a building, how much work is done in lifting just the cable?

3. A cylindrical tank with radius 5 feet and height 10 feet is half-filled with water. How much work must be done to pump all the water over the top rim of of the tank? (The density of water is 62.4 pounds per cubic foot.) [**HINT**: Slice the tank into layers, each of thickness Δx.

The layer of water with its bottom at height x needs to be pumped a distance of $10 - x$ feet.]

4. Is it more work to raise a 60-pound bucket from a 60-foot well or a 50-pound bucket from a 70-foot well? Rope weighs 0.25 pounds per foot.

5. According to Hooke's law, the force $F(x)$ necessary to stretch or compress an ideal spring x units away from its natural length is proportional to x; i.e., $F(x) = kx$. (The value of k, called the spring constant, depends on the spring.)

   A certain spring has a natural length of 18 inches; a force of 10 pounds is enough to compress it to a length of 16 inches.

   (a) What is the value of the spring constant k?

   (b) How much work is done in compressing the spring from 16 inches to 12 inches?

6. Redo Example 2 in this section, but assume that the orbit is 20,000 miles from Earth's center.

FURTHER EXERCISES

7. A cylindrical gasoline tank with radius 4 feet and length 15 feet is buried on its side under a service station. The top of the tank is 10 feet underground; its flat ends are perpendicular to the ground surface. Find the amount of work needed to pump all the gasoline in the tank to a nozzle that is 3 feet above the ground. (Gasoline weighs 42 pounds per cubic foot.)   [**HINT:** Each horizontal cross-section of the tank is a rectangle, one side of which has length 15 feet.]

8. A bucket that weighs 80 pounds when filled with water is lifted from the bottom of a well that is 75 feet deep. However, the bucket has a hole in it, so it weighs only 40 pounds when it reaches the top of the well. Assuming that water leaks from the bucket at a constant rate and that the rope weighs 0.65 pounds per foot, find the work required to lift the bucket from the bottom of the well to the top.

9. It takes $40x$ pounds of force to keep a certain spring stretched x feet from rest.
   (a) Find the work done in stretching the spring 2 feet, starting from rest.
   (b) Find the work done in stretching the same spring s feet from rest. (The answer depends on s.)
   (c) Farmer Ole's aging plow horse, Sven, can do only 10,000 foot-pounds of work on his daily oats ration. Ole hitches Sven to the spring described above. How far can Sven pull?

10. It takes kx pounds of force to keep a spring stretched x feet from rest.

   (a) How much work is done in stretching the spring 10 feet from rest? (The answer depends on k.)
   (b) How much work is done in stretching the spring from $x = a$ feet to $x = a + 10$ feet? (The answer depends on both k and a.)

11. A certain spring exerts $4x$ pounds of force when stretched x feet from rest. One end is fixed to the ceiling. A chain that is 10 feet long and weighs 2 pounds per foot hangs from the other end. The end of the chain just brushes the floor. Find the work done in pulling down on the chain a distance of 2 feet. Don't ignore the weight of the chain.

12. An object weighing k pounds slides without friction along the graph of $y = f(x)$. (The x and y units are feet.) A horizontal force $F(x)$ is applied to move the object from $x = a$ to $x = b$. Physical intuition says that at x, the horizontal force $F(x)$ required is proportional to the slope of the graph. (Right? What should $F(x)$ be if the graph is horizontal at x? Vertical?) In other words, $F(x) = kf'(x)$.
   (a) Find the work done if $k = 10$, $a = 0$, $b = 1$, and $f(x) = x$.
   (b) Find the work done if $k = 10$, $a = 0$, $b = 1$, and $f(x) = x^3$.
   (c) Find the work done if $k = 10$, $a = 0$, $b = 1$, and $f(x) = x^n$, for n any positive integer.
   (d) Explain the relations among your answers to parts (a)–(c).

8.5 Present Value

The Idea: Money Grows

Anyone, given the choice, would prefer a dollar today to a dollar tomorrow. This makes excellent sense. A dollar tomorrow is worth less than a dollar today, because today's dollar, prudently invested, can earn some interest by tomorrow.

But *how much* less is a dollar tomorrow worth than a dollar today? What is an advertised $1 million lottery prize really worth if it's paid in 20 yearly $50,000 installments? Given the choice, should I collect my salary in annual, monthly, or weekly installments, or does it matter? How much should I deposit now to cover 4 years of college tuition starting in 10 years?

All these questions ▶ concern what economists call **present value**, i.e., the value *now* of a future payment or "stream" of payments. Present-value calculations, using integral calculus, are vital in economic decisionmaking, e.g., in choosing among competing investment options.

We'll answer some of them.

The Present Value of One Future Payment

The first and most important example illustrates the simplest case: the present value of *one* future payment. Notice especially that exponential functions are involved and that present value depends on an interest rate.

E X A M P L E 1   A savings bond will pay $1000 on its maturity date, 10 years from today. What is it worth now, assuming that 5% compound interest is available? What if 10% interest were available?

S o l u t i o n   Let $V(t)$ denote the bond's value t years from now. We're given that $V(10) = 1000$, and we want to find $V(0)$.

To start, recall this important fact: Under continuously compounded interest, the value V grows *exponentially*. In other words, $V(t)$ has the form

$$V(t) = V(0)e^{rt},$$

where r is the annual interest rate. (Remember why? Here's a quick refresher. Growth at continuously compounded interest rate r means, in differential equation language, that for all t,

$$V'(t) = rV(t).$$

See for yourself, by differentiation, that $V' = rV$.

It's easy to check that for any constant C, the function $V(t) = Ce^{rt}$ solves this DE. ◄ Since $V(0) = Ce^0 = C$, it follows that $V(0)$ is the "right" value for C.)

Knowing that $V(10) = 1000$, we can solve for $V(0)$. For any interest rate r,

$$V(10) = V(0)e^{10r} = 1000 \implies V(0) = 1000e^{-10r}.$$

Thus, if $r = 0.05$, $V(0) = 1000e^{-0.5} \approx 606.53$; if $r = 0.1$, $V(0) = 1000e^{-1.0} \approx 367.88$. These calculations mean that the **present value** of a single $1000 payment 10 years in the future is about $607, assuming 5% annual interest, or about $368, assuming 10% interest.   ■

The preceding example prompts three general remarks:

A Negative Exponential The present value, PV, of one future payment depends on three things: P, the size of the payment; t, the time until the payment is made; and r, the interest rate. We found that

$$PV = Pe^{-rt}.$$

The general form of this expression will reappear soon. The negative exponent reflects the fact that the present value of a future payment is normally *less* than the future payment itself.

Backward in Time The present value of a future payment P can be thought of as the amount to be deposited now, at interest rate r, to accrue to value P after time t. Equivalently, one can think of following an investment's value *backward* in time, from value P at time t to value PV at time zero.

Simplifying Assumptions Present-value calculations are simplest if we assume that (i) the interest rate r remains constant over time, and (ii) interest is compounded continuously. (The first assumption is just for convenience; the second lets us use calculus.) We make both of these assumptions in most of what follows.

The Discrete Case: Present Value of Several Future Payments

Many cases of interest involve several payments, occurring at several future times. The *total* present value of a sequence of payments is found, naturally, by addition.

EXAMPLE 2   A lottery jackpot, although advertised as $1 million, is paid in 20 annual installments of $50,000. What's the present value of the prize, assuming 6% annual interest? What if 8% interest were available?

Solution   By the preceding formula, the present value of one $50,000 payment k years in the future is $50{,}000e^{-0.06k}$. Thus, the total present value of all 20 payments is

$$50{,}000(e^{-0.06 \times 0} + e^{-0.06 \times 1} + e^{-0.06 \times 2} + \cdots + e^{-0.06 \times 19}) \approx \$599{,}983.$$

That's much less than the advertised million. At 8% interest, the computation is similar; the present value is

$$50{,}000(e^{-0.08 \times 0} + e^{-0.08 \times 1} + e^{-0.08 \times 2} + \cdots + e^{-0.08 \times 19}) \approx \$519{,}033. \quad \blacksquare$$

Several Payments: The General Form.   Example 2 illustrated the general picture. If r is the interest rate, and payments $P_1, P_2, \ldots, P_n$ are made at times $t_1, t_2, \ldots, t_n$, then the present value of all n payments is found by summation:

$$PV = \sum_{i=1}^{n} P_i e^{-rt_i}. \tag{8.2}$$

The Continuous Case: Present Value of an Income Stream

In real life, financial transactions occur at specific moments—the beginning of a month, the end of a pay period, 11:59 P.M. on April 15, and so on. Nevertheless, economists often picture income as an unbroken "stream," flowing continuously in time (hence, for instance, the term "cash flow"). The next example—in which all three graphs are drawn as continuous, unbroken curves—illustrates this point of view. Notice that the graphs show the *rates* of income flow as functions of time.

EXAMPLE 3   The following graphs ▶ show varying rates of daily income, over a 360-day "year," for three idealized workers: a tax consultant, an ice cream vendor, and a factory worker. (Which is which? Day 0 is January 1; U.S. tax returns are due April 15.)

Each one represents an income stream.

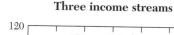

Three income streams

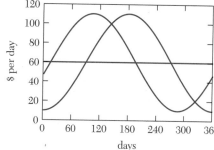

How much does each worker earn over the whole year? Who's best off? Why?

Our fable occurs north of the equator. How do the graphs show this?

Solution   First, which graph is which? Judging from their peaks (around April 15 [day 105] and July 1 [day 180]), the two sinusoidal graphs represent the tax consultant and the ice cream vendor. ◄   The straight-line graph therefore represents the factory worker's constant $60 daily wage. Label the graphs T, I, and F.

The factory worker's total annual income is easiest to calculate. At $60 per day for a 360-day year, the factory worker earns $21,600—the area under the F-graph from $t = 0$ to $t = 360$. The total incomes of the other two workers can also be found as areas, under the T- and I-graphs, respectively. A close look reveals a mild surprise. All three graphs enclose the *same* total area, ◄   so all three workers earn the same total annual income: $21,600.

Take a close look now; convince yourself.

Nevertheless, the workers are not (quite) equally well off. The tax consultant is luckiest. T receives more income early in the year than I and F do. For now that's all we can say. Soon we'll *quantify* T's advantage by calculating the present value of each income stream at $t = 0$. ∎

Fair Assumptions?   We made several simplifying assumptions in Examples 1–3: that interest rates remain constant over time, that interest is compounded continuously, that daily earnings follow simple patterns, and so on. None of these assumptions need be literally true. For instance, few factory workers are paid by the day, including weekends. Ice cream sales rise and fall with temperature, not just with the time of year.

Do such simplifications render our results useless? Certainly not. *All* economic decisions require some combination of calculation and guesswork. To some extent, the future is always unknowable. In practice, making sensible simplifying assumptions is inevitable—and it's standard operating procedure.

Why Treat Income as a Stream?   Treating income as a continuous stream lends itself nicely to calculus; functions, derivatives, and integrals all arise naturally. From this point of view, an income stream is a function p of time: At time t, $p(t)$ is the (instantaneous) *rate* at which income "flows" to the recipient. It follows, then, that the integral $\int_a^b p(t)\, dt$ is the total income received over the time interval $a \le t \le b$. (That's why we looked at areas under the curves in Example 3.)

After stating and using the definition, we'll say why it's sensible.

Present Value as an Integral: The Definition.   Building on the preceding examples, let's define ◄   the present value of a continuous income stream. Three ingredients are involved:

- A time scale t; $t = 0$ corresponds to the *present time*.

In practice, r is not always constant, of course.

- An interest rate r, usually taken as a constant. ◄

That is, the income stream starts at time a and ends at time b.

- A continuous income-stream function $p(t)$, defined for $a \le t \le b$. ◄   At time t, $p(t)$ tells the rate of income flow with respect to time.

Finally, an important caution about compatible units of measurement:

The quantities t, r, and p must use the same time unit.

For example, if t measures days, then r is the *daily* interest rate and $p(t)$ is measured in money units per *day*.

> **Definition** Let t, r, and p be as before. The **present value of the income stream** p is defined by the integral
>
> $$PV = \int_a^b p(t)e^{-rt}\, dt.$$

First we'll use the definition, then defend it as sensible. In the meantime, notice the similarity to Equation 8.2.

EXAMPLE 4   Assuming a 5% annual interest rate, find the present value (at the beginning of the 360-day year) for each of the three workers T, I, and F of Example 3. How different are the results if 10% interest is available?

Solution   Let t denote time in days; then $t = 0$ is the present time, and the income flow lasts from $t = 0$ to $t = 360$. Because t involves days, r should measure the daily, not yearly, interest rate. Here, therefore, $r = 0.05/360$.

The three stream functions—we'll call them p_T, p_I, and p_F—are shown graphically ► on page 479. To find our present values we need formulas for p_T, p_I, and p_F. One formula can be "read" easily from the graphs: $p_F(t) = 60$. The other two graphs look periodic; with some effort we could produce their formulas from trigonometric ingredients. ►   Here we just state them:

Look again.

See the exercises at the end of this section.

$$p_T(t) = 50\cos\left(\pi \cdot \frac{t - 105}{180}\right) + 60; \qquad p_I(t) = 50\cos\left(\pi \cdot \frac{t - 180}{180}\right) + 60.$$

The rest of the solution consists of straightforward (though messy) calculation. Here are the results. ►   With $r = 0.05/360$,

We spare you the calculation—but see the exercises at the end of this section.

$$PV_F = \int_0^{360} 60e^{-0.05t/360}\, dt \approx 21068.89;$$

$$PV_T = \int_0^{360} \left(50\cos\left(\pi \cdot \frac{t - 105}{180}\right) + 60\right) e^{-0.05t/360}\, dt \approx 21203.55;$$

$$PV_I = \int_0^{360} \left(50\cos\left(\pi \cdot \frac{t - 180}{180}\right) + 60\right) e^{-0.05t/360}\, dt \approx 21067.78.$$

As predicted, T comes out on top, ►   about \$135 ahead of the others.

At a 10% annual interest rate—i.e., $r = 0.1/360$—the tax consultant's advantage is a little greater:

Tax consultants usually seem to.

$$PV_F = \int_0^{360} 60e^{-0.1t/360}\, dt \approx 20555.12;$$

$$PV_T = \int_0^{360} \left(50\cos\left(\pi \cdot \frac{t - 105}{180}\right) + 60\right) e^{-0.1t/360}\, dt \approx 20817.26;$$

$$PV_I = \int_0^{360} \left(50\cos\left(\pi \cdot \frac{t - 180}{180}\right) + 60\right) e^{-0.1t/360}\, dt \approx 20550.78. \qquad \blacksquare$$

Why the Integral Definition Makes Sense. The typographical similarity between

$$\sum_{i=1}^{n} P_i e^{-rt_i} \quad \text{and} \quad \int_a^b p(t) e^{-rt}\, dt$$

(the formulas for present value in the discrete and continuous cases, respectively) is no accident. Roughly speaking, the integral is the continuous version of the discrete sum.

Remember partitions? See, e.g., Section 5.4.

To relate the integral and the sum more precisely, we approximate a continuous income stream with a discrete sequence of payments. Suppose, then, that for $a \le t \le b$, $p(t)$ describes a continuous income stream, and let $a = t_0 < t_1 < t_2 < \cdots < t_n = b$ partition the interval $[a, b]$ into n short subintervals. ◄ We approximate the income stream p with a finite sequence of payments $P_1, P_2, \ldots, P_n$, occurring at times $t_1, t_2, \ldots, t_n$, respectively. ◄

Payments P_i occur at the end of each subinterval.

How large should each payment P_i be? Over a (short) subinterval $[t_{i-1}, t_i]$, the stream function p doesn't change much (because p is continuous), so the estimate $p(t) \approx p(t_i)$ is reasonable. ◄ Thus, the *total income* received over the time interval $[t_{i-1}, t_i]$ is approximately $p(t_i)\,\Delta t_i$. In our (approximate) discrete version, therefore, it's natural to use $P_i = p(t_i)\,\Delta t_i$. In summary, we approximate the continuous stream p, defined for $a \le t \le b$, with n separate payments; at time t_i, the payment is $P_i = p(t_i)\Delta t_i$. As n tends to infinity, we expect our approximation to improve.

We pretend, in effect, that $p(t)$ is constant on each subinterval.

We saw earlier that the total present value of these n payments, given interest rate r, is

$$\sum_{i=1}^{n} P_i e^{-rt_i} = \sum_{i=1}^{n} p_i e^{-rt_i}\,\Delta t_i.$$

The key point is that the right side is an approximating sum for $\int_a^b p(t)e^{-rt}\, dt$—just the integral we've been waiting for. Therefore, as n tends to infinity, the approximation tends both to the desired integral and to the desired present value.

BASICS

1. Al and Bob, both 42 years old, hope to retire as millionaires. To that end, they'll deposit money now to accrue to $1 million by the time they're 65.
   (a) How much should Al deposit now if 6% interest is available?
   (b) How much should Al deposit now if 8% interest is available?
   (c) Bob has $100,000 on hand. What interest rate will he need to meet his retirement goal?

2. Christine, now 20, wants to retire at 65 with a $1 million nest egg. How much should she deposit now at 6%? At 8%? How much should she invest in speculative junk bonds that pay 20%?

3. **Real interest rates.** Inflation, the rise of prices over time, reduces the buying power of future money. In effect, inflation imposes negative interest on a deposit. Economists "control for inflation" by defining the real interest rate as

$$\begin{array}{c} \text{Real} \\ \text{interest rate} \end{array} = \begin{array}{c} \text{nominal} \\ \text{interest rate} \end{array} - \text{inflation rate}$$

Predicting future inflation necessarily involves guesswork; so, therefore, must the real interest rate. Historically, real interest rates have tended to hover in the range 2% to 4%.

Anne and Betty, now 42 years old, hope to retire as *real* millionaires, i.e., as millionaires after inflation. They can accomplish this by depositing enough money now to accrue—at real interest rate r—to $1 million by the time they're 65.

(a) How much should Anne deposit now if 2% real interest is available?

(b) How much should Anne deposit now if 4% real interest is available?

(c) Betty has $200,000 on hand. What real interest rate will she need to meet her retirement goal?

4. Christopher, now 20, wants to retire at 65 with a $1 million real nest egg. How much should he deposit now at 2% real interest? At 4%? How much should he invest in speculative junk bonds that pay 15% real interest?

5. A child born in 1993 will typically start college in 2011 and graduate in 2015. Nobody really knows what college tuitions will be then. However, St. Lena College offers an "early decision" guarantee for the class of 2015: Enroll as an infant and lock in yearly tuition payments of $40,000, $42,000, $44,000, and $46,000, payable in 2011, 2012, 2013, and 2014.

How much should new parents deposit in 1993 to cover these future payments if 6% annual interest is available? What if 8% interest is available?

6. Consider again the college tuition scenario in Exercise 5. This time we take a continuous point of view. Over the 4 years 2011–2015, total tuition paid is $172,000. Suppose, now, that this money is paid continuously, at the *constant* rate of $43,000 per year.

How much difference does the continuous point of view make? To decide, answer the same questions as in Exercise 5.

7. Consider carefully the two wavy graphs in Example 3, page 479. We claimed that their formulas are

$$p_T(t) = 50 \cos\left(\pi \frac{t - 105}{180}\right) + 60;$$

$$p_I(t) = 50 \cos\left(\pi \frac{t - 180}{180}\right) + 60.$$

(a) Which graph goes with which formula? Explain briefly, in words, how you know.

(b) The functions $\cos t$, p_T, and p_I are all periodic functions; i.e., they repeat themselves on time intervals of a fixed length. How long is the period of each of these functions?

(c) The formulas for p_T and p_I are "built" from $\cos t$ using various constants: 50, 60, 105, 180, and so on. Briefly discuss the effect each constant has on the corresponding graph.

FURTHER EXERCISES

8. Find the present value of the income stream in parts (a)–(e). In each case, treat income as a continuous stream. (Thus, for instance, think of yearly income of $12,000 as flowing in at the *constant rate* of $12,000 per year.)

(a) A yearly income of $12,000, beginning 10 years in the future and lasting for 10 years. Assume annual interest rate $r = 0.06$.

(b) A yearly income of $12,000, beginning 10 years in the future and lasting for 10 years. Assume annual interest rate $r = 0.08$.

(c) A yearly income of $12,000, beginning 10 years in the future and lasting for 10 years. Assume annual interest rate $r = 0$. (This investor doesn't trust banks; he "deposits" the money under the floorboards.)

(d) At time t years, with $10 \le t \le 20$, income flows at the rate $p(t) = 12,000e^{0.04t}$. Assume annual interest rate $r = 0.08$. [NOTE: In practice, many income streams do increase over time—for example, to correct for inflation. This one rises at a 4% instantaneous rate.]

(e) Repeat part (d), but assume an annual interest rate of $r = 0.1$.

9. Suppose that an investment that pays a continuous return of $5000 per year for 8 years is offered for sale at $30,000. If the current interest rate is 6%, should you invest?

10. Symbolic calculation of the integrals in Example 4, page 481, is a rather messy affair. Instead, use the midpoint rule with 50 subdivisions to estimate the first three integrals there.

11. Symbolic calculation of the integrals in Example 4, page 481, is messy, but it can be done. One way is to use this antiderivative formula:

$$\int e^{ax} \cos(bx)\, dx = \frac{e^{ax}}{a^2 + b^2}\left(a \cos(bx) + b \sin(bx)\right)$$

(a) Verify by differentiation that the antiderivative formula really holds.

(b) Use the integral formula to show symbolically (as claimed in Example 4) that

$$\int_0^{360} \left(50 \cos\left(\pi \cdot \frac{t - 180}{180}\right) + 60\right) e^{-0.1t/360}\, dt$$
$$\approx 20,550.78.$$

12. Suppose that an acquaintance offers to sell you an investment that will produce income at a continuous rate of $1000/(1+t)$ dollars per year for 10 years. The asking price is $2000. If 10-year bank certificates of deposit that pay 5% per year compounded continuously are available, should you purchase this investment from your acquaintance?

13. Suppose that an income stream p is continuously invested at the same continuously compounded interest rate r that is currently available. Show that the total income accrued between $t = 0$ and $t = T$ is

$$e^{rT}\,PV = e^{rT}\int_0^T p(t)e^{-rt}\,dt.$$

8.6 Fourier Polynomials

Approximating Functions with Other Functions

There are many ways to approximate one function with another, and many reasons for doing so. **Taylor polynomials** are the most familiar examples. They are ordinary polynomials chosen to approximate a given function as closely as possible near a given point x_0. (Taylor polynomials are discussed in detail in Section 4.3. This section, however, can be read independently of that one.) For example, the cubic polynomial

$$p_3(x) = 1 + x + \frac{x^2}{2} + \frac{x^3}{6}$$

Convince yourself.

is called the **third-order** Taylor polynomial for $f(x) = e^x$ at $x = 0$, because p_3 and f have the same value and first three derivatives at $x = 0$: ◄

$$f(0) = p_3(0); \qquad f'(0) = p_3'(0); \qquad f''(0) = p_3''(0); \qquad f'''(0) = p_3'''(0).$$

Can you tell which is which?

Because of this degree of agreement, the functions f and p_3 behave very similarly in the vicinity of $x = 0$, as the following graphs show. ◄

Graphs of f and p_3

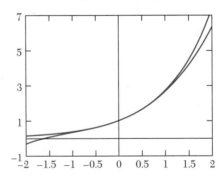

Why would one bother to approximate one type of function with another? One reason is that polynomial functions, such as p_3, are easy to evaluate, even by hand or with crude technology, whereas transcendental functions, such as exponential functions, may require more effort to manipulate.

For other purposes, ordinary polynomials, simple as they are, may not be the best choice for approximating a given "target" function f. It may be known, for instance, that f is a periodic function—one that repeats itself on intervals of fixed

length. Such functions arise often in modeling of physical phenomena, where repeating or rhythmic behaviors are common. In such cases, trigonometric functions—
among the simplest, best-understood, and most-studied periodic functions—are better suited than ordinary polynomials for use as basic modeling elements.

Trigonometric Polynomials

A **trigonometric polynomial** is a function of the form

$$q_n(x) = a_0 + a_1 \cos x + a_2 \cos(2x) + \cdots + a_n \cos(nx)$$
$$+ b_1 \sin x + b_2 \sin(2x) + \cdots + b_n \sin(nx),$$

where n is a nonnegative integer and the coefficients a_k and b_k are real numbers.
Trigonometric polynomials are "linear combinations" of the trigonometric functions

$$1, \cos x, \sin x, \cos(2x), \sin(2x), \cos(3x), \sin(3x), \ldots$$

just as ordinary polynomials are linear combinations of the power functions 1, x, x^2,
$x^3, \ldots$.

By choosing the coefficients a_k and b_k properly, we can construct trigonometric
polynomials to model and approximate a remarkable variety of functions. Here are
several, plotted over the interval $-\pi \le x \le \pi$:

Several trigonometric polynomials

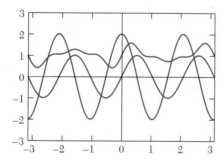

Right now, the particular formulas of the curves don't matter; simply note the variety
of shapes that trigonometric polynomials can assume.

Properties of Trigonometric Polynomials.   All trigonometric polynomials, thanks
to their structure as combinations of sines and cosines, have certain properties in
common. We list several such properties, some of which are reflected in the preceding graphs. Some details are left to exercises.

Periodicity Every trigonometric polynomial is 2π-periodic; i.e., it repeats itself
(perhaps more than once) on every interval of length 2π. That is so because
each summand $a_k \cos(kx)$ and $b_k \sin(kx)$ repeats itself (k times, actually)
on every interval of length 2π. For this reason, we'll work mainly on
the basic interval $[-\pi, \pi]$. On larger intervals, trigonometric polynomials
simply repeat themselves.

Zero Integrals For any positive integer k,

$$\int_{-\pi}^{\pi} \cos(kx)\,dx = 0 \qquad \text{and} \qquad \int_{-\pi}^{\pi} \sin(kx)\,dx = 0.$$

See the exercises at the end of this section, too.

Both integrals are easily done with u-substitution. ◄   Thus, for any trigonometric polynomial $q_n(x) = a_0 + a_1 \cos x + b_1 \sin x + \ldots$, we have

$$\int_{-\pi}^{\pi} q_n(x)\, dx = \int_{-\pi}^{\pi} (a_0 + a_1 \cos x + b_1 \sin x + \ldots)\, dx = 2\pi a_0.$$

More Zero Integrals   Let m and n be any integers. Then

$$\int_{-\pi}^{\pi} \cos(mx) \sin(nx)\, dx = 0.$$

If m and n are *distinct* integers, then

$$\int_{-\pi}^{\pi} \sin(mx) \sin(nx)\, dx = 0; \qquad \int_{-\pi}^{\pi} \cos(mx) \cos(nx)\, dx = 0.$$

All three of the preceding facts can be proved with a combination of integral tables and trigonometric identities; see the exercises for details. Here, we'll simply assume these results.

Some Nonzero Integrals   If $m = n$ in the preceding integrals, we get

$$\int_{-\pi}^{\pi} \cos^2(mx)\, dx = \pi \qquad \text{and} \qquad \int_{-\pi}^{\pi} \sin^2(mx)\, dx = \pi.$$

Both integrals are straightforward calculations; see the exercises for details.

With these facts at hand, we can show how to use trigonometric polynomials to approximate almost any function. First, the main definition:

Definition (Fourier Polynomials) Let $f(x)$ be a continuous function on $(-\pi, \pi)$. The **Fourier polynomial** of degree n for f is the trigonometric polynomial

$$q_n(x) = a_0 + a_1 \cos x + a_2 \cos(2x) + \cdots + a_n \cos(nx)$$
$$+ \; b_1 \sin x + b_2 \sin(2x) + \cdots + b_n \sin(nx),$$

with coefficients given by

$$a_0 = \frac{1}{2\pi} \int_{-\pi}^{\pi} f(x)\, dx;$$

$$a_k = \frac{1}{\pi} \int_{-\pi}^{\pi} f(x) \cos(kx)\, dx \qquad \text{if } k > 0;$$

$$b_k = \frac{1}{\pi} \int_{-\pi}^{\pi} f(x) \sin(kx)\, dx \qquad \text{if } k > 0.$$

Fourier polynomials are named for the French mathematician Joseph Fourier (1768–1830), who used them to study the mathematics of various physical problems, including heat diffusion. Over the last 200 years, a large (and still growing) mathematical theory, harmonic analysis, has grown up around the subject.

For a given function f, the definition tells how to find the coefficients of a trigonometric polynomial that approximates f on the interval $[-\pi, \pi]$. After an example to illustrate *that* Fourier polynomials have this approximation property, we'll suggest *why* they do.

EXAMPLE 1 Find several Fourier polynomials q_n for $f(x) = x$. Plot them together with f. Do the q_n appear to approximate f? Where?

Solution Let's use the definition to find some coefficients by integration. Luckily, there's a simple pattern. ▶ The following observations are key:

Watch carefully. A similar pattern often works.

For all k, $\cos(kx)$ is an even function and $\sin(kx)$ is an odd function.

These properties can save us much labor, for two reasons: ▶

Convince yourself of these basic facts.

1. If g is any odd function, then $\int_{-\pi}^{\pi} g(x)\, dx = 0$.

2. If h is any even function, then $\int_{-\pi}^{\pi} h(x)\, dx = 2\int_{0}^{\pi} h(x)\, dx$.

In this case, the "target" function $f(x) = x$ is odd; thus, for all $k \geq 0$, $x\cos(kx)$ is also odd. Therefore, for all $k \geq 0$,

$$\int_{-\pi}^{\pi} x\,\cos(kx)\, dx = 0 \implies a_k = 0.$$

The b_k are readily found using the table of integrals (a computer could help, too); ▶ the result is

The calculation is an exercise.

$$b_k = \frac{1}{\pi}\int_{-\pi}^{\pi} x\,\sin(kx)\, dx = \frac{-2\cos(k\pi)}{k}.$$

The numerical value depends on k:

$$b_k = \frac{-2}{k} \quad \text{if } k \text{ is even;} \qquad b_k = \frac{2}{k} \quad \text{if } k \text{ is odd.}$$

Thus, for any n, the nth Fourier polynomial is

$$q_n(x) = 2\sin x - 1\sin(2x) + \frac{2}{3}\sin(3x) - \frac{2}{4}\sin(4x) + \cdots + \frac{\pm 2}{n}\sin(nx).$$

Here are graphs of f, q_3, and q_7 on the same axes, on the interval $[-\pi, \pi]$: ▶

Try to see which is which; think about the period of each function.

A function f and two Fourier polynomials

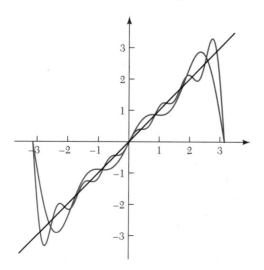

A close look shows that q_7 does a better job than q_3 of approximating f on $[-\pi, \pi]$. Of course, neither q_3 nor q_7 does well in approximating f outside the interval $[-\pi, \pi]$, since q_3 and q_7 are periodic and $f(x) = x$ isn't:

A function f and two Fourier polynomials

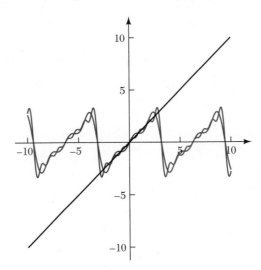

The moral is that Fourier polynomials are effective only for approximating functions that, like Fourier polynomials themselves, repeat on intervals of length 2π. In the present case, we can think of q_3 and q_7 as approximations not to $f(x) = x$ itself but to the "sawtooth" function that's formed by repeating the basic shape of $y = x$ on each interval of length 2π. ■

Why Fourier Approximation Works

Why do Fourier polynomials approximate a given function f on $[-\pi, \pi]$? Why are the coefficient formulas given in the definition "correct"? Rigorous answers are well beyond our scope, but the basic idea is straightforward.

The deep fact is that for *any* continuous function f on $(-\pi, \pi)$, $f(x)$ can be approximated as closely as we like by *some* trigonometric polynomial. In other words,

$$f(x) \approx q(x) = a_0 + a_1 \cos x + b_1 \sin x + \cdots + a_n \cos(nx) + b_n \sin(nx),$$

for some coefficients a_k and b_k. Assuming that, the remaining question is how to find the coefficients. Here's the answer. Since $f(x) \approx q(x)$, we'd expect that, for all k,

$$\int_{-\pi}^{\pi} f(x) \cos(kx)\, dx \approx \int_{-\pi}^{\pi} q(x) \cos(kx)\, dx.$$

The earlier integral identities imply that the right-hand integral is simply πa_k. Therefore,

$$\int_{-\pi}^{\pi} f(x) \cos(kx)\, dx \approx \pi a_k.$$

This explains (for the a_k; the b_k are similar) why the formulas given in the definition are reasonable.

BASICS

1. Let $f(x) = e^x$ and $p_3(x) = 1 + x + x^2/2 + x^3/6$. (The function p_3 is the third-order Taylor polynomial for $f(x)$ at $x = 0$.)
   (a) Show that $f(0) = p_3(0)$, $f'(0) = p_3'(0)$, $f''(0) = p_3''(0)$, and $f'''(0) = p_3'''(0)$.
   (b) Let $p_1(x) = 1 + x$, and let $p_2(x) = 1 + x + x^2/2$. Plot the functions f, p_1, p_2, and p_3 on the same axes.

2. Let $f(x) = \cos(kx)$, and let $g(x) = \sin(kx)$, where k is a nonnegative integer.
   (a) Explain why, for all x, $f(x + 2\pi) = f(x)$ and $g(x + 2\pi) = g(x)$, regardless of the value of k. (This means that f and g are 2π-periodic.)
   (b) Use part (a) to explain why every trigonometric polynomial is 2π-periodic.
   (c) What is the smallest period of the trigonometric polynomial $p(x) = \cos(4x) + \sin(8x)$? Show that p actually has this period. [**HINT**: Plot $p(x)$ first to guess the period.]

3. We said in this section that for any positive integer k,
   $$\int_{-\pi}^{\pi} \cos(kx)\,dx = 0 \qquad \text{and} \qquad \int_{-\pi}^{\pi} \sin(kx)\,dx = 0.$$
   (a) Plot $\cos(kx)$ and $\sin(kx)$ on $[-\pi, \pi]$ for several positive integers k. How do the graphs support the claims?
   (b) Use u-substitution to show that both of the integrals are zero, as claimed.

4. Let $f(x) = \cos x$, and let $p_4(x) = 1 - x^2/2 + x^4/24$. (The function p_4 is the fourth-order Taylor polynomial for $f(x)$ at $x = 0$.)
   (a) Show that $f(0) = p_4(0)$, $f'(0) = p_4'(0)$, $f''(0) = p_4''(0)$, $f'''(0) = p_4'''(0)$, and $f^{(4)}(0) = p_4^{(4)}(0)$. (The notation $f^{(n)}$ refers to the nth derivative.)
   (b) Let $p_2(x) = 1 - x^2/2$. Plot the functions f, p_2, and p_4 on the same axes.

5. Let $p(x) = a_0 + a_1 \cos x + b_1 \sin x + \cdots$ be a Fourier polynomial. Show that
   $$\int_{-\pi}^{\pi} p(x)\,dx = \int_{-\pi}^{\pi} (a_0 + a_1 \cos x + b_1 \sin x + \cdots)\,dx$$
   $$= 2\pi a_0.$$

6. Use an integral table or u-substitution to show:
   (a) $\int_{-\pi}^{\pi} \cos(mx)\sin(mx)\,dx = 0$
   (b) $\int_{-\pi}^{\pi} \cos^2(mx)\,dx = \pi$
   (c) $\int_{-\pi}^{\pi} \sin^2(mx)\,dx = \pi$

7. Show that if m and n are distinct positive integers, then $\int_{-\pi}^{\pi} \cos(mx)\sin(nx)\,dx = 0$. [**HINT**: The integrand is an odd function; explain why this is true and why it matters.]

8. Use the integral table to show that for all positive integers k,
   $$\frac{1}{\pi} \int_{-\pi}^{\pi} x \sin(kx)\,dx = \frac{-2\cos(k\pi)}{k}.$$
   (This gives the coefficients b_k in approximating the function $f(x) = x$ on $[-\pi, \pi]$.)

9. Consider the function $f(x)$ defined by
   $$f(x) = \begin{cases} 1 & \text{if } 2m\pi < x \leq (2m+1)\pi \\ 0 & \text{if } (2m+1)\pi < x \leq (2m+2)\pi \end{cases}$$
   for all integers m. This function is sometimes called a **pulse train** or a **box wave**.
   (a) Plot the function f. Notice that it is discontinuous but repeats on intervals of length 2π.
   (b) Use the definition of Fourier polynomials to find formulas for a_k and b_k. [**HINT**: For any function g, $\int_{-\pi}^{\pi} f(x)g(x)\,dx = \int_0^{\pi} g(x)\,dx$.]
   (c) Find and plot the Fourier polynomials q_1, q_3, q_5, and q_7 on the same axes.

10. Repeat Exercise 9 using the target function that is defined on $[-\pi, \pi]$ by
    $$f(x) = \begin{cases} \pi + x & \text{if } -\pi \leq x < 0 \\ \pi - x & \text{if } 0 \leq x \leq \pi \end{cases}$$
    and repeats itself on intervals of length 2π. This function is sometimes called a **sawtooth wave**.

8.7 Chapter Summary

This chapter surveyed several standard applications of the integral to geometric and physical problems.

Measurement and the Definite Integral. In practical applications a definite integral usually *measures* something: area, volume, the length of a curve, a physical or economic quantity, and so on. This chapter told how to use integrals to measure various quantities.

Usually the hardest task is identifying the integral that measures a given quantity. In choosing the right integral, it's useful to keep two different but equivalent interpretations of the integral in mind: (1) $\int_a^b f(x)\,dx$ as a limit of approximating sums, and (2) $\int_a^b f(x)\,dx$ as the accumulated change in an antiderivative. Section 8.1 described and compared these points of view.

Volumes.   Calculating volumes offers another geometric application of the integral. For complicated three-dimensional figures, such calculations require multivariable calculus, but for many simple solids, ordinary calculus integrals suffice. The principle behind typical volume calculations is **slicing**. A convenient x-axis is chosen; the solid lies along the axis from $x = a$ to $x = b$. The solid is then sliced (in one's imagination, of course) with cuts perpendicular to the axis. If $A(x)$ denotes the cross-sectional area revealed by such a cut at position x along the axis, then the solid's volume is the integral

$$\text{Volume} = \int_a^b A(x)\,dx.$$

As with other applications, the main challenge is usually setting up, rather than evaluating, the integral. For this purpose, **solids of revolution** offer a key advantage: All cross sections perpendicular to the axis of rotation are circles. As a result, finding the integrand $A(x)$ for such solids is usually easy.

Arclength.   The length of a curve C is defined as the limit, as n tends to infinity, of the lengths of "polygonal approximations" to C. If the curve is the graph of $y = f(x)$ from $x = a$ to $x = b$, then each polygonal approximation can be thought of as an approximating sum for the integral $\int_a^b \sqrt{1 + f'(x)^2}\,dx$. The value of this integral therefore gives the length of C.

Work.   Calculating **work** offers a physical application of the integral. Work (in the physicist's sense) is done when a force acts through a distance. In the simplest case, when the force is constant, work is the product of the force and the distance through which it acts. Calculus methods are needed if the force is variable—as it is in most real-life applications.

The force of gravity, for instance, is essentially constant near the surface of the earth but varies significantly, as described by Newton, over astronomical distances. By combining Newton's law of gravitation with elementary integral techniques, one can compute, for example, the work done against gravity in lifting satellites to various levels of orbit.

Present Value.   Suppose that $t = 0$ represents the present time, that a function $p(t)$ represents the rate at which a stream of income flows into an account during a time interval $a \le t \le b$, and that interest is continuously compounded at the constant rate r. Then, as shown in Section 8.5, the integral $\int_a^b p(t)e^{-rt}\,dt$ gives the **present value** of the income stream. The integral takes account of the fact that future payments are worth less than present payments because of interest effects.

Fourier Polynomials.   Trigonometric polynomials are combinations of sines and cosines, much as ordinary polynomials are combinations of powers of x. Taylor polynomials (discussed in Chapter 4) are ordinary polynomials chosen, using *derivatives*, to approximate nonpolynomial functions. Fourier polynomials, discussed in Section 8.6, are trigonometric polynomials chosen, using *integrals*, to approximate other functions. All Fourier polynomials are periodic, with period 2π, so they're especially useful in approximating periodic phenomena, such as physical wave forms.

9

More Antidifferentiation Techniques

9.1 Integration by Parts

Introduction

In this chapter we return to a project begun in Chapter 6: finding, for a given elementary function f, an elementary antiderivative function, i.e., a function F for which $F' = f$. (Recall that an elementary function is one built from the standard basic "elements"—power functions, exponential and logarithm functions, trigonometric functions, and so on.)

See the end of this section for several examples.

As we said in Chapter 6, finding elementary *antiderivatives* is, as a rule, harder than finding *derivatives*. Sometimes it is actually impossible—some elementary functions, even simple-looking ones, have *no* elementary antiderivatives. ◀

Still, many important elementary functions *do* have elementary antiderivatives. Searching for them is a worthwhile goal in its own right and an excellent vehicle for understanding classes of functions.

This chapter presents methods for finding antiderivative formulas for several types of integrands. We make no pretense of exhausting the surprisingly deep and difficult subject of antidifferentiation. However, if we combine the symbolic methods discussed here with a moderate-sized integral table, we can handle most integrals that arise in practical applications. With extra help from modern mathematical software, we can do better still.

Integration by Parts and the Product Rule

Substitution is the reverse version of the *chain rule*. In a similar sense, integration by parts reverses the *product rule*.

The product rule for derivatives says that, for differentiable functions u and v,

$$(u \cdot v)' = u \cdot v' + v \cdot u'.$$

From the *antiderivative* point of view, the product rule says that uv is an antiderivative of $uv' + u'v$. In symbols, ▶

$$\int \big(u(x) \cdot v'(x) + v(x) \cdot u'(x)\big)\, dx = u(x) \cdot v(x).$$

We added the input variable x this time.

Equivalently,

$$\int u(x) \cdot v'(x)\, dx = u(x) \cdot v(x) - \int v(x) \cdot u'(x)\, dx.$$

This last identity is worth a closer look. Its simple but crucial idea—swapping derivatives for antiderivatives ▶ inside an integral sign—is surprisingly useful throughout mathematics. In elementary calculus, the equation is known as the **integration-by-parts formula**. As we'll see, the formula sometimes lets us trade a difficult integration problem for an easier one.

We trade uv' for vu'.

If we write $u = u(x)$, $v = v(x)$, $du = u'(x)\, dx$, and $dv = v'(x)\, dx$, the formula looks simpler still. We state the result as a theorem; it applies to both definite and indefinite integrals. ▶

The FTC explains why it applies to indefinite integrals.

Theorem 1 (Integration by Parts) If u and v are differentiable functions, then

- $$\int u\, dv = uv - \int v\, du.$$

- $$\int_a^b u\, dv = uv \Big]_a^b - \int_a^b v\, du.$$

Here's how the method works under ideal conditions.

EXAMPLE 1 Find $\int x\, e^x\, dx$ and $\int_0^1 x\, e^x\, dx$.

Solution Setting $u = x$ and $dv = e^x\, dx$ "fits" the Theorem's template:

$$\int x\, e^x\, dx = \int u\, dv.$$

To use the formula, we need values for du and v. The first is simple: Since $u = x$, $du = dx$. Finding a suitable v can be hard, ▶ but here it's easy: If $v = e^x$, then $dv = e^x\, dx$, as desired.

Finding v from dv is another antiderivative problem; sometimes that's hard.

Now the formula kicks in:

$$\int x\, e^x\, dx = \int u\, dv = uv - \int v\, du = x\, e^x - \int e^x\, dx = x\, e^x - e^x + C.$$

Checking the answer involves—of all things—the product rule:

$$(x\, e^x - e^x + C)' = e^x + x\, e^x - e^x = x\, e^x.$$

Our "candidate" does what it should.

Evaluating the *definite* integral is now routine:

$$\int_0^1 x\, e^x\, dx = x\, e^x \Big]_0^1 - \int_0^1 e^x\, dx = x\, e^x - e^x \Big]_0^1 = 1.$$ ∎

EXAMPLE 2   Find $\int x^2 \ln x \, dx$.

Solution   If we set $u = \ln x$, then

$$dv = x^2 dx, \qquad du = \frac{1}{x} \, dx, \qquad v = \frac{x^3}{3},$$

and everything works nicely:

$$\int x^2 \ln x \, dx = \int u \, dv = uv - \int v \, du$$

$$= \frac{x^3}{3} \ln x - \int \frac{x^2}{3} \, dx$$

$$= \frac{x^3}{3} \ln x - \frac{x^3}{9} + C.$$

Do so. Which differentiation rule is involved?

It's now a routine matter to check, by differentiation, that the answer is correct. ◄   ■

How to choose u and dv successfully isn't always obvious. Setting $u = \ln x$ and $dv = x^2 \, dx$ in Example 2, for instance, was natural enough, but not inevitable. We might have tried, say, $u = x^2$ and $dv = \ln x \, dx$ instead. (The latter choices, though not incorrect, turn out to be unhelpful.)

Sometimes, surprising choices of u and dv turn out to work.

EXAMPLE 3   Find $\int \ln x \, dx$.

Solution   At first glance, the problem doesn't seem to fit the mold at all. Yet it does fit. If we set $u = \ln x$, then

$$dv = dx, \qquad du = \frac{1}{x} \, dx, \qquad v = x,$$

Verify that everything "fits."

and the formula says that ◄

$$\int \ln x \, dx = x \ln x - \int 1 \, dx = x \ln x - x + C.$$

Convince yourself.

As usual, the answer is easy to check by differentiation, using the product rule. ◄   ■

Tricks of the Trade: Wise and Foolish Choices

Integration by parts trades one indefinite integral expression, $\int u \, dv$, for another, $uv - \int v \, du$. The bargain is worth making, of course, only if the second expression is simpler or more tractable than the first.

When Things Go Wrong

In practice, such satisfaction cannot be guaranteed. On the contrary, choosing u and dv unwisely may make things *worse*, not better.

▓ E X A M P L E 4   (**A Step in the Wrong Direction**) What's wrong ▶ with   *In hindsight, anyway.*
choosing $u = e^x$ and $dv = x\,dx$ for the indefinite integral $\int x\,e^x\,dx$?

▓ S o l u t i o n   Our choices aren't wrong; they just don't help. Setting $u = e^x$,
$dv = x\,dx$, $du = e^x\,dx$, and $v = x^2/2$ gives

$$\int x\,e^x\,dx = uv - \int v\,du = e^x \cdot \frac{x^2}{2} - \int \frac{x^2}{2} \cdot e^x\,dx.$$

The new integral looks harder than the old. This time we struck out.   ▓

Once u and dv are chosen, we need values for du and v. Finding du is *al-ways* straightforward; only differentiation is involved. Finding v, by contrast, is an antidifferentiation problem; it may be hard or even impossible.

▓ E X A M P L E 5   Find $\int 2x\,e^{x^2}\,dx$.

▓ S o l u t i o n   The integrand is certainly a product and, therefore, a candidate for integration by parts. Plunging ahead, let's try $u = 2x$ and $dv = \exp(x^2)\,dx$. Then $du = 2\,dx$, but what's v? The answer: Nothing useful; $\exp(x^2)$ has no elementary antiderivative. We can't even write v down, let alone use it to solve our problem.

Two possible conclusions follow: Either we chose u and dv poorly, or integration by parts was the wrong tool for this job. In this case the latter is correct. The ordinary substitution $u = x^2$, $du = 2x\,dx$ works nicely: ▶   *Yet another moral: Try easy methods—especially substitution—first.*

$$\int 2x\,e^{x^2}\,dx = \int e^u\,du = e^{x^2} + C.$$ ▓

When Things Go Right

Mishaps are possible, but for surprisingly many antiderivatives—even ones that don't appear to be products—integration by parts succeeds. The trick is to choose u and dv successfully. In practice, "successfully" means two things:

- dv can be antidifferentiated to give v;
- $\int v\,du$ is simpler or more familiar than $\int u\,dv$.

As we've seen, not every choice of u and dv succeeds. But if one choice fails, another may work. Intuition for recognizing good and bad choices comes with practice.

> **LIATE: A Mnemonic for Choosing u and dv.**   Whether an attempt at integration by parts succeeds or fails usually depends on the choices of u and dv.
>
> In his article "A Technique for Integration by Parts" (*American Mathematical Monthly* 90, 1983, pp. 210–211), Herbert E. Kasube proposes the LIATE rule for choosing u. (Choosing u is enough; given u, dv must be "everything else.") He observes that most integrands are built from functions of five types: logarithmic (L); inverse (I) trigonometric; algebraic (A); trigonometric (T); and exponential (E). The LIATE rule says:
>
> *Choose u in the order LIATE.*
>
> In other words, u should be a function of the *first* available type in the list L, I, A, T, E. In Example 1, for instance, the choice $u = x$ was of type A (algebraic), because neither logarithmic (L) nor inverse trigonometric (I) functions were present.
>
> Not every indefinite integral succumbs to integration by parts. For those that do, the LIATE method usually seems to work.

Mixing Methods

Some problems require more than one method.

What does LIATE say here?

EXAMPLE 6   Find $\int \arctan x \, dx$. ◄

Short for "integration by parts."

Solution   Let $u = \arctan x$ and $dv = dx$. Then $du = \frac{1}{1+x^2} \, dx$, $v = x$, and, by "parts," ◄

$$\int \arctan x \, dx = uv - \int v \, du = x \arctan x - \int \frac{x}{1+x^2} \, dx.$$

Watch the 2's.

We're nearly finished—the last integral yields to the ordinary substitution $u = 1+x^2$, $du = 2x \, dx$: ◄

$$
\begin{aligned}
x \arctan x - \int \frac{x}{1+x^2} \, dx &= x \arctan x - \frac{1}{2} \int \frac{du}{u} \\
&= x \arctan x - \frac{1}{2} \ln |u| + C \\
&= x \arctan x - \frac{1}{2} \ln \left(1 + x^2\right) + C.
\end{aligned}
$$

■

Integration by Parts for Experts

Several useful variations on the basic theme of integration by parts exist. Two examples follow.

EXAMPLE 7   (**Repeated Integration by Parts**)   Find $\int x^2 e^x \, dx$.

Solution   If $u = x^2$, then $dv = e^x \, dx$, $du = 2x \, dx$, $v = e^x$, and

$$\int x^2 e^x \, dx = x^2 e^x - 2 \int x \, e^x \, dx.$$

With $u = x$, $dv = e^x \, dx$.

Did we get anywhere? Yes—we've already seen that the last integral can be handled by *another* integration by parts. ◄   We won't redo the problem. Here's the result:

$$\int x^2 e^x \, dx = x^2 e^x - 2 \int x \, e^x \, dx = x^2 e^x - 2 \left(x \, e^x - e^x\right) + C.$$

■

EXAMPLE 8   **(Integrate Twice, Then Solve)**   Find $\int e^x \sin x \, dx$.

Solution   If we set $u = \sin x$, then $dv = e^x \, dx$, $du = \cos x \, dx$, $v = e^x$, and ▶

Watch carefully, especially for signs.

$$I = \int e^x \sin x \, dx = e^x \sin x - \int e^x \cos x \, dx.$$

We haven't made much progress yet; let's try parts again on the last integral. If $u = \cos x$, then $dv = e^x \, dx$, $du = -\sin x \, dx$, and $v = e^x$, so

$$I = e^x \sin x - \int e^x \cos x \, dx = e^x \sin x - \left(e^x \cos x + \int e^x \sin x \, dx \right).$$

The original integral, I, has reappeared! Are we chasing our tails? No; we can *solve* the last equation for I:

$$I = e^x \sin x - \left(e^x \cos x + I \right) \implies 2I = e^x \sin x - e^x \cos x$$

$$\implies I = \frac{1}{2} \left(e^x \sin x - e^x \cos x \right) + C. \qquad \blacksquare$$

Reality Check

That "wraparound" trick in the preceding example seems almost too good to be true. Let's check the result graphically. Here, for comparison, are graphs of both $f(x) = e^x \sin x$ and $F(x) = (e^x \sin x - e^x \cos x)/2$, supposedly an antiderivative:

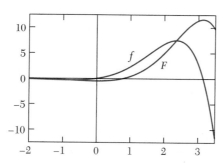

Graphs of $f(x) = e^x \sin x$ and $F(x) = (e^x \sin x - e^x \cos x)/2$

The graphs look promising. Observe:

Asymptotes The integrand $f(x)$ tends quickly to zero as $x \to -\infty$. Properly, therefore, the antiderivative F appears essentially horizontal toward the left of the picture.

Wild Swings in the Long Run As $x \to \infty$, $f(x)$ changes sign at each multiple of π, oscillating between larger and larger positive and negative values. The antiderivative F therefore also oscillates dramatically, with successive peaks and valleys at multiples of π. The graphs suggest this behavior; a larger window would show it more clearly.

Reduction Formulas: Stepping Down to a Solution

It holds for any integer
n > 0.

The equation ◄

$$\int x^n \, e^x \, dx = x^n \, e^x - n \int x^{n-1} \, e^x \, dx$$

is called a **reduction formula**. It expresses the left-hand indefinite integral in terms of another, slightly simpler ◄   integral.

I.e., "reduced."

E X A M P L E 9   Use the preceding reduction formula to find $\int x^3 \, e^x \, dx$.

S o l u t i o n   The reduction formula says, for $n = 3$, that

$$\int x^3 \, e^x \, dx = x^3 \, e^x - 3 \int x^2 \, e^x \, dx.$$

In Example 7 we found a value for the last integral; plugging that result in above gives the answer:

$$\int x^3 \, e^x \, dx = x^3 \, e^x - 3 \int x^2 \, e^x \, dx$$

$$= x^3 \, e^x - 3 \left(x^2 e^x - 2x e^x + 2e^x + C \right)$$

$$= x^3 e^x - 3x^2 e^x + 6x e^x - 6e^x - 3C.$$

Since C is an arbitrary constant, so is $-3C$. Therefore, we write simply

$$\int x^3 \, e^x \, dx = x^3 e^x - 3x^2 e^x + 6x e^x - 6e^x + C.$$

By differentiation.

A routine check ◄   shows that the answer is right.   ■

Example 9 shows that the preceding reduction formula works for $n = 3$, but why does it hold for *every* positive integer n? Integration by parts explains why. If we set $u = x^n$, $dv = e^x \, dx$, $du = nx^{n-1} \, dx$, and $v = e^x$, then the integration-by-parts formula says precisely what we want:

$$\int x^n e^x \, dx = x^n \, e^x - n \int x^{n-1} \, e^x \, dx.$$

See the next section.

Integral tables ◄   contain many reduction formulas; most of them are proved by integration by parts.

Is a Function Antidifferentiable in Closed Form?

As we know well by now, the answer is "Not necessarily." Our favorite examples of functions without elementary antiderivatives are $\sin(x^2)$, $\cos(x^2)$, and $\exp(x^2)$, but there are many more. Integration by parts sometimes helps us recognize such functions.

E X A M P L E 10   For a given nonnegative integer n, does $x^n \exp(x^2)$ have an elementary antiderivative?

Check that it's right!

S o l u t i o n   For $n = 0$, no—in that case, the integrand is $\exp(x^2)$, which we already discussed. For $n = 1$, yes—in that case, the simple substitution $u = x^2$ produces the antiderivative $\exp(x^2)/2$. ◄   We might guess, then, that the answer is yes for odd n and no for even n.

Our guess is correct. To see why, consider the integral

$$\int x^n \, \exp(x^2) \, dx$$

for $n \geq 2$. Integration by parts, with $u = x^{n-1}$ and $dv = x \exp(x^2) \, dx$, leads ▶ to   *Convince yourself.*
the reduction formula

$$\int x^n \, \exp(x^2) \, dx = x^{n-1} \frac{\exp(x^2)}{2} - \frac{n-1}{2} \int x^{n-2} \, e^{x^2} \, dx.$$

Observe two things:

- The left-hand integrand has an elementary antiderivative *if and only if* the one on the right does.

- Applying the reduction formula repeatedly knocks down the power of x by 2 each time. We'll eventually reach either 0 or 1, depending on whether n is odd or even.

These two facts, combined with what we know for $n = 0$ and $n = 1$, show what we claimed:

$x^n \, \exp(x^2)$ *is antidifferentiable in closed form if and only if n is odd.*   ■

BASICS

In Exercises 1–6, find the antiderivative using integration by parts and the suggested u and dv. Check your answers by differentiation.

1. $\int x e^{2x} \, dx$ $u = x, \quad dv = e^{2x} \, dx$

2. $\int x \sin(3x) \, dx$ $u = x, \quad dv = \sin(3x) \, dx$

3. $\int x \sec^2 x \, dx$ $u = x, \quad dv = \sec^2 x \, dx$

4. $\int \sqrt{x} \ln x \, dx$ $u = \ln x, \quad dv = \sqrt{x} \, dx$

5. $\int x\sqrt{1+x} \, dx$ $u = x, \quad dv = \sqrt{1+x}$

6. $\int \arcsin x \, dx$ $u = \arcsin x, \quad dv = dx$

In Exercises 7–12, evaluate the definite integral using integration by parts and check your answer by comparing it to a midpoint rule estimate computed with $n = 2$.

7. $\int_0^\pi x \cos(2x) \, dx$

8. $\int_0^1 x e^{-x} \, dx$

9. $\int_1^e x \ln x \, dx$

10. $\int_{\pi/4}^{\pi/2} x \csc^2 x \, dx$

11. $\int_{-1}^{\sqrt{2}/2} x^2 \arctan x \, dx$

12. $\int_1^4 e^{3x} \cos(2x) \, dx$

FURTHER EXERCISES

13. Evaluate $\int x \cos^2 x \, dx$ using integration by parts. [**HINT:** $\cos^2 x = \frac{1}{2}(1 + \cos(2x))$.]

14. Evaluate $\int x \sin x \cos x \, dx$ using integration by parts. [**HINT:** $\sin^2 x = \frac{1}{2}(1 - \cos(2x))$.]

15. (a) Use integration by parts to show that

$$\int \sin^2 x \, dx = -\sin x \cos x + \int \cos^2 x \, dx.$$

  (b) Use part (a) and the trigonometric identity $\sin^2 x + \cos^2 x = 1$ to show that

$$\int \sin^2 x \, dx = \frac{1}{2}(x - \sin x \cos x).$$

In Exercises 16–31, find the antiderivative and check your result graphically.

16. $\int x^2 \cos x \, dx$

17. $\int \arccos x \, dx$

18. $\int \sqrt{x} \ln \left(\sqrt[3]{x} \right) \, dx$

19. $\int (\ln x)^2 \, dx$

20. $\int \frac{\ln x}{x^2} \, dx$

21. $\int x e^x \sin x \, dx$

22. $\int \arctan(1/x) \, dx$

23. $\int x \arctan x \, dx$

24. $\int e^{\sqrt{x}} \, dx$

25. $\int x^5 \sin \left(x^3 \right) \, dx$

26. $\int \sin(\ln x)\, dx$

27. $\int \sqrt{x}e^{-\sqrt{x}}\, dx$

28. $\int \sin\left(\sqrt{x}\right) dx$

29. $\int \dfrac{\arctan\left(\sqrt{x}\right)}{\sqrt{x}}\, dx$

30. $\int \sqrt{x}\arctan\left(\sqrt{x}\right) dx$

31. $\int x^2 \arcsin x\, dx$

32. Find a function f such that
$$\int f(x)\sin x\, dx = -f(x)\cos x + \int x^3 \cos x\, dx.$$

33. Let n be a positive integer such that $n \ge 1$, and define $I_n = \int x\,(\ln x)^n\, dx$.
   (a) Evaluate I_1.
   (b) Use integration by parts to derive the reduction formula
$$I_n = \frac{x^2}{2}(\ln x)^n - \frac{n}{2}I_{n-1}.$$
   (c) Use the reduction formula in part (b) to evaluate I_2 and I_3.
   (d) Explain why $\left(\int x\,(\ln x)^n\, dx\right)' = x\,(\ln x)^n$.
   (e) Verify the reduction formula in part (b) by showing that
$$x\,(\ln x)^n = \frac{d}{dx}\left(\frac{x^2}{2}(\ln x)^n - \frac{n}{2}\int x\,(\ln x)^{n-1}\, dx\right).$$

34. Let n be a positive integer such that $n \ge 1$.
   (a) Verify the reduction formula
$$\int x^n e^x\, dx = x^n e^x - n\int x^{n-1}e^x\, dx$$
   by differentiation.
   (b) Use integration by parts to derive the reduction formula in part (a).

35. Let n be a positive integer such that $n \ge 1$. Use integration by parts to show that
$$\int (\ln x)^n\, dx = x\,(\ln x)^n - n\int (\ln x)^{n-1}\, dx.$$

36. Suppose that $r \ne -1/2$. Use integration by parts to show that
$$\int \left(x^2 + a^2\right)^r = \frac{x\left(x^2 + a^2\right)^r}{2r+1}$$
$$+ \frac{2ra^2}{2r+1}\int \left(x^2 + a^2\right)^{r-1}\, dx.$$

37. We said in this section that none of the functions $\sin\left(x^2\right)$, $\cos\left(x^2\right)$, and $\exp\left(x^2\right)$ has an elementary antiderivative. Use these facts, if necessary, in the following parts of this exercise.
   (a) The function $x\sin\left(x^2\right)$ has an elementary antiderivative. Find it.
   (b) The function $x^3 \sin\left(x^2\right)$ has an elementary antiderivative. Find it. [**HINT:** Try $u = x^2$, $dv = x\sin(x^2)$.]
   (c) The function $x^3 \cos\left(x^2\right)$ has an elementary antiderivative. Find it.
   (d) Show that the function $x^2 \cos\left(x^2\right)$ has no elementary antiderivative. [**HINT:** Set $u = x$, $dv = x\sin\left(x^2\right)$; then imitate the argument in Example 10, page 498.]

38. In Example 10, page 498, we showed that for positive integers n, $x^n \exp(x^2)$ has an elementary antiderivative if n is odd but has no elementary antiderivative if n is even. Imitate the arguments there to show that the same result holds for the function $x^n \sin(x^2)$.

39. Suppose that f is a continuous function, that $f(0) = 2$, and that $\int_0^\pi f(x)\sin x\, dx + \int_0^\pi f''(x)\sin x\, dx = 6$. Find $f(\pi)$.

40. Suppose that f is a continuous function, that $\int_{-\pi/2}^{3\pi/2} f'(x)\, dx = 1$, and that $\int_{-\pi/2}^{3\pi/2} f''(x)\cos x\, dx = 4$. Evaluate $\int_{-\pi/2}^{3\pi/2} f(x)\cos x\, dx$.

9.2 Partial Fractions

Rational Functions and Their Antiderivatives

Recall that a **rational function** is any function that can be written as the *ratio* $p(x)/q(x)$ of two polynomials. All three of the following expressions define rational functions.

$$\frac{2}{1-x^2}; \qquad \frac{2+5x+3x^2+3x^3}{x(1+x^2)(x+2)}; \qquad \frac{x^3}{1+x^2}.$$

This section is about the problem of antidifferentiating rational functions. Notice first that there *is* a problem. Polynomials themselves are easy to antidifferentiate, but *quotients* of polynomials are another matter entirely. Even derivatives of quotients are sometimes sticky to compute; *anti*derivatives can be worse still.

The method of **partial fractions** is a systematic technique for antidifferentiating rational functions. The basic idea is to "divide and conquer": Using some flashy algebra, we rewrite a given rational function as a sum of simpler rational functions—ones we can easily antidifferentiate.

First Examples: The Method in Action

Before describing the method in full theoretical regalia, we illustrate it with several examples. ▶

Read the examples carefully; they motivate the general discussion that follows.

EXAMPLE 1   Find $\int \dfrac{2}{1-x^2}\,dx$.

Solution   As it stands, the problem looks hard. The integrand doesn't directly resemble any of our standard basic forms, and neither u-substitution nor integration by parts looks promising.

The trick is to rewrite the integrand as a sum of two simpler terms:

$$\frac{2}{1-x^2} = \frac{2}{(1+x)(1-x)} = \frac{1}{1+x} + \frac{1}{1-x}.$$

The last two summands are the **partial fractions** for which the method is named. ▶ We'll explain later how we found them; for the moment, the point is that rewriting the integrand as we just did brightens our outlook considerably. Each partial fraction is easy to antidifferentiate separately; adding the results completes the problem. Here are the details: ▶

Summing the partial fractions gives the total fraction.

Watch the minus signs.

$$\int \frac{2}{1-x^2}\,dx = \int \frac{dx}{1+x} + \int \frac{dx}{1-x} = \ln|1+x| - \ln|1-x| + C.$$

The answer, as always, is easy to check by differentiation. Let's do so:

$$\left(\ln|1+x| - \ln|1-x| + C\right)' = \frac{1}{1+x} - \frac{-1}{1-x} = \frac{2}{1-x^2},$$

as claimed.   ∎

The Gap.   The gap in the preceding solution concerns how we *found* the useful equation

$$\frac{2}{(1+x)(1-x)} = \frac{1}{1+x} + \frac{1}{1-x}.$$

We'll fill this gap soon. Notice, however, that once written down, the equation is easy to *check* algebraically—for example, by finding a common denominator for both terms on the right.

EXAMPLE 2   Find $\int \dfrac{2 + 5x + 3x^2 + 3x^3}{x(1+x^2)(x+2)}\,dx$.

Solution   It's easy to check ▶ that

By manipulating the right-hand side.

$$\frac{2 + 5x + 3x^2 + 3x^3}{x(1+x^2)(x+2)} = \frac{1}{x} + \frac{1}{1+x^2} + \frac{2}{x+2}.$$

With the integrand written as a sum of partial fractions (as in the previous example), the rest is easy:

$$\int \frac{dx}{x} + \int \frac{dx}{1+x^2} + \int \frac{2}{x+2}\, dx = \ln|x| + \arctan x + 2\ln|x+2| + C. \qquad \blacksquare$$

EXAMPLE 3   Find $\displaystyle \int \frac{x^3}{1+x^2}\, dx$.

I.e., the highest power of x.

S o l u t i o n   This time the denominator cannot be factored. However, since the degree ◀ of the numerator is higher than that of the denominator, long division is possible. Here's the result:

$$\frac{x^3}{1+x^2} = x - \frac{x}{1+x^2}.$$

For the second integral, substitute $u = 1 + x^2$.

Once again, rewriting the integrand (as the sum of a polynomial and a partial fraction) makes antidifferentiation comparatively easy: ◀

$$\int x\, dx - \int \frac{x}{1+x^2}\, dx = \frac{x^2}{2} - \frac{1}{2}\ln\left|1+x^2\right| + C. \qquad \blacksquare$$

The Forest for the Trees: Antiderivatives and Their Ingredients

For example, $\sin(x^2)$ doesn't.

As we know, not every elementary function has an elementary antiderivative. ◀ Must every *rational* function have one? We found a nice antiderivative for each of the preceding functions, but that could be just chance. ◀

Authors have been known to rig problems.

Happily, every rational function does have an elementary antiderivative. A rigorous proof of this theoretical fact involves deep mathematics. We won't attempt such a proof, but the key idea is that of partial fractions. In particular, it can be shown that the antiderivative of *any* rational function involves *only* the ingredients seen in the preceding examples: logarithms, arctangents, and rational functions.

Knowing that an antiderivative formula exists is one thing; finding one explicitly is quite another. The hardest problem is usually algebraic—rewriting the given rational function as a sum of partial fractions. Since the method itself is our main interest, we'll arrange problems to keep the algebra simple.

Rational Numbers, Rational Functions, and Partial Fractions

The analogy continues in higher mathematics, as those who study abstract algebra will see.

Rational *functions* are closely akin to rational *numbers*. ◀ Both are quotients of simpler objects; many of the standard ideas and operations that apply to ordinary fractions apply to rational functions, as well.

Proper vs. Improper

A positive rational number p/q is called **improper** if $p \geq q$. Any improper fraction can be written as the sum of an integer and a proper fraction. For example,

$$\frac{29}{6} = 4 + \frac{5}{6}.$$

4 is the quotient; 5 is the remainder.

The numbers 4 and 5 are found by long-dividing 29 by 6. ◀

Rational functions behave similarly. A rational function $r(x) = p(x)/q(x)$ is **proper** if p has lower degree than q; otherwise, r is **improper**. ▶ The analogy with rational numbers continues:

The integrand in Example 3 is improper.

> Any improper rational function can be written as the sum of a polynomial and a proper rational function.

For instance,

$$\frac{x}{x+1} = 1 - \frac{1}{x+1}; \qquad \frac{x^2}{x+1} = x - 1 + \frac{1}{x+1}.$$

As with ordinary fractions, the right sides of these equations can be found by long division. ▶

Convince yourself.

Even Miss Manners Would Approve. Rational functions, no matter how "improper," commit no breach of morals or manners. Using such terminology, quaint as it may be, is part of the fun of mathematics.

Partial Fractions—of Numbers and of Polynomials

Any rational *number* m/n can be written as the sum of fractions of a special type: fractions whose denominators are the prime factors of n or powers thereof. For example, 6 has prime factors 2 and 3, and

$$\frac{5}{6} = \frac{5}{2 \cdot 3} = \frac{1}{2} + \frac{1}{3}; \qquad \frac{5}{12} = \frac{5}{2^2 \cdot 3} = \frac{3}{4} + \frac{-1}{3}.$$

The idea for rational *functions* is similar. Any rational function $p(x)/q(x)$ can be written as the sum of **partial fractions**—i.e., other rational functions whose denominators are the **irreducible factors** of $q(x)$, or powers thereof. We used this idea when we wrote

$$\frac{p(x)}{q(x)} = \frac{2 + 5x + 3x^2 + 3x^3}{x(1 + x^2)(x + 2)} = \frac{1}{x} + \frac{1}{1 + x^2} + \frac{2}{x + 2}.$$

Notice especially the denominators on the right: Each one is an irreducible factor of $q(x)$. (An irreducible polynomial is one that cannot be factored any further.) ▶

Irreducible factors are analogous to prime numbers: neither can be factored.

Partial Fractions: Only Two Possible Forms. All this jargon may sound formidably abstract. Indeed, the theory of polynomials and rational functions *is* a formidable subject. We need to examine only the tip of that iceberg. For us, only the following facts matter:

Factoring Every polynomial can be factored as a product of *linear* and *irreducible quadratic* factors.

What's Irreducible? Linear polynomials are automatically irreducible. A quadratic polynomial $ax^2 + bx + c$ may or may not have linear factors. The quadratic formula tells which is the case: If $b^2 - 4ac < 0$, then no further factoring is possible.

Proper Behavior Any *proper* rational function can be written as a sum of *proper* partial fractions.

The situation boils down to the fact that any proper rational function can be written as a sum of partial fractions of just two basic types:

$$\frac{A}{(ax+b)^n} \quad \text{and} \quad \frac{Ax+B}{(ax^2+bx+c)^n}.$$

(In each case n is a positive integer; a, b, c, A, and B are real constants.) Antidifferentiating *any* rational function, therefore, reduces to antidifferentiating these two basic types.

Antiderivatives of the Basic Forms

Do you agree?

Type I Antiderivatives. Partial fractions of the first type are always easy to antidifferentiate. The substitution $u = ax + b$ shows ◄ that

$$\int \frac{A}{(ax+b)^n}\, dx = \begin{cases} A\dfrac{\ln|ax+b|}{a} & \text{if } n = 1; \\[2ex] \dfrac{A}{a(1-n)(ax+b)^{n-1}} & \text{if } n > 1. \end{cases}$$

The general formula may look forbidding. In most actual cases, though, antidifferentiation is easy.

EXAMPLE 4 Find $\displaystyle\int \frac{6}{2x+1}\, dx$ and $\displaystyle\int \frac{6}{(2x+1)^5}\, dx$.

Check the easy details.

Solution For both integrals, substituting $u = 2x+1$ and $du = 2\, dx$ does the trick: ◄

$$\int \frac{6}{2x+1}\, dx = \int \frac{3}{u}\, du = 3\ln|2x+1| + C;$$

$$\int \frac{6}{(2x+1)^5}\, dx = \int \frac{3}{u^5}\, du = -\frac{3}{4(2x+1)^4} + C. \qquad \blacksquare$$

Type II Antiderivatives. Antidifferentiating partial fractions of the second form,

$$\frac{Ax+B}{(ax^2+bx+c)^n},$$

We'll usually avoid such cases, but see the exercises at the end of the section for an example.

can be quite messy, especially if $n > 1$. In that case, reduction formulas can be used to knock the exponent n down. ◄

For $n = 1$, **completing the square** in the denominator simplifies things considerably. Let's see some examples.

EXAMPLE 5 Find $\displaystyle\int \frac{dx}{x^2+2x+2}$.

The quadratic formula says so.

Solution Because $b^2 - 4ac < 0$, the denominator is irreducible, so we won't bother trying to factor it. ◄ Instead, we'll complete the square:

$$\frac{1}{x^2+2x+2} = \frac{1}{x^2+2x+1+1} = \frac{1}{(x+1)^2+1}.$$

Substituting $u = x + 1$ and $du = dx$ completes the problem:

$$\int \frac{dx}{(x+1)^2 + 1} = \int \frac{du}{u^2 + 1} = \arctan(x+1) + C.$$

■

EXAMPLE 6   Find $\displaystyle\int \frac{2x+3}{x^2 + 2x + 2}\, dx$.

Solution   Completing the square and substituting $u = x + 1$ helps here, too: ▶

Since $u = x + 1$, $x = u - 1$.

$$\int \frac{2x+3}{x^2 + 2x + 2}\, dx = \int \frac{2x+3}{(x+1)^2 + 1}\, dx = \int \frac{2(u-1) + 3}{u^2 + 1}\, du$$

$$= \int \frac{2u+1}{u^2+1}\, du = \int \frac{2u}{u^2+1}\, du + \int \frac{du}{u^2+1}$$

$$= \ln|u^2 + 1| + \arctan u + C$$

$$= \ln|x^2 + 2x + 2| + \arctan(x+1) + C.$$

■

Reality Check: A Graphical Interlude

We've been throwing plenty of symbols around, but does it all make sense? For some reason, antidifferentiating rational functions seems to spawn logarithms and arctangents. When the dust cleared in Example 2, for instance, we concluded that

$$\int \frac{2 + 5x + 3x^2 + 3x^3}{x(1 + x^2)(x+2)}\, dx = \ln|x| + \arctan x + 2\ln|x+2| + C.$$

A look at graphs helps explain the appearance of these ingredients. Here, together, are graphs of the integrand f and an antiderivative F:

**A rational function f
and an antiderivative F**

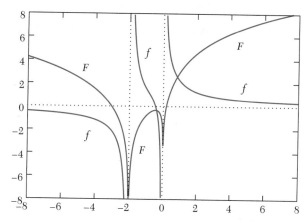

The graphs are complicated (like the functions themselves), but they show the right things:

Vertical Asymptotes   The integrand f has vertical asymptotes at $x = -2$ and $x = 0$. So does the antiderivative F: Near these values of x, the F-graph resembles a logarithm function.

Only Two Vertical Asymptotes   The integrand f has only two vertical asymptotes; the factor $(1 + x^2)$ in the denominator is never zero. For the same reason, the arctangent in F contributes no vertical asymptotes.

In the Long Run   The general shape of the F-graph looks roughly logarithmic as $x \to \pm\infty$. As $x \to \infty$, for instance, the F-graph, like a logarithm, is
Look closely; do you agree? ◀   increasing and concave down. ◀   This is as it should be. Here's an intuitive way to see why. Expanding the denominator of f gives

$$f(x) = \frac{2 + 5x + 3x^2 + 3x^3}{x(1 + x^2)(x + 2)} = \frac{2 + 5x + 3x^2 + 3x^3}{2x + x^2 + 2x^3 + x^4}.$$

When $|x|$ is large, the highest powers in the numerator and denominator "dominate." Thus, for large $|x|$,

$$f(x) = \frac{2 + 5x + 3x^2 + 3x^3}{2x + x^2 + 2x^3 + x^4} \approx \frac{3x^3}{x^4} = \frac{3}{x}.$$

This suggests (and the graph agrees) that for large x, $F(x) \approx 3 \ln|x| + C$.

With our graphical side trip completed, we return to the algebraic mainstream.

Last, Not Least: How to Find the Partial Fractions

We've seen *why* we'd want to write a rational function as a sum of partial fractions,
We saved the best for last.   but not yet *how* to do so. ◀   How, for instance, did we arrive at the equation

$$\frac{2 + 5x + 3x^2 + 3x^3}{x(1 + x^2)(x + 2)} = \frac{1}{x} + \frac{1}{1 + x^2} + \frac{2}{x + 2}$$

in Example 2?

Writing a rational function $p(x)/q(x)$ as a sum of partial fractions takes four steps:

Make It Proper   If $p(x)/q(x)$ is improper, use long division to rewrite it as the sum of a polynomial and a proper rational function.

Often, this step is done for you.   **Factor q**   Write the denominator, $q(x)$, as a product of linear and irreducible quadratic factors; some may be repeated. ◀

Write as a Sum   Write $p(x)/q(x)$ as a sum of partial fractions with **undetermined coefficients** (represented by letters) in their numerators.

Find the Coefficients   Solve one or more algebraic equations to find numerical values for the coefficients.

The last two steps need elaboration. After brief comments, we'll illustrate everything by example.

The General Form of an Answer

As earlier examples show, each partial fraction summand comes from some factor of the denominator of the original rational function. For the rational function

$$\frac{2 + 5x + 3x^2 + 3x^3}{x(1 + x^2)(x + 2)},$$

for instance, we expect a partial fraction sum of the general form

$$\frac{2 + 5x + 3x^2 + 3x^3}{x(1 + x^2)(x + 2)} = \frac{A}{x} + \frac{Bx + C}{1 + x^2} + \frac{D}{x + 2};$$

one partial fraction corresponds to each irreducible factor of the denominator. (Since the middle summand has a *quadratic* denominator, its numerator can have degree no more than one. The other summands, which have *linear* denominators, can have only constant numerators.) ▶

Otherwise, a summand would be improper.

Handling Repeated Factors

If the denominator has a **repeated factor**, that factor contributes several partial fractions—one for each power of the factor—to the general form. For example, the partial fraction version of

$$\frac{1}{x^3(x^2 + 1)^2(x + 1)}$$

has the general form

$$\frac{A}{x} + \frac{B}{x^2} + \frac{C}{x^3} + \frac{Dx + E}{x^2 + 1} + \frac{Fx + G}{(x^2 + 1)^2} + \frac{H}{x + 1}.$$

(Without allowing for these "extra" summands, a partial fraction form might not exist.)

Finding Values for the Coefficients

Finding values for the unknown constants, as we just did, is an algebra problem, not a true calculus problem. Examples of several techniques follow.

EXAMPLE 7 Derive the **partial fraction decomposition** ▶

Math jargon—impress your friends.

$$\frac{2}{(1 + x)(1 - x)} = \frac{1}{1 + x} + \frac{1}{1 - x}.$$

(We used it in Example 1, page 501.)

Solution The denominator's factors show that we need constants A and B such that

$$\frac{2}{(1 + x)(1 - x)} = \frac{A}{1 + x} + \frac{B}{1 - x}.$$

Multiplying both sides by $(1 + x)(1 - x)$ clears out the denominators; we get

$$2 = A(1 - x) + B(1 + x).$$

Now it's easy to find A and B. One way is to expand ▶ the right side and collect powers of x:

I.e., multiply out.

$$2 = A(1 - x) + B(1 + x) = A - Ax + B + Bx = (A + B) + (B - A)x.$$

Equating powers of x on the far left and far right shows that $A + B = 2$ and $B - A = 0$, so $A = B = 1$, as claimed earlier.

That is, values that lead to equations we can solve easily for A and B.

Another, sometimes simpler, route to the same result is to plug judiciously chosen values of x ◀ into the equation

$$2 = A(1 - x) + B(1 + x).$$

If, say, $x = 1$, then $2 = 2B$, so $B = 1$. If $x = -1$, then $2 = 2A$, so $A = 1$.   ■

EXAMPLE 8   Explain why

$$\frac{2 + 5x + 3x^2 + 3x^3}{x(1 + x^2)(x + 2)} = \frac{1}{x} + \frac{1}{1 + x^2} + \frac{2}{x + 2}.$$

Solution   We want constants A, B, C, and D such that

$$\frac{2 + 5x + 3x^2 + 3x^3}{x(1 + x^2)(x + 2)} = \frac{A}{x} + \frac{Bx + C}{1 + x^2} + \frac{D}{x + 2}.$$

We omitted some algebra details.

Multiplying both sides by the denominator on the right and collecting powers of x gives ◀

$$2 + 5x + 3x^2 + 3x^3 = A(1 + x^2)(x + 2) + (Bx + C)x(x + 2) + Dx(1 + x^2)$$

$$= 2A + (A + 2C + D)x + (2A + 2B + C)x^2$$

$$+ (A + B + D)x^3.$$

Equating the coefficients of various powers of x in the first and last expressions gives four equations in four unknowns:

$$2 = 2A; \qquad 5 = A + 2C + D; \qquad 3 = 2A + 2B + C; \qquad 3 = A + B + D.$$

Solving these equations simultaneously gives $A = 1$, $B = 0$, $C = 1$, and $D = 2$.

At last!

That's why ◀

$$\frac{2 + 5x + 3x^2 + 3x^3}{x(1 + x^2)(x + 2)} = \frac{1}{x} + \frac{1}{1 + x^2} + \frac{2}{x + 2},$$

as claimed.   ■

BASICS

1. (a) Show algebraically (by adding the terms on the right) that
$$\frac{5x + 7}{(x + 1)(x + 2)} = \frac{2}{x + 1} + \frac{3}{x + 2}.$$
   (b) Use part (a) to find $\displaystyle\int \frac{5x + 7}{(x + 1)(x + 2)}\, dx.$
   (c) Check your answer to part (b) by differentiation.

2. (a) Show algebraically (by adding the terms on the right) that
$$\frac{2}{x^2 - 1} = \frac{1}{x - 1} - \frac{1}{x + 1}.$$
   (b) Use part (a) to find $\displaystyle\int \frac{2}{x^2 - 1}\, dx.$
   (c) Check your answer to part (b) by differentiation.

3. (a) Find A, B, and C such that
$$\frac{x^2 + 3x - 1}{x(x + 1)(x - 2)} = \frac{A}{x} + \frac{B}{x + 1} + \frac{C}{x - 2}.$$
   (b) Use part (a) to find $\displaystyle\int \frac{x^2 + 3x - 1}{x(x + 1)(x - 2)}\, dx.$

4. (a) Find A, B, and C such that
$$\frac{x^2 - 1}{x(x^2 + 4)} = \frac{A}{x} + \frac{Bx + C}{x^2 + 4}.$$
   (b) Use part (a) to find $\displaystyle\int \frac{x^2 - 1}{x(x^2 + 4)}\, dx.$

5. (a) Find A, B, and C such that
$$\frac{6}{(x-2)\left(x^2-1\right)} = \frac{A}{x-2} + \frac{B}{x-1} + \frac{C}{x+1}.$$
   (b) Use part (a) to find $\int \dfrac{6}{(x-2)\left(x^2-1\right)}\,dx.$

6. (a) Find A, B, and C such that
$$\frac{x^2}{(x+1)^3} = \frac{A}{x+1} + \frac{B}{(x+1)^2} + \frac{C}{(x+1)^3}.$$
   (b) Use part (a) to find $\int \dfrac{x^2}{(x+1)^3}\,dx.$

7. (a) Find A, B, C, and D such that
$$\frac{x^2+x}{(x^2+1)^2} = \frac{Ax+B}{x^2+1} + \frac{Cx+D}{(x^2+1)^2}.$$
   (b) Use part (a) to find $\int \dfrac{x^2+x}{(x^2+1)^2}\,dx.$

8. Evaluate $\displaystyle\int_0^1 \dfrac{dx}{(x-2)(x^2+1)}.$

9. Evaluate $\displaystyle\int_0^2 \dfrac{dx}{x^3+1}.$ [**HINT:** $x^3+1 = (x+1)(x^2-x+1)$.]

FURTHER EXERCISES

In Exercises 10–20, find the antiderivative.

10. $\displaystyle\int \dfrac{2x+1}{(x-2)(x+3)}\,dx$

11. $\displaystyle\int \dfrac{x+1}{(x-1)(x+2)}\,dx$

12. $\displaystyle\int \dfrac{x^2+x}{\left(x^2+4\right)^2}\,dx$

13. $\displaystyle\int \dfrac{5x^2+3x-2}{x^3+2x^2}\,dx$

14. $\displaystyle\int \dfrac{4x^2-3x+2}{x\,(2x-1)^2\,dx}$

15. $\displaystyle\int \dfrac{x}{x^2+2x+6}\,dx$

16. $\displaystyle\int \dfrac{x^4}{x^4-1}\,dx$

17. $\displaystyle\int \dfrac{x^3}{x^2+1}\,dx$

18. $\displaystyle\int \dfrac{x^3}{x^2-1}\,dx$

19. $\displaystyle\int \dfrac{3x^2-1}{(x-1)(x+2)}\,dx$

20. $\displaystyle\int \dfrac{x^2}{(x^2+1)(x+1)^2}\,dx$

21. Use the substitution $u = \sqrt{x+1}$ to evaluate $\displaystyle\int \dfrac{dx}{x\sqrt{x+1}}.$

22. Use the substitution $u = \sqrt{x-1}$ to evaluate $\displaystyle\int \dfrac{dx}{x\sqrt{x-1}}.$

23. Use the substitution $u = 1 + e^x$ to evaluate $\displaystyle\int \dfrac{dx}{1+e^x}.$

24. Suppose that p a quadratic polynomial such that $p(0) = 1$, $p'(0) = 0$, and
$$\int \frac{p(x)}{x^3(x-1)^2}\,dx$$
   is a rational function.
   (a) Show that $p(x) = 1 + Ax^2$, where A is a constant.

   (b) Let q be a polynomial of degree 4 or lower. Then, the partial fraction decomposition of the rational function
$$r(x) = \frac{q(x)}{x^3(x-1)^2}$$
   has the general form
$$\frac{q(x)}{x^3(x-1)^2} = \frac{B}{x} + \frac{C}{x^2} + \frac{D}{x^3} + \frac{E}{x-1} + \frac{F}{(x-1)^2}.$$
   If $\int r(x)\,dx$ is a rational function, explain why $B = E = 0$.

   (c) Let r be the rational function defined in part (b). Show that if $\int r(x)\,dx$ is a rational function, then
   $$q(x) = D + (C-2D)x + (D-2C)x^2 + (C+F)x^3.$$
   (Thus, q is a polynomial of degree at most 3.)

   (d) Use parts (a) and (c) to show that $p(x) = 1 - 3x^2$.

25. Is there a quadratic polynomial $q(x)$ such that
$$\int \frac{q(x)}{(1-x)^2(3+x)}\,dx = \frac{1}{1-x} + \arcsin x$$
$$+ \ln(3+x) + C?$$
   Justify your answer.

26. Let q be a quadratic polynomial. Does
$$\int \frac{q(x)}{(1-x)^2(x+3)}\,dx = \frac{A}{(1-x)^2} + B\ln|x+3| + C,$$
   where A, B, and C are constants?

27. (a) Show that
$$\int \frac{dx}{x+a} = \ln|x+a|.$$
   (b) Let n be an integer such that $n > 1$. Show that
$$\int \frac{dx}{(x+a)^n} = \frac{1}{1-n}\frac{1}{(x+a)^{n-1}}.$$

28. (a) Show that
$$\int \frac{dx}{x^2 + a^2} = \frac{1}{a}\arctan(x/a).$$
(b) Show that
$$\int \frac{x}{x^2 + a^2}\,dx = \frac{1}{2}\ln\left|x^2 + a^2\right|.$$

29. Let n be an integer such that $n > 1$.
(a) Use integration by parts to show that
$$\int \frac{dx}{\left(x^2 + a^2\right)^n} = \frac{x}{\left(x^2 + a^2\right)^n}$$
$$+ 2n\int \frac{x^2}{\left(x^2 + a^2\right)^{n+1}}\,dx.$$
(b) Use part (a) to show that
$$\int \frac{dx}{\left(x^2 + a^2\right)^n} = \frac{x}{\left(x^2 + a^2\right)^n} + 2n\int \frac{dx}{\left(x^2 + a^2\right)^n}$$
$$- 2na^2\int \frac{dx}{\left(x^2 + a^2\right)^{n+1}}.$$

[**HINT:** $x^2 = (x^2 + a^2) - a^2$.]

(c) Use part (b) to show that
$$\int \frac{dx}{\left(x^2 + a^2\right)^{n+1}} = \frac{1}{2na^2}\frac{x}{\left(a^2 + x^2\right)^n}$$
$$+ \frac{2n - 1}{2na^2}\int \frac{dx}{\left(a^2 + x^2\right)^n}.$$

30. Use the reduction formula
$$\int \frac{dx}{\left(ax^2 + bx + c\right)^{n+1}} = \frac{2ax + b}{n\left(4ac - b^2\right)\left(ax^2 + bx + c\right)^n}$$
$$+ \frac{2(2n - 1)a}{n\left(4ac - b^2\right)}\int \frac{dx}{\left(ax^2 + bx + c\right)^n}$$

to evaluate
$$\int \frac{4x^2 + 2x + 1}{\left(4x^2 + 5x + 3\right)^2}\,dx.$$

[**HINT:** Compute the partial fraction decomposition of the integrand.]

9.3 Trigonometric Antiderivatives

In this section we continue our quest for antiderivatives of elementary functions. Although the symbolic methods developed so far—substitution, integration by parts, and partial fractions—are adequate for many classes of functions, some classes remain unexplored. Here we consider functions that involve two special ingredients: powers of trigonometric functions and roots of quadratic expressions. At present, we can *check* ◄ antiderivatives such as

By differentiation—try these.

$$\int \sin^2 x \cos^3 x\,dx = \frac{\sin^3 x}{3} - \frac{\sin^5 x}{5} + C;$$
$$\int \sqrt{-x^2 - 2x}\,dx = \frac{(x + 1)\sqrt{-x^2 - 2x}}{2} + \frac{\arcsin(x + 1)}{2} + C,$$

but we cannot yet *find* them. This section shows how to do both.

This section contains no truly new or different ideas or methods. We explore, mainly through examples, how methods of earlier sections can be combined with properties of trigonometric and algebraic functions to handle new classes of antiderivative problems.

Powers of Trigonometric Functions

Every antiderivative of the forms

$$\int \cos^3 x\,dx; \qquad\qquad \int \cos^3 x \sin^4 x\,dx;$$
$$\int \cos^2 x \sin^3 x \tan^2 x\,dx; \qquad \int \sec^3 x \tan^3 x\,dx$$

(i.e., any product of integer powers of the six trigonometric functions) can be solved in elementary form. All such antiderivatives can be found by combining substitution, integration by parts, and (sometimes) clever applications of trigonometric identities.

Useful Properties of Trigonometric Functions

Before proceeding to antidifferentiation, let's recall some properties of and relations among the trigonometric functions.

Two Main Types   All six trigonometric functions are defined in terms of sines and cosines. It follows that *every* product of integer powers of trigonometric functions can be written in the form $\sin^n x \, \cos^m x$, where n and m are (not necessarily positive) integer powers. For instance,

$$\cos^4 x \, \sin^3 x \, \tan^2 x = \cos^4 x \, \sin^3 x \, \frac{\sin^2 x}{\cos^2 x} = \cos^2 x \, \sin^5 x.$$

With enough work, ▶ every function of the form $\sin^n x \, \cos^m x$ can be antidifferentiated.   *Some examples follow.*

Finding trigonometric antiderivatives tends to be easier when only *nonnegative* powers are involved. If negative powers occur, it may help to rewrite a function in terms of positive powers of secants and tangents. For instance,

$$\cos^{-5} x \, \sin^2 x = \cos^{-3} \frac{\sin^2 x}{\cos^2} x = \sec^3 x \, \tan^2 x.$$

In this section, therefore, we mainly consider integrals of two types:

$$\int \sin^n x \, \cos^m x \, dx \qquad \text{and} \qquad \int \sec^n x \, \tan^m x \, dx,$$

where m and n are nonnegative integers.

Reduction Formulas   If either $n = 0$ or $m = 0$ in the preceding forms, then the integral can be handled—with care—using one of the reduction formulas:

$$\int \sin^n x \, dx = -\frac{\sin^{n-1} x \, \cos x}{n} + \frac{n-1}{n} \int \sin^{n-2} x \, dx; \qquad (n \neq 0)$$

$$\int \cos^n x \, dx = \frac{\cos^{n-1} x \, \sin x}{n} + \frac{n-1}{n} \int \cos^{n-2} x \, dx; \qquad (n \neq 0)$$

$$\int \tan^n x \, dx = \frac{\tan^{n-1} x}{n-1} - \int \tan^{n-2} x \, dx; \qquad (n \neq 1)$$

$$\int \sec^n x \, dx = \frac{\sec^{n-2} x \, \tan x}{n-1} + \frac{n-2}{n-1} \int \sec^{n-2} x \, dx. \qquad (n \neq 1).$$

Large powers may require *repeated* use of the reduction formulas. For an example using tangents, see Example 7, page 417.

Using Trigonometric Identities   Many integrands can be simplified to one of the preceding forms, often by using either the **Pythagorean identities**

$$\sin^2 x + \cos^2 x = 1, \qquad \tan^2 x + 1 = \sec^2 x,$$

Also known as half-angle formulas.

or the **double-angle formulas** ◄

$$\sin^2 x = \frac{1}{2} - \frac{\cos(2x)}{2}, \qquad \cos^2 x = \frac{1}{2} + \frac{\cos(2x)}{2}.$$

The first two identities permit us, in effect, to trade sines for cosines, secants for tangents, and vice versa. The double-angle formulas can be used to convert even *powers* of sines and cosines to cosines of even *multiples* of x.

Some concrete examples will illustrate the use of these facts in finding antiderivatives. We'll usually stop as soon as we've transformed the problem into something manageable.

EXAMPLE 1 Find $\int \cos^3 x \, dx$.

Solution With the substitution $u = \sin x$ in mind, we reserve one power of the cosine for du and convert the other two powers to sines:

$$\int \cos^2 x \cos x \, dx = \int (1 - \sin^2 x) \cos x \, dx$$

$$= \int (1 - u^2) \, du$$

$$= \sin x - \frac{\sin^3 x}{3} + C.$$

Are the two answers comparable? See the exercises.

Alternatively, we could have used the reduction formula for cosines—in this case, with $n = 3$: ◄

$$\int \cos^3 x \, dx = \frac{\cos^2 x \sin x}{3} + \frac{2}{3} \int \cos x \, dx = \frac{\cos^2 x \sin x + 2 \sin x}{3} + C. \qquad ■$$

EXAMPLE 2 Find $\int \sin^4 x \, dx$.

Watch the steps.

Solution We use the double-angle formulas to rewrite $\sin^4 x$ in terms of $\cos(2x)$ and $\cos(4x)$, as follows: ◄

$$\sin^4 x = (\sin^2 x)^2 = \left(\frac{1}{2} - \frac{\cos(2x)}{2} \right)^2 = \frac{1}{4} - \frac{\cos(2x)}{2} + \frac{\cos^2(2x)}{4}.$$

The first two summands are now easy to integrate. To the last summand we apply the double-angle formula *again*:

$$\frac{\cos^2(2x)}{4} = \frac{1}{8} + \frac{\cos(4x)}{8}.$$

We've produced another tractable integrand. Here's the result:

$$\int \sin^4 x \, dx = \frac{x}{4} - \frac{\sin(2x)}{4} + \frac{x}{8} + \frac{\sin(4x)}{32} + C.$$

(Other approaches to the same problem may produce an equivalent—but differentlooking—answer.) ■

EXAMPLE 3   Find $\int \cos^3 x \sin^4 x \, dx$.

Solution   One strategy is to convert all the sines to cosines, and then to attack the result with the preceding cosine reduction formula. Instead, we substitute $u = \sin x$, $du = \cos x \, dx$. We chip off one power of $\cos x$ for du; the remaining two powers of cosine become sines:

$$\int \cos^3 x \sin^4 x \, dx = \int (1 - \sin^2 x)(\sin^4 x) \cos x \, dx = \int (1 - u^2)u^4 \, du.$$

The last integral is simple, so we'll leave it alone.   ■

EXAMPLE 4   Find $\int \sec^3 x \tan^3 x \, dx$.

Solution   If we substitute $u = \sec x$ and $du = \sec x \tan x \, dx$, the remaining tangents can be converted to secants:

$$\int \sec^3 x \tan^3 x \, dx = \int \sec^2 x \tan^2 x (\sec x \tan x) \, dx$$

$$= \int \sec^2 x (\sec^2 x - 1)(\sec x \tan x) \, dx$$

$$= \int u^2 (u^2 - 1) \, du.$$

The rest is easy.   ■

Trigonometric Substitutions

Trigonometric substitutions help us antidifferentiate certain integrands that involve roots of quadratic expressions. The simplest such expressions are these:

$$\sqrt{a^2 - x^2}; \qquad \sqrt{x^2 - a^2}; \qquad \sqrt{a^2 + x^2}.$$

(In each case, a is a positive constant.) The fact that *trigonometric* functions arise at all is at least mildly surprising; so far, there's nothing trigonometric in sight. Yet it turns out that carefully chosen substitutions—with trigonometric ingredients—often reduce integrals of the present type to powers of trigonometric functions. A (relatively) simple example ▶ illustrates the idea and some of the subtleties involved.

Follow the example carefully; it gives the general idea.

EXAMPLE 5   According to our integral table,

$$\int \sqrt{4 - x^2} \, dx = \frac{x\sqrt{4 - x^2}}{2} + 2 \arcsin\left(\frac{x}{2}\right).$$

How was this found?

Solution   If $x = 2 \sin t$, then $dx = 2 \cos t \, dt$. These substitutions and a bit of algebra produce a simpler-looking integral:

$$\int \sqrt{4 - x^2} \, dx = 2 \int \sqrt{4 - 4\sin^2 t} \, \cos t \, dt = 4 \int \cos^2 t \, dt.$$

The last integral is of the type discussed earlier in this section. Using methods developed there (or looking in a table) gives

$$4 \int \cos^2 t \, dt = 2t + \sin(2t) + C.$$

So far so good, but we want an answer involving x, not t. To eliminate t in favor of x, we need several facts. All follow from having set $x = 2 \sin t$, and from the formula $\sin(2t) = 2 \sin t \cos t$.

$$t = \arcsin\left(\frac{x}{2}\right); \qquad 2 \cos t = \sqrt{4 - x^2}; \qquad \sin(2t) = 2 \sin t \cos t = \frac{x\sqrt{4 - x^2}}{2}.$$

Now we can finish our problem:

$$\int \sqrt{4 - x^2} \, dx = 2t + \sin(2t) + C = 2 \arcsin\left(\frac{x}{2}\right) + \frac{x\sqrt{4 - x^2}}{2} + C. \qquad \blacksquare$$

The example needs some amplification:

Inverse Substitution Substituting $x = 2 \sin t$ and $dx = 2 \cos t \, dt$ represents an "inverse" variant of the more usual u-substitution technique. Such substitutions, although legal, require special care.

Which Values of t? The equation $x = 2 \sin t$ makes sense for *all* real t. In order to write $t = \arcsin(x/2)$, however, we must (and hereafter will) assume that t lies in the interval $[-\pi/2, \pi/2]$. ◄

That is, t must lie in the range of the arcsine function.

Absolute Values In the preceding calculation we used, without comment, the fact that

$$\sqrt{4 - x^2} = \sqrt{4 - 4\sin^2 t} = 2 |\cos t| = 2 \cos t.$$

The last equation assumes, in effect, that $\cos t$ is nonnegative. Fortunately, that's true; restricting t to the interval $[-\pi/2, \pi/2]$ guarantees so.

The Bottom Line Whatever the subtleties of the trigonometric substitution, the bottom line remains: The result *is* a suitable antiderivative, as differentiation can verify.

Trigonometric Substitutions: Three Types

Example 5 illustrates one of three types of trigonometric substitutions. Here, telegraphically, are all three:

Trigonometric Substitutions			
Radical Form	**Substitution**	**t-Domain**	**Result**
$\sqrt{a^2 - x^2}$	$x = a \sin t$	$[-\pi/2, \pi/2]$	$\sqrt{a^2 - x^2} = a \cos t$
$\sqrt{a^2 + x^2}$	$x = a \tan t$	$[-\pi/2, \pi/2]$	$\sqrt{a^2 + x^2} = a \sec t$
$\sqrt{x^2 - a^2}$	$x = a \sec t$	$[0, \pi], \, x \neq \pi/2$	$\sqrt{x^2 - a^2} = \pm a \tan t$

These substitutions, combined (if necessary) with completing the square, produce antiderivatives of many functions that involve square roots of quadratic expressions.

From t Back to x: Helpful Pictures. Trigonometric substitutions, if successful, trade a troublesome integral in x for a simpler integral in t. However, the problem of translating back to an expression in x always remains. As the preceding example showed, doing so can be difficult. The following pictures can help jog one's memory: ▶

Pictures aren't proofs. Careful proofs depend on properties of the trigonometric function and their inverses.

Trigonometric substitutions: pictorial aids

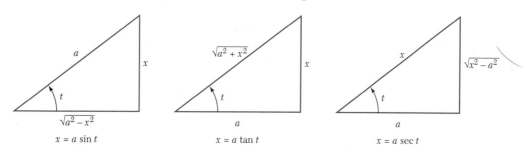

$$x = a \sin t \qquad\qquad x = a \tan t \qquad\qquad x = a \sec t$$

From the middle picture, for instance, one can read such implications as

$$x = a \tan t \implies \sec t = \frac{\sqrt{a^2 + x^2}}{a}.$$

We use such facts in the following examples.

EXAMPLE 6 Find $\int \dfrac{dx}{\sqrt{x^2 + 4}}$.

Solution Let $x = 2 \tan t$; then $dx = 2 \sec^2 t \, dt$ and $\sqrt{x^2 + 4} = \sqrt{4 \tan^2 t + 4} = 2 \sec t$. ▶ Therefore,

In the assumed t-domain (see the preceding table), $\sec t$ is nonnegative.

$$\int \frac{dx}{\sqrt{x^2 + 4}} = \int \frac{2 \sec^2 t}{2 \sec t} \, dt$$

$$= \int \sec t \, dt = \ln |\sec t + \tan t| + C.$$

To reconvert the result to an expression in x, we use the conclusion drawn just before this example:

$$\int \sec t \, dt = \ln \left| \frac{\sqrt{4 + x^2}}{2} + \frac{x}{2} \right| + C. \qquad\qquad ∎$$

EXAMPLE 7 Find $\int_0^2 \dfrac{dx}{\sqrt{x^2 + 4}}$.

Solution We just calculated the antiderivative, so there's nothing but algebra left:

$$\int_0^2 \frac{dx}{\sqrt{x^2 + 4}} = \ln \left| \frac{\sqrt{4 + x^2}}{2} + \frac{x}{2} \right| \Bigg]_0^2 = \ln(\sqrt{2} + 1).$$

Example 6 also suggested another approach to the same goal: substitution in the *definite integral*. Notice that if $x = 2\tan t$, then $x = 0$ when $t = 0$ and $x = 2$ when $t = \pi/4$. Substituting all this into the original integral gives

$$\int_0^2 \frac{dx}{\sqrt{x^2 + 4}} = \int_0^{\pi/4} \sec t \, dt = \ln|\sec t + \tan t| \Big]_0^{\pi/4} = \ln(1 + \sqrt{2}),$$

just as above.

■

EXAMPLE 8   Find $I = \displaystyle\int \frac{dx}{\sqrt{x^2 + 2x + 5}}$.

Solution   Completing the square in the denominator gives

$$\int \frac{dx}{\sqrt{x^2 + 2x + 5}} = \int \frac{dx}{\sqrt{(x + 1)^2 + 4}}.$$

Next, we substitute $u = x + 1$, $du = dx$; the result looks familiar:

$$I = \int \frac{du}{\sqrt{u^2 + 4}}.$$

Work already done in the last two examples completes the problem.

■

BASICS

1. We solved Example 1, page 512, in two different ways and apparently got two different answers. Are the answers really different? Why or why not?

In Exercises 2–15, find the antiderivative.

2. $\int \sin^2(3x) \, dx$

3. $\int \cos^2(x/3) \, dx$

4. $\int \sin^3 x \cos^3 x \, dx$

5. $\int \cos^2 x \sin^3 x \, dx$

6. $\int \cos^3(2x) \sin^2(2x) \, dx$

7. $\int \sin^2 x \cos^2 x \, dx$

8. $\int \cos^4 x \sin^2 x \, dx$

9. $\int \dfrac{\sin^3 x}{\cos x} \, dx$

10. $\int \sqrt{1 - x^2} \, dx$

11. $\int \dfrac{dx}{\left(x^2 + 4\right)^2}$

12. $\int x^2 \sqrt{1 - x^2} \, dx$

13. $\int \dfrac{x^2}{\sqrt{9 - x^2}} \, dx$

14. $\int \dfrac{dx}{x^2 \sqrt{x^2 + 1}}$

15. $\int \dfrac{dx}{\sqrt{1 + x^2}}$

16. Show that $\displaystyle\int \frac{\arctan x}{(1 + x^2)^{3/2}} \, dx = \frac{1 + x \arctan x}{\sqrt{1 + x^2}}$.

   [**HINT:** Start by making the substition $w = \arctan x$.]

FURTHER EXERCISES

In Exercises 17–28, find the antiderivative.

17. $\int \tan^4 x \, dx$

18. $\int \sec^2 x \tan^2 x \, dx$

19. $\int \sec^3 x \tan^2 x \, dx$

20. $\int \sin(2x) \cos^2 x \, dx$

21. $\int \sqrt{\cos x} \sin^5 x \, dx$

22. $\int \sqrt{1 + \sin x} \, dx$

23. $\int \sqrt{1 + x^2} \, dx$

24. $\int \dfrac{dx}{x^2 \sqrt{4 - x^2}}$

25. $\int \dfrac{dx}{x^2 \sqrt{x^2 - 4}}$

26. $\int \dfrac{\sqrt{4-x^2}}{x^2}\,dx$

28. $\int x \arcsin x\,dx$

27. $\int \dfrac{x+2}{x\left(x^2+1\right)}\,dx$

29. (a) Use the addition formula $\cos(x+y)=\cos x \cos y - \sin x \sin y$ to show that $2\cos x \cos y = \cos(x+y)+\cos(x-y)$.
    (b) Let a and b be constants such that $a \neq b$. Use part (a) to evaluate $\int \cos(ax)\cos(bx)\,dx$.
    (c) Let a be a constant. Evaluate $\int \cos(ax)\cos(ax)\,dx$.

30. (a) Use the addition formula $\cos(x+y)=\cos x \cos y - \sin x \sin y$ to show that $2\sin x \sin y = \cos(x-y)-\cos(x+y)$.
    (b) Let a and b be constants such that $a \neq b$. Use part (a) to evaluate $\int \sin(ax)\sin(bx)\,dx$.
    (c) Let a be a constant. Evaluate $\int \sin(ax)\sin(ax)\,dx$.

31. (a) Use the addition formula $\sin(x+y)=\sin x \cos y + \cos x \sin y$ to show that $2\sin x \cos y = \sin(x+y)+\sin(x-y)$. [**HINT:** Cosine is an even function and sine is an odd function (i.e., $\cos(-x)=\cos x$ and $\sin(-x)=-\sin x$.]
    (b) Let a and b be constants such that $a \neq b$. Use part (a) to evaluate $\int \sin(ax)\cos(bx)\,dx$.
    (c) Let a be a constant. Evaluate $\int \sin(ax)\cos(ax)\,dx$.

32. (a) Evaluate $\int_1^2 \dfrac{\sqrt{x^2-1}}{x}\,dx$.
    (b) Evaluate $\int_{-2}^{-1} \dfrac{\sqrt{x^2-1}}{x}\,dx$.

33. (a) Use integration by parts with $u = \sin^{n-1} x$ and $dv = \sin x\,dx$ to show that
$$\int \sin^n x\,dx = -\sin^{n-1} x \cos x$$
$$+\,(n-1)\int \sin^{n-2} x \cos^2 x\,dx.$$
    (b) Use part (a) and the trigonometric identity $\sin^2 x + \cos^2 x = 1$ to show that
$$\int \sin^n x\,dx = -\dfrac{\sin^{n-1} x \cos x}{n}$$
$$+\,\dfrac{n-1}{n}\int \sin^{n-2} x\,dx \ (n \neq 0).$$

34. (a) Use the reduction formula
$$\int \sin^n x\,dx = -\dfrac{\sin^{n-1} x \cos x}{n}$$
$$+\,\dfrac{n-1}{n}\int \sin^{n-2} x\,dx \ (n \neq 0)$$
    to show that if $n>0$ is an odd integer, then
$$\int_0^{\pi/2} \sin^n x\,dx = \dfrac{(n-1)}{n}\dfrac{(n-3)}{n-2}\cdots\dfrac{4}{5}\dfrac{2}{3}.$$
    (b) Use the reduction formula in part (a) to find an expression for $\int_0^{\pi/2} \sin^n x\,dx$ if $n>0$ is an even integer.

35. When using the trigonometric substitution $x = 2\sin t$, the expression $\sqrt{4-x^2}$ becomes $2\cos t$. When using the trigonometric substitution $x = 2\sec t$, the expression $\sqrt{x^2-4}$ becomes $|2\tan t|$. Explain why the absolute value is required for one of these substitutions but not the other.

36. The substitution $x = 2\arctan t$ can be used to transform rational expressions involving trigonometric functions into rational expressions involving the variable t. This exercise explores this substitution.
    (a) Show that
$$\sin x = \dfrac{2t}{1+t^2}.$$
    (b) Show that
$$\cos x = \dfrac{1-t^2}{1+t^2}.$$
    (c) Show that
$$\int \dfrac{dx}{1+\sin x + \cos x} = \ln\left|1+\tan(x/2)\right| + C.$$

37. Use the substitution described in Exercise 36 to evaluate
$$\int \dfrac{dx}{1+\cos x}.$$

9.4 Miscellaneous Exercises

In Exercises 1–80, find the antiderivative.

1. $\int \dfrac{\sin x}{(3+\cos x)^2}\,dx$

2. $\int \dfrac{x^2}{x+1}\,dx$

3. $\int x\left(3+4x^2\right)^5\,dx$

4. $\int \dfrac{dx}{\sqrt{1-x^2}}$

5. $\displaystyle\int \frac{x}{\sqrt[3]{x^2+4}}\,dx$

6. $\displaystyle\int \frac{(\ln x)^2}{x}\,dx$

7. $\displaystyle\int xe^x\,dx$

8. $\displaystyle\int e^x \sin x\,dx$

9. $\displaystyle\int \frac{\ln x}{x}\,dx$

10. $\displaystyle\int x\sqrt{x+2}\,dx$

11. $\displaystyle\int \sin^2(3x)\cos(3x)\,dx$

12. $\displaystyle\int xe^{3x}\,dx$

13. $\displaystyle\int xe^{3x^2}\,dx$

14. $\displaystyle\int \frac{dx}{1+4x^2}$

15. $\displaystyle\int \frac{7-x}{(x+3)(x^2+1)}\,dx$

16. $\displaystyle\int (2-3x)^{10}\,dx$

17. $\displaystyle\int \arctan x\,dx$

18. $\displaystyle\int \frac{\sec^2 x}{3+\tan x}\,dx$

19. $\displaystyle\int x\sin x\,dx$

20. $\displaystyle\int \frac{dx}{(x-1)(x+2)}$

21. $\displaystyle\int x^2 \ln x\,dx$

22. $\displaystyle\int \frac{x+1}{x^2+1}\,dx$

23. $\displaystyle\int \frac{e^x}{\sqrt{1-e^{2x}}}\,dx$

24. $\displaystyle\int \frac{\sin x}{2+\cos x}\,dx$

25. $\displaystyle\int \ln x\,dx$

26. $\displaystyle\int x\cos\left(3x^2\right)\,dx$

27. $\displaystyle\int \arcsin x\,dx$

28. $\displaystyle\int \frac{dx}{x^2+2x+3}$

29. $\displaystyle\int \frac{x}{\sqrt{x-2}}\,dx$

30. $\displaystyle\int \frac{dx}{\sqrt{1-4x^2}}$

31. $\displaystyle\int \frac{x+6}{(x+1)\left(x^2+4\right)}\,dx$

32. $\displaystyle\int x\sin^2 x\cos x\,dx$

33. $\displaystyle\int \frac{x^3}{1+x^2}\,dx$

34. $\displaystyle\int \tan x\,dx$

35. $\displaystyle\int \cos(2x)\,dx$

36. $\displaystyle\int e^{2x}\sqrt{1+e^x}\,dx$

37. $\displaystyle\int \frac{dx}{1+x^2}$

38. $\displaystyle\int \frac{dx}{\sqrt{3-2x-x^2}}$

39. $\displaystyle\int \sin^3 x\cos^3 x\,dx$

40. $\displaystyle\int x^2 \arcsin x\,dx$

41. $\displaystyle\int \frac{dx}{x(\ln x)^2}$

42. $\displaystyle\int x\arctan x\,dx$

43. $\displaystyle\int \frac{dx}{x^3+x}$

44. $\displaystyle\int \tan^4 x\,dx$

45. $\displaystyle\int \frac{x+5}{x^2+3x-4}\,dx$

46. $\displaystyle\int \frac{x^3}{\sqrt{4-x^2}}\,dx$

47. $\displaystyle\int \frac{dx}{\sqrt[3]{x-1}}$

48. $\displaystyle\int \frac{x}{(x-1)(x+1)}\,dx$

49. $\displaystyle\int x^3 e^{x^2}\,dx$

50. $\displaystyle\int \frac{dx}{\sqrt{9+x^2}}$

51. $\displaystyle\int \frac{dx}{2x-x^2}$

52. $\displaystyle\int \frac{x^2}{1-3x}\,dx$

53. $\displaystyle\int \left(x^2+2x+3\right)^{3/2}\,dx$

54. $\displaystyle\int \frac{x}{\left(x^2-1\right)^3}\,dx$

55. $\displaystyle\int e^x e^{2x}\,dx$

56. $\displaystyle\int \sqrt{4x-3}\,dx$

57. $\displaystyle\int \ln(1+x^2)\,dx$

58. $\displaystyle\int \sin(\sqrt{x})\,dx$

59. $\displaystyle\int \frac{x}{9+4x^4}\,dx$

60. $\displaystyle\int \frac{dx}{\left(4-x^2\right)^{3/2}}$

61. $\displaystyle\int \sin(3x)\cos(5x)\,dx$

62. $\displaystyle\int \ln\left(\sqrt{x^2+1}\right)\,dx$

63. $\displaystyle\int x\sqrt{2x+1}\,dx$

64. $\displaystyle\int \frac{dx}{x\left(x+\sqrt[3]{x}\right)}$

65. $\displaystyle\int \frac{\tan x}{\sec^2 x}\,dx$

66. $\displaystyle\int \frac{dx}{x^3+1}$

67. $\displaystyle\int \frac{x}{16+9x^2}\,dx$

68. $\displaystyle\int \frac{dx}{e^x-1}$

69. $\displaystyle\int \frac{dx}{(e^x-e^{-x})^2}$

70. $\displaystyle\int \frac{dx}{\sqrt{2x-x^2}}$

71. $\displaystyle\int x\tan^2 x\,dx$

72. $\displaystyle\int \frac{dx}{1+\sqrt{x}}$

73. $\displaystyle\int \frac{x^3}{\left(x^2+1\right)^2}\,dx$

74. $\displaystyle\int \cos^3 x\,dx$

75. $\displaystyle\int \sin x\sin(2x)\,dx$

76. $\displaystyle\int x^2 \ln(3x)\,dx$

77. $\displaystyle\int \frac{x}{1+x^4}\,dx$

78. $\displaystyle\int \sqrt{x}\ln x\,dx$

79. $\displaystyle\int \sin^5 x\cos^2 x\,dx$

80. $\displaystyle\int x\sec^2 x\,dx$

10

Improper Integrals

10.1 When Is an Integral Improper?

Each of the following expressions is an **improper integral**. ▶

$$\int_1^\infty \frac{dx}{x^2} \qquad \int_1^\infty \frac{dx}{x} \qquad \int_{-\infty}^\infty \frac{dx}{1+x^2}$$

$$\int_0^\infty e^{-x^2}\,dx \qquad \int_0^1 \frac{dx}{x^2} \qquad \int_0^1 \frac{dx}{\sqrt{x}}$$

It will pay to keep these standard examples in mind. They illustrate most of what can go right—and wrong—with improper integrals.

Improper integrals, like improper fractions, commit no breach of morals or manners. ▶ The adjective "improper" is a warning sticker attached to integrals that differ somehow from the ordinary $\int_a^b f(x)\,dx$ variety, in which $[a, b]$ is a finite interval and $f(x)$ is a continuous function. Each of the preceding integrals, examined carefully, should raise some suspicion.

Moreover, they're just as useful as "proper" integrals.

Integrals can commit two types of "impropriety":

- *The interval of integration may be infinite,* as in the first four sample integrals. This is technically illegal, because the formal definition of definite integral relies on partitions of a *finite* interval.

- *The integrand may be unbounded on the interval of integration,* ▶ as in the last two samples. This, too, is technically illegal, because the integrand is not defined at an endpoint of the interval of integration.

The integrand "blows up," in other words.

A few really obstreperous integrals, such as

$$\int_0^\infty \frac{dx}{\sqrt{x}+x^2},$$

commit *both* types of impropriety; they need especially strict handling.

Convergence and Divergence: Basic Ideas and Examples

Some integrals, despite being improper, have a sensible numerical value; they are called **convergent**. For other integrals the impropriety is fatal—these integrals have no sensible finite value, and are called **divergent**. First, some concrete examples; formal definitions come later. ◄

Study the examples carefully; they motivate the formal definitions.

EXAMPLE 1   Make sense of $\int_1^\infty \dfrac{dx}{x^2}$.

Solution   What could such an integral mean? Interpreted geometrically, the integral represents the shaded area in the following graph.

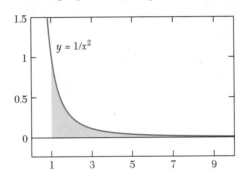

The improper integral $\int_1^\infty dx/x^2$ as area

The shaded region extends infinitely far to the right, but could its *area* be finite? The answer is yes; the reason involves a limit. For any number $t > 1$, consider

$$\int_1^t \frac{dx}{x^2} = \text{area from } x = 1 \text{ to } x = t.$$

Check the easy computation.   We can calculate this area exactly: ◄

$$\int_1^t \frac{dx}{x^2} = \frac{-1}{x}\Bigg]_1^t = 1 - \frac{1}{t}.$$

As $t \to \infty$, this quantity tends to 1. In other words, the total shaded region, although infinitely long, has *finite* area. ◄   In telegraphic summary:

Does this result go against intuition? We'll see many similar results.

$$\lim_{t\to\infty} \int_1^t \frac{dx}{x^2} = \lim_{t\to\infty}\left(1 - \frac{1}{t}\right) = 1.$$

We say that the integral *converges* to 1.   ∎

EXAMPLE 2   Does $\int_1^\infty \dfrac{dx}{x}$ converge or diverge? Why?

Solution   At first glance, the situation *looks* similar to the preceding one.

The improper integral $\int_1^\infty dx/x$ as area

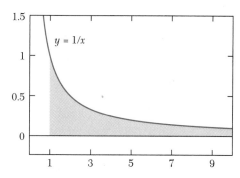

The question, again, is whether the shaded region—unbounded on the right—has finite or infinite area.

The picture alone, being finite, can't tell us the answer. Nor can a comparison with the previous region—the current region has larger area. ▶ From what we know so far, the area could be either finite or infinite. To answer finally, we'll calculate much as before. For $t > 1$, the area from $x = 1$ to $x = t$ is ▶

Look carefully; do you see why?

Check details.

$$\int_1^t \frac{dx}{x} = \ln x \Big]_1^t = \ln t.$$

(The area shown in the second graph, for instance, is $\ln 10 \approx 2.3026$.) As $t \to \infty$, $\ln t \to \infty$. We conclude, therefore, that the current improper integral *diverges* to infinity. In symbolic shorthand,

$$\lim_{t \to \infty} \int_1^t \frac{dx}{x} = \lim_{t \to \infty} (\ln t) = \infty.$$ ■

EXAMPLE 3   Does $\displaystyle\int_{-\infty}^\infty \frac{dx}{1 + x^2}$ converge? If so, to what?

Solution   The integral is improper at both ends, so we break it into two pieces,

$$\int_{-\infty}^\infty \frac{dx}{1 + x^2} = \int_{-\infty}^0 \frac{dx}{1 + x^2} + \int_0^\infty \frac{dx}{1 + x^2},$$

and handle each separately. ▶

First consider the last summand. A straightforward calculation,

One impropriety at a time is plenty!

$$\lim_{t \to \infty} \int_0^t \frac{dx}{1 + x^2} = \lim_{t \to \infty} \arctan x \Big]_0^t = \lim_{t \to \infty} \arctan t = \frac{\pi}{2},$$

shows that the second summand converges to $\pi/2$. Because the integrand is even, ▶ the first summand has the same value as the second. In other words,

Do you see why? What does "even" mean graphically?

$$\int_{-\infty}^0 \frac{dx}{1 + x^2} = \frac{\pi}{2}.$$

The conclusion is now clear: The original integral converges to π. In symbols,

$$\int_{-\infty}^\infty \frac{dx}{1 + x^2} = \int_{-\infty}^0 \frac{dx}{1 + x^2} + \int_0^\infty \frac{dx}{1 + x^2} = \pi.$$ ■

Caveat Grapher

The preceding examples illustrate one of the subtleties inherent in improper integrals. At first glance, the graphs of $1/x^2$ and $1/x$ appear similar: Both approach zero asymptotically as $x \to \infty$. Nevertheless, the first graph bounds only *one* unit of area, while the second graph bounds *infinite* area.

Another Impropriety

The improprieties in the preceding examples involved infinite *intervals*. Almost the same strategy applies to infinite *integrands*.

E X A M P L E 4   Discuss $\displaystyle\int_0^1 \frac{dx}{x^2}$.

S o l u t i o n   This time, the geometric question is whether the *vertically* unbounded region in the following graph has finite or infinite area.

The improper integral $\int_0^1 dx/x^2$ as area

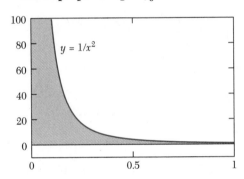

To settle the question, we find another limit of areas—this time as t tends to 0 *from the right*. ◄   For any $t > 0$, the area from $x = t$ to $x = 1$ makes good sense, and it's easy to find: ◄

Rather than as $t \to \infty$.

We calculated almost the same integral before.

$$\int_t^1 \frac{dx}{x^2} = -\frac{1}{x}\Big]_t^1 = \frac{1}{t} - 1.$$

The result shows that, as t tends to 0 from the right, the area in question *tends to infinity*. Symbolically put:

$$\lim_{t \to 0^+} \int_1^t \frac{dx}{x^2} = \lim_{t \to 0^+}\left(\frac{1}{t} - 1\right) = \infty.$$

The original integral diverges to infinity.   ∎

What the Examples Say; Formal Definitions

We applied the same basic idea to each of the preceding improper integrals, regardless of whether the interval or the integrand was unbounded. First we *located* the impropriety at ∞, at $-\infty$, or at a finite endpoint of the interval of integration. (If an integral is improper at more than one place, we rewrite it as a sum of simpler integrals, each with only one impropriety.) Then, given any integral with one improper endpoint, we considered the limit as a variable endpoint tends to the troublesome value from either above or below.

In the case of greatest interest, the fine print reads like this:

Definition Consider the integral $I = \int_a^\infty f(x)\,dx$, where f is continuous for $x \geq a$. If the limit

$$L = \lim_{t \to \infty} \int_a^t f(x)\,dx$$

exists and is finite, then I **converges** to L. Otherwise, I **diverges**.

Graphically speaking, the question concerns the long-term behavior of the following shaded area: As $t \to \infty$, does the area converge or diverge?

Convergence: what happens as $t \to \infty$?

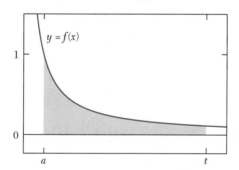

The integrals in Examples 2 and 4 diverge to infinity. The next example illustrates another way in which an integral can diverge.

EXAMPLE 5   Does $I = \int_0^\infty \cos x\,dx$ converge or diverge?

Solution   The definition gives a quick answer. Because

$$\int_0^t \cos x\,dx = \sin x \Big]_0^t = \sin t,$$

it follows that

$$\lim_{t \to \infty} \int_0^t \cos x\,dx = \lim_{t \to \infty} \sin t.$$

The last limit doesn't exist: As $t \to \infty$, $\sin t$ oscillates endlessly between -1 and 1. Hence, I diverges. ∎

A similar but more general definition of convergence applies to *any* improper integral:

> **Definition** Let the integral $I = \int_a^b f(x)\,dx$ be improper either at a or at b. (The cases $a = -\infty$ and $b = \infty$ are allowed.) If either
>
> $$\lim_{t \to a^+} \int_t^b f(x)\,dx \qquad \text{or} \qquad \lim_{t \to b^-} \int_a^t f(x)\,dx$$
>
> exists and has finite value L, then I **converges** to L. Otherwise, I **diverges**.

Notice:

A Limit of Proper Integrals Every convergent *improper* integral is a limit of *proper* integrals. For example,

$$\int_1^\infty f(x)\,dx = \lim_{t \to \infty} \int_1^t f(x)\,dx$$

if the limit exists.

$\int_{-\infty}^\infty dx/(1+x^2)$ *is doubly improper; we already saw that it converges to* π.

More Than One Impropriety If an integral is improper at both ends, ◄ it can be broken into two summands in any convenient manner; the whole thing converges only if *both* summands converge.

EXAMPLE 6   Discuss $I = \int_0^\infty \dfrac{dx}{x^2}$.

Solution   The integral is improper at both ends. To separate the improprieties, we write

$$I = \int_0^\infty \frac{dx}{x^2} = \int_0^1 \frac{dx}{x^2} + \int_1^\infty \frac{dx}{x^2} = I_1 + I_2.$$

As we saw in earlier examples, I_2 *converges* to 1, but I_1 *diverges* to ∞. Hence, I itself diverges.   ■

BASICS

In Exercises 1–6, explain why the integral is improper.

1. $\displaystyle\int_0^\infty x^2 e^{-x^2}\,dx$

2. $\displaystyle\int_0^1 \frac{x}{\sqrt{x^2 - 3x + 2}}\,dx$

3. $\displaystyle\int_1^4 \frac{dx}{x^2 \ln x}$

4. $\displaystyle\int_0^3 \frac{dx}{x^2 - x - 2}$

5. $\displaystyle\int_0^{\pi/2} \tan x\,dx$

6. $\displaystyle\int_0^{2\pi} \frac{\cos x}{\sqrt{1 + \cos x}}\,dx$

7. Show that the improper integral $\displaystyle\int_0^\infty \frac{dx}{x^2}$ diverges.

8. Explain why the integral $\displaystyle\int_0^{\pi/2} \frac{\cos x}{\sqrt{1 - \sin^2 x}}\,dx$ is not improper.

FURTHER EXERCISES

In Exercises 9–16, use an antiderivative to evaluate the improper integral.

9. $\int_0^\infty e^{-x}\,dx$

10. $\int_e^\infty \dfrac{dx}{x\,(\ln x)^2}$

11. $\int_1^\infty \dfrac{dx}{x(1+x)}$

12. $\int_0^4 \dfrac{dx}{\sqrt{x}}$

13. $\int_{-2}^2 \dfrac{2x+1}{\sqrt[3]{x^2+x-6}}\,dx$

14. $\int_\pi^\infty e^{-x}\sin x\,dx$

15. $\int_0^\infty \dfrac{\arctan x}{(1+x^2)^{3/2}}\,dx$

16. $\int_2^4 \dfrac{x}{\sqrt{|x^2-9|}}\,dx$

In Exercises 17–19, find a value for the parameter a that makes the value of the improper integral less than 10^{-5}.

17. $\int_a^\infty e^{-x}\,dx$

18. $\int_a^\infty \dfrac{dx}{x^2+1}$

19. $\int_a^\infty \dfrac{dx}{x\,(\ln x)^3}$

20. (a) Suppose that $f(x) \geq 0$ for all $x \geq 1$ and that $\int_1^\infty f(x)\,dx$ converges. Explain why there is a number a such that $\int_a^\infty f(x)\,dx \leq 10^{-10}$.

(b) Suppose that $g(x) \geq 0$ for all $x \geq 1$ and that $\int_1^\infty g(x)\,dx$ diverges. Explain why there is a number a such that $\int_1^a g(x)\,dx \geq 10^{10}$.

21. Suppose that $I = \int_0^\infty f(x)\,dx$ converges and that $\left|\int_a^\infty f(x)\,dx\right| \leq 0.0001$. Explain why

$$\left| I - \int_0^a f(x)\,dx\right| \leq 0.0001.$$

22. (a) Show that $\lim_{a\to\infty} \int_{-a}^a x\,dx = 0$.

(b) Explain why $\int_{-\infty}^\infty x\,dx$ diverges.

23. (a) Explain why the integral $\int_{-1}^1 x^{-3}\,dx$ is improper.

(b) Does the improper integral in part (a) converge?

In Exercises 24–40, use an antiderivative to determine whether the improper integral converges or diverges. If the integral converges, evaluate it.

24. $\int_0^\infty \dfrac{x}{\sqrt{1+x^2}}\,dx$

25. $\int_0^\infty \dfrac{\arctan x}{1+x^2}\,dx$

26. $\int_e^\infty \dfrac{dx}{x\ln x}$

27. $\int_3^\infty \dfrac{x}{\left(x^2-4\right)^3}\,dx$

28. $\int_{-\infty}^1 e^x\,dx$

29. $\int_0^8 \dfrac{dx}{\sqrt[3]{x}}$

30. $\int_1^3 \dfrac{dx}{\sqrt[3]{x-2}}$

31. $\int_2^3 \dfrac{x}{\sqrt{3-x}}\,dx$

32. $\int_0^2 \dfrac{dx}{\sqrt{4-x^2}}$

33. $\int_1^\infty \dfrac{dx}{x(\ln x)^2}$

34. $\int_0^\infty \dfrac{dx}{(x-1)^2}$

35. $\int_0^\infty \dfrac{dx}{e^x-1}$

36. $\int_{-\infty}^\infty e^{-x}\,dx$

37. $\int_{-\infty}^\infty \dfrac{dx}{e^x+e^{-x}}$

38. $\int_0^1 \dfrac{x}{\sqrt{1-x^2}}\,dx$

39. $\int_0^1 \dfrac{e^{-\sqrt{x}}}{\sqrt{x}}\,dx$

40. $\int_0^{\pi/2} \dfrac{\cos x}{\sqrt{\sin x}}\,dx$

41. (a) Show that $\int_1^\infty \dfrac{dx}{x}$ diverges.

(b) Show that $\int_1^\infty \dfrac{dx}{x^p}$ converges if $p > 1$.

(c) Show that $\int_1^\infty \dfrac{dx}{x^p}$ diverges if $p < 1$.

42. For which values of p does $\int_0^1 \dfrac{dx}{x^p}$ converge?

43. (a) For which values of p does $\int_1^e \dfrac{dx}{x(\ln x)^p}$ converge?

(b) For which values of p does $\int_e^\infty \dfrac{dx}{x(\ln x)^p}$ converge?

44. (a) Suppose that c is a real number and that $f(x) \geq c > 0$ for all $x \geq 0$. Give a graphical argument that explains why $\int_0^\infty f(x)\,dx$ diverges.

(b) Suppose that $0 \leq g(x) \leq x^{-2}$ for all $x \geq 1$. Give a graphical argument that explains why $\int_1^\infty g(x)\,dx$ converges.

(c) Suppose that $h(x) \geq x^{-1}$ for all $x \geq 1$. Give a graphical argument that explains why $\int_1^\infty h(x)\,dx$ diverges.

In Exercises 45–50, evaluate the improper integral for all values of the parameter C for which it converges.

45. $\int_0^\infty \left(\dfrac{2x}{x^2+1} - \dfrac{C}{2x+1}\right)dx$

46. $\int_1^\infty \left(\dfrac{C}{x+1} - \dfrac{3x}{2x^2+C}\right)dx$

47. $\int_1^\infty \left(\dfrac{Cx^2}{x^3+1} - \dfrac{1}{3x+1}\right)dx$

48. $\displaystyle\int_0^\infty \left(\frac{1}{\sqrt{x^2+4}} - \frac{C}{x+2} \right) dx$

49. $\displaystyle\int_0^\infty \left(\frac{x}{x^2+1} - \frac{C}{3x+1} \right) dx$

50. $\displaystyle\int_1^\infty \left(\frac{Cx}{x^2+1} - \frac{1}{2x} \right) dx$

Some improper integrals can be transformed into proper integrals by an appropriate change of variables. In Exercises 51–52, use the specified substitution to transform the improper integral into a proper integral with the same value. (Do not evaluate these integrals.)

51. $\displaystyle\int_1^\infty \frac{x}{x^3+1}\, dx; \quad u = x^{-1}$

52. $\displaystyle\int_0^{\pi/2} \frac{\cos x}{\sqrt{\pi - 2x}}\, dx; \quad u = \sqrt{\pi - 2x}$

53. Show that $\displaystyle\int_0^\infty \frac{dx}{1+x^4} = \int_0^\infty \frac{x^2}{1+x^4}\, dx$.

   [**HINT:** See Exercise 51.]

54. Show that $\displaystyle\int_0^\infty x^3 e^{-x}\, dx = \int_0^1 (-\ln x)^3\, dx$.

   [**HINT:** See Exercise 51.]

55. Evaluate $\displaystyle\int_0^\infty \frac{x \ln x}{1+x^4}\, dx$.

56. Let $\displaystyle I = \int_{\pi/4}^{\pi/2} \sqrt{1 + \tan x}\, dx$.

   (a) Explain why I is an improper integral.
   (b) Use the substitution $\tan x = 1/u^2$ to show that

$$I = 2 \int_0^1 \frac{\sqrt{1+u^2}}{1+u^4}\, du.$$

10.2 Detecting Convergence, Estimating Limits

The preceding section was about the idea of convergence or divergence of an improper integral. When handling an integral of the form $\int_1^\infty f(x)\, dx$, for example, all depends on these questions:

> *Does $\int_1^t f(x)\, dx$ approach a finite limit as t tends to infinity? If so, what is this limit?*

Answering these questions was a routine matter for the relatively simple examples already presented. Each time, we found by antidifferentiation a simple, explicit formula (in t) for $\int_1^t f(x)\, dx$. ◀ Letting t tend to infinity, we could easily see whether a limit existed and, if so, find its value.

For example,
$\int_1^t dx/x^2 = 1 - 1/t.$

Things don't always go that smoothly. Finding a symbolic expression for $\int_1^t f(x)\, dx$ may be difficult, or even impossible. Even if a symbolic expression *is* found, finding a limit—or even deciding whether one exists—may be tricky.

All is not lost. As they do for *proper* definite integrals, numerical methods (such as the trapezoid and midpoint rules) can help us estimate *improper* integrals. This section tells how. For simplicity, we discuss improper integrals of only one type: those with infinite intervals of integration.

Important Examples

Estimating improper integrals numerically takes special care. One problem is obvious: The improper integral $\int_a^\infty f(x)\, dx$ is taken over an infinite interval, but all our error-bound formulas for estimating definite integrals involve the *length* of the interval of integration. How can these formulas apply to infinite intervals?

The following examples suggest strategies for dealing with (or dodging) such difficulties. They illustrate and motivate the theory that follows. ▶

Notice especially how the words "comparison" and "tail" are used. They'll reappear later.

EXAMPLE 1   Does $I = \displaystyle\int_1^\infty \frac{dx}{x^5 + 1}$ converge or diverge?

Solution   The question is easy to *state* symbolically:

$$\text{Does the limit } \lim_{t \to \infty} \left(\int_1^t \frac{dx}{x^5 + 1} \right) \text{ exist? } \blacktriangleright$$

If so, we'll worry later about its value.

Alas, it is harder to *answer* symbolically. The given integrand has a very complicated antiderivative; ▶   even if we wrote it down, finding a limit (if one exists) symbolically would be very hard.

Complicated even by Maple *standards; it fills most of a computer screen.*

It's hard to calculate the preceding limit. Yet, perhaps surprisingly, it's easy to see that *some* finite limit must exist. Because the integrand $1/(x^5 + 1)$ is positive for all $x \geq 1$, the question is whether the following (positive) area "blows up" or remains bounded as t tends to infinity.

Does I converge? What happens as $t \to \infty$?

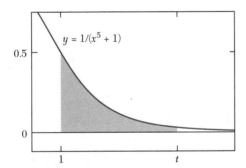

The answer is that the area remains bounded, so I converges. A natural **comparison** of the two similar-looking integrals

$$\int_1^\infty \frac{dx}{x^5 + 1} \quad \text{and} \quad \int_1^\infty \frac{dx}{x^5}$$

shows why. Notice first that the second integral converges to 1/4: ▶

The computation is easy, but check details.

$$\lim_{t \to \infty} \int_1^t \frac{dx}{x^5} = \lim_{t \to \infty} \left(\frac{1}{4} - \frac{1}{4t^4} \right) = \frac{1}{4}.$$

The key idea is that for all $x \geq 1$, ▶

The only x's we care about.

$$\frac{1}{x^5 + 1} < \frac{1}{x^5}.$$

The graphs agree:

Comparing improper areas

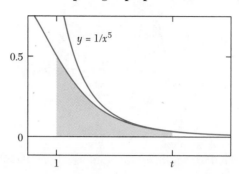

The shaded area, whatever its exact value, is evidently less than the corresponding area under the graph of $y = 1/x^5$. Our conclusion now follows:

The original integral I converges; its limit is less than 1/4.

Briefly, in symbols,

$$\int_1^\infty \frac{dx}{x^5 + 1} < \int_1^\infty \frac{dx}{x^5} = \frac{1}{4}.$$

■

EXAMPLE 2   OK, $I = \displaystyle\int_1^\infty \frac{dx}{x^5 + 1}$ converges. To what value does it converge?

Solution   Symbolic methods aren't promising, for reasons already stated, so we try a numerical approach.

Numerical integration rules apply only to *finite* intervals $[a, b]$. With this restriction in mind, we break I into two parts:

$$I = \int_1^\infty \frac{dx}{x^5 + 1} = \int_1^5 \frac{dx}{x^5 + 1} + \int_5^\infty \frac{dx}{x^5 + 1} = I_1 + I_2.$$

The first integral, I_1, is over the finite interval $[1, 5]$; we'll estimate its value numerically. (There's nothing sacred about $[1, 5]$; we could have used, say, $[1, 10]$.) The second integral, I_2, is called, picturesquely, a **tail** of I; it represents the total area—small, we hope—under the graph to the *right* of $x = 5$.

To start, we apply the midpoint rule with 50 subdivisions to I_1. Here's the result (courtesy of *Maple*):

$$\int_1^5 \frac{dx}{x^5 + 1} \approx M_{50} \approx 0.17991.$$

We expect M_{50} to approximate I_1 closely, but how closely does it approximate I itself? There are two sources of error:

1. The error M_{50} commits in estimating I_1

2. The error due to ignoring the tail I_2

We'll bound each error separately.

Bounding the first type of error is by now a routine matter. The usual midpoint-rule error-bound formula guarantees ▶ that this error is less than 0.003. Bounding the second type of error—known as **bounding the tail**—involves the same comparison as in Example 1. Because the inequality

We spare you the details.

$$\frac{1}{x^5 + 1} < \frac{1}{x^5}$$

holds for all x of interest, it follows that

$$\int_5^\infty \frac{dx}{x^5 + 1} < \int_5^\infty \frac{dx}{x^5}.$$

Zoologically speaking, the first tail is "skinnier" than the second, and therefore has less area. The right-hand integral is easy to calculate:

$$\int_5^\infty \frac{dx}{x^5} = \lim_{t \to \infty} -\frac{1}{4x^4}\bigg]_5^t = \frac{1}{2500} = 0.0004.$$

The result is an upper bound on the second type of error.

Having bounded both types of possible error, we conclude (with some relief) that the estimate $I \approx 0.1781$ commits total error less than $0.003 + 0.0004 = 0.0034$. ∎

Comparing Improper Integrals: Two Theorems

The preceding examples used the comparisons

$$\int_1^\infty \frac{dx}{x^5 + 1} < \int_1^\infty \frac{dx}{x^5} \quad \text{and} \quad \int_5^\infty \frac{dx}{x^5 + 1} < \int_5^\infty \frac{dx}{x^5}.$$

From the first inequality we inferred that the "smaller" integral converges; using the second, we bounded a tail. The following theorem guarantees the legitimacy of what we did. ▶

Not that it was in much doubt.

T h e o r e m 1 (**Comparison for Nonnegative Improper Integrals**) Let f and g be continuous functions. Suppose that for all $x \geq a$,

$$0 \leq f(x) \leq g(x).$$

- If $\int_a^\infty g(x)\, dx$ *converges, then so does* $\int_a^\infty f(x)\, dx$, *and*

$$\int_a^\infty f(x)\, dx \leq \int_a^\infty g(x)\, dx.$$

- If $\int_a^\infty f(x)\, dx$ *diverges, then so does* $\int_a^\infty g(x)\, dx$.

The theorem's two claims, once understood, are certainly plausible. The first claim says:

> If the area under the g-graph is *finite*, then so is the area under the "lower" f-graph.

The second claim says:

> If the area under the f-graph is *infinite*, then so is the area under the "higher" g-graph.

The following picture supports both claims.

What Theorem 1 says, graphically

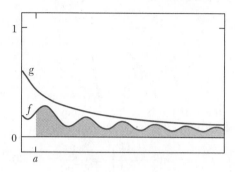

The formal proof would add little, so we omit it.

Using the Theorem

We saw both in the preceding examples.

The theorem is useful in two ways: ◄   for recognizing convergence or divergence and for estimating limits. We illustrate both uses by example.

EXAMPLE 3   Does $I = \displaystyle\int_0^\infty e^{-x^2}\,dx$ converge? If so, to what limit?

Solution   We need to determine the value of the limit

$$\lim_{t\to\infty} \int_0^t e^{-x^2}\,dx,$$

if it exists. Since we can't antidifferentiate the integrand, estimates are called for.

To see that I converges, we compare it to a nicer integral, one that we *know* converges. Notice first that if $x \geq 1$,

$$e^{-x^2} \leq x\,e^{-x^2}.$$

The graphs agree:

Comparing e^{-x^2} to xe^{-x^2}

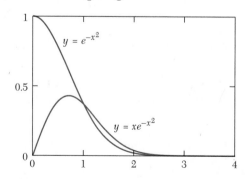

Now the theorem applies; it says that

$$\int_1^\infty e^{-x^2}\, dx < \int_1^\infty x\, e^{-x^2}\, dx.$$

The right-hand integral can be worked out directly:

$$\int_1^\infty xe^{-x^2}\, dx = \lim_{t\to\infty} \int_1^t xe^{-x^2}\, dx = \lim_{t\to\infty} -\frac{e^{-x^2}}{2}\bigg]_1^t = \frac{1}{2e}.$$

Therefore, I itself ▶ converges to some (still unknown) limit.

To estimate I numerically, we'll break it into a convenient finite piece and a tail, and estimate each one separately:

$$I = \int_0^\infty e^{-x^2}\, dx = \int_0^4 e^{-x^2}\, dx + \int_4^\infty e^{-x^2}\, dx = I_1 + I_2.$$

To estimate I_1, we use $M_{100} \approx 0.886226$. The usual error-bound formula shows ▶ that M_{100} commits error less than 0.00054 in the process. The tail integral, I_2, turns out to be even smaller. Using the same comparison as before (with $a = 4$, this time) gives

$$I_2 = \int_4^\infty e^{-x^2}\, dx < \int_4^\infty xe^{-x^2}\, dx = \frac{1}{2e^{16}} \approx 0.000000056.$$

(The value of the right-hand integral is found as before; we omit the calculation.)

The error from both sources is therefore tiny; we conclude that I converges to a limit very close to 0.886226—at least three decimal places are meaningful. ▶ ■

The theorem actually shows that $\int_1^\infty e^{-x^2}\, dx$ (rather than $\int_0^\infty e^{-x^2}\, dx$) converges. That's good enough, since $\int_0^1 e^{-x^2}\, dx$ certainly converges.

We omit the computation.

It can be shown by more advanced methods that the exact value of our integral is $\sqrt{\pi}/2 \approx 0.8862269255$.

Compared to What? Integrals for Reference

The idea of comparison testing is simple enough. The hardest part, in practice, is to decide *which* known integral to compare to a given unknown integral.

Many improper integrals can be successfully compared to one of these "benchmark" integrals (or a constant multiple thereof):

$$\int_0^\infty e^{-x}\, dx = 1$$

$$\int_1^\infty \frac{dx}{x^p} = \frac{1}{p-1} \qquad \text{if } p > 1;$$

$$\int_1^\infty \frac{dx}{x^p} = \infty \qquad \text{if } p \leq 1.$$

(The last equation means that the integral diverges to infinity.)

EXAMPLE 4 Does $I = \displaystyle\int_1^\infty \frac{dx}{x+1}$ converge or diverge? Why?

Solution The obvious comparison is to

$$\int_1^\infty \frac{dx}{x},$$

which we know to diverge; we'd therefore expect I to diverge as well. Unfortunately, the inequality

$$\frac{1}{x+1} \leq \frac{1}{x}$$

runs the wrong way for our purposes. However, the inequality

$$\frac{1}{x+1} \geq \frac{1}{2x}$$

Convince yourself—with graphs, if necessary.

is valid for $x \geq 1$. ◄   Therefore, by Theorem 1,

$$\int_1^\infty \frac{dx}{x+1} \geq \int_1^\infty \frac{dx}{2x} = \frac{1}{2} \int_1^\infty \frac{dx}{x} = \infty,$$

We could also have worked directly, antidifferentiating and taking a limit.

so I diverges, as expected. ◄   ■

EXAMPLE 5   Discuss the convergence of $I = \int_3^\infty \frac{2x^2}{x^4 + 2x + \cos x} \, dx.$

Large values of x determine whether I converges or diverges.

Solution   For large values of x ◄   the denominator behaves like x^4. This observation suggests that we compare I with the known-to-converge integral

$$\int_3^\infty \frac{2}{x^2} \, dx.$$

The comparison *is* valid. For $x \geq 3$,

$$x^4 + 2x + \cos x > x^4 \implies \frac{2x^2}{x^4 + 2x + \cos x} < \frac{2}{x^2}.$$

Now Theorem 1 guarantees that I converges and, by an easy calculation, that

$$I < \int_3^\infty \frac{2}{x^2} \, dx = \frac{2}{3}.$$ ■

Integrands That Change Sign; Absolute Convergence

Does the integral

$$I = \int_1^\infty \frac{\sin x}{x^2} \, dx$$

converge or diverge? Theorem 1 doesn't say—it applies only to *nonnegative* integrands. Nevertheless, Theorem 1 does say something useful. Because

$$\left| \frac{\sin x}{x^2} \right| \leq \frac{1}{x^2}$$

for all $x \geq 1$, it follows that

$$\int_1^\infty \left| \frac{\sin x}{x^2} \right| \, dx \leq \int_1^\infty \frac{dx}{x^2} = 1.$$

We would therefore expect that I itself converges and that

$$|I| = \left| \int_1^\infty \frac{\sin x}{x^2} \, dx \right| \leq \int_1^\infty \left| \frac{\sin x}{x^2} \right| \, dx \leq 1.$$

The next theorem justifies all these speculations.

Theorem 2 (**Absolute Comparison**) Let f and g be continuous functions such that, for all $x \geq a$,

$$0 \leq |f(x)| \leq g(x).$$

Suppose that $\int_a^\infty g(x)\,dx$ converges. Then $\int_a^\infty f(x)\,dx$ also converges, and

$$\left| \int_a^\infty f(x)\,dx \right| \leq \int_a^\infty g(x)\,dx.$$

Absolute Convergence. Theorem 2 says, among other things, that if $\int_a^\infty |f(x)|\,dx$ converges, then so must $\int_a^\infty f(x)\,dx$. The first condition is called **absolute convergence.** ▶

We'll see the same phrase later, for infinite series.

Tips on Tails

In any improper integral—say, of the form $I = \int_1^\infty f(x)\,dx$—some of the most interesting action happens in the tail. Any such integral can be broken apart, perhaps as follows:

$$\int_1^\infty f(x)\,dx = \int_1^{1000} f(x)\,dx + \int_{1000}^\infty f(x)\,dx.$$

The first term is a proper integral; it can be handled either by antidifferentiation or by numerical methods. Whether I converges or diverges depends only on the second term—if the tail converges, then so does I. If, better yet, the tail is small, as in the preceding examples, then the first term closely approximates I.

Comparison, in the sense of the two preceding theorems, is the key to keeping tails small. We illustrate with a final example.

EXAMPLE 6 Both $I = \int_1^\infty e^{-x}\,dx$ and $J = \int_1^\infty \dfrac{\sin x}{x^2}\,dx$ converge. For each, find a tail with absolute value less than 0.001.

Solution For each integral, we need an a such that $\int_a^\infty f(x)\,dx < 0.001$. For I, such an a is relatively easy to find. Notice first that, for any real number a,

$$\int_a^\infty e^{-x}\,dx = \lim_{t \to \infty} -e^{-x}\bigg]_a^t = e^{-a} = \frac{1}{e^a}.$$

Therefore,

$$\int_a^\infty e^{-x}\,dx = \frac{1}{e^a} < 0.001 \iff e^a > 1000 \iff a > \ln 1000 \approx 6.9.$$

To handle J, we need Theorem 2. Because

$$\left| \frac{\sin x}{x^2} \right| \leq \frac{1}{x^2},$$

Theorem 2 guarantees that

$$\left| \int_a^\infty \frac{\sin x}{x^2} \, dx \right| \le \int_a^\infty \frac{1}{x^2} \, dx.$$

Finally, an easy calculation shows that

$$\int_a^\infty \frac{1}{x^2} \, dx = \frac{1}{a} < 0.001 \iff a > 1000.$$ ∎

BASICS

1. (a) Explain why the inequalities $x - 1 \le x + \sin x \le x + 1$ are valid for all $x \in \mathbb{R}$.

   (b) Use the comparison test and one of the inequalities in part (a) to show that the improper integral

   $$\int_2^\infty \frac{dx}{x + \sin x}$$

   diverges.

2. (a) Explain why the inequalities $x^2 \le x^2 + \sqrt{x} \le 2x^2$ are valid for all $x \ge 1$.

   (b) Use the comparison test and one of the inequalities in part (a) to show that the improper integral

   $$\int_1^\infty \frac{dx}{x^2 + \sqrt{x}}$$

   converges. (Do not evaluate the integral.)

   (c) Explain why the inequalities $\sqrt{x} \le x^2 + \sqrt{x} \le 2\sqrt{x}$ are valid if $0 \le x \le 1$.

   (d) Use the comparison test and the inequalities in part (c) to determine whether the improper integral

   $$\int_0^1 \frac{dx}{x^2 + \sqrt{x}}$$

   converges. (Do not evaluate the integral.)

   (e) Does the improper integral

   $$\int_0^\infty \frac{dx}{x^2 + \sqrt{x}}$$

   converge? Justify your answer.

3. (a) Explain why the inequalities $\frac{1}{2}x^2 \le x^2 - \sqrt{x} \le x^2$ are valid for all $x \ge 2$.

   (b) Use the comparison test and one of the inequalities in part (a) to show that the improper integral

   $$\int_3^\infty \frac{dx}{x^2 - \sqrt{x}}$$

   converges. (Do not evaluate the integral.)

4. (a) Show that the inequalities

   $$\frac{1}{2} \le \frac{e^x}{1 + e^x} \le 1$$

   are valid for all $x \ge 0$.

   (b) Use the inequalities from part (a) to determine whether the improper integral

   $$\int_0^\infty \frac{e^x}{1 + e^x} \, dx$$

   converges.

5. (a) Explain why the inequalities

   $$\frac{1}{\sqrt{2x}} \le \frac{x}{\sqrt{1 + x^3}} \le \frac{1}{\sqrt{x}}$$

   are valid for all $x \ge 1$.

   (b) Use the comparison test and the inequalities in part (a) to determine whether the improper integral

   $$\int_0^\infty \frac{x}{\sqrt{1 + x^3}} \, dx$$

   converges. (Do not evaluate the integral.)

6. (a) Explain why the inequalities $\sqrt{x} \le 1 + \sqrt{x} \le 2\sqrt{x}$ are valid for all $x \ge 1$.

   (b) Use the comparison test and the inequalities in part (a) to determine whether the improper integral

   $$\int_0^\infty \frac{dx}{1 + \sqrt{x}}$$

   converges. (Do not evaluate the integral.)

7. (a) Explain why the inequalities $x^{9/2} \leq \sqrt{1+x^9} \leq \sqrt{2}\,x^{9/2}$ are valid for all $x \geq 1$.
   (b) Use the comparison test and the inequalities in part (a) to determine whether the improper integral
   $$\int_0^\infty \frac{dx}{\sqrt{1+x^9}}$$
   converges. (Do not evaluate the integral.)

In Exercises 8–11, find a definite integral whose value approximates that of the given (convergent) improper integral within 10^{-5}. (Do not evaluate the definite integral.)

8. $\displaystyle\int_0^\infty \frac{dx}{x^2 + e^x}$

9. $\displaystyle\int_1^\infty \frac{dx}{x^4\sqrt{2x^3+1}}$

10. $\displaystyle\int_0^\infty \frac{\arctan x}{\left(1+x^2\right)^3}\,dx$

11. $\displaystyle\int_0^\infty \frac{e^{-x}}{2+\cos x}\,dx$

FURTHER EXERCISES

12. Consider the integral $I = \displaystyle\int_e^\infty \frac{dx}{(\ln x)^2}$.
    (a) Explain why I is an improper integral.
    (b) Show that $1 \leq (\ln x)^2 \leq x^2$ for all $x \geq e$. [**HINT:** Start by showing that $1 \leq \ln x \leq x$ for all $x \geq e$.]
    (c) Explain why the inequalities in part (b) aren't helpful for determining whether I converges.
    (d) Show that $1 \leq \ln x \leq \sqrt{x}$ for all $x \geq e$.
    (e) Show that I diverges.

13. (a) Show that $0 \leq \frac{1}{2}x \leq \sin x$ if $0 \leq x \leq 1$. [**HINT:** $\frac{1}{2} \leq \cos x$ if $0 \leq x \leq 1$.]
    (b) Use the inequalities in part (a) to show that the improper integral
    $$\int_0^1 \frac{dx}{\sqrt{\sin x}}$$
    converges.

14. Let $I = \displaystyle\int_0^\infty \frac{dx}{\sqrt{x+x^4}}$.
    (a) Explain why I is improper.
    (b) Show that $0 \leq I \leq 3$. [**HINTS:** If $0 \leq x \leq 1$, then $x \geq x^4$. If $x \geq 1$, then $x^4 \geq x$.]

In Exercises 15–28, use the comparison test to determine whether the improper integral converges or diverges. (Do not evaluate the integral.)

15. $\displaystyle\int_0^\infty \frac{dx}{x^4+1}$

16. $\displaystyle\int_0^\infty \frac{dx}{x^4+x}$

17. $\displaystyle\int_2^\infty \frac{dx}{\sqrt{x}-1}$

18. $\displaystyle\int_0^\infty e^{\sin x}\,dx$

19. $\displaystyle\int_1^\infty \frac{dx}{x\sqrt{1+x}}$

20. $\displaystyle\int_0^\infty \frac{dx}{x+e^x}$

21. $\displaystyle\int_0^\infty \frac{dx}{x+e^{-x}}$

22. $\displaystyle\int_1^\infty \frac{dx}{\sqrt[3]{x^6+x}}$

23. $\displaystyle\int_0^\infty \frac{\sqrt{x}}{x+1}\,dx$

24. $\displaystyle\int_2^\infty \frac{\sin x}{x^2\sqrt{x-1}}\,dx$

25. $\displaystyle\int_0^\infty \frac{dx}{1+\sqrt{x}}$

26. $\displaystyle\int_3^\infty \frac{x}{\ln x}\,dx$

27. $\displaystyle\int_0^\infty \frac{dx}{\sqrt{x}(1+x)}$

28. $\displaystyle\int_0^\infty \frac{\sin^2 x}{(1+x)^2}\,dx$

29. Show that $\displaystyle\int_0^1 \frac{dx}{\sqrt{x+x^3}} = \int_1^\infty \frac{dx}{\sqrt{x+x^3}}$.

30. (a) Show that the improper integral
    $$\int_0^\infty \frac{e^{-x}}{\sqrt{x}}\,dx$$
    converges.
    (b) Find α and β such that the definite integral
    $$\int_\alpha^\beta \frac{e^{-x}}{\sqrt{x}}\,dx$$
    approximates the improper integral in part (a) within 0.005.

31. Let $f(x) = \int_3^x \sqrt{t}e^{-t}\,dt$.
    (a) Show that $\lim_{x\to\infty} f(x)$ exists.
    (b) Find a such that $f(a)$ approximates $\int_3^\infty \sqrt{x}e^{-x}\,dx$ within 0.001.

32. Let $I = \int_0^\infty f(x)\,dx$ where $f(x) = e^{-x^3}\sin^2 x$.
    (a) Show that I converges. [**HINT:** For every $x \geq 0$, $e^{-x^3}\sin^2 x \leq x^2 e^{-x^3}$.]
    (b) Compute an estimate of I that is guaranteed to be correct within 0.005 and explain why your estimate has this accuracy. [**HINT:** For every $x \geq 0$, $-3 < f''(x) < 2$ and $-44 < f^{(4)}(x) < 61$.]

In Exercises 33–36, use numerical integration to estimate the value of the (convergent) improper integral to within 0.005. Explain why your estimate has the desired error bound.

33. $\displaystyle\int_1^\infty \frac{dx}{x^4 + \sqrt{x}}$

34. $\displaystyle\int_0^\infty e^{-x^2} \sin x \, dx$

35. $\displaystyle\int_0^\infty \frac{dx}{e^{x^2} + x}$

36. $\displaystyle\int_1^\infty \frac{\arctan x}{\left(1 + x^2\right)^4} \, dx$

37. (a) Show that $\displaystyle\int_1^\infty \frac{\cos x}{x^2} \, dx$ converges.

(b) Use integration by parts to show that $\displaystyle\int_1^\infty \frac{\sin x}{x} \, dx$ converges.

(c) Use parts (a) and (b) to show that $\displaystyle\int_0^\infty \sin\left(e^x\right) \, dx$ converges. [**HINT**: Use the substitution $u = e^x$.]

10.3 Improper Integrals and Probability

Random Behavior

Many real-world phenomena exhibit forms of random behavior: the number of heads in ten throws of a coin, the height of a male college student chosen at random, the actual weight of a nominally 16-ounce cereal package, and so on. One may want to know (or estimate) the probability that a male student is between 68 and 75 inches tall, or the likelihood of getting less than 15 ounces of Cheerios in a package.

Statisticians call such quantities **random variables**. A random variable that can take only *finitely* many values (such as the number of heads in ten coin tosses) is called **discrete**. A random variable that ranges over an *interval* of real numbers is called **continuous**. Adult height is a good example: Any number in some interval— $(36, 85)$, say—may conceivably be someone's height in inches. We consider only continuous random variables, even though calculus methods can also be applied in the discrete case.

Basics of Continuous Probability

Probability and statistics, fields of study in their own right, draw heavily on calculus ideas—particularly the idea of improper integrals. Here we'll glimpse one such application.

By convention, probabilities are measured on a scale from 0 to 1. An event that is certain to occur has probability 1. A continuous random variable X, such as the height of a college student chosen at random, has infinitely many possible values. It's not interesting, therefore, to ask for the probability that X is *precisely* equal to *That probability is zero.* some number. ◀   A better question concerns the probability that X lies in some *interval*, such as $(68, 75)$. If this probability is, say, 0.78, then 78% of male students are between 68 and 75 inches tall.

Associated with every continuous random variable X is a function f, called the **probability density function**, or **pdf**, of X. The connection between f and X is given by an integral: ▶

This is all the probability theory we'll need!

$$\int_a^b f(x)\, dx = \text{probability that } X \text{ lies in } (a, b).$$

The following shaded region represents this probability geometrically.

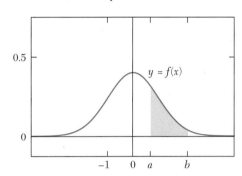

Probability that $a < X < b$

Here are some commonsense facts about f and X:

Domain   The domain of f is the set of possible values of X. (Often f has an infinite domain—hence the need for improper integrals.)

Near, Not Exact   For a given x, $f(x)$ measures the comparative likelihood that the value of X falls near x. The probability that X takes any *precise* value x is zero.

Certainty   Because X must take *some* real value, $\int_{-\infty}^{\infty} f(x)\, dx = 1$.

A Positive Function   For all inputs x, $f(x) \geq 0$.

The Normal Density Function

Many real-life random phenomena can be modeled by a **normal** (also called **Gaussian normal**) density function, i.e., one of the form

$$f(x) = \frac{1}{\sqrt{2\pi}\, s} \exp\left(-\frac{(x - m)^2}{2s^2}\right).$$

(A random variable whose density function has this form is said to be **normally distributed**.) This function looks messy, but in practice it's not too bad. The letters s and m stand for constant parameters called, respectively, the **standard deviation** and the **mean** of the random variable. (Statisticians usually write σ and μ—we'll stick with the English.) Their numerical values depend on the situation being modeled. The mean m, as its name suggests, is a kind of average value of X. The standard deviation s is a measure of how "spread out" X's values are. ▶ The following graphs illustrate the geometric effects of varying m and s; all four

Statisticians use more precise definitions of m and s. For us the general ideas suffice.

graphs bound the same total (unit) area:

Normal density, $m = 0$, $s = 1$

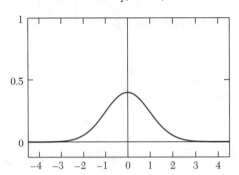

Normal density, $m = 2$, $s = 1$

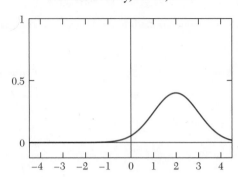

Normal density, $m = 0$, $s = 0.5$

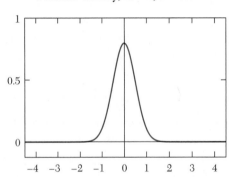

Normal density, $m = 0$, $s = 2$

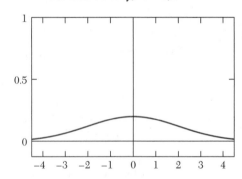

Tips on Integrating the Normal Density—Properly and Improperly

Using the normal density function

$$f(x) = \frac{1}{\sqrt{2\pi}\,s} \exp\left(-\frac{(x-m)^2}{2s^2}\right)$$

requires integrating it over various intervals, either finite or infinite. Because the normal density function has no elementary antiderivative, such integrals must be estimated numerically. The following properties of f help simplify this process.

Convergence For any values of m and s, the improper integral $\int_{-\infty}^{\infty} f(x)\,dx$ converges, and its limit is 1. (Showing that the integral converges is not too difficult; we showed something similar in Example 3, page 530. Estimating a limit is easy, but showing rigorously that the limit is *precisely* 1 is considerably harder, and we omit the proof. For more on such matters, see the exercises at the end of this section.)

Symmetry For any values of m and s, the graph of f is "centered" on the line $x = m$; the graph bounds total area 0.5 to each side of this line. This fact can simplify area computations and improve the accuracy of numerical estimates.

EXAMPLE 1   Suppose that the birth weight (in pounds) of infants born in Northfield, Minnesota, is normally distributed, with mean 7.5 pounds and standard deviation 1 pound. What percentage of infants can be expected to weigh between 6.5 and 8.5 pounds? Over 10 pounds?

Solution   Both answers are areas bounded by the normal density function with parameter values $m = 7.5$ and $s = 1$. Here are both areas, shown shaded: ▶

Observe the tiny shaded area on the far right.

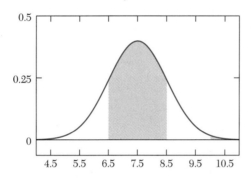

Normal density, m = 7.5, s = 1

Written as integrals, the answers are, respectively, $I_1 = \int_{6.5}^{8.5} f(x)\, dx$ and $I_2 = \int_{10}^{\infty} f(x)\, dx$. Applying the midpoint rule to I_1 gives $M_{20} \approx 0.6829$. ▶   Therefore, given our assumptions, about 68% of babies should weigh within a pound of the mean.

We used Maple. We won't bother estimating the (small) error committed.

The picture shows that I_2 is small. To estimate it numerically, we use the fact that the total area to the right of $x = 7.5$ is ▶   0.5. Therefore,

Exactly!

$$I_2 = \int_{10}^{\infty} f(x)\, dx = 0.5 - \int_{7.5}^{10} f(x)\, dx.$$

Applied to the last integral, the midpoint rule gives $M_{20} \approx 0.4938$, so $I_2 \approx 0.0062$. Therefore, only around 0.6% of babies can be expected to weigh over 10 pounds. ∎

Is Life Normal?   Are real-life phenomena—baby weights, SAT scores, heights of male adults, and so on—really normally distributed?

In one sense, certainly not. One obvious problem is that every normal distribution is infinitely "spread out": The variable can take *any* real value—large or small, positive or negative. Few real-life phenomena behave this way. No baby has negative weight, for instance; yet every normal distribution assigns some tiny but positive probability to this peculiar event.

In another, more practical sense, normal distributions do usefully model many phenomena. "Model" is the key word (here and elsewhere in applied mathematics). In suitable settings, the normal density function provides an effective, albeit imperfect, *approximation* to an underlying reality. Imperfection is no surprise; it comes with the territory of real life. Effectiveness is the surprise.

Historically, this assumption is not far off.

EXAMPLE 2   Possible scores on the math and verbal parts of the Scholastic Aptitude Test (SAT) range from 200 to 800. Assuming that test scores are normally distributed, with mean 500 and standard deviation 100, ◄ what percentage of students score between 400 and 600? Over 750?

Solution   The analysis and geometry are similar to those in Example 1. Setting $m = 500$ and $s = 100$ gives the following picture, with the areas of interest shaded:

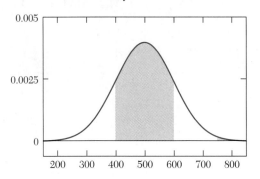

Normal density, $m = 500$, $s = 100$

The shaded areas can be estimated numerically, just as before. We omit details this time, because, for reasons we'll soon see, the shaded regions here have exactly the same areas as those in Example 1. We therefore conclude that about 68% of SAT scores fall between 400 and 600, i.e., within one standard deviation of the mean; only 0.6% of scores are above 750.   ∎

Keeping It Simple: the Standard Normal Density

A graph of this function appears earlier in this section.

Calculations are simplest if $m = 0$ and $s = 1$. With these parameter values, the **standard normal density** function takes the particularly pleasant symbolic form ◄

$$n(x) = \frac{1}{\sqrt{2\pi}} \exp\left(\frac{-x^2}{2}\right).$$

Remarkably, any normal distribution can easily be "standardized"—i.e., reinterpreted as a standard normal distribution. In the SAT case, for instance, we could consider how much scores differ from 500, rather than "raw" test scores themselves. Doing so moves the mean to 0; the standard deviation remains at 100. Dividing *these* results by 100 leaves the mean unchanged at 0 but cuts the standard deviation to 1. The results therefore have a standard normal distribution.

Statisticians call such altered data **Z-scores**. For example, raw SAT scores of 620 and 450 correspond to Z-scores of 1.20 and -0.5, i.e., to scores 1.2 standard deviations above and 0.5 standard deviation below the mean, respectively. Similarly, a 10-pound baby weighs in at 2.5 standard deviations above the mean of 7.5 pounds.

To find the probability of a Z-score between -1 and 1, we integrate the standard normal density:

$$\int_{-1}^{1} \frac{1}{\sqrt{2\pi}} \exp\left(\frac{-x^2}{2}\right) dx.$$

No symbolic antiderivative is available, but numerical integration methods work just fine. The midpoint rule with 20 subdivisions, for instance, gives ▶ $M_{20} \approx 0.6829$. *The error is small; we* This by-now-familiar result offers evidence that Z-scores make mathematical sense. *ignore it.*

Why Z-Scores Work: Substitution in Definite Integrals

What statisticians call Z-scores *we* call Z-substitution ▶ in a definite integral. We *Also known as* illustrate with an example. *u-substitution.*

EXAMPLE 3 Write integrals, with and without Z-scores, for the probability of scoring between 500 and 700 on the SAT. Why do both integrals give the same result?

Solution With raw scores, we use $m = 500$ and $s = 100$; the desired probability is

$$I_1 = \frac{1}{\sqrt{2\pi} \cdot 100} \int_{500}^{700} \exp\left(-\frac{(x-500)^2}{2 \cdot 100^2}\right) dx.$$

With Z-scores we want the probability of a result between 0 and 2, ▶ i.e., the *Raw scores of 500 and 700* integral *give Z-scores of 0 and 2.*

$$I_2 = \frac{1}{\sqrt{2\pi}} \int_0^2 \exp\left(-\frac{x^2}{2}\right) dx.$$

Numerical evidence that $I_1 = I_2$ is easy to find: Applied to either integral, $M_{20} \approx 0.47729$. A symbolic calculation confirms the result. The substitution

$$Z = \frac{x-500}{100}; \qquad dZ = \frac{dx}{100}$$

in I_1 yields I_2. ▶ ■ *Convince yourself; see the exercises at the end of this section.*

Z-Score Tables

In precomputer days, it was difficult to estimate the definite integrals that arise in the study of normal distributions. The only recourse was to numerical tables; part of such a table follows. For a given input x, the table entry is $A(x) = \int_{-\infty}^{x} n(t)\, dt$, the area shown graphically as follows:

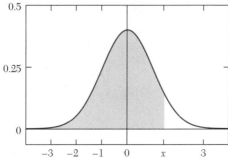

Standard normal density;

area $A(x) = \int_{-\infty}^{x} n(t)\, dt$

Values of $A(x) = \int_{-\infty}^{x} n(t)\,dt$									
x	0.0	0.1	0.2	0.3	0.4	0.5	0.6	0.7	0.8	0.9
$A(x)$	0.5000	0.5398	0.5793	0.6179	0.6554	0.6915	0.7257	0.7580	0.7881	0.8159
x	1.0	1.1	1.2	1.3	1.4	1.5	1.6	1.7	1.8	1.9
$A(x)$	0.8413	0.8643	0.8849	0.9032	0.9192	0.9333	0.9452	0.9554	0.9641	0.9713
x	2.0	2.1	2.2	2.3	2.4	2.5	2.6	2.7	2.8	2.9
$A(x)$	0.9773	0.9822	0.9861	0.9893	0.9918	0.9938	0.9953	0.9965	0.9974	0.9981

The numbers compare nicely with the graph. As we'd expect, for instance, $A(0) = 0.5$ and $A(2.9) \approx 1.00$.

BASICS

1. Sketch a graph of the normal probability density function with the given parameter values. On each graph, shade the area corresponding to the probability that the value of a random variable having this distribution will be within one standard deviation of the mean.
   (a) $m = 1$, $s = 1$ (c) $m = 0$, $s = 1/4$
   (b) $m = 2$, $s = 1/2$

2. (a) Let X be a normally distributed random variable with $m = 0$ and $s = 1$. Explain why the fraction of the possible values of X that lie within three standard deviations of the mean is
   $$\frac{1}{\sqrt{2\pi}} \int_{-3}^{3} \exp\left(-x^2/2\right)\,dx.$$
   (b) Let Y be a normally distributed random variable with $m = 0$ and $s = 2$. Write a definite integral whose value is the fraction of the possible values of Y that lie within three standard deviations of the mean.
   (c) Use the substitution $u = x/2$ to show that the integrals in parts (a) and (b) are equal.
   (d) Use numerical integration to estimate the integral in part (a).

3. Let I_1 and I_2 be the integrals discussed in Example 3. Show that making the substitution $Z = (x - 500)/100$ transforms I_1 into I_2.

4. Use the substitution $z = (x - m)/s$ to show that
   $$\frac{1}{\sqrt{2\pi}\,s} \int_{x_1}^{x_2} \exp\left(-\frac{(x - m)^2}{2s^2}\right)\,dx = \frac{1}{\sqrt{2\pi}} \int_{z_1}^{z_2} e^{-z^2/2}\,dz$$
   where $z_1 = (x_1 - m)/s$ and $z_2 = (x_2 - m)/s$.

5. Throughout this section we used the fact that if m is any real number, s is any positive number, and
   $$f(x) = \frac{1}{\sqrt{2\pi}\,s} \exp\left(-\frac{(x - m)^2}{2s^2}\right),$$
   then $\int_{-\infty}^{\infty} f(x)\,dx = 1$. This exercise explores this fact.

   (a) We showed in Example 3, page 530, that $\int_0^{\infty} e^{-x^2}\,dx$ converges, and we claimed that its limit is $\sqrt{\pi}/2$. Assuming these facts, explain geometrically why $\int_{-\infty}^{\infty} e^{-x^2}\,dx = \sqrt{\pi}$. [**HINT:** The integrand is an even function.]

   (b) If $m = 0$ and $s = 1$, then $f(x) = \dfrac{1}{\sqrt{2\pi}} \exp\left(-x^2/2\right)$.
   Use the substitution $u = x/\sqrt{2}$ to show that $\int_{-\infty}^{\infty} f(x)\,dx = 1$.

   (c) Part (b) shows that
   $$\frac{1}{\sqrt{2\pi}\,s} \int_{-\infty}^{\infty} \exp\left(-\frac{(x - m)^2}{2s^2}\right)\,dx = 1$$
   in the special case that $m = 0$ and $s = 1$. Assuming this, show that the same thing holds for any real number m and positive number s. [**HINT:** Use the substitution $u = (x - m)/s$.]

6. Suppose that an observation of a normally distributed random variable has a Z-score of -1.3.
   (a) Is this observation larger than the mean? Justify your answer.
   (b) How many standard deviations from the mean is this observation?

7. (a) What Z-score corresponds to an SAT score of 600?
   (b) What Z-score corresponds to an SAT score of 450?
   (c) Use part (b) to write an integral whose value is the probability of getting an SAT score greater than 450.
   (d) Use part (a) to write an integral whose value is the probability of getting an SAT score less than 600.
   (e) Use parts (c) and (d) to write an integral whose value is the probability of getting an SAT score between 450 and 600.

8. Let $A(x) = \displaystyle\int_{-\infty}^{x} n(t)\,dt$. Explain why $\displaystyle\int_{z_1}^{z_2} n(t)\,dt = A(z_2) - A(z_1)$.

9. Let $A(x) = \int_{-\infty}^{x} n(t)\,dt$.
   (a) Explain why $A(-z) = 1 - A(z)$.
   (b) Use part (a) and the table on page 542 to estimate $A(-1.2)$.

10. Use the table on page 542 to compute the probability of getting an SAT score
   (a) greater than 600.
   (b) between 350 and 500.

11. (a) Use the table on page 542 to estimate

$$\frac{1}{5\sqrt{2\pi}} \int_{-\infty}^{14} \exp\left(-\frac{(x-10)^2}{50}\right).$$

   (b) Use the table on page 542 to estimate

$$\frac{1}{5\sqrt{2\pi}} \int_{-\infty}^{4} \exp\left(-\frac{(x-10)^2}{50}\right).$$

FURTHER EXERCISES

12. Estimate the integrals I_1 and I_2 in Example 1, using the trapezoid rule with 20 subintervals, and find bounds on the approximation errors. Are these results consistent with those reported in the text? Justify your answer.

13. How high must a student score on the SAT to have a score in the top 10%? In the top 5%? [**HINT:** Use the tabulated values of $A(x)$ on page 542.]

14. Suppose that the average rainfall in May is 5 inches with a standard deviation of 0.6 inches. What is the probability that the rainfall next May will differ from the average by more than 1 inch? (Assume that May rainfall amounts are normally distributed.)

15. During the grand finale of a circus, a performer is shot from a special cannon. The distance the performer travels varies but is normally distributed with a mean of 150 feet and a standard deviation of 10 feet. The landing net is 30 feet long.
   (a) How far away from the cannon should the nearest edge of the net be placed?
   (b) Given the net position in part (a), what is the probability that the performer will miss the landing net?

16. A stamping machine produces can tops whose diameters are normally distributed with a standard deviation of 0.01 inch. At what "nominal" (i.e., mean) diameter should the machine be set so that no more than 10% of the can tops produced have diameters greater than 3 inches.

17. Let $A(x) = \int_{-\infty}^{x} n(t)\,dt$.
   (a) Since 0.75 lies midway between 0.7 and 0.8, it is reasonable to estimate that $A(0.75) = (0.7580 + 0.7881)/2 = 0.77305$. Does this approximation overestimate the exact value? Justify your answer. [Assume that the values of $A(0.7)$ and $A(0.8)$ are exact.]
   (b) Use a tangent line approximation to estimate $A(0.75)$. [**HINT:** $A'(x) = n(x)$.]
   (c) Does the estimate you computed in part (b) overestimate the exact value? Justify your answer.

18. The probability density function for the **log-normal** distribution is

$$g(x) = \begin{cases} \dfrac{1}{\sqrt{2\pi}\,\beta x}\, e^{-(\ln x - \alpha)^2/2\beta^2} & \text{if } x > 0, \\ 0 & \text{if } x \leq 0. \end{cases}$$

   (a) Explain why $\beta > 0$ must be true for g to be a probability density function.
   (b) Use the substitution $u = (\ln x - \alpha)/\beta$ to show that the probability that $a \leq x \leq b$ is

$$\int_{a}^{b} g(x)\,dx = A\left(\frac{\ln b - \alpha}{\beta}\right) - A\left(\frac{\ln a - \alpha}{\beta}\right),$$

   where $A(x) = \int_{-\infty}^{x} n(x)\,dx$.

19. The graph of the probability density function for a random variable X with values in the interval $[0, \infty)$ is shown below. Use this graph to estimate the probability that a random value of X lies in each of the following intervals.

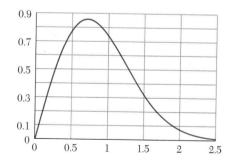

   (a) $[0, 0.5]$
   (b) $[0.5, 1.5]$
   (c) $[0, 1.5]$
   (d) $[1.5, \infty)$

20. Let T be the number of minutes that it takes a randomly selected person to solve a certain puzzle. The probability density function of T is

$$f(x) = \begin{cases} 3/x^4 & \text{if } x \geq 1, \\ 0 & \text{if } x < 1. \end{cases}$$

(a) Show that $\int_{-\infty}^{\infty} f(x)\, dx = 1$.
(b) What is the probability that it takes a person more than 2 minutes to solve this puzzle?
(c) What is the probability that it takes a person less than 1.5 minutes to solve this puzzle?
(d) Evaluate $\int_{-\infty}^{\infty} xf(x)\, dx$. (This is the average time it takes someone to solve this puzzle.)
(e) Find m such that 50% of the people solve the puzzle in m minutes or less. (This is the **median** of the random variable T.)

21. Find k so that $f(x) = k/(1 + x^2)$ is a probability density function.

22. Find k so that

$$f(x) = \begin{cases} k(1 - x^2) & \text{if } 0 \leq x \leq 1, \\ 0 & \text{otherwise,} \end{cases}$$

is a probability density function.

23. Let $\lambda > 0$ be a constant, and let f be the function

$$f(x) = \begin{cases} \lambda e^{-\lambda x} & \text{if } x \geq 0, \\ 0 & \text{if } x < 0. \end{cases}$$

(a) Show that f is a probability density function. [**NOTE:** f is known as the **exponential** density function.]
(b) Evaluate $\int_{-\infty}^{\infty} xf(x)\, dx$. [**NOTE:** If X is a random variable that has f as its probability density function, then the value of this integral is the **mean** of X.]

24. The exponential density function (see Exercise 23) is often used to model the life span of electrical devices. Suppose that 10% of a certain type of device fail after six months.

(a) What value of λ corresponds to the observed life span?
(b) What is the probability that a randomly selected device will still be functioning after 18 months?
(c) What is the mean lifetime of this device?

10.4 l'Hôpital's Rule: Comparing Rates

Indeterminate Forms: Posing the Question

Values of simple limits, such as

$$\lim_{x \to \infty} \frac{1}{x^2} = 0 \quad \text{and} \quad \lim_{x \to 0} \frac{1}{x^2} = \infty,$$

Mathematicians' jargon. are easy to guess "by inspection," ◀ i.e., just by looking at the relative sizes of numerators and denominators.

When Is a Limit Indeterminate?

Other limits are less susceptible to intuition. Consider these:

$$\lim_{x \to \infty} \frac{x^2}{2^x}; \quad \lim_{x \to 0} \frac{\sin(2x)}{x}; \quad \lim_{x \to \infty} xe^{-x}; \quad \lim_{x \to \infty} \frac{x^2 + 1}{2x^2 + 3}.$$

More mathematicians' jargon. Limits such as these are called **indeterminate forms**. ◀ They are "indeterminate" because, in every case, two conflicting tendencies operate. In the first limit, for instance, both numerator and denominator "blow up":

$$x^2 \to \infty \quad \text{and} \quad 2^x \to \infty \quad \text{as } x \to \infty.$$

The ratio is therefore **indeterminate of type** ∞/∞. Thus, the real question is *how fast* the numerator and denominator grow compared with each other.

The second limit is ambiguous for another reason: Both numerator and denominator tend, simultaneously, to zero. Their ratio is therefore called **indeterminate of type 0/0**. Again the question concerns the relative rates at which numerator and denominator tend to zero.

The third limit involves the product $x \cdot e^{-x}$. ► The two factors behave in "opposite" ways—

It isn't a quotient this time.

$$x \to \infty \quad \text{but} \quad e^{-x} \to 0.$$

Therefore the limit is called **indeterminate of type** $\infty \cdot 0$.

The last limit, like the first, is indeterminate of type ∞/∞. Much earlier in this course, we saw an algebraic method for detecting "asymptotic behavior" of rational functions like the one here. Dividing top and bottom by x^2 yields a new, simpler ► limit—and an answer:

The second limit is not indeterminate!

$$\lim_{x \to \infty} \frac{x^2 + 1}{2x^2 + 3} = \lim_{x \to \infty} \frac{1 + 1/x^2}{2 + 3/x^2} = \frac{1}{2}.$$

The result means graphically that the line $y = 1/2$ is a horizontal asymptote for our rational function.

When Algebra Fails: Graphical and Numerical Evidence

The algebraic limit calculation just shown worked—but only because both numerator and denominator happened, by good luck, to be *algebraic* functions (polynomials, in this case). With the other preceding indeterminate forms, algebra alone is ineffectual.

Graphical and numerical approaches, although less precise, are more forgiving. The following graphs ► strongly suggest values for these three limits as $x \to \infty$:

Decide which is which.

$$\frac{x^2}{2^x} \to 0; \qquad xe^{-x} \to 0; \qquad \frac{x^2 + 1}{2x^2 + 3} \to \frac{1}{2}.$$

Three functions' long-run behavior

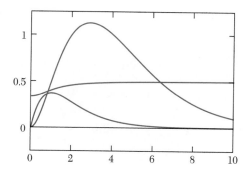

The following table ► suggests just as convincingly that, as $x \to 0$,

We could also have drawn a graph.

$$\frac{\sin(2x)}{x} \to 2.$$

As $x \to 0$, $\sin(2x)/x \to 2$: Numerical Evidence						
x	-0.1	-0.01	-0.0001	$\ldots$	0.0001	0.01	0.1
$\sin(2x)/x$	1.98670	1.99987	1.99999	$\ldots$	1.99999	1.99987	1.98670

l'Hôpital's Rule: Finding Limits by Differentiation

Graphics, numerics, and algebra helped us find (approximately, anyway) plausible values in all four of the preceding cases. l'Hôpital's rule offers another, more powerful, approach to indeterminate forms. It handles all four indeterminate forms discussed, and many more besides.

How It Works: Simplest Examples

We'll discuss them below.

l'Hôpital's rule says that, *under appropriate conditions,* ◄ an indeterminate form can be evaluated by *differentiating the numerator and the denominator separately.* In symbols,

$$\lim_{x \to a} \frac{f(x)}{g(x)} = \lim_{x \to a} \frac{f'(x)}{g'(x)}. \tag{10.1}$$

Deferring legalities for the moment, we illustrate l'Hôpital's method by example.

The numerical data suggest the answer 2.

EXAMPLE 1   Use l'Hôpital's rule to show that ◄ $\lim_{x \to 0} \dfrac{\sin(2x)}{x} = 2$.

Solution   By Equation 10.1,

$$\lim_{x \to 0} \frac{\sin(2x)}{x} = \lim_{x \to 0} \frac{2\cos(2x)}{1}.$$

The right-hand limit is no longer indeterminate; simply plugging in $x = 0$ gives 2, the expected answer. ■

EXAMPLE 2   Determine $\lim_{x \to \infty} \dfrac{x^2}{2^x}$.

Solution   Assuming that Equation 10.1 applies, it says that

$$\lim_{x \to \infty} \frac{x^2}{2^x} = \lim_{x \to \infty} \frac{2x}{2^x \ln 2}.$$

The second limit is *still* indeterminate of type ∞/∞. We apply Equation 10.1 *again*:

$$\lim_{x \to \infty} \frac{x^2}{2^x} = \lim_{x \to \infty} \frac{2x}{2^x \ln 2} = \lim_{x \to \infty} \frac{2}{2^x \ln 2 \cdot \ln 2}.$$

The rightmost limit is, at last, not indeterminate. Because the numerator remains constant while the denominator grows without bound, the last limit must be zero. So, therefore, must the first. ■

But doesn't prove—graphs aren't proofs.

EXAMPLE 3   A graph suggests that $\lim_{x \to \infty} xe^{-x} = 0$. ◄   Does Equation 10.1 agree?

Solution   Yes, but only after a little algebra. To use Equation 10.1, we must first rewrite the expression in *quotient* form:

$$\lim_{x \to \infty} xe^{-x} = \lim_{x \to \infty} \frac{x}{e^x}.$$

Now Equation 10.1 makes good sense:

$$\lim_{x \to \infty} \frac{x}{e^x} = \lim_{x \to \infty} \frac{1}{e^x} = 0,$$

exactly as expected.   ∎

When It Works, When It Doesn't: Careful Statements

Using l'Hôpital's rule requires a little care; thoughtless applications quickly lead one astray. The tempting idea that *every* limit of quotients can be handled à la l'Hôpital ▶ is just plain wrong. For instance,

I.e., by differentiating top and bottom.

$$\lim_{x \to 0} \frac{x + 42}{x + 1} = 42, \qquad \text{but} \qquad \lim_{x \to 0} \frac{(x + 42)'}{(x + 1)'} = \lim_{x \to 0} \frac{1}{1} = 1.$$

We erred in trying to apply the rule where it doesn't apply—to a limit that wasn't indeterminate to start with.

When, exactly, *does* the rule work? ▶   The following theorem gives the answer.

As a rule, it works only for indeterminate forms.

> **T h e o r e m   3** (l'Hôpital's Rule) Let f and g be differentiable functions. Suppose, too, that
>
> (a) as $x \to a$, either (i) $f(x) \to 0$ and $g(x) \to 0$ or (ii) $f(x) \to \pm\infty$ and $g(x) \to \pm\infty$; and
>
> (b) $\displaystyle\lim_{x \to a} \frac{f'(x)}{g'(x)}$ exists.
>
> Then $\displaystyle\lim_{x \to a} \frac{f(x)}{g(x)} = \lim_{x \to a} \frac{f'(x)}{g'(x)}.$

Note:

Limits at Infinity   The values $a = \pm\infty$ are allowed in the theorem. ▶   One-sided limits (e.g., $x \to 0^+$) are also permitted.

Truly Indeterminate   Hypothesis (a) guarantees that $\lim f/g$ is genuinely indeterminate, of either type $0/0$ or type ∞/∞. ▶

Luckily for us—we've already used the result twice with $a = \infty$.

Without this proviso, the theorem would fail.

EXAMPLE 4   Find $\displaystyle\lim_{x \to 0^+} x \ln x$.

Solution   The theorem applies only to quotients. Let's produce one:

$$\lim_{x \to 0^+} x \ln x = \lim_{x \to 0^+} \frac{\ln x}{1/x}.$$

The right-hand form is indeterminate of type ∞/∞; moreover,

$$\lim_{x \to 0^+} \frac{(\ln x)'}{(1/x)'} = \lim_{x \to 0^+} \frac{1/x}{-1/x^2} = \lim_{x \to 0^+} -x = 0.$$

Hence, the theorem applies; the original limit is 0.   ∎

Why It Works: Comparing Rates

The theorem concerns the behavior, as x tends to a, of the ratio $f(x)/g(x)$. If *both* numerator and denominator tend either to zero or to infinity, then the limit—if it exists—depends on the relative *rates* at which f and g tend to their limits. The derivatives $f'(x)$ and $g'(x)$ measure these rates. A formal proof of l'Hôpital's rule ◀ makes this general idea rigorous. Among its ingredients is an appeal to the mean value theorem of differential calculus.

We won't give one.

Fine Print: Pointers Toward a Proof

To see the general idea of a proof, let's suppose that f and g are differentiable functions, with $f(a) = g(a) = 0$. Then, if $x \approx a$, f and g are close to their respective tangent lines at $x = a$. In other words, ◀

Convince yourself that the tangent lines have the claimed formulas.

$$f(x) \approx f(a) + f'(a)(x - a) = f'(a)(x - a);$$

$$g(x) \approx g(a) + g'(a)(x - a) = g'(a)(x - a).$$

If $g'(a) \neq 0$, then

$$\frac{f(x)}{g(x)} \approx \frac{f'(a)}{g'(a)}, \quad \text{so} \quad \lim_{x \to a} \frac{f(x)}{g(x)} = \lim_{x \to a} \frac{f'(x)}{g'(x)} = \frac{f'(a)}{g'(a)}.$$

This is what l'Hôpital's rule says, in its simplest form. A fully general proof requires more care, but is based on the same idea.

EXAMPLE 5   By l'Hôpital's rule, $\displaystyle \lim_{x \to 0} \frac{\sin(2x)}{\sin x} = \lim_{x \to 0} \frac{2\cos(2x)}{\cos x} = 2$. Interpret the result graphically, using tangent lines.

Solution   Let $f(x) = \sin(2x)$ and $g(x) = \sin x$. Graphs of f, g, and their tangent lines at $x = 0$ ◀   follow:

$y = 2x$ and $y = x$, respectively.

Near $x = 0$, $\sin(2x) \approx 2x$, $\sin x \approx x$

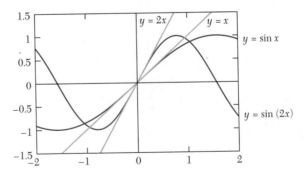

The picture shows that for x near 0, $f(x) \approx 2x$ and $g(x) \approx x$; hence,

$$\frac{f(x)}{g(x)} \approx \frac{2x}{x} = 2,$$

as claimed.

BASICS

1. (a) Using a calculator, fill in the table below.

x	1	10	100	1000
x^2				
2^x				
$(3x+10)^2$				

(b) Use the entries in the table to guess the limits

$$\lim_{x\to\infty} x^2/2^x, \quad \lim_{x\to\infty} x^2/(3x+10)^2, \quad \text{and} \quad \lim_{x\to\infty} 2^x/(3x+10)^2.$$

(c) Use l'Hôpital's rule to verify your guesses in part (b).

In Exercises 2–9, use l'Hôpital's rule to evaluate the limit.

2. $\lim\limits_{x\to 1} \dfrac{x^3+x-2}{x^2-3x+2}$

3. $\lim\limits_{x\to\infty} \dfrac{x^2+1}{2x^2+3}$

4. $\lim\limits_{x\to\infty} \dfrac{\ln x}{x^{2/3}}$

5. $\lim\limits_{x\to 0} \dfrac{5x-\sin x}{x}$

6. $\lim\limits_{x\to 0} \dfrac{1-\cos x}{\sin(2x)}$

7. $\lim\limits_{x\to 0} \dfrac{1-\cos(5x)}{4x+3x^2}$

8. $\lim\limits_{x\to\infty} \dfrac{e^x}{x^2+x}$

9. $\lim\limits_{x\to 0} \dfrac{e^x-x-1}{x^2}$

10. Can l'Hôpital's rule be used to evaluate $\lim\limits_{x\to\pi/2} \dfrac{\tan x}{x-\pi/2}$? Justify your answer.

FURTHER EXERCISES

11. Let f be the function shown. Use the graph to estimate each of the following limits.

Graph of f

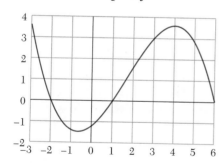

(a) $\lim\limits_{x\to 1} \dfrac{f(x)}{x^2-1}$

(b) $\lim\limits_{x\to 2} \dfrac{f(x)}{x^2-4}$

(c) $\lim\limits_{x\to 1} \dfrac{f(x)}{f(-2x)}$

(d) $\lim\limits_{x\to 1} \dfrac{f(x-3)}{f(x+3)}$

In Exercises 12–40, evaluate the limit.

12. $\lim\limits_{x\to\infty} e^{-x}\ln x$

13. $\lim\limits_{x\to 0} x\cot x$

14. $\lim\limits_{x\to 8} \dfrac{x-8}{\sqrt[3]{x}-2}$

15. $\lim\limits_{x\to 0^+} \dfrac{\sin x}{x+\sqrt{x}}$

16. $\lim\limits_{x\to 0} \dfrac{\sin x}{x-\sin x}$

17. $\lim\limits_{x\to 0} \dfrac{\tan(3x)}{\ln(1+x)}$

18. $\lim\limits_{x\to 0} \dfrac{e^x-1}{x}$

19. $\lim\limits_{x\to 0} \dfrac{\arctan(2x)}{\sin(3x)}$

20. $\lim\limits_{x\to 0} \dfrac{e^x-e^{-x}}{x}$

21. $\lim\limits_{x\to 0} \dfrac{\sin^2 x}{\cos(3x)-1}$

22. $\lim\limits_{x\to 0} \dfrac{1-\cos^2 x}{x^2}$

23. $\lim\limits_{x\to 0} \dfrac{1-x-e^{-x}}{1-\cos x}$

24. $\lim\limits_{x\to 1} \dfrac{\ln x}{x^2-x}$

25. $\lim\limits_{x\to\infty} x^2 e^{-x^2/2}$

26. $\lim\limits_{x\to 1} \dfrac{\sin(\pi x)}{x^2-1}$

27. $\lim\limits_{x\to 0} \dfrac{1-x^2-e^{-x^2}}{x^4}$

28. $\lim\limits_{x\to 1} \dfrac{\cos^3(\pi x/2)}{\sin(\pi x)}$

29. $\lim\limits_{x\to\pi/2} \dfrac{\ln(\sin x)}{(x-\pi/2)^2}$

30. $\lim\limits_{x\to 0^+} x^2 \ln x$

31. $\lim\limits_{x\to\infty} x\sin(1/x)$

32. $\lim\limits_{x\to 0} x^2 \ln(\cos x)$

33. $\lim\limits_{x\to 0^+} \sin x \ln(\sin x)$

34. $\lim\limits_{x\to\infty} x\left(\dfrac{\pi}{2}-\arctan x\right)$

35. $\lim\limits_{x\to\pi/2} \left(\dfrac{\pi}{2}-x\right)\tan x$

36. $\lim\limits_{x\to 0} \left(\dfrac{1}{\sin x}-\dfrac{1}{x}\right)$

37. $\lim\limits_{x\to\infty} \dfrac{\int_1^x \sqrt{1+e^{-3t}}\,dt}{x}$

38. $\lim\limits_{x\to 0} \dfrac{\int_0^x \sin(t^2)\,dt}{x^3}$

39. $\lim\limits_{x\to\infty} e^{x^2}\int_0^x e^{-t^2}\,dt$

40. $\lim\limits_{x\to\infty} e^{-x^2}\int_0^x e^{t^2}\,dt$

41. Let $f(k) = \lim_{x \to 2} \dfrac{x^2 - 5x + 3k}{x^2 - 3x + k}$.   Determine $f(k)$ for all k.

42. Suppose that $f(1) = 1$ and $f'(1) = 2$. Evaluate
$$\lim_{x \to 1} \frac{\big(f(x)\big)^2 - 1}{x^2 - 1}.$$

43. Suppose that f and f' are continuous functions, that $f(0) = 0$, and that
$$\lim_{x \to 0} \frac{f(x)}{\sin(2x)} = 5.$$
Evaluate $f'(0)$.

44. Suppose that f and g are differentiable functions and that $g(0) \neq 0$. Show that
$$\lim_{x \to 0} \frac{x f(x)}{(e^x - 1)\, g(x)} = \frac{f(0)}{g(0)}.$$

45. Suppose that
$$\lim_{x \to 0^+} \frac{f(x)}{g(x)} = -2 \quad \text{and} \quad \lim_{x \to 0^-} \frac{f(x)}{g(x)} = 3.$$
Evaluate $\lim_{x \to \infty} \dfrac{g(1/x)}{f(1/x)}$.

46. Use l'Hôpital's rule to evaluate the (convergent) improper integral $\int_0^\infty x e^{-x}\, dx$.

47. (a) Explain why L'Hôpital's rule can be used to evaluate
$$\lim_{x \to 0} \frac{\sin x}{x} \quad \text{but not} \quad \lim_{x \to 0} \frac{\cos x}{x}.$$

(b) Explain why the integral
$$\int_0^1 \frac{\cos x}{x}\, dx$$
is improper.

(c) Does the improper integral in part (b) converge or diverge? Justify your answer.

(d) Explain why the integral
$$\int_0^1 \frac{\sin x}{x}\, dx$$
is *not* improper. [**HINT**: Use l'Hôpital's rule.]

(e) Show that $0.8 < \int_0^\infty \sin\left(e^{-x}\right) dx < 1$. [**HINT**: Use the substitution $u = e^{-x}$.]

48. Is the integral $\int_0^1 x \ln x\, dx$ improper? Justify your answer.

49. Is the integral
$$\int_0^1 \frac{\arctan x}{\sqrt[3]{x}}\, dx$$
improper? Justify your answer.

50. Is the integral
$$\int_0^1 \frac{\tan x}{\sqrt{x}}\, dx$$
improper? Justify your answer.

51. Let $f(x) = \dfrac{\sin x}{\sqrt{x^3}}$.

(a) Show that the improper integral $\int_0^\infty f(x)\, dx$ converges. [**HINT**: $\sin x \le x$ for all $x \ge 0$.]

(b) Find real numbers α and β such that the definite integral $\int_\alpha^\beta f(x)\, dx$ approximates the improper integral in part (a) within 0.001.

52. (a) Suppose that $\lim_{x \to a} f(x) = A > 0$. Explain why $\lim_{x \to a} \ln\big(f(x)\big) = \ln A$.

(b) Suppose that $\lim_{x \to a} \ln\big(f(x)\big) = B$. Evaluate $\lim_{x \to a} f(x)$.

In Exercises 53–60, evaluate the limit. [**HINT**: See Exercise 52.]

53. $\lim\limits_{x \to 0} \cos\left(\dfrac{\tan x}{x}\right)$

54. $\lim\limits_{x \to 0^+} \arctan(x \ln x)$

55. $\lim\limits_{x \to 0^+} x^{2x}$.

56. $\lim\limits_{x \to \infty} (1 + x)^{1/x}$

57. $\lim\limits_{x \to 0} (1 + x)^{1/x}$

58. $\lim\limits_{x \to 0} (e^x + x)^{1/x}$

59. $\lim\limits_{x \to 1} \big(\ln x\big)^{\sin x}$

60. $\lim\limits_{x \to (\pi/2)^-} \sin\left(x^{\tan x}\right)$

61. Suppose that f is a differentiable function, that f' is continuous on $[1, \infty)$, and that $|f(x)| \le e^{-x} \ln x$ if $x \ge 1$. Show that $\int_1^\infty f'(x)\, dx$ converges.

62. l'Hôpital's rule says that if f is a differentiable function and $f(0) = 0$, then $\lim_{x \to 0} \dfrac{f(x)}{x} = f'(0)$. Why isn't this a surprise?

63. Suppose that P_2 is the second-order Taylor polynomial for f based at a. Also, suppose that $q \neq P_2$ is a quadratic polynomial such that $q(a) = f(a)$ and $q'(a) = f'(a)$.

(a) Show that $\lim\limits_{x \to a} \dfrac{f(x) - P_2(x)}{x - a} = 0$.

(b) Show that $\lim\limits_{x \to a} \dfrac{f(x) - P_2(x)}{(x - a)^2} = 0$.

(c) Show that $\lim\limits_{x \to a} \dfrac{f(x) - q(x)}{(x - a)^2} \neq 0$.

(d) Interpret the results of parts (a)–(c) in terms of the graphs of f, P_2, and q.

(e) Suppose that P_n is the nth-order Taylor polynomial for f based at a. Explain why
$$\lim_{x \to a} \frac{f(x) - P_n(x)}{(x - a)^n} = 0.$$
Does any other nth-order polynomial have this property? Justify your answer.

64. Let $\Gamma(x) = \int_0^\infty t^{x-1} e^{-t}\, dt$.

    (a) Evaluate $\Gamma(1)$.

    (b) Use integration by parts to show that if $x > 1$, $\Gamma(x) = (x-1)\Gamma(x-1)$.   [**HINT:** For every $z \in \mathbb{R}$, $\lim_{t\to\infty} t^z e^{-t} = 0$.]

    (c) Show that $0 < \Gamma\left(\frac{2}{3}\right) < 2$.   [**HINT:** Be careful; the integral has *two* improprieties.]

    (d) It is known that $\int_0^\infty e^{-x^2}\, dx = \frac{1}{2}\sqrt{\pi}$. Use this result to evaluate $\Gamma\left(\frac{1}{2}\right)$.   [**HINT:** Let $u = \sqrt{t}$.]

65. Show that $\int_0^\infty t^z e^{-t}\, dt = \int_0^1 \left(\ln(1/x)\right)^z dx$.
    [**HINT:** Use the substitution $t = \ln(1/x)$.]

66. It is a fact that $\int_0^\infty x^n e^{-x}\, dx = n!$ when $n \geq 0$ is an integer. Use this fact to show that

$$\int_0^1 x^m (\ln x)^n\, dx = \frac{(-1)^n n!}{(m+1)^{n+1}}$$

    when m and n are positive integers.   [**HINT:** Make the substitution $x = e^{-u}$ in the second integral.]

10.5 Chapter Summary

Both of this chapter's main subjects—improper integrals and l'Hôpital's rule—concern the long-term behavior of functions. Both topics laid the groundwork for the later study of infinite series, in which more delicate questions of long-term behavior arise.

Improper Integrals.   The integrals

$$\int_1^\infty \frac{1}{x^2}\, dx \qquad \text{and} \qquad \int_0^1 \frac{1}{x^2}\, dx$$

are both improper, for different reasons. In the first case, the interval of integration is infinite; in the second, the integrand itself "blows up" at the left endpoint of integration. Despite such problems, some improper integrals "converge," in a natural sense, to finite values. Whether a given integral converges or diverges depends subtly on how the function behaves on its domain of integration; which alternative occurs for a given integral can be hard to tell at a glance. The first of the preceding integrals, for instance, converges (to the value 1), but the second one diverges (to ∞).

Improper integrals, like proper ones, are sometimes best handled numerically. The second section of the chapter treated some of the special problems improper integrals raise in this regard.

Probability and Improper Integrals.   Improper integrals have important applications in probability and statistics. Working with the famous "bell-shaped" or **normal distribution**, for example, requires evaluating—numerically, in most cases—improper integrals like this one:

$$\int_{-1}^1 \frac{1}{\sqrt{2\pi}} \exp\left(\frac{-x^2}{2}\right) dx.$$

Doing so enables us to model a large and important variety of random phenomena.

l'Hôpital's Rule.   l'Hôpital's rule is a technique for evaluating certain "indeterminate" limits, often of the form

$$\lim_{x\to\infty} \frac{f(x)}{g(x)},$$

where both $f(x)$ and $g(x)$ tend either to 0 or to ∞. Specialized as such problems may seem at first, they often arise in the context of infinite sequences and infinite series, our next topic.

11

Infinite Series

11.1 Sequences and Their Limits

We'll study them in some detail.

This section, on infinite *sequences*, prepares the ground for the next topic—infinite *series*. Convergent series ◄ are defined in terms of the simpler, more basic idea of convergent sequences. We start with a quick introduction to sequences—what they are, what it means for them to converge or diverge, and how to find their limits.

The Idea and Language of Sequences; Simple Examples

A **sequence** is an infinite list of numbers, of the general form

$$a_1, a_2, a_3, a_4, \ldots, a_k, a_{k+1}, \ldots.$$

Read "a sub three" and "a sub k."

Individual entries are called the **terms** of the sequence; a_3 and a_k, ◄ for instance, are the third term and the kth term, respectively. The full sequence is, technically speaking, an **ordered set**; the standard notation

$$\{a_k\}_{k=1}^{\infty}$$

(or simply $\{a_k\}$) uses set brackets to emphasize this view.

Our main interest in sequences is in their **limits**. For the simplest sequences, limits are evident at a glance. The next three examples are of this type; we include them mainly to illustrate ideas, notations, and terminology.

> **EXAMPLE 1** Discuss the sequence $\{a_k\}_{k=1}^{\infty}$ defined by the formula $a_k = 1/k$, $k = 1, 2, 3, \ldots$. Does this sequence have a limit?

> **Solution** Sampling the first few terms,

$$\frac{1}{1}, \frac{1}{2}, \frac{1}{3}, \ldots, \frac{1}{10}, \frac{1}{11}, \ldots, \frac{1}{100}, \frac{1}{101}, \ldots,$$

shows (to nobody's surprise) that the sequence converges to zero: As k increases, the terms approach zero arbitrarily closely. In symbols,

$$\lim_{k \to \infty} a_k = \lim_{k \to \infty} \frac{1}{k} = 0.$$ ∎

EXAMPLE 2 Suppose that $\{b_j\}_{j=1}^{\infty}$ is defined ▶ by

$$b_j = \frac{(-1)^j}{j}, \qquad j = 1, 2, 3, \dots.$$

This time we used j rather than k; the name of the "index variable" doesn't matter.

What's $\lim_{j\to\infty} b_j$?

Solution Writing out terms shows a similar pattern:

$$-\frac{1}{1}, \frac{1}{2}, -\frac{1}{3}, \dots, \frac{1}{10}, -\frac{1}{11}, \dots, \frac{1}{100}, -\frac{1}{101}, \dots.$$

Although the terms oscillate in *sign*, they approach zero more and more closely. Eventually, all the terms—positive or negative—remain within any specified distance from zero. ▶ Hence, the sequence $\{b_j\}$ also converges to zero:

All terms past b_{1000}, for instance, are within 0.001 of zero.

$$\lim_{j\to\infty} b_j = \lim_{j\to\infty} \frac{(-1)^j}{j} = 0. \qquad ■$$

EXAMPLE 3 Does the sequence $\{c_k\}_{k=0}^{\infty}$ with general term $c_k = (-1)^k$ converge?

Solution No; it diverges. Successive terms have the pattern

$$1, -1, 1, -1, 1, \dots,$$

never settling on a single limit. ■

Fine Points

Sequences have their own notational quirks and conventions. Here are several to watch for.

Where To Start? The sequence in the preceding example began with c_0, not c_1. Other starting points, such as a_2 or even b_{-3}, occasionally arise. In practice, fortunately, the difference is usually unimportant. What usually matters for sequences is their long-run behavior, not the presence or absence of a few initial terms.

Index Names Don't Matter We can define the squaring function by writing either $f(t) = t^2$ or $f(x) = x^2$; the *variable name* makes no difference. In the same way, a sequence's *index name* is arbitrary: $\{a_k\}_{k=1}^{\infty}$ and $\{a_j\}_{j=1}^{\infty}$ mean exactly the same thing.

Reindexing The sequence

$$\frac{1}{1}, \frac{1}{2}, \frac{1}{3}, \dots, \frac{1}{10}, \frac{1}{11}, \dots$$

of Example 1 can be described symbolically in various different-looking, but still equivalent, ways. Here are two:

$$a_k = \frac{1}{k}, \quad k = 1, 2, \dots \quad \text{or} \quad a_j = \frac{1}{j+1}, \quad j = 0, 1, 2, \dots.$$

Depending on the situation, one description or the other may be preferable.

Sequences as Functions

Sequences are closely related to functions, as expressions such as

$$a_k = \frac{\sin k}{k} \quad \text{and} \quad f(x) = \frac{\sin x}{x}$$

illustrate. The formal definition of sequence makes this connection precise.

> **Definition** An **infinite sequence** is a real-valued function that is defined for *positive integer inputs*.

Alternative Views of Sequences. With the definition and the preceding examples, we have several useful ways to think of a sequence.

As a List As an infinite list of numbers: $a_1, a_2, a_3, \ldots$

As a Function As a function $a(n)$, where n takes only positive integer values. ◄ Thus, $a(1) = a_1$, $a(2) = a_2$, $a(3) = a_3$, and so on.

As a Discrete Sample As a discrete sample of values of an ordinary function $f(x)$, defined for real $x \geq 1$. ◄

The function a may or may not make sense for other inputs.

Most of the preceding examples are of this type.

Graphs of Sequences, Graphs of Functions

The graph of a function f consists of all points $\left(x, f(x)\right)$, as x runs through the domain of f. The graph of a sequence $\{a_k\}$ is therefore the set of points (k, a_k), as k runs through positive integer values.

EXAMPLE 4   Let a function f and a sequence $\{a_k\}$ be defined by

$$f(x) = \frac{\sin x}{x}; \qquad a_k = \frac{\sin k}{k}.$$

Plot graphs of both. What do the graphs say about limits?

Solution   Here are the graphs:

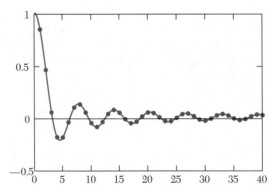

Graphs of $f(x) = \sin(x)/x$ and $a_k = \sin(k)/k$

They show how the sequence $\{a_k\}$ is a "discrete sample" of the continuous function f. They show, too, that as x (or k) tends to infinity, this sequence and function, although oscillating in sign, tend to zero. ∎

Not every sequence comes naturally from a familiar function. Nevertheless, graphs or tables often suggest limits.

EXAMPLE 5   A sequence $\{b_j\}$ has general term

$$b_j = 1 \cdot 2 \cdot 3 \cdot 4 \cdots (j-1) \cdot j = j!$$

(In words, b_j is j **factorial**.) Tabulate $\{b_j\}$; find its limit, if any.

S o l u t i o n   As the table suggests, $\{b_j\}$ diverges—quickly—to infinity:

As $k \to \infty$, $k! \to \infty$: Explosive Numerical Evidence						
k	1	2	4	8	16	32	64
$k!$	1	2	24	40,320	2.092×10^{13}	2.631×10^{35}	1.269×10^{89}

Can any doubt remain? ▶

■ *Graphs don't work well for this sequence. Do you see why not?*

Limits of Sequences, Limits of Functions

Sequences are special sorts of functions; limits of sequences, therefore, are mild variants on limits at infinity. An informal definition will suffice. ▶

A formal definition describes more precisely what "approaches" means.

> **D e f i n i t i o n** (**Limit of a Sequence**) Let $\{a_k\}$ be a sequence and L a real number. If a_k approaches L to within any desired tolerance as k increases without bound, then the sequence converges to L. In symbols,
>
> $$\lim_{k \to \infty} a_k = L.$$
>
> Otherwise, the sequence diverges.

Notice:

Divergence to Infinity   If either $a_k \to \infty$ or $a_k \to -\infty$ as k increases without bound, then the sequence diverges to (positive or negative) infinity. We write, for instance,

$$\lim_{k \to \infty} k! = \infty.$$

Other Divergence Behavior   Example 3 shows that a divergent sequence need not "blow up"; other patterns of "wandering" behavior (or no pattern at all) are possible.

Asymptotes   A sequence, like a function, converges to a finite limit L if and only if its graph has a horizontal asymptote at $y = L$.

Sequences, Functions, and Limits.   Many sequences, as we've said, are "discrete samples" of familiar functions. For such sequences we can often use our knowledge of the underlying function.

> **Fact** Let f be a function defined for $x \geq 1$. If $\lim_{x \to \infty} f(x) = L$, and $a_k = f(k)$ for all $k \geq 1$, then $\lim_{k \to \infty} a_k = L$.

EXAMPLE 6 Does the sequence with general term $a_k = \sin k$ converge or diverge?

Solution Here's the surprising graph.

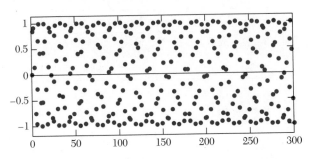

The surprising graph of $a_k = \sin k$:
structure but no convergence

Although full of fascinating shapes, the graph never settles on a single limit, so the sequence diverges. ◄

This example is based on an idea of Gilbert Strang.

Limits of Sequences and l'Hôpital's Rule

If a sequence happens to be given by a symbolic formula, l'Hôpital's rule can often be applied to the formula to find the limit.

EXAMPLE 7 Numerical evidence suggests that if $a_k = 2^k/k^2$, then $\{a_k\}$ diverges to infinity. Use l'Hôpital's rule to show this result symbolically.

Solution l'Hôpital's rule, applied to the continuous function $h(x) = 2^x/x^2$, says that ◄

Differentiate for yourself!

$$\lim_{x \to \infty} \frac{2^x}{x^2} = \lim_{x \to \infty} \frac{2^x \ln 2}{2x} = \lim_{x \to \infty} \frac{2^x \ln 2 \cdot \ln 2}{2} = \infty.$$

(The last equation holds because the numerator blows up as the denominator remains fixed.) Because $h(x) \to \infty$, $a_k \to \infty$ too. ∎

EXAMPLE 8 Show that $\lim_{n \to \infty} n^{1/n} = 1$.

Solution l'Hôpital's rule doesn't apply directly, because no fraction is involved. Applying the natural logarithm produces a fraction:

$$a_n = n^{1/n} \implies \ln(a_n) = \frac{\ln n}{n}.$$

Now l'Hôpital's rule *does* apply:

$$\lim_{n \to \infty} \frac{\ln n}{n} = \lim_{n \to \infty} \frac{1/n}{1} = 0.$$

We've shown that as $n \to \infty$, $\ln(a_n) \to 0$; it follows that $a_n \to 1$. ∎

New Sequence Limits from Old

Limits of sequences, like limits of functions, can be combined in various ways to give *new* limits.

Standard Examples

Before combining limits, we'll need some known limits to combine. The following limits can be thought of as basic building blocks.

$$\lim_{n \to \infty} n^{1/n} = 1$$

$$\lim_{n \to \infty} x^{1/n} = 1 \quad \text{(for all } x > 0\text{)}$$

$$\lim_{n \to \infty} \frac{1}{n^k} = 0 \quad \text{(for all } k > 0\text{)}$$

$$\lim_{n \to \infty} r^n = 0 \quad \text{(if } -1 < r < 1\text{)}$$

(We derived the first of these limits in Example 8, using l'Hôpital's rule. The other limits are easier.)

Algebraic Combinations

Plausible-seeming calculations such as this one—

$$\lim_{k \to \infty} \left(\frac{1}{k} + \frac{3k}{k+1} \right) = \lim_{k \to \infty} \frac{1}{k} + 3 \lim_{k \to \infty} \frac{k}{k+1} = 0 + 3 = 3,$$

rely implicitly on the following theorem. We've already seen it, in virtually identical form, for functions.

Theorem 1 (**Algebra with Limits**) Suppose that as $k \to \infty$,

$$a_k \to L \quad \text{and} \quad b_k \to M,$$

where L and M are finite numbers. Let c be any real constant. Then

$$(ca_k) \to cL, \qquad (a_k \pm b_k) \to L \pm M, \qquad \text{and} \qquad a_k b_k \to LM.$$

If $M \neq 0$, then $\dfrac{a_k}{b_k} \to \dfrac{L}{M}$.

The Squeeze Principle

For sequences as for functions, new, unknown limits can sometimes be found by "squeezing" them between old, known limits.

Theorem 2 (**The Squeeze Principle**) Suppose that for all $k > 0$,

$$a_k \leq b_k \leq c_k \quad \text{and} \quad \lim_{k \to \infty} a_k = \lim_{k \to \infty} c_k = L.$$

Then $\lim_{k \to \infty} b_k = L$.

EXAMPLE 9 Explain legalistically why $\lim\limits_{k \to \infty} \dfrac{\sin k}{k} = 0$.

Solution The "squeeze inequality"

$$-\frac{1}{k} \leq \frac{\sin k}{k} \leq \frac{1}{k}$$

holds for all integers $k > 0$. As $k \to \infty$, both the left and the right sides tend to 0; so, therefore, must the middle expression. ∎

When Must a Sequence Converge? An Existence Theorem

The ideal way to show that a sequence converges is to find a limit. Alas, sometimes that's difficult.

EXAMPLE 10 Does the sequence with general term

$$a_n = \left(1 + \frac{1}{n}\right)^n$$

converge or diverge? Why?

A graph would show the same thing.

Solution The answer isn't obvious at a glance, so let's tabulate some values. ◀

As $n \to \infty$, $(1 + 1/n)^n$ Appears to Converge							
n	16	32	64	128	256	512	1024
$(1 + 1/n)^n$	2.63793	2.67699	2.69734	2.70774	2.71299	2.71563	2.71696

As $n \to \infty$, a_n seems to increase but not blow up. Therefore, apparently, the sequence converges to some limit. ◀ ∎

The number $e \approx 2.71828$ is a reasonable guess.

A Convergence Theorem. The sequence in Example 10 seems to converge for two reasons: (1) it's increasing (i.e., $a_1 \leq a_2 \leq a_3 \ldots$), and (2) it's bounded above. Common sense suggests that any increasing sequence must either converge to a limit—as the preceding sequence appears to do—or blow up to ∞. Similarly, common sense suggests that a decreasing sequence (i.e., a sequence for which $a_1 \geq a_2 \geq a_3 \ldots$) either converges to a limit or diverges to $-\infty$. ◀

Think it through; do you agree?

The following theorem corroborates these commonsense impressions. It applies to any **monotone sequence**, i.e., any sequence that is *either* increasing or decreasing.

> **Theorem 3** Suppose that the sequence $\{a_k\}$ is *increasing* and *bounded above* by a number A. In other words,
>
> $$a_1 \leq a_2 \leq a_3 \leq \ldots a_k \leq a_{k+1} \leq \ldots \leq A.$$
>
> Then $\{a_k\}$ converges to some finite limit a, with $a \leq A$. Similarly, if $\{b_k\}$ is *decreasing* and *bounded below* by a number B, then $\{b_k\}$ converges to a finite limit b, with $b \geq B$.

EXAMPLE 11   Consider the increasing sequence $\{a_k\}$ defined by

$$a_1 = 0; \qquad a_{k+1} = \sqrt{6 + a_k} \qquad \text{if } k \geq 1.$$

Show that $\{a_k\}$ converges and that the limit is less than or equal to 3.

Solution   The first few terms suggest (and we take it for granted) that the sequence is increasing:

$$a_1 = 0; \qquad a_2 = \sqrt{6} \approx 2.45; \qquad a_3 \approx 2.91; \qquad a_4 \approx 2.98; \qquad a_5 \approx 2.997.$$

The sequence also seems to be bounded above by 3. This is easy to show. If $a_k < 3$, then, by definition,

$$a_{k+1} = \sqrt{6 + a_k} < \sqrt{6 + 3} = 3.$$

Thus, *no* term can exceed 3. It follows from the theorem, therefore, that the sequence converges to a limit no greater than 3. (The limit *is* 3, in fact, but that requires further proof.) ■

BASICS

In Exercises 1–4, find a symbolic expression for the general term a_k of the sequence.

1. $1, -\dfrac{1}{3}, \dfrac{1}{9}, -\dfrac{1}{27}, \dfrac{1}{81}, \ldots$

2. $3, 6, 9, 12, 15, \ldots$

3. $\dfrac{1}{2}, \dfrac{2}{4}, \dfrac{3}{8}, \dfrac{4}{16}, \dfrac{5}{32}, \ldots$

4. $\dfrac{1}{2}, \dfrac{1}{5}, \dfrac{1}{10}, \dfrac{1}{17}, \dfrac{1}{26}, \dfrac{1}{37}, \ldots$

In Exercises 5–20, first calculate a_1, a_2, a_5, and a_{10}, then determine $\lim_{k \to \infty} a_k$ (if the limit exists). (Round your answers to four decimal places.)

5. $a_k = (-3/2)^k$

6. $a_k = (-0.8)^k$

7. $a_k = (1.1)^k$

8. $a_k = \left(\sqrt{26}/17\right)^k$

9. $a_k = \left(\dfrac{\pi}{e}\right)^k$

10. $a_k = e^{-k}$

11. $a_k = (\arcsin 1)^k$

12. $a_k = (1/k)^k$

13. $a_k = \sin k$

14. $a_k = \arctan k$

15. $a_k = \cos(1/k)$

16. $a_k = \sin(k\pi)$

17. $a_k = \cos(k\pi)$

18. $a_k = \dfrac{k^2}{k^2 + k + 3}$

19. $a_k = \sqrt{\dfrac{2k}{k + 3}}$

20. $a_k = \dfrac{k}{\sqrt{k} + 10}$

FURTHER EXERCISES

In Exercises 21–26, calculate a_1, a_{10}, and a_{100} (to four decimal places), then find the limit of the sequence. [**HINT:** l'Hôpital's rule may be useful.]

21. $a_n = \dfrac{n + 2}{n^3 + 4}$

22. $a_m = m^2/e^m$

23. $b_j = \ln j / \sqrt[3]{j}$

24. $c_k = \dfrac{\ln(1 + k^2)}{\ln(4 + 3k)}$

25. $d_n = n \sin(1/n)$

26. $a_k = (2^k + 3^k)^{1/k}$

27. Show that $\lim_{n \to \infty} x^{1/n} = 1$ for all $x > 0$.

28. Let $a_k = \left(1 + \dfrac{x}{k}\right)^k$, where x is a real number.

    (a) Show that $\lim_{k \to \infty} \ln(a_k) = x$.

    (b) Use part (a) to evaluate $\lim_{k \to \infty} a_k$.

29. Evaluate $\displaystyle\lim_{n \to \infty} \left(1 - \dfrac{1}{2n}\right)^n$.

In Exercises 30–38, first calculate a_1, a_2, a_5, and a_{10}, then determine $\lim_{k \to \infty} a_k$ (if the limit exists). (Round your answers to four decimal places.)

30. $a_k = \dfrac{k!}{(k+1)!}$

31. $a_k = \ln\left(\dfrac{k}{k+1}\right)$

32. $a_k = \displaystyle\int_0^k e^{-x}\, dx$

33. $a_k = \displaystyle\int_k^\infty \dfrac{dx}{1+x^2}$

34. $a_k = \dfrac{\cos k}{\ln(k+1)}$

35. $a_k = e^{-k} \sin k$

36. $a_k = \sqrt{k^2 + 1} - k$

37. $a_k = \sqrt{k^2 + k} - k$

38. $a_k = 3^{1/k}$

39. Give an example of a sequence that is
   (a) convergent but not monotone.
   (b) bounded but not monotone.
   (c) monotone but not convergent.
   (d) monotone decreasing and unbounded.
   (e) monotone decreasing and convergent.
   (f) unbounded but not monotone.

40. Let $a_n = \displaystyle\sum_{k=1}^n \dfrac{k}{n^2}$.
   (a) Evaluate a_{10}.
   (b) Explain why $\lim_{n \to \infty} a_n = \int_0^1 x\, dx = 1/2$.

   [**HINT:** $k/n^2 = (k/n) \cdot (1/n)$.]

41. Define a sequence $\{a_n\}$ by $a_n = \displaystyle\sum_{k=1}^n \dfrac{1}{n+k}$.
   (a) Show that this is a monotonically increasing sequence (i.e., $a_{n+1} > a_n$).
   (b) Show that $a_n \leq \dfrac{n}{n+1} < 1$.
   (c) What do parts (a) and (b) imply about $\lim_{n \to \infty} a_n$?
   (d) Explain why $\lim_{n \to \infty} a_n > 1/2$. [**HINT:** $a_1 = 1/2$.]
   (e) Show that $\lim_{n \to \infty} a_n = \ln 2$. [**HINT:** $\dfrac{1}{n+k} = \dfrac{1}{1+k/n} \cdot \dfrac{1}{n}$, so a_n is a Riemann-sum approximation to an integral.]

42. Let $a_n = \sum_{k=1}^n k^2/n^3$. Evaluate $\lim_{n \to \infty} a_n$. [**HINT:** Think of a_n as a Riemann-sum approximation to an integral.]

43. Let $a_n = \dfrac{\sqrt[n]{n!}}{n}$.
   (a) Show that $\ln a_n = \dfrac{1}{n}\displaystyle\sum_{k=1}^n \ln k - \dfrac{1}{n}\sum_{k=1}^n \ln n$.
   (b) Use part (a) to show that $\ln a_n$ is a right-sum approximation to $\int_0^1 \ln x\, dx$.
   (c) Use part (b) to show that $\lim_{n \to \infty} a_n = e^{-1}$.

44. Let $a_n = 4^n/n!$.
   (a) Find a number N such that $a_{n+1} \leq a_n$ for all $n \geq N$.
   (b) Use part (a) to explain why $\lim_{n \to \infty} a_n$ exists.
   (c) Evaluate $\lim_{n \to \infty} a_n$.

45. Suppose that $\{a_n\}$ is a sequence with the property $|a_{n+1}/a_n| \leq (n+3)/(2n+1)$ for all $n \geq 1$. Show that $\lim_{n \to \infty} a_n = 0$. [**HINT:** Start by showing that $|a_{n+1}/a_3| \leq (6/7)^{n-2}$ for all $n \geq 3$.]

46. (a) For which values of x does $\lim_{n \to \infty} x^n$ diverge?
   (b) Find all values of x for which $\lim_{n \to \infty} x^n$ converges to 0.
   (c) Are there any values of x for which $\lim_{n \to \infty} x^n$ converges to a number other than 0? For each such x, find the limit.

In Exercises 47–50, determine the values of x for which the sequence converges as $k \to \infty$. Evaluate $\lim_{k \to \infty} a_k$ for these values of x.

47. $a_k = e^{kx}$

48. $a_k = (\ln x)^k$

49. $a_k = (\arcsin x)^k$

50. $a_k = 2^{-k}(\arctan x)^k$

51. Let $a_n = \cos 1 \cdot \cos 2 \cdot \cos 3 \cdot \cos 4 \cdots \cos n$.
   (a) Evaluate a_1, a_2, a_3, and a_5. (Round your answers to four decimal places.)
   (b) Is the sequence $\{a_n\}$ bounded? Justify your answer.
   (c) Is the sequence $\{a_n\}$ monotone? Justify your answer.
   (d) Is the sequence $\{|a_n|\}$ bounded? Justify your answer.
   (e) Is the sequence $\{|a_n|\}$ monotone? Justify your answer.

52. Use Theorem 3 to show that the sequence 0.7, 0.77, 0.777, 0.7777, 0.77777, 0.777777, 0.7777777, ... has a limit.

53. Let $a_n = \dfrac{1 \cdot 3 \cdot 5 \cdots (2n-1)}{2 \cdot 4 \cdot 6 \cdots (2n)}$.
   (a) Calculate a_1, a_2, and a_5.
   (b) Use Theorem 3 to show that $\lim_{n \to \infty} a_n$ is a finite number.

54. Show that $\lim_{n \to \infty} \sin\left(\dfrac{\pi}{2^2}\right) \cdot \sin\left(\dfrac{\pi}{3^2}\right) \cdots \sin\left(\dfrac{\pi}{n^2}\right) = 0$. [**HINT:** $0 < \sin x < x$ when $0 < x < 1$.]

55. Does the sequence defined by $a_1 = 1$, $a_{n+1} = 1 - a_n$ converge? Justify your answer.

56. Does the sequence defined by $a_1 = 1$, $a_{n+1} = a_n/2$ converge? Justify your answer.

57. Consider the sequence defined by
$$a_1 = 1, \qquad a_{n+1} = \left(\dfrac{n}{n+1}\right) a_n.$$
   (a) Show that the sequence converges.
   (b) Find the limit. [**HINT:** Write out the first few terms and look for a pattern.]

58. Suppose that $\lim_{k \to \infty} a_k = L$, where L is a finite number, and that the terms of the sequence $\{b_k\}$ are defined by $b_k = L - a_k$. Explain why $\lim_{k \to \infty} b_k = 0$.

59. For which values of $x \geq 0$ does the sequence defined by $a_1 = x$, $a_{n+1} = \sqrt{a_n}$ converge? Justify your answer.

60. (a) Suppose that f is a differentiable function and that $f(0) = 0$. Show that $\lim_{n \to \infty} nf(1/n) = f'(0)$.
    (b) Use the result in part (a), not l'Hôpital's rule, to evaluate $\lim_{n \to \infty} n \arctan(1/n)$.

61. Let $F_0 = 1$, $F_1 = 1$, and define $F_{n+1} = F_n + F_{n-1}$ for all $n \geq 1$. (The terms of this sequence are the Fibonacci numbers.) Assume that there is a number L such that $\lim_{n \to \infty} F_{n+1}/F_n = L$.

(a) Show that L is a solution of the equation $x^2 = x + 1$.
(b) Suppose that x satisfies the equation $x^2 = x + 1$. Use mathematical induction to show that $x^n = xF_n + F_{n-1}$ for all $n \geq 1$.
(c) Let $r_1 = (1 + \sqrt{5})/2$ and $r_2 = (1 - \sqrt{5})/2$. Use part (c) to show that $F_n = (r_1^n - r_2^n)/\sqrt{5}$.
(d) Use part (d) to evaluate L.

11.2 Infinite Series, Convergence, and Divergence

Introduction

An **infinite series**, or **series** for short, is an infinite sum, i.e., an expression of the form ▶

On the left side we used **sigma notation** *for brevity.*

$$\sum_{k=1}^{\infty} a_k = a_1 + a_2 + a_3 + a_4 + \cdots + a_k + a_{k+1} + \cdots.$$

Thus, a series results from *adding* the terms of a sequence $a_1, a_2, a_3, \ldots$. If, say, $a_k = 1/k^2$, then

$$\sum_{k=1}^{\infty} a_k = \sum_{k=1}^{\infty} \frac{1}{k^2} = \frac{1}{1} + \frac{1}{4} + \frac{1}{9} + \frac{1}{16} + \cdots.$$

If $a_k = k$, then

$$\sum_{k=1}^{\infty} a_k = \sum_{k=1}^{\infty} k = 1 + 2 + 3 + \cdots.$$

The idea of an infinite sum raises natural questions:

- What does it *mean* to add infinitely many numbers?
- Which series add up to a finite number? Which blow up?
- If a series has a finite sum, how can we find (or estimate) it?

The rest of this chapter addresses these questions.

Improper Sums, Improper Integrals

Standard examples of infinite series include

$$\sum_{k=1}^{\infty} \frac{1}{k^2}, \quad \sum_{k=1}^{\infty} \frac{1}{k}, \quad \sum_{k=1}^{\infty} \frac{1}{\sqrt{k}}, \quad \text{and} \quad \sum_{k=1}^{\infty} \frac{1}{2^k}.$$

The typographical resemblance to the improper integrals

$$\int_1^{\infty} \frac{dx}{x^2}, \quad \int_1^{\infty} \frac{dx}{x}, \quad \int_1^{\infty} \frac{dx}{\sqrt{x}}, \quad \text{and} \quad \int_1^{\infty} \frac{dx}{2^x}$$

is no accident; infinite series and improper integrals are similar in many respects. We'll see, in fact, that the first and fourth of the preceding series converge, while the second and third diverge—exactly as for the corresponding integrals. We'll often return to the important analogy between integrals and series.

Why Series Matter: A Preview

We'll see at the end of this chapter why it's true.

Understanding series takes some legwork. As a first glimpse of why the work is worth doing, consider the fact ◄ that for any real number x,

$$\cos x = 1 - \frac{x^2}{2!} + \frac{x^4}{4!} - \frac{x^6}{6!} + \frac{x^8}{8!} - \cdots. \tag{11.1}$$

The "dots" in each equation mean that the pattern continues.

If, say, $x = 1$, then ◄

$$\cos 1 = 1 - \frac{1}{2!} + \frac{1}{4!} - \frac{1}{6!} + \frac{1}{8!} - \cdots. \tag{11.2}$$

Why should we care about such equations? Why write something familiar—the cosine function—in terms of something exotic—a series?

That's why every scientific calculator has a COS key.

For a few special values of x, such as $x = 0$, $x = \pi/6$, and $x = \pi/4$, finding $\cos x$ is easy. For most inputs x, it's genuinely hard.

Ordinary formulas don't have dots in them.

One good answer is that the cosine function, although crucial in applications, ◄ has no *algebraic* formula. Without a concrete formula, finding accurate numerical values of $\cos x$ for given inputs x is a genuine problem. ◄

The preceding equations help solve this problem. Equation 11.1, although not a formula in the ordinary sense, ◄ gives a concrete, computable recipe for approximating $\cos x$: Given an input x, calculate as far out as practically possible in the "infinite polynomial"

$$1 - \frac{x^2}{2!} + \frac{x^4}{4!} - \frac{x^6}{6!} + \frac{x^8}{8!} - \cdots.$$

With luck, the result should closely approximate the true value of $\cos x$.

Try it with your calculator.

Equation 11.2 shows how to approximate $\cos 1$. It's easy to check ◄ that, to seven decimals,

$$1 - \frac{1}{2!} = 0.5000000;$$

$$1 - \frac{1}{2!} + \frac{1}{4!} = 0.5416667;$$

$$1 - \frac{1}{2!} + \frac{1}{4!} - \frac{1}{6!} = 0.5402778;$$

$$1 - \frac{1}{2!} + \frac{1}{4!} - \frac{1}{6!} + \frac{1}{8!} = 0.5403026.$$

The results converge with gratifying speed to the "right" answer—the true value of $\cos 1$ (≈ 0.5403023).

Open Questions. These calculations raise as many questions as they answer.

Where did Equation 11.1 come from? What do all the "dots" really mean? Are similar equations available for other functions—sine, arctangent, logarithm, and so on? How many terms are needed to guarantee accuracy to, say, five decimals?

Questions such as these guide our study of series.

Convergence: Definitions and Terminology

The payoff comes later, in clarity.

Working successfully with series requires some up-front investment in definitions and technical language. ◄ After stating terms and definitions, we show by example why they're reasonable.

Series Language: Terms, Partial Sums, Tails, Convergence, Limit

Let $\sum_{k=1}^{\infty} a_k = a_1 + a_2 + a_3 + \cdots + a_k + a_{k+1} + \cdots$ be an infinite series. ▶ Recall that the summand a_k is called the kth term of the series. The nth **partial sum**, denoted by S_n, is the (finite) sum of all terms *through index n*:

$$S_n = a_1 + a_2 + a_3 + \cdots + a_{n-1} + a_n = \sum_{k=1}^{n} a_k.$$

To keep notation simple, k usually runs from 1 to ∞. Sometimes it's convenient to let k start at 0, 2, or elsewhere.

The nth tail, denoted by R_n, is the (infinite) sum of all terms *beyond index n*:

$$R_n = a_{n+1} + a_{n+2} + a_{n+3} + \cdots = \sum_{k=n+1}^{\infty} a_k.$$

As the notation R_n suggests, the nth tail is the remainder—what's left after the terms through index n are added. In symbols,

$$\sum_{k=1}^{\infty} a_k = (a_1 + a_2 + \cdots + a_n) + (a_{n+1} + a_{n+2} + a_{n+3} + \cdots) = S_n + R_n.$$

The crucial definition ▶ of convergence involves partial sums:

The notation is as before.

> **Definition** If $\lim_{n\to\infty} S_n = S$, for some finite number S, then the series $\sum_{k=1}^{\infty} a_k$ converges to the limit S. Otherwise, the series *diverges*.

Notice the following aspects of the definition.

Divergent Series A divergent series is one for which the sequence of partial sums does *not* converge to a finite limit S. One possibility is that the partial sums S_n blow up to infinity. Another possibility is that the partial sums remain bounded, but never settle on a specific limit.

Improper Integrals, Improper Sums The preceding definition says, in symbols, that

$$\sum_{k=1}^{\infty} a_k = \lim_{n\to\infty} \sum_{k=1}^{n} a_k$$

if the limit exists. Convergence for improper integrals means much the same thing:

$$\int_{x=1}^{\infty} f(x)\, dx = \lim_{n\to\infty} \int_{x=1}^{n} f(x)\, dx$$

if *this* limit exists. ▶ An infinite series is an improper sum in exactly the sense that an integral may be improper.

In each case we take the limit of something proper.

What Convergence Means for Partial Sums and Tails To say that $\sum a_k$ converges to the limit (or **sum**) S means that $S_n \to S$ as $n \to \infty$. Since for all n, $S = S_n + R_n$, it follows that $R_n \to 0$ as $n \to \infty$. In particular, to decide whether a series converges, we need to examine the sequence of partial sums.

Two Sequences: Keep Them Straight Every series $\sum a_k$ involves two sequences: the sequence $\{a_k\}$ of *terms* and the sequence $\{S_n\}$ of *partial*

Many, maybe most, student difficulties arise from confusing them.

sums. Keeping these related but different sequences separate is essential. ◄ We'll take special care to do so in the next examples.

EXAMPLE 1   (**A Geometric Series**)   Does the series

$$\sum_{k=0}^{\infty} \frac{1}{2^k} = 1 + \frac{1}{2} + \frac{1}{4} + \frac{1}{8} + \cdots$$

converge? If so, to what limit?

Solution   In this series, the index k began at 0, not 1. So, therefore, does the sequence of partial sums. Direct calculation yields

$$S_0 = 1 = 2 - 1;$$

$$S_1 = 1 + \frac{1}{2} = \frac{3}{2} = 2 - \frac{1}{2};$$

$$S_2 = 1 + \frac{1}{2} + \frac{1}{4} = \frac{7}{4} = 2 - \frac{1}{4};$$

$$S_3 = 1 + \frac{1}{2} + \frac{1}{4} + \frac{1}{8} = \frac{15}{8} = 2 - \frac{1}{8};$$

$$S_4 = 1 + \frac{1}{2} + \frac{1}{4} + \frac{1}{8} + \frac{1}{16} = \frac{31}{16} = 2 - \frac{1}{16}.$$

The pattern is easy to see: For any $n \geq 1$,

$$S_n = 2 - \frac{1}{2^n}.$$

From this explicit formula for S_n, the conclusion follows: Because $S_n \to 2$ as $n \to \infty$, the series converges to 2.   ■

A table of numerical values gives the same impression, and also illustrates the behavior of tails. Note that, for each n,

$$S_n = \sum_{k=0}^{n} \frac{1}{2^k}; \qquad R_n = \sum_{k=n+1}^{\infty} \frac{1}{2^k} = 2 - \sum_{k=0}^{n} \frac{1}{2^k}.$$

Check some for yourself.   Here are the numbers: ◄

Partial Sums and Tails of $1 + 1/2 + 1/4 + \cdots$										
n	0	1	2	3	4	5	6	7	8	9	10
S_n	1	1.5	1.75	1.875	1.938	1.969	1.984	1.992	1.996	1.998	1.999
R_n	1	0.5	0.25	0.125	0.063	0.031	0.016	0.008	0.004	0.002	0.001

EXAMPLE 2   Does $\sum_{k=1}^{\infty} (-1)^k = -1 + 1 - 1 + 1 - 1 + \cdots$ converge?

Solution   No. Successive partial sums are of the form $-1, 0, -1, 0, \ldots$. Since the sequence $\{S_n\}$ diverges, so does the series.   ■

EXAMPLE 3 Does the **harmonic series** $\sum\limits_{k=1}^{\infty} \dfrac{1}{k}$ converge?

Solution The answer depends, as always, on the partial sums S_n. By definition,

$$S_n = 1 + \frac{1}{2} + \frac{1}{3} + \frac{1}{4} + \frac{1}{5} + \frac{1}{6} + \cdots + \frac{1}{n}.$$

This time, unfortunately, no simple formula for S_n in terms of n comes to mind, so we'll investigate the sequence $\{S_n\}$ numerically. A computer makes quick work of the calculations. Here are some results, to three decimal places:

$$S_1 = 1; \qquad S_{20} = 3.598; \qquad S_{50} = 4.499; \qquad S_{100} = 5.187; \qquad S_{1000} = 7.485.$$

The numerical evidence is ambiguous; the S_n's seem to keep growing, although slowly. Whether the S_n's converge to some finite number or diverge to infinity is not yet clear. ▶

■ *In fact, the series diverges. We'll show this fact soon by comparing the series to the integral $\int_1^\infty 1/x \, dx$.*

When Does a Series (or an Integral) Converge?

Deciding whether a given infinite series converges or diverges is a delicate question; so was the analogous question for improper integrals. Such series as

$$\sum_{k=1}^{\infty} \frac{1}{k} \qquad \text{and} \qquad \sum_{k=1}^{\infty} \frac{1}{k^2}$$

pose the typical dilemma: Although successive terms of both series tend to zero, the *number* of terms is infinite. Convergence or divergence hinges on which of these conflicting tendencies "wins" in the long run.

The same dilemma arose ▶ for the improper integrals

$$\int_1^\infty \frac{1}{x} \, dx \qquad \text{and} \qquad \int_1^\infty \frac{1}{x^2} \, dx.$$

Remember? See Section 10.1.

Although the integrands tend to zero as x grows without bound, the total area bounded by their graphs may or may not converge to a finite limit. Indeed, we discovered earlier that these two integrals have opposite outcomes: The first integral diverges to infinity, and the second converges to one.

Convergence vs. Divergence, Graphically

Deciding whether the preceding two series converge or diverge is a bit harder than for the corresponding integrals, because (as we saw in Example 3) no convenient formulas for S_n are available.

For a series $\sum a_k$, plotting both $\{a_k\}$ and $\{S_n\}$ ▶ on the same axes illustrates the connection between the two—and sometimes suggests whether the series converge

The terms and the partial sums, respectively.

or diverges. The harmonic series $\sum 1/k$ generates this picture:

Terms and partial sums for $\sum 1/k$

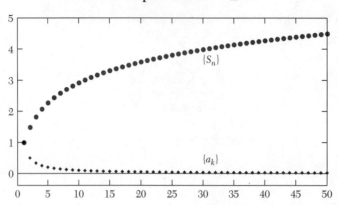

The picture for the series $\sum 1/k^2$ gives a different impression:

Terms and partial sums for $\sum 1/k^2$

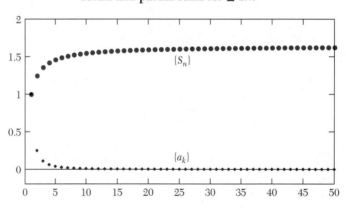

The question for any series is whether the sequence $\{S_n\}$ converges to a limit. The pictures suggest (but don't prove!) that the second of the preceding series converges and the first diverges. (From the second picture, what would you estimate as the limit of the series $\sum_{k=1}^{\infty} 1/k^2$?)

Geometric Series

Geometric series form the simplest and most important class of infinite series. A **geometric series** is one of the form

$$a + ar + ar^2 + ar^3 + ar^4 + \cdots = \sum_{k=0}^{\infty} ar^k \, ;$$

Each term is r times the previous term.

a is called the **leading term**, and r is the **ratio**. ◄ The series

$$1 + \frac{1}{2} + \frac{1}{4} + \frac{1}{8} + \cdots = \sum_{k=0}^{\infty} \left(\frac{1}{2} \right)^k$$

of Example 1 is geometric, with $a = 1$ and $r = 1/2$.

Partial Sums of Geometric Series

Geometric series have one great advantage: It's easy to decide whether they converge or diverge and, if they converge, to find their limits. The reason it's easy is that—unlike for many other series—there's a convenient, explicit formula for an arbitrary partial sum S_n. The formula depends on the fact that if $r \neq 1$ and $n \geq 1$, then

$$1 + r + r^2 + r^3 + \cdots + r^n = \frac{1 - r^{n+1}}{1 - r}.$$

(Multiply both sides by $(1 - r)$ to convince yourself that the formula holds.)

Here, then, is our formula for S_n. If $r \neq 1$ and $n \geq 1$, then

$$S_n = a + ar + ar^2 + ar^3 + \cdots + ar^n = a\frac{1 - r^{n+1}}{1 - r}. \tag{11.3}$$

From this formula follows the whole story of convergence and divergence for geometric series. As for any series, the present question is how the sequence $\{S_n\}$ of partial sums behaves as n tends to infinity. From the facts that

$$\lim_{n \to \infty} r^n = 0 \qquad \text{if} \qquad |r| < 1,$$

$$\lim_{n \to \infty} r^n = 1 \qquad \text{if} \qquad r = 1,$$

$$\lim_{n \to \infty} r^n = \infty \qquad \text{if} \qquad r > 1,$$

$$\lim_{n \to \infty} r^n \text{ does not exist} \qquad \text{if} \qquad r \leq -1,$$

the conclusion follows. It's worth emphasizing:

Fact If $|r| < 1$, the geometric series

$$\sum_{k=0}^{\infty} ar^k = a + ar + ar^2 + ar^3 + \cdots$$

converges to $\dfrac{a}{1 - r}$. If $a \neq 0$ and $|r| \geq 1$, the series diverges.

EXAMPLE 4 The series $\dfrac{1}{3} - \dfrac{1}{6} + \dfrac{1}{12} - \dfrac{1}{24} + \cdots$ converges. To what limit?

Solution The series is geometric, with $a = 1/3$ and $r = -1/2$. It therefore converges to

$$\frac{a}{1 - r} = \frac{1/3}{1 + 1/2} = \frac{2}{9} \approx 0.2222222.$$

A look at partial sums and tails supports the computation.

Partial Sums and Tails of $1/3 - 1/6 + 1/12 - 1/24 + \cdots$										
n	0	1	2	3	4	5	6	7	8	9	10
S_n	0.3333	0.1667	0.2500	0.2083	0.2292	0.2188	0.2240	0.2214	0.2227	0.2220	0.2223
R_n	−0.1111	0.0555	−0.0278	0.0139	−0.0069	0.0035	−0.0017	0.0009	−0.0004	0.0002	−0.0001

Telescoping Series

Geometric series are the most important variety for which partial sums can be found explicitly. Telescoping series offer the same possibility. The next example illustrates how they work and explains the name "telescoping."

EXAMPLE 5   Show that $\displaystyle\sum_{k=1}^{\infty} \frac{1}{k(k+1)}$ converges, and find its limit.

Solution   A bit of algebra lets us rewrite the series in a more helpful form: ◄

To convince yourself that the two sides are equal, find a common denominator on the right side.

$$\sum_{k=1}^{\infty} \frac{1}{k(k+1)} = \sum_{k=1}^{\infty} \left(\frac{1}{k} - \frac{1}{k+1} \right).$$

Writing out a few terms shows the "telescoping" pattern:

$$S_n = \sum_{k=1}^{n} \left(\frac{1}{k} - \frac{1}{k+1} \right)$$

$$= \left(\frac{1}{1} - \frac{1}{2} \right) + \left(\frac{1}{2} - \frac{1}{3} \right) + \left(\frac{1}{3} - \frac{1}{4} \right) + \cdots + \left(\frac{1}{n} - \frac{1}{n+1} \right)$$

$$= 1 - \frac{1}{n+1}.$$

Now it's clear that, as $n \to \infty$, $S_n \to 1$. That's the limit.

Watch the tails go to zero.

Again the numbers agree: ◄

Partial Sums and Tails of $\sum_{k=1}^{\infty} 1/(k^2 + k)$										
n	1	2	3	4	5	6	7	8	9	10	11
S_n	0.5000	0.6667	0.7500	0.8000	0.8333	0.8571	0.8750	0.8889	0.9000	0.9091	0.9167
R_n	0.5000	0.3333	0.2500	0.2000	0.1667	0.1429	0.1250	0.1111	0.1000	0.0909	0.0833

■

Algebra with Series

As with functions and sequences, combining series algebraically produces new series. Combining *convergent* series produces new *convergent* series, with limits related in the expected way.

> **Theorem 4** Suppose that $\sum_{k=1}^{\infty} a_k$ converges to S and that $\sum_{k=1}^{\infty} b_k$ converges to T. Let c be any constant. Then
>
> $$\sum_{k=1}^{\infty} (a_k \pm b_k) \qquad \text{converges to } S \pm T;$$
>
> $$\sum_{k=1}^{\infty} ca_k \qquad \text{converges to } cS.$$

These reasonable-looking properties of convergent *series* follow directly from the analogous properties of convergent *sequences*. ▶

A little series algebra, cleverly applied, can immensely simplify finding the limits of certain series.

After all, the limit of a series is defined as the limit of the sequence of partial sums.

EXAMPLE 6   Evaluate $\displaystyle\sum_{k=0}^{\infty} \frac{4 + 2^k}{3^k}$.

Solution   The given series is the sum of two convergent geometric series. ▶ The theorem says, therefore, that

We can find limits of geometric series.

$$\sum_{k=0}^{\infty} \frac{4 + 2^k}{3^k} = \sum_{k=0}^{\infty} \frac{4}{3^k} + \sum_{k=0}^{\infty} \left(\frac{2}{3}\right)^k = 6 + 3 = 9.$$ ∎

EXAMPLE 7   For the geometric series $\displaystyle\sum_{k=0}^{\infty} \frac{3}{2^k}$, calculate the tail R_{10}.

Solution   A little algebra is all we need:

$$R_{10} = \sum_{k=11}^{\infty} \frac{3}{2^k} = \frac{3}{2^{11}} + \frac{3}{2^{12}} + \frac{3}{2^{13}} + \cdots$$

$$= \frac{3}{2^{11}} \left(1 + \frac{1}{2} + \frac{1}{2^2} + \frac{1}{2^3} + \cdots\right)$$

$$= \frac{3}{2^{11}} \cdot 2 = \frac{3}{2^{10}} = \frac{3}{1024}.$$

(The geometric series in the middle line sums to 2; hence the final answer.)   ∎

Detecting Divergent Series

For us, *convergent* series and their limits are mainly of interest. The preceding theorem, for instance, applies safely only to convergent series. It's important, therefore, to recognize *divergence* when it occurs. The **nth term test** sometimes helps.

A series $\sum a_k$ converges if and only if its partial sums S_n converge to a limit. For this to occur, the difference $S_n - S_{n-1}$ between successive partial sums must tend to zero. ▶   This difference is simply the nth term:

Otherwise the partial sum sequence wouldn't "level off."

$$S_n - S_{n-1} = (a_1 + a_2 + \cdots + a_{n-1} + a_n) - (a_1 + a_2 + \cdots + a_{n-1}) = a_n.$$

Thus, for a series to converge, its *terms* must tend to zero. We restate this fact in its most useful form:

> **Theorem 5** (**The nth Term Test for Divergence**) If $\lim_{n \to \infty} a_n \neq 0$, then $\sum a_n$ diverges.

What the Theorem Doesn't Say.   It's important to notice that Theorem 5 does *not* guarantee that $\sum a_n$ converges if $a_n \to 0$. The harmonic series $\sum 1/k$ illustrates that the terms of a divergent series may tend to zero. The rather blunt nth term test may detect *divergence*, but never convergence. Sharper instruments are needed to detect convergence. We develop several in the next section.

EXAMPLE 8   Does $\displaystyle\sum_{k=1}^{\infty} \frac{k}{k+1000}$ converge?

Solution   The nth term test says no. Because

$$a_n = \frac{n}{n+1000} \to 1 \quad \text{as} \quad n \to \infty,$$

the series diverges.   ∎

EXAMPLE 9   Given that $\displaystyle\sum_{k=0}^{\infty} \frac{3^k}{k!}$ converges, find $\displaystyle\lim_{k\to\infty} \frac{3^k}{k!}$.

Try some.

Solution   The limit is zero. By the preceding theorem, the terms of any convergent series must tend to zero. (Tabulating values ◄ of $3^n/n!$ supports this conclusion.)   ∎

BASICS

1. This exercise is about partial sums of the geometric series $\sum_{k=0}^{\infty} ar^k$. By definition, a partial sum S_n of this series is $S_n = \sum_{k=0}^{n} ar^k = a + ar + ar^2 + ar^3 + \cdots + ar^n$. According to Equation 11.3, page 567, there's a simpler, more explicit formula for S_n:

$$S_n = a\,\frac{1 - r^{n+1}}{1 - r}.$$

   (a) Let $a = 1$, $r = 2$, and $n = 5$. Calculate both sides of the expression. Are they equal?

   (b) Repeat part (a) using $a = 1$, $r = 0.01$, and $n = 5$.

   (c) Suppose that $r = 1$. Explain why the formula does not hold in this case. Then find a formula for S_n when $r = 1$.

   (d) Calculate and simplify the quantity $S_n - rS_n$.

   (e) Use your result from part (d) to show that the formula holds if $r \neq 1$. [**HINT:** $S_n - rS_n = (1-r)S_n$.]

   (f) Find $3 + 6 + 12 + 24 + 48 + 96 + \cdots + 3072$. [**HINT:** Use the formula for S_n.]

2. The series $\displaystyle\sum_{k=0}^{\infty} a_k = \sum_{k=0}^{\infty} \frac{1}{k!}$ converges to $e \approx 2.718282$.

   (a) Evaluate a_1, a_2, a_5, and a_{10}.

   (b) Evaluate S_1, S_2, S_5, and S_{10}.

   (c) Explain why $\{S_n\}$ is an increasing sequence.

   (d) Evaluate R_1, R_2, R_5, and R_{10}.

   (e) Show that $R_n > 0$ for all $n \geq 0$.

   (f) Explain why $\{R_n\}$ is a decreasing sequence.

   (g) Using a calculator or a computer, determine a value of n for which S_n differs from e by less than 0.001.

   (h) Using a calculator or a computer, determine a value of n for which S_n differs from e by less than 10^{-5}.

   (i) Use parts (f) and (h) to show that $R_{50} < 10^{-5}$.

   (j) Explain why $\lim_{n\to\infty} R_n = 0$.

3. The series $\displaystyle\sum_{k=1}^{\infty} a_k = \sum_{k=1}^{\infty} \frac{1}{k^2}$ converges to $\dfrac{\pi^2}{6} \approx 1.64493$.

   (a) Evaluate a_1, a_2, a_5, and a_{10}.
   (b) Evaluate S_1, S_2, S_5, and S_{10}.
   (c) Is $\{S_n\}$ an increasing sequence? Justify your answer.
   (d) Evaluate R_1, R_2, R_5, and R_{10}.
   (e) Is $\{R_n\}$ a decreasing sequence? Justify your answer.
   (f) Show that $0 < R_n < 0.05$ for all $n \geq 20$.
   (g) Evaluate $\lim_{n\to\infty} R_n$.

4. Consider the series $\displaystyle\sum_{k=0}^{\infty} a_k = \sum_{k=0}^{\infty} \frac{1}{5^k}$.

   (a) Evaluate a_1, a_2, a_5, and a_{10}.
   (b) Evaluate S_1, S_2, S_5, and S_{10}.
   (c) Show that the sequence $\{S_n\}$ is increasing and bounded above. What does this imply about the sequence of partial sums?
   (d) Find the sum of the series (i.e., $\lim_{n\to\infty} S_n$).
   (e) Evaluate R_1, R_2, R_5, and R_{10}.
   (f) Show that $\{R_n\}$ is decreasing and bounded below.
   (g) Evaluate $\lim_{n\to\infty} R_n$.

5. Consider the series $\displaystyle\sum_{k=0}^{\infty} a_k = \sum_{k=0}^{\infty} (-0.8)^k$.

   (a) Evaluate a_1, a_2, a_5, and a_{10}.
   (b) Evaluate S_1, S_2, S_5, and S_{10}.
   (c) Find the sum of the series (i.e., $\lim_{n\to\infty} S_n$).
   (d) Evaluate R_1, R_2, R_5, and R_{10}.
   (e) Is the sequence $\{S_n\}$ increasing? Justify your answer.
   (f) Show that the sequence $\{R_n\}$ is neither increasing nor decreasing.
   (g) Show that the sequence $\{|R_n|\}$ is decreasing.
   (h) Evaluate $\lim_{n\to\infty} R_n$.

6. Consider the series $\displaystyle\sum_{k=0}^{\infty} a_k = \sum_{k=0}^{\infty} \frac{1}{k + 2^k}$.

   (a) Evaluate S_1, S_2, S_5, and S_{10}.
   (b) Show that the sequence $\{S_n\}$ is increasing.
   (c) Explain why $a_k \leq 2^{-k}$ for all $k \geq 0$.
   (d) Use part (c) to show that $S_n \leq 2 - 2^{-n} < 2$.
   [**HINT:** $\sum_{k=0}^{n} 2^{-k}$ is a geometric series.]
   (e) Use parts (b) and (d) to show that the series $\sum_{k=0}^{\infty} a_k$ converges (i.e., that $\lim_{n\to\infty} S_n$ exists).
   (f) Show that $\lim_{n\to\infty} R_n = 0$.

7. Consider the series $\displaystyle\sum_{j=0}^{\infty} a_j = \sum_{j=0}^{\infty} \frac{1}{2 + 3^j}$.

   (a) Evaluate S_1, S_2, S_5, and S_{10}.

(b) Show that the sequence $\{S_n\}$ is increasing and bounded above.
(c) Does $\sum_{j=0}^{\infty} a_j$ converge? Justify your answer.

8. It is known that $\displaystyle\sum_{m=1}^{\infty} \frac{1}{m^4} = \frac{\pi^4}{90}$. Use this fact to evaluate

   (a) $\displaystyle\sum_{i=0}^{\infty} \frac{1}{(i + 1)^4}$ (b) $\displaystyle\sum_{k=3}^{\infty} \frac{1}{k^4}$

In Exercises 9–16, find the limit of the series.

9. $\dfrac{1}{16} + \dfrac{1}{32} + \dfrac{1}{64} + \dfrac{1}{128} + \cdots + \dfrac{1}{2^{i+4}} + \cdots$

10. $2 - 5 + 9 + \dfrac{1}{3} + \dfrac{1}{9} + \dfrac{1}{27} + \dfrac{1}{81} + \cdots + \dfrac{1}{3^n} + \cdots$

11. $\displaystyle\sum_{n=0}^{\infty} e^{-n}$ 14. $\displaystyle\sum_{i=10}^{\infty} \left(\frac{2}{3}\right)^i$

12. $\displaystyle\sum_{k=3}^{\infty} \left(\frac{e}{\pi}\right)^k$ 15. $\displaystyle\sum_{j=5}^{\infty} \left(-\frac{1}{2}\right)^j$

13. $\displaystyle\sum_{m=1}^{\infty} (\arctan 1)^m$ 16. $\displaystyle\sum_{j=0}^{\infty} \frac{3^j + 4^j}{5^j}$

17. Show that the series $\displaystyle\sum_{k=1}^{\infty} \frac{1}{2 + \sin k}$ diverges.

18. Use partial sums to explain why the series $\displaystyle\sum_{k=0}^{\infty} (-1)^k$ diverges.

In Exercises 19–26, find an expression for the partial sum S_n of the series. Use this expression to determine whether the series converges and, if so, to find its limit.

19. $\displaystyle\sum_{k=0}^{\infty} \left(\arctan(k + 1) - \arctan k\right)$

20. $\displaystyle\sum_{i=2}^{\infty} \frac{1}{i(i - 1)} = \sum_{i=2}^{\infty} \left(\frac{1}{i - 1} - \frac{1}{i}\right)$

21. $\displaystyle\sum_{k=1}^{\infty} \frac{2}{k^2 + k}$ 22. $\displaystyle\sum_{j=1}^{\infty} \frac{j}{(j + 1)!}$

23. $\displaystyle\sum_{m=1}^{\infty} \left(\frac{1}{\sqrt{m}} - \frac{1}{\sqrt{m + 2}}\right)$

24. $\displaystyle\sum_{j=1}^{\infty} \ln\left(1 + \frac{1}{j}\right) = \sum_{j=1}^{\infty} \left(\ln(j + 1) - \ln j\right)$

25. $\displaystyle\sum_{k=0}^{\infty} \cos(k\pi)$ 26. $\displaystyle\sum_{j=2}^{\infty} (-1)^j j$

FURTHER EXERCISES

27. It is known that

$$\sum_{i=1}^{\infty} \frac{1}{i^2} = 1 + \frac{1}{4} + \frac{1}{9} + \frac{1}{16} + \frac{1}{25} + \cdots = \frac{\pi^2}{6}.$$

(a) Use this fact to evaluate

$$\sum_{j=1}^{\infty} \frac{1}{(2j)^2} = \frac{1}{4} + \frac{1}{16} + \frac{1}{36} + \frac{1}{64} + \frac{1}{100} + \cdots.$$

(b) Use part (a) to show that

$$\sum_{k=0}^{\infty} \frac{1}{(2k+1)^2} = 1 + \frac{1}{9} + \frac{1}{25} + \frac{1}{49} + \cdots = \frac{\pi^2}{8}.$$

(c) Evaluate

$$\sum_{m=1}^{\infty} \frac{(-1)^{m+1}}{m^2} = 1 - \frac{1}{4} + \frac{1}{9} - \frac{1}{16} + \frac{1}{25} - \frac{1}{36} + \cdots.$$

28. Express $\displaystyle\sum_{m=3}^{\infty} \frac{2^{m+4}}{5^m}$ as a rational number.

For each of the series in Exercises 29–37, find all values of x for which the series converges, then state the limit as a simple expression involving x. (Assume that $x^0 = 1$ for all x.)

29. $\displaystyle\sum_{k=0}^{\infty} x^k$

30. $\displaystyle\sum_{m=2}^{\infty} \left(\frac{x}{5}\right)^m$

31. $\displaystyle\sum_{j=5}^{\infty} x^{2j}$

32. $\displaystyle\sum_{k=1}^{\infty} x^{-k}$

33. $\displaystyle\sum_{n=3}^{\infty} (1+x)^n$

34. $\displaystyle\sum_{j=4}^{\infty} \frac{1}{(1-x)^j}$

35. $\displaystyle\sum_{k=11}^{\infty} \left(\frac{\sin x}{2}\right)^k$

36. $\displaystyle\sum_{m=2}^{\infty} (\ln x)^m$

37. $\displaystyle\sum_{n=0}^{\infty} (\arctan x)^n$

38. Find the limit of the sequence defined by $S_1 = 1$, $S_{n+1} = S_n + 1/3^n$. [**HINT:** Write out the first few terms to see the pattern.]

39. Find the limit of the sequence defined by $a_1 = 4$, $a_{n+1} = a_n - 1/2^n$.

In Exercises 40–55, determine whether the series converges or diverges. If a series converges, find its limit. Justify your answers.

40. $2 - 2 + 2 - 2 + 2 - 2 + \cdots$

41. $\dfrac{3}{10} - \dfrac{3}{20} + \dfrac{3}{40} - \dfrac{3}{80} + \dfrac{3}{160} - \dfrac{3}{320} + \cdots$

42. $1 - \dfrac{1}{2} + \dfrac{1}{2} - \dfrac{1}{3} + \dfrac{1}{3} - \cdots$

43. $1 - 1 + 2 - 1 - 1 + 3 - 1 - 1 - 1 + 4 - 1 - 1 - 1 - 1 + \cdots$

44. $\dfrac{4}{7^{10}} + \dfrac{4}{7^{12}} + \dfrac{4}{7^{14}} + \dfrac{4}{7^{16}} + \dfrac{4}{7^{18}} + \cdots$

45. $\displaystyle\sum_{n=0}^{\infty} \frac{n+1}{2n+1}$

46. $\displaystyle\sum_{j=0}^{\infty} (\ln 2)^j$

47. $\displaystyle\sum_{n=2}^{\infty} \frac{2}{n^2 - 1}$

48. $\displaystyle\sum_{k=1}^{\infty} \frac{k^\pi}{k^e}$

49. $\displaystyle\sum_{m=2}^{\infty} \frac{1}{(\ln 3)^m}$

50. $\displaystyle\sum_{j=0}^{\infty} \left(\frac{1}{2^j} + \frac{1}{3^j}\right)^2$

51. $\displaystyle\sum_{k=1}^{\infty} \left(\int_k^{k+1} \frac{dx}{x^2}\right)$

52. $\displaystyle\sum_{m=0}^{\infty} \left(\int_0^m e^{-x^2}\, dx\right)$

53. $\displaystyle\sum_{n=1}^{\infty} \left(1 + \frac{1}{n}\right)^n$

54. $\displaystyle\sum_{j=1}^{\infty} \sqrt[j]{\pi}$

55. $\displaystyle\sum_{n=1}^{\infty} \frac{\ln n}{\ln(3 + n^2)}$

56. A rubber ball rebounds to two-thirds the height from which it falls. If it is dropped from a height of 4 feet and is allowed to continue bouncing indefinitely, what is the total distance it travels?

57. Let $S_n = \displaystyle\sum_{k=1}^{n} \frac{1}{\sqrt{k}}$.

(a) Evaluate $\displaystyle\lim_{k\to\infty} \frac{1}{\sqrt{k}}$.

(b) Show that $S_n \geq \dfrac{n}{\sqrt{n}} = \sqrt{n}$ for all $n \geq 1$.
[**HINT:** If $k \leq n$, then $1/k \geq 1/n$.]

(c) Use part (b) to show that $\displaystyle\sum_{k=1}^{\infty} \frac{1}{\sqrt{k}}$ diverges.

58. Let $\{a_k\}$ be an increasing sequence such that $a_1 > 0$ and $a_k \leq 100$ for all $k \geq 1$.
(a) Does $\lim_{k\to\infty} a_k$ exist? Justify your answer.
(b) Show that $\sum_{k=1}^{\infty} a_k$ diverges.

59. Let $\{a_k\}$ be a sequence of positive terms such that $\sum_{k=1}^{n} a_k \leq 100$ for all $n \geq 1$. Explain why $\lim_{k\to\infty} a_k = 0$ must be true.

60. A certain series $\sum_{k=1}^{\infty} a_k$ has partial sums

$$S_n = \sum_{k=1}^{n} a_k = 5 - \frac{3}{n}.$$

   (a) Evaluate $S_{100} = \sum_{k=1}^{100} a_k$.
   (b) Evaluate $\sum_{k=1}^{\infty} a_k$.
   (c) Evaluate $\lim_{k\to\infty} a_k$.
   (d) Show that $a_k > 0$ for all $k \geq 1$.
   [**HINT**: $a_{n+1} = S_{n+1} - S_n$.]

61. A certain series $\sum_{j=1}^{\infty} b_j$ has partial sums

$$S_n = \sum_{j=1}^{n} b_j = \ln\left(\frac{2n+3}{n+1}\right).$$

   (a) Evaluate $\lim_{n\to\infty} S_n$.
   (b) Does the series converge? Justify your answer.
   (c) Show that $b_j < 0$ for all $j \geq 1$.

62. Suppose that the partial sums of the series $\sum_{k=1}^{\infty} a_k$ satisfy the inequality

$$\frac{6\ln n}{\ln(n^2+1)} < S_n < 3 + ne^{-n}$$

   for all $n \geq 100$.
   (a) Does the series converge? If so, to what limit? Justify your answers.
   (b) What, if anything, can be said about $\lim_{k\to\infty} a_k$? Explain.

63. Let $S_n = \sum_{k=1}^{n} a_k$, and suppose that $0 \leq S_n \leq 100$ for all $n \geq 1$.
   (a) Give an example of a sequence $\{a_k\}$ that satisfies the given conditions but $\sum_{k=1}^{\infty} a_k$ diverges.
   (b) Show that if $a_k > 0$ for all $k \geq 1$, then $\sum_{k=1}^{\infty} a_k$ converges.
   (c) Show that if $a_k > 0$ for all $k \geq 10^6$, then $\sum_{k=1}^{\infty} a_k$ converges.

64. Suppose that $\sum_{k=1}^{\infty} a_k$ diverges.
   (a) Explain why $a_k > 0$ for all $k \geq 1$ implies that $\lim_{n\to\infty} S_n = \infty$.
   (b) Give an example of a divergent series for which $\lim_{n\to\infty} S_n$ does not exist.

65. What is wrong with the following argument?

   Let $S = 1 + 2 + 4 + 8 + \cdots$.

   Then $2S = 2 + 4 + 8 + \cdots = S - 1$, so $S = -1$.

66. (a) Use the trigonometric identity $\tan x = \cot x - 2\cot(2x)$ to show that

$$\sum_{k=1}^{n} \frac{1}{2^k}\tan\left(\frac{x}{2^k}\right) = \frac{1}{2^n}\cot\left(\frac{x}{2^n}\right) - \cot x.$$

   (b) Use part (a) to show that

$$\sum_{k=1}^{\infty} \frac{1}{2^k}\tan\left(\frac{x}{2^k}\right) = \frac{1}{x} - \cot x.$$

67. Let S_n denote the nth partial sum of the harmonic series

$$\sum_{k=1}^{\infty} \frac{1}{k} = 1 + \frac{1}{2} + \frac{1}{3} + \frac{1}{4} + \cdots$$

   (i.e., $S_n = \sum_{k=1}^{n} 1/k$).
   (a) Complete the following table; report answers to three decimal places. In the third row, enter *differences* between successive entries in the second row. A few entries are given.

n	10	20	30	40	50
S_n	2.929	3.598			
ΔS_n	—	0.671			
n	60	70	80	90	100
S_n					5.187
ΔS_n					

   (b) Do the numbers in the table in part (a) suggest clearly whether the harmonic series converges or diverges? Why or why not?
   (c) Complete the following table; report answers to three decimal places. In the third row, enter *differences* between successive entries in the second row. A few entries are given.

n	2	4	8	16	32
S_n	1.500	2.083			
ΔS_n	—	0.583			
n	64	128	256	512	1024
S_n					7.509
ΔS_n					0.693

   (d) Do the numbers in the table in part (c) suggest clearly whether the harmonic series converges or diverges? Why or why not?
   (e) The bottom row of the table in (c) shows that doubling n causes S_n to increase by about 0.693. Use this fact to guess values for S_{2048}, S_{4096}, and S_{8192}. (Don't try to calculate these numbers directly!)

68. Let $H_n = \sum_{k=1}^{n} \frac{1}{k}$, and let $I_n = \int_1^{n+1} \frac{dx}{x}$.
   (a) Let L_n be the left Riemann-sum approximation, with n equal subdivisions, to I_n. Show that $L_n = H_n$.
   (b) Use part (a) to show that the harmonic series diverges. [**HINT**: Start by comparing L_n and I_n.]

69. Let $H_n = \sum_{k=1}^{n} \frac{1}{k}$, and let $a_m = \sum_{j=1}^{2^{m-1}} \frac{1}{2^{m-1}+j}$.   Then

    $H_{2^n} = 1 + \sum_{m=1}^{n} a_m$. [**NOTE:** a_m is the sum of a "block" of 2^{m-1} consecutive terms of the harmonic series—those from $n = 2^{m-1}+1$ through $n = 2^m$.]

    (a) Show that $a_1 = 1/2$, $a_2 = 7/12$, and $a_3 = 533/840$.
    (b) Show that $H_8 = 1 + a_1 + a_2 + a_3 = 761/280$.
    (c) Show that $a_k \geq 1/2$ for all $k \geq 1$.
        [**HINT:**

        $$\frac{1}{2^{m-1}+j} \leq \frac{1}{2^{m-1}+2^{m-1}}$$

        if $1 \leq j \leq 2^{m-1}$.]
    (d) Use part (b) to show that $\lim_{n\to\infty} H_n = \infty$ (i.e., the harmonic series diverges).

70. Let $H_n = \sum_{k=1}^{n} \frac{1}{k}$, and let $I = \int_1^2 \frac{dx}{x}$.

    (a) Let L_n be the left Riemann-sum approximation to I with n equal subdivisions. Show that

    $$L_n = \sum_{k=0}^{n-1} \frac{1}{n+k}$$
    $$= \frac{1}{n} + \frac{1}{n+1} + \frac{1}{n+2} + \cdots + \frac{1}{2n-1}.$$

    (b) Explain why $L_n > \ln 2$ for all $n \geq 1$.
    (c) Show that $H_{2n} - H_n = L_n - \frac{1}{2n}$ for all $n \geq 1$.
    (d) Show that $\lim_{n\to\infty}(H_{2n} - H_n) = \ln 2$.
    (e) Explain why part (d) implies that the harmonic series diverges.
    (f) Explain why $H_{2n} - H_n > \ln 2 - \frac{1}{2}$ for all $n \geq 1$. [**HINT:** Use parts (b) and (c).]
    (g) Show that $H_2 > 1 + \ln 2 - \frac{1}{2}$.
    (h) Show that that $H_{2^m} > 1 + m\left(\ln 2 - \frac{1}{2}\right)$ if $m \geq 1$. [**HINT:** Use parts (f) and (g).]

71. (a) Show that $\sin x \leq x \leq \tan x$ if $0 \leq x < \pi/2$.
    (b) Use part (a) to show that $\cot^2 x \leq 1/x^2 \leq 1 + \cot^2 x$ if $0 \leq x < \pi/2$.
    (c) Use part (b) to show that

    $$\sum_{k=1}^{n} \cot^2\left(\frac{k\pi}{2n+1}\right) \leq \frac{(2n+1)^2}{\pi^2} \sum_{k=1}^{n} \frac{1}{k^2}$$
    $$\leq n + \sum_{k=1}^{n} \cot^2\left(\frac{k\pi}{2n+1}\right).$$

    (d) It can be shown that

    $$\sum_{k=1}^{n} \cot^2\left(\frac{k\pi}{2n+1}\right) = \frac{n(2n-1)}{3}.$$

Thus, part (c) is equivalent to

$$\frac{n(2n-1)}{3} \leq \frac{(2n+1)^2}{\pi^2} \sum_{k=1}^{n} \frac{1}{k^2} \leq n + \frac{n(2n-1)}{3}.$$

Use these inequalities to show that $\sum_{k=1}^{\infty} \frac{1}{k^2} = \frac{\pi^2}{6}$.

72. Let $I_n = \int_0^{\pi/4} \tan^n x \, dx$ for integer $n \geq 0$.
    (a) Show that $\{I_n\}$ is a monotone decreasing sequence (i.e., $I_{n+1} \leq I_n$).
    (b) Show that $I_n + I_{n-2} = \frac{1}{n-1}$.
    (c) Use parts (a) and (b) to show that

    $$\frac{1}{2(n+1)} \leq I_n \leq \frac{1}{2(n-1)}.$$

    [**HINT:** Part (b) implies that $I_{n+2} + I_n = 1/(n+1)$.]
    (d) Show that, if $n \geq 2$,

    $$I_n = \frac{1}{n-1} - \int_0^{\pi/4} \tan^{n-2} x \, dx.$$

    (e) Use part (d) to show that

    $$I_{2n} = (-1)^n \left(\frac{\pi}{4} + \sum_{k=1}^{n} \frac{(-1)^k}{2k-1}\right)$$

    for all integers $n \geq 1$.
    (f) Use parts (c) and (e) to show that $\sum_{k=0}^{\infty} \frac{(-1)^k}{2k+1} = \frac{\pi}{4}$.
    (g) Use part (d) to show that

    $$I_{2n+1} = (-1)^n \left(\frac{1}{2}\ln 2 + \sum_{k=1}^{n} \frac{(-1)^k}{2k}\right)$$

    for all integers $n \geq 1$.
    (h) Use parts (c) and (g) to show that $\sum_{k=1}^{\infty} \frac{(-1)^{k+1}}{k} = \ln 2$.

73. Consider the series $\sum_{k=1}^{\infty} \frac{1}{k^p}$ with $p > 1$. This exercise outlines a proof that this series converges.

    (a) Let $S_n = \sum_{k=1}^{n} \frac{1}{k^p}$. Show that the sequence of partial sums $\{S_n\}$ is increasing.
    (b) Show that $S_{2m+1} = 1 + \sum_{k=1}^{m} \frac{1}{(2k)^p} + \sum_{k=1}^{m} \frac{1}{(2k+1)^p}$.
    (c) Explain why $S_{2m+1} < 1 + 2\sum_{k=1}^{m} \frac{1}{(2k)^p}$.

    [**HINT:** $1/(x+1) < 1/x$ for all $x > 0$.]
    (d) Show that $S_{2m+1} < 1 + 2^{1-p} S_{2m+1}$. [**HINT:** First show that $S_{2m+1} < 1 + 2^{1-p} S_m$.]
    (e) Show that $\{S_n\}$ is bounded above.

11.3 Testing for Convergence; Estimating Limits

Converge or Diverge: What's the Question?

In theory, the question of convergence is simple:

> The series $\sum a_k$ converges to S if the sequence $\{S_n\}$ of partial sums tends to S.

The trouble, in practice, is that a simple, explicit formula for S_n is often unavailable. ▶ It might seem, then, that testing for convergence—let alone finding a limit—would be difficult or impossible. Surprisingly, that isn't so. All the convergence tests of this section and the next (comparison test, integral test, ratio test, and so on) offer clever, indirect ways of testing whether $\{S_n\}$ converges—even without knowing each S_n exactly.

The harmonic series poses this problem.

Nonnegative Series

This apparent sleight-of-hand depends on a surprisingly simple observation: ▶

Convince yourself of this simple but important fact.

> If $a_k \geq 0$ for all k, then the sequence $\{S_n\}$ of partial sums of $\sum a_k$ is nondecreasing:

$$S_1 \leq S_2 \leq S_3 \leq \cdots S_n \leq S_{n+1} \cdots.$$

A series with this convenient property is called **nonnegative**. By Theorem 3 of Section 11.1, a nondecreasing sequence must either converge or diverge to infinity. Therefore, to decide whether a nonnegative series $\sum a_k$ converges, it's enough to answer this key question:

> As $n \to \infty$, do the partial sums $\{S_n\}$ blow up or remain bounded?

If a simple formula for S_n is available (as it is, for instance, for geometric series), the answer may be obvious. If not, our best recourse is to try to compare ▶ the given series to something that is better understood (another series, an integral—anything that works). Sometimes an obvious comparison suggests itself.

Watch for frequent reappearances of the words "compare" and "comparison."

EXAMPLE 1 Does $\displaystyle\sum_{k=0}^{\infty} \frac{1}{2^k + 1}$ converge?

Solution We saw in the preceding section ▶ that the similar series $\sum_{k=0}^{\infty} 1/2^k$ converges to 2. For all k,

See Example 1, page 564.

$$\frac{1}{2^k + 1} < \frac{1}{2^k},$$

so each partial sum of $\sum_{k=0}^{\infty} 1/(2^k + 1)$ is *less than* the corresponding partial sum of $\sum_{k=0}^{\infty} 1/2^k$. Because the latter sums tend to 2, the former must tend to a limit less than 2. Briefly, in symbols,

$$\frac{1}{2^k + 1} < \frac{1}{2^k} \implies \sum_{k=0}^{\infty} \frac{1}{2^k + 1} < \sum_{k=0}^{\infty} \frac{1}{2^k} = 2.$$

Easily found by calculator or computer.

Numerical evidence ◄ agrees. For the original series,

$$S_{10} \approx 1.263523536; \qquad S_{20} \approx 1.264498827; \qquad S_{100} \approx 1.264499780. \qquad \blacksquare$$

The Comparison Test: One Series vs. Another

In Example 1 we showed that one series converges by comparing it to another series that is *known* to converge. The following theorem makes this idea precise, duly noting all necessary hypotheses.

Theorem 6 (Comparison Test for Nonnegative Series) Suppose that for all $k \geq 1$, $0 \leq a_k \leq b_k$. Consider the two series $\sum_{k=1}^{\infty} a_k$ and $\sum_{k=1}^{\infty} b_k$.

- If $\sum_{k=1}^{\infty} b_k$ *converges, so does* $\sum_{k=1}^{\infty} a_k$, and

$$\sum_{k=1}^{\infty} a_k \leq \sum_{k=1}^{\infty} b_k.$$

- If $\sum_{k=1}^{\infty} a_k$ *diverges, so does* $\sum_{k=1}^{\infty} b_k$.

Observe:

What We'd Expect The theorem's assertion is reasonable, and is not difficult to prove. A formal proof depends on the fact that, like the terms themselves, the partial sums of $\sum a_k$ are *all* less than the corresponding partial sums of $\sum b_k$.

Successful Comparisons In a "successful" comparison, either both series converge or both diverge. To use the comparison test, therefore, it's necessary first to *guess* whether the series in question converges or diverges. ◄

We did so implicitly in Example 1.

Comparing Tails We assumed earlier that the comparison inequality $0 \leq a_k \leq b_k$ holds for *all* possible k. This assumption isn't really necessary; if N is any positive integer, and $0 \leq a_k \leq b_k$ for all $k \geq N$, then

$$\sum_{k=N}^{\infty} a_k \leq \sum_{k=N}^{\infty} b_k.$$

We'll use this fact in the next example to estimate a limit numerically.

Sequences are "discrete" versions of functions; series are "discrete" versions of integrals.

Comparing Integrals, Comparing Series An almost identical comparison theorem applies to improper integrals. ◄ Theorem 1, Section 10.2, says that if $0 \leq a(x) \leq b(x)$ for all $x \geq 1$, then

$$\int_1^{\infty} a(x)\, dx \leq \int_1^{\infty} b(x)\, dx.$$

EXAMPLE 2 As we saw, $\sum_{k=0}^{\infty} 1/(2^k + 1)$ converges. How closely does $S_{100} \approx 1.264499781$ approximate the true limit S?

Solution   For any series, the error in using S_{100} to estimate the limit comes from ignoring the tail, R_{100}. In symbols,

$$S = \sum_{k=0}^{\infty} a_k = \sum_{k=0}^{100} a_k + \sum_{k=101}^{\infty} a_k = S_{100} + R_{100}.$$

Because, for our series, the inequality

$$\frac{1}{2^k + 1} < \frac{1}{2^k}$$

holds for all $k \geq 101$, ▶   it follows that

The inequality actually holds for all k; here we care only about $k \geq 101$.

$$\sum_{k=101}^{\infty} \frac{1}{2^k + 1} < \sum_{k=101}^{\infty} \frac{1}{2^k}.$$

We can calculate the right side directly:

$$\sum_{k=101}^{\infty} \frac{1}{2^k} = \frac{1}{2^{101}} + \frac{1}{2^{102}} + \frac{1}{2^{103}} + \cdots = \frac{1}{2^{101}}\left(1 + \frac{1}{2} + \frac{1}{2^2} + \cdots\right) = \frac{1}{2^{100}}.$$

Thus, as expected, we commit *very* little error by ignoring the tail R_{100}; the estimate S_{100} differs from S by less than $1/2^{100} \approx 8 \times 10^{-31}$. ■

Compared to What?

The *idea* of comparison is easy. Harder, in practice, is deciding what to compare a series *to*. Our only reliable "benchmark" series, so far, are geometric series. The **integral test** enlarges our stock of known series considerably.

The Integral Test: Series vs. Integrals

We know that

$$\int_1^{\infty} \frac{1}{x}\,dx \quad \text{diverges,} \qquad \text{but} \qquad \int_1^{\infty} \frac{1}{x^2}\,dx \quad \text{converges.}$$

One might guess, therefore, that

$$\sum_{k=1}^{\infty} \frac{1}{k} \quad \text{diverges,} \qquad \text{but} \qquad \sum_{k=1}^{\infty} \frac{1}{k^2} \quad \text{converges.}$$

These guesses are correct, as graphical evidence has already suggested.

Integrals and Series: Comparing Areas

Thinking of the terms of a positive series $\sum_{k=1}^{\infty} a_k$ as rectangular areas ▶   clarifies the connection with the integral $\int_1^{\infty} a(x)\,dx$. The first picture shows one way of

Each has base 1.

doing so:

If $\int_1^\infty a(x)\,dx$ diverges, so must $\sum\limits_{k=1}^{\infty} a_k$

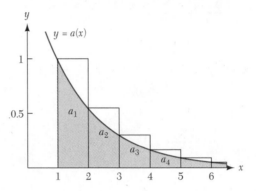

Look closely at the picture.

Total Areas The successive rectangles have *heights* $a(1) = a_1$, $a(2) = a_2$, $a(3) = a_3$, and so on; each *base* is 1. The respective *areas*, therefore, are a_1, a_2, a_3, and so on. Here's the key idea:

> *The series $a_1 + a_2 + a_3 + a_4 + \cdots$ represents the total "left-rule" rectangular area from 1 to ∞.*

An Important Inequality The shaded area—less than the rectangular area—represents the *integral* $\int_1^\infty a(x)\,dx$. Thus, if the integral diverges, so must the series. ◀ Here's the message of the picture, in inequality form:

We collect our results in the next theorem.

$$a_1 + a_2 + a_3 + \cdots \geq \int_1^\infty a(x)\,dx.$$

If the right side diverges to infinity, so must the left.

A Decreasing Integrand The reasoning that led to the preceding inequality requires that the integrand a be *decreasing* for $x \geq 1$, as shown in the picture. We collect such technical hypotheses carefully in the next theorem.

In the next picture, an integral bounds a series from *above*.

If $\int_1^\infty a(x)\,dx$ converges, so must $\sum\limits_{k=1}^{\infty} a_k$

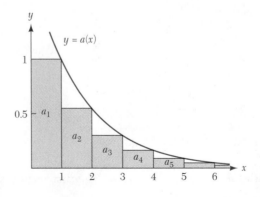

Again, successive rectangles have areas a_1, a_2, a_3, and so on. This time, comparing areas shows that ▶

Convince yourself; notice how the first term is "broken off."

$$a_1 + a_2 + a_3 + \cdots \le a_1 + \int_1^\infty a(x)\,dx.$$

If the right side converges, so must the left.

Combining both of the preceding inequalities gives upper *and* lower bounds for the series:

$$\int_1^\infty a(x)\,dx \le \sum_{k=1}^\infty a_k \le a_1 + \int_1^\infty a(x)\,dx.$$

In particular:

> The integral $\int_1^\infty a(x)\,dx$ and the series $\sum_{k=1}^\infty a_k$ either both converge or both diverge.

The final picture relates the tails of an integral and of a series.

Comparing tails: why $\displaystyle\sum_{k=n+1}^\infty a_k \le \int_n^\infty a(x)\,dx$

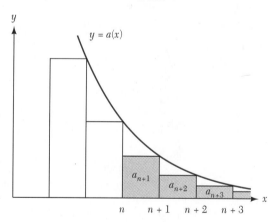

It shows, again by comparing areas, that for any n, the tail R_n satisfies

$$R_n = \sum_{k=n+1}^\infty a_k \le \int_n^\infty a(x)\,dx.$$

The preceding pictures show a particular function $a(x)$, but the conclusions we drew are generic—they hold for *any* function $a(x)$ that is both positive and decreasing. ▶

What could go wrong if $a(x)$ isn't decreasing?

It's time to collect all the foregoing observations in a theorem. As before, $a_k = a(k)$ for all integers $k \ge 1$.

Theorem 7 (Integral Test for Positive Series)   Suppose that, for all $x \geq 1$, the function $a(x)$ is continuous, positive, and decreasing. Consider the series $\sum_{k=1}^{\infty} a_k$ and the integral $\int_1^{\infty} a(x)\,dx$.

- If either diverges, so does the other.
- If either converges, so does the other. In this case,

$$\int_1^{\infty} a(x)\,dx \leq \sum_{k=1}^{\infty} a_k \leq a_1 + \int_1^{\infty} a(x)\,dx.$$

- If the series converges, then

$$R_n = \sum_{k=n+1}^{\infty} a_k \leq \int_n^{\infty} a(x)\,dx.$$

P-Series, Convergent and Divergent

Series of the form $\displaystyle\sum_{k=1}^{\infty} \frac{1}{k^p}$ are called **p-series**. They form an important family of examples, with behavior depending on the value of p.

EXAMPLE 3   Which p-series converge?

Are all the hypotheses really satisfied?

Solution   We found in Chapter 10 that the integral $\int_1^{\infty} dx/x^p$ converges if and only if $p > 1$. Therefore, using the integral test, ◀

> *The p-series $\displaystyle\sum_{k=1}^{\infty} \frac{1}{k^p}$ converges if and only if $p > 1$.*

In particular, the harmonic series $\sum 1/k$ diverges. ■

Harmonic Divergence.   The fact that the harmonic series

$$1 + \frac{1}{2} + \frac{1}{3} + \frac{1}{4} + \frac{1}{5} + \cdots$$

diverges to infinity—even though the terms themselves tend to zero—has fascinated mathematicians for many centuries. One early proof (not the one given here) is attributed to Nicole Oresme, a 14th-century French bishop, scientist, and mathematician.

EXAMPLE 4   The p-series $\sum_{k=1}^{\infty} 1/k^3$ converges, by the integral test, to some limit S. How large must n be to ensure that S_n differs from S by less than 0.0001?

See the "Comparing tails" picture.

Solution   We choose n so that the tail R_n is less than 0.0001. Theorem 7 shows how. By the last part, ◀

$$R_n \leq \int_n^{\infty} \frac{1}{x^3}\,dx.$$

See for yourself.

The right side is easily calculated; we get $1/2n^2$. ◀   This quantity (and hence also R_n) are less than 0.0001 if $n \geq 71$. Hence, $S_{71} \approx 1.20196$ differs from the true limit by less than 0.0001. ■

EXAMPLE 5   Does $\displaystyle\sum_{k=1}^{\infty} \frac{1}{10k+1}$ converge?

Solution   The integral test says no.

$$\int_{1}^{\infty} \frac{1}{10x+1}\, dx = \lim_{n\to\infty} \int_{1}^{n} \frac{1}{10x+1}\, dx = \lim_{n\to\infty} \left. \frac{\ln(10x+1)}{10} \right]_{1}^{n} = \infty.$$

The comparison test *also* says no. Because for all $k \geq 1$,

$$\frac{1}{10k+1} \geq \frac{1}{11k},$$

it follows that

$$\sum_{k=1}^{\infty} \frac{1}{10k+1} \geq \sum_{k=1}^{\infty} \frac{1}{11k} = \frac{1}{11} \sum_{k=1}^{\infty} \frac{1}{k}.$$

Since the last series diverges to infinity, so must the first.   ∎

The Ratio Test: Comparison with a Geometric Series

In a geometric series

$$a + ar + ar^2 + ar^3 + ar^4 + \cdots$$

the ratio of successive terms is r; the series converges if and only if $|r| < 1$. ▶

The **ratio test** is based on the same principle. In the end, it amounts to a lightly disguised form of comparison with a geometric series.

Why? See the Fact on page 567.

Theorem 8 (**Ratio Test for Positive Series**) Suppose that $\sum_{k=1}^{\infty} a_k$ is a positive series and that

$$\lim_{k\to\infty} \frac{a_{k+1}}{a_k} = L.$$

- If $L < 1$, then $\sum a_k$ *converges.*
- If $L > 1$, then $\sum a_k$ *diverges.*
- If $L = 1$, either convergence or divergence is possible. The test is inconclusive.

Notice these possibilities:

When the Ratio Tends to 1   For many series, unfortunately, $a_{k+1}/a_k \to 1$ as $k \to \infty$. In such cases, the ratio test says *nothing*. This happens for every p-series $\sum 1/k^p$, for instance.

When the Ratio Test Works Best   The ratio test works best for such series as $\sum 1/k!$, $\sum r^k$, and $\sum 1/(2^k + 3)$, in which the index k appears in an exponent or a factorial.

The Idea of a Proof

To illustrate the connection between the ratio test and geometric series, and to give the idea of a proof, let's suppose that

$$\lim_{k \to \infty} \frac{a_{k+1}}{a_k} = \frac{1}{2}.$$

Why must $\sum a_k$ converge?

The inequality must hold for all large k because of the limit above.

The idea is that, for large k, $a_{k+1} \approx a_k/2$, so the series behaves similarly to a geometric series. Suppose, for instance, that $a_{k+1} < 0.6a_k$ for all $k \geq 1000$. ◄ Then

$$a_{1001} < (0.6)a_{1000}, \ldots; \qquad a_{1002} < (0.6)a_{1001} < (0.6)^2 a_{1000},$$
$$a_{1003} < (0.6)^3 a_{1000},$$

so

$$\sum_{k=1000}^{\infty} a_k = a_{1000} + a_{1001} + a_{1002} + \cdots < a_{1000}\left(1 + (0.6) + (0.6)^2 + (0.6)^3 + \cdots\right).$$

The last inequality is the point; it shows that $\sum a_k$ converges *by comparison with the geometric series* $\sum a_{1000}(0.6)^k$.

The divergence statement can be proved in a similar way.

EXAMPLE 6   Show that $\displaystyle\sum_{k=0}^{\infty} \frac{1}{k!}$ converges. Guess a limit.

Check details carefully.

Solution   The ratio test works nicely. ◄   Since

$$\lim_{k \to \infty} \frac{a_{k+1}}{a_k} = \lim_{k \to \infty} \frac{k!}{(k+1)!} = \lim_{k \to \infty} \frac{1}{k+1} = 0,$$

the series converges. In fact, it converges very, very fast. Here are some representative partial sums:

$$S_5 \approx 2.716666667;$$
$$S_{10} \approx 2.718281801;$$
$$S_{30} \approx 2.718281828459045235360287471\,35.$$

Is e involved somehow? To 30 decimals,

$$e = 2.718281828459045235360287471\,35.$$

In fact, this series can be shown rigorously to converge to e. We explore this phenomenon further in later sections.   ∎

EXAMPLE 7   Does $\displaystyle\sum_{k=0}^{\infty} \frac{100^k}{k!}$ converge?

Solution   Yes, by the ratio test:

$$\lim_{k \to \infty} \frac{a_{k+1}}{a_k} = \lim_{k \to \infty} \frac{100^{k+1}}{(k+1)!} \cdot \frac{k!}{100^k} = \lim_{k \to \infty} \frac{100}{k+1} = 0.$$

Notice what the result means: Even though 100^k grows very fast, $k!$ grows faster still.   ∎

BASICS

1. Consider the series $\sum_{k=0}^{\infty} a_k$, where $a_k = \dfrac{1}{k + 2^k}$.

   (a) Use the comparison test to show that the series converges. [**HINT:** $a_k \leq 2^{-k}$ for all $k \geq 0$.]
   (b) Show that $0 \leq R_{10} \leq 2^{-10}$.
   (c) Compute an estimate of the limit of the series that is guaranteed to be within 0.001 of the exact value.
   (d) Is your estimate in part (c) an overestimate or an underestimate? Justify your answer.

2. Consider the series $\sum_{j=0}^{\infty} \dfrac{1}{2 + 3^j}$.

   (a) Show that the series converges.
   (b) Estimate the value of the limit of the series within 0.01.
   (c) Is your estimate in part (b) an overestimate or an underestimate? Justify your answer.

In Exercises 3–7, suppose that $a(x)$ is continuous, positive, and decreasing for all $x \geq 1$ and that $a_k = a(k)$ for all integers $k \geq 1$.

3. Rank the values $\int_1^n a(x)\,dx$, $\sum_{k=1}^{n-1} a_k$, and $\sum_{k=2}^{n} a_k$ in increasing order. [**HINT:** Draw a picture.]

4. Rank the values $\int_n^{\infty} a(x)\,dx$, $\sum_{k=n+1}^{\infty} a_k$, and $\int_{n+1}^{\infty} a(x)\,dx$ in increasing order.

5. Draw a carefully annotated picture that shows that $\int_1^{n+1} a(x)\,dx \leq \sum_{k=1}^{n} a_k$.

6. Draw a carefully annotated picture that shows that $\sum_{k=2}^{n} a_k \leq \int_1^n a(x)\,dx$.

7. Draw a carefully annotated picture that shows that

$$\sum_{k=n+1}^{\infty} a_k \leq a_{n+1} + \int_{n+1}^{\infty} a(x)\,dx \leq \int_n^{\infty} a(x)\,dx.$$

In Exercises 8–10, use the integral test to find upper and lower bounds on the limit of the series.

8. $\sum_{k=0}^{\infty} \dfrac{1}{k^2 + 1}$

9. $\sum_{k=1}^{\infty} \dfrac{1}{k\sqrt{k}}$

10. $\sum_{j=1}^{\infty} j e^{-j}$

11. Show that $\sum_{n=2}^{\infty} \dfrac{1}{(\ln n)^2}$ diverges.
    [**HINT:** $\ln x \leq \sqrt{x}$ for all $x \geq 1$.]

12. Show that $\sum_{k=3}^{\infty} \dfrac{1}{(\ln k)^k}$ converges.
    [**HINT:** $\ln k \geq \ln 3 \approx 1.0986$ for all $x \geq 3$.]

13. Consider the series $\sum_{k=1}^{\infty} \dfrac{2 + \sin k}{k^2}$.

    (a) Explain why Theorem 7 cannot be used to prove that this series converges.
    (b) Show that this series converges.

In Exercises 14–17, use the ratio test to show that the series converges.

14. $\sum_{j=0}^{\infty} \dfrac{j^2}{j!}$

15. $\sum_{k=1}^{\infty} \dfrac{2^k}{k!}$

16. $\sum_{n=1}^{\infty} \dfrac{n^2}{2^n}$

17. $\sum_{m=1}^{\infty} \dfrac{m!}{(2m)!}$

FURTHER EXERCISES

18. Suppose that for all $k \geq 1$, $0 \leq a_k \leq b_k$. Let $S_n = \sum_{k=1}^{n} a_k$ and $T_n = \sum_{k=1}^{n} b_k$.

    (a) Suppose that $\sum_{k=1}^{\infty} b_k$ converges. Explain why there is a number M such that $S_n \leq T_n \leq M$ for all $n \geq 1$.
    (b) Explain why $\{S_n\}$ is an increasing sequence.
    (c) Explain why parts (a) and (b) together imply that $\sum_{k=1}^{\infty} a_k$ converges.
    (d) Suppose that $\sum_{k=1}^{\infty} a_k$ diverges. Explain why $\lim_{n \to \infty} S_n = \infty$.
    (e) Suppose that $\sum_{k=1}^{\infty} a_k$ diverges. Use part (d) to show that $\sum_{k=1}^{\infty} b_k$ diverges.

19. Suppose that $a(x)$ is continuous, positive, and decreasing for all $x \geq 1$, that $a_k = a(k)$ for all integers $k \geq 1$, and that $\int_1^{\infty} a(x)\,dx$ converges.

    (a) Explain why the sequence of partial sums $\{S_n\}$ is an increasing sequence.
    (b) Explain why $\int_1^n a(x)\,dx \leq \int_1^{\infty} a(x)\,dx$.
    (c) Use parts (a) and (b) to show that the sequence of partial sums $\{S_n\}$ converges.

20. Suppose that $a_n \geq 0$ for all $n \geq 1$ and that $\sum_{n=1}^{\infty} a_n$ converges. Show that $\sum_{n=1}^{\infty} \sin(a_n)$ converges.
    [**HINT:** $|\sin x| \leq |x|$ for all x.]

21. Does the series

$$1 + \frac{1}{1 \cdot 3} + \frac{1}{1 \cdot 3 \cdot 5} + \frac{1}{1 \cdot 3 \cdot 5 \cdot 7}$$
$$+ \cdots + \frac{1}{1 \cdot 3 \cdot 5 \cdot 7 \cdots (2k+1)} + \cdots$$

converge? Justify your answer.

22. Show that $\displaystyle\sum_{k=3}^{\infty} \frac{1}{(\ln k)^{\ln k}}$ converges.

[**HINT:** $\ln k > e^2$ if $k > 1619$.]

23. Let $H_n = \displaystyle\sum_{k=1}^{n} \frac{1}{k}$ and let $S_n = \displaystyle\sum_{k=0}^{n} \frac{1}{2k+1}$.

(a) Explain why $\lim_{n \to \infty} H_n = \infty$.

(b) Show that $S_n \geq \frac{1}{2} H_n$.

(c) What do the results in parts (a) and (b) imply about $\sum_{k=0}^{\infty} 1/(2k+1)$? Explain.

24. (a) Estimate a lower bound for $n!$ by comparing $\ln(n!)$ and $\int_1^n \ln x \, dx$.

(b) Let a be a positive number. Use part (a) to find an integer N such that $a^N/N! < 1/2$.

25. For which values of p does the series

$$\sum_{n=3}^{\infty} \frac{1}{n(\ln n)^p}$$

converge? Justify your answer.

26. Consider the series $\displaystyle\sum_{k=1}^{\infty} a_k = \sum_{k=1}^{\infty} \frac{\ln k}{k}$.

(a) Use the integral test to show that the series diverges. [**HINT:** The function $\ln x/x$ is monotone on $[3, \infty)$. Start by showing that $\sum_{k=3}^{\infty} a_k$ diverges.]

(b) Use the comparison test to show that the series diverges. [**HINT:** $1 - x^{-1} \leq \ln x$ for all $x > 0$.]

(c) Can the ratio test be used to show that the series diverges? Explain.

27. Consider the series

$$\sum_{k=1}^{\infty} a_k = \frac{1}{2} + \frac{1}{3} + \frac{1}{2^2} + \frac{1}{3^2} + \frac{1}{2^3} + \frac{1}{3^3} + \cdots .$$

(a) Explain why $\lim_{k \to \infty} a_{k+1}/a_k$ does not exist.

(b) What does the ratio test say about the convergence of the series $\sum_{k=1}^{\infty} a_k$?

(c) Show that the series converges, and evaluate its limit. [**HINT:** Rewrite the given series as the sum of two series.]

28. (a) What does the ratio test say about the convergence of the series

$$\frac{1}{2} + \frac{1}{2} + \frac{1}{4} + \frac{1}{4} + \frac{1}{8} + \frac{1}{8} + \cdots ?$$

(b) Does the series in part (a) converge or diverge? Explain.

29. Give an example of a divergent series $\sum a_k$ such that $a_k > 0$ and $a_{k+1}/a_k < 1$ for all $k \geq 1$.

30. Use the ratio test to show that $\sum_{n=1}^{\infty} n^{-n}$ converges.

31. Use the ratio test to show that the series $\displaystyle\sum_{n=1}^{\infty} \frac{n^n}{n!}$ diverges.

In Exercises 32–35, use the comparison test to show that the series converges. Then find an upper bound on the limit of the series.

32. $\displaystyle\sum_{n=1}^{\infty} \frac{1}{n^2 + \sqrt{n}}$

33. $\displaystyle\sum_{j=0}^{\infty} \frac{1}{j + e^j}$

34. $\displaystyle\sum_{m=1}^{\infty} \frac{1}{m\sqrt{1+m^2}}$

35. $\displaystyle\sum_{k=1}^{\infty} \frac{k}{(k^2+1)^2}$

In Exercises 36–56, determine whether the series converges or diverges. If the series converges, find an upper bound on its limit. Justify your answers.

36. $\dfrac{1}{100} + \dfrac{1}{200} + \dfrac{1}{300} + \cdots$

37. $\displaystyle\sum_{n=1}^{\infty} \frac{\arctan n}{1 + n^2}$

38. $\displaystyle\sum_{m=1}^{\infty} \frac{m^3}{m^5 + 3}$

39. $\displaystyle\sum_{j=1}^{\infty} \frac{1}{100 + 5j}$

40. $\displaystyle\sum_{k=2}^{\infty} \frac{1}{k \ln k}$

41. $\displaystyle\sum_{n=1}^{\infty} \frac{1}{n \, 3^n}$

42. $\displaystyle\sum_{k=1}^{\infty} \frac{1}{\ln(10^k)}$

43. $\displaystyle\sum_{j=1}^{\infty} \frac{j}{5^j}$

44. $\displaystyle\sum_{k=0}^{\infty} \frac{k^2}{5k^2 + 3}$

45. $1 - \dfrac{1}{2} - \dfrac{1}{3} - \dfrac{1}{4} - \dfrac{1}{5} - \cdots$

46. $\displaystyle\sum_{j=1}^{\infty} \frac{j}{j^4 + j^2 - 1}$

47. $\displaystyle\sum_{n=2}^{\infty} \frac{1}{\sqrt[3]{n^2 - 1}}$

48. $\displaystyle\sum_{k=1}^{\infty} \frac{\sqrt{k}}{k^2 + k + 1}$

49. $\displaystyle\sum_{m=0}^{\infty} e^{-m^2}$

50. $\displaystyle\sum_{j=0}^{\infty} \frac{j!}{(j+2)!}$

51. $\displaystyle\sum_{n=0}^{\infty} \frac{n!}{(2n)!}$

52. $\displaystyle\sum_{m=1}^{\infty} \frac{m^3}{m^4 - 7}$

53. $\displaystyle\sum_{k=1}^{\infty} \frac{k!}{(k+1)! - 1}$

54. $\displaystyle\sum_{j=2}^{\infty} \frac{\ln j}{j^2}$

55. $\displaystyle\sum_{n=1}^{\infty} \left(\sum_{k=1}^{n} k^{-1} \right)$

56. $1 - \dfrac{1}{2} + \dfrac{1}{2} - \dfrac{1}{4} + \dfrac{1}{3} - \dfrac{1}{6} + \dfrac{1}{4} - \dfrac{1}{8} + \dfrac{1}{5} - \dfrac{1}{10} + \cdots$

In Exercises 57–63, determine whether the series converges or diverges. If the series converges, find a number N such that $n \geq N$ implies that the partial sum S_n approximates the sum of the series within 0.001. If the series diverges, find a number N such that $n \geq N$ implies that $S_n \geq 1000$, or explain why there is no such N.

57. $\displaystyle\sum_{k=0}^{\infty} \frac{1}{k^2 + 3}$

58. $\displaystyle\sum_{m=1}^{\infty} \frac{\arctan m}{m}$

59. $\displaystyle\sum_{j=2}^{\infty} \frac{3^j}{4^{j+1}}$

60. $\displaystyle\sum_{k=0}^{\infty} \frac{1}{2 + \cos k}$

61. $\displaystyle\sum_{m=2}^{\infty} \frac{\ln m}{m^3}$

62. $\displaystyle\sum_{k=0}^{\infty} \frac{k}{k^6 + 17}$

63. $\displaystyle\sum_{k=2}^{\infty} \frac{1}{k (\ln k)^5}$

64. (a) Where in the proof of the integral test (Theorem 7) is the assumption that $a(x)$ is a decreasing function used?

  (b) Suppose that the requirement that $a(x)$ be decreasing for all $x \geq 1$ is replaced by the "weaker" requirement that $a(x)$ be decreasing for all $x \geq 10$. How does this change in assumptions affect the conclusions of Theorem 7?

65. Consider the series $\displaystyle\sum_{k=1}^{\infty} \frac{1}{k!}$.

  (a) Explain why $\dfrac{1}{k!} \leq \dfrac{1}{2^{k-1}}$ for all $k \geq 1$.

  [**HINT:** $n \geq 2 \Longrightarrow 1/n \leq 1/2$.]

  (b) Show that $\dfrac{1}{k!} \leq \dfrac{1}{10! \, 10^{k-10}}$ for all $k \geq 10$.

  (c) Explain why S_{10} underestimates the limit of the series, and find a bound on the magnitude of the approximation error. [**HINT:** Use part (b) to bound R_{10}.]

66. Let $n \geq 2$ be an integer.

  (a) Explain why
  $$\frac{1}{n!} + \frac{1}{(n+1)!} + \frac{1}{(n+2)!} + \cdots$$
  $$< \frac{1}{n!} + \frac{1}{n \cdot n!} + \frac{1}{n^2 \cdot n!} + \cdots.$$

  (b) Use part (a) to show that
  $$\sum_{k=n}^{\infty} \frac{1}{k!} < \frac{1}{n!} \frac{1}{1 - \frac{1}{n}}.$$

  (c) Use part (b) to show that $\displaystyle\sum_{k=1}^{\infty} \frac{1}{k!}$ converges.

  (d) Find an integer N such that the partial sum S_N approximates the sum of the series in part (c) within 0.00001.

67. Consider the series $\displaystyle\sum_{k=1}^{\infty} \left(\frac{1}{k!} \right)^2$.

  (a) Show that this series converges.

  (b) Find an integer N such that the partial sum S_N approximates the sum of this series within 5×10^{-6}.

68. Consider the series $\displaystyle\sum_{k=1}^{\infty} \frac{e^k}{k!}$. Find an integer N such that the partial sum S_N approximates the sum of this series within 5×10^{-6}.

69. Let x be a positive real number, and let N be an integer such that $N \geq x$.

  (a) Show that $\dfrac{x^k}{k!} \leq \dfrac{x^N}{N!} \left(\dfrac{x}{N+1} \right)^{k-N}$ for all $k \geq N$.

  (b) Use part (a) to show that
  $$\sum_{k=N}^{\infty} \frac{x^k}{k!} \leq \frac{x^N}{N!} \cdot \frac{1}{1 - x/(N+1)}.$$

70. Let r be a positive number less than 1. Suppose that $\{a_k\}$ is a sequence of positive terms and that $a_{k+1}/a_k \leq r$ for all $k \geq 1$.

  (a) Show that $a_2 \leq a_1 r$ and that $a_3 \leq a_1 r^2$. (A similar argument shows that $a_{k+1} \leq a_1 r^k$.)

  (b) Use the result mentioned parenthetically in part (a) to show that $\sum_{k=1}^{\infty} a_k$ converges. [**HINT:** Use the comparison test and the formula for the sum of geometric series.]

  (c) Show that $R_n = \displaystyle\sum_{k=n+1}^{\infty} a_k \leq \frac{a_{n+1}}{1 - r}$.

71. Let $a_k = 1/k$. Then $a_{k+1}/a_k < 1$ for all $k \geq 1$. Why can't the ideas outlined in Exercise 70 be used to "prove" that the harmonic series converges?

72. Use Exercise 70 to find an integer N such that the partial sum S_N approximates the sum of the series $\sum_{n=1}^{\infty} n^2/2^n$ within 0.0005.

73. Use Exercise 70 to find an integer N such that the partial sum S_N approximates the sum of the series $\sum_{n=1}^{\infty} (n!)^2/(2n)!$ within 0.0005.

74. Let $H_n = \displaystyle\sum_{k=1}^{n} \frac{1}{k}$ be the nth partial sum of the harmonic series.

  (a) Show that $\ln(n+1) < H_n < 1 + \ln n$. [**HINT:** Use the integral test.]

  (b) Use part (a) to show that $H_N > 10$ implies that $N > 8000$.

  (c) Show that $\displaystyle\lim_{n \to \infty} \frac{H_n}{\ln n} = 1$.

  (d) Show that the sequence defined by $a_n = H_n - \ln n$ is decreasing. [**HINT:** Explain why $\ln(n+1) - \ln n = \int_n^{n+1} x^{-1} \, dx > (n+1)^{-1}$.]

  (e) Use part (d) to show that $\lim_{n \to \infty} a_n$ exists. (This limit, denoted by γ, is called **Euler's constant**; $\gamma \approx 0.57722$.)

75. Let a_n and γ be as in Exercise 74, and let

$$f(x) = \ln(x+1) - \ln x - \frac{1}{x+1}.$$

(a) Show that $f(x) = -\int_x^\infty f'(t)\,dt.$

(b) Use part (a) to show that

$$f(x) > \frac{1}{2(x+1)^2}.$$

[**HINT:** If $x > 0$, then

$$-f'(x) = \frac{1}{x(x+1)^2} > \frac{1}{(x+1)^3}.]$$

(c) Show that $a_n - \gamma = \sum_{k=n}^\infty (a_k - a_{k+1}).$

(d) Use parts (b) and (c) to show that

$$\frac{1}{2(n+1)} < a_n - \gamma < \frac{1}{2n}.$$

[**HINT:** $f(k) = a_k - a_{k+1}.]$

76. **Cauchy condensation theorem.** Let $\sum_{n=1}^\infty a_n$ be a series of positive terms such that $a_{n+1} \le a_n$ for all n.

(a) Let $m \ge 1$ be an integer. Explain why

$$2^{m-1} a_{2^m} \le a_{2^{m-1}+1} + a_{2^{m-1}+2} + \cdots + a_{2^m} \le 2^{m-1} a_{2^{m-1}}.$$

(b) Use part (a) to show that $\dfrac{1}{2} \displaystyle\sum_{k=1}^m 2^k a_{2^k} \le \sum_{k=2}^{2^m} a_k.$

(c) Use part (b) to show that if $\sum_{k=1}^\infty 2^k a_{2^k}$ diverges, then $\sum_{k=1}^\infty a_k$ diverges.

(d) Let m and n be integers such that $n \le 2^m$. Use part (a) to show that $\sum_{k=2}^n a_k \le \sum_{k=1}^m 2^{k-1} a_{2^{k-1}}.$

(e) Use part (d) to show that if $\sum_{k=0}^\infty 2^k a_{2^k}$ converges, then $\sum_{k=1}^\infty a_k$ converges.

77. Use Exercise 76 to prove that the harmonic series diverges.

78. (a) Use the properties of geometric series and Exercise 76 to prove that $\sum_{k=1}^\infty 1/k^p$ converges if $2^{1-p} < 1$ (i.e., $p > 1$) and diverges otherwise.

(b) Use Exercise 76 and part (a) to prove that the series

$$\sum_{k=2}^\infty \frac{1}{k(\ln k)^p}$$

converges if $p > 1$ and diverges otherwise.

11.4 Absolute Convergence; Alternating Series

Not-Necessarily-Positive Series

The integral, comparison, and ratio tests, as stated in the last section, apply only to *nonnegative* series. Some interesting series, however, have both positive and negative terms.

EXAMPLE 1   From numerical and graphical evidence, does the **alternating harmonic series**

$$\sum_{k=1}^\infty \frac{(-1)^{k+1}}{k} = 1 - \frac{1}{2} + \frac{1}{3} - \frac{1}{4} + \frac{1}{5} - \cdots$$

converge or diverge? To what limit?

Solution   As for any series, the question is how partial sums behave. Tabulating some of them ◀ shows a pattern:

Computed by machine, of course.

Partial Sums of $1 - 1/2 + 1/3 - 1/4 + 1/5 - \cdots$										
n	1	2	3	4	5	6	7	8	9	10	$\cdots$
S_n	1	0.500	0.833	0.583	0.783	0.617	0.760	0.635	0.746	0.646	$\cdots$
n	51	52	53	54	55	56	57	58	59	60	$\cdots$
S_n	0.703	0.684	0.702	0.684	0.702	0.684	0.702	0.685	0.702	0.685	$\cdots$

Successive partial sums seem to hop back and forth across some limiting value. Plots of partial sums (together with terms) exhibit the same pattern:

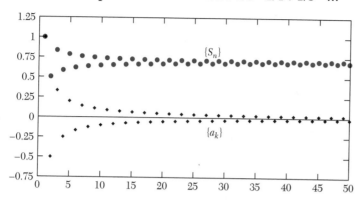

Terms and partial sums for 1 − 1/2 + 1/3 − 1/4 + 1/5 − ...

Because the terms alternate in sign, the partial sums successively rise and fall, alternately overshooting and undershooting the limiting value, which is apparently around 0.69. ▶

■ *It can be shown (with considerable effort) that the exact limit is* $\ln 2 \approx 0.69315$.

In this section we develop tools for handling such not-necessarily-positive series.

Absolute vs. Conditional Convergence

The alternating harmonic series illustrates the phenomenon of **conditional convergence**. Although

$$1 - \frac{1}{2} + \frac{1}{3} - \frac{1}{4} + \frac{1}{5} - \cdots$$

itself converges, as the preceding example leads one to expect, the *ordinary* harmonic series ▶

Obtained from the previous series by taking the absolute value of each term.

$$1 + \frac{1}{2} + \frac{1}{3} + \frac{1}{4} + \frac{1}{5} + \cdots$$

diverges, as we saw from the integral test.

EXAMPLE 2   Does $\displaystyle\sum_{k=1}^{\infty} \frac{\sin k}{k^2}$ converge? Does $\displaystyle\sum_{k=1}^{\infty} \frac{|\sin k|}{k^2}$? Estimate limits.

Solution   The first series, like the alternating harmonic series, has both negative and positive terms—although in no regular order this time. Plotting terms and partial sums suggests (but doesn't prove) that this series, too, converges:

Terms and partial sums for $\sum\limits_{k=1}^{\infty} \sin(k)/k^2$

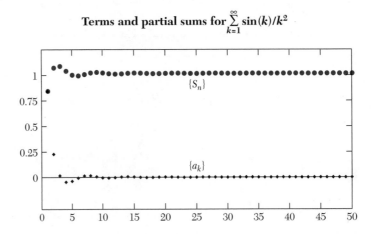

The partial sums wander up and down slightly but still appear to approach a horizontal asymptote, perhaps near $y = 1$. We'll show that this impression is correct. The second series also seems to converge, this time to some limit near 1.25:

Terms and partial sums for $\sum\limits_{k=1}^{\infty} |\sin(k)|/k^2$

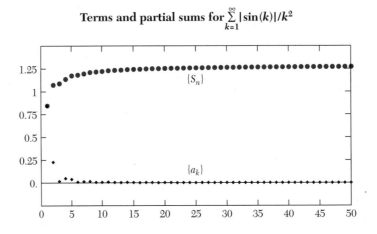

If $\sum a_k$ is the first series, $\sum |a_k|$ is the second.

The second series comes from the first by taking the absolute value of each term. ◄ The new series is nonnegative, so the comparison test applies. Since

$$0 \leq \frac{|\sin k|}{k^2} \leq \frac{1}{k^2}$$

for all $k \geq 1$ and $\sum_{k=1}^{\infty} 1/k^2$ converges, so must $\sum_{k=1}^{\infty} |\sin k|/k^2$. ■

This example illustrates the phenomenon of **absolute convergence**: Not only does the original series $\sum_{k=1}^{\infty} a_k$ converge, but so does its "absolute version" $\sum_{k=1}^{\infty} |a_k|$.

Definition Let $\sum_{k=1}^{\infty} a_k$ be any series.

- If $\sum_{k=1}^{\infty} |a_k|$ diverges but $\sum_{k=1}^{\infty} a_k$ converges, then $\sum_{k=1}^{\infty} a_k$ converges *conditionally*.
- If $\sum_{k=1}^{\infty} |a_k|$ converges, then $\sum_{k=1}^{\infty} a_k$ *converges absolutely*.

The Wacky World of Conditional Convergence.   Conditionally convergent series have some surprising properties. Here is one of the oddest:

Let $\sum a_k$ be conditionally convergent, and let L be any real number. Then the terms of $\sum a_k$ can be reordered in such a way that the resulting series converges to L.

(For more details, see your instructor.)   Notice how drastically this peculiar property of conditionally convergent series upsets the naive hope that addition is commutative.

Pluses and Minuses of Pluses and Minuses

Let $\sum_{k=1}^{\infty} a_k$ be any series. If, perchance, $a_k \geq 0$ for all k, the advantage is simplicity: The partial sums are nondecreasing. The disadvantage, as the harmonic series shows, is that the partial sums may tend to infinity.

Mixing positive and negative terms may cost something in simplicity, but it's an advantage for convergence. As the alternating harmonic series shows, positive and negative terms can offset each other, thus helping in the cause of convergence.

Absolute Convergence Implies Ordinary Convergence

We saw in Chapter 10 for *integrals* that, if $\int_1^{\infty} |f(x)|\,dx$ converges, then so must $\int_1^{\infty} f(x)\,dx$, and

$$\left| \int_1^{\infty} f(x)\,dx \right| \leq \int_1^{\infty} |f(x)|\,dx.$$

The same principle applies to infinite series.

Theorem 9   If $\sum_{k=1}^{\infty} |a_k|$ converges, then so does $\sum_{k=1}^{\infty} a_k$, and

$$\left| \sum_{k=1}^{\infty} a_k \right| \leq \sum_{k=1}^{\infty} |a_k|.$$

The idea of a rigorous proof is to write the original series as a sum of two new series, one entirely positive and the other entirely negative. Using the comparison test, one can show that each of the new series converges.

The theorem shows that, as the picture suggested, the series $\sum_{k=1}^{\infty} \sin(k)/k^2$ of Example 2 does indeed converge, because $\sum_{k=1}^{\infty} |\sin k|/k^2$ does. Our limit estimates are also consistent with the theorem:

$$1 \approx \left| \sum_{k=1}^{\infty} \frac{\sin k}{k^2} \right| \leq \sum_{k=1}^{\infty} \frac{|\sin k|}{k^2} \approx 1.25.$$

EXAMPLE 3 For which values of x does the power series

$$\sum_{k=1}^{\infty} kx^k = x + 2x^2 + 3x^3 + 4x^4 + \cdots$$

A power series is something like an "infinite polynomial." We discuss power series in the next section.

Watch the algebra.

converge? ◄

Solution First use the ratio test to check for absolute convergence. ◄

$$\lim_{k \to \infty} \left| \frac{(k+1)x^{k+1}}{kx^k} \right| = \lim_{k \to \infty} \left| \frac{k+1}{k} \right| \cdot |x| = 1 \cdot |x| = |x|.$$

If $|x| < 1$, the original series converges absolutely. (Therefore, by Theorem 9, it also converges *without* absolute value signs.)

See for yourself.

It's easy to see ◄ that, for $|x| \geq 1$, the series diverges, by the nth term test. ∎

Using the Theorem to Estimate Limits

Estimating a limit for any series—nonnegative or not—depends upon keeping the upper tail small. Theorem 9, combined with earlier estimates, can help.

We used Maple.

EXAMPLE 4 For the series $\sum_{k=1}^{\infty} (-1)^{k+1} k^{-3}$, we find ◄ that $S_{100} \approx 0.901542$. How closely does S_{100} approximate S, the true limit of the series?

Solution Since

$$S = \sum_{k=1}^{\infty} \frac{(-1)^{k+1}}{k^3} = \sum_{k=1}^{100} \frac{(-1)^{k+1}}{k^3} + \sum_{k=101}^{\infty} \frac{(-1)^{k+1}}{k^3} = S_{100} + R_{100},$$

The first inequality uses Theorem 9; the second uses the integral test.

we need only estimate R_{100}. ◄

$$|R_{100}| = \left| \sum_{k=101}^{\infty} \frac{(-1)^{k+1}}{k^3} \right| \leq \sum_{k=101}^{\infty} \frac{1}{k^3} < \int_{100}^{\infty} \frac{1}{x^3} \, dx.$$

Its lower limit is 100, not 101. Theorem 7, page 580, says why.

The last integral ◄ is easy to calculate:

$$\int_{100}^{\infty} \frac{1}{x^3} \, dx = \left. \frac{-1}{2x^2} \right]_{100}^{\infty} = \frac{1}{20,000} = 0.00005.$$

In other words, the estimate $S \approx S_{100} \approx 0.901542$ is good to at least four decimal places. ∎

Alternating Series: Convergence and Estimation

For most series with both positive and negative terms, testing for absolute convergence is usually the best option. In the special (but surprisingly useful) case that the terms alternate *strictly* in sign, we can sometimes do better.

Definition An **alternating series** is one whose terms alternate strictly in sign, i.e., a series of the form

$$c_1 - c_2 + c_3 - c_4 + c_5 - c_6 + \cdots,$$

where each c_i is positive.

The alternating harmonic series ▶

$$1 - \frac{1}{2} + \frac{1}{3} - \frac{1}{4} + \frac{1}{5} - \cdots$$

illustrates the best possibility. Because successive terms alternate in sign and decrease in size, successive partial sums straddle smaller and smaller intervals. If the terms also tend to zero, then the partial sums narrow down on a limit. ▶ The following theorem makes these observations formal and gives a convenient error bound.

See Example 1, especially the picture.

Here, more than ever, the picture is key.

Theorem 10 (Alternating Series Test)   Consider the series

$$\sum_{k=1}^{\infty}(-1)^{k+1}c_k = c_1 - c_2 + c_3 - c_4 + \cdots,$$

where

- $c_1 \geq c_2 \geq c_3 \geq \cdots \geq 0$;
- $\lim_{k \to \infty} c_k = 0$.

Then the series converges, and its limit S lies between any two successive partial sums; that is, for each $n \geq 1$, either $S_n \leq S \leq S_{n+1}$ or $S_{n+1} \leq S \leq S_n$. In particular,

$$|S - S_n| < c_{n+1}.$$

A formal proof is slightly tricky, but the underlying idea is simple. Because the limit S lies between successive partial sums, adding another term to any partial sum always "overshoots" the limit—hence the final inequality.

Although the hypotheses of this theorem seem restrictive, a surprising number of interesting series turn out to satisfy them.

Using the Theorems: Miscellaneous Examples

Combining Theorems 9 and 10 with results from earlier sections, we can handle many not-necessarily-positive series, detecting convergence or divergence and, when possible, estimating limits. The following examples illustrate some useful tricks of this trade.

EXAMPLE 5   (**An Alternating p-Series: Another Look**)   What does Theorem 10 say about $\sum_{k=1}^{\infty}(-1)^{k+1}/k^3$ and its 100th partial sum, $S_{100} \approx 0.9015422$?

Solution   In this context, $c_k = 1/k^3$. Now Theorem 10 says not only that the series converges—which we already knew—but also that

$$|S - S_{100}| < c_{101} = \frac{1}{101^3} \approx 0.000001.$$

Thus, $S_{100} \approx 0.9015422$ ▶   lies within 0.000001 of the true limit S. Equivalently, S lies between $S_{100} \approx 0.9015422$ and $S_{101} \approx 0.9015432$.   ∎

Does S_{100} overshoot or undershoot? Why?

EXAMPLE 6   **(The nth Term Test: Always Available)**

Does $\displaystyle\sum_{j=1}^{\infty} (-1)^j \frac{j}{j+1}$ converge or diverge? Why?

Solution   The alternating series test looks tempting at first glance, but it doesn't apply. The given series *is* alternating, but another important hypothesis isn't satisfied:

$$\lim_{j \to \infty} \frac{j}{j+1} = \lim_{j \to \infty} \frac{1}{1+1/j} = 1,$$

Maybe we should call it the "jth term test" here. In any case, the index name is immaterial.

not zero, as Theorem 10 requires. The nth term test ◄   does apply, however. Since $j/(j+1) \to 1$ as $j \to \infty$, it follows that $(-1)^j j/(j+1)$ has no limit as $j \to \infty$. ◄ It follows that the given series diverges.   ■

Successive terms are alternately near 1 and near −1.

EXAMPLE 7   Does the series

$$1 + 2 + 3 + 4 + 5 - \frac{1}{6} + \frac{1}{7} - \frac{1}{8} + \frac{1}{9} - \cdots$$

converge? If so, find or estimate the limit.

We studied it in Section 11.2.

Solution   The alternating series test doesn't apply right out of the box, because the first five terms break the desired pattern. The problem isn't fatal, however. Basic series algebra ◄   lets us group our terms into two blocks as follows:

$$(1 + 2 + 3 + 4 + 5) - \left(\frac{1}{6} - \frac{1}{7} + \frac{1}{8} - \frac{1}{9} + \cdots \right).$$

The first block is finite, so convergence isn't an issue; its sum is 15. The second block clearly satisfies all hypotheses of the alternating series test and so converges to some limit L. Any partial sum of the second block, moreover, differs from L by less than the magnitude of the next term (by the the last line of Theorem 10).

Adding more terms would approximate S more closely.

The entire series therefore converges to $S = 15 - L$, and any partial sum differs from S by no more than the next term. The partial sum $S_9 = 1 + 2 + \cdots + 1/9 \approx 14.962$, for instance, overshoots the true limit by less than $1/10$. In other words, the exact limit S satisfies $14.862 \le S \le 14.962$. ◄   ■

EXAMPLE 8   Does $\displaystyle\sum_{n=1}^{\infty} \frac{\sin n}{n^3 + n^2 + n + 1 + \cos n}$ converge or diverge? Why?

Solution   The problem is easier than it looks. Hoping for absolute convergence, we start by taking absolute values:

$$\left| \frac{\sin n}{n^3 + n^2 + n + 1 + \cos n} \right| = \frac{|\sin n|}{n^3 + n^2 + n + 1 + \cos n}.$$

Convince yourself of each one.

From the general appearance of numerator and denominator, comparison suggests itself. Some simple inequalities make the job much easier: ◄

$$\frac{|\sin n|}{n^3 + n^2 + n + 1 + \cos n} \le \frac{1}{n^3 + n^2 + n} \le \frac{1}{n^3}.$$

It's a p-series with $p = 3$.

We know that $\sum 1/n^3$ converges; ◄   so, by comparison, must the absolute-value version of the given series. By Theorem 9, the original series must converge, too.
  ■

BASICS

1. We showed in Example 7, page 592, that the series

$$1 + 2 + 3 + 4 + 5 - \frac{1}{6} + \frac{1}{7} - \frac{1}{8} + \frac{1}{9} - \cdots$$

converges to some limit S.
   (a) Does the series converge conditionally or absolutely? Why?
   (b) Calculate S_{15} for this series. Does S_{15} overestimate or underestimate S? How do you know?
   (c) According to *Maple*, $S_{60} \approx 14.902$. Use this result to find good upper and lower bounds for S. Explain your answer.
   (d) We said in this section that the alternating harmonic series can be shown to converge to $\ln 2$. Use this fact to find the limit of the series exactly.

2. In Example 8, page 592, we showed that the series

$$\sum_{n=1}^{\infty} \frac{\sin n}{n^3 + n^2 + n + 1 + \cos n}$$

converges to some limit S, but we didn't find or estimate S.
   (a) Compute S_{50} (not by hand!).
   (b) Explain why $|R_{50}| \leq \int_{50}^{\infty} \frac{dx}{x^3}$.
   (c) Use part (b) to give good upper and lower bounds for S.

3. (a) Suppose that the series $\sum_{k=1}^{\infty} a_k$ converges absolutely. Show that the series $\sum_{k=1}^{\infty} \frac{a_k}{k}$ converges.
   (b) Suppose that the series $\sum_{k=1}^{\infty} \frac{a_k}{k}$ converges. Must the series $\sum_{k=1}^{\infty} a_k$ also converge? Justify your answer.

4. Show that $\sum_{k=2}^{\infty} (-1)^k \frac{k}{k^2 - 1}$ converges conditionally.

In Exercises 5–10, show that the series converges. Then compute an estimate of the limit that is guaranteed to be in error by no more than 0.005.

5. $\sum_{k=1}^{\infty} \frac{(-1)^k}{k^4}$

6. $\sum_{k=1}^{\infty} \frac{(-1)^k}{k^2 + 2^k}$

7. $\sum_{k=0}^{\infty} \frac{(-2)^k}{7^k + k}$

8. $\sum_{k=0}^{\infty} \frac{(-3)^k}{(k^2)!}$

9. $\sum_{k=5}^{\infty} (-1)^k \frac{k^{10}}{10^k}$

10. $\sum_{k=0}^{\infty} \frac{(-1)^k}{(k+1) 2^k}$

FURTHER EXERCISES

11. Does the series $\sum_{n=2}^{\infty} (-1)^n \frac{n}{2n - 1}$ converge? Justify your answer.

12. Let $a_k = \int_k^{\infty} \frac{dx}{2x^2 - 1}$.
   (a) Evaluate $\lim_{k \to \infty} a_k$.
       [**HINT:** $1/2x^2 \leq 1/(2x^2 - 1) \leq 1/x^2$ if $x \geq 1$.]
   (b) Does $\sum_{k=1}^{\infty} (-1)^{k+1} a_k$ converge absolutely?
   (c) Does $\sum_{k=1}^{\infty} (-1)^{k+1} a_k$ converge?

13. (a) For which values of p does the series $\sum_{k=2}^{\infty} \frac{\ln k}{k^p}$ converge?
   (b) For which values of p does the series $\sum_{k=2}^{\infty} (-1)^k \frac{\ln k}{k^p}$ converge?
   (c) For which values of p does the series $\sum_{k=2}^{\infty} (-1)^k \frac{\ln k}{k^p}$ converge absolutely?
   (d) For which values of p does the series $\sum_{k=2}^{\infty} (-1)^k \frac{\ln k}{k^p}$ converge conditionally?

In Exercises 14–24, determine whether the series converges absolutely, converges conditionally, or diverges. If it converges, find upper and lower bounds on its limit. Justify your answers.

14. $\sum_{k=1}^{\infty} \frac{(-1)^k}{\sqrt{k}}$

15. $\sum_{j=1}^{\infty} \frac{(-1)^{j+1}}{j^2}$

16. $\sum_{n=1}^{\infty} \frac{(-3)^n}{n^3}$

17. $\sum_{k=4}^{\infty} (-1)^k \frac{\ln k}{k}$

18. $\sum_{m=8}^{\infty} \frac{\sin m}{m^3}$

19. $\sum_{n=1}^{\infty} \frac{\cos(n\pi)}{n}$

20. $\sum_{k=0}^{\infty} (-1)^k \frac{k}{2k + 1}$

21. $\sum_{m=0}^{\infty} (-1)^m \frac{m^3}{2^m}$

22. $\sum_{k=0}^{\infty} \frac{(-2)^k}{3^k + k}$

23. $\sum_{j=0}^{\infty} (-1)^j \frac{j!}{(j^2)!}$

24. $\sum_{n=1}^{\infty} (-1)^{n+1} \frac{\arctan n}{n}$

25. Consider the series $\sum_{k=1}^{\infty}(-1)^{k+1}a_k$. Suppose that the terms of the sequence $\{a_k\}$ are positive and decreasing for all $k \geq 10^9$ and $\lim_{k\to\infty} a_k = 0$ but that $a_{10^9} > a_1$. Explain why the series converges. [**HINT**: Theorem 10 doesn't apply directly.]

26. Does the infinite series
$$1 - \frac{1}{2^3} + \frac{1}{3^2} - \frac{1}{4^3} + \frac{1}{5^2} - \frac{1}{6^3} + \frac{1}{7^2} - \frac{1}{8^3} + \cdots$$
converge or diverge? Justify your answer.

27. Suppose that $\sum_{j=1}^{\infty} b_j$ converges to a number S and that $b_j \geq 0$ for all $j \geq 1$.
    (a) Show that $\sum_{j=1}^{\infty}(-1)^{j+1}b_j$ converges.
    (b) Suppose that $0 \leq b_{j+1} \leq b_j$ for all $j \geq 1$ and that the partial sum $\sum_{j=1}^{100} b_j$ approximates S within 0.005. Explain why
$$0 \leq \sum_{j=1}^{\infty}(-1)^{j+1}b_j - \sum_{j=1}^{100}(-1)^{j+1}b_j < 0.005.$$

28. Give an example of a convergent series $\sum_{k=1}^{\infty} a_k$ with the property that $\sum_{k=1}^{\infty}(a_k)^2$ diverges.

29. Suppose that $a_k \geq 0$ for all $k \geq 1$. Is it possible that $\sum_{k=1}^{\infty} a_k$ converges conditionally? Explain.

30. Suppose that $\sum_{k=1}^{\infty} a_k$ diverges. Is it possible that $\sum_{k=1}^{\infty} |a_k|$ converges? Justify your answer.

31. (a) Show that $\displaystyle\sum_{k=n+1}^{\infty} \frac{(-1)^{k+1}}{k^2} \leq \sum_{k=n+1}^{\infty} \frac{1}{k^2}$ for any $n \geq 1$.

(b) Explain why the result in part (a) implies that the series $\displaystyle\sum_{k=1}^{\infty} \frac{(-1)^{k+1}}{k^2}$ converges faster than the series $\displaystyle\sum_{k=1}^{\infty} \frac{1}{k^2}$.

32. This exercise outlines a proof of the alternating series test.

Let $S_n = \displaystyle\sum_{k=1}^{n}(-1)^{k+1}c_k$ denote the partial sum of the first n terms of a series satisfying the hypotheses of the alternating series test (Theorem 10).
    (a) Show that the sequence of even partial sums, $S_2, S_4, S_6, S_8, \ldots$, is monotone increasing.
    (b) Show that the sequence of odd partial sums, $S_1, S_3, S_5, S_7, \ldots$, is monotone decreasing.
    (c) Show that $S_{2m} \leq S_{2m-1}$ for any integer $m \geq 1$.
    (d) Use part (c) to show that the sequence of even partial sums and the sequence of odd partial sums both converge. [**NOTE**: Although both sequences converge, we must still show that they converge to the same limit.]
    (e) Show that $\lim_{m\to\infty}(S_{2m+1} - S_{2m}) = 0$. From this it follows that there is a real number S such that $\lim_{n\to\infty} S_n = S$.
    (f) Explain why $0 < S - S_{2m} < c_{2m+1}$ and $0 < S_{2m+1} - S < c_{2m+2}$.

33. Let $a_k = \dfrac{|\sin k|}{k^2}$.
    (a) Show that $\dfrac{a_{k+1}}{a_k} = |\cos 1 + \sin 1 \cot k| \cdot \dfrac{k^2}{(k+1)^2}$.
    (b) Explain why the result in part (a) implies that $\{a_k\}$ never becomes a decreasing sequence.
    (c) Show that $\displaystyle\sum_{k=1}^{\infty}(-1)^{k+1}a_k$ converges.

11.5 Power Series

Basic Ideas and Examples

A **power series** is a series of the form
$$a_0 + a_1x + a_2x^2 + a_3x^3 + a_4x^4 + \cdots + a_nx^n + \cdots = \sum_{k=0}^{\infty} a_kx^k.$$

Each a_k is a constant, called the **coefficient of x^k**. The symbol x denotes a variable. A power series may converge for some values of x and diverge for others.

EXAMPLE 1   (**A Geometric Power Series**)   Among the simplest and most useful power series is

$$S(x) = 1 + x + x^2 + x^3 + x^4 + x^5 + \cdots = \sum_{k=0}^{\infty} x^k.$$

(For all $k \geq 0$, $a_k = 1$.) For which real numbers x does $S(x)$ converge?

S o l u t i o n   Setting $x = 1$ gives the divergent ▶ series

Why is it divergent?

$$S(1) = 1 + 1 + 1^2 + 1^3 + 1^4 + \cdots = 1 + 1 + 1 + 1 + 1 + \cdots.$$

If, say, $x = 1/2$, the series converges to 2: ▶

Check the arithmetic.

$$S(1/2) = 1 + \frac{1}{2} + \frac{1}{4} + \frac{1}{8} + \cdots + \frac{1}{2^n} + \cdots = 2.$$

Indeed, for *any* value of x, $S(x)$ is a geometric series in x, so it converges if and only if $|x| < 1$. We even know the limit: ▶

Recall these properties of geometric series? Section 11.2 has details.

If $|x| < 1$, then $1 + x + x^2 + x^3 + x^4 + \cdots$ converges to $\dfrac{1}{(1-x)}$. ∎

E X A M P L E 2   It's a fact ▶ that, for any number x,

In the next section we'll see why.

$$e^x = 1 + x + \frac{x^2}{2!} + \frac{x^3}{3!} + \frac{x^4}{4!} + \cdots.$$

Interpret the right side in the language of power series.

S o l u t i o n   Writing the series in the form

$$1 + x + \frac{x^2}{2!} + \frac{x^3}{3!} + \cdots = \sum_{k=0}^{\infty} \frac{1}{k!} x^k$$

shows the pattern of coefficients.

For what values of x does the series converge?

We'll use the ratio test to decide. Since the ratio test works only for positive series, we'll use it to check for absolute convergence. For any input x,

$$\lim_{k \to \infty} \frac{|a_{k+1} x^{k+1}|}{|a_k x^k|} = \lim_{k \to \infty} \frac{|x|^{k+1}}{(k+1)!} \cdot \frac{k!}{|x|^k} = \lim_{k \to \infty} \frac{|x|}{k+1} = 0.$$

Because $0 < 1$, the ratio test guarantees that this series converges absolutely—and therefore in the ordinary sense—for *all* values of x. ▶ It can also be shown, though we haven't shown it yet, that the series converges to e^x for any x. If, say, $x = 1$, then the series converges to e, as partial sums suggest numerically:

Recall: If a series converges with absolute value signs, then it converges without. See Theorem 9, page 589.

$$S_{10} \approx 2.718281801; \qquad S_{20} \approx 2.718281828; \qquad S_{30} \approx 2.71828182845905.$$

The last number agrees with e in all 15 decimal places. ∎

Power Series and Polynomials

Power series are, roughly speaking, "unending" or "infinite-degree" polynomials. More precisely:

Terms Are Power Functions   For both polynomials and power series, each summand is of the form $a_k x^k$, with k a nonnegative integer.

Partial Sums Are Ordinary Polynomials   Every partial sum S_n of the power series $\sum_{k=0}^{\infty} a_k x^k$ has the form

$$S_n = a_0 + a_1 x + a_2 x^2 + a_3 x^3 + \cdots + a_n x^n,$$

i.e., an ordinary polynomial of degree n.

An important proviso!

Easy to Use   Polynomials are easy to differentiate and integrate, term by term. With due care taken for convergence, ◄   so are power series. We'll soon return to this theme and to its practical importance.

Choosing Base Points

The polynomial expressions

$$p(x) = x^2 - 2x + 2,$$
$$q(x) = (x-1)^2 + 1,$$
$$r(x) = (x-2)^2 + 2(x-2) + 2$$

Are you convinced? If not, multiply out q and r.

all represent the same function. ◄   The differences have to do with different choices of **base point**. Version q, for instance, is said to be **expanded about** the base point $x = 1$, because q is written in powers of $(x-1)$. Versions p and r are expanded about the base points $x = 0$ and $x = 2$, respectively.

Which version is "best" depends on the problem at hand. Version q, for instance, focuses attention most clearly on the graph's vertex, which occurs at the base point $x = 1$. Finding values and derivatives of the function at the base point is especially easy.

They're different functions this time.

The same choice of base point applies to power series. For example, the two power series ◄

$$\sum_{k=0}^{\infty} 2^k x^k \qquad \text{and} \qquad \sum_{k=0}^{\infty} 2^k (x-1)^k$$

are written in powers of x and powers of $(x-1)$, respectively; their respective base points are $x = 0$ and $x = 1$. Mathematically, the difference is small: The second form amounts only to a "shift" of one unit to the right. We'll usually, but not always, treat power series based at $x = 0$.

Power Series as Functions

Any power series

$$S(x) = \sum_{k=0}^{\infty} a_k x^k = a_0 + a_1 x + a_2 x^2 + a_3 x^3 + \cdots$$

defines, in a natural way, a *function* of x. For a given input x, $S(x)$ is the limit—if one exists—of the power series.

Domains of Power Series

For example, the natural domain of $\sqrt{x}$ is the set of nonnegative numbers.

Any function given by a "formula" in x has a natural domain: the set of x for which the formula makes sense. ◄   Power series are no different. The domain of a function $S(x)$ given by a power series is the set of inputs x for which the series converges—also known as the **interval of convergence**. We saw, for instance, that $\sum_{k=0}^{\infty} x^k$ converges for x in $(-1, 1)$; $\sum_{k=0}^{\infty} x^k / k!$ converges for x in $(-\infty, \infty)$.

EXAMPLE 3   A function $S(x)$ is defined by the power series

$$S(x) = \sum_{k=0}^{\infty} 2^k x^k = 1 + 2x + 4x^2 + 8x^3 + \cdots.$$

What's the domain of S? Is a simpler formula available?

Solution   Think of $S(x)$ as a geometric series $1 + r + r^2 + r^3 + \cdots$, with $r = 2x$: ▶

$$S(x) = 1 + 2x + (2x)^2 + (2x)^3 + \cdots.$$

In other words, $S(x)$ is "geometric in $2x$."

A geometric series converges if $|r| < 1$; *this* one converges, therefore, if $|2x| < 1$, i.e., if $|x| < 1/2$. In that case, the limit is $1/(1-r) = 1/(1-2x)$. To summarize: The power series $S(x)$ converges for x in $(-1/2, 1/2)$; on that domain,

$$S(x) = 1 + 2x + (2x)^2 + (2x)^3 + \cdots = \frac{1}{1-2x}.$$ ∎

Finding the Interval of Convergence

The first task, given a power series, is to find the interval of convergence. For many series, the ratio test is all that's needed. We illustrate with several important examples.

EXAMPLE 4   Show that the series

$$1 + 2x + 3x^2 + 4x^3 + 5x^4 + \cdots = \sum_{k=1}^{\infty} kx^{k-1}$$

converges only for x in $(-1, 1)$. Guess a limit.

Solution   We'll use the ratio test to check for *absolute* convergence. The ratio of successive terms is

$$\frac{|a_{k+1}x^{k+1}|}{|a_k x^k|} = \frac{(k+1)|x|^k}{k|x|^{k-1}} = |x|\frac{k+1}{k}.$$

As $k \to \infty$, this ratio tends to $|x|$. Therefore, the series converges—with or without absolute-value signs—if $|x| < 1$.

If $|x| = 1$, the ratio test is inconclusive. However, it's easy to see that, if $x = \pm 1$,

$$\left|kx^{k-1}\right| = |k| \to \infty \qquad \text{as} \qquad k \to \infty.$$

Thus, by the nth term test, the series diverges if $x = \pm 1$. For the same reason, the series diverges if $|x| > 1$. The convergence interval is therefore $(-1, 1)$, as claimed.

Let's guess a limit. We saw in Example 1 that the equation

$$S(x) = 1 + x + x^2 + x^3 + x^4 + \cdots = \frac{1}{1-x}$$

holds for $|x| < 1$. Differentiating all three quantities suggests a limit for the original series. If $|x| < 1$, then it's reasonable to expect that

$$S'(x) = 1 + 2x + 3x^2 + 4x^3 + \cdots = \frac{1}{(1-x)^2}.$$

If, say, $x = 1/2$, then

$$\sum_{k=1}^{\infty} \frac{k}{2^{k-1}} = 1 + \frac{2}{2} + \frac{3}{4} + \frac{4}{8} + \cdots = \frac{1}{(1 - 1/2)^2} = 4.$$

Numerical evidence suggests that we're right.   For the preceding series, $S_{20} \approx 3.999958038$.   ∎

Our guess is reasonable and correct, but it raises important questions:

- Is it legitimate to differentiate a series term by term?
- On what interval does the resulting series converge?

We address these questions in the next section.

EXAMPLE 5   Antidifferentiating the geometric series $S(x) = 1 + x + x^2 + x^3 + \cdots$ term by term gives the new series

$$T(x) = x + \frac{x^2}{2} + \frac{x^3}{3} + \frac{x^4}{4} + \frac{x^5}{5} + \cdots = \sum_{k=1}^{\infty} \frac{x^k}{k}.$$

Where does $T(x)$ converge? Guess a limit.

Check the algebraic details.

Solution   Because $S(x)$ converges (absolutely) for $|x| < 1$, it's reasonable to expect the same of $T(x)$. The ratio test agrees: ◄

$$\lim_{k \to \infty} \frac{|a_{k+1} x^{k+1}|}{|a_k x^k|} = \lim_{k \to \infty} |x| \cdot \frac{k}{k + 1} = |x|.$$

Therefore, as expected, $T(x)$ converges absolutely on the interval $(-1, 1)$.

The harmonic series and the alternating harmonic series.

What happens at the endpoints, $x = \pm 1$? Setting $x = \pm 1$ in $T(x)$ produces two by-now-familiar series: ◄

$$T(1) = \sum_{k=1}^{\infty} \frac{1}{k}; \qquad T(-1) = \sum_{k=1}^{\infty} \frac{(-1)^k}{k}.$$

As we saw earlier, the first series diverges; the second converges conditionally (by the alternating-series theorem). Thus, T converges for x in $[-1, 1)$.

Since $T(x)$ came from $S(x)$ by antidifferentiation, it's reasonable to guess a similar relationship for limits:

$$1 + x + x^2 + x^3 + \cdots = \frac{1}{1 - x} \implies x + \frac{x^2}{2} + \frac{x^3}{3} + \cdots = -\ln(1 - x).$$

Numerical evidence suggests that we're right.   If $x = 1/2$, the series gives $S_{20} \approx 0.69314714$; that's not far from $-\ln 1/2 \approx 0.69314718$.   ∎

Familiar questions arise again.

- Is it legitimate to antidifferentiate a series term by term?
- On what interval does the resulting series converge?

We address these questions in the next section.

EXAMPLE 6   Where does the power series $\sum_{k=0}^{\infty} k! x^k$ converge?

Solution Like any power series, this one converges at its base point, $x = 0$. But if $x \neq 0$, the ratio test (applied to absolute values) gives

$$\lim_{k \to \infty} \frac{\left|a_{k+1}x^{k+1}\right|}{\left|a_k x^k\right|} = \lim_{k \to \infty} \frac{(k+1)!\,|x|^{k+1}}{k!\,|x|^k} = \lim_{k \to \infty} (k+1) \cdot |x| = \infty.$$

This result amounts to a dramatic violation of the nth term test. Far from tending to zero, successive terms grow larger and larger. This power series has, in a sense, the smallest possible domain: It converges only if $x = 0$. ∎

Power Series Convergence: Lessons from the Examples

The preceding examples illustrate several useful properties of power series and their convergence sets.

The Radius of Convergence. *Every* power series $\sum_{k=0}^{\infty} a_k x^k$ converges for $x = 0$. The real question is this:

> *How far from zero can x be without destroying convergence?*

In every example so far, the convergence set turned out to be an *interval centered at 0*. ▶ This was no accident. The convergence domain for *any* power series is an interval centered at zero; ▶ its radius is called the **radius of convergence** of the power series. The following theorem guarantees all these claims.

> **Theorem 11** Let $S(x) = \sum_{k=0}^{\infty} a_k x^k$ be a power series, and let C be any real number. If $S(x)$ converges for $x = C$, then $S(x)$ also converges for $|x| < |C|$.

In the last example, the power series converged only for $x = 0$. Stretching a point slightly, we'll call the set $\{0\}$ an interval of radius 0.

In particular, the convergence interval is symmetric about the "base point."

The Idea of Proof for $C = 1$. Suppose that $\sum_{k=0}^{\infty} a_k x^k$ converges for $x = 1$. Then $\sum_{k=0}^{\infty} a_k$ converges. By the nth term test, $a_k \to 0$ as $k \to \infty$. Since the a_k's cannot blow up, they must be bounded in absolute value. ▶ In other words, there is a number $M > 0$ such that, for all k,

$$|a_k| \le M.$$

Therefore

$$\left|a_k x^k\right| \le M|x|^k$$

holds for all k. Now the comparison test applies: For $|x| < 1$, the geometric series $\sum_{k=0}^{\infty} M|x|^k$ converges, so $\sum_{k=0}^{\infty} \left|a_k x^k\right|$ must converge, too. That's what we wanted to show. ▶

If the a_k's weren't bounded, they couldn't tend to zero.

A general proof for any value of C isn't much harder.

At Endpoints, Anything Can Happen. In several of the preceding examples, the series converged on the *open* interval $(-1, 1)$; in another example, it converged on $[-1, 1)$. ▶ A series' interval of convergence may include either, both, or neither of its endpoints—any combination is possible. ▶ In practice, what really matters is a series' *radius* of convergence; what happens at the endpoints, although sometimes interesting, is usually less important.

In both cases, the radius of convergence is 1.

See the exercises at the end of this section.

Any Radius of Convergence Is Possible. In several examples, power series turned out to converge on $(-1, 1)$. Actually, any (positive) radius of convergence is possible. Indeed, for any positive constant R, the series $\sum_{k=0}^{\infty} x^k/R^k$ has radius of convergence precisely R.

Power Series Convergence, Graphically

For any $n \geq 0$, the nth partial sum of the power series $S(x) = \sum_{k=0}^{\infty} a_k x^k$ is the polynomial

$$p_n(x) = a_0 + a_1 x + a_2 x^2 + a_3 x^3 + \cdots + a_n x^n.$$

To say that the power series converges for x in $(-R, R)$ means that, for any x in that interval, there's a number $S(x)$ such that $p_n(x) \to S(x)$ as $n \to \infty$.

Sorting out precisely what this means is a worthy challenge. The following picture gives a graphical sense of the situation for the geometric power series $S(x) = \sum_{k=0}^{\infty} x^k$:

On $(-1, 1)$, $\sum\limits_{k=0}^{\infty} x^k$ converges to $1/(1-x)$

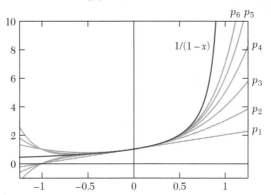

Labeled p_1 through p_6.

In the picture, the polynomial graphs ◄ represent the first six partial sums. Over the interval $(-1, 1)$, they appear to approach the graph of the limiting function more and more closely. Outside that interval, the polynomial graphs appear to diverge, rather than to approach any common limiting function.

BASICS

1. The power series $\sum_{k=1}^{\infty} x^k/k$ has radius of convergence 1. Plot the partial sum polynomials of degree 1, 2, 4, 6, 8, and 10 over the interval $[-2, 2]$. Is the interval of convergence of the series apparent? Explain.

In Exercises 2–7, find the radius of convergence of the power series.

2. $\sum\limits_{j=1}^{\infty} \left(\dfrac{x}{2}\right)^j$

3. $\sum\limits_{k=1}^{\infty} \dfrac{x^k}{k \, 2^k}$

4. $\sum\limits_{k=1}^{\infty} \dfrac{x^k}{\sqrt{k}}$

5. $\sum\limits_{m=1}^{\infty} \dfrac{x^m}{m^2 + 1}$

6. $\sum\limits_{n=1}^{\infty} n^n x^n$

7. $\sum\limits_{n=0}^{\infty} \dfrac{x^n}{n! + n}$

In Exercises 8–15, find the radius and the interval of convergence of the power series.

8. $\displaystyle\sum_{k=1}^{\infty}(3x)^k$

9. $\displaystyle\sum_{m=0}^{\infty}\frac{(3x)^m}{m!}$

10. $\displaystyle\sum_{n=1}^{\infty}\frac{(3x)^n}{n}$

11. $\displaystyle\sum_{j=1}^{\infty}\frac{(3x)^j}{j^2}$

12. $\displaystyle\sum_{n=0}^{\infty}(x-2)^n$

13. $\displaystyle\sum_{n=2}^{\infty}\frac{(x-3)^{2n}}{n^4}$

14. $\displaystyle\sum_{n=2}^{\infty}\frac{(x-5)^n}{n\ln n}$

15. $\displaystyle\sum_{n=1}^{\infty}\frac{(x+1)^n}{n}$

FURTHER EXERCISES

16. Let $R > 0$ be any positive constant.
    (a) Show that $\displaystyle\sum_{k=0}^{\infty}\frac{x^k}{R^k}$ converges on $(-R, R)$.
    (b) Show that $\displaystyle\sum_{k=1}^{\infty}\frac{x^k}{kR^k}$ converges on $[-R, R)$.
    (c) Show that $\displaystyle\sum_{k=1}^{\infty}\frac{x^k}{k^2R^k}$ converges on $[-R, R]$.
    (d) Concoct a power series that converges on $(-R, R]$.

17. For each of the following intervals, give an example of a power series that has the given interval as its interval of convergence. [**HINT**: See Exercise 16.]
    (a) $[-4, 4)$ (d) $(8, 16]$
    (b) $[-1, 5]$ (e) $[-11, -3)$
    (c) $(-4, 0)$

18. Suppose that $\sum_{n=1}^{\infty} a_n$ converges and that $|x| < 1$. Show that $\sum_{n=1}^{\infty} a_n x^n$ converges absolutely.

19. Suppose that the power series $\sum_{k=0}^{\infty} a_k x^k$ converges only if $-2 < x \le 2$.
    (a) Explain why the radius of convergence of this power series is 2.
    (b) Explain why the power series $\sum_{k=0}^{\infty} a_k(x-1)^k$ has radius of convergence 2.
    (c) Show that the interval of convergence of the power series $\sum_{k=0}^{\infty} a_k(x-3)^k$ is $(1, 5]$.
    (d) Find the interval of convergence of the power series $\sum_{k=0}^{\infty} a_k(x+1)^k$.

20. Suppose that the power series $\sum_{k=0}^{\infty} a_k(x-b)^k$ converges only if $-11 \le x < 17$.
    (a) What is the radius of convergence of the power series?
    (b) Determine the value of b.

In Exercises 21–26, find the interval of convergence (endpoint behavior too!) of the power series.

21. $\displaystyle\sum_{m=0}^{\infty}\left(\frac{x-3}{2}\right)^m$

22. $\displaystyle\sum_{j=0}^{\infty}\frac{(x-2)^j}{j!}$

23. $\displaystyle\sum_{k=1}^{\infty}\frac{(x-1)^k}{k4^k}$

24. $\displaystyle\sum_{n=1}^{\infty}\frac{(x-1)^n}{\sqrt{n}}$

25. $\displaystyle\sum_{i=1}^{\infty}\frac{(x+5)^i}{i(i+1)}$

26. $\displaystyle\sum_{m=1}^{\infty}\frac{2^m(x-1)^m}{m}$

27. Suppose that the power series $\sum_{k=0}^{\infty} a_k x^k$ converges if $x = -3$ and diverges if $x = 7$. Indicate which of the following statements *must* be true, which *may* be true, and which *cannot* be true. Justify your answers.
    (a) The power series converges if $x = -10$.
    (b) The power series diverges if $x = 3$.
    (c) The power series converges if $x = 6$.
    (d) The power series diverges if $x = 2$.
    (e) The power series diverges if $x = -7$.
    (f) The power series converges if $x = -4$.

28. Suppose that the power series $\sum_{k=0}^{\infty} a_k(x+2)^k$ converges if $x = -7$ and diverges if $x = 7$. Indicate which of the following statements *must* be true, which *may* be true, and which *cannot* be true. Justify your answers.
    (a) The power series converges if $x = -8$.
    (b) The power series converges if $x = 1$.
    (c) The power series converges if $x = 3$.
    (d) The power series diverges if $x = -11$.
    (e) The power series diverges if $x = -10$.
    (f) The power series diverges if $x = 5$.
    (g) The power series diverges if $x = -5$.

29. Consider a power series of the form $\sum_{k=0}^{\infty} a_k(x-1)^k$. Indicate whether each of the following statements *must* be true, *may* be true, or *cannot* be true. Justify your answers.
    (a) The power series converges only if $|x| > 2$.
    (b) The power series converges for all values of x.
    (c) If the radius of convergence of the power series is 3, the power series converges if $-2 < x < 4$.
    (d) The interval of convergence of the power series is $[-5, 5]$.
    (e) If the interval of convergence of the power series is $(-7, 9)$, the radius of convergence is 7.

30. The power series

$$\sum_{k=0}^{\infty} \frac{1}{k!} x^k = 1 + x + \frac{x^2}{2!} + \frac{x^3}{3!} + \cdots$$

converges, for any x, to e^x.

(a) If $x = -1$, the series converges to $1/e$. For this case, find S_{10}. By how much does S_{10} differ from the *actual* value of $1/e$? (Use a calculator or a computer.)

(b) If $x = -1$, the series is alternating. What does this say about the maximum possible error committed by S_{10} in estimating $1/e$? For which n must S_n estimate $1/e$ with error less than 10^{-10}?

31. Let $f(x) = \displaystyle\sum_{n=0}^{\infty} \frac{2x^n}{3^n + 5}$.

(a) Show that $f(10)$ is undefined (i.e., the series that defines $f(x)$ diverges when $x = 10$).

(b) Which of the numbers 0.5, 1.5, 3, and 6 are in the domain of f?

(c) Estimate $f(1)$ within 0.01 of its exact value.

32. Let $h(x) = \displaystyle\sum_{k=0}^{\infty} \frac{(x-2)^k}{k! + k^3}$.

(a) What is the domain of h (i.e., the set of x for which the series converges)?

(b) Estimate $h(0)$ within 0.005 of its exact value.

(c) Estimate $h(3)$ within 0.005 of its exact value.

33. Let $g(x) = \displaystyle\sum_{n=1}^{\infty} \frac{(x+4)^n}{n^3 5^n}$.

(a) What is the domain of g?

(b) Estimate $g(0)$ within 0.005 of its exact value.

(c) Estimate $g(-5)$ within 0.005 of its exact value.

34. (a) Evaluate $\lim_{x \to 1^-} \sum_{k=0}^{\infty} (-1)^k x^k$.

(b) Explain why the result in part (a) does *not* mean that $\sum_{k=0}^{\infty} (-1)^k$ converges.

11.6 Power Series as Functions

Any power series

$$S(x) = \sum_{k=0}^{\infty} a_k x^k = a_0 + a_1 x + a_2 x^2 + a_3 x^3 + \cdots$$

can be thought of as a function of x; its domain is the series' interval of convergence. As we saw in the last section, this domain is always an interval centered at 0. ◄   In this section we explore the remarkable—and useful—properties of functions defined by power series.

The convergence interval may or may not contain its endpoints; for most purposes, it doesn't matter much either way.

Calculus with Power Series

Given *any* power series $S(x)$, convergent or divergent, it's easy to differentiate or antidifferentiate term by term to produce new series $D(x)$ and $A(x)$:

$$S(x) = \sum_{k=0}^{\infty} a_k x^k = a_0 + a_1 x + a_2 x^2 + a_3 x^3 + \cdots ;$$

$$D(x) = \sum_{k=1}^{\infty} k a_k x^{k-1} = a_1 + 2a_2 x + 3a_3 x^2 + \cdots ;$$

$$A(x) = \sum_{k=0}^{\infty} a_k \frac{x^{k+1}}{k+1} = a_0 x + a_1 \frac{x^2}{2} + a_2 \frac{x^3}{3} + a_3 \frac{x^4}{4} + \cdots .$$

Prudence dictates a measure of caution; important questions remain to be answered.

- If the series S has radius of convergence r, is the same true of D and A?
- Even if we assume that S, D, and A all converge on $(-r, r)$, must D be the derivative of S, in the ordinary calculus sense? Is A necessarily an antiderivative of S?

The following convenient theorem answers all these questions in the affirmative. Happily, everything works just as we'd hope.

Theorem 12 (**Derivatives and Antiderivatives**) Let $S(x)$ be a power series with radius of convergence $r > 0$. Let $D(x)$ and $A(x)$ be defined as before. Then:

- Both D and A have radius of convergence r.
- For $|x| < r$, $D(x) = S'(x)$.
- For $|x| < r$, $A(x) = \int_0^x S(t)\,dt$.

The theorem says, among other things, that a function S given by a power series is differentiable and that its derivative is another power series, S', with the same radius of convergence. The same theorem applies to S', to S'', and so on, to show that S has *infinitely many* derivatives—all available by repeated term-by-term differentiation.

EXAMPLE 1   For any x, $e^x = 1 + x + \dfrac{x^2}{2!} + \dfrac{x^3}{3!} + \cdots$. Explain why.

Solution   Let $S(x)$ represent the preceding series; we saw in the last section that $S(x)$ converges for *all* x. ▶ By Theorem 12, S' can be found by differentiating S term by term. However:

This S has infinite radius of convergence.

    Differentiating S term by term leaves S unchanged.

But *every* differentiable function S for which $S' = S$ has the form $S(x) = Ce^x$. Since $S(0) = 1$, it follows that $C = 1$; hence, $S(x) = e^x$, as claimed. ∎

Writing Known Functions as Power Series

In preceding examples, including the last one, we've written various functions—$1/(1-x)$, e^x, $\ln(1-x)$, and so on—in power series form. It's natural to ask whether *other* functions can be "represented" as power series and, if so, how. Theorem 12 suggests several techniques for doing so. First, let's address an even more basic question.

Why Bother? Examples with the Sine Function

Why write a function as a power series? What good, for instance, is the equation

$$\sin x = x - \frac{x^3}{3!} + \frac{x^5}{5!} - \frac{x^7}{7!} + \frac{x^9}{9!} - \cdots,$$

which holds for all real numbers x? ▶

It is an equation; assume so for now. We'll soon give convincing reasons.

Good Family Values.   Transcendental functions—trigonometric functions, exponential functions, and so on—have no finite algebraic formulas. For such functions, power series are the next best thing. Using them, we can find very accurate, albeit approximate, values of many transcendental functions.

EXAMPLE 2   Use the preceding sine series to approximate sin 1 accurately.

Solution   Substituting $x = 1$ into the sine series gives

$$\sin 1 = 1 - \frac{1}{3!} + \frac{1}{5!} - \frac{1}{7!} + \frac{1}{9!} - \frac{1}{11!} + \cdots,$$

Rapidly!

an alternating series with terms decreasing ◄ to zero. By the alternating series theorem, any partial sum of such a series differs from the limit by no more than the next term. In particular, the partial sum $1 - 1/3! + 1/5! - \cdots - 1/11! \approx 0.8414709846$ differs from sin 1 by no more than $1/13! \approx 2 \times 10^{-10}$. In other words, our estimate is good to at least nine decimal places. This checks out numerically: $\sin 1 \approx 0.8414709848$. Not bad for so little work! ■

Hard Integrals Made Easy.   As we've seen repeatedly, many integrals, even simple-looking ones, cannot be calculated in "closed form," i.e., by elementary antidifferentiation. Numerical methods—such as the midpoint rule—offer one recourse. Infinite series, being easy to integrate, offer another way to transcend such difficulties.

EXAMPLE 3   Find, in series form, an antiderivative for $\sin(x^2)$. Use it to estimate $I = \int_0^1 \sin(x^2)\,dx$. (For comparison, the midpoint rule applied to I gives $M_{50} \approx 0.31025$.)

Solution   Even though the function $\sin(x^2)$ has no *elementary* antiderivative, it's easy to find an antiderivative in series form. Here's how.
Replacing x with x^2 in the sine series gives the new series

$$\sin(x^2) = x^2 - \frac{x^6}{3!} + \frac{x^{10}}{5!} - \frac{x^{14}}{7!} + \cdots.$$

(Because the original series converges for all x, so does this one.) Antidifferentiating term by term gives a new power series:

$$\int_0^x \sin(t^2)\,dt = \frac{x^3}{3} - \frac{x^7}{7 \cdot 3!} + \frac{x^{11}}{11 \cdot 5!} - \frac{x^{15}}{15 \cdot 7!} + \cdots.$$

It converges for all x, by Theorem 12.

The new series ◄ is not an elementary function, but it's a perfectly honest antiderivative for $\sin(x^2)$. We can therefore find our definite integral in the obvious way:

$$\int_0^1 \sin(x^2)\,dx = \frac{x^3}{3} - \frac{x^7}{7 \cdot 3!} + \frac{x^{11}}{11 \cdot 5!} - \frac{x^{15}}{15 \cdot 7!} + \cdots \Big]_0^1$$

$$= \frac{1}{3} - \frac{1}{7 \cdot 3!} + \frac{1}{11 \cdot 5!} - \frac{1}{15 \cdot 7!} + \cdots.$$

The alternating series theorem applies to the last series, so the estimate

$$\int_0^1 \sin(x^2)\,dx \approx \frac{1}{3} - \frac{1}{7 \cdot 3!} + \frac{1}{11 \cdot 5!} - \frac{1}{15 \cdot 7!} \approx 0.3102681578$$

is in error by no more than $1/19 \cdot 9! \approx 1.5 \times 10^{-7}$ (the size of the next term). The result agrees with the midpoint rule estimate through four decimal places. ■

New Series from Old: Help from Algebra and Calculus

For many familiar functions, power series can be found by simple algebra or calculus operations, starting from a few standard known series. Differentiating the sine series, for instance, gives (thanks to Theorem 12)

$$\cos x = 1 - \frac{x^2}{2!} + \frac{x^4}{4!} - \frac{x^6}{6!} + \cdots.$$

The new series, like the old, converges for all x.

We can also start with another famous series,

$$\frac{1}{1-x} = 1 + x + x^2 + x^3 + x^4 + x^5 + \cdots,$$

which converges for $|x| < 1$. With a little algebraic ingenuity we can produce many other useful series, all converging on the same set. Replacing x with $-x$, for instance, gives the alternating series

$$\frac{1}{1+x} = 1 - x + x^2 - x^3 + x^4 - x^5 + \cdots.$$

Replacing x with x^2 gives *another* alternating series:

$$\frac{1}{1+x^2} = 1 - x^2 + x^4 - x^6 + x^8 - x^{10} + \cdots.$$

Integrating *this* series term by term gives an even more striking result:

$$\arctan x = x - \frac{x^3}{3} + \frac{x^5}{5} - \frac{x^7}{7} + \frac{x^9}{9} - \cdots.$$

Setting $x = 1$ in this last series yields a remarkable and beautiful result, one that has been discovered and rediscovered by some of history's greatest mathematicians:

$$\frac{\pi}{4} = 1 - \frac{1}{3} + \frac{1}{5} - \frac{1}{7} + \frac{1}{9} - \cdots.$$

(A little caution is needed, however. These simple arguments show only that the series for $\arctan x$ is valid if $-1 < x < 1$. Showing carefully that the series converges to $\pi/4$ when $x = 1$ requires further argument.) ▶

But it's true!

Multiplying Power Series.   Convergent power series can be multiplied together, something like polynomials, to form new convergent series. As always with series, convergence is a question. Here's the answer: The product of two power series converges wherever both factors converge. We illustrate with an example.

EXAMPLE 4   We already showed that

$$\frac{1}{1-x} = 1 + x + x^2 + x^3 + \cdots$$

and

$$\frac{1}{1+x} = 1 - x + x^2 - x^3 + \cdots.$$

Multiply these series. Where does the new series converge? What familiar function does the result represent?

Solution   Symbolically, the problem looks like this:

$$(1 + x + x^2 + x^3 + \cdots) \cdot (1 - x + x^2 - x^3 + \cdots) = a_0 + a_1 x + a_2 x^2 + a_3 x^3 + \cdots.$$

We want numerical values for the constants on the right.

Both factors have infinitely many summands, so ordinary expansion quickly gets out of hand. To avoid this, we collect like powers right from the start. It's clear, for instance, that $a_0 = 1 \cdot 1 = 1$; no other combination of factors yields a constant result. Similarly, tracking the first and second powers of x gives

$$a_1 = 1 \cdot (-1) + 1 \cdot 1 = 0; \qquad a_2 = 1 \cdot 1 + 1 \cdot (-1) + 1 \cdot 1 = 1.$$

Try the next term or two for yourself. There's no substitute for experience.

Continuing this process ◄   quickly produces a simple pattern:

$$(1 + x + x^2 + x^3 + \cdots) \cdot (1 - x + x^2 - x^3 + \cdots) = 1 + x^2 + x^4 + x^6 + \cdots.$$

The result is a geometric series in powers of x^2; it converges for $|x^2| < 1$, i.e., if $-1 < x < 1$.

What function does the product series represent? Since the two factors represent the functions $1/(1 - x)$ and $1/(1 + x)$, it follows that the product *series* represents the product *function* $1/(1 - x^2)$.   ■

The next example concerns yet another useful descendant of the geometric series.

EXAMPLE 5   Find a power series for $\ln(1 + x)$; use it to estimate $\ln 1.5$ with error less than 0.0001.

Solution   Integrating the geometric series gives

$$\int \frac{1}{1 + x} \, dx = \int \left(1 - x + x^2 - x^3 + \cdots \right) dx$$

$$= x - \frac{x^2}{2} + \frac{x^3}{3} - \frac{x^4}{4} + \cdots = \ln(1 + x).$$

(We used $C = 0$ as the constant of integration, because our "target" function $\ln(1+x)$ has the value 0 when $x = 0$.) The new series, like the old, converges for x in $(-1, 1)$.

To estimate $\ln 1.5$, we plug $x = 0.5$ into our series expression:

$$\ln 1.5 = 0.5 - \frac{0.5^2}{2} + \frac{0.5^3}{3} - \frac{0.5^4}{4} + \cdots = \sum_{k=1}^{\infty} (-1)^{k+1} \frac{0.5^k}{k}.$$

Now the alternating series theorem applies. To achieve our target accuracy, any partial sum S_n for which

$$\frac{0.5^{n+1}}{n + 1} < 0.0001$$

will do. It's easy to see that $n = 10$ works, with room to spare. In fact, $S_{10} \approx 0.405435$; this compares favorably with the "exact" value $\ln 1.5 \approx 0.405465$.   ■

A Brief Atlas of Power Series

For ease of reference, here is a short list of "standard" power series for basic calculus functions. Each series appears with a suitable interval of convergence. In the next section we develop additional tools for showing rigorously that the limits are as stated.

A Power Series Sampler		
Function	**Series**	**Convergence Interval**
$\sin x$	$x - \dfrac{x^3}{3!} + \dfrac{x^5}{5!} - \dfrac{x^7}{7!} + \dfrac{x^9}{9!} - \cdots$	$(-\infty, \infty)$
$\cos x$	$1 - \dfrac{x^2}{2!} + \dfrac{x^4}{4!} - \dfrac{x^6}{6!} + \dfrac{x^8}{8!} - \cdots$	$(-\infty, \infty)$
$\exp x$	$1 + x + \dfrac{x^2}{2!} + \dfrac{x^3}{3!} + \dfrac{x^4}{4!} + \dfrac{x^5}{5!} + \cdots$	$(-\infty, \infty)$
$\dfrac{1}{1-x}$	$1 + x + x^2 + x^3 + x^4 + x^5 + \cdots$	$(-1, 1)$
$\dfrac{1}{1+x}$	$1 - x + x^2 - x^3 + x^4 - x^5 + \cdots$	$(-1, 1)$
$\dfrac{1}{1+x^2}$	$1 - x^2 + x^4 - x^6 + x^8 - x^{10} + \cdots$	$(-1, 1)$
$\arctan x$	$x - \dfrac{x^3}{3} + \dfrac{x^5}{5} - \dfrac{x^7}{7} + \dfrac{x^9}{9} - \cdots$	$[-1, 1]$

What's Next? A Power Series for Any Function

As we've seen, knowing a power series expression for one function can lead, via various manipulations, to power series versions of related functions. A good question remains:

> *Given any function f, how can we find a power series "from scratch," without knowing a related series to begin with?*

We answer this question in the next section.

BASICS

1. Let $f(x) = \displaystyle\sum_{k=0}^{\infty} \left(\dfrac{x}{2}\right)^k$.

   (a) What is the radius of convergence of the power series for f?

   (b) According to Theorem 12, $f'(x) = \displaystyle\sum_{k=1}^{\infty} \dfrac{kx^{k-1}}{2^k}$. What is the radius of convergence of the series for f'?

   (c) According to Theorem 12, $F(x) = \displaystyle\sum_{k=0}^{\infty} \dfrac{x^{k+1}}{(k+1)2^k}$ is an antiderivative of f. What is the radius of convergence of the series for F?

In Exercises 2–5, use the power series representation of the $(1-x)^{-1}$ to produce a power series representation of the function f.

2. $f(x) = \dfrac{x^2}{1+x}$

3. $f(x) = \dfrac{1}{1-x^2}$

4. $f(x) = \dfrac{1}{(1+x)^2}$

5. $f(x) = \dfrac{x}{1-x^4}$

In Exercises 6–9, find a power series representation of the function and the radius of convergence of this power series. Then plot the function and the fifth-order polynomial that is a partial sum of the power series on the same axes. [**HINT**: Write out the first few terms of the series before trying to find the form of the general term.]

6. $f(x) = \arctan(2x)$

7. $f(x) = \cos(x^2)$

8. $f(x) = x^2 \sin x$

9. $f(x) = \ln\left(1 + \sqrt[3]{x}\right)$

10. Let $f(x) = \ln(1+x)$. Show that the power series for f and f' have the same radius of convergence but not the same interval of convergence.

11. Use the partial sum of a series to estimate $1/\sqrt{e}$ with an error less than 0.005.

12. Use the partial sum of a series to estimate $\int_0^{0.2} xe^{-x^3}dx$ with an error less than 10^{-5}.

13. Evaluate $\sum_{n=1}^{\infty} n/2^n$ exactly. [**HINT:** If $f(x) = \sum_{n=0}^{\infty} x^n$, then $f'(x) = \sum_{n=1}^{\infty} nx^{n-1}$.]

14. Use power series to show that $\lim_{x \to 0} \dfrac{(\sin x - x)^3}{x(1 - \cos x)^4} = -\dfrac{2}{27}$.

15. Show that $\lim_{x \to 0^+} \dfrac{x - \sin x}{(x \sin x)^{3/2}} = \dfrac{1}{6}$.

FURTHER EXERCISES

In Exercises 16–25, use power series to evaluate the limit. Check your answer using l'Hôpital's rule.

16. $\lim_{x \to 0} \dfrac{\sin x}{x}$

17. $\lim_{x \to 0} \dfrac{e^x - 1}{x}$

18. $\lim_{x \to 0} \dfrac{1 - \cos x}{x}$

19. $\lim_{x \to 0} \dfrac{1 - \cos x}{x^2}$

20. $\lim_{x \to 0} \dfrac{\arctan x}{x}$

21. $\lim_{x \to 0} \dfrac{e^x - e^{-x}}{x}$

22. $\lim_{x \to 0} \dfrac{\ln(1+x) - x}{x^2}$

23. $\lim_{x \to 0} \dfrac{x - \arctan x}{x^3}$

24. $\lim_{x \to 1} \dfrac{\ln x}{x - 1}$

25. $\lim_{x \to 0} \dfrac{1 - \cos^2 x}{x}$ [**HINT:** $1 - \cos^2 x = \left(1 - \cos(2x)\right)/2$.]

In Exercises 26–35, find a power-series representation of the function and the radius of convergence of this power series.

26. $f(x) = \dfrac{1}{2+x}$

27. $f(x) = \sin\left(\sqrt{x}\right)$

28. $f(x) = \sin x + \cos x$

29. $f(x) = 2^x = e^{x \ln 2}$

30. $f(x) = \ln\left(1 + x^2\right)$

31. $f(x) = (x^2 - 1) \sin x$

32. $f(x) = \ln\left(\dfrac{1+x}{1-x}\right)$

33. $f(x) = \cos^2 x = \frac{1}{2}\left(1 + \cos(2x)\right)$

34. $f(x) = \dfrac{5+x}{x^2 + x - 2} = \dfrac{2}{x - 1} - \dfrac{1}{x + 2}$

35. $f(x) = \sin^3(x) = \frac{1}{4}\left(3 \sin x - \sin(3x)\right)$

36. Show that $\dfrac{1}{x - 1} = \sum_{k=1}^{\infty} \dfrac{1}{x^k}$ if $|x| > 1$.
    [**HINT:** $1/(x - 1) = x/(x - 1) - 1$.]

37. (a) Use the formula for the sum of a geometric series to show that
$$\sum_{k=1}^{\infty} \dfrac{x^k}{k} = -\ln|1 - x|.$$
    (b) What is the interval of convergence of the series in part (a)?

(c) Show that if $N \geq 1$, then
$$0 < \ln 2 - \sum_{k=1}^{N} \dfrac{1}{k\, 2^k} \leq \dfrac{1}{(N+1)2^N}.$$
    [**HINT:** $-\ln(1/2) = \ln 2$.]

38. Use a power series to show that $x - \frac{1}{2}x^2 < \ln(1+x) < x$ for all x in the interval $(0, 1)$.

39. Use power series to show that $1 - \cos x < \ln(1 + x) < \sin x$ if $0 < x < 1$.

40. (a) Does the series $\sum_{n=1}^{\infty} \sin(1/n)$ converge? Justify your answer.
    (b) Does the series $\sum_{n=1}^{\infty} \dfrac{1}{n} \sin(1/n)$ converge? Justify your answer.

41. (a) Does the series $\sum_{n=1}^{\infty} e^{-1/n}$ converge? Justify your answer.
    (b) Does the series $\sum_{n=1}^{\infty} \left(1 - e^{-1/n}\right)$ converge? Justify your answer.

42. Use the fact that $\int_0^{\infty} t^n e^{-t}\, dt = n!$ to show that
$$\int_0^{\infty} e^{-t} \sin(xt)\, dt = \dfrac{x}{1 + x^2} \qquad \text{if } |x| < 1.$$

43. Let $I = \int_0^{\infty} \dfrac{xe^{-x}}{1 - e^{-x}}\, dx$.
    (a) Show that I is a convergent improper integral.
    (b) Use the subsitution $u = 1 - e^{-x}$ to show that
$$I = -\int_0^1 \dfrac{\ln(1 - u)}{u}\, du.$$
    (c) Use part (b) and the series representation of $\ln(1-u)$ to show that $I = \pi^2/6$. [**HINT:** $\sum_{k=1}^{\infty} k^{-2} = \pi^2/6$.]

44. Let $f(x) = \dfrac{1}{1 + x^4}$.
    (a) Find a power series representation of f.
    (b) What is the interval of convergence of the series in part (a)?
    (c) Use the series found in part (a) to evaluate $\int_0^{0.5} f(x)\, dx$ with an error no greater than 0.001.

45. (a) Find the power series representation of an antiderivative of e^{-x^2}.
    (b) Use the result from part (a) to estimate $\int_0^1 e^{-x^2}\,dx$ within 0.005 of its exact value.

46. Estimate $\int_0^1 \cos\left(x^2\right)\,dx$ with an error no greater than 0.005.

47. Estimate $\int_0^1 \sqrt{x}\sin x\,dx$ with an error no greater than 0.001.

48. Estimate $\displaystyle\int_0^{1000} \frac{\exp^{-10x}\sin x}{x}\,dx$ with an error no greater than 5×10^{-5}.

In Exercises 49–54, find the first four nonzero terms in the power series representation of the function.

49. $f(x) = e^{2x}\ln(1+x^3)$

50. $f(x) = \arctan x \sin(4x)$

51. $f(x) = \dfrac{e^x}{1-x}$

52. $f(x) = \tan x = \dfrac{\sin x}{\cos x}$

53. $f(x) = e^{\sin x}$

54. $f(x) = \ln(\cos x)$

55. Use the fact that $\cos x = -\cos(x-\pi)$ to write $\cos x$ as a series in powers of $x - \pi$.

56. Determine the coefficients a_k such that

$$\frac{1}{1-x} = \sum_{k=0}^{\infty} a_k(x-2)^k.$$

[**HINT:** $\dfrac{1}{1-x} = -\dfrac{1}{1+(x-2)}$.]

In Exercises 57–60, find the elementary function represented by the power series by manipulating a more familiar power series (e.g., the series for $\cos x$, $\sin x$, $(1-x)^{-1}$).

57. $\displaystyle\sum_{k=1}^{\infty} kx^{k-1}$

58. $\displaystyle\sum_{k=0}^{\infty} \frac{x^k}{(k+1)!}$

59. $\displaystyle\sum_{k=1}^{\infty}(-1)^{k+1}x^k$

60. $\displaystyle\sum_{k=1}^{\infty} \frac{(2x)^k}{k}$

61. **A proof that e is irrational.** Assume that $e = m/n$, where m and n are positive integers.
    (a) Explain why

$$m!\left| \frac{1}{e} - \sum_{k=0}^{m} \frac{(-1)^k}{k!} \right| \le \frac{m!}{(m+1)!} = \frac{1}{m+1}.$$

    (b) Explain why $m!/e$ is an integer.
    (c) Explain why $m!\displaystyle\sum_{k=0}^{m} \frac{(-1)^k}{k!}$ is an integer.
    [**HINT:** Start by explaining why $m!/k!$ is an integer if k is an integer and $0 \le k \le m$.]

(d) Parts (a)–(c) imply that

$$N = m!\left| \frac{1}{e} - \sum_{k=0}^{m} \frac{(-1)^k}{k!} \right|$$

is an integer that is less than or equal to $1/(m+1)$. Explain why it follows that $N = 0$.

(e) Explain why the conclusion of part (d) is impossible and therefore e cannot be a rational number. [**HINT:** $\sum_{k=m+1}^{\infty}(-1)^k/k! \ne 0$.]

62. Use power series to show that $y = e^x$ is a solution of the differential equation $y' = y$.

63. Use power series to show that $y = 2e^x$ is the solution of the initial value problem $y' = y$, $y(0) = 2$.

64. Use power series to show that $y = e^{3x}$ is the solution of the initial value problem $y' = 3y$, $y(0) = 1$.

65. Use power series to show that $y = \sin x$ is a solution of the differential equation $y'' = -y$.

66. Use power series to show that $y = (1-x)^{-1}$ is the solution of the initial value problem $y' = y^2$, $y(0) = 1$.

67. (a) Show that $f(x) = \tan x$ is the solution of the initial value problem $f'(x) = 1 + \left(f(x)\right)^2$, $f(0) = 0$.
    (b) Use part (a) to find the first four nonzero terms in the power series representation of $\tan x$.

68. Let r be a fixed number, and define the function f by

$$f(x) = 1 + \sum_{n=1}^{\infty} \frac{r(r-1)(r-2)\cdots(r-n+1)}{n!} x^n.$$

    (a) Show that the series defining f converges if $|x| < 1$.
    (b) Show that $(1+x)f'(x) = rf(x)$.
    (c) Let $g(x) = (1+x)^{-r}f(x)$. Show that $g'(x) = 0$.
    (d) Show that part (c) implies that $f(x) = (1+x)^r$.
    [**NOTE:** The power series for f is known as the **binomial series**.]

69. Use the binomial series defined in Exercise 68 to show that

$$\sqrt{1+x} \approx 1 + \frac{1}{2}x - \frac{1}{8}x^2 + \frac{1}{16}x^3$$
$$- \frac{5}{128}x^4 + \frac{7}{256}x^5 \mp \cdots.$$

In Exercises 70–73, use the series in Exercise 68 to find the first four nonzero terms of a power series representation of the given function.

70. $f(x) = \left(1+x^4\right)^3$

71. $f(x) = \sqrt[3]{1-x^2}$

72. $f(x) = \left(1+x^2\right)^{-3/2}$

73. $f(x) = \arcsin x$.

74. Use the partial sum of a series to estimate $\int_0^{0.4} \sqrt{1+x^3}\,dx$ with an error less than 5×10^{-4}.

11.7 Maclaurin and Taylor Series

In the preceding section we saw some of the practical advantages of writing a function as a power series. We also saw how to use a power series for one function to derive power series for related functions, using algebra, calculus, and other devices. In this section we show, given a suitable function, how to find its power series "from scratch," i.e., without starting from a related series. In the process we draw important connections among a function, its derivatives, and its series. First, however, a note of caution.

Does Every Function Have a Power Series?

Mathematical life would be simpler if *every* function $f(x)$ could be written as a power series—ideally, a series that converges to $f(x)$ for all x. Alas, it isn't so. At least two things can go wrong:

Smaller Domains A power series may have a smaller domain than the function it represents. For instance, the series equation

$$\frac{1}{1+x^2} = 1 - x^2 + x^4 - x^6 + x^8 - x^{10} + \cdots$$

holds if—but only if—$|x| < 1$, even though $f(x) = 1/(1+x^2)$ is defined (and well behaved) for *all* real numbers x.

No Series at All A function may have no series at all. Theorem 12 in the preceding section showed that every power series can be differentiated again and again on its interval of convergence. ◄   Thus, any function that *has* a power series must itself be repeatedly differentiable at $x = 0$. This fact rules out functions such as $f(x) = |x|$, which is continuous everywhere but not differentiable at $x = 0$.

Each derivative is another power series.

Despite these cautions, many important functions *can* be written as power series.

Coefficients and Derivatives at Zero

A simple equation relates the coefficients and the derivatives at zero of a power series. If

$$S(x) = a_0 + a_1x + a_2x^2 + a_3x^3 + a_4x^4 + \cdots ,$$

We can differentiate repeatedly, by Theorem 12 of the preceding section.

then, simplest of all, $S(0) = a_0$. Differentiating S repeatedly ◄   gives

$$S(x) = a_0 + a_1x + a_2x^2 + a_3x^3 + a_4x^4 + \cdots ;$$

$$S'(x) = a_1 + 2a_2x + 3a_3x^2 + 4a_4x^3 + \cdots ;$$

$$S''(x) = 2a_2 + 6a_3x + 12a_4x^2 + \cdots ;$$

$$S'''(x) = 6a_3 + 24a_4x + \cdots$$

$$\vdots$$

These equations show that

$$S(0) = a_0; \quad S'(0) = a_1; \quad S''(0) = 2a_2; \quad S'''(0) = 6a_3; \quad \ldots$$

In general, the following holds true.

Fact If $S(x) = \sum_{k=0}^{\infty} a_k x^k$, then $a_k = \dfrac{S^{(k)}(0)}{k!}$ for all $k \geq 0$.

This important fact connects coefficients and derivatives; knowing either, we can find the other. We illustrate one of the uses of the fact with an example.

EXAMPLE 1 Assuming (for now) that $f(x) = \sin x$ *has* a power series, *find* that power series.

Solution To find the desired series $f(x) = a_0 + a_1 x + a_2 x^2 + a_3 x^3 + \cdots$, we need "only" find the coefficients $a_0, a_1, a_2, a_3, \ldots$.

In principle, finding infinitely many of anything sounds difficult. In practice, it's often easy; for the sine function, as for many functions of interest, the coefficients follow a simple pattern. The first few derivatives of f reveal the pattern:

$$a_0 = \frac{f(0)}{0!} = \sin 0 = 0; \qquad\qquad a_1 = \frac{f'(0)}{1!} = \cos 0 = 1;$$

$$a_2 = \frac{f''(0)}{2!} = \frac{-\sin 0}{2} = 0; \qquad a_3 = \frac{f'''(0)}{3!} = \frac{-\cos 0}{6} = -\frac{1}{6};$$

$$a_4 = \frac{f^{(4)}(0)}{4!} = \frac{\sin 0}{4!} = 0; \qquad a_5 = \frac{f^{(5)}(0)}{5!} = \frac{\cos 0}{5!} = \frac{1}{5!}.$$

Therefore, written as a series,

$$\sin x = x - \frac{x^3}{3!} + \frac{x^5}{5!} - \frac{x^7}{7!} + \cdots.$$

It's not hard to show that this series converges for *all* x, as we'd hope. ▶

Better yet, the series converges to $\sin x$. We haven't proved this fact yet, but numerical evidence is readily available. If, say, $x = 1$, $S_5 \approx 0.8414710096$, while $\sin 1 \approx 0.8414709848$. ∎

But it's beside the main point, here.

Maclaurin Series

Using the preceding fact as a recipe for the coefficients, we can write a power series for any function f that has repeated derivatives at $x = 0$. ▶ Here's the formal definition, named for the 17th-century Scottish mathematician Colin Maclaurin:

*The **zeroth derivative** of f is f itself.*

Definition (**Maclaurin Series**) Let f be any function with infinitely many derivatives at $x = 0$. The **Maclaurin series** for f is the series $\sum_{k=0}^{\infty} a_k x^k$, with coefficients given by

$$a_k = \frac{f^{(k)}(0)}{k!}, \qquad k = 0, 1, 2, \ldots.$$

Finding Maclaurin Series: Help from Technology

Computing the necessary derivatives by hand to find a Maclaurin series can be tedious and error-prone. Fortunately, *Maple* and other software programs can find any partial sum of the Maclaurin series quickly and easily. (Such sums are called either **Maclaurin polynomials** or **Taylor polynomials**.) Here's how the process works (using one version of *Maple*) for the sine function:

```
> taylorpoly( sin(x), x=0, 7 );
```

$$x - 1/6\; x^3 + 1/120\; x^5 - 1/5040\; x^7$$

The output is the *seventh-degree Taylor polynomial for* $\sin x$, based at $x = 0$.

Taylor Series: Expansion About $x = a$

A Maclaurin series

$$\sum_{k=0}^{\infty} \frac{f^{(k)}(0)}{k!} x^k$$

is said to be expanded about $x = 0$, because all derivatives of f are calculated there. This choice, although often convenient, isn't necessary. A similar series can be found by expanding about *any* point $x = a$. The **Taylor series** for f, expanded about $x = a$, has the form

$$\sum_{k=0}^{\infty} \frac{f^{(k)}(a)}{k!} (x - a)^k.$$

For some functions, expansion about a point other than zero is convenient.

We'll let Maple *help.*

See for yourself.

EXAMPLE 2   The function $f(x) = \ln x$ isn't defined at $x = 0$. Expand f about $x = 1$. ◄

Solution   Derivatives of $f(x) = \ln x$ are easy for either a human or a machine to calculate. ◄ Here are several:

$$f'(x) = \frac{1}{x}; \qquad f''(x) = -\frac{1}{x^2}; \qquad f'''(x) = \frac{2}{x^3};$$

$$f^{(4)}(x) = -\frac{6}{x^4}; \qquad f^{(5)}(x) = \frac{24}{x^5}; \qquad f^{(6)}(x) = -\frac{120}{x^6}.$$

At $x = 1$, therefore,

$$f'(1) = 1; \qquad f''(1) = -1; \qquad f'''(1) = 2;$$

$$f^{(4)}(1) = -6; \qquad f^{(5)}(1) = 24; \qquad f^{(6)}(1) = -120.$$

The general pattern is now visible. The Taylor series for the function $\ln x$, expanded about $x = 1$, has the form

$$\sum_{k=1}^{\infty} \frac{(-1)^{k+1}}{k} (x - 1)^k = (x - 1) - \frac{(x - 1)^2}{2} + \frac{(x - 1)^3}{3} - \frac{(x - 1)^4}{4} + \cdots.$$

For the record, the computer agrees (to seventh degree, anyway):

```
> taylorpoly(ln(x),x=1,7);
                2                  3                4
  x - 1 - 1/2 (x - 1)  + 1/3 (x - 1)  - 1/4 (x - 1)
                  5                  6                7
      + 1/5 (x - 1) - 1/6 (x - 1)  + 1/7 (x - 1)
```

Converging to the Right Place: Taylor's Theorem

Any function f that's infinitely differentiable at $x = 0$ has a Maclaurin series—ideally, one with a large radius of convergence. One possible problem remains: The series might converge at x, but perhaps to a limit other than $f(x)$.

Taylor's theorem guarantees that this unfortunate event almost never occurs. In the bargain, it lets us predict in advance how closely a Maclaurin polynomial approximates the "target" function f.

Theorem 13 (**Taylor's Theorem**) Suppose that f is repeatedly differentiable on an interval I containing 0. Let $\sum_{k=0}^{\infty} a_k x^k$ be the Maclaurin series of f, and let

$$P_n(x) = \sum_{k=0}^{n} a_k x^k = a_0 + a_1 x + a_2 x^2 + \cdots + a_n x^n$$

be the nth degree Maclaurin polynomial. Suppose also that for all x in I,

$$\left| f^{(n+1)}(x) \right| \leq K_{n+1}.$$

Then

$$|f(x) - P_n(x)| \leq \frac{K_{n+1}}{(n+1)!} |x|^{n+1}.$$

Notice the following features.

- Taylor's theorem estimates the *error* committed by $P_n(x)$ in estimating $f(x)$. Unless K_{n+1} grows very quickly with n, this error tends to 0 as $n \to \infty$, so the series converges to $f(x)$.

- The theorem involves familiar ingredients. Like our numerical integral-error estimates, this one depends on bounding a higher derivative of the function in question.

- A rigorous proof of Taylor's theorem is beyond the scope of this book. The underlying ideas, however, are based firmly on the mean value theorem. Indeed, Taylor's theorem is sometimes thought of as a general form of the mean value theorem. We pursued this idea briefly at the end of Section 4.3, in deriving error bounds for linear approximation.

EXAMPLE 3   For $f(x) = \sin x$, show that the Maclaurin series converges to $\sin x$ for any value of x.

Solution   For $f(x) = \sin x$, *all* derivatives are sines or cosines or their opposites. Thus, for any n, the inequality

$$\left| f^{(n+1)}(x) \right| \le 1$$

holds for all x. By Taylor's theorem,

$$|P_n(x) - \sin x| \le \frac{1 \cdot |x|^{n+1}}{(n+1)!}.$$

The last quantity tends to 0 as n tends to infinity, so the series converges, for any x, to $\sin x$. The following graphs show what it means, geometrically, for the Maclaurin polynomials to converge to $\sin x$:

Graphs of $\sin x$, $P_1(x)$, $P_3(x)$, $P_5(x)$, $P_7(x)$

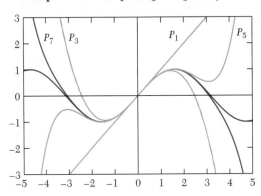

■

BASICS

1. Let $f(x) = x^4 - 12x^3 + 44x^2 + 2x + 1$.
   (a) Find the Maclaurin series representation of f.
   (b) Find the Taylor series representation of f expanded about $x = 3$.

2. Let $p(x) = (1 + x)^n$, where n is a positive integer. Explain why the Maclaurin series for p is the polynomial p itself. [**HINT**: Use Theorem 13.]

3. Let $f(x) = \int_3^x \sqrt{t}\, e^{-t}\, dt$.
   (a) Show that if $x \approx 3$, then

   $$f(x) \approx \sqrt{3}e^{-3}(x-3) - \frac{5}{12}\sqrt{3}e^{-3}(x-3)^2$$
   $$+ \frac{23}{216}\sqrt{3}e^{-3}(x-3)^3.$$

   (b) Use Theorem 13 to bound the error made when $f(3.5)$ is estimated using the polynomial in part (a).

4. Let $f(x) = \sqrt{1+x}$.
   (a) Find the first three nonzero terms in the Maclaurin series for f.

   (b) Use Theorem 13 to bound the approximation error made if the Maclaurin polynomial from part (a) is used to estimate $f(1)$.

5. Explain why the power series

   $$1 - x + \frac{x^2}{2} - \frac{x^4}{8} + \frac{x^5}{15} - \frac{x^6}{240} + \cdots$$

   cannot be the Maclaurin series representation of the function f shown in the graph.

   Graph of f

   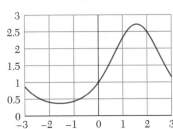

6. (a) Compute the quadratic Maclaurin polynomial for
$f(x) = \sqrt{1+x}$.

   (b) The electrical potential V at a distance r along the axis perpendicular to the center of a uniformly charged disk of radius a and charge density σ is

$$V = 2\pi\sigma\left(\sqrt{r^2 + a^2} - r\right).$$

If r is large in comparison to a, then $a/r \approx 0$. Use part (a) to find an approximation for V.

FURTHER EXERCISES

7. Find numbers a_0, a_1, a_2, a_3, and a_4 such that
$$x^4 - 4x^3 + 5x = a_0 + a_1(x-1) + a_2(x-1)^2 \\ + a_3(x-1)^3 + a_4(x-1)^4.$$

8. Suppose that f is a function such that $f(0) = 1$ and $f'(x) = 1 + \left(f(x)\right)^{10}$ for all x. Find the first four terms of the Maclaurin series for f.

9. Suppose that f is a function that is positive, increasing, and concave down on the interval $[-2, 2]$.
   (a) Is the coefficient of x^2 in the Maclaurin series representation of f positive? Justify your answer.
   (b) Let $g(x) = 1/\sqrt{1 + f(x)}$. Is the coefficient of x^2 in the Maclaurin series representation of g positive? Justify your answer.

10. Let $f(x) = e^{2x}$. What is the coefficient of x^{100} in the Maclaurin series representation of f?

11. Let $f(x) = \dfrac{x}{1 - x^3}$.
    (a) Find the Maclaurin series for $f(x)$.
    (b) What is the interval of convergence of the series in part (a)?
    (c) Use part (a) to find a power series for $f''(x)$.
    (d) Use part (a) to find the Maclaurin series for $\int_0^x f(t)\, dt$.

12. (a) Find the Maclaurin series representation of
$$f(x) = \frac{1}{2 + x}.$$
    [**HINT:** $\frac{1}{2+x} = \frac{1}{2} \cdot \frac{1}{1 + (x/2)}$.]
    (b) Find $f^{(259)}(0)$ exactly.

13. Let
$$f(x) = \begin{cases} \dfrac{1 - \cos x}{x^2} & \text{if } x \neq 0, \\ \dfrac{1}{2} & \text{if } x = 0. \end{cases}$$
Evaluate $f^{(100)}(0)$.

14. Suppose that f is a function such that $f(1) = 1$, $f'(1) = 2$, and $f''(x) = (1 + x^3)^{-1}$ for $x > -1$.
    (a) Estimate $f(1.5)$ using a quadratic Taylor polynomial.
    (b) Find an upper bound on the approximation error made in part (a).

15. Let $f(x) = \begin{cases} x^{-1}\sin x & \text{if } x \neq 0, \\ 1 & \text{if } x = 0. \end{cases}$
    (a) Find the Maclaurin series representation of f.
    (b) What is the interval of convergence of the power series found in part (a)?
    (c) Use the series in part (a) to estimate $f'''(1)$ with an error no greater than 0.005.

16. During an encounter with a friendly extraterrestrial, it is revealed to you that the answer to life, the universe, and everything is $f(1)$ for some function f. It is also revealed that $f(0) = 26$, $f'(0) = 22$, $f''(0) = -16$, $f'''(0) = 12$, and $\left|f^{(4)}(x)\right| \leq 7x^4$ if $|x| \leq 2$.
    (a) Find an upper bound on the value of $f(1)$.
    (b) Find a lower bound on the value of $f(1)$.

17. Use Theorem 13 to show that the Maclaurin series for e^x converges to e^x for all x.

18. Use Theorem 13 to show that the Maclaurin series for $1/(1+x)$ converges to $1/(1+x)$ if $-1/2 < x < 1$. [**HINT:** Consider the cases $-1/2 < x < 0$ and $0 \leq x < 1$ separately.]

19. Suppose that f is a function such that $f^{(n)}$ exists for all $n \geq 1$.
    (a) Explain why the Maclaurin series for f converges to f if $\left|f^{(n)}(x)\right| \leq n$ for all $n \geq 1$.
    (b) Does Theorem 13 guarantee that the Maclaurin series for f converges to f if $\left|f^{(n)}(x)\right| \leq 2^n$ for all $n \geq 1$?

20. Let $f(x) = \begin{cases} e^{-1/x^2} & \text{if } x \neq 0, \\ 0 & \text{if } x = 0. \end{cases}$
    (a) Use the definition of the derivative (and l'Hôpital's rule) to show that $f'(0) = 0$.
    (b) Using methods similar to those in part (a), it can be shown that $f^{(k)}(0) = 0$ for all integers $k \geq 0$. Use this fact to find the Maclaurin series for f.
    (c) What is the radius of convergence of the series in part (b)?
    (d) For which values of x does the series in part (b) converge to $f(x)$?

21. Use the trigonometric identity $\cos(x + y) = \cos x \cos y - \sin x \sin y$ to find the Taylor series representation for $\cos x$ expanded about $a = \pi/3$.

22. Find the Taylor series representation for $\sin x$ expanded about $a = \pi/4$.

23. (a) Use the binomial series (see Exercise 68 in Section 11.6) to show that

$$\arcsin x = \int_0^x \frac{dt}{\sqrt{1 - t^2}}$$

$$= x + \sum_{n=1}^{\infty} \frac{1 \cdot 3 \cdot 5 \cdots (2n - 1)}{2 \cdot 4 \cdot 6 \cdots (2n)} \frac{x^{2n+1}}{2n + 1}.$$

(b) If $n \geq 1$ is an integer, then

$$\int_0^{\pi/2} \sin^{2n+1} x \, dx = \frac{2 \cdot 4 \cdot 6 \cdots (2n)}{3 \cdot 5 \cdot 7 \cdots (2n + 1)}.$$

Use this fact to evaluate $\int_0^1 \frac{x^{2n+1}}{\sqrt{1 - x^2}} dx$.

(c) Use parts (a) and (b) to show that

$$\int_0^1 \frac{\arcsin x}{\sqrt{1 - x^2}} dx = \sum_{k=0}^{\infty} \frac{1}{(2k + 1)^2}.$$

(d) Use the substitution $u = \arcsin x$ to show that

$$\int_0^1 \frac{\arcsin x}{\sqrt{1 - x^2}} dx = \frac{\pi^2}{8}.$$

(e) Use parts (c) and (d) to show that

$$\sum_{k=1}^{\infty} \frac{1}{k^2} = \frac{\pi^2}{6}.$$

24. Let f be a function that has continuous derivatives on an interval containing a and x.
    (a) Explain why $f(x) = f(a) + \int_a^x f'(t) \, dt$.
    (b) Use part (a) and integration by parts to show that

$$f(x) = f(a) + f'(a)(x - a) - \int_a^x (t - x) f''(t) \, dt$$

$$= f(a) + f'(a)(x - a) + \int_a^x (x - t) f''(t) \, dt.$$

[**HINT:** Let $dv = dt$ and $v = t - x$. (This is legitimate since x is not the integration variable.)]
    (c) Use part (b) and integration by parts to show that

$$f(x) = f(a) + f'(a)(x - a) + \frac{1}{2} f''(a)(x - a)^2$$

$$+ \frac{1}{2} \int_a^x (x - t)^2 f'''(t) \, dt.$$

(d) Use part (c) and integration by parts to show that

$$f(x) = f(a) + f'(a)(x - a) + \frac{1}{2} f''(a)(x - a)^2$$

$$+ \frac{1}{3!} f'''(a)(x - a)^3$$

$$+ \frac{1}{3!} \int_a^x (x - t)^3 f^{(4)}(t) \, dt.$$

(e) Show that repeated integration by parts leads to this alternative form of Taylor's theorem:

$$f(x) = f(a) + f'(a)(x - a) + \frac{1}{2} f''(a)(x - a)^2$$

$$+ \frac{1}{3!} f'''(a)(x - a)^3$$

$$+ \cdots + \frac{1}{n!} f^n(a)(x - a)^n$$

$$+ \frac{1}{n!} \int_a^x (x - t)^n f^{(n+1)}(t) \, dt.$$

25. Let $R_n(x) = \frac{1}{n!} \int_a^x (x - t)^n f^{(n+1)}(t) \, dt$. (This is the integral form of the remainder in Taylor's theorem derived Exercise 24.) Show that if $a = 0$ and $\left| f^{(n+1)}(t) \right| \leq K_{n+1}$ for all t, then

$$|f(x) - P_n(x)| = |R_n(x)| \leq \frac{K_{n+1}}{(n + 1)!} |x|^{n+1}.$$

26. Let $f(x) = (1 + x)^r$. In this exercise we prove that the binomial series (see Exercise 68 in Section 11.6) converges to f.
    (a) Use part (e) of Exercise 24 to show that $R_n(x) = f(x) - P_n(x)$ is

$$R_n(x) = \frac{r \cdot (r - 1) \cdot (r - 2) \cdots (r - n)}{n!} \int_0^x \frac{(x - t)^n}{(1 + t)^{n+1-r}} dt.$$

(b) Suppose that $0 \leq t \leq x < 1$. Show that $\frac{|x - t|}{1 + t} \leq |x|$.
    (c) Use part (b) to show that if $0 \leq x < 1$, then

$$|R_n(x)| \leq \frac{|(r - 1) \cdot (r - 2) \cdots (r - n)|}{n!} x^n |(1 + x)^r - 1|.$$

(d) Use part (c) to show that the binomial series converges to f if $0 \leq x < 1$.
    (e) Adapt the reasoning in parts (b)–(d) to show that the binomial series converges to f if $-1 < x < 0$.

11.8 Chapter Summary

In this long and relatively technical chapter, we studied the sophisticated topic of infinite series. Infinite series are formed by adding—in a special sense, and with due concern for convergence—infinitely many numbers or functions. Making sense of such infinite summation requires special care—hence the subtlety of the subject.

Convergence.   An infinite series converges if its partial sums, formed from only finitely many terms, converge as a sequence. Unfortunately, the sequence of partial sums, being so defined, is often hard to understand or handle directly. For **geometric series**, however, partial sums and limits are easy to calculate.

Convergence Tests.   Among the first questions to ask about any series is whether it converges or diverges. As with improper integrals, the answer may be far from obvious at a glance. To address this problem, several tests for convergence and divergence were discussed, including the **integral test**, the **comparison test**, and the **ratio test**. All these tests apply, in raw form, only to series of positive terms. Series that contain both positive and negative terms need special care. Given that care, they too can be tested for convergence.

Estimating Limits.   Many infinite series, even ones known to converge, are difficult to evaluate exactly. The problem of estimating their limits therefore arises naturally. As with integration, technology proves helpful, both for calculating partial sums and for estimating the error committed in using partial sums to approximate series.

Power Series: Series as Functions.   Power series, or "infinite polynomials," are among the most useful and convenient functions of calculus. Moreover, many calculus functions can be written in power series form. Finding such a power series and understanding its functional properties are the subjects of the last two sections of the chapter.

Taylor and Maclaurin Series; Taylor's Theorem.   Taylor's theorem—one of the key theorems of calculus—guarantees that under appropriate conditions a function can be written in series form, usually as a Maclaurin series. The partial sums of a Maclaurin series, being polynomials, are easy to handle, and they offer useful approximations to less convenient functions.

12

Differential Equations

12.1 Differential Equations: The Basics

Ideas, Definitions, and Examples: A Quick Review

(Basic ideas, definitions, and terminology of differential equations were introduced in Chapter 4. In this subsection we present only a quick summary. For more details, definitions, and examples, see Section 4.1.)

DEs and Solutions to DEs

A **differential equation (DE)** is any equation that involves at least one derivative. Each of the equations

$$y' = 0; \qquad y' = y; \qquad y' = 2xy; \qquad \frac{dP}{dt} = \frac{K}{\sqrt{P}}; \qquad y'' = -y$$

is a differential equation.

This bears repeating: Solutions of differential equations are functions, not numbers.

A **solution** of a differential equation is any *function* that satisfies the differential equation. ◄ As the following example illustrates, a DE normally has *many* solution functions. To solve a DE means to find at least one—but preferably all—of these solution functions.

EXAMPLE 1 Find solutions of the DE $y' = y$.

Solution A solution of the DE $y' = y$ is a function $y(x)$ that is *its own derivative*. The exponential function $y(x) = e^x$ is famous for just this reason:

$$y = e^x \implies y' = e^x = y.$$

Thus, $y(x) = e^x$ is a solution of the DE.

C may be positive, negative, or zero.

In fact, *every* function of the form $y(x) = Ce^x$, where C is a constant, ◄ is a solution of the same DE, because

$$y = Ce^x \implies y' = Ce^x = y.$$

This DE, in other words, has an *infinite family of solutions*. Several of these solution functions are shown as follows; each is labeled with its value of C.

Solutions to $y' = y$: $y = Ce^x$ for various values of C

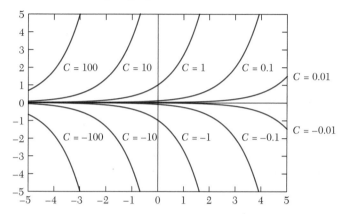

Each graph represents one solution of the DE $y' = y$. Not all solutions are shown, of course. By choosing C correctly, we can find a solution curve passing through *any* point of the xy-plane. (Approximately what value of C would cause the solution curve to pass through the point $(2, 2)$, for example?) ■

IVPs vs. DEs

A typical DE has infinitely many solution functions. Adding an **initial condition** (i.e., a specified "starting value") to a DE usually lets us choose a *single* desired solution. The combination of an initial condition and a DE is called an **initial value problem (IVP)**. The next example illustrates this idea.

EXAMPLE 2   Find a function $y = y(x)$ that satisfies both the DE $y' = y$ and the initial condition $y(0) = -1$.

Solution   We already saw that *every* function of the form $y(x) = Ce^x$ satisfies the DE. By choosing C judiciously, we'll select the one solution that satisfies both the DE and our initial condition. Here's how:

$$-1 = y(0) = Ce^0 = C \Longrightarrow C = -1.$$

Thus, $y = -1e^x$ is our desired solution—and this time, the *only* solution function.

The curve $y = -1e^x$ is among those shown in Example 1. Notice especially that this curve passes through the point $(0, -1)$—as it should, because $y(0) = -1$. ■

DEs and IVPs: Vocabulary and Notation

Every area of mathematics develops its own convenient technical terms and notations. The following list contains several such terms and shorthand forms related to DEs and IVPs.

Order   The **order** of a DE refers to its highest-order derivative. For example, $y' = y$ is a **first-order DE**; $y'' = 0$ is a **second-order DE**.

Omitting Arguments   Consider this typical calculus sentence:

If $f(x) = e^x$, then f is its own derivative.

The function f is mentioned twice—first as $f(x)$ with the **argument** x, then simply as f, without an argument. The terminology of DEs allows the same option. For instance, the two DEs $y'(t) = y(t)$ and $y' = y$ mean the same thing; each describes a function that is its own derivative. Whether to include or to omit the argument (t, in this case) depends on the situation. By convention, DEs are usually written without explicit

For typographical convenience, perhaps.

function arguments. ◄ Omitting understood arguments simplifies a DE's appearance, but it can take some getting used to. The DE

$$y' = y + t,$$

for example, relates variables y and t; y is understood to denote some (unknown) function $y(t)$.

Variable Names   In written DEs, the symbol t, rather than x, often denotes the input variable. This choice is made because the letter t suggests time. A solution y, also known as $y(t)$, is then naturally thought of as a quantity that varies with time. But this is only an aid to intuition. DEs often model physical situations that have nothing to do with time.

DEs, IVPs, and Graphs

One way to interpret the derivatives in DEs and IVPs is as *rates of growth*. From this point of view, a solution to a first-order IVP is a function $y(t)$ that (i) "grows" at a prescribed rate $y'(t)$ (given by the DE) and (ii) has a prescribed value $y(t_0)$ at a prescribed time t_0 (given by the initial condition).

Derivatives can also be thought of in terms of *slope*: For any input x, $y'(x)$ is the slope of the y-graph at the point (x, y). (Because time is no longer of the essence, we revert to using x as the input variable.) From this geometric point of view, a first-order DE prescribes slopes of solution curves, while an initial condition $y(x_0) = y_0$ prescribes a particular point through which a solution curve should pass.

EXAMPLE 3   Interpret the DE $y' = y$ in terms of slope.

Solution   Here's what it means, geometrically, for a function $y(x)$ to satisfy the DE $y' = y$:

At any point (x, y), the y-graph has slope y.

In Example 1, page 618, we plotted several solution functions for this DE. A close look at the picture shows that each of these curves does indeed have the claimed slope property at each of its points.

Consider, for instance, the curve $y = e^x$ as it passes through the point $P = (\ln 2, 2) \approx (0.693, 2)$. Our claim is that the slope at P is 2. To estimate this slope graphically, we can zoom in on P, the point shown bulleted in the following graph.

Zooming in on $y = e^x$:
why $y' = y$ **at** $(\ln 2, 2)$

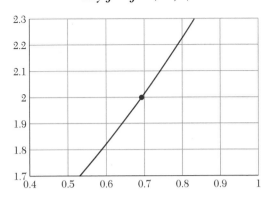

The picture reveals what we expect: At $P = (\ln 2, 2)$, $y = y' = 2$.

Although the picture shows just one solution curve, ▶ *all* solution curves have the same property: At every point (x, y), the solution curve through (x, y) has slope y. ■

It happens to be the solution curve that satisfies $y(0) = 1$

Initial Conditions and Families of Curves. Typically, as we've seen, a first-order DE has infinitely many solutions. Geometrically, the solutions form a family of curves, with one curve passing through any given point in some part of the plane. Geometrically, an initial condition amounts to choosing the one solution curve that passes through a particular point. That curve may be of special interest.

EXAMPLE 4   All five of the following curves are solutions to a certain DE, of the form $y' = k(y - 70)$. ▶ They describe the different temperature "histories" of several forgotten cups of coffee in a 70-degree room.

More details on this DE appear in the exercises at the end of this section.

Five forgotten cups of coffee: how they "cool"

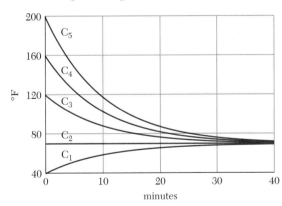

minutes

How long does it take a cup of coffee, initially at 200 degrees, to cool to 160 degrees? When does the temperature reach 120 degrees?

Solution   The initial condition, put symbolically, says that $y(0) = 200$. Only C_5 passes through $(0, 200)$, so it's the solution we want. A look at the picture shows that coffee drops from 200 degrees to 160 degrees in about 3 minutes and to 120 degrees in about 7 minutes.   ∎

Checking Guesses

This section is mainly about the idea and meaning of differential equations, not the mechanics of solving them. (We'll consider the problem of finding solutions soon.) It's worth remarking now that, although *finding* solutions to DEs can be difficult, *checking* candidates for solutions is usually quite easy. This pattern should be no surprise; we've already seen the closely related fact that checking antiderivatives is much easier, as a rule, than finding them. The next example illustrates this principle.

> **Antiderivatives and DEs.**   Finding antiderivatives and solving DEs are closely re-lated problems. Indeed, antiderivatives are nothing more than special types of DEs. The antiderivative statement
>
> $$\int \cos x \, dx = \sin x + C,$$
>
> for example, means, in DE language, that every function of the form $f(x) = \sin x + C$ is a solution of the DE
>
> $$y' = \cos x.$$
>
> More generally, any antiderivative expression $\int f(x)\,dx$ is equivalent to the DE $y' = f(x)$.
>   Some DEs are readily translated into antiderivative problems. DEs with this property are called **separable**; we'll study some in a later section. On the other hand, not all DEs can be solved by antidifferentiation. The DE $y' = \sin y + t^2$, for instance, has no convenient antiderivative version. But all is not lost. As we did with antiderivatives that we couldn't solve symbolically, we will describe graphical and numerical methods for solving almost any first-order DE approximately.

E X A M P L E 5   Is $y = \sin x$ a solution to the second-order DE $y'' = -y$? Is $y = 2\sin x - 3\cos x$ another solution? Why or why not?

Do you agree?

S o l u t i o n   The answer is yes in both cases, as direct calculations show. Here are details for the latter solution: ◄

$$y = 2\sin x - 3\cos x \implies y' = 2\cos x + 3\sin x$$

$$\implies y'' = -2\sin x + 3\cos x = -y.$$

(A similar calculation shows that $y = a\sin x + b\cos x$ solves the DE for *any* constants a and b. Setting $a = 1$ and $b = 0$ shows that $y = \sin x$ is another solution.)   ∎

E X A M P L E 6   Verify that $y = \sqrt{x}$ solves the IVP

$$y' = \frac{1}{2y}; \qquad y(1) = 1.$$

S o l u t i o n   Direct calculation is enough. If $y = \sqrt{x}$, then

$$y' = \frac{1}{2\sqrt{x}} = \frac{1}{2y} \qquad \text{and} \qquad y(1) = \sqrt{1} = 1.$$ ∎

BASICS

1. For any function of the form $y = Ce^x$, $y(0) = Ce^0 = C$ and $y'(0) = Ce^0 = C$.
   (a) What do these conditions mean about the graphs of such functions?
   (b) Graphs of several functions of the form $y = Ce^x$ appear in Example 1. How do the graphs there exhibit the properties you identified in part (a)?

2. Look again at the cooling coffee graphs on page 621. The curves labeled C_1 through C_5 represent five different solutions to a DE of the form
   $$y' = k(y - 70).$$
   (a) Use any convenient graph to estimate the value of k. [**HINT**: Choose any convenient point on a graph; at that point, estimate both the slope (y') and the height (y). Use your results and the DE to solve for k.]
   (b) For each of the curves C_1 through C_5, write an IVP of which the graph is the solution. [**HINT**: Look at each graph to choose an appropriate initial condition.]

3. Verify that $y = x^3$ is a solution of the DE $xy' = 3y$.

4. Verify that $y = 1 - e^{-2x}$ is a solution of the DE $y' + 2y = 2$.

5. A small-town charity fund drive aims to raise $65,000; updated current totals are posted in the town square. According to Alfred E. Neuman's law of cooling of enthusiasm, the rate at which people contribute to such a drive is proportional to the difference between the current total and the announced target amount.

   Let $y(t)$ represent the current total, in thousands of dollars, t weeks after the start of the drive.
   (a) Does Neuman's law of cooling sound reasonable? Why or why not?
   (b) Express Neuman's law of cooling as a DE.
   (c) Why is the name "Neuman's law of cooling" appropriate?

6. It's a fact that for any constants k, T, and A, the function $y(t) = T + Ae^{kt}$ is a solution of the DE $y' = k(y - T)$.
   (a) Show by differentiation that $y(t) = T + Ae^{kt}$ solves the DE $y' = k(y - T)$.

7. Show that $y(x) = x^{-1} - 1$ is a solution of the differential equation $x^3 y'' + x^2 y' - xy = x$.

8. Consider the DE $xy'' - (2x + 1)y' + (x + 1)y = 0$.
   (a) Verify that $y = e^x$ is a solution of this DE.
   (b) Verify that $y = x^2 e^x$ is a solution of this DE.
   (c) Use parts (a) and (b) to show that $3e^x - 7x^2 e^x$ is a solution of this DE.

9. Verify that $y = 1 + C_1 \cos x + C_2 \sin x$ is a solution of the DE $y'' + y = 1$ for any values of the constants C_1 and C_2.

10. Find real numbers a and b so that $y = e^x$ and $y = e^{2x}$ are both solutions of the differential equation $ay'' + by' + y = 0$.

11. Find functions $f(x)$ and $g(x)$ so that $y = x$ and $y = x^2$ are both solutions of the differential equation $f(x)y'' + g(x)y' + y = 0$.

12. Consider the function f shown here.

Graph of f

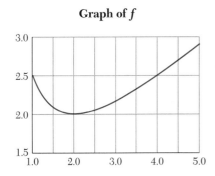

$y = f(x)$ is a solution of one of the following differential equations. Decide which one. Then explain why each of the others is not a solution.

(i) $y' = \dfrac{y - x}{x}$ (ii) $y' = \dfrac{x - y}{x}$ (iii) $y' = \dfrac{x^2 - y}{x}$

(b) Consider the fund drive in Exercise 5. Assume that the drive starts with a $10,000 gift and that after 5 weeks the total is $45,000. How long does it take to raise $60,000? $64,900?

12.2 Slope Fields: Solving DEs Graphically

All derivatives can be understood in terms of slope. The derivatives in DEs and IVPs are no exception. As usual in calculus, the geometric viewpoint is vital to understanding what DEs and IVPs are really "about." Interpreting DEs graphically also leads to very simple but powerful techniques for finding *approximate* solutions of almost any first-order DE or IVP. In the last section, we alluded briefly to the graphical view of DEs; now we take up the subject in more detail.

DEs and Slopes of Solution Curves

Any first-order DE can be understood as a statement about slopes of solution curves. Doing so is simplest if the DE is given in the form

$$y' = \text{an expression involving } y, t, \text{ or both},$$

as is each of the following DEs.

$$y' = y; \qquad y' = 2t - 5; \qquad y' = t + y; \qquad y' = \sin(t^2 - 3y).$$

(Many DEs come ready-made in this form. Of those that don't, most can be rewritten to do so.)

EXAMPLE 1   Discuss the DE $y' = y$ in terms of slope.

Solution   In "slope language," the DE $y'(t) = y(t)$ means the following:

At any point (t, y) on a solution curve $y = y(t)$, the slope $y'(t)$ is equal to y.

In Example 3, page 620, we looked closely at the specific point $(\ln 2, 2)$. We found what the DE requires: at that point, the solution curve has slope 2.

The simple form of the particular DE $y' = y$ says something else about its solution curves. That y' depends *only* on y ◄ (rather than on t or on both y and t) means that all solution curves have the same slope for any given value of y. A close look at several solution curves bears this out:

There's no t on the right-hand side.

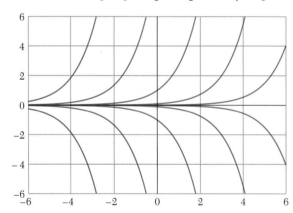

Solutions to $y' = y$: slopes depend only on y

EXAMPLE 2   Interpret the DE $y' = 2t - 5$ graphically.

Solution   The DE says, in slope-talk, that all solution curves must have slope $2t - 5$ at the point (t, y). For this simple DE, it's easy to check directly—by finding solutions—that this slope property holds as claimed. Whether by antidifferentiation or simple guessing, one sees that

$$y'(t) = 2t - 5 \iff y(t) = t^2 - 5t + C.$$

All solution curves therefore have the form $y = t^2 - 5t + C$; C may be any constant. Several such curves follow, each labeled with its C-value.

$y = t^2 - 5t + C$ **for various values of** C:
solutions to $y' = 2t - 5$

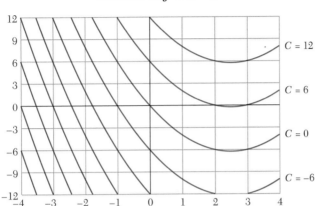

Compare these curves with those in Example 1. For the DE $y' = y$, all solution curves have the same slope at a given y-value. For the DE $y' = 2t - 5$, all solution curves have the same slope at a given t-value. This difference stems from the fact that, for the first DE, y' depends only on y; for the second DE, y' depends only on t. ■

In many DEs, such as $y' = y + t$, y' depends on *both* y and t. The next example shows what this means for solution curves.

EXAMPLE 3   Interpret the DE $y' = t + y$ graphically.

Solution   This time, the DE means that the solution curve through any point (t, y) has slope $t + y$ at that point. In particular, solution curves become steeper as *either* t or y increases. In a moment we'll plot some solution curves and see that their slopes behave this way.

But what *are* some solution functions for the current DE? In the last two examples, the DEs were simpler, and it wasn't hard to *guess* solution functions. With the current DE, guessing a solution is considerably harder. ▶ Therefore, we simply state that solutions have the form   *But don't let that stop you. Try guessing anyway.*

$$y = Ce^t - t - 1,$$

where C is any constant. As usual, *checking* that such functions are solutions is a routine exercise. ▶ Following are graphs of several solution functions, each labeled with its C-value.   *So we leave it as an exercise.*

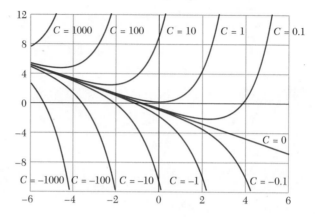

$y = Ce^t - t - 1$ **for various values of** C:
solutions to $y' = y + t$

As we see, slopes of solution curves vary both with y and with t. ■

First-Order IVPs, Graphically

A first-order IVP is a DE together with an initial condition. In symbolic garb, the package usually has the form

$$y' = f(t, y); \qquad y(a) = b,$$

where $f(t, y)$ is an expression involving y or t or both. In graphical terms, the initial condition $y(a) = b$ requires that a solution curve pass through (a, b). In all cases of interest to us, only *one* such curve exists—the solution curve for the IVP. An initial condition, in other words, lets us choose, from among all the solution curves for a DE, the single curve that solves the IVP.

EXAMPLE 4   Among the solution curves shown in Example 3, find one that solves the IVP

$$y' = y + t; \qquad y(0) = -2.$$

For this curve, find $y(2)$, $y(10)$, and $y(-10)$.

Solution   A look at the picture shows that the curve labeled $C = -1$ passes through $(0, -2)$. Therefore, $y = -1e^t - t - 1$ solves the IVP, and

$$y(2) \approx -10.4; \qquad y(10) \approx -22037.5; \qquad y(-10) \approx 8.99995.$$

(Only the first result appears on the graph. The second is far outside the window shown. The third, as the graph suggests, lies near the line $y = -t - 1$.) ■

Slope Fields

In the preceding examples, we illustrated the graphical meaning of DEs by plotting families of solution curves. Doing so requires, of course, that we know (or be told) formulas for the solution functions we plot. Another graphical approach to a first-order DE, called a **slope field**, ◄ uses only the DE itself—no solution

Some authors use the term **direction field**.

formulas are needed. Better yet, slope fields offer a simple and natural method of *approximating* solution curves.

Slope Fields and Solution Curves

A first-order DE $y' = f(t, y)$ ▶ prescribes the slope of a solution curve through the point (t, y) in the ty-plane. A slope field captures this information by showing, centered at each of many "grid points" (t, y), a short line segment with the appropriate slope. These segments, in effect, "point the way" along solution curves. We illustrate the idea of slope fields with three by-now-familiar DEs.

▶ *$f(t, y)$ can be any expression in one or more of the variables t and y, such as y, $2t - 5$, or $y + t$.*

EXAMPLE 5   Draw slope fields for the DEs $y' = y$, $y' = 2t - 5$, and $y' = y + t$. How are the results related to the pictures in Examples 1–3?

Solution   The first slope field is for the DE $y' = y$.

Slope field for the DE $y' = y$

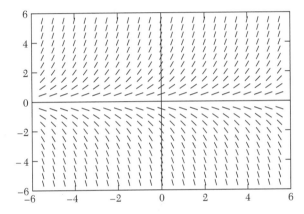

Plotting solution curves and the slope field together shows clearly how the two are related:

Slope field and solution curves for the DE $y' = y$

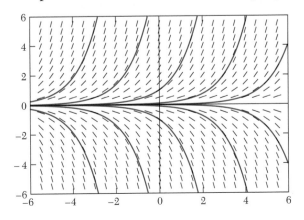

That is, the field "points along" the curve.

Solution curves quite literally "go with the flow" described by a slope field: At each point on a solution curve, the slope field is *tangent* to the curve. ◄

For the DEs $y' = 2t - 5$ and $y' = y + t$, the respective slope fields and solution curves agree in the same sense:

Slope field and solution curves for $y' = 2t - 5$

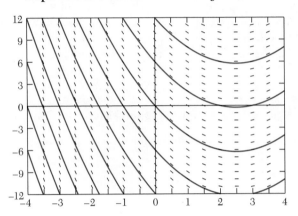

Slope field and solution curves for $y' = y + t$

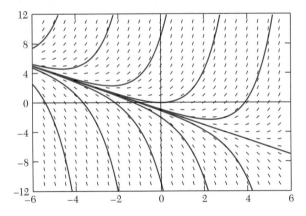

Notice, in particular, how the first slope field varies only with y, the second only with t, and the third with both y and t, as we expected. ∎

Drawing Slope Fields

Given a first-order DE $y' = f(t, y)$ and a rectangle in the ty-plane, how would a person (or a computer) *draw* a slope field? The *idea* is simple.

Step 1 Choose a "grid" of conveniently spaced points within the rectangle. (For each of the preceding pictures, we used a 20×20 grid—400 points in all.)

Short enough not to interfere with the other segments.

Step 2 At each grid point $P(t, y)$, use the DE to calculate y'. Then use the result to draw a short line segment ◄ centered at P, with slope y'.

There's a problem, of course: No sane person wants to repeat Step 2 hundreds of times. Luckily, computers thrive on such repetitive tasks. We therefore use computer-drawn slope fields except in exercises, and then only for simple DEs and "coarse" grids.

Go with the Flow: How To Solve DEs and IVPs Graphically

Solving DEs. The slope field for any first-order DE $y' = f(y, t)$ suggests a natural method of drawing solution curves: Start *anywhere* in the direction field and "go with the flow," always following where the field ticks lead. (Solution curves are usually drawn in *both* directions from the starting point, corresponding both to increasing t—the "future"—and to decreasing t—the "past.") ▶ This step produces one solution curve; to produce others, start somewhere else and repeat the process as often as desired. The result is a collection of solution curves for the given DE— one curve for each starting point. Sometimes it's possible to use such curves to guess formulas for the solution curves. Even without formulas, solution curves can be very useful in practice. ▶

"Future" and "past" should be understood metaphorically; DEs need not always involve time.

See the next example.

Solving IVPs. Given the appropriate slope field, IVPs are even easier to solve (approximately, of course) than are DEs, because only one solution curve needs to be drawn, rather than a whole family. If the IVP has the form

$$y' = f(t, y), \qquad y(a) = b,$$

for instance, then the initial condition ▶ $y(a) = b$ tells *where* in the slope field to start drawing—at the point (a, b). The resulting curve represents *the* solution to the IVP.

A good name, from this point of view.

EXAMPLE 6   Hot coffee in a 70-degree room cools at a rate proportional to the difference between the coffee temperature and room temperature. ▶ It's equivalent to say, in DE language, that $y'(t) = k(y - 70)$, where $y(t)$ denotes coffee temperature at time t. At a certain time, a thermometer showed a coffee temperature of 190 degrees, dropping at the rate of 12 degrees per minute. How much later did the temperature reach 130 degrees? How hot would the coffee have had to be initially to be 130 degrees after 10 minutes?

Newton's law of cooling says so.

Solution   Let $t = 0$ denote the "certain time" just mentioned. The facts (which were lightly disguised) that $y(0) = 190$ and $y'(0) = -12$, and the DE $y' = k(y - 70)$, let us solve for k:

$$-12 = k(190 - 70) = 120k \implies k = -0.1.$$

Putting everything together, we have the IVP

$$y' = -0.1(y - 70); \qquad y(0) = 190.$$

We chose $0 \leq t \leq 20$, $60 \leq y \leq 210$; other choices are possible.

Starting with a slope field on a convenient rectangle, ◄   we draw a plausible solution curve emanating from the initial point (0, 190).

Solving the IVP $y' = -0.1(y - 70)$; $y(0) = 190$

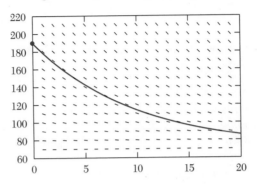

A close look at the solution curve suggests that the temperature reaches 130 degrees at around $t = 7$ minutes.

Coffee at 130 degrees at $t = 10$ suggests another IVP, this time with initial value $y(10) = 130$. The solution curve through (10, 130) looks like this:

Solving the IVP $y' = -0.1(y - 70)$; $y(10) = 130$

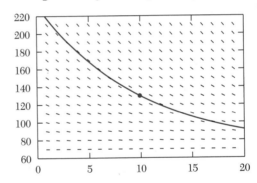

Following the curve backward in time shows the difficulty: At time $t = 0$, the coffee would have to be above 220 degrees—a physical impossibility.   ■

Does Every DE Have Solutions? How Many?

For every slope field we've seen in this section, one solution curve, and only one, passes through any given point in the plane. In other words, every first-order DE we've seen has infinitely many solution functions. Specifying an initial condition narrows the choice to just one solution function.

Geometric intuition suggests (and it can be shown) that the same principle holds for any DE of the form $y' = f(t, y)$, as long as $f(t, y)$ is a well-behaved function of t and y. Given such a DE and any starting point (a, b), the slope field near (a, b) prescribes exactly one curve through (a, b).

The fine points of this theory (some are very fine indeed) constitute the *existence and uniqueness theory of DEs*, a subject of more advanced study in mathematics.

BASICS

1. Look again at Example 1 and its accompanying graphs. It's observed there that *all solution curves for the DE* $y' = y$ *have the same slope for a given value y.*
   (a) How do the shapes of the various solution curves reflect this property? (Answer in a sentence or two.)
   (b) Five of the curves shown in Example 1 pass through the level $y = 3$. Do all five curves appear to have the same slope at that level? Use a ruler and any curve you like to *estimate* this common slope as closely as you can. (The grid should help.) Could your answer have been predicted in advance?
   (c) Repeat part (b), but at the level $y = -4$.
   (d) Repeat part (b), but at the level $y = 0$. [**HINT:** Despite appearances, only one solution curve touches the line $y = 0$.]

2. In Example 3 we claimed (but didn't show) that all functions of the form $y = Ce^t - t - 1$, where C is any constant, are solutions of the DE $y' = y + t$.
   (a) Verify this claim by differentiation.
   (b) Find one function (there is *only* one) that solves the IVP $y' = y + t$; $y(0) = 1$. What's $y(2)$?
   (c) Is the graph of the solution function you found in part (b) one of those shown in Example 3? If so, which one? If not, draw a graph of your solution function.

3. Here are graphs of several functions of the form $y = Ce^t - t - 1$, i.e., solutions of the DE $y' = y + t$. (Compare them to the graphs in Example 3.)

$y = Ce^t - t - 1$: still more solutions to $y' = y + t$

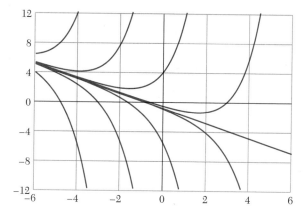

   (a) One of the "curves" is a straight line. Which straight line? Why? Label this curve with the appropriate value of C.
   (b) One of the curves passes through the point $(0, 4)$. What value of C corresponds to this curve? How

do you know? Label this curve with the appropriate value of C.
   (c) Estimate (using the graph) the slope at $(0, 4)$ of the curve mentioned in part (b). Does your answer agree with what the DE predicts?
   (d) Draw—carefully, with a ruler and a sharp pencil—the line $y + t = 0$ (also known as $y = -t$) on the axes. This line crosses four of the solution curves at what points of special interest? Why does this occur? [**HINT:** What does the DE $y' = y + t$ say about points on the line $y + t = 0$?]
   (e) Draw—carefully, with a ruler—the line $y + t = -3$ (also known as $y = -t - 3$) on the axes. This line crosses four of the solution curves. What do these crossing points have in common? Explain your answer.
   (f) Draw—carefully, with a ruler—several lines with slope -1 on the axes. Each such line crosses several solution curves. What can be said about the points at which these crossings occur? Why?
   (f) The curves shown correspond to the C-values $C = \pm 500$, $C = \pm 50$, $C = \pm 5$, and $C = \pm 0.2$. Label each curve with its appropriate C-value.

4. All solution curves shown in Example 3 and in Exercise 3 have one property in common: They seem to converge leftward toward the line $y = -t - 1$. To explain this phenomenon, show that if $y(t)$ is any solution function,
$$\lim_{t \to -\infty} \left(y(t) - (-t - 1) \right) = 0.$$

5. In Example 5 we remarked that "the first slope field varies only with y, the second only with t, and the third with both y and t."
   (a) What visual property of the second slope field corresponds to "varying only with t?" [**HINT:** The word "parallel" should come in handy.]
   (b) What visual property of the first slope field corresponds to "varying only with y?"

6. (a) Suppose that a DE has the form $y' = f(x)$. Explain how you could recognize this fact from the direction field of the DE.
   (b) Suppose that a DE has the form $y' = g(y)$. Explain how you could recognize this fact from the direction field of the DE.

7. Consider the DE $y' = x/y$.
   (a) Plot the direction field for this DE for $-3 \le x \le 3$ and $-3 \le y \le 3$.
   (b) Sketch the solution curve that satisfies the initial condition $y(0) = 1$ on the direction field from part (a).
   (c) Sketch the solution curve that satisfies the initial condition $y(1) = 0$ on the direction field from part (a).

8. Consider the DE $y' = 1 + 2xy$.

   (a) Plot the direction field for this DE for $-2 \le x \le 2$ and $-2 \le y \le 2$.

   (b) Sketch the solution curve that satisfies the initial condition $y(0) = 0.5$ on the direction field from part (a).

   (c) Sketch the solution curve that satisfies the initial condition $y(0) = -1$ on the direction field from part (a).

9. Consider the DE $y' = x + 2xy$.

   (a) Plot the direction field for this DE for $-2 \le x \le 2$ and $-2 \le y \le 2$.

   (b) Sketch the solution curve that satisfies the initial condition $y(0) = 1$ on the direction field from part (a).

   (c) Sketch the solution curve that satisfies the initial condition $y(-1) = -1$ on the direction field from part (a).

12.3 Euler's Method: Solving DEs Numerically

See Example 6, especially the pictures.

In the preceding section we showed ◄   how to solve a first-order IVP

$$y' = f(t, y), \qquad y(a) = b,$$

graphically, using its slope field:

The graph of a solution function.

> *Starting at (a, b), move along the slope field. The result is a solution curve ◄ for the IVP.*

Euler's Method: The Idea by Example

The numerical version of the same idea is known as **Euler's method**. It describes in concrete numerical terms exactly how to move, step by step, through a slope field. We illustrate with another look at the situation of Example 6, Section 12.2.

Time $t = 0$ can denote any convenient reference time.

EXAMPLE 1   Let $y(t)$ denote the temperature (in degrees Fahrenheit) of a cup of coffee at time t (in minutes). ◄   Room temperature is 70 degrees; the coffee starts at 190 degrees. The coffee's temperature is described by the IVP

$$y' = -0.1(y - 70); \qquad y(0) = 190.$$

How hot is the coffee after 10 minutes?

The bulleted starting point represents the initial condition.

Solution   Here again are the slope field and a plausible solution curve: ◄

Solving the IVP $y' = -0.1(y - 70)$; $y(0) = 190$

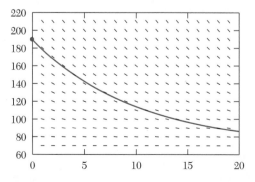

Judging from the solution curve, the temperature after 10 minutes is about 115 degrees.   ■

Drawing a solution curve by "going with the flow" is simple and natural. Any two humans, given the slope field and initial condition just shown, would draw very similar ▶ solution curves and thus arrive at similar estimates for $y(10)$.

But not identical—no two humans draw exactly alike.

This graphical approach, for all its virtues, has two main flaws: (1) It gives inexact answers, and (2) "drawing a curve" isn't precisely defined as a mathematical process. The second flaw is more serious than the first. Approximate answers are perfectly respectable; sometimes they're the best available. To have any idea how *accurate* estimates are, however, we must first describe precisely, in mathematical terms, how we *produce* the estimates. A precise description is also essential if we want (as we do) to use technology to produce and improve our estimates.

Integration Revisited. The old problem of estimating definite integrals, such as

$$I = \int_0^1 \sin(x^2)\, dx,$$

illustrates the pros and cons of graphical and numerical methods. To say that I is the following shaded area

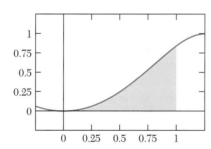

is helpful intuitively, but it doesn't generate practical estimates for I. A precisely described numerical method, such as the left rule,

$$L_{100} = 0.30607,$$

is easier to use and more amenable to accuracy checks.

Let's describe (in terms even a computer could understand) one sensible way to move through the slope field, starting from $(0, 190)$ and ending at $t = 10$. We'll take *ten straight-line steps, each of horizontal length 1.* ▶ The slope of each linear step is determined by the slope field at its *left* end. To make a "curve," ▶ we join successive linear steps. The graphical result follows, with dots at successive **Euler points**.

Each step corresponds to a time interval of 1 second.

Look closely; the "curve" is really ten straight segments.

Ten Euler steps through a slope field

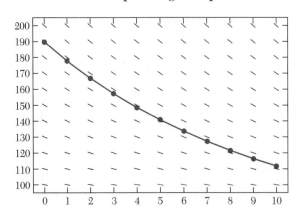

Here's how we found the successive Euler points:

Step 1   The first linear step starts at $(0, 190)$ and ends one unit to the right, at $(1, 178)$. Its slope is given by the DE. If $t = 0$ and $y = 190$, then

$$y'(0) = -0.10(y(0) - 70) = -0.10(190 - 70) = -12.$$

Our estimate, therefore, is

$$y(1) \approx y(0) + y'(0) \cdot 1 = 190 - 12 \cdot 1 = 178.$$

Step 2   The second step begins at $(1, 178)$ and moves one unit to the right. As before, its slope is determined by the DE, but with the updated y-value. Since $y = 178$,

$$y'(1) = -0.10(178 - 70) = -10.8.$$

Therefore,

$$y(2) \approx y(1) + y'(1) \cdot 1 \approx 178 - 10.8 \cdot 1 = 167.2.$$

Each step just described is called an **Euler step with step size 1**. It took ten Euler steps to reach $t = 10$. The following table (with entries rounded to two decimals) summarizes our numerical work:

From $t = 0$ to $t = 10$ in Ten Euler Steps										
Step	0	1	2	3	4	5	6	7	8	9	10
t	0	1	2	3	4	5	6	7	8	9	10
y'	-19.00	-10.80	-9.72	-8.75	-7.87	-7.09	-6.34	-5.74	-5.17	-4.65	-4.18
y	190.00	178.00	167.20	157.48	148.73	140.86	133.77	127.40	121.66	116.49	111.84

We estimate, therefore, that the coffee is about 112 degrees after 10 minutes.

How Good Is Euler's Method?

The two are closely related. See the exercises at the end of this section.

Both Euler's method and the left rule for definite integrals ◄ are systematic and precisely defined, but both find only approximate answers. Moreover, the sources of error in the two cases are exactly the same. Whereas the left rule "pretends" that an *integrand* remains constant over short intervals, Euler's method "pretends" that the *slope of a solution curve* remains constant over short intervals. Neither of these assumptions is usually true.

In the coffee example, we can tell *exactly* how "good" our Euler's method results are. ▶ An easy calculation shows that the function $y(t) = 70 + 120e^{-0.1t}$ is an exact solution of the IVP $y' = -0.1(y - 70)$, $y(0) = 190$. In particular,

We couldn't do this for a harder IVP.

$$y(10) = 70 + 120e^{-1} \approx 114.15;$$

the result is significantly different from what Euler's method predicted. Plotting both the Euler "curve" and the exact solution curve shows how Euler errors accumulate over the time interval.

Exact *vs.* Euler solutions

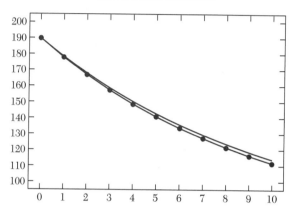

Shorter Steps: Improving Euler's Method

We saw in earlier work that the left rule's accuracy generally improves ▶ as the number of subdivisions increases, i.e., as the left rule's step size Δx decreases. Euler's method behaves similarly: Smaller steps usually produce more accurate results. There is a price, of course: Smaller steps mean more work, for person or for machine. ▶

As a rule—exceptions exist.

For very small steps, roundoff errors may also accumulate.

What happens in the coffee situation if we use steps of size 0.1 rather than 1? Carrying out 100 steps by hand would be tedious, but computers don't complain. Following are some sample results. *Exact* values of y appear in the last column, for comparison.

From $t = 0$ to $t = 10$ in 100 Euler Steps								
Step	0	1	2	3	10	20	30	90	100
t	0	0.1	0.2	0.3	1.0	2.0	3.0	9.0	10.0
y'	−12.00	−11.89	−11.76	−11.64	−10.85	−9.82	−8.88	−4.86	−4.39
y_{Euler}	190.00	188.80	187.61	186.43	178.53	168.15	158.76	118.57	113.92
y_{exact}	190.00	188.81	187.62	186.45	178.58	168.25	158.90	118.79	114.15

The Euler estimates are better this time than with only 10 steps.

We might, in theory, use even smaller step sizes, with thousands or millions of Euler steps. As with numerical integration, however, this is computationally foolish. For one thing, roundoff error eventually becomes significant. For another thing, better and more efficient numerical strategies are available for DEs. We won't pursue such "smart" strategies except to mention that one of the most popular, the **Runge-Kutta method**, is related to Simpson's rule in the same way as Euler's method is related to the left rule.

Euler's Method in General

Suppose that we're given a first-order DE $y' = f(t, y)$ and an initial condition $y(a) = y_0$, and that we use Euler's method with n steps on the interval $a \le t \le b$. What exactly does Euler's method give? The answer depends on our point of view.

Math-speak for a "curve" built from linear pieces.

Graphically Speaking   Euler's method says how to move step by step through the DE's slope field, starting at (a, y_0) and ending when $t = b$. Every Euler step has *horizontal* length $\Delta t = (b - a)/n$, called the **step size**. Each step goes in the direction determined by the slope field at the *beginning point* of the step. Joining successive **Euler points** with line segments (connecting the dots, in other words) produces a piecewise-linear curve. ◄   This curve is the graph of an approximate solution function $Y(t)$ for the DE. (We'll reserve the lowercase $y(t)$ for the exact solution.)

Numerically Speaking   Euler's method produces a list

$$Y(t_0), \ Y(t_1), \ Y(t_2), \ Y(t_3), \ \ldots, \ Y(t_n)$$

of numerical values of an approximate solution function $Y(t)$ for equally spaced inputs

$$a = t_0 < t_1 < t_2 < t_3 < \cdots < t_n = b;$$

each t_i is Δt units greater than its predecessor. Plotting these values and joining successive points with line segments produces the graph just mentioned.

What Euler's Method Doesn't Produce: Formulas

Euler's method uses—and produces—only numerical information. In other words, we can use Euler's method to produce a graph or a table of values but not a formula for a solution function. (The same principle holds if the left rule is applied to $\int_a^b f$. Because the left rule "sees" only isolated values of f, ◄   it can produce only numbers, not formulas.)

It doesn't "see" the formula for f.

This property of Euler's method (and of the left rule) is a mixed bag. To its credit, Euler's method can be used with almost any DE, no matter how complicated. To its debit, Euler's method produces only approximate results, even for simple DEs that might otherwise be solved exactly.

Euler's Method Step by Step

How, in general, does Euler's method produce the successive values

$$Y(t_0), \ Y(t_1), \ Y(t_2), \ Y(t_3), \ \ldots, Y(t_n)?$$

Step 1 The starting point $Y(t_0) = y_0$ comes from the initial condition. By the DE, a solution curve through (t_0, y_0) has slope $m_0 = f(t_0, y_0)$, so the first Euler step has slope m_0. Its rise (or fall) is therefore $m_0 \cdot \Delta t$ vertical units. In other words,

$$Y(t_1) = Y(t_0) + m_0 \Delta t.$$

Step 2 The second Euler step starts at $(t_1, Y(t_1))$, where the previous step ended. At $(t_1, Y(t_1))$, says the DE, a solution curve has slope $m_1 = f(t_1, Y(t_1))$. Thus, the second Euler step rises (or falls) $m_1 \cdot \Delta t$ vertical units:

$$Y(t_2) = Y(t_1) + m_1 \Delta t.$$

Continue The pattern should now be clear. At each step, we (1) use the DE to "update" the slope m_i, and (2) move with slope m_i from $(t_i, Y(t_i))$ to $(t_{i+1}, Y(t_{i+1}))$. After n steps, we arrive at $t_n = b$.

Reduced entirely to symbols, Euler's recipe for assembling an approximate solution $Y(t)$ looks like this:

$$Y(t_0) = y_0;$$
$$Y(t_1) = Y(t_0) + f(t_0, Y(t_0)) \cdot \Delta t;$$
$$Y(t_2) = Y(t_1) + f(t_1, Y(t_1)) \cdot \Delta t;$$
$$\vdots$$
$$Y(t_{i+1}) = Y(t_i) + f(t_i, Y(t_i)) \cdot \Delta t;$$
$$\vdots$$
$$Y(t_n) = Y(t_{n-1}) + f(t_{n-1}, Y(t_{n-1})) \cdot \Delta t.$$

> **Simple But Powerful.** It is striking that Euler's method—the idea of which could hardly be simpler—should be named after one of history's greatest mathematicians, Leonhard Euler (1707–1783). One lesson may be the surprising power of simple ideas. Another lesson may be that powerful ideas, properly understood, *become* simple.

Modeling Logistic Population Growth: Euler's Method and the Wise Flies

Euler's method looks best in action. To show it off, we return to again model the prudent fruit fly population of Section 4.2. ▶ We assume that the fly population grows **logistically**. In other words:

Look back if necessary, but what follows is self-contained.

The population's growth rate is proportional both to the population itself and to the difference between the carrying capacity and the population.

In symbols:

$$P' = kP(C - P),$$

where P represents the population, P' the population's growth rate, and C the carrying capacity of the environment; k is a constant of proportionality. ▶

P and P' vary with time; k and C are constants.

EXAMPLE 2   Through empirical measurements, researchers determine that their captive fly population, initially 1000 members strong, satisfies the logistic IVP

$$P' = 0.00000556P(10,000 - P); \qquad P(0) = 1000.$$

(Here t measures time, in days, since the original measurement; $P = P(t)$ is the population at time t; and $P' = P'(t)$ is the growth rate, in flies per day, at time t.)

Biologists like to say "hence."

What will the population be 10 days hence? ◀   How will the population evolve?

Follow along with a calculator.

S o l u t i o n   To estimate $P(10)$, we take ten (1-day) Euler steps. The numbers are clumsy, so we get technological help. ◀

Step 1   When $t = 0$, $P = 1000$, so

$$P'(0) = 0.00000556 \cdot 1000 \cdot 9000 \approx 50.04 \text{ flies per day.}$$

After one day, therefore, we estimate a population of $P(1) \approx 1000 + 50.04 = 1050.04$ flies.

Step 2   When $t = 1$, $P \approx 1050.04$, so

$$P'(1) = 0.00000556 \cdot 1050.04 \cdot (10000 - 1050.04) \approx 52.25 \text{ flies per day.}$$

Therefore, we estimate

$$P(2) \approx P(1) + P'(1) \cdot 1 \approx 1050.04 + 52.25 = 1102.29 \text{ flies.}$$

Numbers rounded to two decimal places.

The idea should now be clear, so we let the computer do the rest. Here are the results: ◀

Flies: From $t = 0$ to $t = 10$ in Ten Euler Steps										
Step	0	1	2	3	4	5	6	7	8	9	10
t	0	1	2	3	4	5	6	7	8	9	10
P'	50.04	52.25	54.53	56.88	59.29	61.77	64.31	66.91	69.56	72.27	75.02
P	1000.00	1050.04	1102.29	1156.82	1213.70	1272.99	1334.76	1399.07	1465.97	1535.53	1607.80

Thus we'd guess a population of about 1608 flies on day 10. ∎

The program happens to be written in ISETL.

A Computer Program (Optional).   Following is the computer program that produced the preceding table of values. Language details ◀   don't matter, but try to understand how the program works—it shows clearly how Euler's method proceeds.

A technical remark may help. The statement starting with `for` indicates a **loop**. The next three commands (up to `end for;`) are repeated ten times, once for each `t` from 1 to 10.

```
program flies;
    population := 1000.0;
    rate := 50.04;
    t := 0;
    writeln t, rate, population;
    for t in [1..10] do
        rate := 0.00000556*population*(10000-population);
        population := population + rate*1;
        writeln t, rate, population;
    end for;
end program;
```

Caveats. The tabulated estimate $P(10) \approx 1607.80$ must be viewed cautiously.

- Those decimals look impressive, but flies don't come in fractions. Our model only approximates reality.

- The figure 1607.80 *underestimates* the true population at day 10. The reason has to do with step size: Taking 1-day Euler steps amounts, in effect, to pretending that the growth rate remains constant over 1-day periods. In reality, the growth rate *increases* along with the population.

- A smaller step size would update the growth rate more often, and so it would better estimate the population on day 10. With, say, 100 Euler steps, ▶ the computer gives

$$P(0.1) \approx 1005.00, \qquad P(0.2) \approx 1010.03, \qquad \ldots \qquad P(10) \approx 1621.43.$$

The last estimate is *still* low but is better than before. ▶

Each step is 0.1 day long.

Flies in the Long Run. With machine help, it's easy to use Euler's method ▶ to project the fly population far into the future—remembering, of course, that errors accumulate over time.

We did so for a 150-day period, using step size 0.1 day. The following graph shows our results. ▶ Each dot represents 1 day; between every two dots, the machine took ten Euler steps. ▶ Reassuringly, the graph's shape looks reasonable.

With other techniques it can be shown that $P(10) = 1638.09$. See the next section.

A slight variant of the program would work.

Tabulating full results would take a lot of space!

The machine worked very hard.

Flies: how they thrive

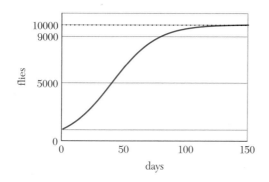

Observe:

- On day 100, the population is over 9500; room is fast running out. Day 200 isn't shown, but the graph's shape suggests that at $t = 200$ the population will still be just below 10,000.

- In about 40 days the population reaches 5000, half the upper limit. In about 80 days, the population reaches 9000—90% of capacity. Around day 130, 99% capacity is reached.

Do you agree? Look carefully.

- The fly population seems to increase *fastest* around day 40, when the population is 5000. ◄ Intuition says that under logistic conditions a population increases slowly at first (when there are few breeding members), more quickly as the population rises, and then more slowly as the upper limit nears. Thus, we might expect what we see on the graph: A population grows fastest when it is *halfway* to its carrying capacity.

BASICS

1. Let $f(t) = \sin t$. Consider the IVP $y' = f(t)$ and $y(0) = 0$.

    (a) Carry out Euler's method with four steps to estimate $y(1)$. Do all steps by hand.

    (b) Use the left rule with four subdivisions to estimate $I = \int_0^1 f(t)\, dt$ by hand.

    (c) Explain why $y(t) = y(0) + \int_0^t f(x)\, dx$ is the solution of the IVP.

    (d) Find the exact value of $y(1)$.

    (e) Find $\int_0^1 f(t)\, dt$ exactly. How much error did L_4 commit?

    (f) Let $y(t)$ be the *exact* solution function, and let $Y(t)$ be the *approximate* solution function constructed by Euler's method. Complete the following table.

t	0.00	0.25	0.50	0.75	1.00
$Y(t)$					
$y(t)$					

    (g) Plot both $y(t)$ and $Y(t)$ on the same axes. What do you see?

2. Repeat Exercise 1 using $f(t) = e^t$.

3. Here's a very simple IVP: $y' = 3$, $y(0) = 0$.
    (a) Just by guessing, find an exact solution $y(t)$ to the IVP. Then list $y(1)$, $y(2)$, $y(3)$, $y(4)$, $y(5)$. [**HINT**: What functions $y(t)$ have constant derivatives?]
    (b) Use Euler's method, with five steps of size 1, to estimate $y(1)$, $y(2)$, $y(3)$, $y(4)$, $y(5)$. How close are your estimates?
    (c) Explain what you found in part (b). When would you expect Euler's method to behave this way?

FURTHER EXERCISES

4. The number e is sometimes defined as $y(1)$, where $y(t)$ is the solution to the IVP $y' = y$, $y(0) = 1$.

    (a) Use Euler's method with one subdivision to estimate e. Is your answer an underestimate or an overestimate? Can you tell even without knowing the value of e?

    (b) Repeat part (a) with four subdivisions.

    (c) Repeat part (a) with 100 subdivisions.

    (d) Imagine that Euler's method, with 10,000 subdivisions, is to be carried out on the IVP to estimate

$y(1)$. Write out the results of the first three steps. Convince yourself that the result after 10,000 steps is $(1.0001)^{10,000}$. Use a calculator to estimate this number. Does it look familiar?

5. On page 635 we said:

An easy calculation shows that the function $y(t) = 70 + 120e^{-0.1t}$ is an exact solution of the IVP $y' = -0.1(y - 70)$; $y(0) = 190$.

Do this "easy calculation." What's $y(5)$?

6. Another population of flies breeds according to the same (logistic) DE as in Example 2,

$$P' = 0.00000556 P (10{,}000 - P),$$

but it starts with only 100 members so $P(0) = 100$.
(a) Carry out by hand the first three Euler steps; use step size 1 day.
(b) The original fly population was about 9600 on day 100 (see the graph on page 639). Use step size 1 to estimate *this* fly population on day 100.

7. Consider again the flies' logistic DE,

$$P' = 0.00000556 P (10{,}000 - P),$$

but this time with the new initial condition $P(0) = 0$.
(a) Carry out by hand three Euler steps; use step size 1 day. Estimate the population on day 200.
(b) Explain in biological terms what's going on.

8. Consider the IVP $y' = 1 + t - y$, $y(0) = 0$.
(a) Use Euler's method with five steps of size 0.4 to estimate $y(2)$.
(b) Use Euler's method with ten steps of size 0.2 to estimate $y(2)$.
(c) Show that the exact solution of the IVP is $y(t) = t$.
(d) Compare the exact value of $y(2)$ with the estimates you computed in parts (a) and (b).

9. Consider the IVP $y' = y - 2$, $y(0) = 1$.
(a) Use Euler's method with ten steps of size 0.1 to estimate $y(1)$.
(b) Use Euler's method with 20 steps of size 0.05 to estimate $y(1)$.
(c) Show that the exact solution of the IVP is $y(t) = 2 - e^t$.
(d) Compare the exact value of $y(1)$ with the estimates you computed in parts (a) and (b).

10. Consider the IVP $y' = xy$, $y(0) = 1$.
(a) Use Euler's method with one step of size 1 to estimate $y(1)$.
(b) Use Euler's method with two steps of size 0.5 to estimate $y(1)$.
(c) Use Euler's method with four steps of size 0.25 to estimate $y(1)$.
(d) Use Euler's method with eight steps of size 0.125 to estimate $y(1)$.
(e) Show that the exact solution of the IVP is $y(x) = e^{x^2/2}$.
(f) Compare the exact value of $y(1)$ with the estimates you computed in parts (a)–(d).
(g) Show that the results in part (f) suggest that the error made by Euler's method is proportional to $1/n$, where n is the number of steps.

11. Consider the IVP $y' = y^2$, $y(0) = 1$.
(a) Use Euler's method with four steps of size 0.2 to estimate $y(0.8)$.
(b) Show that the exact solution of the IVP is $y(t) = (1 - t)^{-1}$.
(c) Explain why Euler's method doesn't provide a good estimate of $y(0.8)$.

12. (a) Consider the IVP $y' = f(t)$, $y(0) = 0$. Explain the relationship between the left-sum estimate of $\int_0^x f(t)\,dt$ and the Euler's-method estimate of $y(x)$, if both are computed using the same number of subdivisions.
(b) Consider the IVP $y' = f(t)$, $y(t_0) = y_0$. Explain the relationship between the left-sum estimate of $\int_{t_0}^x f(t)\,dt$ and the Euler's-method estimate of $y(x)$ if both are computed using the same number of subdivisions.

12.4 Separating Variables: Solving DEs Symbolically

In earlier sections we approached DEs and IVPs graphically, with slope fields, and numerically, with Euler's method. In this section we solve DEs *symbolically*, by **separating variables**. Before describing the method in general terms, we illustrate how it works—when it works—with another look at the situation of Example 1, page 632.

The Idea—An Example and Remarks

EXAMPLE 1   Let $y(t)$ denote (again!) the temperature (in degrees Fahrenheit) of a cup of coffee at time t (in minutes). ▶ Room temperature is 70 degrees;

Time $t = 0$ can denote any convenient reference time.

the coffee starts at 190 degrees. The coffee's temperature is described by the IVP

$$y' = -0.1(y - 70); \qquad y(0) = 190.$$

How hot is the coffee after 10 minutes? How hot is the coffee at *any* time t?

Using the Leibniz notation for y'.

Solution   First we rewrite the DE slightly: ◄

$$\frac{dy}{dt} = -0.1(y - 70).$$

Hence the name of the technique.

Next comes the key step: ◄

Separate the y's and dy's from the t's and dt's.

Here's the result:

$$\frac{dy}{y - 70} = -0.1\,dt.$$

Antidifferentiating both sides gives a new equation:

$$\int \frac{dy}{y - 70} = \int -0.1\,dt.$$

Convince yourself that the antiderivatives are correct.

Both antiderivatives are straightforward; ◄   a simple calculation gives

$$\ln|y - 70| + C_1 = -0.1t + C_2.$$

Because both C_1 and C_2 are arbitrary constants, it's convenient (and legal!) to combine them into one:

$$\ln|y - 70| = -0.1t + C.$$

Check each step.

To solve for y, we exponentiate both sides: ◄

$$\ln|y - 70| = -0.1t + C \Longrightarrow |y - 70| = e^{-0.1t+C} \Longrightarrow y = \pm e^C e^{-0.1t} + 70.$$

One last simplification helps. Because C is an arbitrary constant, so is $K = \pm e^C$, so y has the form

$$y = Ke^{-0.1t} + 70.$$

The last step is to use the initial condition to assign K a numerical value:

$$y(0) = 190 \quad \Longrightarrow \quad 190 = Ke^0 + 70 \quad \Longrightarrow \quad K = 120.$$

Our problem is now completely solved; the function

$$y(t) = 120e^{-0.1t} + 70$$

See $y(t)$ plotted on page 632.

solves the IVP *exactly*, for any time t. ◄   At $t = 10$, for example,

$$y(10) = 120e^{-1} + 70 \approx 114.146 \text{ degrees.} \qquad \blacksquare$$

A Legal Separation?

By differentiation; try it!

It's easy to check ◄   that $y(t) = 120e^{-0.1t} + 70$ *does* solve the preceding IVP. But does the method really make sense? Can we really "separate" dy from dt as we just did?

That's a good question; dy/dt usually denotes *one* quantity—the derivative function—not the quotient of two separate quantities. Care with symbols *is* wise, but things work out all right here. The precise reasons are mathematically subtle. Roughly speaking, the antiderivative notations $\int f(t)\,dt$ and $\int g(y)\,dy$ are designed to ensure that dt and dy can be handled separately.

Our best defense of the separation method, however, is that it works—*and can be verified to work* by an easy, direct check. The same important principle applies to solving DEs and to finding antiderivatives: ▶ *Finding* answers may be difficult, but *checking* answers is easy.

As we saw in Chapters 6 and 9.

Symbolic Solutions: From One, Many

Solving the preceding DE by symbolic antidifferentiation was easy. Solving more complicated DEs symbolically can be *much* harder, if not impossible. Symbolic methods—when they work—have two important advantages over numerical and graphical approaches:

1. They produce *exact* solutions, not approximations.

2. They handle DEs that include parameters ▶ as well as numerical constants.

Parameters are symbols that can stand for any constant.

The second property means that solving *one* DE symbolically can amount, in effect, to solving whole families of DEs. The next example illustrates this advantage.

EXAMPLE 2 Let $y(t)$ denote the temperature (in degrees Fahrenheit) of a cup of coffee at time t (in minutes). ▶ Room temperature is T_r degrees; at time $t = 0$ the coffee is at T_0 degrees. The coffee's temperature is described by the IVP

Time $t = 0$ can denote any convenient reference time.

$$y' = -0.1(y - T_r); \qquad y(0) = T_0.$$

How hot is the coffee after 10 minutes? How hot is it at *any* time t?

Solution (Notice first that neither slope fields nor Euler's method would be any use here, since both require numerical data.)

This example and the last are virtual clones. So are their solutions; we need only replace 70 with T_r, 190 with T_0, and 0 with t_0 wherever they appear in the preceding solution. Here's the result: ▶

Check that this is the result.

$$y(t) = (T_0 - T_r)e^{-0.1t} + T_r.$$

The beauty of such a solution is that it applies for *all* values of the parameters T_0 and T_r. Several solution curves, all with $T_r = 70$ but with various values of T_0,

See the exercises for more on these curves.

appear on page 621. Another set of solution curves, each with $T_0 = 190$ but with various values of T_r, is shown following. ◄

Nine cups of coffee

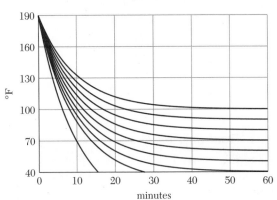

The Method in General: When It Works, When It Doesn't

Separable and Inseparable DEs.   Separating variables can "work" only with separable DEs—those in which the t and y variables *can* be separated. Following are several separable DEs, each with its "separated" form:

$$\frac{dy}{dt} = ty \quad \Longrightarrow \quad \frac{dy}{y} = t\,dt;$$

$$\frac{dy}{dt} = \frac{\sin t}{y} \quad \Longrightarrow \quad y\,dy = \sin t\,dt;$$

$$\frac{dy}{dt} = \frac{\sin(t^2)}{y} \quad \Longrightarrow \quad y\,dy = \sin(t^2)\,dt.$$

Separable DEs often have, or can be written in, one of the forms

$$\frac{dy}{dt} = f(t)g(y) \qquad \text{or} \qquad \frac{dy}{dt} = \frac{f(t)}{g(y)},$$

so that either

$$\frac{dy}{g(y)} = f(t)\,dt \qquad \text{or} \qquad g(y)\,dy = f(t)\,dt.$$

In some DEs (called **inseparable**), the t and y variables *cannot* be separated. Here are three examples:

$$\frac{dy}{dt} = t + y; \qquad \frac{dy}{dt} = \sin(t + y); \qquad \frac{dy}{dt} = \frac{\sec(y^2) + \sin(t + y)}{\sqrt{t^2 + y + \cos y}}.$$

Antiderivative Problems.   Even if t and y can be separated, the antiderivative problems

$$\int \frac{dy}{g(y)} = \int f(t)\,dt \qquad \text{or} \qquad \int g(y)\,dy = \int f(t)\,dt$$

remain. Again, trouble may loom: Some elementary functions cannot be antidifferentiated in elementary symbolic form. One of the separable DEs seen earlier poses just this problem; the right-hand antiderivative

$$\int y \, dy = \int \sin(t^2) \, dt$$

has no elementary solution.

The Big Picture and Two Morals. Separation of variables is only one symbolic method; it applies only to separable DEs. DEs can take many other symbolic forms, and many other symbolic methods exist to solve them. For example, many practically useful DEs have the form

$$y' = a(t)y + b(t);$$

they're called **first-order linear DEs.** General methods ▶ exist for solving first-order (and higher-order) linear DEs. If $a(t)$ and $b(t)$ are suitably simple functions, solutions can be found explicitly in terms of elementary functions.

We won't study them systematically; that's done in DE courses.

For example, the DE $y' = y + t$ is linear but not separable. ▶ With linear DE methods, one can produce the solution $y = Ce^t - t - 1$; C can be any constant.

What's $a(t)$? What's $b(t)$?

There are two main morals. First, if separating variables fails, other methods may succeed. Second, solving DEs symbolically, like finding antiderivatives symbolically, is art as well as science; there is no fail-safe route to a solution.

Rumors, Separable DEs, and Logistic Growth

In earlier sections we used the logistic DE

$$\frac{dP}{dt} = kP(C - P) = kP(t)\big(C - P(t)\big)$$

to model populations that grow logistically. ▶ Both constants k and C can be interpreted in biological terms: C represents the environment's long-run carrying capacity, and k measures the population's reproduction rate.

Flies in a lab, for example; see Section 12.3.

Biological populations come first to mind, but the logistic DE works elsewhere, too. Translating the logistic DE into words gives the key property of *any* quantity that grows logistically toward a long-run value C:

> *The quantity P grows at a rate that is proportional both to P itself and to $C - P$, the "room available" for further growth.*

Solving the Logistic DE

For later use, let's solve the logistic DE in general form. Separating variables sets up the antiderivative problem:

$$\frac{dP}{dt} = kP(C - P) \Longrightarrow \int \frac{dP}{P(C - P)} = \int k \, dt.$$

The right side is easy. The left side is stickier, but it can be found in an integral table. Do-it-yourselfers may prefer the following approach, based on splitting the integrand into partial fractions.

$$\int \frac{dP}{P(C - P)} = \frac{1}{C} \int \left(\frac{1}{P} + \frac{1}{C - P} \right) dP = \frac{1}{C} \ln \left| \frac{P}{C - P} \right|.$$

D is a constant of integration; watch for it.

However it's obtained, the antiderivative shown *is* correct, as differentiation shows. (We can drop the absolute-value signs if we assume—as we will—that $0 < P < C$.) Putting everything together and solving for P solves the logistic DE: ◀

$$\frac{dP}{dt} = kP(C - P) \implies \int \frac{dP}{P(C - P)} = \int k\,dt$$

$$\implies \ln \frac{P}{C - P} = Ckt + CD$$

$$\implies \frac{P}{C - P} = e^{Ckt} e^{CD}$$

$$\implies \frac{C - P}{P} = e^{-Ckt} e^{-CD}$$

$$\implies \frac{C}{P} = e^{-Ckt} e^{-CD} + 1$$

$$\implies P = \frac{C}{de^{-Ckt} + 1}.$$

(Because D is an arbitrary constant, so is $d = e^{-CD}$.) The result is prettier than the calculation.

Fact For any positive constants C, d, and k, the function

$$P(t) = \frac{C}{de^{-Ckt} + 1}$$

solves the logistic DE $P' = kP(C - P)$.

To apply this result, we need only choose appropriate values for the constants.

Rumors and Logistic Growth

A rumor spreads as "tellers" pass it on to "hearers." (Once told, a hearer becomes a teller.) The rumor spreads slowly at first, when tellers are few. It spreads faster when both tellers and hearers are plentiful but slows down again as hearers become scarce, and it stops when everyone knows the rumor. The teller population, in other words, grows as shown:

Logistic rumor-spreading

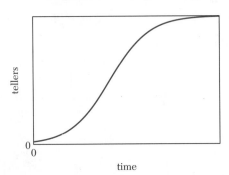

tellers

time

As the shape of the graph suggests, logistic growth is one ▶ plausible model for rumor-spreading.

Not the only one, as we'll see.

EXAMPLE 3   Riverdale High has 1000 students. On day 0, Archie, Jughead, Betty, Veronica, and 16 of their friends start a rumor, which spreads logistically. A day later, 50 students know it. What happens over the next 10 days? When is the rumor spreading fastest?

Solution   Let $P(t)$ denote the number of tellers (people who know the rumor) after t days. Then $P(t)$ satisfies the logistic DE and two additional conditions:

$$\frac{dP}{dt} = kP(1000 - P); \qquad P(0) = 20; \qquad P(1) = 50.$$

By the preceding Fact, ▶

Set $C = 1000$.

$$P(t) = \frac{1000}{de^{-1000kt} + 1}$$

for appropriate constants d and k. The other conditions ▶ let us evaluate d and k:

$P(0) = 20$ is an initial condition.

$$P(0) = \frac{1000}{d+1} = 20 \quad \Longrightarrow \quad d = 49;$$

$$P(1) = \frac{1000}{49e^{-1000k} + 1} = 50 \quad \Longrightarrow \quad k \approx 0.000947.$$

We've found our solution function:

$$P(t) = \frac{1000}{49e^{-0.947t} + 1}.$$

Here's the graph:

Two rumors: logistics at Riverdale

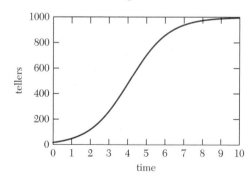

By 10 days, almost everyone knows the rumor. The rumor seems to spread fastest, moreover, when $P = 500$, just after day 4. ▶ (With elementary calculus we can find this time exactly.) ▶

The graph is steepest there.

The answer is $t \approx 4.11$—see the exercises.

EXAMPLE 4   Chagrined that bad news travels so fast, Archie, Jughead, Betty, and Veronica start the next rumor all by themselves. It travels through Riverdale with the same "transmission coefficient" $k = 0.000947$. ▶ What happens this time?

See this section's exercises for more on this terminology.

Solution   Except for the new initial condition, the situation is exactly as in the preceding example. Thus,

$$P(0) = \frac{1000}{d+1} = 4 \quad \Longrightarrow \quad d = 249 \quad \Longrightarrow \quad P(t) = \frac{1000}{249e^{-0.947t} + 1}.$$

Here is the new rumor graph; the old one appears for comparison.

Two rumors: logistics at Riverdale

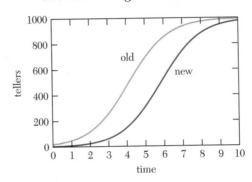

Yesterday's News: Non-logistic Rumors

Not all rumors spread logistically. One possible flaw in the logistic model $P' = kP(C - P)$ concerns the "transmission coefficient" k. In the logistic model, k remains constant over time. In practice, however, k probably *shrinks* over time; as a rumor ages, people care less and less.

A more realistic rumor model might somehow reflect this "staling" process. One approach is to replace the constant k in the logistic DE with a *decreasing function of t*. Here's one possibility. If

$$k(t) = \frac{K}{1+t},$$

where K is any positive constant, then (as we'd hope) $k(t) \to 0$ as $t \to \infty$. Then the new differential equation would have the form

$$\frac{dP}{dt} = k(t)P(t)\big(C - P(t)\big) = \frac{K}{1+t}P(C - P).$$

EXAMPLE 5   Still another Riverdale rumor spreads according to the preceding DE. When the principal hears the rumor at time 0, 100 students know the rumor, and it's spreading at the rate of 100 students per day. What happens as time goes on?

Solution   As before, let $P(t)$ denote the number of students who know the rumor at time t. The new DE, like the old, is separable:

$$\frac{dP}{dt} = \frac{K}{1+t}P(1000 - P) \Longrightarrow \int \frac{dP}{P(1000 - P)} = \int \frac{K}{1+t}\,dt.$$

It takes some effort.   Both sides can be antidifferentiated and the resulting equation solved for P. ◄

When the dust settles, the result is

$$P(t) = \frac{1000}{d(1+t)^{-1000K} + 1}.$$

The additional information lets us find the constants K and d. From $P(0) = 100$ it follows ▶ that $d = 9$. To find K we use the DE itself: *Check it!*

$$P'(0) = 100 = \frac{K}{1+0}100(1000 - 100) \implies K = \frac{1}{900}.$$

All the numbers are now in place. Here's the formula:

$$P(t) = \frac{1000}{9(1+t)^{-10/9} + 1}.$$

Here's the graph, over a 100-day interval:

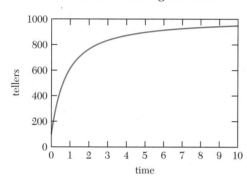

Stale news: a non-logistic rumor

Stale news travels slowly.   ■

BASICS

1. Consider the DE $y' = xy$.
   (a) Find functions f and g such that this DE can be written as $f(y)y' = g(x)$.
   (b) Find functions F and G such that $F' = f$ and $G' = g$.
   (c) Explain why $F(y) = G(x) + C$ implicitly defines a solution y of the DE.
   (d) Use part (c) to show that $y(x) = Ce^{x^2/2}$ is a solution of the DE.

2. (a) Explain why $y(x) = C + \int_a^x g(t)\, dt$ is a solution of the IVP $y' = g(x)$, $y(a) = C$.
   (b) Consider the IVP $f(y)y' = g(x)$, $y(a) = C$, and let F be an antiderivative of f. Explain why

   $$F(y) = F(C) + \int_a^x g(t)\, dt$$

   implicitly defines a solution of this IVP.

Find a solution of each of the following separable differential equations.

3. $y' = 1 + y^2$
4. $y' = y^2$
5. $y' = \sqrt{1 - y^2}$
6. $y' = 2e^{-y}$
7. $y' = 1 - y$
8. $y' = 4 - y^2$
9. $y' = y \ln|y|$
10. $y' = x^2 y$
11. $y' = \dfrac{x}{1 + y^2}$
12. $y' = x^2 \cos y$
13. $y' - xy^2 = 0$
14. $y' = x/y$

15. Solve the flies' IVP

    $$P' = 0.00000556 P(10{,}000 - P), \quad P(0) = 1000,$$

    exactly, by separating variables. Find $P(10)$. How does the answer compare with the one we found in Section 12.3, using Euler's method?

16. Consider the nine coffee curves shown in Example 2, page 644. Label them C_1 through C_9, from top to bottom. Each curve describes the temperature "evolution" of a different cup of coffee.
    (a) All curves start at $(0, 190)$. What does this mean about the cups of coffee?
    (b) Each curve corresponds to a different value of T_r. What does this mean in terms of the coffee situation?
    (c) What values of T_r correspond to C_1, C_4, and C_9?
    (d) Which cup of coffee is cooling fastest at $t = 10$? Why?
    (e) Curve C_4 solves the IVP $y' = -0.1(y - 70)$, $y(0) = 190$. What IVP does curve C_1 solve? What IVP does curve C_9 solve?
    (f) Use the curve C_1 to estimate the rate at which coffee cup C_1 is cooling at time $t = 10$.
    (g) Use the DE for C_1 to find the rate at which coffee cup C_1 is cooling at time $t = 10$.
    (h) Use the curve C_9 to estimate the rate at which coffee cup C_9 is cooling at time $t = 10$.
    (i) Use the DE for C_9 to estimate the rate at which coffee cup C_9 is cooling at time $t = 10$.
    (j) Still another cup of coffee, initially at 190 degrees, is at 100 degrees after 20 minutes. Use the graphs to estimate the room temperature.

17. In Example 2, page 643, we solved the IVP $y' = -0.1(y - T_r)$, $y(0) = T_0$. We found the general solution $y(t) = (T_0 - T_r)e^{-0.1t} + T_r$.
    (a) Check (by differentiation) that
    $$y(t) = (T_0 - T_r)e^{-0.1t} + T_r$$
    is a solution of the IVP.
    (b) Suppose that $T_0 = 200$ and $y(10) = 100$. Find T_r. Then find $y(20)$. Interpret everything in coffee terms.
    (c) Suppose that $T_r = 80$ and $y(10) = 120$. Find T_0. Then find $y(20)$. Interpret everything in coffee terms.
    (d) Suppose that $y(10) = 100$ and $y(20) = 80$. How hot was the coffee at time $t = 0$? What's the room temperature? How hot will the coffee be at $t = 40$?
    (e) It's true that $\lim_{t \to \infty} e^{-0.1t} = 0$. Use this fact to find $\lim_{t \to \infty} y(t)$. Interpret the result in coffee terms.

18. A population P that satisfies a DE of the form $P'(t) = kP(t)\big(C - P(t)\big)$, where k and C are constants, is said to grow logistically.
    (a) If $P < C$, then (says the DE) $P' > 0$. What does this mean in population terms?
    (b) If $P > C$, then (says the DE) $P' < 0$. What does this mean in population terms?
    (c) If $P = C$, then (says the DE) $P' = 0$. What does this mean in population terms?

19. We showed in this section that nonzero solutions to the DE $P'(t) = kP(t)\big(C - P(t)\big)$, where k and C are constants, have the form
    $$P(t) = \frac{C}{Ke^{-Ckt} + 1}$$
    for some constant K. Assuming that C and k are positive constants, show that $P(t) \to C$ as $t \to \infty$. What does this mean in biological terms?

20. In Example 3, page 647, we said that
    $$\frac{1000}{49e^{-1000k} + 1} = 50 \implies k \approx 0.00947.$$
    Verify this claim. (First solve for k exactly; then find a decimal approximation.)

21. In Example 3, page 647, we said that (i) the rumor spreads fastest when the population is 500; (ii) this occurs just after day 4.
    (a) The DE $P' = kP(C - P)$ gives the slope of the P-graph as a function of P. By differentiating the right side with respect to P, show that P' is maximum when $P = C/2$. (In the present case, this means $P = 500$.)
    (b) For what value of t is $P(t) = 500$? [**HINT**: Solve the equation for t.]

22. Here are plots of several variants on the logistic-rumor graph of Example 4, page 648. In each case, 50 people know the rumor on day 0.

    Four rumors: logistics at Riverdale

    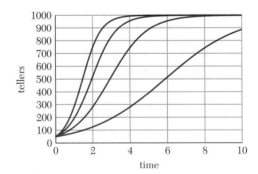

    (a) Call the curves P_1 through P_4, reading from top to bottom on the picture. Each curve's equation is of the form $P(t) = 1000/(19e^{-kt} + 1)$ for some value of k. Use the graph to estimate a value of k for each curve P_1 through P_4.

(b) In this context, the parameter k can be thought of as the rumor's "transmission coefficient." It measures how likely a given teller is to pass the rumor on to a given hearer. Alternatively, k can be thought of as a measure of how "interesting" the rumor may be. Which rumor curve corresponds to the hottest rumor? Which corresponds to the dullest rumor? Why?

23. Throughout this exercise, assume that rumors spread through a 1000-student school in the manner described in Example 5. In each part, use the given information to calculate an appropriate population function $P(t)$. [**HINT**: Find appropriate values for K and d.]
    (a) $P(0) = 100$ and $P'(0) = 50$
    (b) $P(0) = 50$ and $P'(0) = 100$
    (c) $P(0) = 800$ and $P'(0) = 50$
    (d) $P(0) = 800$ and $P'(0) = 100$
    (e) On the same axes, plot your results from parts (a)–(d). (Use a machine!) Label your graphs.
    (f) Discuss the differences among the rumors in parts (a)–(d) using such terms as "hot," "cool," "long-running," and "short-running."

24. Find an equation of a curve $y = f(x)$ that passes through the point $(0, 1)$ and has the property that the length of any segment of the graph is the same as the area under that segment of the graph (ignoring units).

25. Two students, Amy and Joan, order an after-dinner coffee. Amy immediately adds cream to her coffee to cool it, but Joan waits 10 minutes before she adds the same amount of cream. Neither drinks any of her coffee until Joan has added cream to hers. Who drinks the hotter coffee?

26. **First-order linear equations.** Suppose that $F'(x) = f(x)$.
    (a) Show that $\left(e^{F(x)} y(x)\right)' = e^{F(x)}\left(y'(x) + f(x)y(x)\right)$.

(b) Show that
$$y(x) = e^{-F(x)}\left(C + \int_a^x e^{F(t)} g(t)\, dt\right)$$
is a solution of the IVP $y'(x) + f(x)y(x) = g(x)$, $y(a) = Ce^{-F(a)}$. [**HINT**: Multiply both sides of the DE by $e^{F(x)}$.]

(c) Use part (b) to derive $y(x) = (C + 1)e^x - x - 1$ as the solution of each of the IVP $y' = y + x$, $y(0) = C$.

27. Use part (b) of Exercise 26 to find a solution of the IVP $y' + 2xy = x$, $y(0) = 0$.

Use part (b) of Exercise 26 to find the solution of each of the IVPs in Exercises 28–31.

28. $y' - 2xy = x$, $y(0) = 1$

29. $y' + y = e^{2x}$, $y(1) = 2$

30. $y' + (\cos x)y = \cos x$, $y(\pi) = 0$

31. $xy' + 2y = \sin x$, $y(\pi/2) = 1$

32. **Bernoulli equations.** Consider the DE $y' + f(x)y = g(x)y^n$, where $n \geq 2$ is an integer.
    (a) Let $z(x) = \left(y(x)\right)^{1-n}$. Show that z satisfies the DE
    $$\frac{1}{1-n} z' + f(x)z = g(x)$$
    (b) Use part (a) of this exercise and part (b) of Exercise 26 to find the solution of the logistic equation $P' = kCP - kP^2$ ($k > 0$ and $C > 0$).
    (c) Find a solution of the DE $y' - y = y^3$.

12.5 Chapter Summary

Differential equations—equations that contain derivatives—permit the most important and useful applications of calculus. This chapter illustrates the meaning, uses, and techniques of differential equations.

Differential Equations: The Idea. We first broached the idea of a DE in Chapter 4, where we used DEs to model growth phenomena. Key to understanding DEs is the notion of a **solution**. A function—not a number—is a solution to a DE if it satisfies the DE's condition. Every function of the form $y = Ce^x$, for instance, solves the DE $y' = y$. Adding an **initial condition** to a DE produces an **initial value problem**, or **IVP**. Here's one:

$$y' = y; \qquad y(0) = 1.$$

The same functions $y = Ce^x$ still solve the DE, but only one function—$y = 1e^x$—solves the IVP.

Solving Differential Equations Graphically: Slope Fields.   A **slope field** is associated with every first-order DE $y' = f(t, y)$; at each point of a rectangle in the ty-plane, a small segment points in the direction of a solution curve. Given a slope field, a solution curve (i.e., the graph of a solution function) can be drawn through any initial point by "going with the flow."

Solving Differential Equations Numerically: Euler's Method.   Euler's method is the numerical version of the graphical technique just described. To solve a DE $y' = f(t, y)$ with initial condition (t_0, y_0), Euler's method starts at the initial point (t_0, y_0) and proceeds, step by step, toward an estimate for $y(t)$ at later values of t.

Euler's method, like the left rule (which it closely resembles), solves DEs only approximately. The error Euler's method commits (like that of the left rule) can usually be reduced by using a smaller Euler step size.

Solving Differential Equations Symbolically: Separation of Variables.   Some DEs can be solved symbolically by antidifferentiation. Among the simplest of these are **separable DEs**, those that can be written in the form

$$\frac{dy}{dt} = \frac{f(t)}{g(y)}.$$

Separating the t's and dt's from the y's and dy's produces a new equation, one that involves antiderivatives:

$$\int g(y)\,dy = \int f(t)\,dt.$$

Solving the new equation by antidifferentiation solves the original DE.

13

Polar Coordinates

13.1 Polar Coordinates and Polar Curves

Any point P in the xy-plane has a familiar and natural "address": its **rectangular** (or **Cartesian**) **coordinates**. The point P in the left hand picture, for instance, has rectangular coordinates $(4, 3)$.

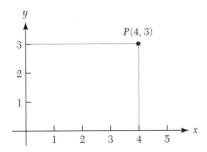

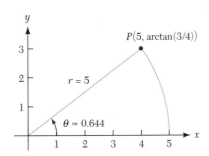

The x- and y-coordinates of P measure directed ▶ distances from P to the two perpendicular coordinate axes. To reach $P(4, 3)$ from the origin, one moves 4 units right and 3 units up.

That is, either positive or negative, depending on direction.

Polar coordinates offer another way of locating a point in the plane. In the **polar coordinate system**, a point P has coordinates r and θ; they tell, respectively, the *distance* from the origin O to P and the *angle* (in radians!) from the positive x-axis to the ray from O to P. The picture on the right shows that the

Convince yourself that
$\theta = \arctan(3/4)$.

point P with rectangular coordinates $(4, 3)$ has polar coordinates $(5, \arctan(3/4))$ $\approx$ $(5, 0.644)$. ◄

Polar Coordinate Systems

But not invariably.

A **rectangular coordinate system** in the Euclidean plane starts with an origin O and two perpendicular coordinate axes. Usually ◄ the x-axis is horizontal and the y-axis is vertical; x-coordinates increase to the right, and y-coordinates increase upward.

A polar coordinate system starts with different ingredients: an origin O, called the **pole**, and a ray (i.e., a half-line) beginning at the origin, called the **polar axis**. The polar axis normally points to the right, along the positive x-axis. With these ingredients and a unit for measuring distance, we can assign polar coordinates (r, θ) to any point P:

> r *is the distance from O to P; θ is any angle from the polar axis to the segment $\overline{OP}$.*

Check carefully that each point's coordinates are correct.

The following picture shows several points with their polar coordinates, plotted on a polar grid. ◄

Points on a polar grid

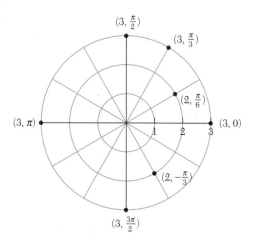

Polar vs. Rectangular Grids.   A rectangular coordinate system leads naturally to a rectangular grid, with vertical lines $x = a$ and horizontal lines $y = b$. In a polar system, holding the coordinates r and θ constant produces, respectively, concentric circles and radial lines. The result is a weblike polar grid. Here are grids of both types:

A rectangular grid

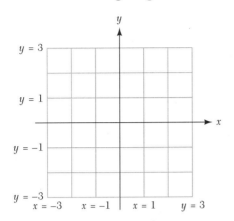

A polar grid

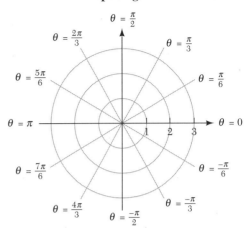

On the Scales.   In a rectangular coordinate system, the two axes often have different scales of measurement. As a result, on a graph, a vertical inch and a horizontal inch may represent different distances. In particular, "circles" may look far from round.

A polar coordinate system, by contrast, has just one axis—the polar axis. As a result, distance does not depend on direction, and circles look round.

Polar Coordinates: Not Unique.   A point in the plane—$P(4, 3)$, for instance—has just one possible pair of rectangular coordinates. Different rectangular coordinate pairs (x_1, y_1) and (x_2, y_2) correspond to different points in the plane.

Polar coordinates, by contrast, are *not* unique. Every point in the plane has many possible pairs of polar coordinates. For example, all of the polar coordinate pairs

$$\left(2, \frac{\pi}{4}\right), \quad \left(2, \frac{9\pi}{4}\right), \quad \left(2, -\frac{7\pi}{4}\right), \quad \left(2, -\frac{15\pi}{4}\right), \quad \left(-2, \frac{5\pi}{4}\right), \quad \left(-2, -\frac{3\pi}{4}\right),$$

(and many others) represent the same point—the one with rectangular coordinates $(\sqrt{2}, \sqrt{2})$. Notice especially the last two pairs. A negative r-coordinate means that, to locate the point P, one moves r units in the direction *opposite* the θ-direction. The point $(-2, 5\pi/4)$, for instance, lies 2 units from the origin on the ray $\theta = \pi/4$. ▶ The origin O allows even more freedom: It's represented by *any* pair of the form $(0, \theta)$, regardless of θ.

This ambiguity of polar coordinates arises for a simple reason. All angles that differ by integer multiples of 2π ▶ determine the same direction. In practice, this ambiguity can be annoying but is seldom a serious problem. Two simple rules help. ▶

Multiples of 2π   For any r and θ, the pairs (r, θ) and $(r, \ \theta + 2\pi)$ describe the same point.

Negative r   For any r and θ, the pairs (r, θ) and $(-r, \ \theta + \pi)$ describe the same point.

That's the ray opposite $\theta = \pi/4$.

For example, $\theta = \pi/4$, $\theta = 9\pi/4$, $\theta = -7\pi/4$, and so on.

See the exercises at the end of this section for more on these rules.

> **Polar Coordinates on Earth.**   The polar grid somewhat resembles an overhead view of earth, looking "down" at the North Pole. In cartographers' language, lines of **longitude** (or **meridians**) converge at the pole; the concentric circles are lines of **latitude** (or **parallels**). For hundreds of years, the **prime meridian** (polar axis, in calculus language), for which $\theta = 0$, has been taken to be the line of longitude that passes through the Greenwich Observatory, just east of London, England. For the same reason, Greenwich Mean Time (the time of day along the prime meridian) is used worldwide as a reference point.
>
>   Why is Greenwich "prime" rather than, say, India or Arabia, where navigation and timekeeping flourished even in antiquity? Nothing intrinsic; Greenwich just happened to be a center of attention when the terms were defined—an early (and quite literal) instance of Eurocentrism.
>
>   Polar coordinates in the *plane*, it should be said, aren't perfectly suited to measuring the (almost) spherical earth. In practice, geographers use a related system called *spherical* coordinates.

Polar Graphs

The ordinary graph of an equation in x and y is the set of points (x, y) whose coordinates satisfy the equation. The graph of $x^2 + y^2 = 1$, for instance, is the circle of radius 1 about the origin. The point $(2, 3)$ does not lie on this graph, because $2^2 + 3^2 \neq 1$.

Because $3 = 2 + \cos 0$.

   The idea of a **polar graph** is similar but not quite identical. The graph of an equation in r and θ is the set of points whose polar coordinates r and θ satisfy the equation. For instance, the polar point $(3, 0)$ lies on the graph of $r = 2 + \cos\theta$, ◄ but the polar point $(2, \pi)$ does not.

A Warning.   The fact that a point in the plane has more than one pair of polar coordinates means that polar plotting requires extra care. At first glance, for instance, the point P with polar coordinates $(-3, \pi)$ seems *not* to satisfy the polar equation $r = 2 + \cos\theta$. A closer look, however, shows that P can also be written with polar coordinates $(3, 0)$—which *do* satisfy the given equation. Here's the moral:

   A point P lies on the graph of a polar equation if P has any pair of polar coordinates that satisfy the equation.

Drawing Polar Graphs

The simplest polar graphs come from functions, usually of the form $r = f(\theta)$. Given such a function and a specific θ-domain, it's a routine matter to tabulate points and then plot them. We illustrate by example.

EXAMPLE 1   Plot the equation $r = 2 + \cos\theta$ for $0 \le \theta \le 2\pi$.

Solution   Let $f(\theta) = 2 + \cos\theta$. We want the r-θ graph of f. First we tabulate some values.

\multicolumn{13}{c}{**Values of $r = f(\theta) = 2 + \cos\theta$**}
θ	0	$\dfrac{\pi}{6}$	$\dfrac{\pi}{3}$	$\dfrac{\pi}{2}$	$\dfrac{2\pi}{3}$	$\dfrac{5\pi}{6}$	π	$\dfrac{7\pi}{6}$	$\dfrac{4\pi}{3}$	$\dfrac{3\pi}{2}$	$\dfrac{5\pi}{3}$	$\dfrac{11\pi}{6}$	2π
r	3	2.87	2.5	2	1.5	1.14	1	1.14	1.5	2	2.5	2.87	3

Next we plot the data (polar "graph paper" makes the job easier) and fill in the gaps smoothly. Here's the result:

A polar graph: $r = 2 + \cos\theta$

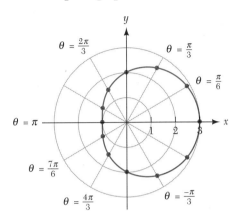

Polar Graphs: A Sampler

Several polar graphs follow. The simplest polar graphs of all—those of the equations $r = a$ and $\theta = b$—we've seen already.

Cardioids and Limaçons. Graphs of the form $r = a \pm b\cos\theta$ and $r = a \pm b\sin\theta$, where a and b are positive numbers, are called **limaçons**; if $a = b$, the term **cardioid** ("heartlike") is used. (The graph in Example 1 is a limaçon.) The following graphs illustrate the variety of limaçons and show the effects of the constants a and b.

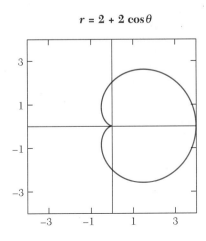

$r = 2 + 2\cos\theta$

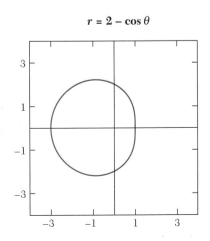

$r = 2 - \cos\theta$

$$r = 1 + 2\cos\theta$$

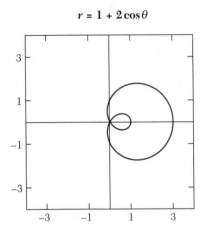

$$r = 2 + 2\sin\theta$$

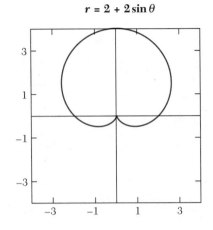

$$r = 2 - \sin\theta$$

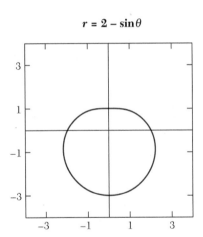

$$r = 1 + 2\sin\theta$$

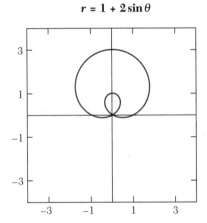

Notice the following features of the graphs.

We could have used the interval $-\pi \le t \le \pi$ instead.

What θ-range? The graphs were drawn by letting θ vary through the interval $[0, 2\pi]$. Since all functions involved are 2π-periodic, any other interval of length 2π ◄ would produce the same result.

Symmetry Three of the preceding limaçons—those that involve the cosine function—are symmetric about the x-axis, i.e., the line $\theta = 0$. (The other three are symmetric about the y-axis.) This symmetry occurs because the cosine function is *even*: For any θ, $\cos\theta = \cos(-\theta)$. The other graphs are symmetric about the y-axis because the sine function is odd. (For more on symmetry, see the exercises at the end of this section.)

Inner Loops   Each of the limaçons $r = 1 + 2\cos\theta$ and $r = 1 + 2\sin\theta$ has an inner loop. A close look at the graphs and the formulas reveals that these loops correspond to *negative* values of r. For $r = 1 + 2\cos\theta$, for instance, we have $r = 0$ when $\theta = 2\pi/3$ or $\theta = 4\pi/3$, and $r < 0$ for $2\pi/3 < \theta < 4\pi/3$. For these θ-values, therefore, the curve is drawn on the opposite side of the origin.

Roses.   Equations of the form $r = a\cos(k\theta)$ and $r = a\sin(k\theta)$, where a is a constant and k is a positive integer, produce graphs called **roses**. The following pictures explain the name.

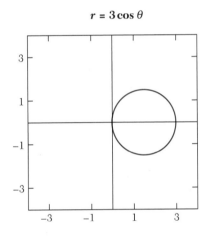

$r = 3\cos\theta$

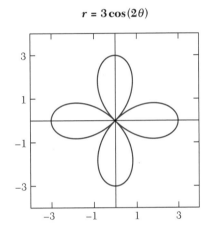

$r = 3\cos(2\theta)$

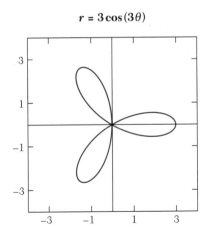

$r = 3\cos(3\theta)$

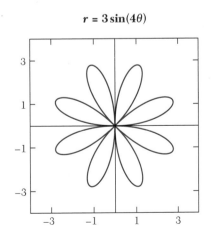

$r = 3\sin(4\theta)$

$$r = 3\sin(5\theta)$$

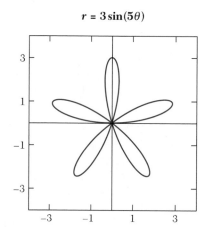

$$r = 3\sin(6\theta)$$

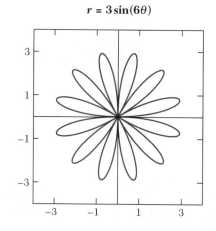

Notice the following features of the graphs.

Symmetry Like limaçons (and for the same reason), all roses are symmetric about an axis—"cosine roses" about the x-axis and "sine roses" about the y-axis.

The Rose's Radius The coefficient a in $r = a\cos(k\theta)$ and $r = a\sin(k\theta)$ determines the rose's "radius."

How Many Petals? The coefficient k in $r = a\cos(k\theta)$ and $r = a\sin(k\theta)$ determines the number of "petals": k if k is odd, $2k$ if k is even. But here's a subtlety, best revealed by plotting some roses by hand:

If k is odd, then each petal is traversed *twice* for $0 \leq \theta \leq 2\pi$.

In other words, for odd k, the rose $r = a\cos(k\theta)$ (or $r = a\sin(k\theta)$) has k *double* petals.

Trading Polar and Rectangular Coordinates

How are the polar coordinates (r, θ) of a point P related to the rectangular coordinates (x, y) of the same point? How can either type of coordinates be found from the other? Following is a useful picture.

Relating polar and rectangular coordinates

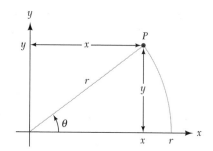

The picture illustrates many relations ▶ among x, y, r, and θ. Among the simplest   *Convince yourself of each.*
are

$$x = r\cos\theta; \qquad y = r\sin\theta; \qquad r^2 = x^2 + y^2; \qquad \tan\theta = \frac{y}{x}.$$

(The last equation holds only if $x \neq 0$.)

These relations let us convert from one type of coordinates to the other. Equations in x and y, for instance, are easy to rewrite in terms of r and θ, as the next two examples illustrate.

EXAMPLE 2   Find a polar equation for the straight line $y = mx + b$.

Solution   Substituting $x = r\cos\theta$ and $y = r\sin\theta$ into the equation for the line gives

$$y = mx + b \iff r\sin\theta = m(r\cos\theta) + b \iff r = \frac{b}{\sin\theta - m\cos\theta}.$$

(One moral is that rectangular coordinates are better suited to straight lines than are polar coordinates!)   ■

EXAMPLE 3   The graph of $r = 3\cos\theta$ *looks* like a circle. Is it a circle? Which circle?

Solution   It *is* a circle. Changing to rectangular coordinates shows why:

$$r = 3\cos\theta \implies r^2 = 3r\cos\theta \implies x^2 + y^2 = 3x.$$

As expected, the last equation does define a circle. To decide *which* circle, complete the square: ▶   *Check details.*

$$x^2 + y^2 = 3x \implies x^2 - 3x + y^2 = 0 \implies \left(x - \frac{3}{2}\right)^2 + y^2 = \frac{9}{4}.$$

The circle therefore has radius 3/2 and center at (3/2, 0)—just as the picture suggests.   ■

BASICS

In Exercises 1–6, a point is given in rectangular coordinates. Plot the given point, then give three different pairs of *polar* coordinates for the same point. For at least one pair, r should be negative. [**NOTE:** Answers involve familiar angles θ, so give exact answers, rather than decimal approximations.]

1. $(1, 1)$

2. $(-1, 1)$

3. $(1, \sqrt{3})$

4. $(-\sqrt{3}, 1)$

5. $(\pi, 0)$

6. $(0, \pi)$

In Exercises 7–12, a point is given in rectangular coordinates. Plot the given point, then give three different pairs of *polar* coordinates for the same point. For at least one pair, r should be negative. [**NOTE:** Angles in this problem are not necessarily familiar, so use a calculator in radian mode. Round answers to three decimal places.]

7. $(1, 2)$

8. $(-1, 2)$

9. $(1, 4)$

10. $(1, 100)$

11. $(0.356, 0.478)$

12. $(-0.356, -0.478)$

In Exercises 13–18, a point is given in polar coordinates. Plot the given point, then give *rectangular* coordinates for the same point. [**NOTE:** Answers involve familiar angles θ, so give exact answers rather than decimal approximations.]

13. $(2, \pi/4)$

14. $(-2, 5\pi/4)$

15. $(1, 13\pi/6)$

16. $(42, 0)$

17. $(a, 0)$ (*a* any positive number)

18. $(-a, 0)$ (*a* any positive number)

In Exercises 19–24, a point is given in polar coordinates. Plot the given point, then give *rectangular* coordinates for the same point. [**NOTE:** Angles are not necessarily familiar, so use a calculator in radian mode. Round answers to three decimal places.]

19. $(1, 1)$

20. $(-1, 1)$

21. $(2, 2)$

22. $(2, -2)$

23. $(1, \arctan 1)$

24. $(1, \arctan 2)$

25. The point $P(1, 0)$ has identical rectangular and polar coordinates. What other points have this property?

26. Imitate Example 1, page 656, to plot the limaçon $r = 2 + \sin\theta$. Proceed as follows:
    (a) Make a table of values like that in Example 1—let θ range from 0 to 2π in steps of $\pi/6$. Round r-values to two decimals.
    (b) Copy the polar grid shown in Example 1. On it, plot the points calculated in part (a). Join them with a smooth curve.
    (c) Discuss the symmetry of the resulting limaçon. What is its axis of symmetry?
    (d) Look carefully at the table of values in Example 1; notice how the r-values are symmetric about $\theta = \pi$. (The graph in Example 1 is also symmetric about $\theta = \pi$.) What similar type of symmetry does your table from part (a) show? Does your graph agree?

27. Imitate Example 1, page 656, to plot the cardioid $r = 1 + \cos\theta$. Proceed as follows:
    (a) Make a table of values like that in Example 1—let θ range from 0 to 2π in steps of $\pi/6$. Round r-values to two decimals.
    (b) Copy the polar grid shown in Example 1. On it, plot the points calculated in part (a). Join them with a smooth curve. Why does the name "cardioid" fit?
    (c) Discuss the symmetry of this cardioid. What is its axis of symmetry? How does the table of values in part (a) reflect the cardioid's symmetry?

28. Plot and discuss the limaçon $r = 1 - 2\cos\theta$, as follows:
    (a) Make a table of values like that in Example 1—let θ range from 0 to 2π in steps of $\pi/6$. Round r-values

to two decimals. Plot the points, and join them with a smooth curve.
    (b) For what values of θ is $r = 0$? How do these values appear on the graph?
    (c) On what θ-interval is $r < 0$? How does this interval show up on the graph?

In Exercises 29–34, draw a plot of the given polar equation. To avoid tedious point plotting, use the models of cardioids and limaçons shown on page 657–658. (A calculator or computer is OK but shouldn't be necessary. Be sure to label your graphs with appropriate units.)

29. $r = 3 + 3\cos\theta$

30. $r = 3 - \cos\theta$

31. $r = 1 + \sqrt{3}\cos\theta$

32. $r = 4 + 4\sin\theta$

33. $r = 4 - 2\sin\theta$

34. $r = 2 - 4\sin\theta$

In Exercises 35–40, sketch the given polar rose. To avoid tedious point plotting, use the models of roses shown on page 659–660. (A calculator or computer is OK but shouldn't be necessary. Be sure to label your graphs with appropriate units.)

35. $r = 2\sin\theta$

36. $r = 2\sin(2\theta)$

37. $r = 2\sin(3\theta)$

38. $r = 2\cos(4\theta)$

39. $r = 2\cos(5\theta)$

40. $r = 2\cos(1001\theta)$
    (rough is OK!)

41. We claimed in this section that for any numbers r and θ the pairs (r, θ), $(r, \theta + 2\pi)$, and $(-r, \theta + \pi)$ all describe the same point in the plane.
    (a) What does the claim say if $r = 1$ and $\theta = 0$? Is it true? Why or why not? Explain in your own words.
    (b) What does the claim say if $r = -1$ and $\theta = \pi/4$? Is it true? Why or why not? Explain in your own words.
    (c) The point with rectangular coordinates $(1, 0)$ can be written in polar coordinates as $(1, 2k\pi)$, where k is any integer, or as $(-1, (2k-1)\pi)$, where k is any integer. In the same sense, describe all the possible polar coordinates of the point with rectangular coordinates $(1, 1)$.

In Exercises 42–45, change the given equation in r and θ to an equivalent equation in x and y. Then plot the result, using whichever form seems simpler.

42. $r = 2\sec\theta$

43. $r = 4$

44. $\tan\theta = 1$

45. $r = 2\sin\theta$

In Exercises 46–49, change the given equation in x and y to an equivalent equation in r and θ. Then plot the result, using whichever form seems simpler.

46. $x^2 + y^2 = 9$

47. $y = 4$

48. $y = 2x$

49. $(x - 1)^2 + y^2 = 1$

50. (Use a graphing calculator or other polar plotting tool.) This exercise concerns limaçons of the form $r = 1 + a\cos\theta$, where a is any real constant. The following questions are open-ended. Answer them by experimenting with plots for various values of a: positive, negative, large, small, etc.

    (a) For which positive values of a does the graph have an inner loop?

    (b) What happens at $a = 1$?

    (c) What happens as $a \to 0$?

    (d) How are the graphs for a and $-a$ (e.g., $r = 1 + 0.5\cos\theta$ and $r = 1 - 0.5\cos\theta$) related to each other?

    (e) What happens as $a \to \infty$?

In Exercises 51–54, the graph is some sort of spiral. Plot each one, either by hand or by calculator (if the latter, be sure to adjust the θ-range appropriately).

51. $r = \theta$ for $0 \le \theta \le 2\pi$ (an ordinary spiral)

52. $r = \theta$ for $-2\pi \le \theta \le 2\pi$ (another ordinary spiral)

53. $r = \ln(\theta)$ for $1 \le \theta \le 4\pi$ (a logarithmic spiral)

54. $r = e^\theta$ for $-\pi \le \theta \le \pi$ (an exponential spiral)

13.2 Calculus in Polar Coordinates

In the last section we explored the polar coordinate system, polar equations, and the geometry of polar curves. This section is about calculus with polar curves. Two main problems stand out for polar curves, as for rectangular curves: finding the slope at a point and finding the area enclosed.

To visualize these problems, consider the following pictures.

Interesting points on a cardioid:
$r = 1 + \cos\theta$

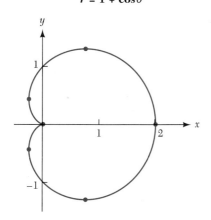

How large a rose? $r = \cos(2\theta)$

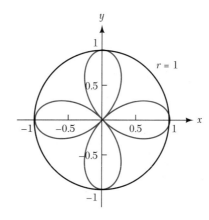

Among the questions we'll address are these:

Horizontal and Vertical Tangents Where—exactly—does the cardioid have horizontal and vertical tangent lines? The points marked with dots on the cardioid seem to be of interest, but what are their coordinates? ▶

Make a guess. Record it here in the margin.

Slope at Any Point   What's the slope of either curve at *any* point?

Areas   How much area does one petal of the rose enclose? What fraction of the circular area does the rose enclose? ◄

Make a guess. Record it here in the margin.

Slopes on Polar Curves

Both curves just shown are of this type.

Let's consider curves of the form $r = f(\theta)$, where f is a differentiable function. ◄ For given θ_0, we want the slope at the polar point (r_0, θ_0).

An Obvious—But Wrong—Guess.   Let's acknowledge, just for the record, that the desired slope is not $f'(\theta_0)$. A look at either of the preceding graphs quickly confirms this. For $f(\theta) = 1 + \cos\theta$, for instance, $f'(\theta) = -\sin\theta$, which varies only between -1 and 1. Yet it's clear at a glance that slopes on the cardioid take *all* real values; at three points, moreover, the cardioid has vertical tangent lines. Thus, $f'(\theta)$ is not the slope we seek, although it will play a role.

Polar Curves and Parametric Equations

We showed in earlier work how to find the slope at a point on a curve defined by parametric equations. Here is the result, for reference:

> **Fact** (**Slope of a Parametric Curve**)   Let a smooth curve C be given by parametric equations $x = f(t)$, $y = g(t)$, with $a \le t \le b$. If $f'(t) \ne 0$, then the slope dy/dx of C at (x, y) is given by
> $$\frac{dy}{dx} = \frac{g'(t)}{f'(t)} = \frac{dy/dt}{dx/dt}.$$

We explained them in the last section.

To *use* the result, we need first to write a polar curve $r = f(\theta)$ in parametric form. That is surprisingly easy to do. The key facts ◄ are the conversion formulas from polar to rectangular coordinates:

$$x = r\cos\theta; \qquad y = r\sin\theta.$$

For points on our polar curve $r = f(\theta)$, therefore,

$$x = r\cos\theta = f(\theta)\cos\theta; \qquad y = r\sin\theta = f(\theta)\sin\theta.$$

These are the desired parametric equations. We've rewritten the curve, originally given in r-θ form, as a pair of parametric equations, with θ as parameter. All that's left is to apply the preceding slope formula. We need $dx/d\theta$ and $dy/d\theta$; let's compute them now. By the product rule,

$$x = f(\theta)\cos\theta \implies \frac{dx}{d\theta} = f'(\theta)\cos\theta - f(\theta)\sin\theta;$$

$$y = f(\theta)\sin\theta \implies \frac{dy}{d\theta} = f'(\theta)\sin\theta + f(\theta)\cos\theta.$$

Everything is in place. Here's the result, stated with appropriate technical hypotheses: ▶

See the exercises at the end of this section for more on these hypotheses.

Fact (**Slope of a Polar Curve**) Let a curve C be given in polar coordinates by a function $r = f(\theta)$, $\alpha \le \theta \le \beta$, where f and f' are continuous on (α, β), and not simultaneously zero. Then, for θ in (α, β), the slope of C at $(r, \theta) = (f(\theta), \theta)$ is given by

$$\frac{dy}{dx} = \frac{dy/d\theta}{dx/d\theta} = \frac{f'(\theta)\sin\theta + f(\theta)\cos\theta}{f'(\theta)\cos\theta - f(\theta)\sin\theta}$$

wherever the denominator is not zero.

Using this Fact, we can answer the questions on tangents raised at the beginning of this section. The next example shows how. It also illustrates some of the caution needed in working with polar coordinates.

EXAMPLE 1   Consider the cardioid $r = 1 + \cos\theta$ shown on page 663. ▶ Where is the curve horizontal? Where is it vertical? What happens at the origin?

Take a close look.

Solution   Here $f(\theta) = 1 + \cos\theta$ and $f'(\theta) = -\sin\theta$, so

$$\frac{dy/d\theta}{dx/d\theta} = \frac{f'(\theta)\sin\theta + f(\theta)\cos\theta}{f'(\theta)\cos\theta - f(\theta)\sin\theta} = \frac{(1 - 2\cos\theta)(1 + \cos\theta)}{(2\cos\theta + 1)\sin\theta}.$$

(We used the preceding Fact and a little algebra.) ▶

Check our work.

The cardioid can have a horizontal tangent line only if the numerator $dy/d\theta$ is zero, i.e., if

$$(1 - 2\cos\theta)(1 + \cos\theta) = 0 \iff \cos\theta = \frac{1}{2} \quad \text{or} \quad \cos\theta = -1.$$

For θ in $[0, \pi]$, one condition or the other holds only if $\theta = \pi/3$ or $\theta = \pi$. (By symmetry, it's enough to look only on the upper half of the cardioid.)

The picture shows that, in fact, the upper half of the cardioid is horizontal *only* at $\theta = \pi/3$, i.e., at the point with polar coordinates $(3/2, \pi/3)$ and rectangular coordinates $(3/4, \ 3\sqrt{3}/4) \approx (0.75, 1.30)$. Symmetry dictates that the cardioid is also horizontal at $\theta = 5\pi/3$.

What happens at $\theta = \pi$, where the cardioid has a "cusp"? At $\theta = \pi$, both $dx/d\theta$ and $dy/d\theta$ are zero, and so the slope expression is undefined. It can be shown using l'Hôpital's rule, however, that the expression tends to zero as θ tends to π. Thus the cardioid can reasonably be said to have *zero* slope at $\theta = \pi$.

A vertical tangent line can occur only where the denominator $dx/d\theta$ is zero, i.e., if

$$\sin\theta \,(2\cos\theta + 1) = 0 \iff \sin\theta = 0 \quad \text{or} \quad \cos\theta = -\frac{1}{2}.$$

What are their rectangular coordinates?

For θ in $[0, \pi]$, one or the other holds if $\theta = 0$ or $\theta = 2\pi/3$. As the picture shows, both $\theta = 0$ and $\theta = 2\pi/3$ correspond to points of vertical tangency on the cardioid. These points have polar coordinates $(2, 0)$ and $(1/2, \ 2\pi/3)$, respectively. ◀

Superimposing the cardioid on a polar grid supports all the preceding calculations.

Horizontal and vertical points on a cardioid:
$$r = 1 + \cos\theta$$

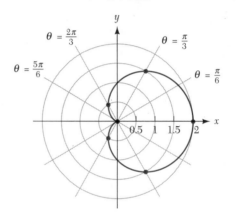

EXAMPLE 2   Let $a > 0$ be any positive constant. What does the slope formula say about the circle $r = a$?

Solution   For the circle $r = a$, we have $f(\theta) = a$ and $f'(\theta) = 0$. Now the Fact says that

$$\frac{dy}{dx} = \frac{f'(\theta)\sin\theta + f(\theta)\cos\theta}{f'(\theta)\cos\theta - f(\theta)\sin\theta}$$

$$= \frac{a\cos\theta}{-a\sin\theta} = -\cot\theta = -\frac{1}{\tan\theta}.$$

Why perpendicular? Because the slopes are negative reciprocals.

Note what the last expression means: The tangent line to the circle at any point $P(a, \theta)$ is perpendicular to the ray from the origin to P. ◀

EXAMPLE 3   Let $r = f(\theta)$ describe a polar curve C; suppose that $f(\theta_0) = 0$. What does the formula

$$\frac{dy}{dx} = \frac{f'(\theta)\sin\theta + f(\theta)\cos\theta}{f'(\theta)\cos\theta - f(\theta)\sin\theta}$$

say about the slope of C at the point $(0, \theta_0)$?

Solution   The slope formula becomes much simpler if $f(\theta_0) = 0$. In that case, ▶

Remember that f and f' aren't simultaneously zero.

$$\frac{dy}{dx} = \frac{f'(\theta_0) \sin \theta_0}{f'(\theta_0) \cos \theta_0} = \tan \theta_0.$$

Note what this means: If a smooth polar curve passes through the origin at $\theta = \theta_0$, its tangent line is simply the ray $\theta = \theta_0$. For the rose $r = \cos(2\theta)$, for example, $r = 0$ when $\theta = \pi/4$, $\theta = 3\pi/4$, $\theta = 5\pi/4$, and $\theta = 7\pi/4$. As the following picture shows, the curve passes through the origin in just these directions.

The rose $r = \cos(2\theta)$:
tangent lines at the origin

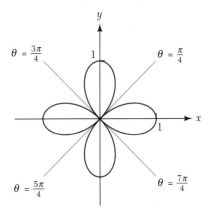

Finding Area in Polar Coordinates

The standard area problem in rectangular coordinates concerns the area defined by an ordinary $y = f(x)$-style graph for $a \le x \le b$. In polar coordinates, the standard area problem is a little different: to find the area bounded by a polar curve $r = f(\theta)$ for $\alpha \le \theta \le \beta$. The following pictures illustrate the "generic" situations; the areas in question are shaded.

Area in rectangular coordinates

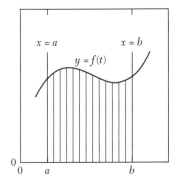

Area in polar coordinates

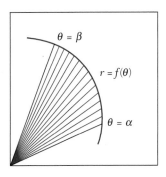

Areas by Integration: Rectangular vs. Polar

The different styles of shading in the two preceding figures—one vertical, the other radial—were chosen intentionally. They reflect two slightly different approaches to finding areas. In order to emphasize both similarities and differences between the rectangular and polar situations, the next two subsections are written in "parallel."

Area, Cartesian Style.   In rectangular coordinates, area is approximated by subdividing the region into thin *vertical strips*, each one based on the x-axis. Each strip corresponds to a small subinterval—say, $[x_i, \ x_i + \Delta x]$—obtained by partitioning the domain interval $a \leq x \leq b$ into n equal pieces. If Δx is small, then ◄   the strip is approximately a rectangle, with base Δx and height $f(x_i)$. Therefore,

We've done this often before, so we omit some details here.

$$\text{Area of one strip} \approx f(x_i) \, \Delta x.$$

Adding all n strips gives

$$\text{Total area} \approx \sum_{i=1}^{n} f(x_i) \, \Delta x.$$

Therefore, taking the limit as $\Delta x \to 0$,

$$\text{Total area} = \lim_{n \to \infty} \sum_{i=1}^{n} f(x_i) \, \Delta x = \int_{a}^{b} f(x) \, dx.$$

Area, Polar Style.   In polar coordinates, by contrast, area is approximated by subdividing the region into thin *pie-shaped wedges*, each with its vertex at the origin. ◄   Each wedge corresponds to a small subinterval—say, $[\theta_i, \ \theta_i + \Delta\theta]$—obtained by partitioning the domain interval $\alpha \leq \theta \leq \beta$ into n equal pieces. If $\Delta\theta$ is small, then each wedge is approximately a sector of a circle with radius $f(\theta_i)$; ◄ the wedge makes the angle $\Delta\theta$ at the origin.

Look carefully at the polar picture: it shows 20 radial wedges.

Remember: $r = f(\theta)$ gives the radius for given θ.

Now, a brief detour. Just ahead, we'll need to know the area of a **circular sector** as just described. The answer, although perhaps less familiar than the area of a rectangle, is easy to find. A full circle of radius R (for which θ runs from 0 to 2π) encloses total area πR^2. A wedge making angle $\Delta\theta$ at the origin represents the fraction $(\Delta\theta)/(2\pi)$ of the total circle, so

$$\text{Area of wedge} = \pi R^2 \times \frac{\Delta\theta}{2\pi} = \frac{R^2}{2} \, \Delta\theta.$$

If, say, $R = f(\theta_i)$, then a circular wedge has area $f(\theta_i)^2 \Delta\theta / 2$.

Back to the main road. Our detour showed that in the "generic" case, illustrated earlier,

$$\text{Area of one wedge} \approx \frac{f(\theta_i)^2}{2} \, \Delta\theta.$$

Adding all n wedges gives

$$\text{Total area} \approx \sum_{i=1}^{n} \frac{f(\theta_i)^2}{2} \, \Delta\theta.$$

Taking the limit as $\Delta\theta \to 0$,

$$\text{Total area} = \lim_{n\to\infty} \sum_{i=1}^{n} \frac{f(\theta_i)^2}{2} \Delta\theta = \int_{\alpha}^{\beta} \frac{f(\theta)^2}{2} \, d\theta.$$

The formula is worth remembering.

Fact (Area in Polar Coordinates) Let f be a continuous function, and let R be the region in the xy-plane bounded by the polar curve $r = f(\theta)$ and the rays $\theta = \alpha$ and $\theta = \beta$. Then

$$\text{Area of } R = \int_{\alpha}^{\beta} \frac{f(\theta)^2}{2} \, d\theta.$$

EXAMPLE 4 What's the area of one leaf of the polar rose $r = f(\theta) = \cos(2\theta)$? What fraction of the circle $r = 1$ does the entire rose cover?

Solution As the picture on page 667 shows, the eastward-pointing leaf lies between $\theta = -\pi/4$ and $\theta = \pi/4$. To ease calculations, we integrate from $\theta = 0$ to $\theta = \pi/4$ and double the result. Here is the integral calculation (a table of integrals helps along the way): ▶

But check our work.

$$\int_0^{\pi/4} \frac{\cos^2(2\theta)}{2} \, d\theta = \frac{\theta}{4} + \frac{\sin(4\theta)}{16} \bigg]_0^{\pi/4} = \frac{\pi}{16}.$$

Thus, *one* leaf has area $\pi/8$, and all four leaves have area $\pi/2$—exactly half the area of the circle $r = 1$. ■

A Rose with Any Other n ... A more general (and more surprising) result than that of the last example is true: ▶

See the exercises for more on this.

For any even n, the rose $r = \cos(n\theta)$ encloses total area $\pi/2$—half the area of the enclosing circle.

EXAMPLE 5 How much area does the cardioid $r = 1 + \cos\theta$ enclose?

Solution With the integral formula, the answer is easy:

$$\text{Area} = \frac{1}{2} \int_0^{2\pi} f(\theta)^2 \, d\theta$$

$$= \frac{1}{2} \int_0^{2\pi} \left(1 + 2\cos\theta + \cos^2\theta\right) d\theta$$

$$= \frac{1}{2} \left(\theta + 2\sin\theta + \frac{\theta}{2} + \frac{\sin(2\theta)}{4}\right) \bigg]_0^{2\pi}$$

$$= \frac{3\pi}{2} \approx 4.71. \quad ■$$

BASICS

1. We claimed in this section that

$$x = f(\theta)\cos\theta \implies \frac{dx}{d\theta} = f'(\theta)\cos\theta - f(\theta)\sin\theta;$$

$$y = f(\theta)\sin\theta \implies \frac{dy}{d\theta} = f'(\theta)\sin\theta + f(\theta)\cos\theta.$$

Use the product rule to show that these claims are valid.

2. We claimed in Example 1 that on the cardioid $r = 1 + \cos\theta$, the slope dy/dx at (r, θ) is

$$\frac{(1 - 2\cos\theta)(1 + \cos\theta)}{(2\cos\theta + 1)\sin\theta}.$$

Use the Fact on page 665 to verify this.

3. Consider the spiral $r = \theta$, for $0 \le \theta \le 4\pi$.
   (a) Draw the spiral.
   (b) At what points does the spiral have horizontal tangent lines? Vertical tangent lines?
   (c) Write an equation in rectangular form for the line tangent to the spiral at the polar point $(1, 1)$. Draw this tangent line on your graph.

4. Consider the limaçon $r = 1 + 2\sin\theta$.
   (a) Draw the limaçon.
   (b) Write an equation in rectangular form for the line tangent to the limaçon at the polar point $(1, 0)$. Draw this tangent line on your graph.
   (c) Find the slope of the line tangent to the limaçon at the polar point $(0, 7\pi/6)$. Draw this tangent line on your graph.
   (d) Find the slope of the line tangent to the limaçon at the polar point $(0, 11\pi/6)$. Draw this tangent line on your graph.
   (e) At what points does the limaçon have horizontal tangent lines?

5. This open-ended problem is about the family of limaçons of the form $r = 1 + a\cos\theta$.
   (a) After plotting example limaçons for various values of a, determine which ones have a "dimple" (i.e., a pushed-in section on either the right or the left) and which ones don't. [**HINT**: Any limaçon with a dimple has three vertical tangent lines.]
   (b) The answer to part (a) is that there's no dimple if $|a| \le 1/2$. Show this using derivatives.
   [**HINT**: Which values of a lead to three vertical tangent lines.]

In Exercises 6–9, draw the region bounded by the given polar curves, and find its area.

6. $r = 1$, $\theta = 0$, $\theta = \pi$
7. $r = 3$, $\theta = 0$, $\theta = \pi$
8. $r = a$, $\theta = 0$, $\theta = \pi$
9. $r = 1$, $\theta = 0$, $\theta = \beta$

In Exercises 10–13, draw the given region and find its area.

10. The outer loop of the limaçon $r = 1 + 2\sin\theta$.
11. The inner loop of the limaçon $r = 1 + 2\sin\theta$.
12. The region bounded by $r = \sec\theta$, $\theta = 0$, $\theta = \pi/4$.
13. The region bounded by $r = \sec\theta$, $\theta = 0$, $\theta = \arctan m$.

14. Consider the $2n$-leafed rose $r = \cos(n\theta)$, where n is even.
    (a) Find the area of one leaf.
    (b) Find the area of all $2n$ leaves. What fraction of the circle $r = 1$ does the entire rose fill up?

15. Consider the n-leafed rose $r = \cos(n\theta)$, where n is odd.
    (a) Find the area of one leaf.
    (b) Find the area of n leaves. What fraction of the circle $r = 1$ does the entire rose fill up?

16. Draw and then calculate the area bounded by one "turn" of the spiral $r = \theta$, i.e., from $\theta = 0$ to $\theta = 2\pi$.

17. Draw and then calculate the area bounded by one "turn" of the exponential spiral $r = e^{\theta}$, i.e., from $\theta = 0$ to $\theta = 2\pi$.

18. Draw and then calculate (or estimate numerically) the area bounded by one "turn" of the logarithmic spiral $r = \ln\theta$, i.e., from $\theta = 2\pi$ to $\theta = 4\pi$.

19. Use polar coordinates to find the area of the region inside the circle $r = 1$ and to the right of $x = 1/2$.

20. Let $0 < a < 1$. Use polar coordinates to find the area of the region inside the circle $r = 1$ and to the right of $x = a$.

In this section we used the idea that a polar curve can also be thought of as a parametric curve. Specifically, the curve defined by the polar function $r = f(\theta)$, for $\alpha \le \theta \le \beta$, can also be defined by the parametric equations $x = f(\theta)\cos\theta$, $y = f(\theta)\sin\theta$, $\alpha \le \theta \le \beta$. Consider, for example, the unit circle $r = 1$, $0 \le \theta \le 2\pi$. Then $f(\theta) = 1$, so the parametric form is simply $x = f(\theta)\cos\theta = \cos\theta$, $y = f(\theta)\sin\theta = \sin\theta$, $0 \le \theta \le 2\pi$.

In Exercises 21–28, a curve is given in polar form. First rewrite the curve in parametric form, then plot it. (Use a graphing calculator or computer.) Do you see the "expected" results?

21. $r = 2$, $0 \le \theta \le 2\pi$

22. $r = 2$, $0 \le \theta \le \pi$

23. $r = \sec\theta$, $-\pi/4 \le \theta \le \pi/4$

24. $r = \csc\theta$, $\pi/4 \le \theta \le 3\pi/4$

25. $r = \theta, \quad 0 \le \theta \le 2\pi$

26. $r = \cos\theta, \quad 0 \le \theta \le \pi$

27. $r = \cos(2\theta), \quad 0 \le \theta \le 2\pi$

28. $r = 1 + \cos\theta, \quad 0 \le \theta \le 2\pi$

29. The formula for the slope of a polar curve requires that f and f' not be simultaneously zero. Show that if this condition holds, then $dy/d\theta$ and $dx/d\theta$ are not simultaneously zero. [**HINT:** Recall that $dy/d\theta = f'(\theta)\sin\theta + f(\theta)\cos\theta$ and $dx/d\theta = f'(\theta)\cos\theta - f(\theta)\sin\theta$. Look at $(dy/d\theta)^2 + (dx/d\theta)^2$.]

14

Multivariable Calculus: A First Look

14.1 Three-Dimensional Space

Single-variable calculus is done mainly in the two-dimensional xy-plane. The Euclidean plane, also known as $\mathbb{R}^2$, is the natural home of such familiar calculus objects as the graph $y = f(x)$, tangent lines to that graph at various points, and the various regions whose areas we might measure by integration.

To do *multivariable* calculus, we need more room. The graph of $z = f(x, y)$, where f is a function of two input variables, lives in *three*-dimensional xyz-space. With three dimensions to work in, we will "see" not only this graph but a variety of multivariable analogues of derivatives and integrals. This section explores three-dimensional Euclidean space, or $\mathbb{R}^3$, where we'll spend most of our time.

In another sense, of course, we spend *all* of our time in $\mathbb{R}^3$. In everyday usage, "space" connotes three physical dimensions. The intuition we gain from living in three spatial dimensions is often useful in mentally picturing and manipulating the objects of multivariable calculus. Familiar as it is, however, three-dimensional space poses special problems for visualization. Two-dimensional pictures (on paper or on a computer screen) of three-dimensional objects are *always* more or less distorted or incomplete. Minimizing such problems is an active science (and an art) in its own right; doing so means carefully controlling viewpoint, perspective, shading, lighting, and other factors.

This chapter's main subject is multivariable calculus, not computer graphics (although we sometimes mention computer graphics) or technical drawing, so we draw pictures to illustrate ideas as simply as possible, not necessarily to look as lifelike as possible. It's worth remarking, however, that many of the basic tools and methods of computer graphics draw directly on the ideas we'll develop in this chapter.

Cartesian Coordinates in Three Dimensions

The *idea* of Cartesian coordinates is the same in both two and three dimensions, but the pictures look a little different. Compare these:

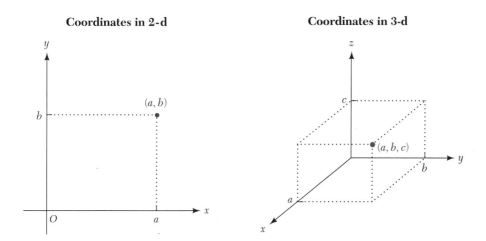

Coordinates in 2-d   **Coordinates in 3-d**

Recall the formalities in the xy-plane. A Cartesian coordinate system consists of an origin, labeled O, and horizontal and vertical coordinate axes, labeled x and y, ▶ passing through O. On each axis we choose a positive direction (usually "east" and "north") and a unit of measurement (not necessarily the same on both axes).

We occasionally use other axis labels.

Given such a coordinate system, every point P in the plane corresponds to one and only one ordered pair (a, b) of real numbers, called the Cartesian coordinates of P. The pair (a, b) can be thought of as P's "Cartesian address." To reach P from the origin, move a units in the positive x-direction and b units in the positive y-direction. ▶

If a or b is negative, go the other way.

Coordinates in three-dimensional xyz-space work the same way—but with *three* coordinate axes, labeled x, y, and z. Each axis is perpendicular to the other two. ▶ To reach the point $P(a, b, c)$ from the origin, go a units in the positive x-direction, b units in the positive y-direction, and c units in the positive z-direction. As the preceding figure illustrates, the resulting point $P(a, b, c)$ can also be thought of as a corner (the one opposite the origin) of a rectangular solid with dimensions $|a|$, $|b|$, and $|c|$.

There's "room" in $\mathbb{R}^3$ for three mutually perpendicular axes. $\mathbb{R}^2$ has room for only two.

Quadrants and Octants. The two axes divide the xy-plane into four **quadrants**, defined by the pattern of positive or negative x- and y-coordinates. The analogous regions in xyz-space are called **octants**. The first octant, for instance, consists of all points (x, y, z) with all three coordinates positive. In the preceding picture, only the first octant is visible. The next picture gives another view. There are eight octants in all in xyz-space—one for each of the possible patterns of signs of the three coordinates: $(+, +, +)$, $(-, +, +)$, ..., $(-, -, -)$. ▶

Are you certain that 8 is the right number? See the exercises at the end of this section for more.

Coordinate Planes. One can think of the first octant as a room, with the origin at the lower left corner of the front wall. At this point three walls meet, all at right angles. These "walls" are known as the **coordinate planes**: the yz-plane (the front wall), the xy-plane (the floor), and the xz-plane (the left wall). ▶ The coordinate

Be sure you agree; look carefully at the standard picture.

planes correspond to simple equations in the variables x, y, and z. The yz-plane, for example, is the graph of the equation $x = 0$, i.e., the set of all points (x, y, z) that satisfy this equation. Similarly, the xy- and xz-planes are graphs of the equations $z = 0$ and $y = 0$, respectively.

Many Possible Views.   The xy-plane, being "flat," is relatively easy to draw. Simulating three-dimensional space on a flat page or computer screen is much harder, and there is always some price to be paid in distortion. For example, in 3-d reality, the x-, y-, and z-axes are all perpendicular to each other, but no flat picture can really show this. The following axes, for instance, don't make right angles on the page. ◄

Study this picture carefully; it's worth the effort.

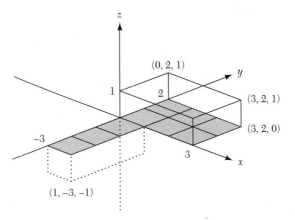

This picture's view of xyz-space is somewhat different from that of the last picture. Observe the following features.

Horizontal and Vertical   The xy-plane (in which the shaded "floor tiles" lie) is drawn to appear horizontal. The z-axis is vertical; the positive direction is up. This is a standard convention, which we'll follow consistently.

Hidden Lines   The dashed lines in the picture lie "below" the xy-plane. They would be hidden from view if the xy-plane (the "floor") were opaque. How much of xyz-space is considered to be visible is a matter of choice. Sometimes only the first octant is shown.

Positive Directions   An arrow on each axis indicates the positive direction. The 3×2 block of shaded squares lies in the first quadrant of the xy-plane. The other shaded squares lie in the plane's fourth quadrant.

Plotting Points: Positive and Negative Coordinates   Any point $P(a, b, c)$ is plotted the same way: From the origin, move a, b, and c units in the positive x-, y-, and z-directions, respectively. Negative coordinates cause no special problem; just move the other way.

Where's the Viewer?   The picture is drawn as though the viewer were floating somewhere above the *fourth* quadrant of the xy-plane. ◄   In the earlier 3-d picture, by contrast, the viewer floats somewhere above the *first* quadrant. There's nothing sacred about *either* viewing angle; we'll use a variety of viewpoints as we go along. For that matter, so do the various computer plotting packages readers may have at hand.

Think about this. Do you agree?

No Perspective To a human viewer, rectangular boxes like the preceding ones would appear in perspective; the sides would taper toward a vanishing point. The closer the viewer, the more pronounced those effects would be. For the sake of simplicity, we ignore perspective effects in the picture. In effect, the viewer is assumed to be *very* far from the origin, perhaps looking through a telescope.

There is no single "best" picture of a 3-d object; choosing a good or convenient view may depend on properties of the object, what needs emphasis, or even the drawing technology at hand. ▶

Computer, calculator, pencil, sharp stick, ...

Distance and Midpoints

Let $P(x_1, y_1)$ and $Q(x_2, y_2)$ be any two points in the xy-plane. Recall that the distance from P to Q ▶ is given by the familiar Pythagorean formula

Or from Q to P—it doesn't matter.

$$d(P, Q) = \sqrt{(x_2 - x_1)^2 + (y_2 - y_1)^2}$$

and that the midpoint M of the segment joining P to Q has these "averaged" coordinates:

$$M = \left(\frac{x_1 + x_2}{2}, \frac{y_1 + y_2}{2} \right).$$

The formulas in three dimensions aren't much different.

Definition The distance between $P(x_1, y_1, z_1)$ and $Q(x_2, y_2, z_2)$ is
$$d(P, Q) = \sqrt{(x_2 - x_1)^2 + (y_2 - y_1)^2 + (z_2 - z_1)^2}.$$
The midpoint of the segment joining P and Q has coordinates
$$M\left(\frac{x_1 + x_2}{2}, \frac{y_1 + y_2}{2}, \frac{z_1 + z_2}{2} \right).$$

Both definitions are simply three-dimensional versions of the corresponding formulas in the xy-plane. In both two and three dimensions, for example, distance is computed as the square root of the sum of the squared differences in coordinates. ▶

In either two or three dimensions, the distance formula reflects the Pythagorean rule. See the exercises for more details.

EXAMPLE 1 Consider the points $P(0, 0, 0)$ and $Q(2, 4, 6)$. Find the distance from P to Q and the midpoint M of the segment joining them. How far is M from P and from Q?

Solution By the distance formula,

$$d(P, Q) = \sqrt{(2 - 0)^2 + (4 - 0)^2 + (6 - 0)^2} = \sqrt{56} \approx 7.483.$$

According to the formula, the midpoint is $M(1, 2, 3)$; each coordinate of M splits the difference between the corresponding coordinates of P and Q. To see why M

deserves the name "midpoint," notice that

$$d(P, M) = \sqrt{(1-0)^2 + (2-0)^2 + (3-0)^2}$$
$$= \sqrt{(2-1)^2 + (4-2)^2 + (6-3)^2} = \sqrt{14} \approx 3.742.$$

Thus, M lies halfway between P and Q, as a midpoint should.   ■

Equations and Their Graphs

The graph of an equation in x and y is the set of all points (x, y) that satisfy the equation. The graph of $x^2 + y^2 = 1$, for instance, is a circle of radius 1 in the xy-plane, centered at the origin. The graph of the equation $x = 0$ is the y-axis. ◄

The graph of an equation may or may not be the graph of a function. The unit circle is not a function graph.

The same idea applies for three variables: The graph of an equation in x, y, and z is the set of points (x, y, z) in space that satisfy the equation. Here, graphically, are three simple examples (only the first octant is shown).

Three simple graphs

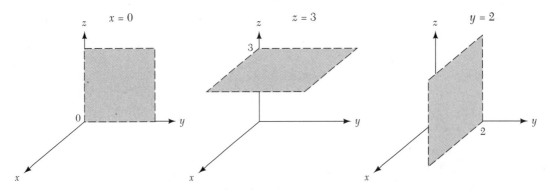

Solutions of the equation $x = 0$ are points of the form $(0, y, z)$, so the graph is the yz-plane. Similarly, solutions of $z = 3$ are all points of the form $(x, y, 3)$, so the graph is a horizontal plane, floating 3 units above the xy-plane. The graph of $y = 2$ is parallel to the xz-plane but moved 2 units in the positive y-direction.

Notice that in each case, the graph of an equation in x, y, and z is a plane—a *two-dimensional* object. In contrast, the graph of one equation in x and y is usually a curve or a line—a *one-dimensional* object. The pattern is the same in both cases:

Watch for this pattern as we go along.

The graph of an equation has dimension 1 less than the number of variables. ◄

A few special types of graphs in xyz-space deserve special mention.

Planes

A **linear equation** is one of the form $ax + by + cz = d$, where a, b, c, and d are constants and at least one of a, b, and c is nonzero. ◄   All three of the equations just plotted are linear. They illustrate an important general fact:

What goes wrong if $a = b = c = 0$?

The graph of any linear equation is a plane.

(We explain this fact carefully in a later section.) To draw planes in xyz-space, we use the fact that a plane is uniquely determined by three points (unless the points happen to be on a straight line).

EXAMPLE 2   Plot the linear equation $x + 2y + 3z = 3$ in the first octant.

Solution   First we'll find some points (x, y, z) that satisfy $x + 2y + 3z = 3$. If anything, this is too easy—there are infinitely many possibilities. Given *any* values for x and y, the equation determines a corresponding value for z. If, say, $x = 1$ and $y = 1$, then $x + 2y + 3z = 3$ can hold only if $z = 0$. Similarly, setting $y = 2$ and $z = 3$ forces $x = -10$. ▶ Among all possible solutions, three of the simplest are

Check this calculation.

$$P(3, 0, 0); \qquad Q\left(0, \frac{3}{2}, 0\right); \qquad R(0, 0, 1).$$

These solutions are both easy to find (set two coordinates to zero and solve for the third) and easy to plot (they lie on the coordinate axes). Now we can plot the plane:

A plane in the first octant:
$x + 2y + 3z = 3$

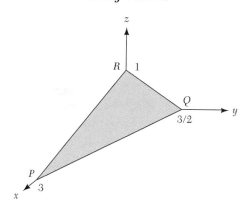

Notice the following features.

Intercepts   A typical *line* in the xy-plane has x- and y-intercepts where the line intersects the coordinate axes. In a similar sense, a typical *plane* in xyz-space has x-, y-, and z-intercepts. In the picture, the intercepts are P, Q, and R. ▶

Traces   If we "slice" a surface in xyz-space with a plane, the intersection of the surface with the plane is called the **trace** of the surface in that plane. The plane p shown in the picture meets each of the three coordinate planes in a straight line. Those three lines are therefore the traces of the surface $x + 2y + 3z = 3$ in the xy-plane, the xz-plane, and the yz-plane, respectively.

  It's easy to find equations for these traces. For example, a point (x, y, z) lies both in the plane p *and* in the xy-plane if and only if it satisfies both $x + 2y + 3z = 3$ and $z = 0$. Setting $z = 0$ in the first equation gives $x + 2y = 3$—as expected, the equation of a line in the xy-plane. This line is therefore the trace of p in the xy-plane. ▶   ■

Not every line in the xy-plane intercepts both axes; not every plane in space intercepts all three axes. See the exercises at the end of the section for more on this.

Do you see this line in the picture?

Spheres

In the plane, a circle of radius $r > 0$ and center $C(a, b)$ is the set of points $P(x, y)$ at distance r from (a, b). Translating this description into symbolic language produces the familiar formula for a circle in the plane:

$$d(P, C) = \sqrt{(x - a)^2 + (y - b)^2} = r, \quad \text{or} \quad (x - a)^2 + (y - b)^2 = r^2.$$

(Squaring both sides does no harm and simplifies the equation's appearance.)

The object in space that is analogous to a circle in the plane is a **sphere** of radius r. Like a circle, a sphere is "hollow," similar to an empty orange skin. Adding the interior (the edible part of the orange) produces a ball. Like a circle, a sphere is the set of points at some fixed distance—the radius—from a fixed center point. Given a radius $r > 0$ and a center point $C(a, b, c)$, the sphere of radius r centered at C is the set of points (x, y, z) such that

$$d(P, C) = \sqrt{(x - a)^2 + (y - b)^2 + (z - c)^2} = r,$$
$$\text{or} \quad (x - a)^2 + (y - b)^2 + (z - c)^2 = r^2.$$

The simplest example, the **unit sphere**, has center $(0, 0, 0)$ and radius 1. Its equation reduces to this simple form:

$$x^2 + y^2 + z^2 = 1.$$

For starters, circles in space don't always look circular. Depending on the viewing angle, they may look like ellipses.

Drawing circles in the xy-plane is easy, even by hand. Drawing spheres (or any "curved" objects, for that matter) convincingly by hand is much harder. ◄ Fortunately, rough sketches usually suffice.

Completing the square may reveal an equation's spherical form.

EXAMPLE 3   Is the graph of $x^2 - 2x + y^2 - 4y + z^2 - 6z = 0$ a sphere? Which one?

Check each step.

Solution   Completing the square in each variable separately gives ◄

$$x^2 - 2x + y^2 - 4y + z^2 - 6z = 0 \iff$$
$$(x^2 - 2x + 1) + (y^2 - 4y + 4) + (z^2 - 6z + 9) = 1 + 4 + 9 \iff$$
$$(x - 1)^2 + (y - 2)^2 + (z - 3)^2 = 14.$$

The last form shows that our equation describes the sphere of radius $\sqrt{14}$ centered at $(1, 2, 3)$. As the equation shows, this sphere passes through the origin. The picture shows this, too:

Graph of $x^2 - 2x + y^2 - 4y + z^2 - 6z = 0$

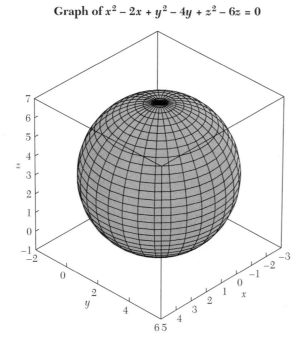

Cylinders

What is the graph of the equation $y = x^2$? The answer depends on where we're working. In the xy-plane, the graph is the familiar parabola—all points of the form (x, x^2). Here's the graph of the same equation in xyz-space:

Graph of $y = x^2$ in xyz-space

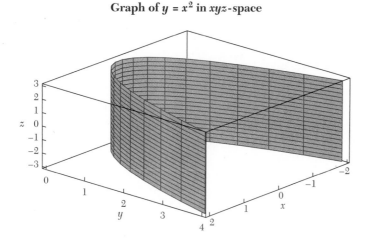

Observe:

A Missing Variable The graph is *unrestricted in the z-direction*—it contains all points that lie directly above or below the graph of $y = x^2$ in the xy-plane. ▶ The graph has this property because z is "missing" in the equation $y = x^2$. This means that if (x, y) satisfies the equation, then so does *every* point (x, y, z), regardless of the value of z.

In other words, the graph has "vertical walls."

*"Simple" is relative.
Drawing anything in
xyz-space poses certain
challenges.*

What's a Cylinder? Graphs like this one, in which at least one of the variables is unrestricted, are called **cylinders**. Any equation that omits one or more variables—$y = z$, say—has a cylindrical graph. Plotting cylinders is comparatively simple. ◄   If the equation involves only y and z, for instance, we first plot the equation in the yz-plane and then "extend" the graph in the x-direction.

In everyday speech, "cylinder" usually means a circular tube. As the next example illustrates, the mathematical idea of a cylinder is much more general.

EXAMPLE 4   Discuss the graph in xyz-space of the equation $z = 2 + \sin y$. Interpret the result as a cylinder.

*The surface, like many
graphs, continues forever; a
picture shows only part of
the graph.*

Solution   There's no variable x in the equation, so the graph is unrestricted in the x-direction; i.e., it's a cylinder in x. Here's a representative view: ◄

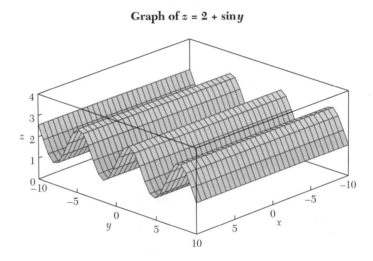

Graph of $z = 2 + \sin y$

The graph resembles the surface of an idealized ocean, with regular waves moving parallel to the y-axis. Each wave is infinitely long, with its trough and crest parallel to the x-axis. Notice especially how the surface meets the yz-plane. The curve of intersection—i.e., the trace of the surface in the yz-plane—is the ordinary sine curve $z = 2 + \sin y$. In fact, the entire surface can be thought of as infinitely many identical copies of this curve, one for each value of x.   ∎

BASICS

1. We said in this section that the graph of $y = x^2$ is a cylinder in xyz-space, unrestricted in the z-direction. (See the picture on page 679.)

   (a) Plot the equation $z = y^2$ in xyz-space. What is the unrestricted direction?   [**HINT:** Start in yz-space.]

   (b) Plot the equation $z = x^2$ in xyz-space. What is the unrestricted direction?

2. Plot the equation $x^2 + y^2 = 1$, first in xy-space, then in xyz-space. (The second graph should resemble a vertical pipe—a cylinder, in the everyday sense of the word.)

3. In each part, plot the given equation in xyz-space (a rough sketch is fine), then describe the graph in words.
   (a) $y^2 - z^2 = 0$ (c) $y^2 + z = -1$
   (b) $y^2 + z^2 = 0$ (d) $y^2 + z^2 = -1$
   [**HINT:** Does the graph contain any points?]

4. Find an equation in x, y, and z for the graph in xyz-space of
   (a) a sphere of radius 2, centered at the origin
   (b) a sphere of radius 1, centered at $(1, 1, 1)$
   (c) a circular cylinder of radius 1, centered along the y-axis
   (d) a circular cylinder of radius 2, centered along the z-axis
   (e) a cylindrical surface that resembles an ocean, with waves rolling in the x-direction (see Example 4, page 680).

5. The equation $z = 3$ omits two variables. Therefore, its graph in xyz-space should be a cylinder in both the x-direction and the y-direction. Is it? What is the trace of the graph in each of the coordinate planes?

6. The graph of $x^2 + y^2 - 6y + z^2 - 4z = 0$ is a sphere. Find the center and radius, then draw the sphere.

7. Consider the unit sphere S, with equation $x^2 + y^2 + z^2 = 1$. If we set $z = 0$ in this equation, we get $x^2 + y^2 + z^2 = x^2 + y^2 = 1$. This means, geometrically, that S intersects the xy-plane in the unit circle $x^2 + y^2 = 1$. In mathematical language, the unit circle is the trace of S in the xy-plane. (To put it another way, the unit circle is the "equator" of the unit sphere.)
   (a) Set $x = 0$ in the original equation to find the equation of the intersection of S and the yz-plane.
   (b) What is the intersection of S and the xz-plane? Describe the answer geometrically.
   (c) Use the results in parts (a) and (b) to sketch the part of S that lies in the first octant. [**HINTS**: First draw a set of coordinate axes. Then draw the traces of S in each of the three coordinate planes.]
   (d) Set $z = 1/2$ in the original equation to show that S intersects the plane $z = 1/2$ in the circle of radius $\sqrt{3}/2 \approx 0.87$, centered at $(0, 0)$.
   (e) What is the trace of S in the plane $z = 0.9$? In the plane $z = 1$? In the plane $z = 2$? Explain your answers.

8. The distance formula in xyz-space can be thought of as just another instance of the Pythagorean rule for right triangles. (The square of the hypotenuse is the sum of the squares of the sides.) This exercise illustrates why.
   (a) Plot and label the points $O(0, 0, 0)$, $P(1, 0, 0)$, $Q(1, 2, 0)$, and $R(1, 2, 3)$ in an xyz-coordinate system. Observe that the triangles $\triangle OPQ$ and $\triangle OQR$ are both right triangles. Mark the sides OP, PQ, and QR with their lengths. (The lengths should be obvious from the picture.)
   (b) Use the Pythagorean rule (not the distance formula) on the triangle $\triangle OPQ$ to find the length of OQ.
   (c) Use the Pythagorean rule (not the distance formula) on the triangle $\triangle OQR$ to find the length of OR.

(d) For comparison, use the distance formula to compute the lengths of OQ and OR.

9. Any reasonable formula for distance should satisfy some commonsense requirements. For example, the distance $d(P, P)$ from any point P to *itself* should certainly be zero. So it is. If $P(x, y, z)$ is any point, then the distance formula says·

$$d(P, P) = \sqrt{(x - x)^2 + (y - y)^2 + (z - z)^2} = 0.$$

In the same spirit, use the distance formula to verify the following commonsense properties. Throughout, use the points $P(x, y, z)$ and $Q(a, b, c)$.
   (a) If $P \neq Q$, then $d(P, Q) > 0$.
   (b) $d(P, Q) = d(Q, P)$.
   (c) If M is the midpoint of P and Q, then $d(P, M) = d(M, Q) = d(P, Q)/2$.

10. In this section we said that xyz-space contains eight different octants. List eight points, all with coordinates ± 1, one in each octant. Draw a picture showing all eight points.

11. Any line in the xy-plane (even a vertical line) has an equation of the form $Ax + By = C$ for some constants A, B, and C. For each of the following lines, state an equation in this form.
   (a) The line $y = 3x + 5$
   (b) The line $y = mx + b$, where m and b are any constants
   (c) The horizontal line through $(2, 3)$
   (d) The vertical line through $(2, 3)$

12. Consider the linear equation $Ax + By = C$ and its graph (a line) in the xy-plane. Here A, B, and C are constants, and we assume that A and B aren't both zero.
   (a) We assumed that A and B aren't both zero. What goes wrong if $A = B = 0$?
   (b) Find the slope of the line $Ax + By = C$. Which lines have undefined slope?
   (c) Find the y-intercept of the line $Ax + By = C$. Which lines have no y-intercept?
   (d) Find the x-intercept of the line $Ax + By = C$. Which lines have no x-intercept?

13. Consider the linear equation $Ax + By + Cz = D$ and its graph (a plane) in xyz-space. Here A, B, C, and D are all constants, and we assume that A, B, and C aren't all zero.
   (a) We assumed that A, B, and C aren't all zero. What goes wrong if $A = B = C = 0$?
   (b) Find (if possible) an x-intercept of the plane $Ax + By + Cz = D$. (Set $y = 0$ and $z = 0$, then solve for x.) Give an example of a plane with no x-intercept.
   (c) Find (if possible) a z-intercept of the plane $Ax + By + Cz = D$. Give an example of a plane with no z-intercept.

14. The linear equation $x + 2y + 3z = 3$ defines a plane p in xyz-space. (See the picture on page 677.)

    (a) Find the equation of the trace of p in the xz-plane. Where does the trace intercept the x- and z-axes?

    (b) Find the equation of the trace of p in the yz-plane. Where does the trace intercept the y- and z-axes?

    (c) Find the equation of the trace of p in the plane $x = 1$.

15. Consider the plane p with equation $4x + 2y + z = 4$.

    (a) Find the x-, y-, and z-intercepts of p. Use them to draw p in the first octant.

    (b) Find the traces of p in each of the three coordinate planes. How do your answers appear in the picture in part (a)?

16. We said in this section that, as a rule, a line in the xy-plane intercepts both coordinate axes, and a plane in xyz-space intercepts all three coordinate axes. But exceptions are possible, as this exercise explores.

    (a) Give an example of a line in the xy-plane that intercepts the x-axis but not the y-axis. Write an equation for your line in the form $ax + by = c$.

    (b) Consider the plane $x = 1$ in xyz-space. Find all possible intercepts with the three coordinate axes.

    (c) Consider the plane $x + 2y = 1$ in xyz-space. Find all possible intercepts with the three coordinate axes.

    (d) Give the equation of a plane in xyz-space that intersects the y-axis and the z-axis but not the x-axis.

17. Suppose that $P_1(x_1, y_1, z_1)$ and $P_2(x_2, y_2, z_2)$ both lie on the plane with equation $Ax + By + Cz = D$. Show that the midpoint of P_1 and P_2 also lies on this plane.

14.2 Functions of Several Variables

Functions of one variable are the basic objects of single-variable calculus. Functions of two (or more) variables play a similar role in multivariable calculus. In this section we meet such functions and consider some of their rudimentary properties.

Functions of One or More Variables

The squaring function, defined for all real numbers x by $f(x) = x^2$, is typical of the functions of beginning calculus: f accepts one number, x, as input, and assigns another number, x^2, as output. If, say, $x = 2$, then $f(2) = 4$.

Consider, by contrast, the function g defined by $g(x, y) = x^2 + y^2$. Unlike f, g accepts a *pair* (x, y) of real numbers as inputs. The output is a third real number, $x^2 + y^2$. If, say, $x = 2$ and $y = 1$, then $g(2, 1) = 4 + 1 = 5$.

"One" and "two" count the input variables to f and g.

Naturally enough, f is called a function of one variable, and g a function of two variables. ◄ The difference has to do with domains: The domain of f is the one-dimensional real number line, while the domain of g is the two-dimensional xy-plane.

The function f corresponds to the equation $y = x^2$; x is the **independent variable**, and y is the **dependent variable**. The function g corresponds to the equation $z = x^2 + y^2$; now both x and y are independent variables, and z is the dependent variable.

We use f and g to illustrate various similarities and differences between functions of one and two variables. Notice, however, that there's nothing sacred about *two* variables—we can (and do) discuss such functions as

$$h(x, y, z) = x^2 + y^2 + z^2 \quad \text{and} \quad k(x, y, z, w) = x^2 + y^2 + z^2 + w^2,$$

which accept three or more variables. For now, however, we'll keep things simple, by sticking mainly to functions of two variables.

Multivariable Functions: Why Bother?

Single-variable calculus is challenging enough. Why complicate things by adding more variables?

It's a fair question. One good practical answer ▶ is that functions of several variables are essential for describing and predicting phenomena we care about, both natural and human-made. In economics, for instance, a manufacturer's profit depends on many input variables: labor costs, distance to markets, tax rates, and so on. In physics, a satellite's motion through space depends on a variety of forces. In biology, populations rise and fall with variations in climate, food supply, predation, and other factors. The weather varies with both longitude and latitude. In short, our world is multidimensional, and modeling it successfully requires multivariable tools.

There are purely mathematical answers, too.

The National Weather Service plots multivariable functions every day. ▶ Here, for instance, are noon surface temperatures on a relatively warm (by Upper Midwest standards) winter day:

See the back page of your newspaper.

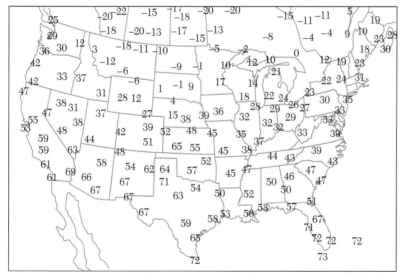

Plot of Surface Temperature (°F)

The map shows how the temperature function varies across its domain, the continental United States.

Functional Vocabulary and Notation. Most of the basic words and notations for functions of several variables are similar to those for functions of one variable. Roughly speaking, a function is a "machine" that accepts inputs and assigns outputs.

A bit more formally:

Domain The domain of a function is the set of permissible inputs. The preceding function g, for instance, accepts *any* 2-tuple (x, y) as input, so the domain is the set $\mathbb{R}^2$.

Range The range of a function is the set of possible outputs. For g, the range is the set $[0, \infty)$ of nonnegative real numbers.

Rule The rule of a function is the method for assigning inputs to outputs. For g, the rule is given by the algebraic formula $g(x, y) = x^2 + y^2$. Not every function has a simple symbolic rule. We often encounter functions given by tables, by graphs, or in other ways.

"Arrow" Notations We used the notation $g(x, y) = x^2 + y^2$ to describe a certain function of two variables. Other notations, involving arrows, are sometimes convenient. The notation

$$g : \mathbb{R}^2 \to \mathbb{R}$$

says that g is a function that accepts *two* real numbers as input and produces *one* real number as output. If we want to emphasize the rule by which g sends inputs to outputs, we can write

$$g : (x, y) \to x^2 + y^2.$$

These notations remind us that a function begins with an input (a 2-tuple, in this case) and ends with an output (a single number, in this case).

Graphs in One and Several Variables

The graph of f is the set of all points (x, y) for which $y = f(x) = x^2$. For example, $f(2) = 4$, so the point $(2, 4)$ lies on the graph. Geometrically, this graph is a curve—a parabola, in this case—in the the xy-plane. Like a straight line, the curve $y = x^2$ is a *one*-dimensional object ◄ that lives in a *two*-dimensional space—the xy-plane. Here's part of the familiar graph:

To an ant walking along it, the parabola looks like a (one-dimensional) straight line.

Graph of $f(x) = x^2$

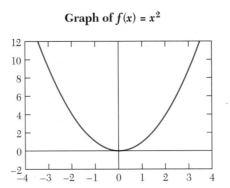

The graph of g is the set of all points (x, y, z) for which $z = g(x, y) = x^2 + y^2$. For example, since $g(2, 1) = 5$, the point $(2, 1, 5)$ lies on the g-graph. Here's a portion of the g-graph:

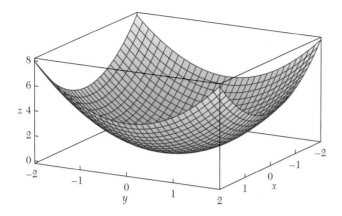

This graph, called a **paraboloid**, is quite different from the parabola shown earlier. Notice especially the dimensions involved: The graph of g is a two-dimensional surface ▶ hovering in three-dimensional xyz-space. As a rule, the *dimension* of any graph is the number of input variables the function accepts. (This rule of thumb applies to almost every function seen in calculus courses. For the record, however, some very ill-behaved functions do violate this rule.)

To an ant walking along it, the g-graph looks like a (two-dimensional) flat plane, just as earth's surface looks flat to a human.

In other respects, the graphs of f and g are quite similar to each other. For both functions, the *height* of the graph above a given domain point (a typical input is x_0 for f and (x_0, y_0) for g) tells the corresponding output value. (Typical outputs are $y_0 = f(x_0)$ and $z_0 = g(x_0, y_0)$, respectively.)

Multivariable Graphs: Beware.   Graphs are at least as important in multivariable calculus as in elementary calculus, but multivariable graphs are usually more complicated; they need a little extra care in handling. Choosing a "good" viewing window, for example, takes some care even for functions of one variable, and the problem can be stickier still for functions of two variables. The fact that multivariable graphs "live" naturally in three-dimensional (or even higher-dimensional) space—not on a flat page or a computer screen—only adds to the problem. For this and other reasons, we try to look at functions from as many points of view as possible.

Level Curves and Contour Maps

Let $f(x, y)$ be a function of two variables, and let c be a number in the range of f. Then $f(x, y) = c$ is an equation in x and y; its graph is (usually) a curve in the xy-plane. ▶ Such curves have a special name.

Occasionally this curve is just a point.

> **Definition**   Let $f(x, y)$ be a function, and let c be a constant. The set of all (x, y) for which $f(x, y) = c$ is called a **level curve** of f. A collection of level curves drawn together is called a **contour map** of f.

Observe:

Why "Level"?   The word "level" makes good sense here, because $f(x, y)$ has the same value, namely c, at each point along the level curve. In other words, the graph of f is level above the level curve $f = c$.

Which Level Curves To Draw?   Each number c in the range of f has its own level curve. Since the range of a function is usually infinite, we cannot possibly draw all the level curves. In practice, we draw some convenient selection of curves, corresponding to *evenly spaced* values of c. The spacing between curves reflects how fast the function increases or decreases.

Labels   We label level curves with their corresponding output values. Therefore, typical labels might be of the form $z = c$, $f = c$, or even simply c.

EXAMPLE 1   Here's a sample of level curves for the function $g(x, y) = x^2 + y^2$. Label each level curve with the appropriate output value. Recall that the graph of g is a paraboloid. How is this shape reflected in the level curves?

Level curves of $g(x, y) = x^2 + y^2$

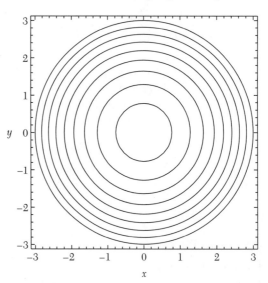

Solution   Each level curve is a circle about the origin; each is the graph of an equation $x^2 + y^2 = c$ for some c. The circles, from smallest to largest, correspond to the levels $z = 1$, $z = 2$, ..., $z = 9$.

Notice that level curves of g get closer and closer together as we move outward from the origin. This reflects the fact that the g-graph is a paraboloid—it gets steeper and steeper as we move away from the origin.   ∎

EXAMPLE 2   Drawing level curves on a temperature map makes the map easier to read and interpret. (Newspapers usually do this.) Following is another version of the preceding map. The boundaries of shaded regions correspond to the level curves of the temperature function; they're called **isotherms**. ▶

Compare this temperature map to the other one.

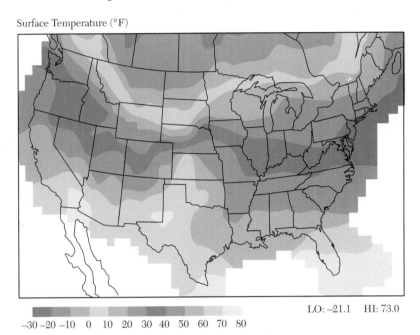

Surface Temperature (°F)

LO: –21.1   HI: 73.0

–30 –20 –10   0   10 20 30 40 50 60 70 80

Linear Functions

A linear function of one variable is one that can be written in the form $L(x) = a + bx$, where a and b are constants. A linear function of several variables has a similar algebraic form.

> **Definition** A **linear function** of two variables is one that can be written in the form
>
> $$L(x, y) = a + bx + cy,$$
>
> where a, b, and c are any constants.

Linear functions of three (or more) variables are similar. A linear function of three variables has the form $L(x, y, z) = a + bx + cy + dz$, where a, b, c, and d are constants.

Why "Linear"?   Why do linear functions deserve that name? In the xy-plane, there's no mystery—the graph of a linear function $y = a + bx$ is a straight line. In xyz-space, however, the graph of a linear function $z = a + bx + cy$ is not a line but a plane. (In $xyzw$-space, the graph of a linear function $w = a + bx + cy + dz$ is even less like a line. It's a three-dimensional solid, called a **hypersurface**.) Nevertheless, we call any function *linear* that involves only constants and the first

The function
$f(x, y) = x^2 + xy + y^2$ *is*
quadratic in this sense.

power of each variable, regardless of the number of variables. In the same spirit, we call a function *quadratic* (no matter how many variables are involved) if the variables appear in nonnegative integer powers, none higher than two. ◄

Readers who have studied linear algebra will recall that the word "linear" (as in "linear transformation") is used in still another way in that subject. Like other very useful English words, "linear" seems to have spawned a whole family of related but not identical meanings.

Why Linear Functions Matter.   Linear functions are simple, useful, and easy to work with. Most important for us, linear functions are prototypes, or models, for *all* differentiable functions. Indeed, any differentiable function, in any number of variables, can be called "almost linear" or "locally linear" in much the same sense that an ordinary calculus function $y = f(x)$ looks like a straight line if we zoom in repeatedly on a typical point on its graph.

BASICS

1. Find the domain and range of each function.
   (a) $g(x, y) = x^2 + y^2$
   (b) $h(x, y) = x^2 + y^2 + 3$
   (c) $j(x, y) = 1/(x^2 + y^2)$
   (d) $k(x, y) = x^2 - y^2$
   (e) $m(x, y) = \sqrt{1 - x^2 - y^2}$

2. Let $f(x, y) = y - x^2$, and let $g(x, y) = x - y^2$.
   (a) In the rectangle $[-3, 3] \times [-3, 3]$, draw and label the level curves of f that correspond to $z = -3$, $z = -2, \ldots, z = 2$, and $z = 3$. What is the shape of each level curve?
   (b) In the rectangle $[-3, 3] \times [-3, 3]$, draw and label the level curves of g that correspond to $z = -3$, $z = -2$, $\ldots, z = 2$, and $z = 3$. What is the shape of each level curve?
   (c) How are the results of parts (a) and (b) similar? How are they different?
   (d) Use technology to plot the graphs $z = f(x, y)$ and $z = g(x, y)$, for (x, y) in $[-3, 3] \times [-3, 3]$. Describe briefly, in words, how the two graphs are related to each other.

3. Let $f(x, y) = x^2 + y^2$, and let $g(x, y) = x^2 + y^2 + 1$.
   (a) In the rectangle $[-3, 3] \times [-3, 3]$, draw and label the level curves of f that correspond to $z = 0$, $z = 2$, $z = 4$, $z = 6$, and $z = 8$. What is the shape of each level curve?
   (b) In the rectangle $[-3, 3] \times [-3, 3]$, draw and label the level curves of g that correspond to $z = 1$, $z = 3$, $z = 5$, $z = 7$, and $z = 9$. What is the shape of each level curve?
   (c) How are the results of parts (a) and (b) similar? How are they different?

(d) Use technology to plot the graphs $z = f(x, y)$ and $z = g(x, y)$ for (x, y) in $[-3, 3] \times [-3, 3]$. Describe briefly, in words, how the two graphs are related to each other.

4. Let f and g be the linear functions $f(x, y) = 2x - 3y$ and $g(x, y) = -2x + 3y$.
   (a) In the rectangle $[-3, 3] \times [-3, 3]$, draw and label the level curves of f that correspond to $z = -5$, $z = -3$, $z = -1$, $z = 1$, $z = 3$, and $z = 5$. What is the shape of each level curve?
   (b) In the rectangle $[-3, 3] \times [-3, 3]$, draw and label the level curves of g that correspond to $z = -5$, $z = -3$, $z = -1$, $z = 1$, $z = 3$, and $z = 5$. What is the shape of each level curve?
   (c) How are the results of parts (a) and (b) similar? How are they different?
   (d) What special properties do the level curves of a *linear* function have?
   (e) Use technology to plot the graphs $z = f(x, y)$ and $z = g(x, y)$ for (x, y) in $[-3, 3] \times [-3, 3]$. Describe briefly, in words, how the two graphs are related to each other.

5. Let f and g be the functions $f(x, y) = 2 + x^2$ and $g(x, y) = 2 + y^2$.
   (a) In the rectangle $[-3, 3] \times [-3, 3]$, draw and label the level curves of f that correspond to $z = 0$, $z = 2$, $z = 4$, $z = 6$, and $z = 8$. What is the shape of each level curve?
   (b) In the rectangle $[-3, 3] \times [-3, 3]$, draw and label the level curves of g that correspond to $z = 0$, $z = 2$, $z = 4$, $z = 6$, and $z = 8$. What is the shape of each level curve?

(c) How are the results of parts (a) and (b) similar? How are they different?

(d) Use technology to plot the graphs $z = f(x, y)$ and $z = g(x, y)$ for (x, y) in $[-3, 3] \times [-3, 3]$. Describe briefly, in words, how the two graphs are related to each other.

(e) The graphs of f and g are both cylinders (in the sense defined in the text). How do the contour maps of f and g reflect this fact?

6. Imagine a map of the United States in the usual position. The positive x-direction is east, and the positive y-direction is north. Suppose that the units of x and y are miles and that Los Angeles, California, has coordinates $(0, 0)$. (Several approximations are involved here. The earth's surface is not flat, and Los Angeles occupies more than a single point.) Let $T(x, y)$ be the temperature, in degrees Celsius, at the location (x, y) at noon, Central Standard Time, on January 1, 1996.

(a) What does it mean in weather language to say that $T(0, 0) = 15$?

(b) What do the level curves of T mean in weather language? As a rule, would you expect level curves of T to run north and south or east and west? Why?

(c) International Falls, Minnesota, is about 1400 miles east and 1100 miles north of Los Angeles. The noon temperature in International Falls on January 1, 1996, was $-15°$ C. What does this mean about $T(x, y)$?

(d) Suppose that International Falls was the coldest spot in the country at the time in question. How would you expect the level curves to look near International Falls?

7. For any point (x, y) in the xy-plane, let $f(x, y)$ be the distance from (x, y) to the origin. Then f has the formula $f(x, y) = \sqrt{x^2 + y^2}$.

(a) Find the domain and range of f.

(b) Plot f. Describe the graph in words.

(c) Draw the level curve of f that passes through $(3, 4)$.

(d) All level curves of f have the same shape. What is it?

8. For any point (x, y) in the xy-plane, let $f(x, y)$ be the distance from (x, y) to the line $x = 1$.

(a) Find a formula for $f(x, y)$.

(b) Plot f. Describe the graph in words.

(c) Draw the level curve that passes through $(3, 4)$.

(d) All level curves of f have the same shape. What is it?

9. Below are some values of a linear function $L(x, y)$. (No explicit symbolic formula for L is given.) Use the table to answer the following questions.

(a) All the level curves of L are straight lines. Using this fact, draw (all in the rectangle $[-3, 3] \times [-3, 3]$) and label the level curves $z = -12$, $z = -8$, $z = -4$, $z = 0$, $z = 4$, $z = 8$, $z = 12$.

(b) Find an equation in x and y for the level line $z = 0$.

(c) Because L is a linear function, its formula has the form $L(x, y) = a + bx + cy$ for some constants a, b, and c. Find numerical values for a, b, and c. [**HINT:** The table says that $L(0, 0) = 0$. Therefore, $L(0, 0) = a + b \cdot 0 + c \cdot 0 = 0$, so $a = 0$. Use similar reasoning to find values for b and c.]

(d) Use technology to plot $L(x, y)$ over the rectangle $[-3, 3] \times [-3, 3]$. Is the shape of the graph consistent with the level curves you plotted in part (a)?

Values of $L(x, y)$							
y x	**−3**	**−2**	**−1**	**0**	**1**	**2**	**3**
3	−15	−12	−9	−6	−3	0	3
2	−13	−10	−7	−4	−1	2	5
1	−11	−8	−5	−2	1	4	7
0	−9	−6	−3	0	3	6	9
−1	−7	−4	−1	2	5	8	11
−2	−5	−2	1	4	7	10	13
−3	−3	0	3	6	9	12	15

14.3 Partial Derivatives

Derivatives in One Variable—Interpretations

Let f be an ordinary function of one variable. Recall some familiar properties of the derivative function f': ▶

To avoid distractions, we assume that f and f' are continuous functions.

Slope For any fixed input $x = x_0$, the derivative $f'(x_0)$ tells the *slope* of the f-graph at the point $(x_0, f(x_0))$. The sign of $f'(x_0)$, in particular, tells whether f is increasing or decreasing (rising or falling, in everyday speech) at $x = x_0$.

Rate of Change   The derivative f' can also be interpreted as the *rate function* associated to f, as follows: For any input x_0, $f'(x_0)$ tells the instantaneous rate of change of $f(x)$ with respect to x. If, say, $f(x)$ gives the *position* of a moving object at time x, then $f'(x)$ gives the corresponding *velocity* at time x. ◀

In a specific example, we would need to specify appropriate units for everything.

Limit   The derivative $f'(x_0)$ is defined as a limit of difference quotients:

$$f'(x_0) = \lim_{h \to 0} \frac{f(x_0 + h) - f(x_0)}{h}.$$

Similar interpretations hold if $h < 0$, but $h = 0$ is taboo.

For any $h > 0$, the difference quotient can be thought of either as the average rate of change of f over the interval $[x_0, x_0 + h]$ or as the slope of a secant line on the f-graph, over the same interval. ◀ Taking the limit as $h \to 0$ corresponds to finding the instantaneous rate of change of f or, equivalently, the slope of the tangent line to the f-graph at $x = x_0$. ◀

See the exercises at the end of this section for a brief refresher on these ideas.

Linear Approximation   At a point $\big(x_0, f(x_0)\big)$ on the curve $y = f(x)$, the tangent line has slope $f'(x_0)$. This tangent line is the graph of the linear function L, with equation

$$y = L(x) = f(x_0) + f'(x_0)(x - x_0).$$

The function L is called the **linear approximation** to f at x_0. The name makes sense because the graphs of f and L are close together near x_0. In symbols,

$$L(x) \approx f(x) \qquad \text{when} \qquad x \approx x_0.$$

Derivatives in Several Variables

Derivatives are just as important in multivariable calculus as in one-variable calculus, but—not surprisingly—the idea is more complicated for functions of several variables.

Take slope, for instance. The graph of a one-variable function $y = f(x)$ is a curve in the xy-plane. The slope at (x_0, y_0)—a single number—completely describes the graph's direction at (x_0, y_0). The graph of a two-variable function $z = f(x, y)$, by contrast, is a *surface*, which has no single slope at a point. A surface's steepness at a point depends on the direction (uphill, downhill, along the "contour," and so on) taken from the point. ◀ To put it another way, the graph of a one-variable function can be approximated near a given point by a one-dimensional tangent *line*. The graph of a two-variable function can be approximated near a fixed point by a two-dimensional tangent *plane*.

Every hiker knows this. How steep a mountain "feels" depends on the direction of the trail.

Here's the moral: To suit *multivariable* calculus, our notion of derivative must go beyond the simple idea of slope. The idea of linear approximation, not slope, turns out to be the key to extending the idea of derivative to functions of more than one variable.

In this section we start to extend the derivative idea to functions of several variables. **Partial derivatives** are the simplest multivariable analogues of ordinary derivatives. ◀

"Partial" suggests correctly that there's more to the derivative story.

Partial Derivatives: The Idea. The basic idea of a partial derivative is to differentiate with respect to *one* variable, holding all the others constant. The following easy example illustrates the idea and introduces some useful terminology and notation.

EXAMPLE 1 Let $f(x, y) = x^2 - 3xy + 6$. Find $f_x(x, y)$ and $f_y(x, y)$, the partial derivatives of f with respect to x and y, respectively. Find the numerical values $f_x(2, 1)$ and $f_y(2, 1)$.

Solution To find f_x, we differentiate $f(x, y) = x^2 - 3xy + 6$ with respect to x, *treating y as a constant:* ▶

Check this calculation and the next carefully. Do you agree?

$$f_x(x, y) = 2x - 3y.$$

To find f_y, we *treat x as a constant:*

$$f_y(x, y) = -3x.$$

Setting $x = 2$ and $y = 1$ in these formulas gives

$$f_x(2, 1) = 2 \cdot 2 - 3 \cdot 1 = 1; \qquad f_y(2, 1) = -3 \cdot 2 = -6. \qquad ■$$

The calculations were easy, ▶ but what do the results mean? What do they say about how f behaves near $(x, y) = (2, 1)$? Can we interpret the results graphically and numerically? Understanding multivariable derivatives fully is a long-term proposition, but here are some starters.

Partial derivatives are usually easy to calculate. What they mean is more important.

Holding Variables Constant. To find the partial derivative f_x of a function f with respect to x, we treat all the other variables as constants. This produces a function of just one variable, x, which we differentiate in the usual way. For example, suppose we fix $y = 3$ in $f(x, y) = x^2 - 3xy + 6$. Then $f(x, y) = f(x, 3) = x^2 - 9x + 6$, and

$$\frac{d}{dx}\big(f(x, 3)\big) = f_x(x, 3) = 2x - 9.$$

This agrees, as it should, with the general formula $f_x(x, y) = 2x - 3y$ found earlier.

Directional Rates of Change. Partial derivatives (like ordinary derivatives) can be interpreted as rates or slopes—but with an important proviso about directions. For a function $f(x, y)$ and a point (x_0, y_0) in its domain, the partial derivative $f_x(x_0, y_0)$ tells the rate of change of $f(x, y)$ with respect to x, ▶ i.e., how fast $f(x, y)$ increases as the input (x, y) moves away from (x_0, y_0) *in the positive x-direction*. The other partial derivative, $f_y(x_0, y_0)$, tells how fast f increases near (x_0, y_0) as y increases. The next example illustrates what this means numerically.

x is the only variable that's free to move.

We saw this already.

EXAMPLE 2   The function $f(x, y) = x^2 - 3xy + 6$ has partial derivatives $f_x(x, y) = 2x - 3y$ and $f_y(x, y) = -3x$. ◄ What does this say about rates of change of f at $(x, y) = (2, 1)$? At $(x, y) = (1, 2)$?

Solution   We start at $(2, 1)$. The formulas give $f_x(2, 1) = 1$ and $f_y(2, 1) = -6$. These numbers represent rates of change of f with respect to x and y, respectively, at $(2, 1)$. A table of f-values centered at $(2, 1)$ shows what this means. ◄

The boxed row and column meet at $(2, 1)$, our target point.

Values of $f(x, y) = x^2 - 3xy + 6$ Near $(2, 1)$

y \ x	1.97	1.98	1.99	2.00	2.01	2.02	2.03
1.03	3.7936	3.8022	3.8110	3.8200	3.8292	3.8386	3.8482
1.02	3.8527	3.8616	3.8707	3.8800	3.8895	3.8992	3.9091
1.01	3.9118	3.9210	3.9304	3.9400	3.9498	3.9598	3.9700
1.00	3.9709	3.9804	3.9901	4.0000	4.0101	4.0204	4.0309
0.99	4.0300	4.0398	4.0498	4.0600	4.0704	4.0810	4.0918
0.98	4.0891	4.0992	4.1095	4.1200	4.1307	4.1416	4.1527
0.97	4.1482	4.1586	4.1692	4.1800	4.1910	4.2022	4.2136

Reading *up* the boxed column (at each step, y increases by 0.01) shows successive corresponding values of f *decreasing* by about 0.06. Thus, -6 is the rate of change of f with respect to y at $(2, 1)$; equivalently, $f_y(2, 1) = -6$. Similarly, the fact that $f_x(2, 1) = 1$ suggests that values of $f(x, 1)$ should increase at about the same rate as x if $x \approx 2$. Reading across the boxed row confirms this expectation. ◄

Convince yourself of this.

Similar reasoning applies at $(1, 2)$. The values $f_x(1, 2) = -4$ and $f_y(1, 2) = -3$ ◄ mean that f decreases (at different rates) with respect to both x and y near $(1, 2)$. Numerical f-values bear this out:

Use the formulas to check these values.

Values of $f(x, y) = x^2 - 3xy + 6$ Near $(1, 2)$

y \ x	0.98	0.99	1.00	1.01	1.02
2.02	1.0216	0.9807	0.9400	0.8995	0.8592
2.01	1.0510	1.0104	0.9700	0.9298	0.8898
2.00	1.0804	1.0401	1.0000	0.9601	0.9204
1.99	1.1098	1.0698	1.0300	0.9904	0.9510
1.98	1.1392	1.0995	1.0600	1.0207	0.9816

Reading either across or up shows f-values decreasing at rates of about -4 and -3, respectively.   ■

Formal Definitions

Partial derivatives are defined formally as limits, much as ordinary derivatives are.

> **Definition** Let $f(x, y)$ be a function of two variables. The partial derivative with respect to x of f at (x_0, y_0), denoted by $f_x(x_0, y_0)$, is defined by
>
> $$f_x(x_0, y_0) = \lim_{h \to 0} \frac{f(x_0 + h, y_0) - f(x_0, y_0)}{h}$$
>
> if the limit exists. The partial derivative with respect to y at (x_0, y_0), denoted by $f_y(x_0, y_0)$, is defined by
>
> $$f_y(x_0, y_0) = \lim_{h \to 0} \frac{f(x_0, y_0 + h) - f(x_0, y_0)}{h}$$
>
> if the limit exists.

The definition says, in effect, that $f_x(x_0, y_0)$ is the ordinary derivative at $x = x_0$ of an ordinary function of one variable—namely, the function given by the rule $x \to f(x, y_0)$. Following are some further comments and observations.

Do They Exist?   If either limit does not exist, then neither does the corresponding partial derivative. It's possible, for instance, that $f_x(0, 0)$ exists but $f_y(0, 0)$ doesn't. ▶   For most functions in this chapter, however, both partial derivatives *do* exist.

The exercises at the end of the section explore this a little further.

About Domains   In order for the preceding limits to exist, $f(x, y)$ must be defined at (x_0, y_0) and at nearby points (x, y). In practice, this condition seldom causes trouble. In particular, there's no problem if (x_0, y_0) lies in the interior of the domain of f, i.e., if f is defined both at and near (x_0, y_0). (For the record, trouble is likeliest if (x_0, y_0) lies on the edge of the domain of f.)

Other Notations   As with ordinary derivatives, various notations are used to denote partial derivatives. If, say, $z = f(x, y) = x \sin y$, then all of the expressions

$$f_x, \quad \frac{\partial f}{\partial x}, \quad \frac{\partial z}{\partial x}, \quad \text{and} \quad \frac{\partial}{\partial x}(x \sin y)$$

mean the same thing, as do

$$f_x(x_0, y_0), \quad \frac{\partial f}{\partial x}(x_0, y_0), \quad \frac{\partial z}{\partial x}\bigg|_{(x_0, y_0)}, \quad \text{and} \quad \frac{\partial}{\partial x}(x \sin y)\bigg|_{(x_0, y_0)}.$$

Partial derivatives with respect to y are expressed with similar notations. Notice especially the "curly-d" symbol ∂. It's usually read aloud as "partial."

Partial Derivatives and Contour Maps

We saw in Example 2, page 692, how to estimate partial derivatives from a table of function values. Contour maps can be used for the same purpose. Thinking of $f_x(x_0, y_0)$ and $f_y(x_0, y_0)$ as rates suggests how: Use the contour map to measure how fast $z = f(x, y)$ rises or falls near (x_0, y_0) as either x or y increases. ◄

On a topographic map oriented the usual way, the partial derivatives f_x and f_y describe the steepness of the terrain in the eastward and northward directions, respectively.

EXAMPLE 3   Following is a contour map of the function $f(x, y) = x^2 - 3xy + 6$, centered at $(2, 1)$. Use the contour map to estimate the partial derivatives $f_x(2, 1)$ and $f_y(2, 1)$.

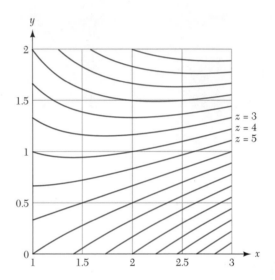

Take a close look at the contour map to convince yourself of these claims.

Solution   The level curves represent successive integer values of $z = f(x, y)$. In general, values of $f(x, y)$ increase as (x, y) moves toward the lower right. ◄

First we estimate $f_x(2, 1)$. A close look at the picture suggests that $f(2.1, 1) \approx 4.1$. Increasing x by 0.1 increases f by 0.1; this suggests that $f_x(2, 1) \approx 1$. Similarly, $f(2, 1.1) \approx 3.4$, so increasing y by 0.1 increases f by -0.6, and so $f_y(2, 1) \approx -6$. ∎

Partial Derivatives and Linear Approximation

For a differentiable function $y = f(x)$ of one variable, the ordinary derivative can be interpreted in terms of linear approximation. For any point (x_0, y_0) on the f-graph, there's a certain line through this point—the tangent line—that best "fits" the graph near $x = x_0$. The derivative $f'(x_0)$ gives the slope of this tangent line. Knowing the slope, it's easy to find an equation for the tangent line in point-slope form:

$$y - y_0 = f'(x_0)(x - x_0).$$

Equivalently, we can think of the tangent line as the graph of the linear function L defined by ▶

Recall that $f(x_0) = y_0$.

$$y = L(x) = y_0 + f'(x_0)(x - x_0).$$

The function L is called the linear approximation to f at x_0. The name is appropriate for three good reasons:

(i) L is linear;   (ii) $L(x_0) = f(x_0)$;   (iii) $L'(x_0) = f'(x_0)$.

In short, L agrees with f at x_0 as closely as any linear function can—both functions have the same value and the same (first) derivative. ▶

The idea of quadratic approximation is similar, except that we'd require agreement in the second derivative, too.

Linear Approximation in Several Variables.   The idea of linear approximation is essentially the same for functions of two (or more) variables: Given a function $f(x, y)$ and a point (x_0, y_0) in the domain of f, we look for a linear function $L(x, y)$ that has the same value and partial derivatives as does f at (x_0, y_0). In other words, we want a linear function L such that

$$L(x_0, y_0) = f(x_0, y_0), \ L_x(x_0, y_0) = f_x(x_0, y_0), \ \text{and } L_y(x_0, y_0) = f_y(x_0, y_0).$$

In the one-variable case, the graph of a linear approximation function is called a tangent line. For a function of two variables, the graph of the linear approximation function is called a **tangent plane**. We illustrate and summarize these ideas in the next example.

EXAMPLE 4   Find the linear approximation function L to $f(x, y) = x^2 + y^2$ at $(x_0, y_0) = (2, 1)$. How are graphs of f and L related?

Solution   The partial derivatives of f are $f_x(x, y) = 2x$ and $f_y(x, y) = 2y$. Thus, at our base point, $(x_0, y_0) = (2, 1)$:

$$f(2, 1) = 5; \qquad f_x(2, 1) = 4; \qquad f_y(2, 1) = 2.$$

Let's find a linear function L to match these values.

It's easiest to start by writing L in the convenient form ▶

We'll soon see why this form is convenient.

$$L(x, y) = a(x - x_0) + b(y - y_0) + c = a(x - 2) + b(y - 1) + c,$$

and then to choose appropriate values for a, b, and c. Since $L(x, y) = a(x - 2) + b(y - 1) + c$, it's easy to see ▶ that

But check for yourself, especially for the derivatives.

$$L(2, 1) = c; \qquad L_x(2, 1) = a; \qquad L_y(2, 1) = b.$$

To match f, we must have $c = 5$, $a = 4$, and $b = 2$, so

$$L(x, y) = a(x - 2) + b(y - 1) + c = 4(x - 2) + 2(y - 1) + 5.$$

Here are views of f and L together:

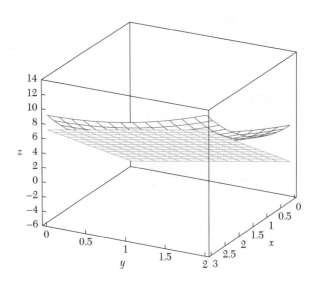

The picture illustrates the phrases "tangent plane" and "linear approximation": The plane $z = L(x, y)$ touches the surface $z = f(x, y)$ at $(2, 1, 5)$; at this point, moreover, the flat plane fits the curved surface as well as possible. A closer look at both functions near $(2, 1)$, this time using contour maps, suggests how good the fit really is: ◄

The dashed lines are contours of L.

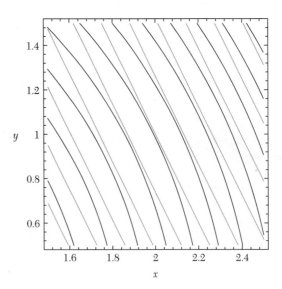

The two contour maps are almost identical near $(2, 1)$.

Linear Approximation: The General Formula. The procedure in Example 4 works the same way for any function of several variables, as long as the necessary partial derivatives exist. Following are definitions for two and three variables. (The same idea works for any number of variables.)

Definition (Linear Approximation) Let $f(x, y)$ and $g(x, y, z)$ be functions, and suppose that all the following partial derivatives exist. The linear approximation to f at (x_0, y_0) is the function

$$L(x, y) = f(x_0, y_0) + f_x(x_0, y_0)(x - x_0) + f_y(x_0, y_0)(y - y_0).$$

The linear approximation to g at (x_0, y_0, z_0) is the function

$$L(x, y, z) = g(x_0, y_0, z_0) + g_x(x_0, y_0, z_0)(x - x_0)$$
$$+ g_y(x_0, y_0, z_0)(y - y_0) + g_z(x_0, y_0, z_0)(z - z_0).$$

EXAMPLE 5 Find the linear approximation to $g(x, y, z) = x + yz^2$ at $(1, 2, 3)$. Does the answer make numerical or graphical sense?

Solution Easy calculations give

$$g(1, 2, 3) = 19; \quad g_x(1, 2, 3) = 1; \quad g_y(1, 2, 3) = 9; \quad g_z(1, 2, 3) = 12.$$

The linear approximation function therefore has the form

$$L(x, y, z) = g(1, 2, 3) + g_x(1, 2, 3)(x - 1) + g_y(1, 2, 3)(y - 2) + g_z(1, 2, 3)(z - 3)$$
$$= 19 + 1(x - 1) + 9(y - 2) + 12(z - 3).$$

To see the situation numerically, we tabulate some values of each function:

Values of L and g Near $(1, 2, 3)$						
(x, y, z)	$(1, 2, 3)$	$(1.1, 2, 3)$	$(1, 2.1, 3)$	$(1, 2, 3.1)$	$(1.1, 2.1, 3.1)$	$(3, 4, 5)$
$g(x, y, z)$	19	19.1	19.9	20.22	21.281	103
$L(x, y, z)$	19	19.1	19.9	20.2	21.2	63

As the numbers illustrate, $L(x, y, z)$ and $g(x, y, z)$ are close together if, but only if, (x, y, z) is near $(1, 2, 3)$. ▶

To plot ordinary graphs of g and L would require four dimensions. ▶ Instead we plot, for comparison, the level surfaces $L(x, y, z) = 19$ and $g(x, y, z) = 19$,

The last column illustrates what happens far from $(1, 2, 3)$.

One for each input variable and one for the output.

both of which pass through the base point $(1, 2, 3)$. (A **level surface** is like a level curve—it's a set of inputs along which a function has constant output value.) The pictures suggest again how similarly $g(x, y, z)$ and $L(x, y, z)$ behave when $(x, y, z) \approx (1, 2, 3)$.

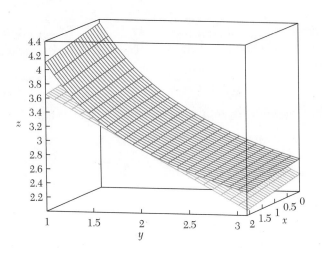

BASICS

1. For each of the following functions, find the partial derivative with respect to each variable.

   (a) $f(x, y) = x^2 - y^2$

   (b) $f(x, y) = x^2 y^2$

   (c) $f(x, y) = \dfrac{x^2}{y^2}$

   (d) $f(x, y) = \cos(xy)$

   (e) $f(x, y) = \cos(x) \cos(y)$

   (f) $f(x, y) = \dfrac{\cos(x)}{\cos(y)}$

   (g) $f(x, y, z) = xy^2 z^3$

   (h) $f(x, y, z) = \cos(xyz)$

2. Let $f(x) = x^2$, and let $x_0 = 3$.

   (a) Let L be the linear approximation to f at x_0. Show that $L(x) = 9 + 6(x - 3)$.

   (b) Plot L and f on the same axes. (Choose your own plotting window.) Supposedly, "L linearly approximates f near x_0." How do the graphs illustrate this?

   (c) Find an interval $a \le x \le b$ on which $|f(x) - L(x)| < 0.01$. (On this interval, $L(x)$ approximates $f(x)$ within 0.01.) [**HINT:** This can be done either graphically, by zooming, or symbolically, by solving inequalities.]

3. Redo Exercise 2, but work with $f(x) = \sqrt{x}$ and $x_0 = 9$.

4. Let $f(x) = x^2$.

   (a) Since $f'(x) = 2x$, $f'(3) = 6$. What does this mean about the graph of f? (Use a graph of f for $2.5 \le x \le 3.5$ to illustrate your answer.)

   (b) Plot f over the interval $2.5 \le x \le 3.5$. (It's OK to use a calculator or computer, but copy the graph onto paper.) On your graph, draw the secant lines from $x = 3$ to $x = 3.5$ and from $x = 3$ to $x = 3.1$. Find their slopes.

   (c) Find the average rate of change $\Delta y / \Delta x$ of f over the intervals $[3, 3.5]$ and $[3, 3.1]$. [**HINT:** The answers should be familiar from part (b).]

   (d) Find the limit $\displaystyle\lim_{h \to 0} \dfrac{f(3 + h) - f(3)}{h}$. Interpret the answer as a derivative. Does the answer agree with other information already given?

5. Redo Exercise 4, but use $f(x) = x^2 - x$.

6. Here are some values of a function $g(x, y)$. (No explicit symbolic formula for g is given.) Use the table to answer the following questions.

Values of $g(x, y)$				
y \ x	**−0.0100**	**0.0000**	**0.0100**	**0.0200**
1.02	2.0603	2.0604	2.0603	2.0600
1.01	2.0300	2.0301	2.0300	2.0297
1.00	1.9999	2.0000	1.9999	1.9996
0.99	1.9700	1.9701	1.9700	1.9697
$\vdots$	$\vdots$	$\vdots$	$\vdots$	$\vdots$
0.02	0.0203	0.0204	0.0203	0.0200
0.01	0.0100	0.0101	0.0100	0.0097
0.00	−0.0001	0.0000	−0.0001	−0.0004
−0.01	−0.0100	−0.0099	−0.0100	−0.0103

Values of $g(x, y)$				
y \ x	**0.9900**	**1.0000**	**1.0100**	**1.0200**
1.02	1.0803	1.0604	1.0403	1.0200
1.01	1.0500	1.0301	1.0100	0.9897
1.00	1.0199	1.0000	0.9799	0.9596
0.99	0.9900	0.9701	0.9500	0.9297
$\vdots$	$\vdots$	$\vdots$	$\vdots$	$\vdots$
0.02	−0.9597	−0.9796	−0.9997	−1.0200
0.01	−0.9700	−0.9899	−1.0100	−1.0303
0.00	−0.9801	−1.0000	−1.0201	−1.0404
−0.01	−0.9900	−1.0099	−1.0300	−1.0503

(a) Use the table to estimate the partial derivatives $g_x(1, 1)$ and $g_y(1, 1)$.

(b) It's true that $g_x(0, 0) = 0$ and $g_y(0, 0) = 1$. How do the table entries reflect these facts?

(c) Consider the linear function $L(x, y) = 0 + 0x + 1y = y$. Show that $L_x(0, 0) = g_x(0, 0) = 0$, $L_y(0, 0) = g_y(0, 0) = 1$, and $L(0, 0) = g(0, 0) = 0$.

(d) Fill in the following table of values for the function L from part (c).

Values of $L(x, y)$				
y \ x	**−0.02**	**−0.01**	**0.00**	**0.01**	**0.02**
0.02					
0.01					
0.00					
−0.01					
−0.02					

Compare your results with the tabulated values of g. (The results show how L linearly approximates g near $(0, 0)$.)

(e) Find the linear function $M(x, y)$ such that (i) $M(1, 1) = g(1, 1)$, (ii) $M_x(1, 1) = g_x(1, 1)$, and (iii) $M_y(1, 1) = g_y(1, 1)$. [**HINT:** One approach is to write $M(x, y) = a + b(x - 1) + c(y - 1)$ and then use the conditions to find values for a, b, and c.]

7. Let $f(x, y) = \sin y + 2$. (The formula is independent of x.)

(a) Plot f; use the domain $-5 \leq x \leq 5$, $-5 \leq y \leq 5$. How does the shape of the graph reflect the fact that f is independent of x? (In Section 14.1 we called such graphs cylinders.)

(b) Find $f_x(x, y)$ and $f_y(x, y)$. How do the answers reflect the fact that f is independent of x?

(c) Find the linear approximation function L for f at the point $(0, 0)$. How does its form reflect the fact that f is independent of x?

8. This exercise is about the situation described in Example 3 (page 694). Use the function f and the contour map given there.

(a) Use the contour map to estimate the partial derivatives $f_x(1.5, 1.5)$ and $f_y(1.5, 1.5)$.

(b) Use the formula $f(x, y) = x^2 - 3xy + 6$ to find $f_x(1.5, 1.5)$ and $f_y(1.5, 1.5)$ exactly.

(c) Use results of part (b) to find the linear approximation $L(x, y)$ to $f(x, y)$ at $(1, 5, 1.5)$.

9. Let $f(x, y) = \sin(x)$.

(a) Draw a contour map of f in the rectangle $-\pi \leq x \leq \pi$, $-2 \leq y \leq 2$. Show the level curves that correspond to $z = \pm 1$, $z = \pm 0.75$, $z = \pm 0.5$, $z = \pm 0.25$, and $z = 0$.

(b) Use the level-curve diagram to estimate $f_x(0, 0)$ and $f_y(0, 0)$.

(c) Use the level-curve diagram to estimate $f_x(\pi/2, 0)$ and $f_y(\pi/2, 0)$.

(d) The formula shows that $f_y(x, y) = 0$ for all (x, y). How does the contour map reflect this fact?

(e) The formula shows that $f_x(x, y)$ is independent of y. How does the contour map of f reflect this fact?

10. Let $f(x, y) = \cos(y)$.
    (a) Draw a contour map of f in the rectangle $-\pi \le x \le \pi$, $-2 \le y \le 2$. Show the level curves that correspond to $z = \pm 1$, $z = \pm 0.75$, $z = \pm 0.5$, $z = \pm 0.25$, and $z = 0$.
    (b) Use the level-curve diagram to estimate $f_x(0, 0)$ and $f_y(0, 0)$.
    (c) Use the level-curve diagram to estimate $f_x(\pi/2, 0)$ and $f_y(\pi/2, 0)$.
    (d) The formula for f shows that $f_x(x, y) = 0$ for all (x, y). How does the contour map of f reflect this fact?
    (e) The formula for f shows that $f_y(x, y)$ is independent of x. How does the contour map of f reflect this fact?

11. Let $f(x, y) = 2x - 3y$.
    (a) Draw a contour map of f in the rectangle $[-3, 3] \times [-3, 3]$. Show the level curves that correspond to $z = -5$, $z = -4$, $z = -3$, ..., $z = 4$, and $z = 5$.
    (b) Use your contour map (not the formula) to find $f_x(0, 0)$ and $f_y(0, 0)$.
    (c) The formula for f implies that both f_x and f_y are constant functions. How does the contour map of f reflect this fact?
    (d) The formula for f implies that for any (x, y), $f_x(x, y) = 2$ and $f_y(x, y) = -3$. How does the contour map reflect the fact that $f_x(x, y)$ is positive but $f_y(x, y)$ is negative?

12. Let $f(x, y) = 2y - x$.
    (a) Draw a contour map of f in the rectangle $[-3, 3] \times [-3, 3]$. Show the level curves that correspond to $z = -5$, $z = -4$, $z = -3$, ..., $z = 4$, and $z = 5$.
    (b) Use your contour map (not the formula) to find $f_x(0, 0)$ and $f_y(0, 0)$.
    (c) The formula for f implies that $f_x(x, y) = -1$ and $f_y(x, y) = 2$ for all (x, y). How does the contour map reflect these facts? In particular, how does the contour map show that $f_x(x, y)$ is negative but $f_y(x, y)$ is positive?

13. Let $f(x, y) = xy$.
    (a) Find the linear approximation function L to f at $(x_0, y_0) = (2, 1)$.
    (b) (Do this part by hand.) On one set of xy-axes, draw the level curves $L(x, y) = k$ for $k = 1, 2, 3, 4, 5$. On another set of axes, draw the level curves $f(x, y) = k$ for $k = 1, 2, 3, 4, 5$. (In each case, draw the curves into the square $[0, 3] \times [0, 3]$.)
    (c) How do the contour maps in part (b) reflect the fact that L is the linear approximation to f at the point $(2, 1)$? Explain briefly in words.

(d) Use technology to plot contour maps of f and L in the window $1.8 \le x \le 2.2$, $0.8 \le y \le 1.2$. (This small window is centered at $(2, 1)$.) Explain what you see.

14. Repeat Exercise 13 using the function $f(x, y) = x^2 - y^2$.

15. Let $f(x, y) = x^2 + y^2$. (See the contour map in Section 14.2.)
    (a) Use the contour map of f to estimate the partial derivatives $f_x(1, 2)$ and $f_y(1, 2)$.
    (b) Check your answers to part (a) by symbolic differentiation.
    (c) Use your answers from part (a) to find the linear approximation $L(x, y)$ to $f(x, y)$ at $(1, 2)$.
    (d) On one set of axes, plot the level curves $L(x, y) = k$ and $f(x, y) = k$ for $k = 3, 4, 5, 6, 7$. (Use the window $[0, 3] \times [0, 3]$.) What's special about the point $(1, 2)$?

16. Let $f(x, y) = \sin(x) + 2y + xy$.
    (a) Find the partial derivatives $f_x(x, y)$ and $f_y(x, y)$; then evaluate $f_x(0, 0)$ and $f_y(0, 0)$.
    (b) Find a linear function $L(x, y) = a + bx + cy$ such that $L_x(0, 0) = f_x(0, 0)$, $L_y(0, 0) = f_y(0, 0)$, and $L(0, 0) = f(0, 0)$.
    (c) Complete the following table (report answers to four decimals).

(x, y)	$(0, 0)$	$(0.01, 0.01)$	$(0.1, 0.1)$	$(1, 1)$
$f(x, y)$				
$L(x, y)$				

How do the answers reflect the fact that L approximates f closely near $(0, 0)$?
    (d) Use technology to draw contour plots of both f and L on the rectangle $-1 \le x \le 1$, $-1 \le y \le 1$. Label several contours on each. How do the pictures reflect the fact that L approximates f closely near $(0, 0)$?

17. For each function f, find the linear function L that linearly approximates f at the given point (x_0, y_0). (If possible, check your answers graphically by plotting both f and L near (x_0, y_0).)
    (a) $f(x, y) = x^2 + y^2$; $(x_0, y_0) = (2, 1)$.
    (b) $f(x, y) = x^2 + y^2$; $(x_0, y_0) = (0, 0)$.
    (c) $f(x, y) = \sin(x) + \sin(y)$; $(x_0, y_0) = (0, 0)$.
    (d) $f(x, y) = \sin(x)\sin(y)$; $(x_0, y_0) = (0, 0)$.

18. Let $f(x, y)$ be a differentiable function of two variables, let (x_0, y_0) be any point in its domain, and let $L(x, y)$ be the linear approximation to f at (x_0, y_0). Show that if f is independent of one of the variables—say, x—then so is L.

19. Suppose we know that for a certain function f, $f(3, 4) = 25$, $f_x(3, 4) = 6$, $f_y(3, 4) = 8$, and $f(4, 5) = 41$.
    (a) Find a linear function $L(x, y)$ that approximates f as well as possible near $(3, 4)$.
    (b) Use L to estimate $f(2.9, 3.9)$, $f(3.1, 4.1)$, and $f(4, 5)$.
    (c) Could f itself be a linear function? Why or why not?

20. Suppose we know that for a certain function g, $g(3, 4) = 5$, $g_x(3, 4) = 3/5$, $g_y(3, 4) = 4/5$, and $g(4, 5) = \sqrt{41}$.
    (a) Find a linear function $L(x, y)$ that approximates g as well as possible near $(3, 4)$.
    (b) Use L to estimate $g(2.9, 4.1)$ and $g(4, 5)$.
    (c) Could g be a linear function? Why or why not?

21. Let $f(x, y) = |y| \cos x$. This exercise explores the fact that the partial derivatives of a function may or may not exist at a given point.
    (a) Use technology to plot $z = f(x, y)$ over the rectangle $[-5, 5] \times [-5, 5]$. The graph suggests that there may be trouble with partial derivatives where $y = 0$, i.e., along the x-axis. How does the graph suggest this? Which partial derivative (f_x or f_y) seems to be in trouble?

    (b) Use the definition to show that $f_y(0, 0)$ does not exist. In other words, explain why the limit
    $$\lim_{h \to 0} \frac{f(h, 0) - f(0, 0)}{h}$$
    does not exist.
    (c) Show that $f_x(0, 0)$ does exist; find its value. How does the result appear on the graph?
    (d) Use the limit definition to show that $f_y(0, \pi/2)$ does exist; find its value.
    (e) How does the graph reflect the result of part (d)? (You may need to do some experimenting with the graph to answer this.)
    (f) Find a function $g(x, y)$ for which $g_y(0, 0)$ exists but $g_x(0, 0)$ does not. Use technology to plot its graph.

22. Let $f(x, y) = |x| y$.
    (a) By experimenting with graphs (use technology!), try to guess where $f_x(x, y)$ and $f_y(x, y)$ do exist and where they don't. (No proofs are needed.)
    (b) Find $f_x(0, 0)$.
    (c) Explain why $f_x(0, 1)$ does not exist.

14.4 Optimization and Partial Derivatives: A First Look

Optimization—finding maximum and minimum values of a function—is as important for multivariable functions as for one-variable functions. Functions of several variables are more complicated, but derivatives are still the crucial tool.

A One-Variable Review

For a one-variable differentiable function $y = f(x)$ on an interval I, finding maximum and minimum values is relatively straightforward. Maxima and minima are found only at stationary points—where $f'(x) = 0$—or at the endpoints (if any) of the interval I. Usually, only a few such "candidate" points exist, and we can check directly which one produces, say, the largest value of f.

A simple idea lies behind all talk of derivatives and extrema: ▶ At a local maximum or minimum point x_0, the graph of a differentiable function f must be "flat," so $f'(x_0) = 0$. However, a stationary point x_0 might be (i) a local maximum point; (ii) a local minimum point, or (iii) neither. (The function $f(x) = x^3$ at $x = 0$ illustrates the "neither" case.)

For simple one-variable functions, deciding which of (i)–(iii) actually holds is easy. One strategy is to check the sign of $f''(x_0)$. If, say, $f''(x_0) < 0$, then f is

"Extremum" (singular) means "either maximum or minimum"; "extrema" is the plural.

Without technology, plotting f may be hard.

concave down at x_0, so f has a local maximum there. Alternatively, we might just plot f and see directly how it behaves near x_0. ◄

Local Talk. When is a maximum or minimum "local"? When is it "global"? The following graph shows the difference.

Graph of f: local vs. global extrema

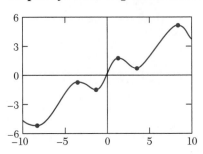

All six bulleted points (three maximum points and three minimum points) correspond to *local* extrema. At each bulleted point $(x_0, f(x_0))$, $f(x_0)$ is either highest or lowest among *nearby* points $(x, f(x))$. (For present purposes, exactly *how* nearby doesn't matter.) Only the first and last points are *global* extrema, because only they represent the largest and smallest values of $f(x)$ among all possibilities shown in the picture. On a larger domain interval—say, $-15 \leq x \leq 15$—these extrema might not be global.

  Local maxima and minima are conceptually simpler than the global variety, are readily recognizable on graphs, and are often convenient to locate symbolically, using derivatives. With all these advantages, local extrema are usually the main tools used to solve optimization problems.

In Several Variables

If a one-variable function has a local maximum or minimum at x_0, then the ordinary derivative (if it exists) must be zero at x_0. If a two-variable function $f(x, y)$ has a local maximum at (x_0, y_0), then *both* partial derivatives (if they exist) must vanish there:

$$f_x(x_0, y_0) = 0; \qquad f_y(x_0, y_0) = 0.$$

This is the natural extension of the one-variable idea.

In words, (x_0, y_0) is a stationary point of f. ◄

  The result sounds plausible, but why is it true? Consider the case of a local maximum. Geometrically, the surface $z = f(x, y)$ has a "peak" above the domain point (x_0, y_0). If we slice the surface with any vertical plane, say the plane $y = y_0$, then the resulting curve—i.e., the trace of the surface $z = f(x, y)$ in the plane $y = y_0$—has the equation $z = f(x, y_0)$, and this curve must have a peak at $x = x_0$. We know from one-variable calculus that if the function $x \to f(x, y_0)$ is differentiable, then its derivative must be zero at x_0. In other words,

$$\frac{dz}{dx}(x_0) = f_x(x_0, y_0) = 0.$$

For similar reasons, $f_y(x_0, y_0) = 0$. Here's the general fact:

> **Theorem 1** (**Extreme Points and Partial Derivatives**) Suppose that $f(x, y)$ has a local maximum or a local minimum at (x_0, y_0). If both partial derivatives exist, then
>
> $$f_x(x_0, y_0) = 0 \quad \text{and} \quad f_y(x_0, y_0) = 0.$$

Caution. The theorem is useful, but what it *doesn't* say is also important. It does *not* guarantee, in particular, that at a stationary point, f must assume either a local maximum or a local minimum value. The next example illustrates both sides of this coin.

EXAMPLE 1 Let $f(x, y) = x^2 + y^2$, and let $g(x, y) = xy$. Find all the stationary points of f and g. What happens at each one?

Solution Finding the partial derivatives is easy:

$$f_x(x, y) = 2x; \quad f_y(x, y) = 2y; \quad g_x(x, y) = y; \quad g_y(x, y) = x.$$

Both f and g, therefore, are stationary only at the origin $(0, 0)$. For f the origin is a minimum point, since $f(x, y) = x^2 + y^2 \geq 0$ for all (x, y). For g, however, the origin is neither a maximum nor a minimum point, since $g(x, y)$ assumes both positive and negative values near $(0, 0)$. (For example, $g(0.1, -0.1) < 0$, but $g(0.1, 0.1) > 0$.) Contour maps of f and g illustrate their very different behavior near the stationary point. (In the following pictures, the level curves àre the edges of the shaded regions. Note the key to the right of each contour map.) Here's f:

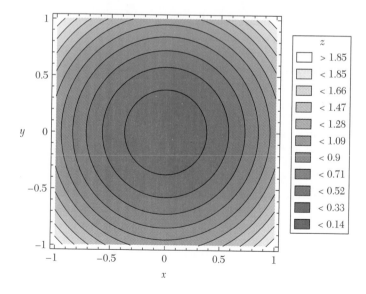

Level curves of f are circles centered on the "basin" at $(0, 0)$. Thus, the picture suggests ▶ that f has a local minimum at the stationary point $(x, y) = (0, 0)$. The

But take a careful look. Notice, in particular, that darker regions are "lower."

suggestion is correct: $f(0, 0) = 0$, and $f(x, y) \geq 0$ for all (x, y), so f assumes a local (and even global) minimum value at $(0, 0)$.

Now look at g near its stationary point:

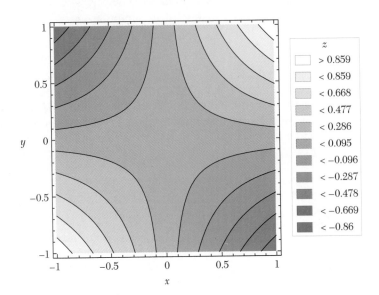

Think about this carefully. Can you see the saddle in the contour map?

Level curves of g show a "saddle" at $(0, 0)$. (If the surface were literally a saddle, the horse would be walking either "northeast" or "southwest.") ◄ The surface rises above the first and third quadrants and falls below the second and fourth quadrants. ∎

Different Types of Stationary Points. One moral of the preceding example is that, although the basic strategy for optimizing a function—find the stationary points and analyze them—is exactly the same for functions of one and of several variables, the situation is usually more complicated for functions of several variables. For one thing, finding stationary points may be harder; for another, functions of several variables can behave in more complicated ways near a stationary point. This makes multivariable optimization harder but also more interesting.

As we did for functions of one variable, we identify three main types of stationary points for a function $f(x, y)$. (The definitions for a three-variable function $g(x, y, z)$ are almost identical.)

Local Minimum Point A stationary point (x_0, y_0) is a **local minimum point** for f if $f(x, y) \geq f(x_0, y_0)$ for all (x, y) near (x_0, y_0). (A little more formally, $f(x, y) \geq f(x_0, y_0)$ for all (x, y) in some rectangle surrounding (x_0, y_0).) In this case, we say that f *assumes a local minimum value* at (x_0, y_0). In Example 1, $(0, 0)$ is a local minimum point for $f(x, y) = x^2 + y^2$.

Local Maximum Point A stationary point (x_0, y_0) is a **local maximum point** for f if $f(x, y) \leq f(x_0, y_0)$ for all (x, y) near (x_0, y_0). In this case, we say that f *assumes a local maximum value* at (x_0, y_0). The following contour map illustrates a local maximum point for the function $h(x, y) = 1 - x^2 - y^2$.

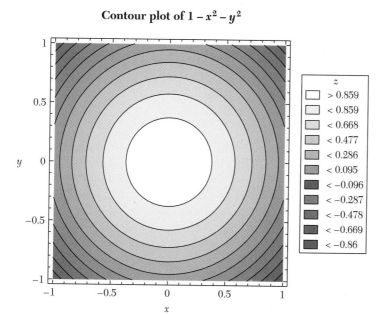

Contour plot of $1 - x^2 - y^2$

Saddle Point A stationary point (x_0, y_0) is a **saddle point** for f if f assumes neither a local maximum nor a local minimum at (x_0, y_0). In Example 1, $(0, 0)$ is a saddle point for $g(x, y) = xy$.

As the following picture shows, all three possibilities can coexist in close quarters. The function in question is $f(x, y) = \cos(x) \sin(y)$. ▶

Look in the picture for a stationary point of each type just listed.

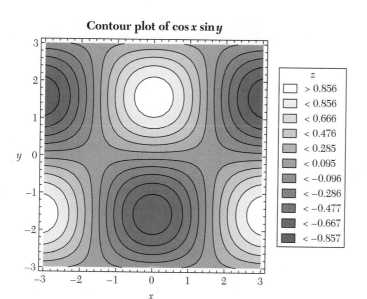

Contour plot of $\cos x \sin y$

Optimization: More to the Story

There's much more to the story of optimizing functions of several variables; we've taken only a first, short look. Following are two questions that arise in multivariable calculus courses.

Second Derivative Tests.   In single-variable calculus, the second derivative f'' is sometimes used to classify stationary points as maxima, minima, or neither. For example, if $f'(x_0) = 0$ and $f''(x_0) > 0$, then x_0 is a local minimum point for f. A similar but (necessarily) more sophisticated approach is possible for functions of two or more variables.

Extremes on the Boundary.   Recall what happens in elementary calculus for a differentiable function $f(x)$ defined on a closed interval $[a, b]$: f may assume its maximum and minimum values either at a stationary point (where $f'(x) = 0$) or at either of the endpoints $x = a$ and $x = b$.

The situation is similar for a function of two variables defined on a region, such as a rectangle or a circle, that has a definite "edge," or boundary: $f(x, y)$ may assume its maximum and minimum either at a stationary point or somewhere on the boundary of the region. We illustrate the situation, and one way to approach it, with a simple example.

EXAMPLE 2   Where on the rectangle $R = [-1, 1] \times [-1, 1]$ does $g(x, y)$ assume its minimum and maximum values?

Solution   We saw in Example 1, page 703, that g has only one stationary point, a saddle point, in the interior of R. Therefore, the maximum and minimum values of g must occur somewhere on the boundary of R. A look at the contour plot of g (notice the symmetry) shows that it's enough to look along *any* boundary edge of R, such as the right edge. On this edge we have $x = 1$, so g behaves like a function of just one variable: $g(x, y) = g(1, y) = y$. Clearly, $g(1, y) = y$ is largest at $y = 1$ and smallest at $y = -1$. We therefore conclude that $g(1, 1) = 1$ and $g(1, -1) = -1$ are, respectively, maximum and minimum values of g on R.   ■

BASICS

1. Let $g(x, y) = xy$. (Its contour map is shown in Example 1.) To an ant walking along the surface $z = g(x, y)$ from lower left to upper right, the origin seems to be a low spot; another ant walking from upper left to lower right would experience the origin as a high spot.

   (a) An ant walks along the surface from $(0, -1)$ to $(0, 1)$. How does the ant's altitude change along the way?

   (b) Another ant walks along the surface from $(0.5, -1)$ to $(0.5, 1)$. How does the ant's altitude change along the way? Where is the ant highest? How high is the ant there?

2. See the contour map of $f(x, y) = \cos(x)\sin(y)$ on page 705.

   (a) The surface $z = f(x, y)$ resembles an egg carton. Where do the eggs go?

   (b) From the picture alone, estimate the coordinates of a local minimum point, a local maximum point, and a saddle point.

   (c) Use the formula $f(x, y) = \cos(x)\sin(y)$ to find (exactly) all the stationary points of f in the rectangle $R = [-3, 3] \times [-3, 3]$.

   (d) Find the maximum and minimum values of f in the rectangle $R = [-3, 3] \times [-3, 3]$.

3. Consider the function $f(x, y) = x(x - 2)\sin(y)$. Here's a contour map:

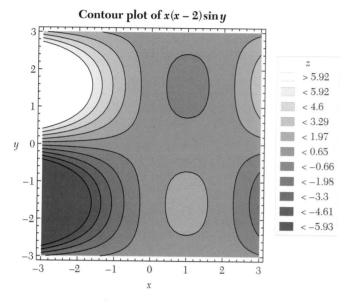

Contour plot of $x(x - 2)\sin y$

z	
	> 5.92
	< 5.92
	< 4.6
	< 3.29
	< 1.97
	< 0.65
	< −0.66
	< −1.98
	< −3.3
	< −4.61
	< −5.93

(a) The function f has two stationary points along the line $x = 1$. Use the picture to estimate their coordinates. What type is each one?

(b) There are four stationary points inside the rectangle $R = [-3, 3] \times [-3, 3]$. Use the formula for f to find all four.

(c) The contour plot shows that f assumes its maximum and minimum values on $R = [-3, 3] \times [-3, 3]$ somewhere along the left boundary—i.e., where $x = -3$. Find these maximum and minimum values.
[**HINT:** If $x = -3$, then $f(x, y) = f(-3, y) = -3\sin(y)$. This is a function of one variable, defined for $-3 \le y \le 3$.]

4. For each of the following functions, use the formula to find all stationary points. Then use technology (e.g., a properly chosen contour plot or surface plot) to decide what type of stationary point each one is.
   (a) $f(x, y) = -x^2 - y^2$
   (b) $f(x, y) = x^2 - y^2$
   (c) $f(x, y) = 3x^2 + 2y^2$
   (d) $f(x, y) = xy - y - 2x + 2$

5. Consider the linear function $L(x, y) = 1 + 2x + 3y$. Does L have any stationary points? If so, what type are they? If not, why not?

6. Consider the linear function $L(x, y) = a + bx + cy$, where a, b, and c are any constants.
   (a) The graph of L is a plane. Which planes have stationary points? For these planes, where are the stationary points?
   (b) Under what conditions on a, b, and c will L have stationary points? In this case, where are the stationary points? Reconcile your answers with those in part (a).

7. Let $f(x, y) = x^2$. The graph of f is a cylinder, unrestricted in the y-direction.
   (a) Use technology to plot the surface $z = f(x, y)$. Where in the xy-plane are the stationary points? What type are they?   [**HINT:** There's a whole line of stationary points.]
   (b) Use partial derivatives of f to find all the stationary points. Reconcile your answer with part (a).

8. Give an example as described in each part.
   [**HINTS:** (1) See Exercise 7 for ideas. (2) Check your answers by plotting.]
   (a) A function $g(x, y)$ for which every point on the x-axis is a local minimum point
   (b) A function $h(x, y)$ for which every point on the line $x = 1$ is a local maximum point
   (c) A nonconstant function $k(x, y)$ that has a local minimum at $(3, 4)$

14.5 Multiple Integrals and Approximating Sums

The last two sections introduced *derivatives* of functions of several variables and a few of their most basic properties. The next few sections, on *integrals*, continue our flying tour of the basic calculus of functions of two or three variables. This section concerns mainly what multiple integrals are and what they mean. In the following two sections we'll consider more systematically how to calculate multiple integrals.

Integrals and Approximating Sums

All integrals—single, double, triple, or whatever—are defined to be certain limits of approximating sums (sometimes known as Riemann sums). This important idea

Remember?

is always studied in single-variable calculus, ◄   but it may be quickly (and perhaps gratefully) forgotten.   Readers whose memories are vague on this score have an excellent excuse: Although integrals are *defined* as limits of approximating sums, they are often *calculated* in an entirely different way, using antiderivatives. Here's

There's not a Riemann sum in sight.

a typical calculation: ◄

$$\int_0^1 x^2 \, dx = \frac{x^3}{3}\bigg]_0^1 = \frac{1}{3}.$$

This method of evaluating an integral—find an antiderivative for the integrand and plug in the endpoints—works just fine, thanks to the fundamental theorem of calculus.

So why bother with approximating sums? Here are two good reasons:

Antiderivative Trouble   The antiderivative method depends on finding a convenient antiderivative of the integrand. Unfortunately, not every function, even in single-variable calculus, *has* an "elementary" antiderivative, i.e., an antiderivative with a symbolic formula built from the usual ingredients.

Spend a moment looking for an antiderivative formula. Nothing works.

The simple-looking function $f(x) = \sin(x^2)$ is an example. ◄   The best we can do with the integral

$$I = \int_0^1 \sin(x^2) \, dx,$$

therefore, is to approximate it with some sort of sum, using, say, the left rule, the midpoint rule, or the trapezoid rule. (For the record, approximating I with a trapezoid-rule sum with 10 subdivisions gives $I \approx 0.311$.)

What Integrals Mean   The fundamental theorem (when it works) makes *calculating* integrals easy, but approximating sums may illustrate more clearly what the answers mean. The following picture, for instance, illustrates the sense in which a midpoint-rule sum with four subdivisions approximates the area bounded by the curve $y = x^2$ from $x = 0$ to $x = 1$.

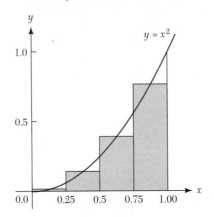

A midpoint sum estimate to $\int_0^1 x^2 \, dx$:
$M_4 = 63/192 \approx 0.328$

The basic idea of an integral as a limit of approximating sums is much the same for functions of two or more variables as for functions of one variable.

Integrals and Sums: A Review

Let's review the main single-variable objects and notations that arise on the way to defining the integral of a function f over an interval $[a, b]$. The first step is to form Riemann sums.

> **Definition** Let $[a, b]$ be partitioned into n subintervals by any $n+1$ points
>
> $$a = x_0 < x_1 < x_2 < \cdots < x_{n-1} < x_n = b;$$
>
> let $\Delta x_i = x_i - x_{i-1}$ denote the width of the ith subinterval. Within each subinterval $[x_{i-1}, x_i]$, choose any point c_i. The sum
>
> $$\sum_{i=1}^{n} f(c_i)\, \Delta x_i = f(c_1)\, \Delta x_1 + f(c_2)\, \Delta x_2 + \cdots + f(c_n)\, \Delta x_n$$
>
> is a Riemann sum with n subdivisions for f on $[a, b]$.

Left-rule, right-rule, and midpoint-rule approximating sums all fit this definition. Each of these sums is built from a partition of $[a, b]$ into subintervals of equal length and some consistent scheme for choosing the sampling points c_i. (For left, right, and midpoint sums, respectively, we choose each c_i as the left endpoint, the right endpoint, or the midpoint of the ith subinterval.)

Graphical and numerical intuition suggest that all of these approximating sums (and others) should converge to some fixed number. Geometrically speaking, this number measures the signed area bounded by the f-graph from $x = a$ to $x = b$.

The limit definition of integral makes these ideas precise.

> **Definition** Let the function f be defined on the interval $[a, b]$. The integral of f over $[a, b]$, denoted by $\int_a^b f(x)\, dx$, is the number to which all Riemann sums S_n tend as n tends to infinity and as the widths of all subdivisions tend to zero. In symbols,
>
> $$\int_a^b f(x)\, dx = \lim_{n \to \infty} S_n = \lim_{n \to \infty} \sum_{i=1}^{n} f(c_i)\, \Delta x_i$$
>
> if the limit exists.

Honesty dictates a brief admission: The limit in the definition, taken at face value, is a slippery customer. Understanding every ramification of permitting arbitrary partitions and sampling points, for example, can be tricky. Fortunately, these issues need not trouble us for the usual, well-behaved functions (e.g., continuous functions) of single-variable and multivariable calculus. For such functions, almost any respectable sort of approximating sum does what we'd expect—approaches the true value of the integral as n tends to infinity.

Two Variables: Double Integrals

Most of the differences between single-variable integrals and multivariable integrals are technical rather than theoretical. Indeed, the definitions of

$$\iint_R f(x, y)\, dA \qquad \text{and} \qquad \int_a^b f(x)\, dx$$

are almost identical. Now f is a function of two variables, and R is a region—a rectangle, in the simplest case—in the xy-plane. (The mechanics of evaluating these two types of integrals by antidifferentiation are quite different, on the other hand.) ◄

We'll get to that in the next section.

First, let's list the ingredients that go into defining the **double integral** $\iint_R f(x, y)\, dA$. In each case, look for similarities to and differences from the one-variable situation.

R, the Region of Integration   In one variable, the region of integration is always an interval $[a, b]$ in the domain of f; this is implicit in the notation $\int_a^b f(x)\, dx$. In two variables, by contrast, the region of integration, denoted by R, may be almost *any* two-dimensional subset of the plane. In the simplest cases, ◄ R is a rectangle $[a, b] \times [c, d]$, and we sometimes write

We emphasize them.

$$\int_a^b \int_c^d f(x, y)\, dy\, dx, \qquad \text{not} \qquad \iint_R f(x, y)\, dA.$$

(The first notation suggests a fact we'll see in the next section—double integrals can sometimes be calculated by integrating "one variable at a time.")

Partitions   In one variable, we partition an interval $[a, b]$ by cutting it, perhaps unevenly, into smaller intervals with endpoints $a = x_0 < x_1 < x_2 < \cdots < x_n = b$. The "size" ◄   of the ith subinterval is simply its length, Δx_i. In two variables, we do much the same thing: We chop the plane region R into m smaller regions $R_1, R_2, R_3, \ldots, R_m$, perhaps of different sizes and shapes. The "size" of a subregion R_i is now taken to be its *area*, denoted by ΔA_i.

We use "size" informally; "measure" is sometimes used in a more technical sense.

In practice—whatever the number of variables—it's usually convenient to choose the partition in some consistent way. In one variable, using equal-length subintervals is simplest. An analogous procedure in two variables, if R is a rectangle $[a, b] \times [c, d]$, is to cut R by an n-by-n grid in each direction, producing n^2 rectangular subregions in all. (This isn't the only possibility. Another alternative is to cut R into small squares.)

Approximating Sums   In one variable, an approximating sum has the form

$$f(c_1)\,\Delta x_1 + f(c_2)\,\Delta x_2 + \cdots + f(c_n)\,\Delta x_n = \sum_{i=1}^{n} f(c_i)\,\Delta x_i,$$

where each c_i is a sample point chosen from the ith subinterval. A two-variable approximating sum is similar. From each subregion R_i we choose a sampling point $P_i(x_i, y_i)$ and then form the approximating sum

$$S_m = f(P_1)\,\Delta A_1 + f(P_2)\,\Delta A_2 + \cdots + f(P_m)\,\Delta A_m = \sum_{i=1}^{m} f(P_i)\,\Delta A_i,$$

where ΔA_i is the area of R_i.

The following picture illustrates the situation for the function $f(x, y) = x + y$ on the rectangle $R = [0, 4] \times [0, 4]$, with $n^2 = 4^2 = 16$ subdivisions. In each subinterval, sampling is done at the corner closest to the origin.

The surface $z = x + y$ and a Riemann approximation

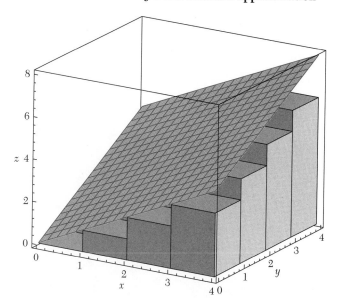

The picture shows (among other things) that the approximating sum gives an estimate to the volume bounded above by the surface $z = f(x, y)$ and below by the rectangle $R = [0, 4] \times [0, 4]$. A close look and some back-of-the-envelope calculations show that the approximating sum shown adds up to 48.

Calculated with help from technology, of course.

It's reasonable to expect these estimates to converge to the "true" volume as n (the number of subdivisions in each direction) increases to infinity. A table of values ◄ supports this expectation:

Approximating Sums for Various m	
Number of Subdivisions (m)	Approximating Sum (S_m)
2^2	32.00
4^2	48.00
8^2	56.00
12^2	58.67
16^2	60.00
20^2	60.80
24^2	61.33
28^2	61.71
32^2	62.00

The next picture shows another approximating sum for the same function over the same interval. This time, however, sampling is done at the *midpoint* of each subrectangle.

The surface $z = x + y$ and a midpoint Riemann approximation

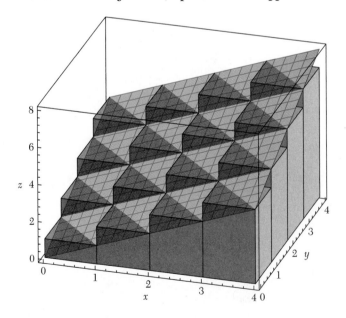

This time the approximating sum adds up to 64.

What's dA?   The symbol "dA" in the double integral resembles the "dx" that appears in single integrals. The "A" reminds us of *area*.

EXAMPLE 1   Consider the double integral $I = \iint_R f(x, y)\, dA$, where $f(x, y) = x + y$ and $R = [0, 4] \times [0, 4]$. Calculate, by hand, an approximating sum S_4 with four equal subdivisions (two in each direction). In each square subregion, evaluate f at the corner nearest to the origin.

Solution   All four subregions are squares with edge length 2 and therefore area 4; all corners have integer coordinates. The sampling points just described are $P_1 = (0, 0)$, $P_2 = (2, 0)$, $P_3 = (0, 2)$, and $P_4 = (2, 2)$. The desired approximating sum is therefore

$$S_4 = \sum_{i=1}^{4} f(P_i)\, \Delta A_i = 0 \cdot 4 + 2 \cdot 4 + 2 \cdot 4 + 4 \cdot 4 = 32.$$ ∎

The Integral as a Limit

We defined the one-variable integral $\int_a^b f(x)\, dx$ as a limit of approximating sums. The double integral $\iint_R f(x, y)\, dA$ can be defined in a similar way. The following definition is adequate for well-behaved functions $f(x, y)$. ▶

Some additional subtleties—which we ignore—may arise for very ill-behaved functions.

> **Definition** Let the function $f(x, y)$ be defined on the region R, and let S_m be an approximating sum with m subdivisions. Let I be a number such that S_m tends to I whenever m tends to infinity and the diameter of all subdivisions tends to zero. Then I is the double integral of f over R, and we write
>
> $$I = \iint_R f(x, y)\, dA = \lim_{m \to \infty} \sum_{i=1}^{m} f(P_i)\, \Delta A_i.$$

As in the one-variable setting, the limit definition—although crucial to understanding integrals and often useful for approximating them—almost never lends itself to calculating integrals *exactly*. Fortunately, there are methods based on antidifferentiation for this purpose. We'll see some soon.

Triple Sums and Triple Integrals

The idea of integral can be extended to dimension three (and even higher dimensions). We will consider only the simplest case: the **triple integral** of a function $g(x, y, z)$ over a rectangular parallelepiped ▶ $R = [a, b] \times [c, d] \times [e, f]$, denoted by either

A "brick," to put it more humbly.

$$\iiint_R g(x, y, z)\, dV \qquad \text{or} \qquad \int_a^b \int_c^d \int_e^f g(x, y, z)\, dz\, dy\, dx.$$

As for double integrals, the second notation suggests, correctly, that such integrals can sometimes be calculated one variable at a time.

Triple integrals, just like single and double integrals, are defined formally as limits of approximating sums. An approximating sum in three dimensions is formed by subdividing a rectangular solid region R into m smaller rectangular subregions R_i, each with volume ΔV_i; choosing a sampling point P_i in each subregion; and then evaluating the sum

$$\sum_{i=1}^{m} f(P_i)\,\Delta V_i.$$

The triple integral is then defined as the limit of such sums as the diameter of all subregions tends to zero.

EXAMPLE 2   Consider the triple integral $I = \iiint_R g(x, y, z)\,dV$, where $g(x, y, z) = x+y+z$ and $R = [0, 2] \times [0, 2] \times [0, 2]$. Calculate an approximating sum S_8 with eight equal subdivisions (two in each direction). In each cubical subregion, evaluate g at the corner nearest to the origin.

Solution   All eight subregions are cubes with edge length 1 and (therefore) volume 1; all the corners have integer coordinates. The sampling points just described are $P_1 = (0, 0, 0)$, $P_2 = (1, 0, 0)$, $P_3 = (0, 1, 0)$, $P_4 = (1, 1, 0)$, $P_5 = (0, 0, 1)$, $P_6 = (1, 0, 1)$, $P_7 = (0, 1, 1)$, and $P_8 = (1, 1, 1)$. The approximating sum is therefore

$$S_8 = \sum_{i=1}^{8} g(P_i)\,\Delta V_i = 0 + 1 + 1 + 2 + 1 + 2 + 2 + 3 = 12.$$ ■

Interpreting Multiple Integrals

Integrals can be interpreted geometrically, physically, or in other ways. Following is a sampler of possibilities.

Double Integrals and Volume.   The standard geometric interpretation of a single-variable integral $\int_a^b f(x)\,dx$ is in terms of *area*: If $f(x) \geq 0$, then $\int_a^b f(x)\,dx$ is the area of the region bounded above by the curve $y = f(x)$, bounded below by the interval $[a, b]$ in the x-axis, and having vertical sides. In a similar vein, as we've already seen from pictures, if $f(x, y) \geq 0$ for (x, y) in R, then the double integral $\iint_R f(x, y)\,dA$ measures the *volume* of the three-dimensional solid bounded above by the surface $z = f(x, y)$, bounded below by the region R in the xy-plane, and having sides perpendicular to the xy-plane.

Double Integrals and Area.   If R is a region in the xy-plane and g is the constant function $g(x, y) = 1$, then (as the preceding paragraph says) the integral $\iint_R 1\,dA$ represents the volume of the solid S bounded below by R and having vertical sides and *constant height* 1. Recall, however, that the volume of any such "cylindrical" solid S is the *area* of the base times the height. Therefore, in this special case, the volume of S happens to be the same as its area. Thus, for any plane region R,

$$\iint_R 1\,dA = \text{area of } R.$$

Surprisingly, this fact is often of practical use in calculating areas of plane regions. The next section contains examples.

Triple Integrals and Volume.   Triple integrals pose a special problem having to do with dimensions. There isn't "room" in three-dimensional space even to plot a function $w = g(x, y, z)$—*four* variables would be needed. For this reason, interpreting triple integrals geometrically is, as a rule, difficult or impossible.

There's one important exception to this rule. For reasons similar to those explained in the preceding paragraph, integrating the constant function $g(x, y, z) = 1$ over a solid region R in xyz-space gives the volume of the region R. In symbols:

$$\iiint_R 1 \, dV = \text{volume of } R.$$

Density, Mass, and Multiple Integrals. Both double and triple integrals can often be interpreted physically in the language of density and mass. (This view has the special advantage of making sense for both double and triple integrals.)

For a double integral $\iint_R f(x, y) \, dA$, one thinks of the plane region R as a flat plate with variable density; at any point (x, y), $f(x, y)$ gives the density, measured in appropriate units (grams per square centimeter, say). From this viewpoint, the double integral $\iint_R f(x, y) \, dA$ is the total mass, in grams, of the plate R.

For a triple integral $\iiint_R g(x, y, z) \, dV$, one imagines a solid region R with variable density; at any point (x, y), $g(x, y, z)$ is the solid's density, measured in appropriate units (grams per cubic centimeter, say). From this viewpoint, the triple integral $\iiint_R g(x, y, z) \, dV$ is the total mass, in grams, of the solid R.

BASICS

Important Notes. *Maple* (or a similar utility) is essential for some parts of these exercises. In all cases, approximating sums are evaluated at the midpoints of subintervals or subregions.

1. Calculate by hand (without technology) the midpoint sum with four subdivisions for $\int_0^1 x^2 \, dx$. Then check that *Maple* agrees. Finally, compare *Maple*'s answer for 100 subdivisions.

2. Calculate by hand (without technology) the double midpoint sum with $n = 3$ (i.e., nine subdivisions in all) for the integral $\iint_R \sin(x) \sin(y) \, dA$ over the rectangle $R = [0, 1] \times [0, 1]$. Check your answer using *Maple*. Finally, compare *Maple*'s answer for $n = 10$, i.e., 100 subdivisions in all.

3. Calculate by hand (without technology) the triple midpoint sum with $n = 2$ (i.e., eight subdivisions in all) for the triple integral $\iiint_R xyz \, dV$ over the cube $R = [0, 4] \times [0, 4] \times [0, 4]$. Check your answer using *Maple*. Finally, compare *Maple*'s answer for $n = 4$, i.e., $4^3 = 64$ subdivisions in all.

4. Let $f(x, y) = x + y$, let $R = [0, 4] \times [0, 4]$, and let $I = \iint_R f(x, y) \, dA$. Here is a contour plot of f:

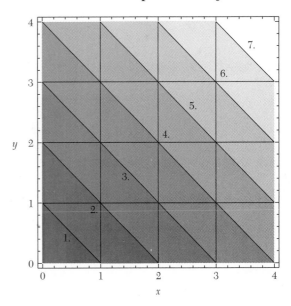

Contour plot of $z = x + y$

(a) Use the contour plot to evaluate a double midpoint sum for I with $n = 4$ (16 subdivisions in all).

(b) Use *Maple* to check your answer from part (a).

(c) Your answer in part (a) is, in fact, the exact value of the integral I. How does the symmetry of the contour map show this?

5. Let $f(x, y) = x^2 + y^2$, let $R = [0, 4] \times [0, 4]$, and let $I = \iint_R f(x, y) \, dA$. Here is a contour plot of f:

Contour plot of $z = x^2 + y^2$

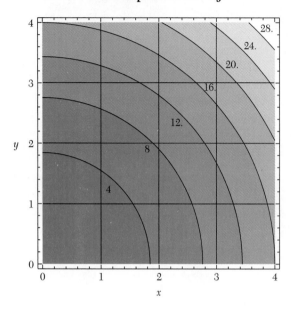

(a) Use the contour plot to evaluate a double midpoint sum for I with $n = 4$ (16 subdivisions in all).
(b) Use *Maple* to check your answer from part (a).
(c) Would you expect your answer from part (a) to overestimate or underestimate the true value of I? How can you tell?

14.6 Calculating Integrals by Iteration

The preceding section was about *defining* multivariable integrals as limits of approximating sums. This section is about *calculating* multivariable integrals, using antidifferentiation. Approximating sums are conceptually simple and (with technology) easy to calculate. But approximating sums are only approximate; to evaluate integrals *exactly*, we'd like an appropriate version of the single-variable antiderivative method, ◄ suitably modified for multivariable use.

It works thanks to the fundamental theorem of calculus.

Iteration: How It Works

The key idea is to integrate a multivariable function *one variable at a time*, treating other variables as constants. The process is called **iterated integration**. ◄ To start, here's an example to show *how* it works; we'll see *why* it works in a moment.

In math-speak, "iterate" means "repeat."

EXAMPLE 1   Let $f(x, y) = x + y$, and let $R = [0, 4] \times [0, 4]$. Find $\iint_R f(x, y) \, dA$ by iterated integration. (We studied this integral in Section 14.5; see the pictures and the table of values starting on page 711.)

Solution   We integrate first in x, treating y as a constant. Watch each step carefully. ▶

Attaching variable names to the limits of integration is optional, but it can help remind us which variable is involved.

$$\iint_R f(x, y)\, dA = \int_{y=0}^{y=4} \left(\int_{x=0}^{x=4} (x + y)\, dx \right) dy$$

$$= \int_{y=0}^{y=4} \left(\frac{x^2}{2} + xy \Big]_{x=0}^{x=4} \right) dy$$

$$= \int_{y=0}^{y=4} (8 + 4y)\, dy$$

$$= 8y + 2y^2 \Big]_{y=0}^{y=4} = 64.$$

Observe:

The Same Answer   The final answer, 64, should be familiar—it's what we estimated in Section 14.5, using a midpoint approximation.

The Answer as Volume   Interpreted as a volume, the answer says that the solid bounded below by $R = [0, 4] \times [0, 4]$, bounded above by the plane $z = x + y$, and having straight vertical walls has volume 64 cubic units. Here's a picture of the solid in question:

A solid bounded above by $z = x + y$

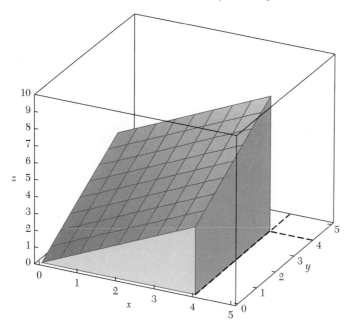

Checking Answers with Technology   Is the answer 64 geometrically reasonable? Is it symbolically correct? Technology (e.g., *Maple*) can help with both questions. A plot like the preceding one suggests that the answer 64 is at least in the ball park. As a further check, here is *Maple*'s version of the symbolic calculation:

```
> int( int( x+y, x=0..4), y=0..4 );
```

$$64$$

Work from Inside Out . Iterated integrals are calculated from the inside out. The "inner" integral (the x-integral in the preceding calculation) is found first.

Either Order Works   There's nothing sacred about integrating first in x and then in y. For well-behaved functions we can integrate in either order, and both orders give the same answer. We'll return to this question.   ■

Cross-Sectional Area: An Intermediate Function.   In the preceding example, the inner integral was calculated with respect to x, with y treated as constant. The result was an intermediate function, g, of y alone; the formula is

$$g(y) = \int_0^4 (x + y)\, dx = 8 + 4y.$$

We integrated $g(y)$ with respect to y to get the final answer.

The function g has a nice geometric meaning: For any fixed y_0 in $[0, 4]$, $g(y_0)$ is the area under the curve $z = f(x, y_0)$ and above the xy-plane. In other words, $g(y)$ is the *area* of the cross section of the solid obtained by slicing with the plane $y = y_0$. Here's the picture for $y = 2$:

The cross-section with the plane $y = 2$

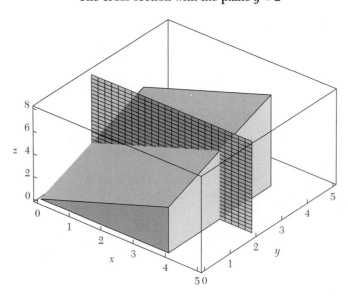

Since $g(y) = 8 + 4y$, we get $g(2) = 16$; this is the area of the part of the plane inside the solid. As y runs from $y = 0$ to $y = 4$, $g(y)$ measures the area "swept out" by planes parallel to the one shown. ◄

This area increases with y, as the picture shows.

As a matter of fact, the idea of a cross-sectional-area function is not new; we used the same idea in single-variable calculus when we calculated volumes by integration.

Back then, we put it like this: ▶

*In the preceding example,
g(y) played the role that
A(x) will play in what
follows.*

> **Fact** Suppose that a solid lies with its base on the xy-plane, between
> the vertical planes $x = a$ and $x = b$. For all x in $[a, b]$, let $A(x)$ denote
> the area of the cross section at x, perpendicular to the x-axis. If $A(x)$ is
> a continuous function, then
>
> $$\text{Volume} = \int_a^b A(x)\, dx.$$

Iteration: Why It Works

Why does the iteration method just demonstrated work? The preceding Fact gives
some geometric feeling for the matter, at least when an integral can be thought of
as the volume of a three-dimensional solid.

But not all integrals can or should be thought of in this way. A better reason why iteration works—a reason that makes sense in *any* dimension—is based
on approximating sums. We describe the idea in two dimensions, but everything
transfers readily to three (or even higher) dimensions. The main idea, in a nutshell, is that an approximating sum for a double integral can be grouped either
along "rows" or along "columns." We first give an example and then a more general
argument.

EXAMPLE 2   Let $f(x, y) = 2 - x^2 y$, let $R = [0, 1] \times [0, 1]$, and let $I =
\iint_R f(x, y)\, dA$. Here's a picture of the solid whose volume is given by I:

The solid under $z = 2 - x^2 y$, over $[0, 1] \times [0, 1]$

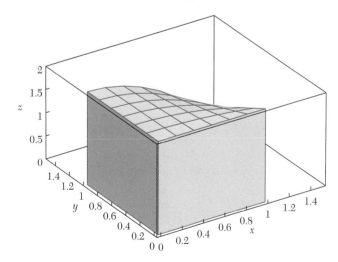

Compute a midpoint approximating sum with 100 equal subdivisions (10 in each
direction) for the integral $I = \iint_R f(x, y)\, dA$. Compare the result to the exact
value of I, found by iteration.

But check our work.

Solution   It's easy to calculate I exactly, by iteration. ◄   This time, for variety, we integrate first in y:

$$I = \int_0^1 \left(\int_0^1 (2 - x^2 y) \, dy \right) dx = \int_0^1 \left(2y - \frac{x^2 y^2}{2} \Big]_0^1 \right) dx$$

$$= \int_0^1 \left(2 - \frac{x^2}{2} \right) dx = \frac{11}{6}.$$

They're rounded to one decimal place.

To find an approximating sum, we first tabulate values (rounded to two decimals) of f at the midpoints of all 100 subrectangles of $[0, 1] \times [0, 1]$. For later use, column sums are at the bottom. ◄

y \ x	0.05	0.15	0.25	0.35	0.45	0.55	0.65	0.75	0.85	0.95
Values of $f(x, y) = 2 - x^2 y$										
0.05	2.00	2.00	2.00	1.99	1.99	1.98	1.98	1.97	1.96	1.95
0.15	2.00	2.00	1.99	1.98	1.97	1.95	1.94	1.92	1.89	1.86
0.25	2.00	1.99	1.98	1.97	1.95	1.92	1.89	1.86	1.82	1.77
0.35	2.00	1.99	1.98	1.96	1.93	1.89	1.85	1.80	1.75	1.68
0.45	2.00	1.99	1.97	1.94	1.91	1.86	1.81	1.75	1.67	1.59
0.55	2.00	1.99	1.97	1.93	1.89	1.83	1.77	1.69	1.60	1.50
0.65	2.00	1.99	1.96	1.92	1.87	1.80	1.73	1.63	1.53	1.41
0.75	2.00	1.98	1.95	1.91	1.85	1.77	1.68	1.58	1.46	1.32
0.85	2.00	1.98	1.95	1.90	1.83	1.74	1.64	1.52	1.39	1.23
0.95	2.00	1.98	1.94	1.88	1.81	1.71	1.60	1.47	1.31	1.14
Sum	20.0	19.9	19.7	19.4	19.0	18.4	17.9	17.2	16.4	15.5

A computer helped!

The approximating sum is found by totaling all the function values and multiplying by 1/100, the area of each subrectangle. (In this case, therefore, the approximating sum is the average of the 100 table entries.) The result ◄ is 1.83; it compares nicely with the exact answer, $11/6 \approx 1.8333$.

Now consider any *column* in the table. The last column, for instance, contains numbers of the form $f(0.95, y)$, for ten equally-spaced values of y, with $\Delta y = 0.1$. Therefore, the column sum multiplied by 0.1 (the answer is 1.55) is a Riemann sum for the integral $\int_0^1 f(0.95, y) \, dy$. We can calculate this last integral directly:

$$\int_0^1 f(0.95, y) \, dy = \int_0^1 (2 - 0.95^2 y) \, dy = 1.54875.$$

Integrals are rounded to two decimals.

That's not far from the Riemann sum. Tabulating results for the other columns produces the same pattern. The Riemann sum associated with each column closely approximates the corresponding y-integral. ◄

Column Sums and y-Integrals										
x	0.05	0.15	0.25	0.35	0.45	0.55	0.65	0.75	0.85	0.95
Column sum	20.0	19.9	19.7	19.4	19.0	18.4	17.9	17.2	16.4	15.5
$\int_0^1 f(x,y)\,dy$	2.00	1.99	1.97	1.94	1.90	1.85	1.79	1.72	1.64	1.55

The calculations are complicated, but the moral is simple: Whether adding up approximating sums or calculating integrals, it's OK to work with one variable at a time. ∎

A General Argument. Let's summarize the ideas of Example 2 in more general terms. To do so, suppose that we're given a rectangle $[a, b] \times [c, d]$ in the xy-plane, and a function f defined on R. We'll see why the integral $I = \int\int_R f(x, y)\,dA$ can reasonably be calculated by iteration.

To begin, let's write an approximating sum for I in the sense of the preceding section. First we subdivide both $[a, b]$ and $[c, d]$ into n equal subintervals. Their lengths are $\Delta x = (b - a)/n$ and $\Delta y = (d - c)/n$, respectively. This produces a grid of n^2 subrectangles R_{ij}, where $1 \le i, j \le n$; each rectangle has area $\Delta x\,\Delta y$. Let (x_i, y_j) be the midpoint of the (i, j)th rectangle; we'll use these midpoints as sampling points for an approximating sum S_{n^2} for I.

With all ingredients now in place, we can write the approximating sum:

$$S_{n^2} = \sum_{i,j=1}^{n} f(x_i, y_j)\,\Delta x\,\Delta y = f(x_1, y_1)\,\Delta x\,\Delta y + \cdots + f(x_n, y_n)\,\Delta x\,\Delta y.$$

(The subscript on the "Σ" means that we sum over all possible values of *both* i and j from 1 to n.) ▶

We can group and add the summands in any order or pattern we like. Here's one convenient pattern: ▶

There are n^2 summands in all.

It's OK to factor Δy out of each row, because it's common to all summands.

$$
\begin{aligned}
S_{n^2} \;=\;& \left(f(x_1, y_1)\,\Delta x + f(x_2, y_1)\,\Delta x + \cdots + f(x_n, y_1)\,\Delta x\right)\Delta y \\
&+ \left(f(x_1, y_2)\,\Delta x + f(x_2, y_2)\,\Delta x + \cdots + f(x_n, y_2)\,\Delta x\right)\Delta y \\
&+ \ldots \\
&+ \left(f(x_1, y_j)\,\Delta x + f(x_2, y_j)\,\Delta x + \cdots + f(x_n, y_j)\,\Delta x\right)\Delta y \\
&+ \ldots \\
&+ \left(f(x_1, y_n)\,\Delta x + f(x_2, y_n)\,\Delta x + \cdots + f(x_n, y_n)\,\Delta x\right)\Delta y.
\end{aligned}
$$

Here's the first of two key points:

The sum inside parentheses on each line above is a Riemann sum with n subdivisions for a single-variable integral in x.

Specifically, the sum on the first line is a Riemann sum for the integral $\int_a^b f(x, y_1)\,dx$; the sum on the second line approximates $\int_a^b f(x, y_2)\,dx$, and so on. Now if n is large, then all of these sums are close to the integrals they approximate. Thus, for large n,

$$S_{n^2} \approx \int_a^b f(x, y_1)\,dx\,\Delta y + \int_a^b f(x, y_2)\,dx\,\Delta y + \cdots + \int_a^b f(x, y_n)\,dx\,\Delta y.$$

Here's the second key observation:

The sum on the right above is a Riemann sum with n subdivisions for the integral $\int_c^d g(y)\,dy$, where $g(y) = \int_a^b f(x, y)\,dx$.

This should sound reasonable, and it is indeed true for the well-behaved functions in this chapter. But a rigorous proof is quite subtle; it depends on technical properties of the integrand function.

For large n this sum, too, is near the integral it approximates. ◄   In other words,

$$S_{n^2} \approx \sum_{j=1}^n g(y)\,\Delta y \approx \int_c^d g(y)\,dy = \int_c^d \left(\int_a^b f(x, y)\,dx \right) dy.$$

This shows (informally) what we hoped to show: For large n,

$$I \approx S_{n^2} \approx \int_c^d \left(\int_a^b f(x, y)\,dx \right) dy.$$

We conclude that we can indeed evaluate a double integral I over a rectangle R by integrating first in x and then in y. Nor does the order matter—we could just as well have reversed the roles of x and y throughout the preceding argument.

Iterated Integrals in Three Variables

Iteration works exactly the same way for a triple integral defined on a cube in xyz-space.

EXAMPLE 3   Let $f(x, y, z) = x + y + z$, let $R = [0, 2] \times [0, 2] \times [0, 2]$, and let $I = \iiint_R f(x, y, z)\,dV$. Calculate I exactly, by iteration. (Compare Example 2, page 714, where we calculated a crude approximating sum for I.) What could the answer *mean*?

Solution   We integrate in each of the three variables in turn:

$$\iiint_R f(x, y, z)\,dV = \int_0^2 \left(\int_0^2 \left(\int_0^2 (x + y + z)\,dx \right) dy \right) dz$$

$$= \int_0^2 \left(\int_0^2 \left(\frac{x^2}{2} + xy + xz \right]_0^2 \right) dy \right) dz$$

$$= \int_0^2 \left(\int_0^2 (2 + 2y + 2z)\,dy \right) dz$$

$$= \int_0^2 \left(2y + y^2 + 2yz \right]_0^2 \right) dz$$

$$= \int_0^2 (8 + 4z)\,dz$$

$$= 24.$$

Maple can do the same thing in one fell swoop:

```
> int( int( int( x+y+z, x=0..2), y=0..2), z=0..2);
```

$$24$$

What the answer means depends on our point of view. If we think of the integrand $f(x, y, z)$ as the density (in, say, grams per cubic centimeter) of the solid R at the point (x, y, z), then the integral tells the *mass* of the solid (in grams). ∎

Integrals Over Nonrectangular Regions

Not every integral of interest is taken over a rectangular domain of integration. It's sometimes useful to integrate over regions with curved boundaries. Much the same process of iteration applies in this case as in that of integrals over rectangles and cubes, but some extra care is needed. We illustrate the process with an example.

EXAMPLE 4   Find $I = \iint_R (2-x)\, dA$, where R is the plane region between the curves $y = 0$ and $y = x$, and $x = 1$. ▶

Solution   The integral gives the volume of the following solid figure:

It's essential to draw the domain; use an ordinary pair of xy-axes.

A surface with non-rectangular base

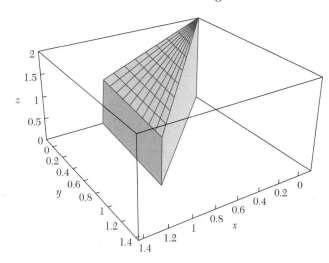

We can think of the domain R as bounded by the straight lines $x = 0$ and $x = 1$ on the left and right, and by the curves $y = 0$ and $y = x$ on the bottom and top. As in earlier cases, we can integrate by iteration. The *outer* integral, in x, runs from $x = 0$ to $x = 1$. The upper and lower limits of the *inner* integral, however, depend on x, as the following computation shows.

$$I = \iint_R (2 - x)\, dA = \int_{x=0}^{x=1} \left(\int_{y=0}^{y=x} (2 - x)\, dy \right) dx$$

$$= \int_{x=0}^{x=1} \left(2y - xy \right]_{y=0}^{y=x} \right) dx$$

$$= \int_{x=0}^{x=1} \left(2x - x^2 \right) dx = \frac{2}{3}.$$

Again, *Maple* can do the whole thing at once:

```
> int( int( 2-x, y=0..x), x=0..1);
```

In fact, this integral is not much different from the preceding ones, which were over rectangular domains. The variable limits of integration in the inner integral simply reflect the fact that the domain of integration has varying "heights," depending on x.

Changing the Order of Integration

For integrals over rectangles or cubes, we could integrate in any order we liked. A similar result holds in the present case, too—as long as the region has an appropriate shape. We illustrate by redoing the preceding example, but integrating the variables in the opposite order.

EXAMPLE 5   Redo the integral $I = \iint_R (2-x)\, dA$ of Example 4, but now integrate first in x and then in y.

Solution   This time we think of R as lying between the lines $y = 0$ and $y = 1$. For given y, R starts at the line $x = y$ and ends at the line $x = 1$. Now the calculation is similar to the preceding one:

$$
I = \iint_R (2-x)\, dA = \int_{y=0}^{y=1} \left(\int_{x=y}^{x=1} (2-x)\, dx \right) dy
$$

$$
= \int_{y=0}^{y=1} \left(2x - \frac{x^2}{2} \right]_{x=y}^{x=1} \right) dy
$$

$$
= \int_{y=0}^{y=1} \left(\frac{3}{2} - 2y + \frac{y^2}{2} \right) dy = \frac{2}{3}.
$$

Here's *Maple*'s version:

```
> int( int( 2-x, x=y..1), y=0..1);
                              2/3
```

Not Always So Simple.   Things aren't always quite so simple. Some domains of integration lend themselves more naturally to one order of integration than to another; the following exercises give examples.

BASICS

Note. *Maple* (or a similar utility) will be helpful in several of these exercises.

1. Use iteration to calculate each of the following integrals by hand (without technology). Then check your answers symbolically, using *Maple*. Finally, plot a 3-d surface over an appropriate domain to see that your answer is reasonable. [**NOTE:** Most of the following integrals appear among the exercises in Section 14.5.]

   (a) $\displaystyle\iint_R \sin(x)\sin(y)\, dA$; $R = [0, 1] \times [0, 1]$

   (b) $\displaystyle\iint_R \sin(x+y)\, dA$; $R = [0, 1] \times [0, 1]$

   (c) $\displaystyle\iint_R (x^2 + y^2)\, dA$; $R = [0, 4] \times [0, 4]$

   (d) $\displaystyle\iiint_R x\, dV$; $V = [0, 1] \times [0, 2] \times [0, 3]$

   (e) $\displaystyle\iiint_R y\, dV$; $V = [0, 1] \times [0, 2] \times [0, 3]$

2. Use iteration to calculate each of the following nonrectangular integrals by hand (i.e., without technology). In each case, the inner integral should be in y and the outer integral in x. Check your answers symbolically, using *Maple*.

(a) $\iint_R (x+y)\,dA$; R the region bounded by the curves $y = x$ and $y = x^2$

(b) $\iint_R x\,dA$; R the region bounded by the curves $y = x^2$ and $y = \sqrt{x}$

(c) $\iint_R 1\,dA$; R the first quadrant part of the circle $x^2 + y^2 \le 1$

3. Redo Exercise 2, but integrate first in x and then in y.

4. Consider the integral $I = \iint_R (x+y)\,dA$, where R is the region bounded by the curves $y = x^2$ and $y = 1$.
   (a) Calculate I by integrating first in y and then in x.
   (b) Calculate I by integrating first in x and then in y.

5. Let $f(x, y) = x$, and let R be the plane region bounded by the curves $y = e^x$, $y = 0$, $x = 0$, and $x = 1$.
   (a) Calculate $I = \iint_R f(x, y)\,dA$ by integrating first in y and then in x.

(b) Calculate $I = \iint_R f(x, y)\,dA$ by integrating first in x and then in y.   [**HINT:** First split the region R into two simpler pieces; each simpler piece should be bounded on the left by one curve and on the right by another.]

6. Let $y = f(x)$ be a function, with $f(x) \ge 0$ if $a \le x \le b$; let R be the plane region bounded by the curves $y = f(x)$, $y = 0$, $x = a$, and $x = b$.

(a) What does single-variable calculus say about the area of R?

(b) According to Section 14.5, the double integral $I = \iint_R 1\,dA$ gives the area of R. Use an iterated integral to reconcile this formula with the one in part (a).

7. Let $x = g(y)$ be a function with $g(y) \ge 0$ if $c \le y \le d$; let R be the plane region bounded by the curves $x = g(y)$, $x = 0$, $y = c$, and $y = d$.

(a) What does single-variable calculus say about the area of R?

(b) According to Section 14.5, the double integral $I = \iint_R 1\,dA$ gives the area of R. Use an iterated integral to reconcile this formula with the one in part (a).

14.7 Double Integrals in Polar Coordinates

Easy Integrals and Hard Integrals

What makes a double integral $I = \iint_R f(x, y)\,dA$ hard to calculate? Both f and R can play a role: If either one is complicated or messy to describe, or both, then I may be correspondingly ugly. Let's see examples of both "good" and "bad" integrals.

EXAMPLE 1   Discuss $I_1 = \iint_{R_1} x^2\,dA$ and $I_2 = \iint_{R_2} \sqrt{x^2 + y^2}\,dA$, where R_1 is the rectangle $[0, 1] \times [0, 2\pi]$, and R_2 is the region inside the unit circle $x^2 + y^2 = 1$. ▶

Draw R_1 and R_2 for yourself.

Solution   The first integral is easy:

$$I_1 = \int_{x=0}^{x=1} \left(\int_{y=0}^{y=2\pi} x^2\,dy \right) dx = \int_0^1 \left(x^2 y \right]_0^{2\pi} \right) dx = \int_0^1 2\pi x^2\,dx = \frac{2\pi}{3}.$$

The ingredients of I_2 are more complicated to describe. ▶   The circular region R_2 can be thought of as bounded by the curves $y = \sqrt{1-x^2}$ and $y = -\sqrt{1-x^2}$ on the top and bottom, and by the lines $x = -1$ and $x = 1$ on the left and right. Now we can write I_2 in iterated form:

In xy-coordinates, at least. Polar coordinates will make things simpler.

$$I_2 = \int_{x=1}^{x=-1} \left(\int_{y=-\sqrt{1-x^2}}^{y=\sqrt{1-x^2}} \sqrt{x^2 + y^2}\,dy \right) dx.$$

The integral looks—and is—complicated. ◄   Just to get started on the inner integral, we'd need the antiderivative formula ◄

$$\int \sqrt{x^2 + p^2}\, dx = \frac{1}{2}\left(x\sqrt{x^2 + p^2} + p^2 \ln\left| x + \sqrt{x^2 + p^2} \right| \right).$$

Faced with this prospect, we retreat. But only temporarily—we shall return to I_2 soon. ∎

What Went Wrong, and What To Do.   The integral I_2 in Example 1 led to an unpleasant calculation in x and y for two reasons:

(i) The integrand, $\sqrt{x^2 + y^2}$, has a complicated antiderivative in x or y. ◄

(ii) The domain of integration, although geometrically simple, is messy to describe in rectangular coordinates. ◄

In polar coordinates, on the other hand, both the integrand and the domain have simple, uncluttered formulas. The integrand is

$$f(x, y) = \sqrt{x^2 + y^2} = r.$$

In polar language, the domain of integration is a lot like a rectangle; it's defined by the inequalities

$$0 \le r \le 1 \quad \text{and} \quad 0 \le \theta \le 2\pi.$$

In this case, apparently, both the integrand f and the domain R "deserve" to be described in polar coordinates, not in rectangular coordinates. It seems reasonable, therefore, that the double integral I_2 should also be calculated using polar rather than rectangular coordinates.

That hunch is correct. We demonstrate in this section how to calculate double integrals in polar form, using r and θ, as opposed to Cartesian or rectangular form, using x and y. Integrals such as I_2, in which the integrand, the domain of integration, or both are simplest in polar form, are natural candidates for polar treatment.

Polar "Rectangles"

A rectangle in Cartesian coordinates is defined by two inequalities, of the form

$$a \le x \le b \quad \text{and} \quad c \le y \le d;$$

each of the coordinates x and y ranges through an interval. A **polar rectangle** is defined by two similar inequalities:

$$a \le r \le b; \qquad \alpha \le \theta \le \beta;$$

again, each of the coordinates r and θ ranges through an interval. Following are generic pictures of both types of rectangles.

A Cartesian rectangle

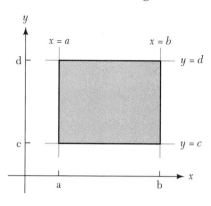

A polar rectangle

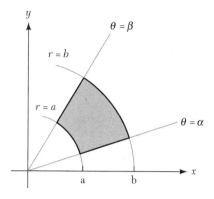

For polar integrals, as for Cartesian ones, rectangles (in the appropriate sense) are the simplest regions over which to integrate.

Polar Integration—How It Works

A double integral $I = \iint_R f(x, y)\, dA$ in rectangular coordinates, where $R = [a, b] \times [c, d]$ is an ordinary rectangle, is written in iterated form ▶ as

We integrate first in y this time.

$$\iint_R f(x, y)\, dA = \int_a^b \int_c^d f(x, y)\, dy\, dx.$$

(Here and later, we omit some parentheses. It's always understood that the inner integral is done first.)

Now suppose we're given a double integral $I = \iint_R g(r, \theta)\, dA$ in polar coordinates, where R is now a *polar* rectangle, defined by inequalities

$$a \le r \le b \quad \text{and} \quad \alpha \le \theta \le \beta,$$

and $g(r, \theta)$ is a function defined on R. The following Fact gives the appropriate integral formula.

Fact (Double Integrals in Polar Coordinates) Let g and R be as before. Then
$$\iint_R g(r, \theta)\, dA = \int_{\theta=\alpha}^{\theta=\beta} \int_{r=a}^{r=b} g(r, \theta)\, r\, dr\, d\theta.$$

The formula prompts several important observations:

Trading x and y for r and θ Any function $f(x, y)$ can be "traded" for an equivalent function $g(r, \theta)$, using the relations

$$x = r \cos \theta \quad \text{and} \quad y = r \sin \theta.$$

The same method works for *equations* in x and y. The equation $x = y$, for example, says in polar coordinates that $r \cos\theta = r \sin\theta$ or, equivalently, that $\tan\theta = 1$. This polar equation describes the same line as the original Cartesian equation. We'll use these principles when evaluating polar integrals.

A Useful Mnemonic   Compare the preceding formulas for integrating in rectangular and polar coordinates. An important difference between the two has to do with the "dA" expression. The full mathematical story is much deeper, ◄ but as a quick aid to memory, the following formulas are very handy.

We won't go into great depth, but we'll give some informal justification soon.

$$dA = dx\,dy \qquad \text{for Cartesian coordinates;}$$
$$dA = r\,dr\,d\theta \qquad \text{for polar coordinates.}$$

That Extra Factor of r   What's that mysterious extra r doing in the polar formula $dA = r\,dr\,d\theta$? Why not just $dA = dr\,d\theta$? We certainly owe the reader an explanation. We'll honor that debt in a moment, when we discuss *why* the formula works. First, however, let's see *that* it works.

EXAMPLE 2   Let R_2 be the region inside the unit circle. Use polar coordinates to calculate that troublesome integral $I_2 = \iint_{R_2} \sqrt{x^2 + y^2}\,dA$, from Example 1, page 725.

This is always possible.

Solution   First we write all the data in polar form. ◄ For the integrand, we have $f(x, y) = \sqrt{x^2 + y^2} = r = g(r, \theta)$. For the domain of integration, we translate the Cartesian equation $x^2 + y^2 = 1$ into its (simpler!) polar form, $r = 1$. The rest is easy: ◄

Watch each step; work from inside out.

$$
\begin{aligned}
\iint_{R_2} \sqrt{x^2 + y^2}\,dA &= \int_{\theta=0}^{\theta=2\pi} \int_{r=0}^{r=1} r\,dA \\
&= \int_{\theta=0}^{\theta=2\pi} \int_{r=0}^{r=1} r^2\,dr\,d\theta \\
&= \int_{\theta=0}^{\theta=2\pi} \frac{r^3}{3}\Bigg]_0^1 d\theta \\
&= \int_{\theta=0}^{\theta=2\pi} \frac{1}{3}\,d\theta = \frac{2\pi}{3}.
\end{aligned}
$$

Notice especially the similarity to the integral I_1 of Example 1—I_1 and I_2 turned out to have the same value. This is no accident. After rewriting in polar coordinates, I_2 turned out to be the same integral in r and θ as I_1 is in x and y.   ∎

Polar Integrals—What They Mean

We discussed some possible interpretations in Section 14.5.

Polar integrals have exactly the same interpretations as any other double integrals. ◄ Depending on the situation and on our point of view, an integral might represent a volume, the area of a plane region, the mass of a thin plate, or many other things.

For example, both the integrals I_1 and I_2 of Example 1, page 725, can be interpreted as volumes of solids. For I_2, the solid lies above the unit disk and below the surface $z = \sqrt{x^2 + y^2}$, as shown.

A polar solid

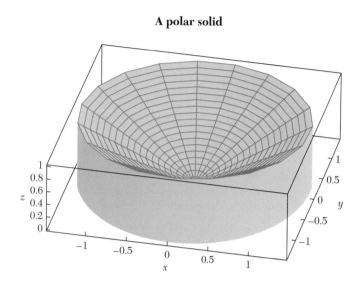

As we just calculated, this figure has volume $2\pi/3 \approx 2.094$ cubic units. ▶

Does this seem reasonable from the picture, given the general size of units?

Polar Integration—Why It Works

Why does the polar integration formula work? Where, especially, does the r in $dA = r \, dr \, d\theta$ come from?

All properties of integrals—whether in Cartesian, polar, or any other form—stem ultimately from properties of the approximating sums that are used to define integrals. For any function f defined on a region R, we have

$$\iint_R f \, dA = \lim_{m \to \infty} \sum_{i=1}^{m} f(P_i) \Delta A_i,$$

where ΔA_i is the area of the ith subregion of R, and P_i is a sampling point chosen inside this subregion.

If $R = [a, b] \times [c, d]$ is a Cartesian rectangle, then it's natural to subdivide R into smaller rectangles, each with sides Δx and Δy. Any such rectangle has area $\Delta A_i = \Delta x \, \Delta y$. In the limit that defines the integral, therefore, $dA = dx \, dy$.

If R is a polar rectangle, the picture is a little different. In this case, a "polar grid" is the natural way to subdivide R. In the following picture, each bulleted point represents a "centered" sampling point in its respective subdivision.

A polar grid on a polar rectangle

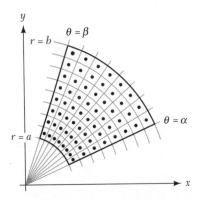

Notice the following features of the grid.

Similar Subregions All subregions correspond to the same $\Delta\theta$ (the angle between any two adjacent dotted radial lines) and the same Δr (the radial distance from one arc to the next).

But Not Identical Similar as they are, the subregions shown are not identical. Here's the key point:

> *In a polar grid, the subregions have different sizes. The area depends not only on $\Delta\theta$ and Δr but also on r: Larger values of r produce larger subregions.*

This fact explains the difference between polar and Cartesian integrals, and hints at why that extra r is needed.

The Area of One Subregion. Each of the preceding subregions is a small polar rectangle with polar dimensions $\Delta\theta$ and Δr, inner radius r, and outer radius $r + \Delta r$. It's an important fact that for such a polar rectangle,

$$\text{Area} = \Delta A_i = \frac{r + r + \Delta r}{2}\, \Delta r\, \Delta\theta.$$

(We'll leave verification of this straightforward fact to the exercises.) The first factor on the right is crucial—it represents the *average* radius of the given subregion, i.e., the r-coordinate of the ith bulleted midpoint (r_i, θ_i) in the illustration. Therefore,

$$\Delta A_i = r_i\, \Delta r\, \Delta\theta.$$

This is just what we've been waiting for. It shows that, for a polar rectangular region, a midpoint approximating sum has the form

$$\sum_{i=1}^{m} f(r_i, \theta_i)\, \Delta A_i = \sum_{i=1}^{m} f(r_i, \theta_i)\, r_i\, \Delta r\, \Delta\theta.$$

The integral itself therefore has the limiting form $\iint_R f(r, \theta)\, r\, dr\, d\theta$.

Polar Integrals Over Nonrectangular Regions

Polar integrals, like Cartesian integrals, can be taken over nonrectangular regions. The method is similar.

Fact Let R be the region bounded by the radial lines $\theta = \alpha$ and $\theta = \beta$, by an inner curve $r = r_1(\theta)$, and by an outer curve $r = r_2(\theta)$. ("Inner" and "outer" are understood relative to the origin.) Let $g(r, \theta)$ be a function defined on R. Then

$$\iint_R g \, dA = \int_{\theta=\alpha}^{\theta=\beta} \int_{r=r_1(\theta)}^{r=r_2(\theta)} g(r, \theta) \, r \, dr \, d\theta.$$

We illustrate with an example.

EXAMPLE 3   Use a polar integral to find the area of the region R inside the cardioid $r = 1 + \cos\theta$.

Solution   We use the familiar principle ▶ that

It applies equally well in polar coordinates!

$$\text{Area of } R = \iint_R 1 \, dA,$$

but we calculate the integral in polar form, as follows:

$$\iint_R 1 \, dA = \int_{\theta=0}^{\theta=2\pi} \int_{r=0}^{r=1+\cos\theta} 1r \, dr \, d\theta$$

$$= \int_{\theta=0}^{\theta=2\pi} \left. \frac{r^2}{2} \right]_0^{1+\cos\theta} d\theta$$

$$= \int_{\theta=0}^{\theta=2\pi} \frac{(1+\cos\theta)^2}{2} \, d\theta.$$

The last integral takes a little doing by hand. ▶ *Maple* has no trouble:

But it's not really difficult.

```
> int( (1+cos(t))^2, t=0 .. 2*Pi );
```

$$3*Pi/2$$

■

BASICS

1. Let R be the polar rectangle defined by $a \le r \le b$ and $\alpha \le \theta \le \beta$.

   (a) Show that the area of R is $\dfrac{a+b}{2}(b-a)(\beta-\alpha)$.
   [**HINT:** Appendix B discusses the area of a circular sector, or "wedge."]

   (b) Use part (a) to show that a polar rectangle with dimensions Δr and $\Delta \theta$ and inner radius r has area $\dfrac{r+r+\Delta r}{2} \Delta r \, \Delta \theta$.

2. Let $f(x, y) = y$, let R be the upper half of the region inside the unit circle $x^2 + y^2 = 1$, and let $I = \iint_R f \, dA$.

   (a) Calculate I as an iterated integral in rectangular coordinates, with the inner integral in y.

   (b) Calculate I as an iterated integral in rectangular coordinates, with the inner integral in x.

   (c) Calculate I as an iterated integral in polar coordinates.

3. In this section, we used polar coordinates to calculate that $I_2 = \iint_{R_2} \sqrt{x^2 + y^2} \, dA = 2\pi/3$, where R_2 is the region inside the unit circle $r = 1$. (See the picture on page 729.) Use formulas for the volumes of cones and cylinders to find the same answer by elementary means.

4. This exercise is about the integral I_1 of Example 1 (page 725).

   (a) Draw (use technology if necessary, but try without it) the solid whose volume is given by I_1.

   (b) Evaluate I_1 again but with the inner integral in x, not y.

5. Use a polar double integral in each of the following parts. (Draw each region first.)

   (a) Find the area of the region inside the car-break dioid $r = 1 + \sin\theta$.

   (b) Find the area of the region bounded by $y = x$, $y = 0$, and $x = 1$. Could you find the answer another way? [**HINT**: First write the boundary equations in polar form.]

   (c) Find the area of the region bounded by the circle of radius 1/2, centered at $(0, 1/2)$. Could you find the answer another way? [**HINT**: One approach is to first write a Cartesian equation for the circle, then change it to polar form.]

6. Use polar coordinates to calculate each of the following quantities.

   (a) Find

   $$\iint_R \frac{1}{\sqrt{x^2 + y^2}}\, dA,$$

   where R is the region inside the cardioid $r = 1 + \sin\theta$ and above the x-axis.

   (b) Find the volume of the solid under the surface $z = 1 - x^2 - y^2$ and above the xy-plane. [**HINT**: First decide where the surface intersects the xy-plane.]

   (c) Find the volume of the conical solid under the surface $z = 1 - \sqrt{x^2 + y^2}$ and above the xy-plane. [**HINT**: First decide where the surface intersects the xy-plane.]

Real Numbers and the Coordinate Plane

The real numbers might seem, at first glance, almost *too* familiar. What is there to say? Surprisingly, a great deal—far more than a calculus course could cover.

Here's just one hint at the subtleties: It's been known since Pythagoras (*ca.* 500 B.C.E.) that some numbers, such as $\sqrt{2}$, are **irrational**—that is, they can't be written as *ratios* of integers. But how thickly scattered are the irrationals among the reals? Are "most" real numbers rational or irrational? The answers are remarkable, even mysterious: "Most" real numbers are irrational; a dart thrown randomly at a number line will almost always hit an irrational. Yet plenty of rational numbers exist; in any interval, no matter how small, there are *infinitely many* rational numbers! Resolving these tantalizing mysteries must wait for a later mathematics course. For us, a brief, brisk, selective review of the properties of real numbers will suffice.

> **Newton Managed.** Isaac Newton himself did quite well without a fully rigorous development of the real numbers. The first mathematically satisfactory definition of the real numbers was given by Richard Dedekind, over 130 years after Newton's death.

The Real Number Line

The real numbers are really no more *real*, in an existential or philosophical sense, than the integers, the rationals, or even the complex numbers. The best reason for the adjective "real" may be that the set of real numbers can be thought of naturally as an unbroken, unending line. Once an origin and a unit of distance are fixed, every point has its own number and every number its own point. Here is the usual picture:

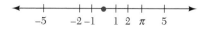

Order.   The line shows nicely how the real numbers are ordered: *As we move to the right on the line, real numbers increase.* This is important, so we restate it as an inequality:

$$a > b \text{ if and only if } a \text{ is to the right of } b.$$

At the risk of redundancy, we'll say it one last time: For real numbers, "greater than" means "to the right of." Thus, for example, $-3 > -4$, even though -4 has greater *magnitude* than -3.

Subsets of the Reals, Subsets of the Line

The set of real numbers, like any infinite set, has a gigantic infinity of possible subsets. Some subsets have fiendishly complicated structures. ("Fractals," for example, are sets that are irregular or ragged at any scale. The most spectacular fractal sets are usually shown in two dimensions, but many fractals have one-dimensional forms.) In calculus, luckily, most subsets of interest are easily understood.

Set Notation: Brackets and More Brackets.   Sets and subsets pop up everywhere in mathematics. By using set notation efficiently, we can avoid tedious and sometimes confusing verbal descriptions.

In calculus, for instance, we might ask:

Let $f(x) = x^3 - x$. For which real numbers x is $f(x) < 0$?

Plot the curve, say for $-2 \le x \le 2$, to see for yourself.

Plotting $y = f(x)$ makes the answer—the **solution set**—apparent. ◄   The solution set, expressed (a bit long-windedly) in words, is

the set of real numbers x such that either x is less than -1 or x lies between 0 and 1.

That's a mouthful, even for this relatively simple problem. In set notation, the same answer is much more concise: $\{ x \mid x < -1 \text{ or } 0 < x < 1 \}$. Read aloud, the symbols say exactly the same thing as the words above. In particular, the vertical line $\mid$ means "such that." It's understood here, moreover, that x represents a real number.

One caution: Three types of brackets ($\{\}$, $()$, and $[\,]$) are commonly used to describe sets. Each type has a different meaning; watch carefully for them below.

Calculus inevitably requires that some mathematical language (closed, open, infinite, bounded, etc.) be learned, and then used carefully. The payoff in efficiency can be great.

Intervals.   Intervals are "pieces" chopped from the real line. They are the simplest and most useful subsets of the reals. Intervals come in various styles: They can be bounded or infinitely long, and they may or may not include their endpoints. (**Closed intervals** contain their endpoints, **open intervals** don't.) ◄   The following table of examples illustrates the full taxonomy. It illustrates, too, how intervals can be defined by one or more inequalities:

<table>
<tr><th colspan="3">Taxonomy of Intervals</th></tr>
<tr><th>Interval</th><th>Description</th><th>Inequality Form</th></tr>
<tr><td>$(-2, 3)$</td><td>an open interval</td><td>$\{\,x \mid -2 < x < 3\,\}$</td></tr>
<tr><td>$[a, b]$</td><td>a closed interval</td><td>$\{\,x \mid a \le x \le b\,\}$</td></tr>
<tr><td>$[2, b)$</td><td>a half-open interval</td><td>$\{\,x \mid 2 \le x < b\,\}$</td></tr>
<tr><td>$(5, \infty)$</td><td>an infinite interval</td><td>$\{\,x \mid x > 5\,\}$</td></tr>
<tr><td>$(-\infty, \infty)$</td><td>another infinite interval</td><td></td></tr>
</table>

More Subsets of $\mathbb{R}$.   Below are more examples of subsets of the reals. Notice carefully our use of various mathematical notations, especially $\mathbb{R}$, $\mathbb{Q}$, and $\mathbb{Z}$ for the sets of reals, rationals, and integers, and the $\cup$ (union) and $\cap$ (intersection) symbols for combining sets.   These notations are standard in mathematical writing; we will use them often.

<table>
<tr><th colspan="3">Subsets of the Reals</th></tr>
<tr><th>Set</th><th>Explanation</th><th>Alternative Notation(s)</th></tr>
<tr><td>$\mathbb{R}$</td><td>the real numbers</td><td>$(-\infty, \infty)$</td></tr>
<tr><td>$\mathbb{N}$</td><td>the natural numbers</td><td>$\{1, 2, 3, \ldots\}$</td></tr>
<tr><td>$\mathbb{Z}$</td><td>the integers</td><td>$\{0, \pm 1, \pm 2, \pm 3, \ldots\}$</td></tr>
<tr><td>$\mathbb{Q}$</td><td>the rationals</td><td></td></tr>
<tr><td>$\{0, 2, 5\}$</td><td>a finite set</td><td>$\{\,x \mid x = 0 \;\text{ or }\; x = 2 \;\text{ or }\; x = 5\,\}$
or $\{\,x \mid x(x-2)(x-5) = 0\,\}$</td></tr>
<tr><td>$(1, 4) \cup (5, 6)$</td><td>union of two intervals</td><td>$\{\,x \mid 1 < x < 4 \;\text{ or }\; 5 < x < 6\,\}$</td></tr>
<tr><td>$(1, 6) \cap [4, 7]$</td><td>intersection of two intervals</td><td>$\{\,x \mid 1 < x < 6 \;\text{ and }\; 4 \le x \le 7\,\}$
or $\{\,x \mid 4 \le x < 6\,\}$ or $[4, 6)$</td></tr>
</table>

Pictures of Subsets.   Uncomplicated subsets of the reals are easy to "draw" on the real line. Intervals are nicest of all: They appear as line segments, including some, all, or none of their endpoints. Errors of approximation are inevitable ▶ in *all* pictures, but for most sets they do no harm. Several straightforward examples appear below; a *closed* dot indicates an *included* point; an *open* dot indicates an *excluded* point.

For example, because ideal "points" on a line, unlike pencil points, have zero thickness. Errors of omission are also inevitable; no picture can show everything.

Drawing Sets	
Set	**Picture**
$(-2, 3)$	
$[3, \infty)$	
$(-3, 2]$	
$\mathbb{Z}$	
$[-4, -2) \cup (1, 3)$	

The Symbol ∞: Not a Number

Every real number corresponds to a unique point on the real line. The symbol ∞ does not; ∞ *is not a real number*.

What *is* ∞, if not a number? That's a thorny question indeed, in a league with "What is truth?" A fully rigorous answer is far beyond the scope of this book. A good enough answer for us (and for most working mathematicians, most of the time) is that the symbol ∞ represents the idea of unboundedness, or endlessness. When we say that the real line has *infinite* length, for instance, we mean that the line goes on forever.

The symbol ∞ is a useful, convenient, and suggestive shorthand for the idea of "without bound." Use it sparingly and with care; ∞ is *not* a number and seldom behaves like one. In particular, "algebra" with the infinity symbol takes special care. ◄

For more on "algebra with ∞," see Section 2.7.

Absolute Value, Distance, and Inequalities

The idea of **absolute value** (or **magnitude**) should be familiar from long experience. Its use in simple expressions like these—

$$|-3| = 3; \qquad |\sin x| \le 1; \qquad \sqrt{x^2} = |x|$$

will cause no trouble. What absolute value means is usually clear in context, but sometimes a formal definition is helpful. Here it is:

$$|x| = \begin{cases} x & \text{if } x \ge 0 \\ -x & \text{if } x < 0 \end{cases}$$

If you learn nothing else about absolute values, learn these facts.

Perhaps we should say the distance between a and b, since the distance in both directions is the same. The equation $|a - b| = |b - a|$ says the same thing.

Familiar as all this is, absolute values can be sticky in practice—especially if inequalities are involved. The following two tricks of the trade often help: ◄

Interpreting Absolute Value as Distance   For *any* numbers a and b, $|a - b|$ can (and should!) be thought of as the *distance from a to b*. ◄   Thinking of absolute value as distance helps us visualize algebraic expressions geometrically. For example, the absolute value equation

$$|x + 2| = 3$$

isn't hard to solve algebraically, but the algebraic solution doesn't give us much insight. Distance brings the situation to life. Because

$$|x + 2| = |x - (-2)| = \text{distance from } x \text{ to } -2,$$

the equation $|x + 2| = 3$ says that x *lies 3 units away from* -2. Clearly, $x = -5$ and $x = 1$ are the only solutions of the equation.

Trading Absolute Values for Inequalities: Two for One   In calculus, inequalities often involve absolute values. Here's one:

$$|x + 2| < 3$$

Solving such inequalities ▶ requires care. Here's the key idea:

> One absolute value inequality corresponds to two ordinary inequalities.

To solve an inequality involving a variable x means to find the solution set and describe it concisely.

The inequality above is equivalent to these two:

$$-3 < x + 2 \qquad \text{and} \qquad x + 2 < 3$$

or, more succinctly, $-3 < x + 2 < 3$. Now it's easy to find the solution set: $-5 < x < 1$. In other words, the inequality $|x + 2| < 3$ holds for those x that lie in the interval $(-5, 1)$, i.e., the set of points that lie within 3 units of -2.

More generally, for *any* quantities x and b:

$$|x| < b \qquad \text{if and only if} \qquad -b < x < b$$
$$|x| > b \qquad \text{if and only if} \qquad -b > x \text{ or } x > b$$

(Similar rules hold for $\leq$ and $\geq$ inequalities.)

EXAMPLE 1   Any open or closed finite interval can be described in terms of its **midpoint** and **radius**; absolute value inequalities help. The open interval $(-3, 5)$, for instance, has midpoint $x = 1$ and radius 4 because the distance from every point in the interval to 1 is less than 4. Therefore, $(-3, 5)$ corresponds naturally to the inequality $|x - 1| < 4$. In formal language:

$$(-3, 5) = \{ x \mid |x - 1| < 4 \} \qquad ∎$$

EXAMPLE 2   Solve the inequality $|2x + 3| \leq 5$.

Solution   By the second fact above, the inequality

$$|2x + 3| \leq 5$$

is equivalent to

$$-5 \leq 2x + 3 \leq 5.$$

Now the usual algebraic rules for inequalities apply:

$$-5 \leq 2x + 3 \leq 5 \iff -8 \leq 2x \leq 2 \iff -4 \leq x \leq 1.$$

(The symbol $\iff$ means "if and only if"; it indicates that each side implies the other.) We conclude that the inequality holds if and only if x lies in the interval $[-4, 1]$. ∎

EXAMPLE 3   Solve $|x - 2| > |x + 3|$.

Solution   Having absolute values on *both sides* of the inequality complicates things. A workable—but tedious—strategy is to look separately at three cases: $x < -3$, $-3 \le x < 2$, and $x \ge 2$.

Convince yourself of this—it's easy but important.

Draw a picture!

An easier approach is to interpret both sides as distance. If we do, the given inequality says: ◄ *x is farther from 2 than from* -3. Now the solution set is easy to visualize; it's everything to the *left* of $-1/2$, the point halfway between 2 and -3. ◄   In interval notation, the solution set is $(-\infty, -1/2)$.   ■

Most absolute value inequalities arise in richer contexts. Here is one in its natural habitat.

EXAMPLE 4   Solve the inequality $|x - \sin(x)| < 0.1$.

Solution   How might such an inequality arise? Why does an answer matter? Consider these graphs of two functions:

Two graphs: $y = x$ and $y = \sin(x)$

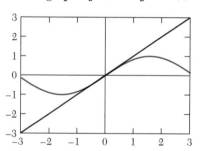

The picture shows that $\sin(x) \approx x$ if $x \approx 0$. This approximation has practical value, but only if we know something about *how* closely x approximates $\sin(x)$. Thus we might ask: For which values of x do x and $\sin(x)$ differ (in either direction) by less than 0.1? That's exactly what our original inequality, $|x - \sin(x)| < 0.1$, asks.

We'll solve our inequality graphically. Here's a piece of the graph of $y = x - \sin(x)$:

Graph of $y = x - \sin(x)$

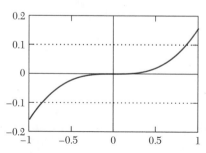

For what values of x does the inequality $-0.1 < x - \sin(x) < 0.1$ hold? The graph shows ► that this occurs for $-0.8 < x < 0.8$.

Approximately, of course.

Further graphical insight is available. Here's a closer look at the two original graphs:

Two graphs: $y = x$ and $y = \sin(x)$

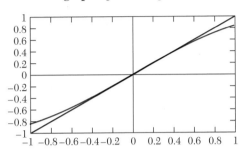

In this setting the inequality $|x - \sin(x)| < 0.1$ holds if and only if the graphs of $y = x$ and $y = \sin(x)$ are less than 0.1 unit apart, vertically. As we've already seen graphically, this occurs for $-0.8 < x < 0.8$. ► ■

The numbers are approximate.

The Coordinate Plane

The real numbers correspond to points on a *line*. In much the same way, pairs of real numbers correspond to points in the **coordinate plane**.

Coordinates and Why They Matter

The idea of attaching numerical coordinates to points in the plane is credited to the French philosopher René Descartes (1596–1650). It is no exaggeration to describe Descartes's idea, natural as it seems in hindsight, as one of the most powerful in intellectual history.

Descartes's idea is powerful because it links geometry with algebra and analysis. Geometry is the study of figures—lines, curves, circles, etc.—in the plane. Algebra studies operations—addition, multiplication, etc.—on numbers and on functions. Analysis is harder to define, but it includes the main ideas of calculus, such as function, derivative, and integral.

Before Descartes, geometry was relatively isolated from other branches of mathematics; geometers used compasses and other drawing tools to construct and measure plane figures. Figures *not* based on circles and lines were harder to describe and manipulate. The parabola, for instance, was described geometrically as the set of points equidistant from a fixed point and a fixed line. Properties of the parabola (such as the **reflection property**: All rays parallel to the axis of symmetry "bounce" to the same point, the **focus**) were hard to understand and prove with purely geometric tools and ways of thinking.

The great advantage of Cartesian coordinates is that they allow *algebraic* descriptions of geometric figures in the plane. The algebraic equation $y = x^2$, for example, describes a parabola in a far simpler and, for many purposes, more convenient fashion than its geometric properties.

When geometric figures are described algebraically the powerful tools and ideas of algebra and analysis become available. The connection, moreover, goes both ways: Geometric intuition sheds light on algebraic and analytic problems.

This link is the essence of the field of mathematics known as **analytic geometry**. In its most familiar form, analytic geometry studies properties of lines, circles, ellipses, parabolas, and hyperbolas. Collectively, these figures are called **conic sections**, because each can be thought of, geometrically, as a slice (i.e., "section") of a cone. (The cone in question is hollow, and may be "double," as in the case of a hyperbola.)

The key idea for us is that the conic sections can be described *algebraically*, as graphs of quadratic equations. Thus, properties of quadratic equations are, from another point of view, properties of conic sections, and vice versa. Knowledge of either tells us something of the other.

Throughout calculus, linking geometric, algebraic, and analytic ideas is crucial to understanding; it is the most important theme of this book.

Coordinate Systems

A **rectangular coordinate system** in the plane has three parts: two perpendicular axes, with scales, and an origin at the point where they intersect. By convention, the x-axis is horizontal and numbered from left to right; the y-axis is vertical and numbered from bottom to top. (These are *only* conventions. "Tilted" axes, other numbering schemes, and other variable names are all possible.)

With a coordinate system in place, every point's coordinates are defined. Here is the familiar picture; several points have coordinates labeled:

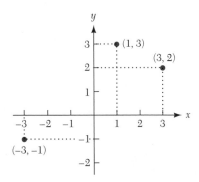

A coordinate system defines a one-to-one correspondence between points in the plane and pairs of real numbers: Each point corresponds uniquely to its two coordinates.

Cartesian Addresses.   The coordinates of a point are, in a sense, the point's "address"—they locate the point in the plane. A point's Cartesian address, moreover, gives explicit directions for getting there: To reach (a, b), from the origin, go a units "east" and b units "north." This is more than can be said for typical street addresses—123 Elm Street, for instance. (Some cities, of course, do use Cartesian-style numerical addresses.)

Calculating with Coordinates: Distance and Midpoint

Using Cartesian coordinates, many questions about points in the plane can be reduced to straightforward, direct calculations.

The Midpoint Formula.   Recall the situation in *one* dimension. If x_1 and x_2 are
points on the real *line*, their midpoint, m, is just their average: $m = (x_1 + x_2)/2$.

A similar situation holds in the plane. To find the midpoint, M, of the segment
joining points $P = (x_1, y_1)$ and $Q = (x_2, y_2)$, we average their x- and y-coordinates
separately:

$$M = \left(\frac{x_1 + x_2}{2}, \frac{y_1 + y_2}{2} \right)$$

A picture makes everything plausible:

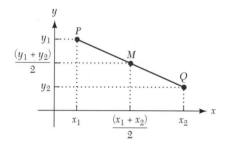

In particular, the coordinates of M are, respectively, midpoints of intervals on the
two axes.

Distance.   On the line the situation is simple: The distance between x_1 and x_2
is $|x_2 - x_1|$. The distance formula in the plane, like the midpoint formula, derives
from the one-dimensional situation.

What's the distance between the points $P = (x_1, y_1)$ and $Q = (x_2, y_2)$? Here's
the picture:

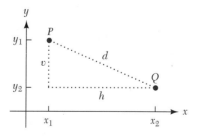

As the picture shows, the distance from P to Q (labeled d) is the hypotenuse of
a right triangle, with sides (labeled h and v) parallel to the axes. We'll use the
Pythagorean theorem to find d, but first we need values for h and v. By the *one-
dimensional* formula:

$$h = |x_2 - x_1| \qquad \text{and} \qquad v = |y_2 - y_1|.$$

Notice where (and why!) the
absolute value signs
conveniently vanished.

The distance formula in the plane now follows easily: ◄

$$d = \sqrt{h^2 + v^2} = \sqrt{|x_2 - x_1|^2 + |y_2 - y_1|^2} = \sqrt{(x_2 - x_1)^2 + (y_2 - y_1)^2}$$

The last form is standard. To summarize:

> **Fact** The distance from $P = (x_1, y_1)$ to $Q = (x_2, y_2)$ is given by
> $d(P, Q) = \sqrt{(x_2 - x_1)^2 + (y_2 - y_1)^2}$.

Elementary properties of distance in the plane follow from this formula:

The Same Distance Either Way.   Intuition says that the distance from P to Q should be equal to the distance from Q to P. All is well (but check the equation carefully to see why):

$$\begin{aligned} d(P, Q) &= \sqrt{(x_2 - x_1)^2 + (y_2 - y_1)^2} \\ &= \sqrt{(x_1 - x_2)^2 + (y_1 - y_2)^2} \\ &= d(Q, P). \end{aligned}$$

Finding the Midpoint.   It stands to reason that the midpoint M between $P = (x_1, y_1)$ and $Q = (x_2, y_2)$ should be equidistant from both. An ugly computation (check the details, as an algebra exercise!) shows that this is indeed so. Squaring everything in sight helps a little:

$$\begin{aligned} d(P, M)^2 &= \left(\frac{x_1 + x_2}{2} - x_1\right)^2 + \left(\frac{y_1 + y_2}{2} - y_1\right)^2 \\ &= \frac{(x_2 - x_1)^2}{4} + \frac{(y_2 - y_1)^2}{4} \\ &= \frac{(x_1 - x_2)^2}{4} + \frac{(y_1 - y_2)^2}{4} \\ &= \left(\frac{x_1 + x_2}{2} - x_2\right)^2 + \left(\frac{y_1 + y_2}{2} - y_2\right)^2 \\ &= d(M, Q)^2. \end{aligned}$$

Using the Distance and Midpoint Formulas

Simply finding distances and midpoints is neither very edifying nor, fortunately, very common in calculus. Such calculations usually arise in more interesting contexts.

Circles.   A circle is, by definition, the set of points at some fixed distance, say r, from a center, say $C = (a, b)$. The distance formula puts all this conveniently and simply, in algebraic form:

The point $P = (x, y)$ lies on the circle if and only if $d(P, C) = \sqrt{(x - a)^2 + (y - b)^2} = r$.

It's equivalent to say that the circle has equation $(x - a)^2 + (y - b)^2 = r^2$.

EXAMPLE 5   Find an algebraic equation for the circle of radius 5, centered at $C = (3, 2)$.

Solution   A point $P = (x, y)$ lies on this circle if and only if $d(P, C) = 5$. Thus,

$$\sqrt{(x - 3)^2 + (y - 2)^2} = 5$$

is an algebraic equation for this circle. We're done; we've found a perfectly acceptable algebraic equation. For anyone squeamish about square roots, there are other, equivalent, forms: ▶

The $\iff$ symbol means "if and only if."

$$\sqrt{(x - 3)^2 + (y - 2)^2} = 5 \iff (x - 3)^2 + (y - 2)^2 = 25$$
$$\iff x^2 + y^2 - 6x - 4y - 12 = 0$$
$$\iff x^2 + y^2 - 6x - 4y = 12 \qquad ■$$

EXAMPLE 6   Is $x^2 + 4x + y^2 - 2y = 10$ the equation of a circle? If so, which one?

Solution   Let's **complete the square** in both x and y:

$$x^2 + 4x + y^2 - 2y = 10 \iff (x^2 + 4x + 4) + (y^2 - 2y + 1) = 10 + 4 + 1$$
$$\iff (x + 2)^2 + (y - 1)^2 = 15$$

The answer can now be read off: The equation describes a circle of radius $\sqrt{15}$, centered at $(-2, 1)$.   ■

EXAMPLE 7   It's well known among those who know such things that *the segment joining the midpoints of two sides of a triangle is parallel to the third side, and half its length.* Classical geometers would have proved this "constructively," with drawing tools. With coordinates, the proof is an easy calculation.

The picture follows. We arranged things to simplify calculations, but convince yourself that we did nothing illegal:

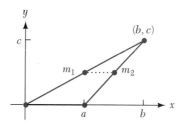

To prove our first claim—that the segment joining m_1 and m_2 is horizontal—it's enough to show that m_1 and m_2 have the same "height." By the midpoint formula,

$$m_1 = (b/2, c/2) \qquad \text{and} \qquad m_2 = \big((a + b)/2, c/2\big).$$

Thus m_1 and m_2 have the same y-coordinate, as claimed. We leave the claim about lengths as an exercise.   ■

BASICS

1. Express each of the following intervals using absolute value inequalities.
   (a) $(0, 10)$ (d) $(2, 8)$
   (b) $[-\pi, \pi]$ (e) $[-3, 7]$
   (c) $(-2, 4)$ (f) $[-10, -2]$

2. Express each of the following sets using absolute value inequalities.
   (a) $\left\{ x \mid x < -3 \quad \text{or} \quad x > 3 \right\}$
   (b) $\left\{ x \mid x < -5 \quad \text{or} \quad x > 3 \right\}$
   (c) $\left\{ x \mid -7 < x \quad \text{and} \quad x < 7 \right\}$
   (d) $\left\{ x \mid -7 < x \quad \text{and} \quad x < 5 \right\}$

3. Express each of the following sets using interval notation.
   (a) $\left\{ x \mid -10 < x \le 15 \right\}$ (e) $\left\{ x \mid |x - 2| \le 5 \right\}$
   (b) $\left\{ x \mid -6 \le x < 4 \right\}$ (f) $\left\{ x \mid |x + 3| \le 4 \right\}$
   (c) $\left\{ x \mid -13 < x < 17 \right\}$ (g) $\left\{ x \mid |x - a| < b \right\}$
   (d) $\left\{ x \mid |x| \le 3 \right\}$ (h) $\left\{ x \mid |x - a| < a \right\}$

4. Find the distance between each of the following pairs of points.
   (a) $(0, 0)$ and $(2, 3)$ (c) $(-4, 2)$ and $(-7, -4)$
   (b) $(1, 2)$ and $(3, 4)$ (d) $(-5, -2)$ and $(2, 5)$

5. Explain why
$$\sqrt{|x_2 - x_1|^2 + |y_2 - y_1|^2} = \sqrt{(x_2 - x_1)^2 + (y_2 - y_1)^2}.$$

6. Find the points in the xy-plane that are at distance 13 from the point $(1, 2)$ and have y-coordinate equal to -3. [**HINT**: Draw a picture.]

7. Find an algebraic equation for the circle of radius 4 centered at $(-3, 7)$.

8. Is $x^2 + y^2 - 6x - 8y = -9$ the equation of a circle? If so, which one?

9. Show that $3x^2 + 3y^2 + 4y = 7$ is the equation of a circle. Find the radius and center of this circle.

10. Show that the line segment joining the midpoint of two sides of a triangle is half the length of the third side.

11. Explain why each of the following statements is true for every real number r such that $3 < r < 7$.
    (a) $|r| < 7$ (c) $|r - 5| < 2$
    (b) $|r| > 3$

12. Suppose that s is a real number and that $-2 \le s \le 1$.
    (a) Must $|s| \le 1$ be true? Justify your answer.
    (b) Must $|s| \le 2$ be true? Justify your answer.
    (c) Must $|s| < 10$ be true? Justify your answer.

13. Which of the following statements is true for every real number x such that $-3 \le x \le 11$? Justify your answers.
    (a) $-3 \le x \le 9$ (c) $-7 \le x \le -4$
    (b) $2 \le x \le 10$ (d) $5.9999 \le x \le 6.0001$

14. Suppose that $L \le x \le U$. Show that $\left| x - \frac{1}{2}(L + U) \right| \le \frac{1}{2}(U - L)$. [**HINT**: Draw a picture.]

15. Suppose that $|x - 3| \le 0.005$ and $|y - 2| \le 0.003$. Find real numbers A and B such that $|(x + y) - A| \le B$. [**HINT**: Draw a picture.]

16. Express each of the following English sentences in mathematical notation using absolute values.
    (a) The temperature outside is within $5°$ (Fahrenheit) of freezing.
    (b) The temperature in that room is closer to $100°$F than to freezing.
    (c) The high temperature today is closer to yesterday's high temperature than yesterday's high was to the high temperature two days ago.

B

Lines and Linear Functions

Manipulating lines and linear functions is important throughout calculus. In this appendix we review the rudiments of lines: algebra, definitions, and techniques.

Slope

Slope measures a line's *steepness* (i.e., the rate at which a line rises or falls). Before reviewing the formal definition, study the picture below. Several lines appear, each labeled ▶ with its slope.

The letter m almost invariably stands for slope. Perhaps from the French monter—to climb.

Lines with various slopes

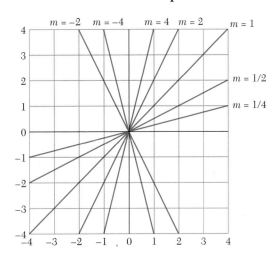

The picture illustrates the general situation:

> *Lines with positive slope rise (from left to right); lines with negative slope fall. The steeper the line, the larger the absolute value of its slope.*

> **Definition** The **slope** of a line is the ratio
>
> $$\frac{y_2 - y_1}{x_2 - x_1} = \frac{\Delta y}{\Delta x}$$
>
> where (x_1, y_1) and (x_2, y_2) are any two points on the line.

Notice:

Same Slope Everywhere A line has the same slope everywhere. *Any* two points can be used to calculate the slope. *Order* doesn't matter either—the slope from P to Q is the same as the slope from Q to P. In algebraic language, ◀

Keeping the same order in numerator and denominator does matter, of course.

$$\frac{y_2 - y_1}{x_2 - x_1} = \frac{y_1 - y_2}{x_1 - x_2}.$$

Parallel and Perpendicular Lines with equal slope are parallel. The slopes of perpendicular lines are negative reciprocals (i.e., their product is -1.)

Horizontal and Vertical Horizontal lines have slope 0; slope is undefined for *vertical* lines. The definition shows why.

How Steep? Scale Effects It's easy to see that the line $y = x$ has slope 1. How steep the line *looks*, on the other hand, depends strongly on the size of units (i.e., the **scales**) on the x- and y-axes. To see the effect of scales, consider these three graphs of the line $y = x$:

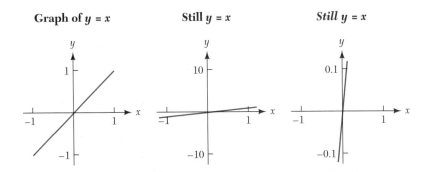

Graph of $y = x$ **Still $y = x$**   *Still $y = x$*

If—but only if—identical scales are used on both axes, the line $y = x$ makes a $45°$ angle with the x-axis.

Rise and Run The formula for slope is sometimes written as $\Delta y / \Delta x$, or as *rise/run*. ◀ The Δ-notation is a convenient shorthand. Invariably, Δ*anything* refers to a change in anything; it's usually calculated as the difference between a final value and an initial value. *Which* values these are must be made clear from context. For lines, things are simple: Δx is a *run*, or horizontal change; Δy is a *rise*, or vertical change. Either or both can be negative.

Convince yourself that these alternative versions really say the same thing.

Slope as a Rate Slope is defined as the ratio $\Delta y / \Delta x$. Slope, therefore, is a rate of change—the rate of change of y with respect to x. The straightness of a line means that this rate is constant.

Lines, Equations, and Functions

A line is completely determined by two bits of information. It's enough to know either (1) two points on the line, or (2) just one point and the slope. With either type of information, several simple algebraic descriptions of a line are possible. Here are two:

> **Definition** (**Point-Slope Form**) The line through (a, b) with slope m has equation $y = m(x - a) + b$.
>
> (**Slope-Intercept Form**) The line with slope m and y-intercept b has equation $y = mx + b$.

EXAMPLE 1   Consider the lines

ℓ_1: through $(0, 1)$ and $(2, 5)$;

ℓ_2: through $(-1, 3)$, with slope -2.

Describe each in point-slope form and in slope-intercept form.

Solution   The line ℓ_1 has slope $(5 - 1)/(2 - 0) = 2$. Now we know both the slope and (at least) one point on each line. Hence, in point-slope form:

$$\ell_1 : y = 2(x - 2) + 5 \qquad \text{and} \qquad \ell_2 : y = -2(x + 1) + 3$$

Now for slope-intercept forms. By inspection, ▶ ℓ_1 has slope 2 and y-intercept 1. Thus, ℓ_1 has slope-intercept equation $y = 2x + 1$. How about ℓ_2? Its slope is -2, so the slope-intercept equation looks like $y = -2x + b$. But what's b? Well, since ℓ_2 passes through $(-1, 3)$, those coordinates must satisfy the equation. In other words,

$$3 = -2 \cdot (-1) + b = 2 + b \, .$$

Clearly, $b = 1$. Therefore, the slope-intercept equation is $y = -2x + 1$. ∎

"By inspection" is the mathematician's grandiloquent way of saying "just by looking."

> **Unique or Not?**   The slope-intercept form of a line is *unique*—there's only one way to write a line in slope-intercept form. Moreover, the slope-intercept form works for any line that *has* slope (i.e., for any line that isn't vertical). Notice that the point-slope form of a line is not unique—*any* point on the line can appear in the point-slope equation. Since, for example, $(10, 21)$ lies on ℓ_1, $y = 2(x - 10) + 21$ is another point-slope form for ℓ_1.

EXAMPLE 2   Write an equation in slope-intercept form for the line ℓ through two given points (a, b) and (c, d). Where does ℓ intersect each axis?

Solution   If $a = c$, then ℓ is vertical; its equation is just $x = a$.

If $a \neq c$, ℓ has slope $(d - b)/(c - a)$, and the point-slope form applies:

$$y = \left(\frac{d - b}{c - a} \right)(x - a) + b \, .$$

More algebra yields this (slightly clumsy) equation for ℓ in slope-intercept form:

$$y = \left(\frac{d - b}{c - a} \right) x + \left(b - a \left(\frac{d - b}{c - a} \right) \right) .$$

Check it, please.

The y-intercept can now be read off; it's everything after the $+$ sign. To find the x-intercept, we set $y = 0$ and solve for x. The result: ◄

$$x = a - b\left(\frac{c-a}{d-b}\right).$$
■

EXAMPLE 3   How many lines do the equations

$$y + 2x = 6, \qquad y = -2(x+2) + 6, \qquad \text{and} \qquad y = -2(x-5) - 8$$

represent? Why?

Solution   Putting everything in slope-intercept form gives

$$y = -2x + 6, \qquad y = -2x + 2, \qquad y = -2x + 2.$$

This shows that the *three* equations represent *two* different lines.   ■

In the next example, lines appear in a typical calculus-style setting.

EXAMPLE 4   Find an equation in point-slope form for the line ℓ that meets the graph of $y = \sin x$ at $x = 1.3$ and at $x = 1.5$. (A line that "cuts" a graph at two prescribed points is called a **secant line**.) ◄   Here's a picture:

From the Latin, secare: to cut.

A secant line to the sine graph

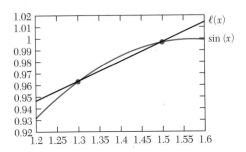

Solution   First let's find the slope of ℓ. Any two points on ℓ will do. As the graph shows, ℓ passes through $(1.3, \sin(1.3))$ and $(1.5, \sin(1.5))$. The slope of ℓ is, therefore,

$$m = \frac{\sin 1.5 - \sin 1.3}{1.5 - 1.3} \approx \frac{0.9975 - 0.9636}{1.5 - 1.3} = 0.1695.$$

Hence an equation for ℓ is

$$y = m(x - 1.3) + \sin(1.3) \approx 0.1695(x - 1.3) + 0.9636.$$
■

Linear Functions

The line with equation $y = 5x + 2$ can be thought of in various ways: as a geometric figure, as an algebraic equation, or—more typically in calculus—as the function $f(x) = 5x + 2$. Functions whose graphs are lines are called, naturally enough, **linear ► functions**. Such functions have an especially simple algebraic form:

Unfortunately, the word "linear" is used in more than one way in mathematics; confusion can result.

> **Definition** A linear function is one whose rule can be written in the form
>
> $$f(x) = Ax + B$$
>
> where A and B are constants.

A crucial property of linear functions, from our point of view, is that they "grow" at *constant* rates. In geometric terms, graphs of linear functions are straight. The numerical interpretation is also important.

EXAMPLE 5   The graph of the linear function $f(x) = 5x + 2$ is a line with slope 5. Interpret this slope *numerically*, as a rate of change.

Solution   Here's a sampler of x- and y-values:

Sample Values: $y = f(x) = 5x + 2$											
x	1.2	1.3	1.4	1.5	1.6	...	−3.4	−3.3	−3.2	−3.1	−3.0
y	8.0	8.5	9.0	9.5	10.0	...	−15.0	−14.5	−14.0	−13.5	−13.0

Notice what the numbers say: Whenever x increases by 0.1, y increases by 0.5 (i.e., 5 times as much). In symbols: $\Delta y = 5 \Delta x$. ► In words:

The Δ symbol denotes a change, or increment, in a variable. We'll use it repeatedly.

> *Along the line $y = 5x + 2$, y increases 5 times as fast as x.*

This rate, moreover, is *constant*—it applies for all values of x. ■

EXAMPLE 6   Its curved graph shows that the sine function is *not* linear. How do numerical values show the same thing?

Solution   Here's another sampler, this time of values of the sine function:

Sample Values: $y = \sin x$										
x	1.2	1.3	1.4	1.5	1.6	...	−3.4	−3.3	−3.2	−3.1
y	0.932	0.964	0.985	0.997	1.00	...	0.256	0.158	0.0584	−0.0416

The nonlinearity of the sine function shows clearly in the numbers. This time ► there's no regular "growth" pattern. Increasing x by 0.1 produces varying increases— or even decreases!—in y. ■

As opposed to the previous table.

Lines and Linear Functions in Calculus

Lines are important throughout calculus—perhaps surprisingly so, given that most functions of interest aren't linear.

An idea we'll return to repeatedly.

We'll return repeatedly to this subtle but important idea. Rough synonyms for "locally straight" include "almost linear" and "locally linear."

To choose a domain point at random.

Local Straightness.   The key idea, ◄   and the link to linear functions, is that a typical calculus function is "locally straight," i.e., it "looks straight at small scale" near every point of its domain. ◄   The next (important) example explores what local straightness means.

EXAMPLE 7   In what *graphical* sense is the sine function "locally straight" near $x = 1.4$? ◄   *Which* line does the sine graph resemble near $x = 1.4$? What does all this mean *numerically*?

Solution   "Locally straight" means that the sine graph looks straight near the point $(1.4, \sin 1.4)$ if viewed in a small enough window—i.e., after enough "zooming in." One view of the graph appears in Example 4. Here are two successively closer views: ◄

Zooming is easy with a graphing calculator. As always, a graph's appearance depends on the x- and y-scales.

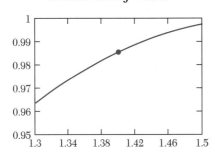

A closer view: *y* = sin x

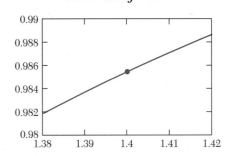

Closer still: *y* = sin x

The last view makes our point: the sine graph is all but indistinguishable from a straight line.

To estimate slope accurately, read values from the graph at $x = 1.38$ and $x = 1.42$.

Allowing for some estimation.

   Which straight line?   Because our graph passes through $(1.4, \sin 1.4)$ with slope ≈ 0.17, ◄   the line we want has equation ◄

$$y = \ell(x) = \sin 1.4 + 0.17(x - 1.4) = 0.17x + 0.7474.$$

What all this means is that $\sin x$, although *not* a linear function, behaves similarly to $\ell(x)$ for x near 1.4. Numerical data support this conclusion:

\multicolumn{10}{c}{**If $x \approx 1.4$, $\sin x \approx \ell(x)$**}									
x	1.380	1.385	1.390	1.395	1.400	1.405	1.410	1.415	1.420
$\sin x$	0.9819	0.9828	0.9837	0.9846	0.9854	0.9863	0.9871	0.9879	0.9887
$\ell(x)$	0.9820	0.9829	0.9837	0.9846	0.9854	0.9863	0.9871	0.9880	0.9888

The next example illustrates an entirely different use of linear functions.

EXAMPLE 8   Are hardness of water and central nervous system (CNS) disorders in infants related, as some health researchers suspect? Below are some relevant data; hardness is measured in parts per million, the disorder rate in malformations per 1000 births. [These data come from *Biostatistics: A Foundation for Analysis in the Health Sciences*, 2nd ed., Wayne W. Daniel (Wiley, 1978).]

Hard Water: Good or Bad?										
Hardness	50	25	15	75	100	150	180	250	275	220
Disorder Rate	7.2	8.1	11.1	9.3	9.4	5.0	5.8	3.3	3.6	4.8
Hardness	160	50	45	60	100	155	200	240	40	65
Disorder Rate	6.3	12.5	15.0	6.5	8.0	10.0	5.3	4.9	7.2	11.9

Is there pattern in the data? It's hard to tell from the table. Let's plot the data:

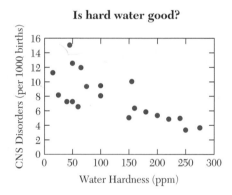

Is hard water good?

What *does* the picture show? The data don't lie on any obvious curve or line, but they suggest a general tendency—with harder water, the CNS disorder rate seems to decrease.

Can we describe this situation mathematically? One possibility is to "fit" a reasonable line to the data, perhaps something like this:

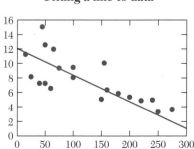

Fitting a line to data

The picture raises obvious questions: How was the line chosen? Is it a "good" fit? Might another line fit better? What does the line mean about CNS disorders? Does *any* line really fit this situation?

We chose the line entirely by appearance—it just *looked* right. Having confessed this, let's find an equation for the line. The y-intercept, 12, can be read directly from the graph. Reading from upper left to lower right shows a rise of -11, a run of 300, and hence a slope of $-11/300 \approx -0.0367$. A suitable equation, therefore, is

$$y = -\frac{11}{300}x + 12.$$

What good is such an equation? If $x = 125$, then

$$y = -\frac{11}{300} \cdot 125 + 12 \approx 7.4.$$

This means, in context, that in an area with water hardness 125 ppm we'd predict about 7.4 CNS disorders per 1000 births.

How trustworthy is such a prediction? The answer depends both on the philosophy and on the mathematics of mathematical modeling.   ■

BASICS

1. A line L passes through the points $(1, 2)$ and $(-2, 0)$.
   (a) Find an equation for the line L in slope-intercept form.
   (b) Does the line L pass through the point $(\pi, 2\pi)$? Explain your answer.

2. (a) Find an equation of the line through $(2, 1)$ that has slope 2/3.
   (b) Find an equation of the line through $(2, 1)$ that is perpendicular to the line in part (a).

3. Find equations in slope-intercept form for
   (a) the line that is parallel to the x-axis and passes through $(-3, 2)$.
   (b) the line that is parallel to the y-axis and passes through $(2, 4)$.
   (c) the line that is parallel to the line $y = x$ and passes through $(-3, 1)$.
   (d) the line through $(-3, 1)$ that is perpendicular to the line $y = 5x + 7$.

4. Find the x- and y-intercepts of the following lines
   (a) $3x + 4y = -12$
   (b) $y = m(x - a) + b$

5. For each condition below, find *all* values of k, if any, for which the line $2x + ky = -4k$ satisfies the condition.
   (a) The line has slope 3.
   (b) The line has slope $m \neq 0$.
   (c) The line passes through the point $(0, 4)$.
   (d) The line is horizontal.
   (e) The line is vertical.

6. Consider the curve shown below.

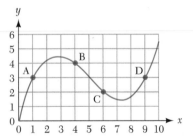

For each of the following pairs of points, find the slope of the secant line passing through the points and an equation of this line.   [**HINT:** Draw the secant lines on this page.]
   (a) **A, B**   (c) **B, D**
   (b) **A, C**   (d) **C, D**

7. The graph of the cosine function is "locally straight" near $x = 1$. Which line does the cosine graph resemble near $x = 1$?

8. The graph $y = \sqrt{x}$ is "locally straight" near $x = 4$. Which line does this graph resemble near $x = 4$?

9. The graph $y = x^3$ is "locally straight" near $x = 2$. Which line does this graph resemble near $x = 2$?

10. The graph $y = 1/x$ is "locally straight" near $x = 1$. Which line does this graph resemble near $x = 1$?

Polynomial Algebra: A Brisk Review

Polynomials

Polynomials are among the simplest and most useful calculus functions. In this appendix we offer a brisk review of the *precalculus* of polynomials: basic ideas, terminology, techniques, and algebra. First, the definition:

> **Definition** A **polynomial** in x is an expression of the form
> $$a_0 + a_1 x + a_2 x^2 + a_3 x^3 + \cdots + a_n x^n,$$
> that is, a sum of *constant* multiples of *nonnegative integer* powers of x.

EXAMPLE 1 Each of the following expressions
$$5x + 2, \qquad -x + 5x^{17}, \qquad \text{and} \qquad (x+3)(\pi x^{17} + 7)$$
is a polynomial in x. ► (*Linear* functions, in particular, are polynomials—among the simplest ones.) None of
$$3 - \frac{1}{x}, \qquad \sqrt{x}, \qquad \text{and} \qquad \frac{x+3}{\pi x^{17} + 7}$$
is a polynomial. ■

Use algebra to write the last example in standard form.

Basic Facts, Terms, and Techniques. To help recall basic polynomial properties and concepts, here is a short glossary of polynomial language:

Degree The **degree** of a polynomial is its largest power. Constant polynomials, for instance, have degree zero. The polynomials $x^3 + 5x^2 + 1$, $-x + 5x^{17}$, and $(x+3)(\pi x^{17} + 7)$ have degree 3, 17, and 18, respectively.

Polynomials of small degree have useful, suggestive nicknames. ► Polynomial functions of degree one are called **linear**, those of degree two **quadratic**, or **parabolic**. The naming sequence continues: cubic, quartic, quintic, ...

Learn and use such names; they're part of mathematical culture.

Factoring and Expanding   To **factor** a polynomial is to write it as a product of lower-degree polynomials. To **expand** is to multiply out. The following table shows polynomials in both **factored form** and **expanded form**:

Polynomials in Two Forms	
Expanded Form	**Factored Form**
$x^2 + 3x + 2$	$(x+1)(x+2)$
$x^4 - 1$	$(x+1)(x-1)(x^2+1)$
$x^4 + 6x^3 + 11x^2 + 6x$	$x(x+1)(x+2)(x+3)$
$x^4 + 6x^3 + 11x^2 + 6x$	$x(x+1)(x+2)(x+3)$
$x^2 + 2ax + a^2$	$(x+a)^2$
$x^3 + 3ax^2 + 3a^2x + a^3$	$(x+a)^3$
$x^3 + 3ax^2 + 3a^2x + a^3$	$(x+a)^3$
$x^3 - a^3$	$(x-a)(x^2+ax+a^2)$

The idea of root makes sense for any function f. The number r is a root of f if $f(r) = 0$. Finding roots of arbitrary functions is an important, and sometimes difficult, problem. We'll return to it.

Roots   A number r is a **root** of a polynomial ◄ p if $p(r) = 0$. Graphically speaking, real roots and x-intercepts are the same thing.

Graphs show nicely the connections among real roots, intercepts, and factors. See the following graph for the polynomial expression $y = x^5 - x = x(x+1)(x-1)(x^2+1)$.

Graph of $y = x^5 - x$

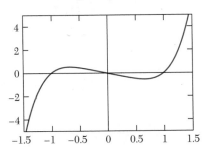

Notice, in particular, the *number* of real roots. Each linear factor contributes one root, in the obvious place. The remaining quadratic factor, $x^2 + 1$, is always positive, so it contributes no real roots.

Roots and Factors   Roots and factors have an important algebraic relationship:

> $(x - r)$ *is a factor of the polynomial $p(x)$ if and only if r is a root of p.*

The connection between roots and factors leads to a simple strategy for factoring polynomials: Somehow find (graphically, by guessing—whatever works) a root; then divide out the corresponding linear factor. ▶   It's easy to see, for instance, that $x = a$ is a root of the polynomial $p(x) = x^3 - a^3$. It follows that $(x - a)$ is a factor of $p(x)$. ▶

How Many Roots?   A polynomial of degree n can have at most n real roots. Geometrically, this means that the graph of a polynomial of degree n crosses the x-axis at most n times. ▶

As we've seen, some polynomials of degree n have fewer than n real roots; *any* number up to n is possible. Want a polynomial of degree 43 with 27 real roots? Here's one (of many):

$$(x - 1)(x - 2) \cdots (x - 27)(x^2 + 1)^8 .$$

Can Every Polynomial Be Factored?   No.   For linear polynomials there's nothing to do; they're already factored. Factoring a quadratic polynomial p is either impossible or easy, depending on whether p has real roots. The **quadratic formula**

$$r = \frac{-b \pm \sqrt{b^2 - 4ac}}{2a}$$

finds the roots—if any exist—of the quadratic function $ax^2 + bx + c$.

Polynomials of degree three and higher are stickier. For the record, abstract theory (well beyond this course) says that *every* polynomial of degree three or higher has at least some lower-degree factors. In other words, every nonconstant polynomial can be written as a product of *linear* and *quadratic* factors; finding these factors, however, can be hard.

Completing the Square   It's sometimes convenient to rewrite a quadratic polynomial in the general form $a(x + b)^2 + c$; doing so is called **completing the square**. The following example illustrates the technique.

EXAMPLE 2   Let $p(x) = 2x^2 + 8x + 1$. Complete the square; use the result to find all roots of p.

Solution   We want to write $p(x)$ in the form $p(x) = 2(x + b)^2 + c$. Here goes—follow each algebra step closely:

$$2x^2 + 8x + 1 = 2(x^2 + 4x) + 1 = 2(x^2 + 4x + 4) + 1 - 8 = 2(x + 2)^2 - 7.$$

The third equation explains the name. Adding 4 to $x^2 + 4x$ "completed" the perfect square $(x + 2)^2$. ▶

The roots of p can now be read off:

$$p(x) = 0 \iff 2(x + 2)^2 - 7 = 0 \iff 2(x + 2)^2 = 7 \iff (x + 2)^2 = \frac{7}{2}$$

$$\iff x + 2 = \pm\sqrt{7/2} \iff x = -2 \pm \sqrt{7/2}.$$

(It's easy to check ▶   that the quadratic formula gives the same roots.)   ■

Polynomials as Functions.   The polynomial *expression* $2x^3 + 5x$ corresponds naturally to the *function* $p(x) = 2x^3 + 5x$. *Any* function defined by a polynomial expression is called a **polynomial function**. In practice the technical distinction between polynomial and polynomial function is seldom important. We'll use the terms interchangeably.

This strategy seldom works in real life. In calculus textbooks, it seldom fails—the problems are rigged to make it so.

See the last line of the table on page 442.

Here's one consequence: The sine function isn't a polynomial.

We "paid" for doing so by subtracting 8 at the end.

Do so!

Domain and Range.   Because polynomials involve only multiplication and addition, every polynomial function has the same domain: all real numbers.

Polynomials' ranges vary drastically. The range of $p(x) = 3$ is $\{3\}$, a set with just one member. By contrast, the range of $q(x) = x^2 + 3$ is $[3, \infty)$, an infinite interval. The range of $r(x) = x^3 - 7x^2$ may not be obvious from the formula. Here's a graph:

Graph of $y = x^3 - 7x^2$

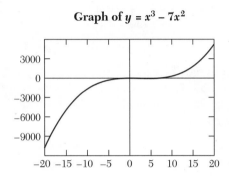

The graph shows how r behaves in the long run. It "blows up" (tends to ∞) as $x \to +\infty$ and "blows down" (tends to $-\infty$) as $x \to -\infty$. Therefore, apparently, r "hits" all real values; its range is $(-\infty, \infty)$.

Polynomial-Speak.   Polynomials have their own specialized vocabulary. Watch carefully in the following discussion for such words as **root**, **degree**, **factor**, **expand**, and **quadratic formula**.

EXAMPLE 3   Find the roots of the fourth-degree polynomial $p(x) = x^4 + 6x^3 + 11x^2 + 6x$; interpret results both algebraically and graphically.

Solution   In its expanded form, the formula for $p(x)$ is unhelpful. But the factored form reveals much more:

$$p(x) = x^4 + 6x^3 + 11x^2 + 6x = x(x + 1)(x + 2)(x + 3).$$

Now it's clear, algebraically, that p has *four* real roots—one where each of the four factors is zero. The following picture tells the same story:

Graph of $y = x(x + 1)(x + 2)(x + 3)$

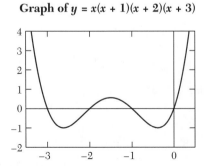

Whether a factored or expanded form of a polynomial is "better" depends on what we're doing. Here the factored form was preferable; the expanded form, on the other hand, may be handier for such symbolic operations as differentiation and integration. ▶

■   *As we'll see!*

Some polynomials have *fewer* real roots than their degrees allow.

EXAMPLE 4   Consider the cubic polynomial polynomial $p(x) = x^3 - 2x^2 + x - 2$. Factor p as much as possible; find all of p's real roots.

Solution   Here's a graph of p:

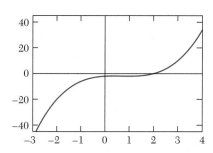

Graph of $y = x^3 - 2x^2 + x - 2$

The graph suggests that 2 is a root; a simple check confirms that, indeed, $p(2) = 0$. Therefore $(x - 2)$ is a factor of p. Indeed, long dividing $(x - 2)$ into $x^3 - 2x + x - 2$ shows ▶ that

$$p(x) = x^3 - 2x^2 + x - 2 = (x - 2)(x^2 + 1).$$

Check the equality by multiplying out the product on the right.

Is further factorization possible? No. Additional factors, if any, would come from $x^2 + 1$. But $x^2 + 1$ has no real roots ▶, and hence no linear factors.

Why not?

Note that the factored form $p(x) = (x - 2)(x^2 + 1)$ tells us something new: $p(x) = 0$ *only if* $x = 2$. In other words, 2 is the only real root. No graph could tell us this, because we can never see all of any polynomial's graph. ■

Why Polynomials Matter

Polynomial functions are simple but surprisingly useful. They lurk in calculus problems ranging from the sublime:

Show that among all rectangles of given perimeter, the square encloses the most area.

to the ridiculous:

Design the most spacious possible rectangular pigpen, using 100 feet of fence. One side abuts a river ....

A subtler but equally important use of polynomials is in *approximating* less convenient, nonpolynomial functions. For example, the following graphs show that, for x near 0, the polynomial function $x - x^3/6$ does a good job of approximating the sine function, which *isn't* a polynomial. ◄

Can you tell which is which?

Two graphs: $y = x - x^3/6$, $y = \sin(x)$

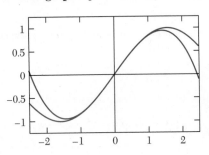

This is worth knowing. It suggests how a calculator might compute numerical values for nonalgebraic functions, such as the sine function.

Rational Functions

Is this really obvious? Think about it; try some examples.

Adding or multiplying two polynomials creates a new polynomial. ◄ *Dividing* two polynomials doesn't (usually) give another polynomial; it gives a **rational function**. Formally:

> **Definition** A rational function is a function that can be written as the quotient, or "ratio," of two polynomials.

EXAMPLE 5   Each function below is rational:

$$f(x) = \frac{1}{x} \qquad\qquad g(x) = \frac{x^2}{x^2 + 3}$$

$$h(x) = \frac{x^7 + x^6 + x^5 - x^4}{x + 1} \qquad k(x) = \frac{2}{x} + \frac{3}{x + 1}$$

The function $m(x) = \sqrt{x} = x^{1/2}$ is not rational because it involves a *fractional* power of x.

The function k may look suspicious. It's certainly the *sum* of two rational functions, but is k itself rational? The answer is yes, because we can rewrite $k(x)$

in the form:

$$k(x) = \frac{2}{x} + \frac{3}{x+1} = \frac{2(x+1)}{x(x+1)} + \frac{3x}{x(x+1)} = \frac{5x+2}{x(x+1)}.$$ ∎

This computation illustrates a more general fact:

> **Fact** Let r and s be rational functions. The sum, difference, product, and quotient of r and s are all rational functions—each can be written as the ratio of polynomials.

Graphs of Rational Functions

Rational functions ▶ have an important new feature: the possibility of **horizontal** and **vertical asymptotes**. Asymptotes are straight lines toward which a graph "tends."

Unlike polynomials.

EXAMPLE 6 Consider the rational function

$$r(x) = \frac{x^2}{x^2-1} = \frac{x^2}{(x-1)(x+1)}$$

and its graph:

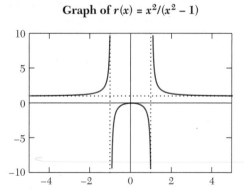

Graph of $r(x) = x^2/(x^2 - 1)$

Technicalities aside, it's evident graphically that this function has three asymptotes: the *vertical* lines $x = -1$ and $x = 1$, and the *horizontal* line $y = 1$.

In Section 1.5, page 54, we show these facts carefully, relating them to infinite limits and limits at infinity. ∎

BASICS

1. Let p, q, r, and t be the polynomials

$$p(x) = 2x^3 + 5x^2 - x$$

$$q(x) = 4x^5 - 6x^3$$

$$r(x) = 5x^{12} - 123x^8 + 47$$

$$t(x) = 17x^{123} - 26x^{19} + 71x^5$$

   What is the degree of each polynomial?

2. Find a polynomial of degree 5 with $x = -2$, $x = 1$, and $x = 3$ as its *only* real roots. Sketch its graph. (There are many possible answers.)

3. The following graphs show (i) a quadratic polynomial with two real roots, (ii) a quadratic polynomial with no real roots, (iii) a cubic polynomial with only one real root, and (iv) a cubic polynomial with three real roots.
   (a) Which graph is which? Label the graphs (i)–(iv).
   (b) Find an equation for the quadratic polynomial with two real roots.
   (c) Find an equation for the cubic polynomial with three real roots.
   (d) The cubic polynomial with only one real root can be written in the form $p(x) = (x + r)(x^2 + s)$. Find the values of r and s.
   (e) Find an equation for the quadratic polynomial with no real roots.

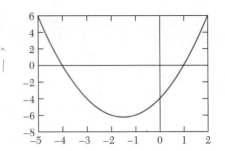

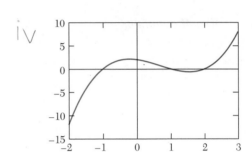

IV

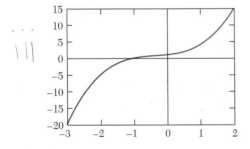

I I I

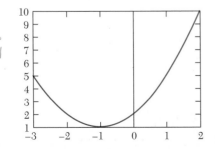

I I

4. Let a be any constant. The table on page A-22 gives the formula $x^3 - a^3 = (x - a)(x^2 + ax + a^2)$.
   (a) Verify the formula by multiplying out the right side.
   (b) If $a \neq 0$, then the second factor (i.e., $x^2 + ax + a^2$) cannot be factored any further. Show this either by completing the square or by using the quadratic formula.
   (c) Use the table on page A-22 to find a factored formula for $x^3 + b^3$. [**HINT:** Let $a = -b$.]

5. The table on page A-22 gives the formula $(x + a)^3 = x^3 + 3ax^2 + 3a^2x + a^3$.
   (a) Verify the formula by multiplying out the left side.
   (b) Let $p(x) = x^3 + 3x^2 + 3x + 1$. Check that $x = -1$ is a root of $p(x)$.
   (c) Since $x = -1$ is a root of $p(x) = x^3 + 3x^2 + 3x + 1$, $(x + 1)$ must be a factor of $p(x)$. Divide $p(x)$ by $(x + 1)$ to get a quadratic polynomial.
   (d) Factor $p(x) = x^3 + 3x^2 + 3x + 1$ as a product of linear factors.

6. Match each of the following functions
   (i) $f(x) = (x-1)(x+2)$
   (ii) $g(x) = \dfrac{x+3}{(x-1)^2}$
   (iii) $h(x) = \dfrac{2x^2 + 3x - 1}{x^2 - 2x - 8}$

to one of the following graphs. Then determine all values of x for which each function is positive.

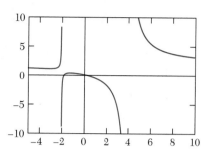

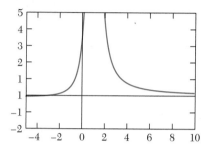

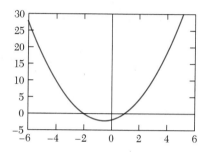

Real-World Calculus: From Words to Mathematics

Translation. Calculus is more than a brilliant theoretical invention. It is also a practical tool—among the most powerful ever discovered—for posing and solving important real-world problems. Real-world problems seldom arrive as neat mathematical packages. More common are untidy bundles of questions, assumptions, information, and (sometimes) disinformation. Important parts may be missing; useless extras may be included.

The first (and sometimes hardest) step in tackling such problems is to translate them, as concisely as possible, into mathematical language. Since calculus operates mainly on functions, "translation" in calculus usually means interpreting a problem in *functional language.*

To understand real phenomena—the Consumer Price Index, gravitational forces, population growth, planetary motion, and so on—we must be able to *predict* their behavior. Useful predictions, moreover, must be *quantitative*, not merely qualitative. Mathematics is the route to quantitative results.

Different areas of mathematics are best suited to describe and predict different sorts of phenomena. Describing phenomena of change is the special genius of calculus; calculus offers both language in which to describe changing quantities, and rules with which to predict their behavior.

Story Problems: You May Feel a Little Discomfort .... Mathematics has been compared (in jest, of course) to dentistry. "Story problems," in particular, sometimes elicit emotions normally associated with the whine of a high-speed drill.

Such fear and loathing can be alleviated—without pharmaceutical intervention. This is not to deny the difficulty. Translation of any sort is expert work, combining both technical and "literary" ◄ expertise.

In our case, mathematical.

Learn the Words by Heart. The essential point is that progress is possible. Indeed, the brevity, bluntness, and precision of mathematical language make mathematical translation simpler and more straightforward than it would otherwise be. For the same reason, mastering basic mathematical vocabulary and "grammar" is especially important. Calculus has its own recurring key words and phrases— **proportional**, **parameter**, **domain**, **maximum**, **increasing**, and many others. An important goal of this appendix is to review several important mathematical terms and techniques and to illustrate their use in calculus-style applications.

EXAMPLE 1   As a rocket rises, the force earth's gravity exerts on it decreases. This *qualitative* observation, although perfectly true, is utterly useless in practice.

It doesn't take a rocket scientist to see why. Without more information, we can't compute anything: how much fuel to load, how much thrust is needed, how long anything will take.

Newton's law of universal gravitation ▶ describes gravity's effect *quantitatively*. It says that for a 1000-lb rocket d miles from the center of the earth, the approximate force (in pounds) of earth's gravity is given by the function

We discuss it more fully elsewhere. See p. A-38.

$$F(d) = \frac{1.6 \times 10^{10}}{d^2}.$$

With the formula, predictions are possible. If, say, $d = 20{,}000$ miles, ▶ then the force of gravity is only

A reasonable distance for a communications satellite.

$$F(20{,}000) = \frac{1.6 \times 10^{10}}{20{,}000^2} = 40 \text{ pounds.} \qquad \blacksquare$$

Garden-Variety Optimization

Many real-life problems boil down to finding the largest or smallest value of some function. Finding the right function hidden in the verbiage may be the main challenge.

EXAMPLE 2   Gardener Alpha plans a rectangular lettuce patch; 100 feet of rabbit-proof fencing is available. What are the best dimensions for Alpha's lettuce patch?

Could "best" possibly mean anything else? Probably so. For real lettuce patches, "best" would involve harvesting convenience, ease of watering, etc.

Translating Alpha's Problem. "Best" can mean different things; we'll take it to mean "largest in area." ▶ The problem, then, is to find the largest rectangle with perimeter 100 feet.

The answer—a square—isn't surprising; translation corroborates our intuition. Thus, let x and y represent the lengths of the "vertical" and "horizontal" sides of the garden:

Alpha's garden

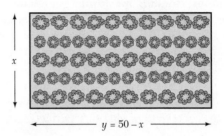

Equivalently, that $y = 50 - x$.

(That 100 feet of fence is available says, algebraically, that $100 = 2x + 2y$.) ◄   The crucial point is that the area A, given by

$$A = \text{length} \times \text{width} = xy = x(50 - x),$$

is a function of the variable x.

The rule $A(x) = x(50 - x)$ makes *algebraic* sense for any real number x. In context, though, both x and $50 - x$ must be nonnegative, because each represents a physical length. Thus the "feasible" input set, or domain, is the x-interval $[0, 50]$.

We've arrived at an optimization problem:

Find the maximum value, and where it occurs, of the function

$$A(x) = x(50 - x)$$

on the interval $[0, 50]$.

Solution   The area A is a quadratic function of x, so its graph is (part of) a parabola:

Graph of $A(x) = x(50 - x)$

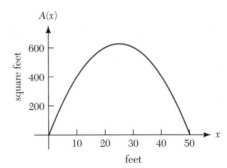

The graph suggests that the maximum value of A occurs at $x = 25$, where the area is $A(25) = 625$ square feet. Graphical guesses are often imperfect, but this time we're right: symmetry guarantees that the vertex of a parabola lies *exactly* midway between the two roots.

In the original language of the problem: *Alpha's lettuce patch should be square, 25 feet on a side, with area 625 square feet.* ■

EXAMPLE 3   Beta, too, has 100 feet of fence. Beta's barn will form one side of a rectangular patch. (The barn is long enough ▶ so that its length isn't a constraint.) Which dimensions are best now?

Over 100 feet, say.

Solution   This time, the answer *isn't* obvious; posing the problem mathematically will help us solve it.

Another "generic" picture applies:

Beta's garden

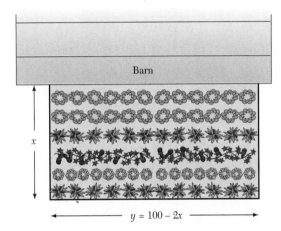

As before, area is a quadratic function of the side we called x: ▶

$$A(x) = xy = x(100 - 2x) = 100x - 2x^2 = 2x(50 - x).$$

Convince yourself that $y = 100 - 2x$.

The feasible domain is (again) the interval $[0, 50]$. ▶ In functional language, the problem is to *maximize the function A over its domain.*

Because x and y represent physical lengths.

The area function A is again quadratic, so its maximum value occurs midway between the two roots, at $x = 25$. The graph of A agrees:

Graph of A(x) = 2x(50 − x)

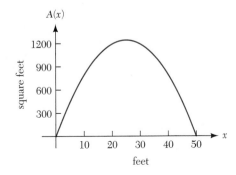

In barnyard language: *The largest feasible garden has dimensions $25' \times 50'$ and area $A(25) = 1250$ square feet.* ■

EXAMPLE 4   Gardener Gamma plans to fence exactly 1000 square feet of land. *Any* rectangular shape will do. For convenience in watering, one side will lie along a (straight) river. How much fencing will Gamma need for the other three sides?

Notice the new twist: Alpha and Beta aimed to *maximize area, for a fixed perimeter.* Gamma wants to *minimize perimeter, for a fixed area.* Optimization problems often come in such pairs. ◄

Paired optimization problems are known as "duals." This problem could be called a fencing dual.

Solution   A familiar picture applies:

Gamma's garden

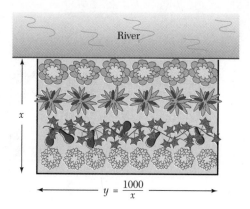

Using it, let's find our function. Let x and y be the lengths shown, and let L denote the total length of the fence. By assumption, $xy = 1000$, so

$$L = 2x + y = 2x + \frac{1000}{x}.$$

There's our function.

The Domain of L: A Subtlety.   What is L's "natural" domain? Technically, *any* positive number is a feasible input—a fence a mile long and 2.27 inches wide makes mathematical sense. Since no such ridiculously oblong design could possibly be most economical, we can safely ignore very large and very small positive inputs x. Given the general size of the numbers, the domain $5 \le x \le 50$ leaves plenty of room. Here, finally, is our problem:

Find the minimum value, and where it occurs, of the function $L(x) = 2x + 1000/x$ on the interval $[5, 50]$.

Solving the problem without calculus is hard, so we'll work approximately, with graphics. Here's a look:

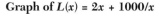

Graph of L(x) = 2x + 1000/x

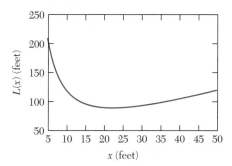

The graph shows that L attains its minimum value (about 90 feet) around $x = 22$. ▶ In agricultural terms: *Gamma needs 90 feet of fence; the garden should have approximate dimensions 22.5′ × 45′.* ∎

With a bit of calculus, we can find exact answers: $x = \sqrt{500} \approx 22.3607$ and $L(\sqrt{500}) \approx 89.4427$.

Parameters

We assumed earlier that Alpha and Beta each had 100 feet of fencing; our answers, naturally, reflected that assumption. But was the number 100 essential in either problem? Hardly. With 200 feet of fence we'd *still* design a square garden for Alpha, and a rectangular one, with proportions 1:2, for Beta.

Such "accidental" aspects of a problem are conveniently handled using **parameters**—symbols that stand for *any* constant value. The next example illustrates a simple use of parameters.

EXAMPLE 5   Delta and Epsilon, horticulturally incompatible, agree to "divorce" their gardens as shown:

Dividing the estate

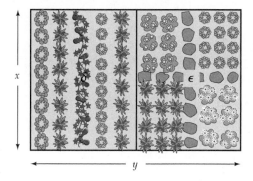

They have L feet of fence. What shape is best?

Solution   Notice two aspects of the problem:

Sorry; it just worked out that way.

- In this problem, the parameter is perimeter. ◀

- Before proceeding, *guess* an answer. Should the total garden be square? Vertically oblong? Horizontally oblong?

Solve this equation for y to get y = (L − 3x)/2.

Let x and y be as shown. As before, the area A satisfies the equation $A = xy$. The garden's peculiar design dictates that the total length of fence needed is $L = 3x + 2y$, so ◀

$$A(x) = xy = \frac{x(L - 3x)}{2}.$$

In context, both x and $L - 3x$ must be nonnegative, so $0 \leq x \leq L/3$ must hold. Our problem, then, is to maximize A for x in the interval $[0, L/3]$.

The presence of the parameter L now complicates life slightly. We'll show two possible strategies.

Maximizing A with Algebra.   For *any* nonnegative number L the function

$$A(x) = \frac{x(L - 3x)}{2}$$

is quadratic in x. Its graph is a parabola, opening *downward*. Symmetry guarantees that the vertex of a parabola lies midway between its two x-intercepts. From the formula above it's clear that $A(x) = 0$ if $x = 0$ or $x = L/3$. Therefore, the largest value of A occurs halfway between, at $x = L/6$.

Maximizing A with Graphics.   Another strategy is to plot $A(x)$ for *several* reasonable values of the parameter, and look for a pattern:

Graphs of A for various lengths L

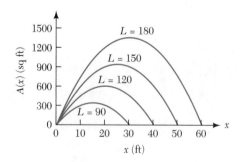

The picture shows that for each value of L, the maximum area occurs at $x = L/6$, the *midpoint* of the domain $[0, L/3]$.

Conclusion.   The most spacious double garden uses $L/6$ feet of fence for each of the three "vertical" runs. The remaining $L/2$ feet is divided equally for the two "horizontal" runs. ▶ The result, drawn proportionately to scale, looks like this:

Again, half the fence runs in each direction. Is this coincidence, or is something happening here?

Biggest gardens with perimeter L

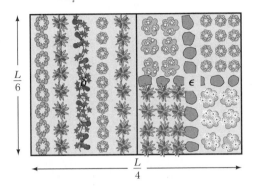

Taking Leave of Our Fences.   Fencing problems, although unglamorous, illustrate nicely the standard and important process of defining a pertinent function on a sensible domain, finding maximum or minimum values, and then interpreting the results.

Proportionality

Proportionality and inverse proportionality are relationships between varying quantities:

> **Definition** Let A and B be quantities which vary together. A is **proportional** to B if, for some nonzero constant k, the equation
>
> $$A = k\,B$$
>
> holds for *all* values of A and B. A and B are **inversely proportional** if
>
> $$A = \frac{k}{B}.$$

EXAMPLE 6   The formula ▶

$$C = \pi\,d$$

says that the circumference of a circle is proportional to the diameter, with **proportionality constant** π. The formula

$$A = \pi\,r^2$$

shows that the area of a circle is proportional to the *square* of the radius, but not to the radius itself. ∎

Although familiar, this formula is not trivial. That the same constant, π, appears here and in the area formula is a remarkable fact.

Gravity—Applying Proportionality

It's well known that objects weigh less on the moon than on Earth. How much less? Why?

Our words—Newton wrote in Latin.

Newton's law of universal gravitation describes the relationships among quantities involved in gravitation: ◄

The attractive force of gravity exerted on one object by another is proportional to the mass of each, and inversely proportional to the square of the distance between them. ◄

In each case, all other quantities are held constant.

In algebraic form, Newton's law says that

$$F = k\frac{M_1 M_2}{d^2},$$

For a sphere the center of gravity is the ordinary (geometric) center.

where F represents gravitational *force*, M_1 and M_2 are the objects' masses, d is the distance between the objects' centers of gravity, ◄   and k is a constant of proportionality. The numerical value of k depends on the units chosen. In the metric system (using kilograms for mass, meters for distance, and newtons for force), $k \approx 6.7 \times 10^{-11}$.

EXAMPLE 7   How much less does an object weigh on the moon than on Earth? What would the same object weigh on Jupiter?

Solution   An object's weight is defined as the force gravity exerts on it. According to Newton's law, the "Earth-weight" F_e of an object with mass M, at the surface of Earth, is

$$F_e = k\frac{M M_e}{R_e{}^2},$$

where M_e is Earth's *mass*, and R_e is Earth's *radius*. Similarly, the same object's "moon-weight" and "Jupiter-weight," are, respectively,

$$F_m = k\frac{M M_m}{R_m{}^2} \quad \text{and} \quad F_j = k\frac{M M_j}{R_j{}^2},$$

where M_m and M_j are the masses, and R_m and R_j are the radii, of the moon and Jupiter, respectively.

Our problem is about the ratios F_m/F_e and F_j/F_e. By easy algebra,

$$\frac{F_m}{F_e} = \frac{M_m R_e{}^2}{M_e R_m{}^2} \quad \text{and} \quad \frac{F_j}{F_e} = \frac{M_j R_e{}^2}{M_e R_j{}^2}.$$

Remarkably enough, every constant above is known—physics and astronomy textbooks have tables like this one:

Celestial Constants		
Object	**Mass (kg)**	**Radius (km)**
Moon	7.35×10^{22}	1740
Earth	5.98×10^{24}	6378
Jupiter	1.90×10^{27}	71398

Now only simple arithmetic remains:

$$\frac{F_m}{F_e} \approx 0.165; \qquad \frac{F_j}{F_e} \approx 2.535.$$

In human terms, an astronaut weighing 75 kg (about 165 pounds) on Earth weighs only $0.165 \cdot 75 \approx 12.38$ kg (about 27 pounds) on the moon. On Jupiter, the same astronaut weighs a staggering 190.13 kg (over 418 pounds).   ■

Mass or Weight?   Physicists distinguish between an object's *mass* and its *weight.* An object's weight can vary (as our example shows) but its mass is the same everywhere. In the metric system, the standard unit of force (and therefore of weight) is the **newton**. At the surface of the earth an object with mass 1 kilogram has weight 9.8 newtons. (Actually, since Earth is not perfectly spherical, an object's weight is slightly different at different points on Earth's surface.)

BASICS

1. Zeta plans to fence off a rectangular field along the bank of a river; no fence is needed along the river. Fencing costs $2 per foot for the ends and $3 per foot for the side parallel to the river. Zeta's budget is $900.

   (a) Let x be the length of the side of the field that lies along the river. Explain why the formula $A(x) = x(225 - 0.75x)$ describes the area of Zeta's field.

   (b) What is the domain of A? Why?

   (c) What are the dimensions of Zeta's *largest* field?

   (d) If given twice as much money, can Zeta enclose twice as much area? Why or why not?

2. Water is flowing into a cubical tank. Each side of the tank is ten feet.

   (a) Write a formula for V, the volume of water in the tank, as a function d, the depth of water in the tank.

   (b) What is the domain of V? What is the range of V?

   (c) Write a formula for d as a function of V.

3. Guido's pizza is delivered in boxes formed from 50-cm by 100-cm rectangular pieces of cardboard. The boxes are made by cutting out six squares of equal size, three from each of the 100-cm edges (one square at each corner, and one in the middle of the edge), then folding in the obvious way.

   (a) Find the volume of the box as a function of the length of the side of the square cutouts.

   (b) How large should the cutouts be to maximize the volume of the box?

4. Suppose that A and B are *directly* proportional, but A and C are *inversely* proportional. Suppose, further, that if $A = 1$, then $B = 2$ and $C = 3$.

   (a) What are the values of B and C if $A = 10$?

   (b) Are B and C directly proportional? Inversely proportional? Explain.

   (c) Suppose that $B = 10$. Find values for A and C.

5. Is the volume of a sphere proportional to its radius? Explain.

6. The weight of an object is inversely proportional to the square of its distance from the center of Earth. A satellite weighs 200 pounds when 10,000 miles from the center of Earth.

   (a) Express the weight of the satellite as a function of d, its distance from the center of Earth.

   (b) At what distance from the center of Earth does the satellite weigh 100 pounds?

7. An open box with a capacity of 36,000 cubic inches is needed. The box must be twice as long as it is wide. The material out of which the box is to be constructed costs $0.10 per square foot. Find the cost of the material for the box as a function of the width of the box.

8. It's 9:45 P.M. Your snowmobile is out of gas and you are 3 miles due south of a major highway in a field. The nearest service station on the highway is 6 miles east of your position. You can walk 4 miles per hour on roads, but only 3 miles per hour through snowy fields.

   Express the time to reach the service station as a function of the distance r you walk along the road. [**HINT:** You could walk straight through the fields, or partway through the fields, then on the road.]

Algebra of Exponentials

Where $e \approx 2.71828$.

Exponential functions are among the most important in both the theory and the applications of calculus. Indeed the exponential function $f(x) = e^x$ ◄ has been called the most important function in all of mathematics. To understand and use such functions effectively requires a basic familiarity with exponential algebra.

Exponential Expressions

An **exponential expression** is any algebraic expression that contains exponents. Each of the following expressions is exponential:

$$2^x, \quad 3^{x+h}, \quad b^3, \quad 2^{x^2}, \quad \frac{2^x}{3^x}, \quad \left(\frac{2}{3}\right)^x, \quad \left(1 + \frac{1}{n}\right)^n.$$

In any expression of the form b^x, b is the **base** and x is the **exponent**.

The Natural Exponential Function

The base-e exponential function, defined by $f(x) = e^x$, is called the **natural exponential function** or simply **the exponential function**. (The definite article in the latter name is slightly unfortunate. Expressions with *other* bases—2^x, 10^x, and so on—also deserve the name "exponential.") Various notations are used for the natural exponential function:

$$\exp(x), \qquad \exp x, \qquad \text{and} \qquad e^x$$

all mean the same thing.

Basics of Exponential Algebra

The basic algebraic rules for exponential expressions are few and simple:

$$b^x b^y = b^{x+y} \tag{E.1}$$

$$\frac{b^x}{b^y} = b^{x-y} \tag{E.2}$$

$$\left(b^x\right)^y = b^{xy} \tag{E.3}$$

$$a^x b^x = (ab)^x \tag{E.4}$$

$$b^0 = 1 \tag{E.5}$$

$$b^{-x} = \frac{1}{b^x} \tag{E.6}$$

In each case, b can be any *positive* number; x and y can take any real values. ▶

We'll see below why $b > 0$ matters.

EXAMPLE 1   Explain Rule (E.1): ▶ $b^x b^y = b^{x+y}$, assuming that x and y are positive integers.

The "addition rule."

Solution   If x and y are positive integers, then

$$b^x = \overbrace{b \cdot b \cdot b \cdots b}^{x\text{-times}} \quad \text{and} \quad b^y = \overbrace{b \cdot b \cdot b \cdots b}^{y\text{-times}}.$$

Therefore,

$$b^x b^y = \overbrace{b \cdot b \cdot b \cdots b}^{x} \cdot \overbrace{b \cdot b \cdot b \cdots b}^{y} = \overbrace{b \cdot b \cdot b \cdots b}^{x+y} = b^{x+y}.$$ ∎

EXAMPLE 2   Make sense of the "zeroth power" rule: $b^0 = 1$.

Solution   Rule(E.5) follows directly from the addition rule: If b is any positive number, then

$$b \cdot b^0 = b^1 \cdot b^0 = b^{1+0} = b^1 = b.$$

Thus $b \cdot b^0 = b$; dividing through by b gives $b^0 = 1$.   ∎

EXAMPLE 3   Rule (E.6) tells how to switch the sign of an exponent. It helps with calculations like this:

$$\left(\frac{1}{2}\right)^{-17} = \frac{1}{2^{-17}} = 2^{17}.$$ ∎

EXAMPLE 4   Derive Rule (E.6) from the other rules.

Solution   Use Rules (E.1) and (E.5) to convince yourself of each equality below:

$$b^{-x} \cdot b^x = b^{x-x} = b^0 = 1.$$

Dividing the first and last quantities by b^x gives Rule (E.6).   ∎

This example is algebraically tricky, but it shows one way exponential algebra is really used in calculus.

EXAMPLE 5 Let k be any constant, and define the function f by $f(x) = 3^{kx}$. Show ◄ that

$$\frac{f(x+h) - f(x)}{h} = f(x) \cdot \frac{f(h) - f(0)}{h}.$$

Solution Watch carefully—we'll use several of the rules cited previously:

$$\frac{f(x+h) - f(x)}{h} = \frac{3^{k(x+h)} - 3^{kx}}{h} = \frac{3^{kx+kh} - 3^{kx}}{h}$$

$$= \frac{3^{kx} \cdot 3^{kh} - 3^{kx}}{h} = 3^{kx} \cdot \frac{3^{kh} - 1}{h}$$

$$= 3^{kx} \cdot \frac{3^{kh} - 3^0}{h} = f(x) \cdot \frac{f(h) - f(0)}{h}. \blacksquare$$

Thinking of Positive Powers: Why $b > 0$ Is Best

Exponential expressions with negative bases can cause trouble. To see how, consider the expression $(-8)^x$ for various values of x:

$$(-8)^2 \qquad (-8)^{-2} \qquad (-8)^{1/2} \qquad (-8)^{1/3} \qquad (-8)^{3/2} \qquad (-8)^{4/3}$$

The first two make good sense:

$$(-8)^2 = 64 \qquad \text{and} \qquad (-8)^{-2} = \frac{1}{64}.$$

What is this value?

The third, $\sqrt{-8}$, isn't a real number, but the fourth, $(-8)^{1/3}$, has a real value. ◄ The fifth expression is meaningless (it involves $\sqrt{-8}$) but the sixth makes good sense:

$$(-8)^{4/3} = (-8) \cdot (-8)^{1/3} = (-8) \cdot (-2) = 16.$$

Faced with such difficulties, we retreat. In calculus, we'll take the base b of every exponential expression b^x to be positive.

Transcendental Meditations (Optional)

The *exponential* function $f(x) = 2^x$ differs from the *power* function $g(x) = x^2$ far more than typography would suggest. One important difference is that the expression x^2 is an algebraic formula (i.e., an explicit recipe for computing output values). If, say, $x = 2.3$, it's easy, even without a calculator, to compute the output: $2.3^2 = 5.29$.

Do you see why? Try it.

A curious word. It suggests that exponential functions somehow float above, or transcend, the mundane world of algebra.

The expression 2^x, by contrast, is not algebraic. Finding the output value $2^{2.3}$ by hand is almost impossible. ◄ For this reason, exponential functions (and other "nonalgebraic" functions) are referred to as **transcendental**. ◄ We explore some ramifications of the nonalgebraic character of exponential functions that follow.

Fractional and Irrational Exponents: Abuse of Powers?

What does b^x really mean if the exponent x isn't an integer? What if x is irrational?

EXAMPLE 6   Find (or estimate) decimal values for

$$2^{1/3}, \qquad 1.6^{11/13}, \qquad \text{and} \qquad \pi^{-2/3}.$$

Solution $2^{1/3}$ is the *cube root* of 2. According to any scientific calculator,

$$\sqrt[3]{2} \approx 1.25992104989.$$

For *any* integer $n > 0$,

$$b^{1/n} = \sqrt[n]{b}.$$

The connection between roots ▶ and powers suggests how to estimate $1.6^{11/13}$ numerically:

$$1.6^{11/13} = \left(1.6^{1/13}\right)^{11} \approx 1.03681563373^{11} \approx 1.4883905652.$$

Is it obvious that every positive number b has a positive nth root? How would one estimate, say, the 11th root of 13?

Similarly,

$$\pi^{-2/3} = \frac{1}{\pi^{2/3}} = \frac{1}{\sqrt[3]{\pi^2}} \approx \frac{1}{\sqrt[3]{9.869604401089}}$$

$$\approx \frac{1}{2.14502939711} \approx 0.466194077036. \qquad ■$$

So far so good: The examples show how to handle *rational* exponents. How about *irrational* exponents?

EXAMPLE 7   Estimate the value of $\pi^{\sqrt{2}}$.

Solution   The "obvious" approach—to approximate everything in sight with rational numbers—works: ▶

$$\pi^{\sqrt{2}} \approx 3.1415^{1.4142} \approx 5.0472. \qquad ■$$

Decimal numbers are rational, aren't they?

Fair Questions.   Does all this approximation really work? Is it really OK to approximate *both* b and x with convenient decimal numbers before computing b^x? The short answer to all of these good questions is "yes." In mathematical language, the fortunate fact is that

$$\text{if } a \approx b \text{ and } x \approx y, \text{ then } b^x \approx a^y.$$

Reasonable as this seems, a rigorous proof is subtle; we'll omit it.

BASICS

1. Evaluate 64^x if x equals
   (a) 0
   (b) 1
   (c) 1/2
   (d) $-1/2$
   (e) 2/3
   (f) 5/6

2. Express each of the following numbers as 4^x for a suitable exponent x.
   (a) 16
   (b) 2
   (c) 8
   (d) $1/\sqrt{2}$
   (e) $4\sqrt{2}$
   (f) 16^{35}

3. Let a be a positive number. Rewrite each of the following expressions in the form a^x.
   (a) $(\sqrt{a})^2$
   (b) $1/a^2$
   (c) 1
   (d) a^3/a^4
   (e) $(a^3/a^5)^{10}$
   (f) $a/\sqrt[3]{a}$

4. Here's further evidence for the fact that $b^0 = 1$:
   $$b^0 \cdot b^0 = b^{0+0} = b^0.$$
   Therefore, b^0 *is its own square*. Does it follow that b^0 must be 1? Explain.

5. Let f be the function $f(x) = 5^{2x}$. Show that
   $$\frac{f(x+h) - f(x)}{h} = f(x) \cdot \frac{f(h) - f(0)}{h}.$$

In Exercises 6–8, mimic the arguments given for the "addition" rule in Example 1 to show that the rule is "obvious" for positive integer powers.

6. Rule (E.2)

7. Rule (E.3)

8. Rule (E.4)

9. Let $g(x) = Ae^{0.2x}$ where A is a real number.
   (a) Suppose that $g(0) = 5$. Find A.
   (b) Suppose that $g(1) = 5$. Find A.
   (c) Suppose that $g(-1) = 5$. Find A.

10. Describe how the graph of each of the following functions is geometrically related to the graph of $y = 2^x$. [**EXAMPLE:** The graph of $y = 2^{-x}$ is the reflection of the graph of $y = 2^x$ about the y-axis.]
    (a) $(1/2)^x$
    (b) $(1/2)^{-x}$
    (c) -2^{-x}
    (d) 2^{x-1}
    (e) $2 - 2^x$
    (f) $2 \cdot 2^{x-1}$

11. Find an equation for the graph that results from each of the following operations on the graph of $y = e^x$.
    (a) Reflection about the x-axis.
    (b) Reflection about the y-axis.
    (c) Reflection about the line $y = 4$.
    (d) Reflection about the line $x = -1$.
    (e) Reflection about the y-axis, then a reflection of the resulting graph about the x-axis.

12. (a) Explain how to obtain the graph of $y = 2e^{x+3}$ from the graph of $y = e^x$ using *only* vertical stretching.
    (b) Explain how to obtain the graph of $y = 2e^{x+3}$ from the graph of $y = e^x$ using *only* horizontal translation.
    (c) Can the graph of $y = 2e^{4x+3}$ be obtained from the graph of $y = e^x$ using only the operations of vertical stretching and horizontal translation? Explain.

Algebra of Logarithms

Logarithm functions (i.e., functions defined by logarithmic expressions) are important in both the theory and the applications of calculus. To understand and use such functions effectively requires a basic familiarity with the algebra of such expressions.

Logarithms and Exponentials: How They're Related

Logarithmic expressions like these—

$$\log_2 x, \qquad \log_{10} t, \qquad \log_e z, \qquad \ln w, \qquad \text{and} \qquad \log_b x$$

may be mystifying at first glance. What does "$\log_b$" *mean*? What is the role of b? What is "ln?" Some mystification is natural: Like "sin," "cos," and "exp," "$\log_b$" is the *name* of a function—not a recipe for computation.

Demystifying logarithms means, first and foremost, understanding how they're related to exponentials. For us, this relation *defines* the idea of logarithm.

> **Definition** For any suitable **base** b and positive number x, the expression $\log_b x$ is defined by the condition
> $$y = \log_b x \iff x = b^y.$$
> (NOTE: "Suitable" means $b > 0$ and $b \neq 1$.)

EXAMPLE 1 Use the definition to simplify the expressions

$$\log_2 8, \qquad \log_{10} 100, \qquad \log_e e^{25}, \qquad \text{and} \qquad \log_b \frac{1}{b^3}.$$

Solution   By definition,

$$y = \log_2 8 \quad \Longleftrightarrow \quad 8 = 2^y.$$

It's clear from the second equation that y must be 3; hence $\log_2 8 = 3$. By similar reasoning,

$$y = \log_{10} 100 \iff 10^y = 100 \iff y = 2;$$

$$y = \log_e e^{25} \iff e^y = e^{25} \iff y = 25;$$

$$y = \log_b \frac{1}{b^3} \iff b^y = b^{-3} \iff y = -3.$$ ■

Logarithm and Exponential Functions as Inverses

The relation

$$y = \log_b x \iff x = b^y$$

means that the logarithm function $\log_b x$ and the exponential function b^x are *inverses*. For any two functions f and g to be inverses means that each function "undoes" the other. In other words, the equations

$$f\big(g(x)\big) = x \qquad \text{and} \qquad g\big(f(x)\big) = x$$

hold for all sensible inputs x. This general fact, applied specifically to the logarithm and exponential functions with base b, looks like this:

$$\log_b b^x = x \qquad \text{and} \qquad b^{\log_b x} = x.$$

For more on domains, see Section 1.5.

These equations hold for all x in the respective domains of $\log_b$ and b^x. ◀

Basics of Logarithmic Algebra

A few algebraic rules for logarithmic expressions go a long way. Here are four to know:

$$\log_b xy = \log_b x + \log_b y \tag{F.1}$$

$$\log_b x/y = \log_b x - \log_b y \tag{F.2}$$

$$\log_b x^r = r \log_b x \tag{F.3}$$

$$\log_b 1/x = -\log_b x \tag{F.4}$$

(Note: $b > 0$ and $b \neq 1$.)

Using the Rules

The rules simplify computations that involve logarithms—and even some that don't seem to.

EXAMPLE 2   Just before it died, your calculator said that $\log_{10} 2 \approx 0.301$ and $\log_{10} 3 \approx 0.477$. You *really* need to know

$$\log_{10} 6, \qquad \log_{10} 1/18, \qquad \log_{10} 2000, \qquad \text{and} \qquad \log_{10} 0.003 \, .$$

What can you do?

Solution   With the rules there's no problem: ▶

$$\log_{10} 6 = \log(3 \cdot 2) = \log_{10} 3 + \log_{10} 2 \approx 0.301 + 0.477 = 0.778$$

$$\log_{10} 1/18 = -\log_{10}(2 \cdot 3^2) = -\log_{10} 2 - 2\log_{10} 3 \approx -1.255$$

$$\log_{10} 2000 = \log_{10} 1000 + \log_{10} 2 = 3 + \log_{10} 2 \approx 3.301$$

$$\log_{10} 0.003 = \log_{10} 0.001 + \log_{10} 3 = -3 + \log_{10} 3 \approx -2.523 \qquad ■$$

But watch closely, and check the results with your own (live) calculator.

EXAMPLE 3   Solve (for x) each of the equations

$$200 \cdot 1.05^x = 375 \qquad \text{and} \qquad \ln x + \ln x^2 = 5 \, .$$

Solution   For the first equation we'll take the natural logarithm of both sides, and then use Rules (F.1) and (F.3):

$$200 \cdot 1.05^x = 375 \iff \ln 200 + \ln(1.05^x) = \ln 375$$

$$\iff x \ln 1.05 = \ln 375 - \ln 200$$

$$\iff x = \frac{\ln 375 - \ln 200}{\ln 1.05} \approx 12.8839$$

To solve the second equation we'll first use Rules (F.1) (in "reverse") and (F.3), and then apply an exponential function to undo the logarithm. Notice first that

$$\ln x + \ln x^2 = \ln x^3 = 3 \ln x \, .$$

Therefore,

$$\ln x + \ln x^2 = 5 \iff 3 \ln x = 5$$

$$\iff \ln x = 5/3$$

$$\iff x = e^{5/3} \approx 5.2945 \, . \qquad ■$$

Why the Rules Hold: Translating between Logarithms and Exponentials

The familiar properties of logarithms listed above are, in a sense, disguised versions of similar rules for exponentials. Properties of logarithms can be "translated" into properties of exponentials, and vice versa.

EXAMPLE 4   Explain, in terms of exponentials, Rule (F.1):

$$\log_b xy = \log_b x + \log_b y \, .$$

Solution   What does Rule (F.1) really say? The equation $\log_b xy = z$—where z is *any* real number—means that $b^z = xy$. Therefore Rule (F.1) really says that

$$b^{\log_b x + \log_b y} = xy \, .$$

Now exponential algebra comes in:

$$b^{\log_b x + \log_b y} = b^{\log_b x} \cdot b^{\log_b y} = xy,$$

Why does the last equation hold? Because logarithms and exponentials are inverses.

just as we wanted. ◄

Which Base?

The general ideas of logarithm and exponential functions are the same for *any* base b. In practice, though, either $b = 10$ or $b = e \approx 2.71828$ is almost invariably used. Logarithms with these bases are known as the **common logarithm** and the **natural logarithm**, respectively. The natural log function, usually written as $\ln x$, is for many good reasons ◄ the favorite in calculus. We'll use base e almost exclusively.

Several good reasons appear in Section 3.3.

Moving Around the Bases: Two Key Identities

To exploit the advantages of base e—for both logarithm and exponential functions—we'll need to move easily from *any* base b to the special base e. Two key identities tell how. For any base b and input x,

$$b^x = e^{x \cdot \ln(b)};$$

$$\log_b(x) = \frac{\ln(x)}{\ln(b)}.$$

Why do these equations hold? Both follow from the fact that logarithm and exponential functions are inverses, and from basic properties of logarithms. For the first:

$$b^x = \left(e^{\ln(b)}\right)^x = e^{x \cdot \ln(b)},$$

Why does $b = e^{\ln(b)}$?

as claimed. ◄ The second identity is similar:

$$x = b^{\log_b x} \implies \ln(x) = \ln\left(b^{\log_b x}\right) = \log_b x \, \ln b$$

$$\implies \log_b(x) = \frac{\ln(x)}{\ln(b)}.$$

EXAMPLE 5 Rewrite the exponential function 3^x in base e.

Solution From the first identity above,

$$3^x = e^{x \ln(3)} \approx e^{1.09861228867x}.$$

EXAMPLE 6 Rewrite $\log_5 x$ in terms of the *natural* logarithm.

Solution From the second identity above,

$$\log_5 x = \frac{\ln x}{\ln 5} \approx \frac{\ln x}{1.60943791243}.$$

BASICS

1. Translate each of the following equations into the language of logarithms.
   (a) $8^{2/3} = 4$
   (b) $10^3 = 1000$
   (c) $10^{-4} = 0.0001$
   (d) $(1/2)^2 = 4$

2. Translate each of the following equations into the language of exponentials.
   (a) $\log_2 7 = x$
   (b) $\log_5 1/25 = -2$
   (c) $\log_{10} 1000 = 3$
   (d) $\log_{64} 128 = 7/6$
   (e) $\log_{1/2} (1/4) = 2$

3. Evaluate each of the following expressions *exactly*.
   (a) $2^{\log_2 16}$
   (b) $2^{\log_2 1/2}$
   (c) $10^{\log_{10} 7}$
   (d) $\log_6 \sqrt{6}$
   (e) $\log_5 5^3$
   (f) $\log_3 (1/27)$

4. If $\log_4 A = 2.1$, evaluate each of the following expressions *exactly*.
   (a) $\log_4 A^2$
   (b) $\log_4 (1/A)$
   (c) $\log_4 16A$
   (d) $\log_2 A$
   (e) $\log_{16} A$

5. Interpret Rules (F.2) and (F.3) in exponential language.

6. Explain why $\ln(x^2 - x - 2) = \ln(x - 2) + \ln(x + 1)$.

7. (a) Explain how to obtain the graph of $y = 2 + \ln(3x)$ from the graph of $y = \ln x$ using *only* vertical translation.
   (b) Explain how to obtain the graph of $y = 2 + \ln(3x)$ from the graph of $y = \ln x$ using *only* horizontal stretching.
   (c) Can the graph of $y = 2 + \ln(3x + 4)$ be obtained from the graph of $y = \ln x$ using *only* the operations of vertical translation and horizontal stretching? Explain.

8. Let $f(x) = 20e^{0.1x}$.
   (a) Find an x for which $f(x) = 20$.
   (b) Find an x for which $f(x) = 40$.
   (c) Find an x for which $f(x) = 80$.
   (d) Find an x for which $f(x) = 10$.

9. Let $h(x) = 20e^{kx}$ where k is a real number.
   (a) Suppose that $h(1) = 20$. Find k.
   (b) Suppose that $h(1) = 40$. Find k.
   (c) Suppose that $h(1) = 10$. Find k.

10. Let $j(x) = Ae^{kx}$ where A and k are real numbers.
    (a) Suppose that $j(0) = 5$ and $j(3) = 10$. Find A and k.
    (b) Suppose that $j(1) = 5$ and $j(3) = 10$. Find A and k.

11. In each part below, plot *both* graphs on the same axes. Explain what you see, referring to properties of exponential and logarithm functions.
    (a) $y = 2^x$, $y = 4^{x/2}$
    (b) $y = \ln(\exp(x^2))$, $y = x^2$
    (c) $y = \ln((\exp x)^2)$, $y = 2x$
    (d) $y = \exp(\ln x)$, $y = x$
    (e) $y = \exp(2 \ln x)$, $y = x^2$
    (f) $y = \exp(\frac{1}{2} \ln x)$, $y = \sqrt{x}$

Trigonometric Functions

The six trigonometric functions are introduced and defined in Chapter 1, Section 5. Start there for definitions and basic properties. In this appendix we collect a variety of information, in more detail, on trigonometric functions, their properties, their graphs, and their importance in calculus.

Trigonometric Function Notation

And a few others.

Trigonometric functions ◄ are often typeset without the customary $f(x)$-style parentheses. It's standard to write

$$\sin t, \qquad \cos t, \qquad \text{and} \qquad \tan t$$

rather than

$$\sin(t), \qquad \cos(t), \qquad \text{and} \qquad \tan(t).$$

But not slavishly.

We'll usually ◄ follow this convention, although a bit reluctantly—parentheses remind us that *functions* are involved. Whenever confusion seems possible (e.g., with the expression $\sin(t + 1)$) or we want to emphasize the functional point of view, we'll use parentheses.

Other quirks of trigonometric notation involve *powers* of trigonometric functions. It's conventional, for instance, to write

Other powers are handled similarly—e.g., $\sin^{17} x$ means $(\sin x)^{17}$.

$$\sin^2 t + \cos^2 t = 1 \qquad \text{instead of} \qquad \big(\sin(t)\big)^2 + \big(\cos(t)\big)^2 = 1,$$

although both mean the same thing. ◄ Neither of these equations should *ever*, of course, be confused with

$$\sin t^2 + \cos t^2 = 1 \quad \text{or} \quad \sin(t^2) + \cos(t^2) = 1,$$

which are equivalent to each other, but *false*.

Sines, Cosines, Angles, and Triangles

Any real number is a legitimate input to the sine or cosine function. We've interpreted such an input as the *signed length of an arc*. (For more on this interpretation, see the definitions on page 62.) Thus $\cos(2) \approx -0.42$, because -0.42 is (approximately) the x-coordinate of the point 2 *units of distance* counterclockwise from $(1, 0)$ on the unit circle. (The same units of distance are used both on the circle and on the axes.) Here's the picture for $\cos(2)$: ▶

Convince yourself that all the quantities shown are plausible.

Why cos(2) ≈ –0.42

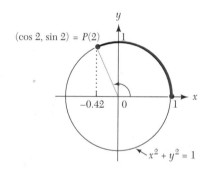

Sometimes we prefer to interpret inputs to the sine and cosine functions as *angles*, not arclengths. Any arc on the unit circle determines an obvious angle: the one between the two line segments joining the origin to the ends of the arc. (See the picture above.) From this point of view, $\cos 2$ is the cosine of the *angle* determined by an arc of length 2.

From Radians to Degrees and Back.   Two standard units (**radians** and **degrees**) are used to measure angles. Radian measure will prove far more convenient for us; we'll use it consistently. ▶ To convert between radians and degrees, use these formulas:

So should you. So should your calculator. Lock it in radian mode and throw away the key.

Radians ⇄ Degrees	
radian measure	$= \dfrac{\pi}{180} \times$ degree measure
degree measure	$= \dfrac{180}{\pi} \times$ radian measure

(Both follow from the fact that 2π radians $= 360°$.) In particular, the angle shown is

$$2 \text{ radians} = 360/\pi \approx 114.6°.$$

Right Triangles. "Trigonometry" refers, etymologically, to *measuring triangles*. Our "circular" viewpoint on trigonometric functions is easily linked to right triangles; doing so sometimes aids intuition.

If $P(t)$ lies in the first quadrant, i.e., if $0 \le t \le \pi/2$, then $P(t)$ naturally determines, as shown below, a right triangle with unit hypotenuse, one of whose angles measures t radians:

**Triangles and
trigonometric functions**

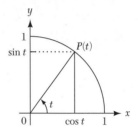

It's easy to see that this triangle has the "right" trigonometric properties. For example:

$$\sin t = \frac{\text{opposite}}{\text{hypotenuse}} = \frac{y}{1} = y; \qquad \cos t = \frac{\text{adjacent}}{\text{hypotenuse}} = \frac{x}{1} = x.$$

Famous Sines and Cosines

*If it isn't, review the
definitions to convince
yourself. It may also help to
review the definition of
"evident."*

*We've already violated our
prohibition against degree
measure! Maybe we should
have said $\pi/6$–$\pi/3$–$\pi/2$ and
$\pi/4$–$\pi/4$–$\pi/2$.*

It's evident ◄   from the definitions of the sine and cosine functions that $\sin 0 = 0$ and $\cos(\pi/2) = 0$. Other important sines and cosines, such as $\sin(\pi/4) = \sqrt{2}/2$ and $\cos(\pi/3) = 1/2$ are well known but less obvious. All the well-known values can be computed as ratios of sides of 30–60–90 and 45–45–90 right triangles. ◄   Know these important values:

Celebrated Sines and Cosines					
t	0	$\pi/6$	$\pi/4$	$\pi/3$	$\pi/2$
$\cos t$	1	$\sqrt{3}/2$	$\sqrt{2}/2$	$1/2$	0
$\sin t$	0	$1/2$	$\sqrt{2}/2$	$\sqrt{3}/2$	1

In daily practice one needs to know sines and cosines of *all* integer multiples of $\pi/4$ and $\pi/6$. This is easier than it seems. Using the values tabulated above for angles in the first quadrant we can use the symmetry of the unit circle to find sines and cosines of many other angles.

EXAMPLE 1   Use symmetry to find $\sin(7\pi/6)$ and $\cos(7\pi/6)$.

Solution   The following diagram

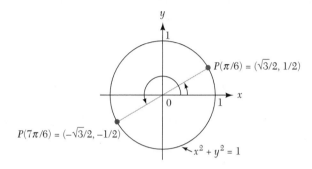

Exploiting symmetry

shows that

$$\sin\left(\frac{7\pi}{6}\right) = -\sin\left(\frac{\pi}{6}\right) = -\frac{1}{2} \quad \text{and} \quad \cos\left(\frac{7\pi}{6}\right) = -\cos\left(\frac{\pi}{6}\right) = -\frac{\sqrt{3}}{2}. \quad \blacksquare$$

Obscure Sines and Cosines

For most inputs, precise values of the trigonometric functions are far from obvious. For example, what are $\sin(1)$ and $\cos(13)$? We could estimate values graphically using a unit circle, but, as it turns out, there is *no* simple algebraic formula for computing *exact* values of trigonometric functions. Fortunately, calculus offers methods of *approximating* values of these functions.

Other Trigonometric Functions

Four additional trigonometric functions are defined in terms of the sine and cosine functions.

Definition For real numbers t,

$$\tan t = \frac{\sin t}{\cos t} \qquad \cot t = \frac{\cos t}{\sin t}$$

$$\sec t = \frac{1}{\cos t} \qquad \csc t = \frac{1}{\sin t}$$

Graphs of all these functions appear on page 64, in Section 1.5.

How Many Trigonometric Functions?   Cosine, sine, tangent, secant, cosecant, and cotangent make six. From another point of view, there are only two—sine and cosine—from which all the others are derived.

If we were serious about this numbers game, we could count at least *ten* trigonometric functions: The familiar six and their (justifiably) obscure relatives *exsecant*, *versine*, *coversine*, and *haversine*. Taking the opposite tack, we might argue for only *one* trigonometric function, say cosine, from which sine, and hence all the others, can be derived.

In the end, the number of trigonometric functions is a matter more philosophical than mathematical. What's important is that, whatever their number, *all* trigonometric functions are closely related to each other and to the unit circle.

More on the Tangent Function

Among the "other" trigonometric functions, the tangent function arises often enough to deserve special attention. To start, here's a close look at its graph. Notice all the asymptotes, shown dotted:

Graph of $y = \tan x$

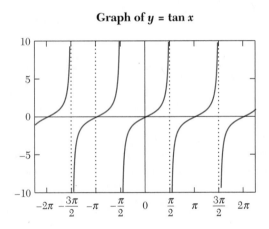

Many properties of the tangent function follow easily from its definition:

Domain and Range   The tangent function accepts all real inputs t *except* those for which $\cos t = 0$ (i.e., $t = \pm\pi/2, \pm3\pi/2, \pm5\pi/2$, and so on). The domain of the tangent function, therefore, consists of all real numbers *other* than these obvious offenders. The graph reflects these troubles. The range is the full set of real numbers. ◄

The graph shows this, too.

Periodicity   The tangent function (like the sine and cosine functions) repeats itself on intervals of length 2π. Here's why:

$$\tan(t + 2\pi) = \frac{\sin(t + 2\pi)}{\cos(t + 2\pi)} = \frac{\sin t}{\cos t} = \tan t .$$

Look at it. Do you agree?

The graph reveals the same property. ◄

More on Periodicity—A Surprise Look again. The tangent function's graph repeats its basic shape on intervals of length π—*twice as often* as the computation above suggests. In other words, the tangent function appears to be π-periodic. Is something wrong?

No. Every π-periodic function is *also* 2π-periodic, automatically. The tangent function repeats itself on intervals of length π for the following reason:

$$\tan(t + \pi) = \frac{\sin(t + \pi)}{\cos(t + \pi)} = \frac{-\sin t}{-\cos t} = \tan t.$$

The second equality is the key. It holds because $\sin(t + \pi) = -\sin t$ and $\cos(t + \pi) = -\cos t$. ▶

Are these equalities obvious? If not, draw the points $P(t)$ and $P(t + \pi)$ on the unit circle. How are they related?

Trigonometric Identities

Equations like

$$\sin^2 t + \cos^2 t = 1, \qquad \sin(t + \pi/2) = \cos t, \qquad \text{and} \quad \tan^2 t + 1 = \sec^2 t,$$

that relate various trigonometric functions to each other are known as **trigonometric identities**. A great many such relations exist—not surprisingly, because all six trigonometric functions spring from the same unit circle idea.

Some trigonometric identities, like the first one above, follow directly from the definitions. Others are far subtler. Memorizing every trigonometric identity is out of the question. It's wiser to memorize a few truly basic identities. ▶

Other—less obvious but still useful—trigonometric identities exist; one should know *that* they exist and *where* to find them.   Below are several such second-rank identities.

The three above enjoy this exalted status.

Fact (Addition Formulas)

$$\sin(s + t) = \sin s \cos t + \cos s \sin t$$

$$\cos(s + t) = \cos s \cos t - \sin s \sin t$$

(**Double-Angle Formulas**)

$$\sin(2t) = 2 \sin t \cos t \qquad\qquad \cos(2t) = \cos^2 t - \sin^2 t$$

$$\sin^2 t = \frac{1}{2}\big(1 - \cos(2t)\big) \qquad \cos^2 t = \frac{1}{2}\big(1 + \cos(2t)\big)$$

The double-angle formulas are junior-grade versions of the addition formulas; the former are readily derived from the latter.

EXAMPLE 2   Derive the formula for $\cos(2t)$.

Solution   Using the addition formula,

$$\cos(2t) = \cos(t + t) = \cos t \cos t - \sin t \sin t = \cos^2 t - \sin^2 t.$$ ■

BASICS

1. The angle drawn within the unit circle shown below measures 1.2 radians. The tick marks around the circumference of the circle indicate tenths of a radian.

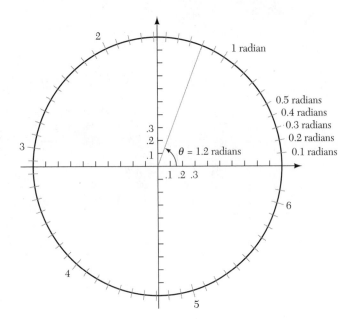

Use this diagram to estimate each of the following values:
(a) $\cos(1.2)$ (d) $\cos(0.2)$
(b) $\sin(0.5)$ (e) $\sin(4.5)$
(c) $\tan(2.8)$ (f) $\cos(13)$

2. Use the diagram above to estimate each of the following. (Then check your answers with a calculator.)
(a) Two values of x such that $\sin x = 1/2$.
(b) Two values of x such that $\tan x = 5$.

3. Let $P(t)$ be the point t units counterclockwise from $(1, 0)$ on the unit circle. Explain in words:
(a) How $P(t)$ and $P(t + \pi)$ are related.
(b) Why $\cos(t + \pi) = -\cos(t)$.
(c) Why $\sin(t + \pi) = -\sin(t)$.
(d) Why $\tan(t + \pi) = \tan(t)$.

4. Find the domain of each of the following functions.
(a) $\cot x$ (b) $\sec x$

5. Graphs of $y = \sin x$ and $y = \cos x$ are shown below.
(a) Which graph is which? Explain.
(b) It's a fact that $\cos x = \sin(x + \pi/2) \approx \sin(x + 1.57)$. How do the graphs reflect this fact?
(c) For which values of x is $\tan x = 0$? How can you tell from the graphs above?
(d) For which values of x is $\tan x = 1$? How can you tell from the graphs above?

(e) Draw the graphs of $y = \sin x$ and $y = 3 \sin x$ over the interval $[-2\pi, 2\pi]$ on the same axes.

Graphs of $y = \sin x$ and $y = \cos x$

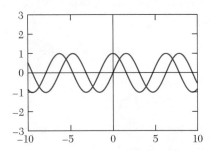

6. Suppose that $\sin a = b$ and $\cos c = 0.3$. Evaluate each of the following expressions. (Your answers should involve only real numbers and the constant b.)
(a) $\sin(-a)$ (d) $\sin c$
(b) $-\sin a$ (e) $\sin^2 a$
(c) $\cos(-c)$ (f) $\sin(2a)$

7. Evaluate each of the following expressions *exactly*. (No decimal approximations allowed!)
(a) $\cos(5\pi/6)$ (e) $\sec(5\pi/3)$
(b) $\sin(2\pi/3)$ (f) $\csc(7\pi/4)$
(c) $\tan(3\pi/4)$ (g) $\sin(257\pi/3)$
(d) $\cot(4\pi/3)$ (h) $\tan(-2000\pi/3)$

8. Use the addition formulas for sine and cosine to evaluate each of the following expressions *exactly*.
(a) $\cos\left(\dfrac{5\pi}{12}\right) = \cos\left(\dfrac{\pi}{4} + \dfrac{\pi}{6}\right)$

(b) $\sin\left(\dfrac{19\pi}{12}\right) = \sin\left(\dfrac{3\pi}{4} + \dfrac{5\pi}{6}\right)$

(c) $\sin\left(\dfrac{\pi}{12}\right) = \sin\left(\dfrac{\pi}{4} - \dfrac{\pi}{6}\right)$

(d) $\tan\left(\dfrac{7\pi}{12}\right) = \tan\left(\dfrac{5\pi}{4} - \dfrac{2\pi}{3}\right)$

9. Which of the following pairs of expressions define the same function?
(a) $\sec^2 x$ and $1 + \tan^2 x$
(b) $\cos^2 x$ and $1 - \sin^2 x$
(c) $\cos x$ and $\sin(x - \pi/2)$
(d) $\sin x$ and $\cos(x + \pi/2)$

10. Find subtraction formulas for sine and cosine similar to the addition formulas.
(a) $\sin(s - t) = ???$ (b) $\cos(s - t) = ???$

11. Derive the formula $\sin(2t) = 2 \sin t \cos t$.

12. Derive the formula $\cos^2 t = \dfrac{1}{2}\left(1 + \cos(2t)\right)$.

Selected Proofs

This appendix contains careful proofs of various results, more or less in order of subtlety. The proofs are intended to be read "as needed," as particular questions arise. Alternatively, some readers may enjoy reading proofs for their own sake, as an introduction to the formal ideas and methods of analytic mathematics.

A Challenging Limit

Many limits that arise in calculus courses can be evaluated either by "inspection" (i.e., at a glance) or with elementary algebra. All of the following limits are of these types:

$$\lim_{x \to \infty} \frac{1}{x} = 0; \qquad \lim_{x \to 3}(x + 2) = 5; \qquad \lim_{x \to 2} \frac{x^2 - 4}{x - 2} = 4.$$

Only the last one requires any argument, and simple algebra is all that's needed:

$$\lim_{x \to 2} \frac{x^2 - 4}{x - 2} = \lim_{x \to 2} \frac{(x - 2)(x + 2)}{x - 2} = \lim_{x \to 2}(x + 2) = 4.$$

The important ▶ limit

$$\lim_{t \to 0} \frac{\sin t}{t} = 1$$

One of the 2 or 3 most important in calculus.

is different. Because $\sin t$ is *not* an algebraic expression, the usual proof tricks—factoring, expansion, and so on—are useless. A rigorous proof must take another tack.

Before beginning a rigorous proof, we remark that overwhelming circumstantial evidence—in graphical and numerical forms—that $\sin t / t \to 1$ is available. (See, e.g., Example 4, page 152.) Still, pictures and numbers aren't proofs. On the contrary, what a computer or calculator "understands" about the sine function probably depends implicitly on some knowledge of the very limit in question.

A Proof. Because $\sin t$ and t are both *odd* functions of t, their ratio is *even*. Thus it's enough to consider only positive values of t, i.e., to show that

$$\lim_{t \to 0^+} \frac{\sin t}{t} = 1.$$

The following unit-circle diagram describes the situation:

Why sin t ≈ t

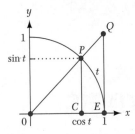

The diagram supports three key claims:

i. $\sin t \le t$; (the segment CP is shorter than the arc EP)

ii. $t \le \sin t + 1 - \cos t$; (the arc EP is shorter than EC+CP—the route traveling *west*, then *north*)

iii. as $t \to 0^+$, $\dfrac{1 - \cos t}{t} \to 0$.

Claim (iii) requires some support. Here's an an airtight proof, based on (i):

$$\frac{1 - \cos t}{t} = \frac{1 - \cos t}{t} \cdot \frac{1 + \cos t}{1 + \cos t} = \frac{1 - \cos^2 t}{t(1 + \cos t)} = \frac{\sin^2 t}{t(1 + \cos t)}$$

$$= \frac{\sin t}{t} \cdot \frac{\sin t}{1 + \cos t} \le \frac{\sin t}{1 + \cos t}.$$

Now it follows that

$$0 \le \frac{1 - \cos t}{t} \le \frac{\sin t}{1 + \cos t}.$$

As $t \to 0^+$, the right side tends to 0. By the squeeze principle, (iii) holds as claimed.

To finish our proof we'll use the squeeze principle again. From (i) and (ii) we have

$$t - 1 + \cos t \le \sin t \le t.$$

Dividing through by t gives

$$\frac{t - 1 + \cos t}{t} = 1 - \frac{1 - \cos t}{t} \le \frac{\sin t}{t} \le 1.$$

Taking limits gives

$$1 - \lim_{t \to 0^+} \frac{1 - \cos t}{t} \le \lim_{t \to 0^+} \frac{\sin t}{t} \le 1.$$

From (iii), the first limit is 1; we're done.　■

Another Proof.　Another "squeeze" proof of the same limit can be based on areas. The same picture shows the following area relation:

$$\text{area(OCP)} \le \text{area(OEP)} \le \text{area(OEQ)}.$$

Here OCP and OEQ are triangles; OEP is a circular sector. Calculating these three areas gives

$$\text{area}(OCP) = \frac{\text{base} \times \text{height}}{2} = \frac{\cos t \ \sin t}{2};$$

$$\text{area}(OEP) = \pi \times \text{fraction of circle} = \pi \times \frac{t}{2\pi} = \frac{t}{2};$$

$$\text{area}(OEQ) = \frac{\text{base} \times \text{height}}{2} = \frac{\tan t}{2}.$$

Thus,

$$\frac{\cos t \ \sin t}{2} \le \frac{t}{2} \le \frac{\tan t}{2}.$$

If $t > 0$, then multiplying through by $2/\sin t$ ▶ gives   *A positive number!*

$$\cos t \le \frac{t}{\sin t} \le \frac{1}{\cos t}.$$

As $t \to 0^+$ the left and right quantities tend to 1; so, therefore, does the middle.   ■

Proving the Product Rule

Recall the product rule:

Fact (**The Product Rule**) Suppose that u and v are differentiable at x. If $p(x) = u(x) \cdot v(x)$, then p is differentiable at x, and

$$p'(x) = \big(u(x) \cdot v(x)\big)' = u'(x) \cdot v(x) + u(x) \cdot v'(x).$$

In Section 3.5 we argued the case informally. Below is a formal proof; notice that it shows—rather than assumes—that the product uv is differentiable.

Proof.   Let u and v be differentiable functions. By definition,

$$(uv)'(x) = \lim_{h \to 0} \frac{u(x+h)v(x+h) - u(x)v(x)}{h}.$$

We need to show that this limit exists, and equals $u'(x)v(x) + u(x)v'(x)$.

An algebraic trick—subtracting and adding the same quantity in the numerator—does the job: ▶   *Check each step.*

$$(uv)' = \lim_{h \to 0} \frac{u(x+h)v(x+h) - u(x)v(x)}{h}$$

$$= \lim_{h \to 0} \frac{u(x+h)v(x+h) - u(x)v(x+h) + u(x)v(x+h) - u(x)v(x)}{h}$$

$$= \lim_{h \to 0} \frac{u(x+h)v(x+h) - u(x)v(x+h)}{h} + \lim_{h \to 0} \frac{u(x)v(x+h) - u(x)v(x)}{h}$$

$$= \lim_{h \to 0} \frac{u(x+h) - u(x)}{h} \cdot \lim_{h \to 0} v(x+h) + u(x) \cdot \lim_{h \to 0} \frac{v(x+h) - v(x)}{h}$$

$$= u'(x)v(x) + u(x)v'(x).$$

We're done.   ■

An $\epsilon - \delta$ Limit Proof

Recall the $\epsilon - \delta$ definition of limit:

> **Definition** Suppose that for every positive number ϵ, no matter how small, there is a positive number δ so that
>
> $$|f(x) - L| < \epsilon \qquad \text{whenever} \qquad 0 < |x - a| < \delta.$$
>
> Then
>
> $$\lim_{x \to a} f(x) = L.$$

Other limit properties, in the same spirit, appear in Theorem 1, page 168.

Recall, too, this plausible-seeming property: ◀

> **Fact** Suppose that
>
> $$\lim_{x \to a} f(x) = L \qquad \text{and} \qquad \lim_{x \to a} g(x) = M,$$
>
> where L and M are finite numbers. Then
>
> $$\lim_{x \to a} [f(x) + g(x)] = L + M.$$

A rigorous proof must use the rigorous definition. Here's how:

Proof. We need to show that for any number $\epsilon > 0$ there's a corresponding number $\delta > 0$ such that

$$0 < |x - a| < \delta \implies |(f(x) + g(x)) - (f(a) + g(a))| < \epsilon.$$

Let $\epsilon > 0$ be given. To find the desired δ, we'll first set $\epsilon_1 = \epsilon_2 = \epsilon/2$. Since ϵ_1 and ϵ_2 are positive, there exist, by our assumption, other positive numbers δ_1 and δ_2 such that

$$0 < |x - a| < \delta_1 \implies |f(x) - f(a)| < \epsilon/2;$$
$$0 < |x - a| < \delta_2 \implies |g(x) - g(a)| < \epsilon/2.$$

If we now set δ to be the smaller of δ_1 and δ_2, then we have what we want:

$$0 < |x - a| < \delta \implies 0 < |x - a| < \delta_1 \qquad \text{and} \qquad 0 < |x - a| < \delta_2,$$

so

$$|(f(x) + g(x)) - (f(a) + g(a))| \leq |f(x) - f(a)| + |g(x) - g(a)|$$
$$< \epsilon/2 + \epsilon/2$$
$$= \epsilon. \qquad \blacksquare$$

I

A Graphical Glossary of Functions

Graphical reasoning—representing functions and their derivatives graphically—is an invaluable skill in calculus. This appendix offers a brief "atlas" of representatives of the most important classes of calculus functions. Each function is shown together with its derivative.

Graph of $f(x) = x$

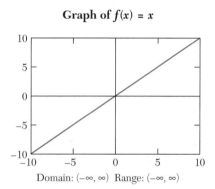

Domain: $(-\infty, \infty)$ Range: $(-\infty, \infty)$

Graph of $f'(x) = 1$

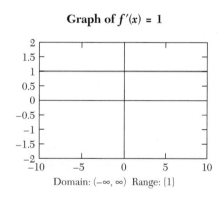

Domain: $(-\infty, \infty)$ Range: $\{1\}$

Notes. The derivative of a linear function is a constant function.

Graph of $f(x) = x^2$

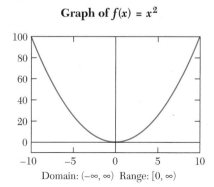

Domain: $(-\infty, \infty)$ Range: $[0, \infty)$

Graph of $f'(x) = 2x$

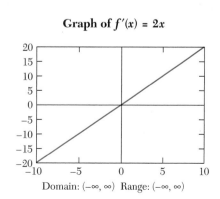

Domain: $(-\infty, \infty)$ Range: $(-\infty, \infty)$

Notes. The derivative of a quadratic function is a linear function. The quadratic function's vertex corresponds to a root of the derivative.

Graph of $f(x) = \dfrac{x}{x^2 - 9}$

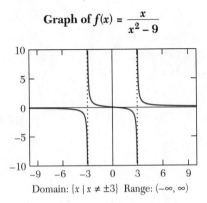

Domain: $\{x \mid x \neq \pm 3\}$ Range: $(-\infty, \infty)$

Graph of $f'(x) = -\dfrac{x^2 - 9}{(x^2 - 9)^2}$

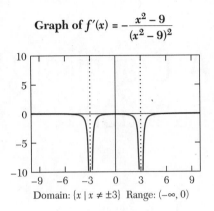

Domain: $\{x \mid x \neq \pm 3\}$ Range: $(-\infty, 0)$

Notes. The derivative of a rational function is another rational function. Notice the behavior of asymptotes: If f has a *vertical* asymptote at $x = a$, then so does f'. If f has a *horizontal* asymptote at $y = b$, then f' has a horizontal asymptote at $y = 0$.

Graph of $f(x) = \sin(x)$

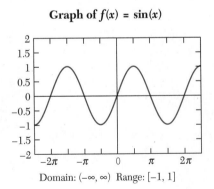

Domain: $(-\infty, \infty)$ Range: $[-1, 1]$

Graph of $f'(x) = \cos(x)$

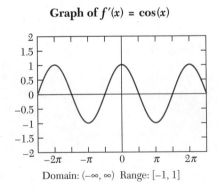

Domain: $(-\infty, \infty)$ Range: $[-1, 1]$

Notes. The sine function is 2π-periodic; so, therefore, is its derivative. Both the sine function and its derivative oscillate between -1 and 1. At the origin, the sine graph has slope 1.

Graph of $f(x) = \cos(x)$

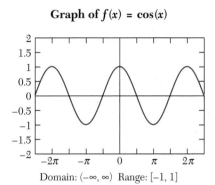

Domain: $(-\infty, \infty)$ Range: $[-1, 1]$

Graph of $f'(x) = -\sin(x)$

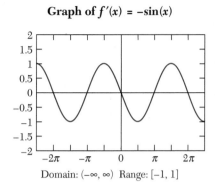

Domain: $(-\infty, \infty)$ Range: $[-1, 1]$

Notes.   The cosine function is 2π-periodic; so, therefore, is its derivative. Both the sine function and its derivative oscillate between -1 and 1. At the origin, the cosine graph is horizontal.

Graph of $f(x) = \tan(x)$

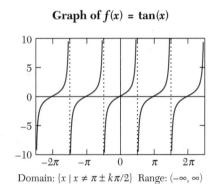

Domain: $\{x \mid x \neq \pi \pm k\pi/2\}$ Range: $(-\infty, \infty)$

Graph of $f'(x) = \sec^2(x)$

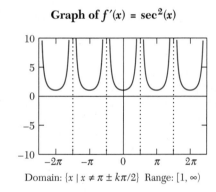

Domain: $\{x \mid x \neq \pi \pm k\pi/2\}$ Range: $[1, \infty)$

Notes.   The tangent function is π-periodic; so is its derivative. Unlike the sine and cosine functions, the tangent function and its derivative have vertical asymptotes. At the origin, the tangent graph has slope 1.

Graphs of $f(x) = e^x$ and $g(x) = 2^x$

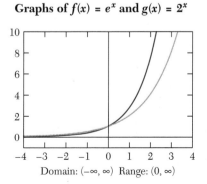

Domain: $(-\infty, \infty)$ Range: $(0, \infty)$

Graphs of $f'(x) = e^x$ and $g'(x) = 2^x \ln 2$

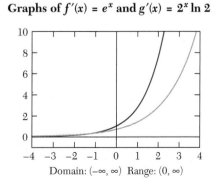

Domain: $(-\infty, \infty)$ Range: $(0, \infty)$

Notes. The natural exponential function $y = e^x$ is its own derivative. Other exponential functions, such as $y = 2^x$, are *proportional* to their own derivatives. The $y = e^x$ graph has slope 1 at $x = 0$; the $y = 2^x$ graph has slope $\ln 2$ at $x = 0$. All exponential functions have horizontal asymptotes; so do their derivatives.

Graph of $f(x) = \ln x$

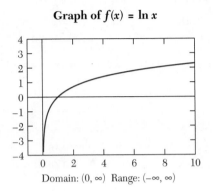

Domain: $(0, \infty)$ Range: $(-\infty, \infty)$

Graph of $f'(x) = 1/x$

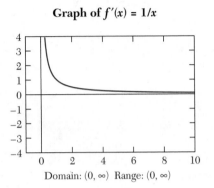

Domain: $(0, \infty)$ Range: $(0, \infty)$

Notes. The natural logarithm function $y = \ln x$ is the inverse function to $y = e^x$. It is defined only for positive inputs. The graph of $y = \ln x$ has slope 1 at $x = 1$.

Index

Right Triangle

$c^2 = a^2 + b^2$ [Pythagorean Theorem]

area $= \frac{1}{2}ab$

$\sin\theta = \dfrac{b}{c} = \dfrac{\text{opp}}{\text{hyp}}$

$\cos\theta = \dfrac{a}{c} = \dfrac{\text{adj}}{\text{hyp}}$

$\tan\theta = \dfrac{\sin\theta}{\cos\theta} = \dfrac{b}{a} = \dfrac{\text{opp}}{\text{adj}}$

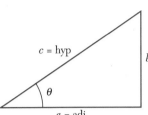

Circular Sector

area $= \frac{1}{2}r^2\theta$

arclength $= s = r\theta$

Any Triangle

area $= \frac{1}{2}bh = \frac{1}{2}ab\sin C$

$a^2 = b^2 + c^2 - 2bc\cos A$

$b^2 = a^2 + c^2 - 2ac\cos B$

$c^2 = a^2 + b^2 - 2ab\cos C$

$\dfrac{\sin A}{a} = \dfrac{\sin B}{b} = \dfrac{\sin C}{c}$

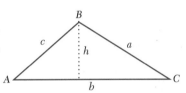

Circular Cylinder

volume $= \pi r^2 h$

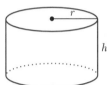

Trapezoid

area $= \frac{1}{2}b(h_1 + h_2)$

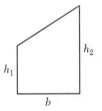

Circular Cone

volume $= \frac{1}{3}\pi r^2 h$

Circle

area $= \pi r^2$

circumference $= 2\pi r$

Sphere

volume $= \frac{4}{3}\pi r^3$

surface area $= 4\pi r^2$

Trigonometric Identities

$\sin^2 x + \cos^2 x = 1$

$\sec^2 x = 1 + \tan^2 x$

$\sin(2x) = 2 \sin x \cos x$

$\cos(2x) = \cos^2 x - \sin^2 x = 2\cos^2 x - 1 = 1 - 2\sin^2 x$

$\sin^2 x = \frac{1}{2}(1 - \cos(2x))$

$\cos^2 x = \frac{1}{2}(1 + \cos(2x))$

$\sin(x \pm y) = \sin x \cos y \pm \cos x \sin y$

$\cos(x \pm y) = \cos x \cos y \mp \sin x \sin y$

$\tan(x \pm y) = \dfrac{\tan x \pm \tan y}{1 \mp \tan x \tan y}$

$\sin x \sin y = \frac{1}{2}\big(\cos(x - y) - \cos(x + y)\big)$

$\cos x \cos y = \frac{1}{2}\big(\cos(x - y) + \cos(x + y)\big)$

$\sin x \cos y = \frac{1}{2}\big(\sin(x + y) + \sin(x - y)\big)$

Algebraic Identities

$ax^2 + bx + c = 0 \iff x = \dfrac{-b \pm \sqrt{b^2 - 4ac}}{2a}$ [Quadratic Formula]

$a^2 - b^2 = (a + b)(a - b)$

$a^3 + b^3 = (a + b)(a^2 - ab + b^2)$

$a^3 - b^3 = (a - b)(a^2 + ab + b^2)$

$(a + b)^n = a^n + na^{n-1}b + \cdots + \binom{n}{k} a^{n-k}b^k + \cdots + b^n; \; \binom{n}{k} = \dfrac{n!}{(n - k)!\,k!}$ [Binomial Theorem]

Derivative Formulas

$\dfrac{d}{dx}c = (c)' = 0$

$\dfrac{d}{dx}x = (x)' = 1$

$\dfrac{d}{dx}x^n = (x^n)' = nx^{n-1}$

$\dfrac{d}{dx}e^x = (e^x)' = e^x$

$\dfrac{d}{dx}\ln x = (\ln x)' = \dfrac{1}{x}$

$\dfrac{d}{dx}\sin x = (\sin x)' = \cos x$

$\dfrac{d}{dx}\cos x = (\cos x)' = -\sin x$

$\dfrac{d}{dx}\tan x = (\tan x)' = \sec^2 x$

$\dfrac{d}{dx}\sec x = (\sec x)' = \sec x \tan x$

$\dfrac{d}{dx}\arcsin x = (\arcsin x)' = \dfrac{1}{\sqrt{1 - x^2}}$

$\dfrac{d}{dx}\arctan x = (\arctan x)' = \dfrac{1}{1 + x^2}$

Derivative Rules

$\big(f(x) + g(x)\big)' = f'(x) + g'(x)$

$\big(f(x)g(x)\big)' = f'(x)g(x) + f(x)g'(x)$

$\left(\dfrac{f'(x)}{g(x)}\right)' = \dfrac{f'(x)g(x) - f(x)g'(x)}{\big(g(x)\big)^2}$

$\big((f \circ g)(x)\big)' = \big(f(g(x))\big)' = f'\big(g(x)\big)g'(x)$

Table of Integrals

Basic Forms

1. $\int x^n dx = \dfrac{x^{n+1}}{n+1}, n \neq -1$

2. $\int \dfrac{dx}{x} = \ln|x|$

3. $\int e^x dx = e^x$

4. $\int b^x dx = \dfrac{1}{\ln b} b^x$

5. $\int \sin x \, dx = -\cos x$

6. $\int \cos x \, dx = \sin x$

7. $\int \tan x \, dx = \ln|\sec x|$
$= -\ln|\cos x|$

8. $\int \cot x \, dx = \ln|\sin x|$
$= -\ln|\csc x|$

9. $\int \sec^2 x \, dx = \tan x$

10. $\int \csc^2 x \, dx = -\cot x$

11. $\int \sec x \tan x \, dx = \sec x$

12. $\int \csc x \cot x \, dx = -\csc x$

13. $\int \dfrac{dx}{x^2 + a^2} = \dfrac{1}{a} \arctan\left(\dfrac{x}{a}\right), \; a \neq 0$

14. $\int \dfrac{dx}{x^2 - a^2} = \dfrac{1}{2a} \ln\left|\dfrac{x-a}{x+a}\right|$

15. $\int \dfrac{dx}{\sqrt{a^2 - x^2}} = \arcsin\left(\dfrac{x}{a}\right), \; a > 0$

16. $\int \ln x \, dx = x(\ln x - 1)$

17. $\int \sec x \, dx = \ln|\sec x + \tan x| = \ln\left|\tan\left(\dfrac{x}{2} + \dfrac{\pi}{4}\right)\right|$

18. $\int \csc x \, dx = \ln|\csc x - \cot x| = \ln\left|\tan\left(\dfrac{x}{2}\right)\right|$

Expressions Containing $ax + b$

19. $\int (ax+b)^n dx = \dfrac{(ax+b)^{n+1}}{a(n+1)}, n \neq -1$

20. $\int \dfrac{dx}{ax+b} = \dfrac{1}{a} \ln|ax+b|$

21. $\int \dfrac{x}{ax+b} dx = \dfrac{x}{a} - \dfrac{b}{a^2} \ln|ax+b|$

22. $\int \dfrac{x}{(ax+b)^2} dx = \dfrac{b}{a^2(ax+b)} + \dfrac{1}{a^2} \ln|ax+b|$

23. $\int \dfrac{dx}{x(ax+b)} = \dfrac{1}{b} \ln\left|\dfrac{x}{ax+b}\right|$

24. $\int \dfrac{dx}{x^2(ax+b)} = -\dfrac{1}{bx} + \dfrac{a}{b^2} \ln\left|\dfrac{ax+b}{x}\right|$

25. $\int \sqrt{ax+b} \, dx = \dfrac{2}{3a} \sqrt{(ax+b)^3}$

26. $\int x\sqrt{ax+b} \, dx = \dfrac{2(3ax-2b)}{15a^2} \sqrt{(ax+b)^3}$

27. $\int \dfrac{dx}{\sqrt{ax+b}} = \dfrac{2\sqrt{ax+b}}{a}$

28. $\int \dfrac{dx}{x\sqrt{ax+b}} = \dfrac{1}{\sqrt{b}} \ln\left|\dfrac{\sqrt{ax+b} - \sqrt{b}}{\sqrt{ax+b} + \sqrt{b}}\right|, \; b > 0$

29. $\int \dfrac{dx}{x\sqrt{ax-b}} = \dfrac{2}{\sqrt{b}} \arctan\sqrt{\dfrac{ax-b}{b}}, \; b > 0$

Expressions Containing $ax^2 + c$, $x^2 \pm p^2$, and $p^2 - x^2, p > 0$

30. $\int \dfrac{dx}{p^2 - x^2} = \dfrac{1}{2p} \ln\left|\dfrac{p+x}{p-x}\right|$

31. $\int \dfrac{dx}{ax^2 + c} = \dfrac{1}{\sqrt{ac}} \arctan\left(x\sqrt{\dfrac{a}{c}}\right), \; a > 0, c > 0$

32. $\int \dfrac{dx}{ax^2 - c} = \dfrac{1}{2\sqrt{ac}} \ln\left|\dfrac{x\sqrt{a} - \sqrt{c}}{x\sqrt{a} + \sqrt{c}}\right|, \; a > 0, c > 0$

33. $\int \dfrac{dx}{(ax^2 + c)^n} = \dfrac{1}{2(n-1)c} \dfrac{x}{(ax^2 + c)^{n-1}} + \dfrac{2n-3}{2(n-1)c} \int \dfrac{dx}{(ax^2 + c)^{n-1}}, n > 1$

34. $\int x(ax^2 + c)^n \, dx = \dfrac{1}{2a} \dfrac{(ax^2 + c)^{n+1}}{n+1}, n \neq -1$

35. $\int \dfrac{x}{ax^2 + c} dx = \dfrac{1}{2a} \ln|ax^2 + c|$

36. $\int \sqrt{x^2 \pm p^2} \, dx = \dfrac{1}{2}\left(x\sqrt{x^2 \pm p^2} \pm p^2 \ln\left|x + \sqrt{x^2 \pm p^2}\right|\right)$

37. $\int \sqrt{p^2 - x^2} \, dx = \dfrac{1}{2}\left(x\sqrt{p^2 - x^2} + p^2\arcsin\left(\dfrac{x}{p}\right)\right), p > 0$

38. $\int \dfrac{dx}{\sqrt{x^2 \pm p^2}} = \ln\left|x + \sqrt{x^2 \pm p^2}\right|$

Expressions Containing Trigonometric Functions

39. $\displaystyle\int \sin^2(ax)\, dx = \frac{x}{2} - \frac{\sin(2ax)}{4a}$

40. $\displaystyle\int \sin^n(ax)\, dx = -\frac{\sin^{n-1}(ax)\cos(ax)}{na} +$
$\displaystyle\frac{n-1}{n}\int \sin^{n-2}(ax)\, dx,\, n > 0$

41. $\displaystyle\int \cos^2(ax)\, dx = \frac{x}{2} + \frac{\sin(2ax)}{4a}$

42. $\displaystyle\int \cos^n(ax)\, dx = \frac{\cos^{n-1}(ax)\sin(ax)}{na} +$
$\displaystyle\frac{n-1}{n}\int \cos^{n-2}(ax)\, dx$

43. $\displaystyle\int \sin(ax)\cos(bx)\, dx = -\frac{\cos\big((a-b)x\big)}{2(a-b)} -$
$\displaystyle\frac{\cos\big((a+b)x\big)}{2(a+b)},\, a^2 \neq b^2$

44. $\displaystyle\int \sin(ax)\sin(bx)\, dx = \frac{\sin\big((a-b)x\big)}{2(a-b)} -$
$\displaystyle\frac{\sin\big((a+b)x\big)}{2(a+b)},\, a^2 \neq b^2$

45. $\displaystyle\int \cos(ax)\cos(bx)\, dx = \frac{\sin\big((a-b)x\big)}{2(a-b)} +$
$\displaystyle\frac{\sin\big((a+b)x\big)}{2(a+b)},\, a^2 \neq b^2$

46. $\displaystyle\int x\sin(ax)\, dx = \frac{1}{a^2}\sin(ax) - \frac{x}{a}\cos(ax)$

47. $\displaystyle\int x\cos(ax)\, dx = \frac{1}{a^2}\cos(ax) + \frac{x}{a}\sin(ax)$

48. $\displaystyle\int x^n\sin(ax)\, dx = -\frac{x^n}{a}\cos(ax) +$
$\displaystyle\frac{n}{a}\int x^{n-1}\cos(ax)\, dx,\, n > 0$

49. $\displaystyle\int x^n\cos(ax)\, dx = \frac{x^n}{a}\sin(ax) -$
$\displaystyle\frac{n}{a}\int x^{n-1}\sin(ax)\, dx,\, n > 0$

50. $\displaystyle\int \tan^n(ax)\, dx = \frac{\tan^{n-1}(ax)}{a(n-1)} -$
$\displaystyle\int \tan^{n-2}(ax)\, dx,\, n \neq 1$

51. $\displaystyle\int \sec^n(ax)\, dx = \frac{\sec^{n-2}(ax)\tan(ax)}{a(n-1)} + \frac{n-2}{n-1}\int \sec^{n-2}(ax)\, dx,\, n \neq 1$

Expressions Containing Exponential and Logarithm Functions

52. $\displaystyle\int xe^{ax}\, dx = \frac{e^{ax}}{a^2}(ax-1)$

53. $\displaystyle\int x^n e^{ax}\, dx = \frac{1}{a}x^n e^{ax} - \frac{n}{a}\int x^{n-1}e^{ax}\, dx,\, n > 0$

54. $\displaystyle\int e^{ax}\sin(bx)\, dx = \frac{e^{ax}}{a^2+b^2}\big(a\sin(bx) - b\cos(bx)\big)$

55. $\displaystyle\int e^{ax}\cos(bx)\, dx = \frac{e^{ax}}{a^2+b2}\big(a\cos(bx) + b\sin(bx)\big)$

56. $\displaystyle\int x^n\ \ln(ax)\, dx = x^{n+1}\left(\frac{\ln(ax)}{n+1} - \frac{1}{(n+1)^2}\right),$
$n \neq -1$

57. $\displaystyle\int (\ln x)^n\, dx = x(\ln x)^n - n\int (\ln x)^{n-1}\, dx$

58. $\displaystyle\int \frac{dx}{a+be^{px}} = \frac{x}{a} - \frac{1}{ap}\ln|a+be^{px}|$

Expressions Containing Inverse Trigonometric Functions

59. $\displaystyle\int \arcsin(ax)\, dx = x\arcsin(ax) + \frac{1}{a}\sqrt{1-a^2x^2}$

60. $\displaystyle\int \arccos(ax)\, dx = x\arccos(ax) - \frac{1}{a}\sqrt{1-a^2x^2}$

61. $\displaystyle\int \arctan(ax)\, dx = x\arctan(ax) - \frac{1}{2a}\ln(1+a^2x^2)$

62. $\displaystyle\int \text{arccot}(ax)\, dx = x\,\text{arccot}(ax) + \frac{1}{2a}\ln(1+a^2x^2)$

63. $\displaystyle\int \text{arccsc}(ax)\, dx = x\,\text{arccsc}(ax) + \frac{1}{a}\ln\left|ax + \sqrt{a^2x^2 - 1}\right|$

64. $\displaystyle\int \text{arcsec}(ax)\, dx = x\,\text{arcsec}(ax) - \frac{1}{a}\ln\left|ax + \sqrt{a^2x^2 - 1}\right|$